An Introduction to System Safety Engineering

An Introduction to System Safety Engineering

Nancy G. Leveson

The MIT Press
Cambridge, Massachusetts
London, England

The MIT Press would like to thank the anonymous peer reviewers who provided comments on drafts of this book. The generous work of academic experts is essential for establishing the authority and quality of our publications. We acknowledge with gratitude the contributions of these otherwise uncredited readers.

This book was set in Times New Roman by Westchester Publishing Services. Printed and bound in the United States of America.

Library of Congress Cataloging-in-Publication Data

Names: Leveson, Nancy, author.
Title: An introduction to system safety engineering / Nancy G. Leveson.
Description: Cambridge, Massachusetts : The MIT Press, [2023] |
Includes bibliographical references and index.
Identifiers: LCCN 2022061535 (print) | LCCN 2022061536 (ebook) |
ISBN 9780262546881 | ISBN 9780262376761 (epub) | ISBN 9780262376754 (pdf)
Subjects: LCSH: Industrial safety. | System safety.
Classification: LCC T55 .L46625 2023 (print) | LCC T55 (ebook) |
DDC 363.11—dc23/eng/20230124
LC record available at https://lccn.loc.gov/2022061535
LC ebook record available at https://lccn.loc.gov/2022061536

10 9 8 7 6 5 4 3 2

Whoever destroys a single life is as guilty as though he has destroyed the whole world, and whoever rescues a single life earns as much merit as though he had rescued the entire world.
—The Talmud
(Sanhedrin 37:a)

A life without adventure is likely to be unsatisfying, but a life in which adventure is allowed to take any form it will, is sure to be short.
—Bertrand Russell
Authority and the Individual, 1949

If you think safety is expensive, try an accident.
—Trevor Kletz

Contents

Preface xiii

1 Historical and Industrial Perspectives on Safety Engineering 1

1.1 Differences between Workplace Safety and Product/System Safety 2
1.2 A Brief Legal View of the History of Safety 2
1.3 A Technical View of the History of Safety 4
1.4 Workplace Safety Today: An Engineer's View 8
1.5 Product/System Safety Today 10
1.5.1 Commercial Aviation 12
1.5.2 Nuclear Power 14
1.5.3 The Chemical Industry 16
1.5.4 Defense and "System Safety" 17
1.5.5 SUBSAFE: The US Nuclear Submarine Safety Program 22
1.5.6 Astronautics and Space 23
1.5.7 Healthcare/Hospital Safety 23
1.6 Summary 24
Exercises 25

2 Risk in Modern Society 27

2.1 Changing Attitudes toward Risk 28
2.2 Changing Risk Factors 30
2.2.1 The Appearance of New Hazards 30
2.2.2 Increasing Complexity 30
2.2.3 Increasing Exposure 32
2.2.4 Increasing Amounts of Energy 32
2.2.5 Increasing Automation of Manual Operations 32
2.2.6 Increasing Centralization and Scale 33
2.2.7 Increasing Pace of Technological Change 34
2.3 How Safe Is Safe Enough? 35
2.3.1 Risk–Benefit Analysis and the Alternatives 36
2.3.2 Trans-Scientific Questions 38
Exercises 40

3 Fundamental Concepts and Definitions 43

3.1 Definitions of Safety and Risk 43
3.2 Hazards and Hazard Analysis 45
3.3 Defining Safety Requirements and Constraints 47

3.4 Safety versus Reliability 49
3.5 What Is a System? 51
3.5.1 Assumptions Underlying the Concept of a System 52
3.5.2 Sociotechnical Systems 54
3.6 Defining Complexity 55
3.7 Approaches to Dealing with Complexity 56
3.7.1 Analytic Decomposition 56
3.7.2 Statistics 59
3.7.3 Systems Thinking and Systems Theory 59
3.7.4 Systems Theory Fundamentals 61
3.8 Summary 68
Exercises 69

4. Why Accidents Occur 73
4.1 The Traditional Conception of Causality 73
4.2 Subjectivity in Ascribing Causality 75
4.3 Oversimplification in Determining Causality 76
4.3.1 The Legal Approach to Causality 76
4.3.2 Human Error as the Cause of Accidents 77
4.3.3 Technical Failures as the Cause of Accidents 77
4.3.4 Organizational Factors as the Cause of Accidents 78
4.4 Multifactorial Explanations of Accidents 79
4.5 Systemic Causes of Accidents 81
4.5.1 Social Dynamics and Organizational Culture 82
4.5.1.1 Overconfidence and complacency 83
4.5.1.2 Low priority given to safety 93
4.5.1.3 Flawed resolution of conflicting goals 95
4.5.1.4 Confusing safety with other system properties 97
4.5.2 Management Decision-Making Structure 100
4.5.2.1 Ill-defined and diffused responsibility, authority, and accountability 100
4.5.2.2 Lack of independence and low-level status of safety personnel 102
4.5.2.3 Limited communication channels and poor information flow 102
4.5.3 Operational Processes and Practices 104
4.5.3.1 Superficial, isolated, or misdirected safety efforts during operations 104
4.5.3.2 Inadequate feedback and learning from events 104
4.5.3.3 Poorly defined operating procedures 106
4.5.3.4 Inadequate training and emergency management 107
4.5.3.5 Inadequate management of change 110
4.5.3.6 Poor maintenance practices 112
4.5.4 Government and Professional Society Oversight 113
4.5.5 Engineering Processes and Practices 116
4.5.5.1 Superficial safety efforts 116
4.5.5.2 Overreliance on redundancy and protection systems 118
4.5.5.3 Ineffective risk control 121
4.5.6 Safety Information System Deficiencies 124
4.6 Summary 126
Exercises 127

5 The Role of Software in Safety 129
5.1 The Use of Software in Systems Today 129
5.2 Understanding the Problem 132
5.3 Why Does Software Present Unique Difficulties? 134
5.3.1 Software Myths 135

5.3.2 Why Software Engineering Is Difficult 140
5.3.2.1 Analog versus discrete state systems 140
5.3.2.2 The curse of flexibility 141
5.3.2.3 Complexity and invisible interfaces 143
5.3.2.4 Lack of historical usage information 144
5.3.3 The Reality We Face 145
5.4 The Way Forward 145
Exercises 146

6 The Role of Humans in Safety 147
6.1 Why Replace Humans with Machines? 148
6.2 Do Human Operators Cause Most Accidents? 149
6.3 The Need for Humans in Automated Systems 155
6.4 Human Error as Human–Task Mismatch 157
6.4.1 Skill-Based Behavior 158
6.4.2 Rule-Based Behavior 159
6.4.3 Knowledge-Based Behavior 159
6.4.4 The Relationship between Experimentation and Error 160
6.5 The Role of Mental Models in Safety 161
6.6 What Is the Appropriate Role for Humans in Complex Systems? 163
6.6.1 The Human as Monitor 164
6.6.2 The Human as Backup 170
6.6.3 The Human as Partner 173
6.7 Conclusions 175
Exercises 176

7 Accident Causality Models 179
7.1 Energy Models 180
7.2 Linear Chain-of-Failure Events Models 181
7.2.1 The Domino Model 184
7.2.2 The Swiss Cheese Model 186
7.2.3 The Functional Resonance Model 187
7.2.4 Limitations of Linear Chain-of-Events Models 188
7.3 Epidemiological Models 190
7.4 More Sophisticated Models of Causality 191
7.5 The STAMP Model of Causality 193
7.6 Looking Ahead 199
Exercises 199

8 Accident Analysis and Learning from Events 201
8.1 Why Are We Not Learning Enough from Accidents? 201
8.1.1 Oversimplification and Root Cause Seduction 202
8.1.2 Hindsight Bias 203
8.1.3 Misunderstanding the Role of Humans in Accidents 205
8.1.4 Focusing on Blame: Blame Is the Enemy of Safety 209
8.2 Goals for Improved Accident Analysis 213
8.3 Example: The Zeebrugge Ferry Accident 214
8.4 Generating Recommendations 235
8.5 Implementing Long-Term Learning 236
8.6 The Cost of Thorough Accident Investigation 236
8.7 Summary 237
Exercises 237

9 Hazard Analysis: Basic Concepts 241
9.1 What Is Hazard Analysis? 241
9.2 The Hazard Analysis Process 243
9.2.1 The Overall Process 244
9.2.2 Detailed Steps 245
9.3 Types of System Models 253
9.4 General Types of Analysis 253
9.4.1 Forward and Backward Searches 254
9.4.2 Top-Down and Bottom-Up Searches 255
9.4.3 Combined Searches 256
9.5 Who Should Do Hazard Analysis? 257
9.6 Limitations and Criticisms of Hazard Analysis 257
9.7 Analysis versus Assessment 259
Exercises 260

10 Hazard Analysis Techniques 261
10.1 Energy Model Techniques: Hazard Indices 262
10.2 Techniques Based on the Chain-of-Failure-Events Causality Model 263
10.2.1 Failure Modes and Effects Criticality Analysis 263
10.2.2 Fault Hazard Analysis 266
10.2.3 Fault Tree Analysis 268
10.2.4 Event Tree Analysis 277
10.2.5 Combinations of Analysis Techniques 281
10.2.6 Hazards and Operability Analysis (HAZOP) 284
10.2.7 Miscellaneous Techniques 289
10.3 STPA: A Technique Based on STAMP 292
10.4 Task and Human Error Analysis Techniques 303
10.4.1 Qualitative Techniques 303
10.4.2 Quantitative Techniques 305
10.5 Conclusions 311
Exercises 312

11 Design for Safety 315
11.1 The Design Process 317
11.1.1 Standards and Codes of Practice 317
11.1.2 Design Guided by Hazard Analysis 318
11.2 Types of Design Techniques and Precedence 320
11.3 Hazard Elimination 323
11.3.1 Substitution 323
11.3.2 Simplification 325
11.3.3 Decoupling 330
11.3.4 Elimination of Specific Human Errors 331
11.3.5 Reduction of Hazardous Materials or Conditions 332
11.4 Hazard Occurrence Reduction 333
11.4.1 Design for Controllability 334
11.4.2 Barriers 335
11.4.3 Monitoring 341
11.4.4 Failure Minimization 343
11.5 Hazard Control 350
11.5.1 Limiting Exposure 351
11.5.2 Isolation and Containment 351
11.5.3 Protection Systems and Fail-Safe Design 352

11.6 Damage Reduction 356
11.7 Design Modification and Maintenance 357
Exercises 357

12 Human Factors in System Design 359
12.1 Determining What Should Be Automated 360
12.2 The Need for Wide Participation in Design Activities 361
12.3 Safety versus Usability and Other Common Goals 362
12.4 Reducing Safety-Critical Human Errors through System Design 363
12.4.1 Safety in the Design of Operator Controls 366
12.4.2 Designing Feedback for Safety 368
12.4.2.1 The role of feedback and independent information 369
12.4.2.2 Alarms 372
12.4.3 Identifying and Designing the Activities and Functions Provided by Humans 376
12.4.3.1 Combating lack of alertness 377
12.4.3.2 Designing for error tolerance 378
12.4.3.3 Task allocation 381
12.4.4 Design of Displays for Safety 384
12.4.4.1 Tailoring the display for human cognitive processing 386
12.4.4.2 Ease of interpretation 387
12.4.4.3 Preparing for failure 387
12.4.4.4 Displaying critical information in a way easy for humans to process 387
12.4.4.5 Feedforward assistance and decision aids 392
12.5 Training and Maintaining Skills 393
12.5.1 Teaching about Safety Features 393
12.5.2 Training for Emergencies 394
Exercises 395

13 Assurance, Assessment, and Certification 397
13.1 Assurance of Safety 397
13.1.1 Limitations of Traditional Assurance Activities When Used for Safety 399
13.2 Hazard and Risk Assessment 404
13.2.1 Qualitative and Quantitative Hazard and Risk Assessment 405
13.2.2 Limitations of Hazard and Risk Assessment 408
13.2.3 Probabilistic Risk Analysis 412
13.3 Certification 415
13.3.1 Types of Certification Approaches 416
13.3.2 National and Industry Practices in Certification 418
13.3.3 Providing Evidence in Performance-Based Regulation and Safety Cases 421
13.3.4 Designing a Certification Program 423
13.4 Some General Conclusions 427
Exercises 427

14 Designing a Safety Management System 429
14.1 Social Dynamics and Organizational Culture 431
14.1.1 Modeling Desired Behavior 433
14.1.2 Documenting Values and Policies 434
14.2 Organizational Structure 435
14.2.1 Assigning Responsibility, Authority, and Accountability 436
14.2.2 Location of System Safety Activities 438
14.2.3 Communication, Coordination, and Information Flow 440
14.3 Management of Safety-Critical System Development 443

14.4 Management of Operational Processes and Practices 445
14.4.1 Providing a Shared and Accurate Perception of Risk 447
14.4.2 Feedback and Learning from Events 448
14.4.3 Creating and Updating Operating Procedures 453
14.4.4 Training and Contingency Management 454
14.4.5 Managing Change 457
14.4.6 Maintenance 460
14.5 Creating an Effective Safety Information System 461
14.6 Summary 465
Exercises 469

Epilogue: Looking Forward 471

Appendix A. Medical Devices: The Therac-25 473

Appendix B. Space: The *Challenger* and *Columbia* Space Shuttle Losses 503

Appendix C. Petrochemicals: Seveso, Flixborough, Bhopal, Texas City, and *Deepwater Horizon* 529

Appendix D. Nuclear Power: Three Mile Island, Chernobyl, and Fukushima Daiichi 609

References 659
Index 675

Preface

These days we adopt innovations in large numbers, and put them to extensive use faster than we can even hope to know their consequences . . . which tragically removes our ability to control the course of events.
—Patrick Lagadec
Major Technological Risk

Don't panic.
—Douglas Adams
The Hitchhiker's Guide to the Galaxy

While this book has some overlap with my first *Safeware* book, I have decided not to write a second edition. In the almost thirty years since I wrote the first book, my knowledge and experience with real systems has led to major changes in what I believe is needed to build safer systems.

This book reflects those changes. The basic principles that underlie system safety still apply, but I've learned much more about how to achieve the goals. That knowledge is what I have tried to include here. Also, my original *Safeware* was heavily oriented toward software and its role in safety. While software is an important and problematic part of safety engineering, this book is intended to provide a textbook on the systems approach to safety engineering as a whole.

There are currently few classes on safety engineering in universities. I am hoping that the existence of this book will encourage more faculty to teach them. I am also hoping that those currently working in this field in industry, most of whom "fell into it" and learned on the job without much academic preparation, can use this book to augment their knowledge and provide the foundation they need to help create safer systems.

Existing textbooks on this field are almost all out of date, as is my first version of *Safeware*. They are also often written like cookbooks, where students are taught to perform specific analysis techniques and fill in endless forms without learning *why* they are doing it or, more important, without learning what they need to know in order to evaluate what they are doing. Instead, I have tried to describe the fundamental principles for creating and operating safety-critical systems today as well as the state of the field as it exists and enough theoretical information to guide those who will advance this field in the future. I have used versions of this book in classes for undergraduates, graduates, and professionals; it is appropriate for all of these groups.

This book is written to provide the technical background for learning about system safety engineering but not to teach specific techniques. Other means are available to achieve that goal and can be used to supplement this book. The changes in the world of engineering are continual and, arguably, are increasing exponentially. Students, of course, need to be able to use various analysis techniques, but including instruction on specific types of analysis in this book will make it obsolete quickly. My goal is to create a foundational textbook that will last a significant amount of time. The use of additional handbooks and other material will allow classes to be kept up to date on new types of analysis while teaching basic concepts that change more slowly.

This book is also not a successor to my second book, *Engineering a Safer World.* That book describes one view of safety engineering, one that I obviously believe is the best. But that also limits its applicability as a general safety engineering textbook. Some may choose to use it to supplement this book, but both stand alone. I tried not to provide too much duplication between the two, but a little repetition was inevitable, particularly the information on managing and operating safety-critical systems.

To create a general textbook, I have tried not to bias the presentation toward my own systems approach. That is difficult, however, because I believe so strongly that this approach is the only viable way forward for the field. Without changes in what has been a basically static field for the past fifty years, it will die. Technology is advancing quickly, and safety engineering has to advance at the same rate or fall behind in usefulness. There is, therefore, some inherent bias in this book, but I do not know how to get rid of it without producing a book that offends nobody but does not help the field advance.

It may be a surprise to many people, but the systems approach to safety is not new. In fact, it was the standard approach to safety engineering developed by the defense industry starting in the 1950s and by some in aviation safety at the same time. That this approach has not spread beyond this community over the years is a shame. Some concepts and practices are being lost even in the defense community due to various pressures to conform to the practices in other industries. Most of what may appear to be my own personal biases have, in fact, been the standard practice and beliefs in this community for more than sixty years.

Unlike the first *Safeware*, exercises are provided to assist in using the book for classes. The goal of the exercises is to make the students think about what they are doing, not just to check whether they can perform a specific analysis. I want the book to help produce engineers and not technicians. Engineers need to understand why they are using the approaches they use, not just blindly apply what they have been taught to do. As new techniques and ideas are created, engineers need to be able to evaluate them as to their relevance, usefulness, and originality. We too often reinvent what did not work in the past, giving it new names and then believing that the renamed techniques somehow will work better after relabeling.

To assist those who are just starting to teach this subject, a repository of teaching materials has been created. Hopefully others will contribute to it. It contains such things as PowerPoint slides I use in my classes, examples of class projects, tests, and supplementary readings. Learning how to prevent accidents requires understanding why they have occurred in the past. The appendices of this book contain detailed descriptions and causal factors for major accidents in a variety of fields. Other accident analyses will be added to the repository in the future.

For the accidents described in the appendices, I provide basic information about the design of the system and the safety features for those not in that industry, the events that occurred, and some of the most important causes of those events. While the official accident reports are the major source of information about the accidents, additional information is related to what I learned by participating in the official investigations and is not necessarily included in the official reports. Students need to learn that accidents are always complicated and multifactorial. I hope they will also discover that many of the same mistakes are made over and over. We can and must learn from them instead of simply repeating them.

My forty years in system safety engineering have involved an incredible journey. Because of the dearth of system safety engineering classes and research in universities, most of my knowledge has come from interacting with great engineers in industry, by working on real systems, and, unfortunately, participating in accident investigations that represent great tragedies such as the Therac-25 radiation overdoses, the Texas City refinery explosion, the *Deepwater Horizon/Macondo* blowout, and the *Columbia* Space Shuttle loss.

During this journey, I have experienced great frustration with the inadequacy of safety engineering as usually practiced, but I have also had the good fortune to work with some of the best. Grady Lee is a shining example of the latter, as is Dr. John Thomas. Most of the early great safety engineers whom I learned the most from are no long working. In this book, I am trying to pass on to future engineers what I learned from those experts, such as C. O. Miller, Willie Hammer, Trevor Kletz, and Jens Rasmussen. They were all ahead of their time and were leaders in introducing new systems thinking paradigms. The systems approach to safety engineering described in this book is an attempt to return to their original vision, but updated to today's technology.

I have had the pleasure of working with hundreds of brilliant students, who have taught me as much as I taught them. I have also learned from those working in industry, especially pilots and others who are responsible for operating our sophisticated systems such as Captain Shem Malmquist. There are too many to mention names here, but know that I cherish the time spent with each of you, including the few who fought me and my systems view of safety engineering every step of the way. A few were never convinced, but the vast majority are or will serve as the future leaders and innovators in this field. Some, such as Col. Bill "Dollar" Young, forced me to consider applying the same systems approach to other system qualities, such as security, and even to system engineering in general. But that view is relegated to *Engineering a Safer World* and to any future books and papers that I or my students and followers may get the chance to write. In this book, I concentrate on the basics of a systems approach to safety engineering.

1

Historical and Industrial Perspectives on Safety Engineering

We must respect the past and mistrust the present if we wish to provide for the safety of the future.
—Joseph Joubert

We seem not to trust one another as much as would be desirable. In lieu of trusting each other, are we putting too much trust in our technology? . . . Perhaps we are not educating our children sufficiently well to understand the reasonable uses and limits of technology.
—T. B. Sheridan,
Trustworthiness of Command and Control Systems [349]

The world of engineering is changing quickly. These changes include greatly increasing complexity, the extensive and growing use of computers and other forms of new technology, and a changing role of humans in complex systems. Advances will be needed in all the forms of safety engineering to keep it relevant for the future as our technology and society change.

The goal of this book is to provide a foundation on which new approaches and advances can be devised without having to slow progress by reinventing things that have been found to be unsuccessful in the past or by taking engineering into byways that do not build on what is already known. To make progress, we need to understand the basic scientific foundations of safety engineering and to understand what works for what types of systems, what does not work and why, and how we should move forward to improve all aspects of safety.

There is no one monolithic approach in our attempts to reduce losses: the approach to safety varies with each industry and sometimes individual companies in an industry. This chapter provides a brief description of the field and its historical development.

The biggest difference is between workplace safety and product or system safety, so let's start there and then survey the state of the practice in each. While workplace safety is mostly practiced the same in most industries, product or system safety is tackled very differently within different types of industries.

1.1 Differences between Workplace Safety and Product/System Safety

Workplace safety and product/system safety have very little in common today. Workplace safety is sometimes also called occupational health and safety, industrial safety, or industrial hygiene. It focuses on injuries to workers during, and as a result of, their job responsibilities, such as exposure to industrial chemicals and toxins, human injury due to the strain or stress involved in their job (ergonomics), or other workplace injuries such as falling off ladders.

A great emphasis in workplace safety has been placed on psychology—particularly behavioral psychology—to change worker behavior to prevent workplace injuries. Standard approaches use training, reward and punishment, and various other types of persuasion to encourage workers to follow rules and procedures and to wear PPE (personal protective equipment). The foundational assumption usually is that worker injuries are due to the workers themselves: if they just were more careful and followed instructions, injuries would not occur.

Much less emphasis has been on changing the design of the system in which the humans work, although guards or barriers and other types of simple devices are often used in the design of assembly and production line automation and have been a part of workplace safety since at least the nineteenth century.

When an accident does occur, workers are usually blamed, with the most common cause cited being noncompliance with specified procedures.

In contrast to workplace safety, product safety, system safety, or safety engineering (all of which are terms used in various contexts) focuses on the dangers of the products or systems we design and use. There is clearly some overlap with workplace safety—for example, workers may be injured due to hazards in the engineered products they use, such as machines or robots in factories. In general, however, workplace safety concentrates on how employees or workers perform their work while product safety focuses on the characteristics of engineered products.

Historically there has been little overlap of practices or practitioners in these two areas of safety after they diverged almost one hundred years ago, even when the problem is that of workers being injured by the use of engineered products in the workplace.

To understand the differences, the next section first describes the history of both fields and their overlaps and divergences. Then the differences in how each is practiced today are outlined. While the primary focus in this book is on product or system safety, the same principles and practices are theoretically applicable in occupational safety.

We start with the common history the two fields share before the practices started to develop separately. Two relevant historic narratives exist: one focusing on the legal aspects and one on the technical ones.

1.2 A Brief Legal View of the History of Safety

Humans have always been concerned about their safety. Prior to the industrial age, natural disasters provided the biggest challenge. Things started to change in the early part of the Industrial Revolution in Europe and the United States. Workers in factories were considered expendable and were often treated worse than slaves: slaves cost a great deal of money, and owners wanted to protect their investment, but workers cost nothing to hire or replace [115]. The prevailing attitude was that when people accepted employment, they also accepted the risks involved in the job and should be smart enough to avoid danger. At the same time, factories were filled with potentially dangerous equipment, such as unguarded machines, flying shuttles, and open belt drives, as well as unsafe conditions, such as open holes. There were no fire escapes, and the lighting was inadequate. Hardly a day went by in most countries without some worker being maimed or killed [115].

Without workers' compensation laws, employees in the Western world had to sue and collect damages for injuries under common law. In the United States, the employer almost could not lose because common law precedents established that an employer did not have to pay injured employees if:

1. The employee contributed at all to the cause of the accident: *Contributory negligence* held that if the employee was responsible or even partly responsible for an accident, the employer was not liable.
2. Another employee contributed to the accident: The *fellow-servant doctrine* held that an employer was not responsible if an employee's injury resulted from the negligence of a coworker.
3. The employee knew of the hazards involved in the accident before the injury and still agreed to work in the conditions for pay: The *assumption-of-risk* doctrine held that an injured employee presumably knew of and accepted the risks associated with the job before accepting the position.

It is not, therefore, surprising that most investigations of accidents, which were conducted by management, found that the workers were responsible.

The horrific workplace conditions existing at this time led to social revolt by activists and union leaders. Miners, railroad workers, and others became concerned about the hazards of their jobs and began to agitate for better conditions. Voluntary safety organizations, such as the American Public Health Association, the National Fire Protection Association, and Underwriters Laboratories, were formed in the late nineteenth century and were active in setting standards. The first efforts focused on health rather than safety: accidents were seen as fortuitous events over which we had little control [100].

Concern over worker safety started in Europe. Otto von Bismarck established workers' compensation and security insurance, paid for by the employees, in Germany during the 1880s. Bismarck sought to undercut the socialists by demonstrating to the German working class that its government was in favor of social reform [115]. Soon, most countries in Europe followed Bismarck's lead.

Other types of safety legislation were also passed in Europe. For example, the Factory and Workshop Act of 1844 in Great Britain was the first legislation for protection from accidents involving shafts, belts, pulleys, and gears used to transmit power from a water wheel or a stationary steam engine throughout the factory. Later, laws setting standards for equipment such as high-pressure steam engines and high-voltage electrical devices were enacted.

The textile industry was one of the first to become mechanized and also contributed more than its share of injuries to workers. The abuses in the mills led to protective legislation and codes in many countries by the end of the nineteenth century. The same was true for industries that employed metal and woodworking machines. Unfortunately, many safety devices, added only grudgingly, were poorly designed or ineffective [330].

In the United States, employers remained indifferent to the many workers being killed and maimed on the job. Eventually, social revolt and agitation by unions against poor working conditions led to social reforms and government intervention to protect employees and the public.

Individual state laws preceded federal legislation. The first successful regulatory legislation in the United States was enacted in 1852, when Congress passed a law to regulate steamboat boilers. This law resulted from public pressure, enhanced by a series of marine disasters that had killed thousands of people.

In 1908, the state of New York passed the first workers' compensation law, which, in effect, required management to pay for injuries that occurred on the job regardless of fault. The New York law was held to be unconstitutional, but a comparable law, passed in Wisconsin in 1911, withstood judicial scrutiny. All other US states and many other countries have now enacted similar laws; the last was in 1947.

When management found that it had to pay for injuries on the job, effort was put into preventing such injuries, and the organized industrial safety movement was born. Owners and managers also began to realize that accidents cost money in terms of lower productivity and started to take safety seriously. The first industrial safety department in a company was established in the early 1900s. The heavy industries, such as steel, in which workplace accidents occurred frequently, began to set safety standards.

This legal history provided an important impetus to technical developments. So now let's examine the historical narrative from a technical viewpoint.

1.3 A Technical View of the History of Safety

Safety has been an issue since early times; the need for safe design is even mentioned in the Bible. In more modern times, however, a few engineers began to recognize the need to prevent hazards early in the industrial era, when the machines they were designing and building began to kill people. James Watt warned about the dangers of high-pressure steam engines in the early 1800s. The Davey Safety Lamp, invented around the same time, helped decrease some of the danger of methane gas explosions in mines. In 1869, George Westinghouse developed a brake based on compressed air, which made railroad travel vastly safer for riders and crew. In 1868, the first patent was granted by the US Patent Office for a machine safety device—an interlocking guard used in a machine for filling bottles with carbonated water. Other patents soon appeared for guards for printing presses, two-hand controls, and circular saws and other woodworking machines [330].

Near the end of the nineteenth century, engineers began to consider safety, as well as functionality, in their designs instead of simply trying to add it on in the form of guards. While the use of guards was prevalent then and still is in workplace safety, much more sophisticated engineering solutions have been found and are used in complex systems, as described in later chapters of this book.

Many of the dangers involved in workplace accidents were due to the increasing use of engineered devices in the workplace, so there was much overlap at this early time in the history of safety engineering and workplace safety. A divergence between the two fields occurred later, as you will see.

One of the first organizations to study accidents was the Society for Prevention of Accidents in Factories, called the Mulhouse Society, so named because it was founded in the town of Mulhouse in Alsace-Lorraine, in the area bordering France and Germany. The Mulhouse Society held annual meetings to exchange ideas about safety improvements in factories and published an encyclopedia of techniques for injury prevention in 1889 and 1895 [274]. By the early part of the twentieth century, a German engineering society was established for the prevention of accidents [330].

At about the same time, the engineering technical literature started to acknowledge that safety should be built into a design. The first paper dealing with the safe design of machinery was presented to the American Society of Mechanical Engineers by John H. Cooper in 1891: "*It is an easy task to formulate a plan of accident-preventing devices after the harm is done, but the wiser engineer foresees the possible weakness, as well as the dangers ahead, which are involved in his new enterprise, and at once embodies all the necessary means of safety in his original design*" [67, p. 250].

In 1899, John Calder published a book in England, *The Prevention of Factory Accidents*, that provided accident statistics and described safety devices in detail. The book emphasized the need for anticipating accidents and building in safety and argued that legislation to compel manufacturers to provide safe products would probably not be necessary; free market forces would provide a stronger incentive: *"Safeguarding by the user, of some kind, can in the long run be compelled by statute, but the author's experience is that, in the case of the multitude of occupiers of small factories, with no mechanical facilities or aptitude, nothing can take the place of good fencing fitted by the makers, and all accidents thereby prevented by being anticipated*" (quoted in [330, p. 89]).

Calder later moved to the United States and changed his mind: By 1911, he was calling for legislation to force manufacturers to provide safe products [56].

Trade journals started to editorialize about the need for designers to eliminate hazards from machinery: "*We reiterate that the time to safeguard machinery is when it is on the drawing board; and designers should awaken fully to a sense of their responsibility in this respect. They should consider safety as of equal importance with operating efficiency, for if the machines unprotected, are not safe to work, they are failures, no matter how efficient they may be as producers*" (quoted in [330, p. 90]).

The first American technical journal devoted solely to accident prevention, *The Journal of Industrial Safety*, began publication in 1911, and individual states began to hold safety conferences. The first American National Safety Congress, organized by the Association of Iron and Steel Electrical Engineers, was held in Milwaukee in 1912.

In 1914, the first safety standards published in the United States, the Universal Safety Standards, were compiled under the direction of Carl Hansen [134]. These standards required first defining hazards and then finding ways to eliminate them by design. This approach is the essence of safety engineering today; notice how far back it goes.

Engineers at this time started to study safety as an independent topic. A study examined the widely held conviction that safer machines with guards were inefficient and resulted in reduced production. The study, conducted by a group that included the major engineering societies at the time, involved employees of twenty industries and sixty product groups who had a combined exposure of over 50 billion hours. The final report confirmed the hypothesis that production increased as safety increased—a lesson still to be learned by many people today.

The study also explained the historical increase in accidents despite industrial safety efforts to reduce them. The increase was found to be related to the tremendous increase in the rate at which American industry was becoming mechanized. Mechanization affected safety in three ways: (1) it displaced the use of hand tools, (2) it increased the exposure of maintenance personnel to machine hazards, and (3) it allowed increased operating and material-feed speeds. The primary conclusion of the study was that "*while [. . .] there has been this recent increase in the hazard of industry per man-hour, production per man-hour has increased so much more rapidly that the hazard in terms of production has decreased*" [330].

Up to this point in time, there are two things to note:

1. Workplace safety and product safety were related and intertwined, primarily because many of the new engineered devices were used in workplaces and were created to reduce workplace injuries. At the same time, however, designing safety into more generally used hazardous products, such as the steam engine, that affected public safety and not just the workplace, was also a focus of study and engineering practices.
2. Post-accident investigation was a very small part of safety engineering in the past (and now). Much more concern was placed on identifying hazards and designing safety into products from the beginning of development.

Around the 1930s, workplace and product/system safety diverged and started to develop separately.

On the workplace safety side, in 1929, H. W. Heinrich, who worked for an insurance company and thus had access to a large number of accident reports, published a study of 50,000 industrial accidents, concluding that for every serious injury that occurred in the workplace, there were twenty minor injuries and 300 near misses without reportable injury. He also suggested that thousands of unsafe acts were occurring as well—although it is not clear how he got this information from the accident reports. This hypothesis became known as Heinrich's pyramid or triangle, shown in figure 1.1 [141].

Heinrich's hypothesis about ratios between serious accidents and less serious incidents does not apply to complex systems and product safety, although that has not deterred many people from applying it anyway. In complex, engineered systems, accidents may result—and often do—from very different things than incidents; that is, the causality may not be related [17]. Even some in occupational safety today are starting to doubt the validity of Heinrich's Triangle.

Heinrich was also the first to propose a conceptual model of accident causation in industrial safety, called the *domino model*, which focuses on human error as the cause of accidents [141]. The domino model is described in chapter 5.

Seizing on these arguments, opponents of mechanical (i.e., engineering) solutions to workplace safety began to direct attention away from unsafe machinery and toward unsafe user acts. Claims were made that accident-prone workers and carelessness were responsible for 85 to 90 percent of all industrial accidents. These arguments were based

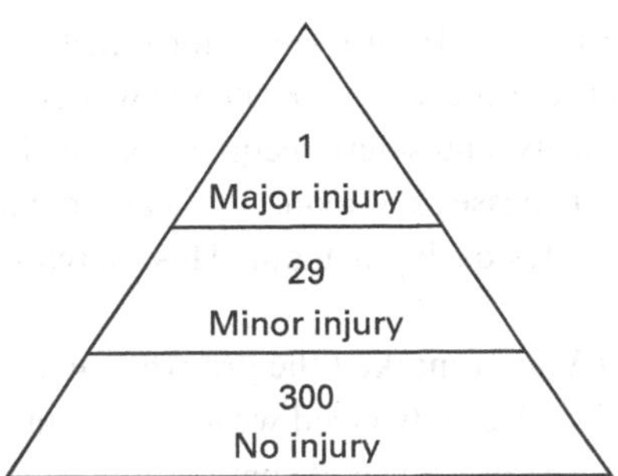

Figure 1.1
Heinrich's triangle.

on the fact that accident investigations concluded that workers were responsible for most accidents and that accidents were occurring despite the use of machine guards.

These arguments ignored the facts, of course, that (1) finding workers responsible for accidents was to the advantage of the companies conducting the investigations as that made them not responsible for the worker deaths and injuries, and that (2) many accidents such as slips and falls or injuries from falling objects or from lifting were not machine-related and thus could not be prevented by any type of guarding [330]. More recent data shows that humans are more likely to be blamed for accidents than are unsafe conditions, even when unsafe conditions make human error almost inevitable [217].

A more important question than assigning blame or cause may be how to eliminate accidents in the future. As Hansen wrote in 1915: "*Forgetfulness, for example, is not a crime deserving of capital punishment; it is one of the universal frailties of humanity. The problem is, therefore, to destroy as far as possible the interrelationship between safety and the universal shortcomings, which can be done by designing the safeguards on machines and equipment so that if a man's acts are essential to safety, it becomes mechanically necessary for him to perform this act before proceeding with his task*" [135, p. 14].

This design principle, technically called an *interlock*, is one of many today and is common to workplace and product safety. Safety design principles are described in more detail in chapter 11.

The Depression, followed by World War II, temporarily diverted attention away from designing safety as well as function into products. In the long run, however, World War II led to even greater safety efforts and the divergence of workplace safety and product/system safety.

Along with greatly increased war manufacturing—the greatest manufacturing effort of all time—came a tremendous increase in accident and injury rates. The rates were so bad that the Allies were actually in danger of losing the war because of them. More people were killed in industrial accidents than on the battlefield. As Ferry writes, "*Not so well known is that for a while, accidents in the workplace were nearly negating the increased production. We were losing twice as many aircraft in training accidents as in combat, worldwide. In nearly all theaters of operation, ground and air accidents were three times those of the combat losses*" [100].

With the realization that accidents at home were hurting military efforts, an enormous safety and health program was initiated, and thousands of people were trained for industrial safety work. We learned from those efforts that most accidents could be prevented and began to take appropriate action. When the war ended, health and safety activities no longer had a high priority, and safety records worsened again. As normal times and prosperity returned, many businesses and industries learned that safety was good business, and the accident and injury rates declined again. This increase in safety efforts, however, was not universal.

At the same time, World War II marked the era when extremely dangerous new technology started to be created such as advanced weapon systems, nuclear power, and highly dangerous chemicals. Increases in the use of some products, such as commercial aircraft, also contributed to public and governmental concern about product/system safety. It is at this point that workplace safety and product/system safety completely diverged, and almost no interaction between them has existed since.

Each of the industries creating the different types of products also developed their own approaches to ensuring safety, based on technological, social, and cultural differences. The result is described later in this chapter.

With this brief introduction to the history of safety completed, let's look at the state of workplace safety and then product/system safety today.

1.4 Workplace Safety Today: An Engineer's View

As seen in the previous historical narratives, workers before the industrial revolution were expected to provide their own tools, to understand the risks associated with their trade, and to accept personal responsibility for their own safety. This attitude was justified partly by the fact that workers devoted their entire careers to the manufacturing of one or two products [331]. They could, therefore, thoroughly understand their jobs and had control over how they performed their tasks.

The world has changed drastically since that time, although the same attitude about personal and workplace safety still exists today in many companies or industries. The workers are often blamed for any accidents although they no longer have the freedom to control their own work situations: for the most part, workers do not have total control over their tools, the way they do their jobs, and the environment in which they work. Because of these changes, responsibility for workplace safety has shifted, or should have shifted, to those creating the workplace environment and tools; that is, from the employee to the employer. The recognition of this shift did not come easily and still is not totally accepted everywhere.

In some cases, the government has had to step in to protect workers. As an example, in the nineteenth century, one of the main causes of injury and death to railroad workers occurred when workers were required to couple and decouple train cars. In the seven years between 1888 and 1894, 16,000 railroad workers were killed in coupling accidents and 170,000 were crippled. Managers claimed that such accidents were due only to worker error and negligence, and therefore nothing could be done aside from telling workers to be more careful. Because of the huge number of accidents, the government finally stepped in and required that automatic couplers be installed. As a result, fatalities and injuries dropped sharply. Three years after Congress acted on the problem, the June 1896 issue of *Scientific American* contained the following:

Few battles in history show so ghastly a fatality. A large percentage of these deaths were caused by the use of imperfect equipment by the railroad companies; twenty years ago it was practically demonstrated that cars could be automatically coupled, and that it was no longer necessary for a railroad employee to imperil his life by stepping between two cars about to be connected. In response to appeals from all over, the US Congress passed the Safety Appliance Act in March 1893. It has or will cost the railroads $50,000,000 to fully comply with the provisions of the law. Such progress has already been made that the death rate has dropped by 35 per cent.

In most industrialized countries today, employers are expected to provide a safe working environment and the necessary tools and equipment to maintain that environment.

One of the important features of workplace safety has been its focus on behaviorism. Behavioral psychology is a theory of learning, advanced by Watson, Skinner, and others, that emphasizes behaviors that are acquired through *conditioning*. Conditioning occurs through interaction with the environment. Strict behaviorists believe that any person can potentially be trained to perform any task, regardless of genetic background, personality traits, and internal thoughts (within the limits of their physical capabilities). It only requires the right conditioning. Simply put, strict behaviorists believe that all behaviors are the result of experience. Behaviorism was the dominant school of thought in psychology from about 1920 through the mid-1950s, when it began to be replaced by other theories. Behavioristic thinking, however, remains in workplace safety even today.

There are two types of conditioning, but the one most relevant to workplace safety is *operant conditioning*. Here learning is thought to occur through reinforcements and punishments. Basically, an association is made between a behavior and a consequence for that behavior. Most people have heard about Pavlov and his dogs. In the case of workplace safety, behaviors deemed to promote safety, such as wearing personal protective equipment like hardhats or following specified procedures, are rewarded and those considered to be unsafe, such as not wearing PPE or acting recklessly, are punished.

Accordingly, in workplace safety, *compliance* with specified procedures is usually emphasized. Compliance may be ensured through safety reviews and audits by industrial safety divisions or safety committees within a company. Lessons learned from accidents are incorporated into standards, both company and government standards, and much of the emphasis in the design of new plants and work rules is in implementing these standards. Often the standards are enforced by the government through occupational safety and health legislation and government agencies focused on this area.

As described in this chapter, workplace safety and product/system safety had many overlaps historically, but at least seventy years ago they diverged and have taken totally different paths since that time with almost no interaction between the two. There are different educational programs, different certification requirements, different oversight agencies with very different approaches to preventing losses, different regulations, different management structures within an organization—Environmental Health and Safety or EHS groups usually oversee workplace safety while product/system safety is part of product development and engineering—different conferences and textbooks/journals, and even different models of causality or beliefs about how and why accidents occur (which is discussed later). Most important, they use very different practices to achieve safety goals. Basically, they are two different fields with very little, if any, overlap.

A few companies mix workplace and product safety in one management structure, but the result has actually contributed to major accidents. Examples are the Texas City Refinery explosion and the *Deepwater Horizon* blowout. In most industries, the training, practices, and management of these two fields are very different, and there is recognition that experts in one are not experts in the other. Mixing them may lead to neither being done well. The two groups within a company usually have few or no interactions nor do they have common practices.

The introduction of more dangerous equipment in workplaces today, such as robots and autonomous vehicles, should have produced more integration of workplace and product/system safety approaches, but so far has primarily led to inappropriate workplace safety practices being applied to this new engineered product domain.

The primary focus of this book is on product or system safety engineering, so let's turn to examining how it has been practiced for the past seventy years, after splitting from workplace safety.

1.5 Product/System Safety Today

Engineers focus on eliminating, preventing, and responding to *hazards* rather than focusing on the behavior or actions of humans. Hazards are, informally, states of the system that can lead to losses, not the losses themselves. Losses are usually labeled as accidents, mishaps, or, in some industries, adverse events or incidents. More formal definitions can be found in chapter 3.

Product/system safety encompasses a wide variety of practices. To try to reduce it to one simple stereotypical approach would be misleading: there is no monolithic approach to our attempts to reduce losses. The approach to safety varies with each industry and sometimes each company in an industry. There are some variations between countries, but for the most part, the few geographic differences are swamped by those between industries. Those differences usually arise due to the nature of the hazards in those industries, but they also arise from different history and traditions in a particular industry and from its general safety culture.

Safety engineering or product safety is usually taught as part of a particular engineering discipline and not called out as a separate topic. It is based on the same scientific and mathematical foundations as other aspects of engineering: The procedures used to improve safety are those used in engineering, primarily modeling and analysis, but in this case applied to hazardous system states. The goal is to identify hazards and then to design, manufacture, operate, and manage systems such that the identified hazards are eliminated or mitigated to prevent losses. In addition, within engineering, the term "safety" is usually broadly defined to include not just human injury (as in workplace safety) but also damage to physical equipment—such as an aircraft or spacecraft—and even mission losses in situations where the mission itself is critical.

Despite the differences, some activities are common to most industries, although they may be performed very differently. These common practices include:

Hazard analysis Hazard analysis is used to identify hazardous states during system design and development and to identify how the system can get into the hazardous states. Many different techniques have been developed for performing hazard analysis, but all start from the hazards identified by the stakeholders. Using basic scientific knowledge about the technology involved along with what is known or assumed about how accidents occur in that industry, hazard analysis identifies the scenarios or behaviors of the system components that can lead to the hazardous states.

Design for safety The goal here is to use the information obtained from the hazard analysis to engineer or "design out" accidents before they occur; that is, to prevent hazards or, if not possible, to prevent hazards from leading to accidents or losses. This goal is accomplished, in general, by design processes that are unique to various industries and the characteristics of the hazards in those industries.

The highest priority when designing for safety is to create designs that cannot get into the identified hazardous states; that is, the potential for hazards is eliminated by, for example, substituting nonflammable materials for flammable ones if the identified hazard is fire. Achieving this goal may require unacceptable tradeoffs with some of the system goals. If hazards cannot be eliminated from the design, then (in priority order) an attempt is made to design to prevent hazardous states from occurring, to reduce their occurrence, and to respond if they do occur and move the system back into a safe or safer state. In the worst case, systems can be designed to reduce potential losses if hazards do occur, such as in the use of seat belts, sirens and alerts, or evacuation. Chapter 11 describes design for safety in more detail.

Human factors engineering and human-centered design In human factors engineering, psychological concepts are applied to engineering designs to prevent human errors and provide human operators with the ability to safely control the system. More focused human-centered design concepts started to be developed in the 1980s and were first applied in the aviation community. In this approach to engineering design, the role of humans in the control of systems is the focus from the beginning of the system concept development.

Operations Systems must not only be designed to be safe, but they must also be operated safely. Operational safety involves such activities as considering operability in the original design and then managing operations to ensure proper training of operators, identification and handling of leading and lagging indicators of risk, management of change procedures, maintenance procedures, and so on. Data collection and analysis during operations has played an important role in improving design and operational safety.

Management and policy Emphasis on the design of Safety Management Systems is a relatively recent emphasis in safety engineering, dating back to the middle of the last century. These efforts involve creating effective safety cultures, information systems, and safety management structures. Effective management of safety includes such things as setting policy and defining goals; planning tasks and procedures; defining responsibility, granting authority, and fixing accountability; documenting and tracking hazards and their resolution (establishing audit trails); maintaining information systems and documentation; establishing reporting and information channels; and so on.

Accident investigation and analysis Every industry investigates accidents, but that effort usually encompasses a small part of the overall safety engineering effort. Learning and continual improvement are, of course, necessary and important, and an accident provides important information about how to improve our prevention efforts. But the consequences of accidents in critical systems today are so great that most of the effort needs to be applied to preventing losses in the first place.

Regulation and licensing Regulation may involve rules enforced by an oversight agency, voluntary standards, or certification of new systems. Regulation usually involves some type of approval of new systems before they are allowed to be used. It also almost always includes oversight into the operation of the systems to ensure that assumptions about operation (such as maintenance assumptions) made during analysis, design, and certification of the system hold in the operational environment and that changes over time are not leading to increasing levels of risk. If dangerous changes and conditions are caught in time, accidents can be prevented. Examples of ways that oversight agencies collect information during operations include licensee event reports in nuclear power plants, aviation safety reporting systems, and auditing of airline and airport operations.

In summary, the primary emphasis in almost all product/system safety fields is on the engineering of systems to prevent accidents through design and analysis activities during product development and by controls and feedback during operations. Almost all industries and companies also investigate accidents—it would be irresponsible not to do so: the information obtained is invaluable in improving prevention activities and our basic engineering knowledge, but no industry relies solely on accident investigation to prevent losses.

Because technology and the basic hazards differ in different industries, it is not possible to describe one approach that is used in all of them. There are many more differences than similarities. A few examples of industry-specific approaches are provided here to give a more complete picture of how safety engineering has been done for the past century.

1.5.1 Commercial Aviation

Aircraft accident rates are very low. But this was not always the case. Aircraft accident rates peaked in the 1970s and have been driven down in the last fifty years, even with a dramatic increase in the number of flights. Aircraft accident rates were so high in the 1950s and 1960s that only 20 percent of Americans were willing to fly [104]. Bill Boeing realized that if he was to build a successful industry, accident rates needed to improve dramatically.

The ensuing effort has been remarkably successful. That success was a result of focusing on all aspects of reducing risk. Aircraft operate in widely differing and very challenging environments. At the same time, high risk is simply unacceptable. The approach used today for safety in commercial aircraft recognizes this reality.

Conservatism and a slow pace in basic design changes has allowed learning from experience and developing the science underlying aeronautics. The industry has emphasized the use of well-understood and debugged designs to maintain and enhance safety. Occasional major technological developments, such as the jet engine, the introduction of computers into control, and structures made of composites, introduced new hazards but also eliminated others. For example, jet engines are much simpler than older types of engines in terms of needing fewer controls and fewer components. In addition, they can sustain large amounts of damage before complete failure. Improved jet aircraft technology provided increased speed, efficiency, and safety. Also, improved cockpit interfaces and standardized training practices and operational procedures have had major influences on safety.

On balance, such developments usually reduced accident rates or at least particular types of accidents. Except for the disruptive effects of these new technologies, the hazards have not changed significantly. Until relatively recently, accidents were assumed to be caused by failures of the aircraft components or pilot errors. The pace of technological change is increasing, however, and new design features are being employed such as the use of fly-by-wire systems, automated flight controls, and some unstable aircraft features. The increased pace of change is partially driven by competitive pressures to introduce advanced control features and reduce fuel consumption. One result is that the causes of aviation accidents are also changing.

When a new aircraft design is created, it is analyzed to identify failure scenarios that could lead to a hazardous state and then those scenarios are used to guide the designers in attempting to eliminate losses. Because there is often no safe state to move the system into (in contrast to nuclear power where the nuclear reaction can be shut down), focus has been on preventing and tolerating failures and errors using:

- design integrity and quality
- component reliability enhancement
- safety margins or factors
- redundancy and backup systems
- design for checkability and inspectability
- error tolerance or designing such that common human errors are easily identified and corrected or do not have catastrophic results—for example, in the design of cockpit layouts and switches or the use of color coding or different shape connectors in wiring
- failure warning and indications to human operators as well as providing information about the current state of the aircraft
- failure containment, which involves isolation of components or subsystems so that failure in one will not affect others
- damage tolerance or designing so that surrounding failures are tolerated in case the failures cannot be contained
- designed failure paths when high energy failure cannot be contained or tolerated, but it can be directed to a safe path—an example is the use of structural fuses in pylons so that the engine will fall off before it damages the structure
- flight crew training and specified flight crew procedures to prevent common errors and for use in situations where there is not enough time or information for human operators to figure out what to do themselves
- incorporating human factors engineering as an important part of design, particularly the design of cockpits and of piloting and maintenance procedures

Beyond these general design strategies, there are also specific airworthiness criteria including the design of seat belts, oxygen masks and systems in case of rapid cabin depressurization, escape and evacuation procedures and equipment, life vests and rafts, and so on. Many of these designs have been improved after accidents showed some of the assumptions used in the design were flawed.

As underlying scientific knowledge about flight, materials, and propulsion was developed, this knowledge was used to design safety into new aircraft and to prevent physical failures and design errors.

In addition to basic design techniques, operability and safety during operations are major considerations in the design process, such as creating practical functional check procedures that can be used to determine a component's condition and its ability to perform its intended function. Such design for operability and maintainability includes periodic inspections, for example, of the aircraft structure and engine rotor disks for fatigue damage.

But the design of the aircraft itself is only a small part of the reason for low accident rates in this industry. Another important factor is government and international regulatory agencies as well as independent industry and user associations, such as airline pilot associations, the Flight Safety Foundation, the NASA aviation safety program, airline passenger associations, and so on. Nearly all aspects of operations are regulated and audited. Programs for reporting near misses and deviations from defined procedures have been enormously successful in getting feedback before accidents occur.

Investigation of accidents has been an important component of reducing accidents in commercial aviation, particularly now that accident rates have been driven down so low. Investigations are remarkably thorough with information gathered about every component of the accident. New components have been added to aircraft to assist in investigations, such as black box recorders, cockpit voice recorders, and other types of aircraft instrumentation.

There was a time, very early in the development of aviation, when the approach known as *fly-fix-fly* was common. But that was before the development of aerodynamics and the science behind flight. Before much is known scientifically about a new technology, the only way to learn is through experimentation. But it is not the primary approach today except for very new concepts. Such experimentation is not done while endangering the public's lives but instead may be done using test pilots and various types of experimental rigs, such as wind tunnels.

The effort put into investigation of accidents today is a result of the success of all the other things that have been done to increase safety in this industry. Accidents now occur because of changes in the design, operations, and environment of aircraft and when the assumptions underlying the original design and even oversight activities no longer hold. We can try to anticipate that the assumptions are changing, but when the change is gradual and not planned, we are often left to gather this information after a tragedy.

1.5.2 Nuclear Power

Nuclear power has in its short history developed a very different approach to safety, at least with respect to design, than that used in aviation. In addition to the general safety engineering problems of all new and potentially dangerous technologies, the nuclear power industry has a relatively unique problem with public relations and has had to put a great deal of energy into convincing government regulators and the public that the plants are safe in order to get approval to build and operate them. This requirement, in turn, has resulted in a greater emphasis on probabilistic methods of risk assessment: the time required for empirical evaluation and measurement of safety is orders of magnitude greater than is practical.

Another result is that nuclear engineering has, until recently, used a few designs for which a lot of past experience can be accumulated and has been very conservative about introducing new technology, such as digital systems. That conservatism, however, is giving way to greatly increased use of digital instrumentation and control and other new technology.

In an effort to promote the development of nuclear power in its early days and also because the hazards were only partly understood, the industry was exempted from the requirements of full third-party insurance in some countries, such as the Price-Anderson Act in the United States [375]. Instead, government regulation and certification were substituted as a means of enforcing safety practices in the industry.

The first nuclear power plant designs and sizes were also limited, although this has changed somewhat over the years as confidence in the designs and protection mechanisms has grown. The primary approach to safety in the nuclear industry is *defense-in-depth.* Defense-in-depth includes [277]:

- a succession of barriers to a propagation of malfunctions, including the fuel itself; the cladding in which the fuel is contained; the closed reactor coolant system; the reactor vessel; any additional biological shield (such as concrete); and the containment building (usually including elaborate spray and filter systems)
- primary engineered safety features to prevent any adverse consequences in the event of a malfunction
- careful design and construction, involving review and licensing in many stages
- training and licensing of operating personnel
- assurance that ultimate safety does not depend on correct personnel conduct in case of an accident
- secondary safety measures designed to mitigate the consequences of conceivable accidents
- oversight of operations and licensee event reports

Licensing is based on identification of the hazards, design to control these hazards under normal circumstances, and backup or shutdown systems that function in abnormal circumstances to further control the hazards. Most of the emphasis, with respect to safety, is placed on the shutdown system that is brought into operation if a hazard occurs. The backup system designs are based on the use of multiple independent barriers, a high degree of single element integrity for passive features, and the provision that no single failure of any active component will disable any barrier. Note that siting nuclear power plants in remote locations is a form of barrier in which the separation is enforced by physical isolation. Because of the difficulty in isolating plants, emergency planning has gotten more attention since the Three Mile Island and other early nuclear power plant accidents.

The dependence on defense-in-depth using barriers and reversion to a safe state, which does not usually exist in aviation and other fields, is unique to nuclear power. While specific details may differ between countries, redundancy and high-integrity parts are commonly used, and engineers have emphasized high component reliability and protection against component failure. Such component failures can lead to failed barriers and inability to reach a safe state.

Because many people outside the United States are most familiar with nuclear power as the most common, highly safety-critical industry in their region, there have sometimes been assumptions that this is the only approach to safety engineering, but the design and general safety engineering approaches used in aerospace and defense are very different.

Using the nuclear power defense-in-depth approach to safety, an accident requires a disturbance in the process, a protection system that fails, and inadequate or failed physical barriers. These events are assumed to be statistically independent because of differences in their underlying physical principles: A very low calculated probability of an accident can be obtained as a result of the independence assumption [318]. This assumption is debatable, however.

Recovery after a problem has occurred depends on the reliability and availability of the shutdown systems and the physical barriers. A major emphasis in building such systems, then, is on how to increase this reliability, usually through redundancy of some sort.

Certification of nuclear power plants has emphasized probabilistic risk assessment that includes estimating the reliability of the barriers and protection systems. Because empirical measurement of probability is not practical for nuclear power designs, most approaches build a model of accident events and construct an analytical risk assessment based on the model. Such models include the events leading up to an uncontrolled release as well as data on factors relating to the potential consequences of the release, such as weather conditions and population density. With the increasing use of nonstochastic components such as computers and digital systems in nuclear power plants, such probabilistic assessments may have decreasing relevance.

In addition to probabilistic risk assessment in certification of nuclear plants, operational concerns are addressed through extensive incident and error reporting, an emphasis on creating a strong safety culture, external oversight and regulation, and in-depth incident investigation.

1.5.3 The Chemical Industry

The approach to safety in the chemical industry was originally driven by insurance needs. The term usually used in the industry, *loss prevention*, reflects these origins.

In chemical and petrochemicals, the three major hazards are fire, explosion, and toxic release. Of these three, fire is the most common; explosions are less common but cause greater losses; and toxic release is relatively rare but has been the cause of the largest losses. Because loss of containment is a precursor for all of these hazards, much of the emphasis in loss prevention is, like the nuclear industry (which is usually considered part of the process industry), on avoiding the escape of explosive or toxic materials through leaks, ruptures, explosions, and so on.

The three major hazards related to the chemical industry have remained virtually unchanged in their nature for decades. Design and operating procedures to eliminate or control these hazards have evolved and been incorporated into codes and standards produced by the chemical industry professional societies and others.

This approach sufficed before World War II because the industry operated on a relatively small scale and development was slow enough to learn by experience. After World War II, however, the chemical and petrochemical industries began to grow in complexity, size, and new technology at a tremendous rate. The potential for major hazards grew at a corresponding rate. The operation of chemical plants also has increased in difficulty, and startup and shutdown have especially become complex and dangerous.

The effect of the changes has been to increase the consequences of accidents, to increase environmental concerns such as pollution and noise, to make the control of hazards more difficult, and to reduce the opportunity to learn by trial and error. At the same time, the social context has been changing. In the past, safety efforts in the process industries were primarily voluntary and based on self-interest and economic considerations. However, pollution has become of increasing concern to the public and to the government. Major accidents, such as the *Deepwater Horizon* oil spill, have drawn enormous publicity and generated political pressure for legislation to prevent similar accidents in the future. Most of this legislation requires a qualitative hazard analysis, including (1) identification of the most serious hazards and their contributing factors and (2) modeling of the most significant accident potentials [335].

These factors led the industry to increase its proactive efforts to analyze potential hazards more carefully and to reduce emissions and noise. The accidental release of MIC (methyl isocyanate) at Bhopal, however, demonstrated that technical hazard analysis by itself is not enough and that management practices may be even more important.

Applying hazard analysis in the chemical industry has special complications compared to other industries, which makes the modeling of causal sequences especially difficult [362]. Although the industry does use some standard reliability and hazard analysis techniques, the unique aspects of the industry have led to the development of industry-specific techniques. For example, the hazardous features of many chemicals are well understood and have been cataloged in indexes that are used in the design and operations of plants.

A hazard analysis technique developed for and used primarily in the chemical and petrochemical industries is called HAZOP (HAZard and Operability Analysis). This technique is a systematic approach to examining each item in a plant to determine the causes and consequences of deviations from normal plant operating conditions. The information about hazards obtained in this study is used to make changes in design, operating, and maintenance procedures. HAZOP dates from the mid-1960s. Newer hazard analysis techniques can be performed earlier, before the design is complete, and the results used to build simpler, cheaper, and safer plants by avoiding the use of hazardous materials, using fewer of them, or using them at lower temperatures or pressures [217].

Accident investigation is of course done if losses occur, but proactive analysis and design efforts are much more heavily emphasized.

1.5.4 Defense and "System Safety"

After World War II, the United States Department of Defense created a unique approach to safety that they called *System Safety* [263]. The United States started to build nuclear weapons, early warning and launch on warning systems, and intercontinental ballistic missiles. The consequences of accidents in these systems were so enormous that safety had a very high priority. Some defense systems, such as atomic weapons, can potentially have such catastrophic consequences that they must not be allowed to behave in an unsafe fashion under any circumstances. Accidents were unthinkable. At the same time, the complexity of these systems was orders of magnitude greater than previously attempted and the inadequacy of previous approaches to safety engineering was dramatically illustrated in close calls. Something new was needed.

Much of the early development of System Safety as a separate discipline in the defense industry began with flight engineers after World War II. The Air Force had long had problems with too many aircraft accidents. For example, from 1952 to 1966, it lost 7,715 aircraft, in which 8,547 people, including 3,822 pilots, were killed [133]. Most of these accidents were blamed on pilots. Some industry aeronautical engineers, however, did not believe the cause was so simple. They argued that safety must be designed and built into aircraft just as are performance, stability, and structural integrity [263; 357]. This was the beginning of a unique systems approach to safety.

Seminars were conducted by the Flight Safety Foundation, headed by Jerome Lederer (who would later head the NASA Apollo safety program), that brought together engineering, operations, and management personnel. It was in 1954, at one of these seminars, that the term *System Safety* may have first been used—in a paper by one of the aviation safety pioneers, C. O. Miller. Around the same time, the Air Force began holding symposiums that fostered a professional approach to safety in propulsion, electrical, flight control, and other aircraft subsystems, but they did not at first treat safety as a system problem.

When the Air Force began to develop ICBMs (intercontinental ballistic missiles), there were no pilots to blame for accidents, yet the liquid-propellant missiles blew up frequently and with devastating results. The missiles used cryogenic liquids with temperatures as low as -320°F and pressurized gases at 6,000 lbs per square inch, which meant the potential safety problems could not be ignored: the highly toxic and reactive propellants were sometimes more lethal than the poison gases used in World War II, more violently destructive than many explosives, and more corrosive than most materials used in industrial processes [133]. The Department of Defense and the Atomic Energy Commission (AEC) were also facing the problems of building and handling nuclear weapons and finding it necessary to establish rigid controls and requirements on nuclear materials and weapon design. Most of the development of System Safety as a unique discipline grew out of these nuclear weapon and ballistic missile programs.

In that same period, the AEC (and later the Nuclear Regulatory Commission or NRC) was engaged in a public debate about the safety of nuclear power. Similarly, civil aviation was attempting to reduce accidents in order to convince a skeptical public to fly, and the chemical industry was coping with larger plants and increasingly lethal chemicals. These parallel activities resulted in different approaches to handling safety issues, as described previously. Little cross-fertilization between these approaches occurred, partly because many of the features of System Safety were not known outside the small group of engineers and companies that worked on these very secret Cold War systems. The System Safety approach developed for them is no longer secret, but it is largely unknown outside the United States or even by many safety practitioners in other industries within the United States.

In the fifties, when the first missile systems such as the Atlas and Titan ICBMs were being developed, standard safety approaches were used. Intense political pressure was focused on building a nuclear warhead with delivery capability as a deterrent to nuclear war. In an attempt to shorten the time between initial concept definition and operational status, a concurrent engineering approach was developed and adopted. In this approach, the missiles and facilities in which they were to be maintained ready for launch were built at the same time that tests of the missiles and training of personnel were proceeding. The Air Force recognized that this approach would lead to many modifications and retrofits that would cost more money, but with nuclear war viewed as the alternative, they concluded that additional money was a cheap way to buy time. A tremendous effort was exerted to make the concurrent approach work [331].

On these first missile projects, system safety was not identified and assigned as a specific responsibility. Instead, as was usual at the time, each designer, manager, and engineer was assigned responsibility for safety. These projects, however, involved advanced technology and much greater complexity than had previously been attempted, and the drawbacks of the standard approach to safety became clear when many interface problems went unnoticed until it was too late.

Within eighteen months after the fleet of seventy-one Atlas F missiles became operational, four blew up in their silos during operational testing. The missiles also had an extremely low launch success rate. An Air Force manual describes several of these accidents:

> *An ICBM silo was destroyed because the counterweights, used to balance the silo elevator on the way up and down in the silo, were designed with consideration only to raising a fueled missile to the surface for firing. There was no consideration that, when you were not firing in anger, you had to bring the fueled missile back down to defuel. The first operation with a fueled missile was nearly successful. The drive mechanism held it for all but the last five feet when gravity took over and the missile dropped back. Very suddenly, the 40-foot diameter silo was altered to about 100-foot diameter.*
>
> *During operational tests on another silo, the decision was made to continue a test against the safety engineer's advice when all indications were that, because of the high oxygen concentrations in the silo, a catastrophe was imminent. The resulting fire destroyed a missile and caused extensive silo damage. In another accident, five people were killed when a single-point of failure in a hydraulic system caused a 120-ton door to fall.* [13]

In general, launch failures were caused by inadequate requirements, design flaws, construction errors (for example, reversed gyros and reversed electrical plugs), bypass of procedural steps, and management decisions to continue in spite of contrary indications because of schedule pressures [13].

Not only were the losses themselves costly, but the resulting investigations detected serious deficiencies in the systems that would require extensive modifications to correct. In fact, the cost of the modifications would have been so high that a decision was made to retire the entire weapon system and accelerate deployment of the Minuteman missile system. Thus, a major weapon system, originally designed to be used for a minimum of ten years, was in service for less than two years primarily because of safety deficiencies [331].

When the early aerospace accidents were investigated, it became apparent that the causes of a large percentage of them could be traced to deficiencies in design, operations, and management. The emphasis for such systems began to be on building them to be safe from the beginning. In this approach, hazards are identified in the early stages of system development and then eliminated in the system design.

A fly-fix-fly approach was clearly not adequate. In this approach, investigations are conducted to resurrect the causes of accidents, action is taken to prevent or minimize the recurrence of accidents with the same cause, and eventually, these preventive actions are incorporated into standards, codes of practice, and regulations. Although the fly-fix-fly approach was effective in reducing the repetition of accidents with identical causes, it became clear to the Department of Defense (DoD), and later others, that it was too costly and, in the case of nuclear weapons, unacceptable. This recognition led to the adoption of system safety approaches sixty to seventy years ago to try to prevent accidents before they occur the first time.

The first military specification on System Safety was published by the Air Force Ballistic Systems Division in 1962, and the Minuteman ICBM became the first weapon system to have a contractual, formal, disciplined System Safety program. From that time on, System Safety received increasing attention, especially in the Air Force missile programs where testing was limited and accident consequences serious.

The Army soon adopted the System Safety programs developed by the Air Force because of the many personnel it was losing in helicopter accidents, and the Navy followed suit. In 1966, the DoD issued a single directive requiring System Safety programs on all development or modification contracts.

At first, there were few techniques that could be used on these complex systems. But step-by-step scientific, technical, and management techniques were developed or adapted from other activities. Contractors were required to establish and maintain a System Safety program that was planned and integrated into all phases of system development, production, and operation. A formal plan was required that ensured that

1. safety, consistent with mission requirements, is designed into the system;
2. hazards associated with the system, each of the subsystems, and the equipment are identified, evaluated, and eliminated or controlled to an acceptable level;
3. control over hazards that cannot be eliminated is established to protect personnel, equipment, and property;
4. minimum risk is involved in the acceptance and use of new materials and new production and testing techniques;
5. retrofit actions required to improve safety are minimized through the timely inclusion of safety engineering activities during the acquisition of a system; and
6. the historical safety data generated by similar programs are considered and used where appropriate.

To summarize all of this, the primary concern of System Safety is the management of hazards: their identification, evaluation, elimination, and control through analysis, design, and management procedures. Losses in general, not just human death or injury, are considered. Such losses may include destruction of property, loss of mission, and environmental harm. The key point is that the loss is considered serious enough that time, effort, and resources must be put into preventing it. How much of an investment is considered worthwhile to avoid the loss or its effects will depend on social, political, and economic factors.

System Safety activities start in the earliest concept development stages of a project and continue through design, production, testing, operational use, and disposal. The primary emphasis is on the early identification and classification of hazards so that corrective action can be taken to eliminate or minimize their impact before final design decisions are made.

To understand the unique aspects of the System Safety approach and differentiate it from the other approaches to safety developed in parallel but independently for such industries as civil aviation and nuclear power, a few basic unique concepts can be identified:

- *System Safety emphasizes building in safety, not adding it on to a completed design or trying to assure it after the design is complete.*

  Seventy to 90 percent of the design decisions that affect safety will be made in the concept development project phase [109]. The degree to which it is economically feasible to eliminate a hazard rather than to control it depends on the stage in system development at which the hazard is identified and considered. Early integration of safety considerations into the system development process allows maximum safety with minimal negative impacts. The alternative is to design the system or product, identify the hazards, and then add on protective equipment to control the hazards when they occur—which is usually more expensive and less effective. Waiting until operations and then expecting human operators to deal with hazards is the most dangerous approach.

- *System Safety deals with systems as a whole rather than with subsystems or components.*

  Safety is a system property, not a component property. The hazards associated with an individual component, such as sharp edges, are very different than the hazards that arise when all the components operate together.

- *System Safety takes a larger view of hazards than just failures.*

  Serious accidents can occur while system components are all functioning exactly as specified—that is, without failure. If failures only are considered in a hazard analysis, many potential accident scenarios will be missed. In addition, the engineering approaches to preventing failures (increasing reliability) and preventing hazards (increasing safety) are different and sometimes conflict.

- *System Safety emphasizes analysis rather than past experience and standards.*

  Standards and codes of practice incorporate experience and knowledge about how to reduce hazards, usually accumulated over long periods of time and resulting from previous mistakes. While such standards and learning from experience are essential in all aspects of engineering, including safety, the pace of change today does not allow for such experience to accumulate and for proven designs to be used. System Safety analysis attempts to anticipate and prevent accidents and near-misses before they occur.

- *System Safety emphasizes qualitative rather than quantitative approaches.*

  System Safety places major emphasis on identifying hazards as early as possible in the design stage and then designing to eliminate or control those hazards. In these early stages, quantitative information usually does not exist. In addition, our technology and innovations are proceeding so fast that historical information may not exist or may not be useful. Some components of high technology systems may be new or may not have been produced and used in sufficient quantity to provide an accurate probabilistic history of failure.

The accuracy of quantitative analyses is also questionable. The majority of factors in accidents cannot be evaluated in numerical terms, and those factors that can be quantified will often receive undue weighting in decisions based on absolute measures while those that cannot be quantified are ignored.

Finally, quantitative evaluations usually are based on unrealistic assumptions that are often unstated, such as accidents are caused by failures; failures are random; testing is perfect; failures and errors are statistically independent; and the system is designed, constructed, operated, maintained, and managed according to good engineering standards. Surprisingly few scientific experiments have been performed to determine the accuracy of probabilistic risk assessment, particularly considering the length of time they have been used. The results of the few scientific experiments that have been done, however, have not been encouraging.

- *System Safety recognizes the importance of tradeoffs and conflicts in system design.*

  Nothing is absolutely safe, and safety is not the only or usually even the primary goal in building systems. Most of the time, safety acts as a constraint on how the system goals (mission) can be achieved and on the possible system designs. Safety may conflict with other goals, such as operational effectiveness, performance, ease of use, time, and cost. System Safety techniques focus on providing information for decision-making about risk management tradeoffs.

- *System Safety is more than just system engineering.*

  System Safety concerns extend beyond the traditional boundaries of engineering to include such things as political and social processes, the interests and attitudes of management, attitudes and motivations of designers and operators, human factors and cognitive psychology, the effects of the legal system on accident investigations and free exchange of information, certification and licensing of critical employees and systems, and public sentiment [205].

The systems approach to safety engineering described in this book and in *Engineering a Safer World* (217) incorporates and extends this traditional defense System Safety approach.

1.5.5 SUBSAFE: The US Nuclear Submarine Safety Program

While US nuclear submarines are engineered using standard Department of Defense System Safety engineering, a special program, called SUBSAFE, to deal with some aspects of nuclear submarine safety was created after the loss of the USS Thresher in 1963. Before that time, the United States lost one submarine (in peacetime) about every two to three years. After SUBSAFE was created, there has not been a loss of a submarine that was in the SUBSAFE program in the past sixty years. Only the highlights are described here. For a more detailed description, see Leveson [217].

SUBSAFE focuses on one hazard: the inability of critical systems to control and recover from flooding, namely, loss of hull integrity. The other hazards are handled using standard System Safety as described previously.

The emphasis in SUBSAFE is on

- establishing and maintaining a strong safety culture;
- sophisticated risk management without the use of untestable or unverifiable arguments based on probabilistic risk assessment;
- certification and performance audits;
- sophisticated safety management;
- education and training;
- specification and documentation; and
- continuous improvement.

1.5.6 Astronautics and Space

The space program was the second major application area after defense to apply the System Safety approach, developed by the US Air Force, in a disciplined fashion. Until the *Apollo 204* launch rehearsal fire in 1967 at Cape Kennedy, in which three astronauts were killed, NASA safety efforts had focused on workplace safety. The accident woke up NASA, and they developed policies and procedures that became models for civilian aerospace activities [264]. Jerome Lederer, one of the pioneers in aviation safety, was hired to head manned space flight safety and, later, all NASA safety efforts. Through his efforts, an extensive program of System Safety was set up for space projects, much of it patterned after the Air Force and DoD programs. Many of the same engineers and companies that had established formal System Safety programs for DoD contracts also were involved in space programs, and the technology and management activities were transferred to these new applications.

More recently, particularly after the *Columbia* Space Shuttle loss, NASA hired safety engineers from the nuclear power community and many of the System Safety approaches and standards were replaced with the probabilistic risk assessment approach used in nuclear power. The creation of new space companies without extensive prior experience in defense and aerospace systems has accelerated the changes in the NASA flight safety program.

1.5.7 Healthcare/Hospital Safety

Health care is, in many ways, closer to a service than a product. Much of safety in health care involves the treatment of patients, not the engineering of medical equipment, although engineered medical devices are playing an increasing role in health care. There are also special factors such as new hazards and diseases appearing all the time.

Workplace safety and patient safety are more closely intertwined in health care than in other industries, although there are still significant differences. While it is normal for goals and hazards to conflict, health care is somewhat unique in that many of the hazards—and thus constraints on how the goals can be achieved—conflict. That often makes finding acceptable solutions to safety problems more difficult. For example, consider the following two hazards:

1. Patients get a treatment that negatively impacts their health.
2. Patients do not get the treatment they need, which results in death or serious negative health impacts.

There are times when taking extraordinary risks may be justified in this context where either choice may involve serious losses. At the same time, much effort is spent in preventing unnecessary losses.

Other unique factors in health care are that it is information-intensive and human-intensive, complex, imprecise, interdisciplinary and constantly changing. Drucker says that "*healthcare institutions are complex, barely manageable places. . . . Large healthcare institutions may be the most complex organizations in human history*" [83].

An engineering approach to patient safety has never been emphasized. Patients are not "engineered," and important biological and health information may be only partially known. But there are lots of engineered or designed components and processes in hospitals and health care facilities including electronic health records; medical devices; pharmacy procedures; and special procedures used to eliminate or control known hazards such as patient handoffs, communication protocols, checklists, and identification devices such as RFID (Radio-Frequency Identification). Procedures such as hand washing are used to control infections, while other interventions are used to reduce wrong site surgery, and so on. In recent years, there has been a tremendous increase in technology in health care. There are also management structures in most hospitals to manage quality and safety, as well as a large number of patient safety organizations devoted to promoting safety by designing healthcare systems to prevent *adverse events*, the common term for accidents or mishaps in health care.

Safety and resilience in medical care and hospitals clearly depend on more than the behavior of frontline workers including individual doctors, nurses, and technicians, which has often been the emphasis in healthcare safety. Design of processes, facilities, and equipment are all important as well as training of personnel, the hospital culture, and management structures for safety and quality in hospitals.

One limitation of the current approaches in healthcare and hospital safety is that they lack a holistic, systems standpoint, and the attempts to improve safety, while sincere, have been largely piecemeal and disjointed. Like the proverbial blind men and the elephant, each is focused on one part of the elephant but misses the whole. Too much emphasis and responsibility have been placed on individual people, such as doctors and nurses, who have only limited control over the operations of the hospital and over the healthcare system as a whole. A more comprehensive approach could be achieved by using a more holistic systems approach.

1.6 Summary

The emphasis on designing safety into systems and products from the beginning is at least one hundred years old. The urgency increased, however, after World War II as companies and industries found themselves faced with similar problems: that is, products and manufacturing facilities that are becoming increasingly complex and increasingly dangerous. Liability concerns have soared. Each industry has created its own approach to safety engineering that reflects the specific hazards and culture involved. But all are now facing similar new challenges in dealing with rapidly increasing complexity in basic designs and technology and with social pressures arising from the growing environmental and human costs at stake.

Exercises

1. Look up the history of safety in your particular field or one that interests you. What are the major changes in technology in that field for the past fifty years? What changes have occurred in the way safety is handled over this time?
2. For your industry or for a specific organization or company, investigate the approach used in safety engineering with respect to: hazard analysis, design for safety practices, human factors in engineering, safety during operation, management and safety policy, accident investigation and analysis, regulation, and licensing. How is workplace safety treated? Compare it to the process for handling engineered systems safety in your industry or company/organization.
3. Have you ever worked on a safety-critical system project? If so, how was safety handled? Do you think it was handled well? If so, why? If not, what would you change?
4. What are some of the major changes in technology in your industry in the past fifty years? Has the approach used in safety changed during this time? How is new technology, such as software, handled in system safety activities in your company or industry? Is it treated the same as hardware or differently? What about human factors? How has complexity increased (if it has) in your industry over the past fifty years?
5. Are there assumptions of the traditional approaches to safety that you think are no longer true or additional goals that might be required for an improved approach to safety today?
6. Look up Environmental Health and Safety on the web or look at how your company or industry regulatory agencies define it. What types of activities are included? Then do the same for system safety or safety engineering (or a regulatory agency that oversees it). What are the differences in terms of what is considered and how it is handled?
7. Look at the OSHA, FAA, and FDA websites and compare how these agencies control safety. What are the differences? Similarities?
8. Look up what are considered the main causes of accidents in your industry for workplace safety and then for safety engineering. Are there differences? Similarities?
9. Does your industry or a specific product or service that you have worked with require liability insurance? How does the insurance company decide whether they will insure the product or service? Do they impose requirements on the manufacturer?

2

Risk in Modern Society

Living in a technological society is like riding a bucking bronco. I don't believe we can afford to get off, and I doubt that someone will magically appear who can lead it about on a leash. The question is: how do we become better bronco busters?

—William Ruckelshaus,
Risk in a Free Society [339, p. 159]

My company has had a safety program for 150 years. The program was instituted as a result of a French law requiring an explosives manufacturer to live on the premises with his family.

—Crawford Greenwalt,
former president of Dupont
quoted in Johnson [166]

Risk is not a new problem. Humans have always had to face risk from their environment, although the risks have changed as society and the natural environment have changed. In the past (and currently in rural areas of developing countries), the greatest concerns were natural geological and climatological disasters such as floods, drought, earthquakes, and tropical storms.

Today, industrialization is adding new human-created hazards to those rooted in nature as well as increasing some environmental hazards through climate change. Damage to our ecosystem is difficult to assess, but danger signals include a high rate of species extinction and high concentrations of toxic chemicals in the environment. Although industrialized nations have used technology to control many natural hazards, a strict distinction between natural and man-made disasters is an oversimplification. Human tampering with the environment has exacerbated and sometimes caused natural disasters, such as the destruction of watersheds leading to flooding or the release of chemicals into the atmosphere affecting both climate and crops. Flood damage in the United States, for example, has increased as expenditures on flood control have increased [185]. Areas prone to flooding were not developed until we introduced flood prevention measures. When an exceptional flood overwhelms these measures or they do not achieve their design goals, the damage is much worse than it would have been before prevention measures were introduced.

Similarly, the amount of damage caused by technological accidents may be affected by natural phenomena. The result of an accidental release of radiation or chemicals, for example, may be dependent on weather factors such as wind direction and strength or population distribution.

Thus, the distinction between natural and technological hazards is often useful but is not completely accurate. In fact, all hazards are affected by complex interactions among technological, ecological, sociopolitical, and cultural systems [39; 317; 379]. Attempts to control risk by treating it simply as a technical problem or only as a social issue are doomed to fail or to be less effective than possible.

Our attitudes toward risk and our approaches to dealing with it have changed significantly over time, as described in the previous chapter. To discover effective solutions to these problems, we must understand the factors underlying the risks we face and identify the most effective ways to deal with risk in general. This book examines what is currently known and used to control technologically related risks in the complex systems that characterize modern society. But it is not possible to understand how and, particularly, why we tackle risk the way we do without considering our general attitudes toward risk and how risks have changed over time. Then we can better determine how to tackle risk in the future.

2.1 Changing Attitudes toward Risk

All human activity involves risk; there is no such thing as a risk-free life. Safety matches and safety razors, for example, are not *safe*, only *safer* than their alternatives; they present a reduced level of risk while preserving the benefits of the devices they replace. No aircraft could fly, no automobile could move, and no ship could be put out to sea if all hazards had to be eliminated first [131].

Progress demands taking some risks, and despite great advances in technology, we are unable to eliminate risk altogether. However, much more so now than in the past, humans are demanding that risks be known and controlled to the extent possible and practical. Societies are recognizing the right of workers, consumers, and the general public to know what risks they face and are making worker and public safety a responsibility of the employer and manufacturer [115; 206].

The shift from purely personal to organizational or public responsibility for risk is a recent phenomenon, as described in the previous chapter. Until the early part of the last century, workers were expected to provide their own tools, to understand the risks associated with their trade, and to accept personal responsibility for their own safety. This attitude was justified partly by the fact that workers devoted their entire careers to the manufacturing of one or two products; they could thoroughly understand their jobs and had control over the way they performed their tasks [331].

Today, workers are more at the mercy of their employers in terms of safety, and, accordingly, responsibility has shifted from the employee to the employer. In most industrialized countries, employers are expected to provide a safe working environment and the necessary tools and equipment to maintain that environment. In addition, changes in legal liability and responsibility have led to product safety programs that are concerned with consumer safety as well as employee safety.

In matters of risk, today's complex, technological society requires that the general public place its trust in expert knowledge [40]. Accordingly, responsibility for detection of and protection from hazards has been transferred from the public to the state, corporate management, engineers, safety experts, and other professionals. Complete abdication of personal responsibility, however, is not always wise. In some instances—such as the Bhopal accident—the public has completely trusted others to plan for and respond effectively to an emergency, with tragic results. The Bhopal Union Carbide plant was run in a way that almost guaranteed a serious accident would occur (see appendix C). In addition, emergency and evacuation planning, training, and equipment were inadequate. The surrounding population was not warned of the hazards, nor were they told—either before or during the chemical release—of the simple measures, such as closing their eyes and putting wet cloths over their faces, that could have saved their lives. Such incidents have aroused the public to become more involved in risks they face.

In turn, public involvement in issues that past generations took for granted—such as the hazards related to medicine, transportation, and industry—has led to government regulation and the creation of public interest groups to control hazards that were previously tolerated [137; 206]. System safety engineers, who used to focus on uncontrolled-energy accidents such as explosions or the inadvertent firing of a weapon, are now being asked to control new hazards such as air and ground pollution and to ensure that the systems we build do not produce, contain, or decompose into hazardous materials.

In writing about the Bhopal tragedy, Bogard expresses this new attitude:

> *We are not safe from the risks posed by hazardous technologies, and any choice of technology carries with it possible worst-case scenarios that we must take into account in any implementation decision. The public has the right to know precisely what these worst-case scenarios are and participate in all decisions that directly or indirectly affect their future health and well-being. In many cases, we must accept the fact that the result of employing such criteria may be a decision to forego the implementation of a hazardous technology altogether.* [39, p. 109]

Increased regional and national concern about safety has expanded in the past decade to include international issues. The greenhouse effect, acid rain, and accidents such as the release of radiation at Chernobyl do not recognize national borders. The global economy and the vast potential destructiveness of some of our technological achievements are forcing us to recognize that risk can have international implications and that control of risk requires cooperative approaches.

Because risk reduction can be expensive and often requires tradeoffs with other desirable goals, it is important to consider whether the increased public and governmental concern about technological risk is justified and whether engineers should be worrying about these problems at all. Is risk increasing in our modern society as a result of new technological achievements, or, on the other hand, are we experiencing a new and unjustified form of Luddism?

2.2 Changing Risk Factors

The major changes occurring in the post–World War II era make most long-term historical risk data inapplicable to today's world. Past experience does not allow us to predict the future when the risk factors in the present and future differ from those in the past. Examining these changes will help us understand the problems we face.

Risk is usually defined as the combination of the *likelihood* of an accident and the *severity* of the potential consequences. Risk increases if either the likelihood or the magnitude of loss increases, as long as the other component does not decrease proportionally. Different factors may affect these two components of risk.

Some factors that are particularly relevant today are the appearance of new hazards and the increasing complexity, exposure, energy, automation, centralization, scale, and pace of technological change in the systems we are attempting to build.

2.2.1 The Appearance of New Hazards

Before the Industrial Revolution, accidents were the result of natural causes or involved a few relatively well-understood, simple technological devices. In the twentieth century, scientific and technological advances have reduced or eliminated many risks that were once commonplace. Modern medicine, for example, has provided cures for previously fatal diseases and eliminated some scourges, such as smallpox, altogether.

At the same time, science and technology have also created new hazards. Misuse and overuse of antibiotics has given rise to resistant microbes. Children no longer work in coal mines or as chimney sweeps, but they are now exposed to man-made chemicals and pesticides in their food or water and to increased environmental pollution [57]. The harnessing of the atom has increased the potential for death and injury from radiation exposure.

Some new hazards are easier to control than those they have replaced. More often, the new hazards are more pervasive and harder to find and eliminate than the ones we eliminated or reduced in the past [57]. In addition, we have no previous experience to guide us in handling these new hazards. Much of what has been learned from past accidents is passed down through codes and standards of practice. But appropriate codes and standards have not yet been developed for many new engineering specializations and technologies. Sometimes lessons learned over centuries are lost when older technologies are replaced by newer ones, for example when digital computers are substituted for mechanical devices.

Many of the approaches that worked on the simpler technologies of the past—such as replication of components to protect against individual component failure—are ineffective in controlling today's complex risks. Although redundancy provides protection against accidents caused by individual component failure, it is not effective against hazards that arise from the interactions among components in the increasingly complex systems being engineered today. In fact, redundancy may increase complexity to the point where the redundancy itself contributes to accidents.

2.2.2 Increasing Complexity

Many of the new hazards are related to increased complexity (both product and process) in the systems we are building. Not only are new hazards created by the complexity, but the complexity itself makes identifying them more difficult. In this chapter the term complexity is used informally, but it is more carefully defined in chapter 3.

High-technology systems are often made up of networks of closely related subsystems. Conditions leading to hazards emerge in the interfaces between subsystems, and disturbances progress from one component to another. As an example of this increasingly common form of complexity, modern petrochemical plants often combine several separate chemical processes into one continuous production, without the intermediate storage that would decouple the subsystems [315].

In fact, analyses of major industrial accidents invariably reveal highly complex sequences of events leading up to accidents, rather than single component failures. Whereas before World War II, component failure was cited as the major factor in accidents, since that time accidents are increasingly resulting from dangerous design characteristics and interactions among components [133; 217].

The operation of some systems today is so complex that it defies the understanding of all but a few experts. Increased complexity makes it difficult for the designer to consider all the hazards, or even the most important ones, or for the operators to handle all normal and abnormal situations and disturbances safely [360].

Not only does functional complexity make the designer's task more difficult, but the complexity and scope of the projects require large numbers of people and teams to work together. The development of a military jet today, for example, may involve eight to ten thousand engineers. The anonymity of team projects dilutes individual responsibility [206]. Moreover, many new specializations do not have standards of individual responsibility and ethics that are as well developed as those of older professions.

Kletz points out the paradox that people are willing to spend money on complexity but not on simplicity [188]. Consider the following accident that occurred in a British chemical plant. In this plant, a pump and various pipelines had several different uses, which included transferring methanol from a road tanker to storage, charging it in the plant, and moving recovered methanol back from the plant. A computer set the various valves, monitored their positions, and switched the transfer pump on and off. On this particular occasion, a tank truck was being emptied:

> *The pump had been started from the control panel but had been stopped by means of a local button. The next job was to transfer some methanol from storage to the plant. The computer set the valves, but as the pump had been stopped manually it had to be started manually. When the transfer was complete the computer told the pump to stop, but because it had been started manually it did not stop and a spillage occurred.* [189, p. 225]

In this case, a simpler design—independent pipelines for different functions, which were actually installed after the spill—makes errors much less likely and may not be more expensive over the lifetime of the equipment.

Computers often allow more interactive, tightly coupled, and error-prone designs to be built, and thus may encourage the introduction of unnecessary and dangerous complexity. Kletz notes that *"programmable electronic systems have not introduced new forms of error, but by increasing the complexity of the processes that can be controlled, they have increased the scope for the introduction of conventional errors"* [189].

Brooks distinguishes between accidental and necessary complexity [53]. Accidental complexity is added when solving a problem and therefore theoretically can be removed. Necessary or essential complexity is in the problem itself and cannot be removed through changing the solution method. It is easy to add accidental complexity when using computers and software to control dangerous systems.

2.2.3 Increasing Exposure

Not only is our *technology* becoming more complex, but our *society* has also become more complex, interdependent, and vulnerable [206]. The consequences of an accident depend not only on the hazard itself but also on the exposure of the hazard—that is, the length of time and the environment within which the hazard exists.

More people may be exposed to a given hazard today than were previously. Passenger capacity in aircraft, for example, is increasing to satisfy economic concerns. The siting of dangerous facilities near large populations is increasing as more people move to cities and larger plants need larger workforces within commuting distance. Interdependence and complexity can cause ripple effects beyond the immediate exposure area of the hazard, magnifying the potential consequences of accidents.

2.2.4 Increasing Amounts of Energy

Another factor related to increased risk is the discovery and use of high-energy sources—such as exotic fuels, high-pressure systems, and atomic fission—which have increased the magnitude of the potential losses. Other new systems use more conventional energy sources, but they involve technology that requires larger amounts of energy than was required in the past.

The larger amounts of energy increase both the surrounding area potentially affected by an accident and the amount of damage possible. New hazards that can cause genetic damage and environmental contamination introduce the potential for affecting not only the current generation but our descendants as well.

2.2.5 Increasing Automation of Manual Operations

Although it might seem that automation would decrease the risk of operator error, the truth is that automation does not remove people from systems—it merely moves them to maintenance and repair functions and to higher-level supervisory control and decision-making [314]. The effects of human decisions and actions can then be extremely serious. At the same time, the increased system complexity makes the human decision-making process more difficult.

Automation often removes the operator from the immediate control of the energies of the system [180]. Consequently, an individual moving within the physical system may find it difficult to anticipate the possible energy flows—for example, the physical behavior of an industrial robot. This difficulty is enhanced in systems that use exotic chemical and physical processes that are not well understood. Automated control systems have become so complex that operators may not be able to understand their behavior, leading to delayed or incorrect responses to disturbances.

In addition, operators in automated systems are often relegated to central control rooms, where they must rely on indirect information about the system state: this information can be misleading. In 1977, New York City had a massive and costly power blackout [303]. When the operator followed prescribed procedures to handle the initial symptoms, the electrical system was brought to a complete halt. The operator did not know that there had been two relay failures—one that would lead to a high flow of current over a line that normally carried little or no current (and thus would have alerted the operator to the real problem) and a second relay failure that blocked the flow over the line, making it appear normal.

The operator was unaware of the particular set of circumstances that made the zero-current reading abnormal and treated it as normal. Lack of direct information increases the probability of such faulty hypotheses. Operators then become scapegoats when accidents are blamed on human error, even though the "error" was induced by features of the automated system.

These problems will only get worse with the current trend toward decentralization of control. Microprocessors embedded in the plant or system are taking over most control functions, with only high-level information being fed back to the central control room or control point. This design limits even further the operator's options and hinders broad comprehension of the system state.

A control loop, by its very nature, masks the occurrence and subsequent development of a malfunction precisely because it copes with the immediate effects of the problem, at least for a time [85]. But this masking does not continue indefinitely: when the malfunction is finally discovered, it may by then be more difficult to control, or the symptoms may be hidden or distorted. By the time a human gets involved, the symptoms may be referred forward or backward by the major loops in the overall process. For example,

> *In 1985, a China Airlines 747 suffered a slow loss of power from its outer right engine. This would have caused the plane to yaw to the right, but the autopilot compensated, until it finally reached the limit of its compensatory abilities and could no longer keep the plane stable. At that point, the crew did not have enough time to determine the cause of the problem and to take action: the plane rolled and went into a vertical dive of 31,500 feet before it could be recovered. The aircraft was severely damaged and recovery was much in doubt.* [286, p. 138]

The energy-saving systems incorporated in many process plants, particularly during the 1970s, to conserve energy and to improve thermal economy complicate this problem further [86]. For example, heat generated by a process might be recovered through a complex exchange system. As is often the case, multiple goals—in this case safety and economy—lead to conflicts. The energy-saving systems introduce component interactions that make the functioning of the system less transparent to its designers and operators [318]. From the designer's standpoint, systematic analysis and prediction of events that could lead to accidents become increasingly difficult. From the operator's standpoint, the extra complexity makes diagnosing problems more difficult and again leads to masking and referral of symptoms: the place in the plant where signs of trouble first emerge may not be where the problems occurred [86].

2.2.6 Increasing Centralization and Scale

Increasing automation has been accompanied by centralization of industrial production in very large plants and the potential for great loss and damage to people, equipment, and the environment. For many decades, plant size has been increasing, resulting in a change of scale in production: devices and whole processes are being extrapolated into untested areas. In nuclear power, for example, Bupp and Derian observed that by 1968, manufacturers were taking orders for plants six times larger than the largest one then in operation: "*And this was in an industry which had previously operated on the belief that extrapolations of two to one over operating experience were at the outer boundary of acceptable risk*" [55, p. 73].

The Browns Ferry Nuclear Power Plant, which was the site of a serious accident in 1975, was ten times the size of any plant already in operation at the time its construction began in 1966. In fact, it was to become one of the world's largest electrical generating facilities [405].

Ocean shipping is another industry experiencing enormous changes in scale that are alien to its previous caution and conservatism. To maximize profits, supertankers are being built without the sound design and redundant systems (such as double hulls) they once had. Mostert writes about these superships: "The gigantic scale of vessels creates an abstract environment in which crews are far removed from direct experience of the sea's unforgiving qualities and potentially hostile environment. Heavy automation undermines much of the old-fashioned vigilance and induces engineers to lose their occupational instincts—qualities that in earlier days of shipping were an invaluable safety factor" [273].

2.2.7 Increasing Pace of Technological Change

A final risk factor is the increasing pace of technological change in the last hundred years. The average time required to translate a basic technical discovery into a commercial product has decreased from thirty years in the early part of the 1900s to five or fewer years today. In addition, the number of new products or processes is increasing exponentially [331]. Beginning in the twentieth century and continuing into the present, there has been a major acceleration in the growth of new industries stemming from scientific and technological innovation. Dangerous substances are being handled on an unprecedented scale, and economic pressures often militate against extensive testing [206].

The time to market for new products has greatly decreased, and strong pressures exist to decrease this time even further. The average time to translate a basic technical discovery into a commercial product in the early part of the last century was thirty years. Today our technologies get to market in two to three years and may be obsolete in five. We no longer have the luxury of carefully evaluating systems and designs to understand all the potential behaviors and risks before commercial or scientific use.

The increased pace of change lessens opportunity to learn from experience. Small-scale and relatively nonhazardous systems can evolve gradually by trial and error. But learning by trial and error is not possible for many modern products and processes because the pace of change is too fast and the penalties of failure too great. Design and operating procedures must be right the first time when there is potential for a major fire, explosion, or release of toxic materials. Nuclear energy is just one example. Christopher Hinton, who was in charge of the first British atomic energy installations, said in 1957: "All other engineering technologies have advanced not on the basis of their successes but on the basis of their failures. The bridges that collapsed . . . have added more to our knowledge of bridge design than the ones which held; the boilers that exploded more than the ones that had no accidents. . . . Atomic energy, however, must forego this advantage of progressing on the basis of knowledge gained by failure" (quoted in [99]).

As a result, empirical design rules and equipment standards are being replaced by reliance on modeling and simulation and by attempts to build ultrareliable systems that never fail. The feasibility of accomplishing either of these goals and the amount of protection they afford are unknown in comparison to the older methods involving learning by experience and using well-tested standards and guidelines.

Given the difficulty that industry is having in coping with all the changes, it is not surprising that government agencies charged with the licensing and safety monitoring of these industries are also having problems keeping pace. New technology, such as self-driving automobiles and other types of software-enabled autonomy, requires new standards, regulatory procedures, and expertise that take time to develop and perfect. Industry and society are unwilling to slow progress while these agencies catch up.

2.3 How Safe Is Safe Enough?

All of these changes in risk factors leads to the question of how safe is safe enough. Because of the recent changes and unique conditions for which historical experience does not apply, the nonoccurrence of particular types of accidents in the past is no guarantee that they will not occur in the future or that they will not increase in number or potential losses. In addition, if we must learn from accidents and if failure teaches us more than success, as Petroski [305] and others have argued, then what can we do about systems in which a single accident can have such tragic consequences that the process of learning from accidents is unacceptable to society?

The system safety approach to reducing risk is to anticipate accidents and their causes in before-the-fact hazard analyses, rather than relying on after-the-fact accident investigations, and to eliminate or control hazards as much as possible throughout the life of a system. The goal is to understand and manage risk in order to eliminate accidents or to reduce their consequences. Frola and Miller [109] claim that system safety investment has reduced losses where it has been applied rigorously in military and aerospace programs.

Unfortunately, hazards will never be eliminated completely from all systems. In addition to the technical difficulty of anticipating and reducing risks, there is the basic problem of conflicting goals. Desirable qualities may conflict with each other, requiring design tradeoffs between safety, performance, and other goals. A large industrial robot arm, for example, carrying a heavy load at high speed cannot be stopped quickly in emergencies without damaging the arm. The longer it takes the arm to stop in response to a deadman switch or other safety device, the less the wear on the arm but also the more likely that the arm will hit something before it stops [296]. Likewise, human–machine interfaces that are designed to be easy to use often are less safe. For example, computer input errors can be reduced if operators are required to repeat operations, but that requires extra time and seemingly wasted effort.

Designing a system to protect against a variety of hazards is clearly possible, but designing a system to protect against all hazards, no matter how perverse or remote, might require making so many compromises in functionality and other goals that the system is not worth building at all [369]. Finding the right balance is difficult.

Wolf [415] suggests that we may be encountering a contradiction in technological culture where we can neither prevent potentially disastrous accidents nor accept their consequences. But even if the number of accidents and their associated losses increase, we are unlikely to abandon the new and risky systems that represent technological progress. The benefits of technology usually come with disadvantages, and society is unwilling to live without many of those benefits. If this assumption is correct, then the process for determining exactly which systems to build and what new technology to introduce into them becomes critical.

Several ways of making this decision have been suggested. At one extreme is an anti-technology stance that blames accidents on new technology alone and concludes that advanced technology should not be used in dangerous systems. Such a simple, negative stance is not the solution to complex engineering and ethical problems. At the opposite extreme is the more common pro-technology position that all new technology is good: if it can be built, then it should be. Those who hold this position often put the blame for accidents on human operators and assume that risks will be reduced by replacing human operators with computers. Again, this position is overly simplistic.

The prevailing position in our society is the utilitarian view that the only reasonable way to make technology and risk decisions is to use risk–benefit analysis. This belief is so widespread that we often accept risk–benefit analysis as the *only* way to make technology and risk decisions, without realizing that there are alternatives.

2.3.1 Risk–Benefit Analysis and the Alternatives

From a utilitarian viewpoint, catastrophic accidents (such as Bhopal) are one of the risks of high technology, which, in the long run, are outweighed by the technology's overall benefits. Decisions can be made by comparing these risks and benefits.

To apply this approach, we must be able (1) to measure risk, and (2) to choose an appropriate level for decision-making. Unfortunately, it is impossible to measure risk accurately, especially before a system is built: systems must be designed and built while knowledge of their risk is incomplete or nonexistent. Even if past experience could be used, it might not be a valid predictor of future risk unless the system and its environment remain static, which is unlikely. Small changes may substantially alter the risk involved [101]. We create new systems and new designs precisely because the old ones no longer satisfy our goals.

Risk assessment tries to solve this dilemma. The goal is to quantify risk, including both the likelihood and severity of a loss, before historical data is available. The accuracy of such risk assessments is controversial. William Ruckelshaus, former head of the US Environmental Protection Agency, argues that the current use of risk assessment data is a kind of pretense: "*To avoid paralysis resulting from waiting for definitive data, we assume we have greater knowledge than scientists actually possess and make decisions based on those assumptions*" [340, p. 110].

In fact, risk assessment may never be able to provide definitive answers to these types of risk questions. Estimates for extremely unlikely events, such as a serious nuclear reactor accident, can never have the same scientific validity as estimates of events for which there are abundant statistics. Because the required probabilities are so small (such as 10^{-7} per reactor per year), there is no practical possibility of determining this failure rate directly—that is, by building 1,000 reactors, operating them for 10,000 years, and tabulating their operating histories [406].

Instead, probabilities of serious accidents are calculated by constructing models of the interaction of events that can lead to the accident. In practice, the only events included are those that can be measured. At the same time, the causal factors involved in most major accidents are almost all unmeasurable. The difficulty of performing risk assessments is discussed in later chapters of this book. In brief, the technique is controversial, and the results are far from universally accepted as meaningful.

Even if risk could be measured, there is still the problem of choosing the level of risk to use in decision-making. The most common criterion is that of *acceptable risk*: a threshold level is selected below which risk will be tolerated. But who determines what level of risk is acceptable in comparison to the benefits?

Often, those getting the benefits are not those assuming the risks. The people who are negatively affected are rarely asked their opinion, especially when they are not represented by an influential lobby or trade association [252]. The attitude of some decision-makers is reflected in a statement by the director of Electricite de France who in 1989 explained French secrecy about nuclear power this way: "You don't tell the frogs when you are draining the swamp" [252, p. 156]. A similar attitude was reflected in the recorded statement by the head of the Nuclear Regulatory Commission at the time of the Three Mile Island accident: "Which amendment [of the Constitution] guarantees the freedom of the press? I am against it" (quoted from *Le Monde*, April 14, 1979, in [200]). Hence, along with technical problems, utilitarianism and risk–benefit analysis present many philosophical and ethical dilemmas.

The moral implication of risk–benefit analysis is epitomized by the Ford Pinto gas tank case. Reportedly, Ford knew of the danger of explosion on impact. But after doing a risk analysis weighing the cost of fixing the gas tank against the incidence of rear-end collisions and the damages usually assessed in wrongful death lawsuits at that time, the company decided that it would be cheaper to settle lawsuits after explosions than to fix the gas tank design. Here the financial benefits went to Ford while the (nonmonetary) risks were unknowingly assumed by the drivers and passengers of the Pinto. Admiral Bobby Inman argued in the wake of the *Challenger* accident that "there is a difference between risks taken with the unknown and risks taken to save on costs" [175, p. 58]. Perrow [303] suggests that the issue ultimately is not risk but power—the power to impose risks on the many for the benefit of the few.

Another moral difficulty with risk–benefit analysis involves selecting a common unit of measurement to compare losses and benefits. Usually, dollars are chosen. This choice raises the question of whether human suffering should simply be regarded as a cost and assigned a dollar value.

Moreover, there is the problem of how to do this assignment. The most common approach is to use the amount of money the person would have earned from the point of death to his or her statistical life expectancy. Thus, a young, healthy, high-earning person would be worth more than a young, low-earning person or an older person close to retirement. The moral difficulties with this approach to assigning dollar values to human life are obvious, especially if you consider what dollar value you personally would place on the life of your child, your spouse, or yourself.

Alternatives to the use of acceptable risk have been proposed. *Optimal risk* involves a tradeoff that minimizes the sum of all undesirable consequences [120]. Optimal risk is achieved when the incremental or marginal cost of risk reduction equals the marginal reduction in societal costs—that is, where the sum of the cost of risk abatement and the expected losses from the risk is at a minimum. Estimating expected losses, however, still requires the ability to make probabilistic likelihood estimates of accidents and losses.

An alternative to optimal risk is to require that accident rates not be increased when new technology is introduced. If, for example, accidents have occurred in a specific type of system at a certain historical rate, then the new technology should only be required to achieve an equivalent rate to be considered acceptable. This approach is based on the belief that if the public currently accepts a technology with a particular accident rate, then they will continue to accept this level of risk.

Aside from the technical problem of how to determine whether the accident rate in the new system will be equivalent, difficult moral problems again arise if the new technology has the potential to reduce the accident rate, but this reduction requires tradeoffs and increased costs. From an ethical standpoint, equivalent safety in this case may not be adequate. Consider air bags and other improvements in automobile safety: by the "acceptable risk is what has been accepted by the public previously" argument, such safety improvements are unnecessary because people apparently accepted the risk prior to these improvements by their willingness to drive.

The use of computers, in particular, may offer a potential increase in safety, but, at the same time, allow a decrease in safety margins; their use, therefore, provides the possibility of economic or productivity benefits along with the same historic level of safety. Should equivalent risk levels be accepted when reduced risk is possible? Even worse, do computers really reduce risk as much as is assumed when these types of tradeoff decisions are made?

It appears that there are no entirely satisfactory methods for making these decisions. Part of the explanation for this lack of mathematical and engineering solutions is that the decisions involve deep philosophical and moral questions—not simply technical choices.

2.3.2 Trans-Scientific Questions

A basic problem with utilitarian approaches is that they attempt to use scientific methods and arguments to answer what are fundamentally *not* scientific questions. Alvin Weinberg, former head of Oak Ridge National Laboratory, writes:

> *Many of the issues that lie between science and politics involve questions that can be stated in scientific terms but that are in principle beyond the proficiency of science to answer. . . . I proposed the term "trans-scientific" for such questions. . . . Though they are, epistemologically speaking, questions of fact and can be stated in the language of science, they are unanswerable by science; they transcend science. . . . In the current attempts to weigh the benefits of technology against its risks, the protagonists often ask for the impossible: scientific answers to questions that are trans-scientific.* [406]

Even though scientists cannot provide definitive answers to such risk questions, they still have an important role to play: to provide what scientific information they can about the question at hand and to make clear where science ends and trans-science begins. Weinberg contends that the debate on risks versus benefits would be more fruitful if we recognized those limits.

Making decisions such as how safe is safe enough involves addressing moral, ethical, philosophical and political questions that cannot be answered fully by differential equations or probabilistic evaluations. "*Scientific truth is established in the traditional methods of peer review: only what has value in the intellectual marketplace survives. By contrast, where trans-science is involved, wisdom (rather than truth) must be arrived at by some other mechanism*" [406].

Although scientists and engineers can legitimately disagree about the extent and reliability of their expertise, they often appear reluctant to concede limits on their ability to answer what are essentially trans-scientific questions. Answering such questions involves moral and aesthetic judgments: they deal not with what is *true* but rather with what is *valuable*. As such, where there can be no consensus on these values, the decisions must be made by political processes.

Unfortunately, conflicts of value present special difficulties for the predominantly scientific and technocratic modes of rationality in Western society. Cotgrove [68] argues that beliefs about risk are embedded in complex belief and value systems that constitute distinct cultures: the way individuals see the world and evaluate risk is part of this culture. Until recently, the master value of our industrial society has been wealth creation: the overall goal for society was taken for granted to be maximizing economic growth and the production of goods and services. This dominant value system is rejected by some members of current society who accept what might be called an *environmental paradigm*. Table 2.1 outlines some of the features of these competing social paradigms.

For groups that hold such different paradigms, communication is nearly impossible: they essentially inhabit different worlds. What is rational and reasonable from one perspective is irrational from another. Each side is unable to comprehend alternative viewpoints, which requires, in Kuhn's terms [196], a *paradigm shift*. If the goal is maximizing output, for example, then not only are nuclear risks justified, but it would be unreasonable not to take them. From a different perspective about how the world works and what kind of society is desirable, to take even the possibly small—but in practice incalculable—risks for future generations stimulates moral indignation. Perhaps the explanation behind scientists' often-expressed frustration with the "irrational" way many people evaluate risk is just that different groups evaluate it from very different cultural viewpoints, all of which are rational within their different contexts.

Because of the futility of attempting to change deep-rooted cultural beliefs, trans-scientific questions are not covered in this book. However, it is important to point out which issues are

Table 2.1
Alternative cultural values

Technological Paradigm	Environmental Paradigm
Core value of material/economic growth	Core value of nonmaterial, self-actualized growth
Nature valued as a resource	Nature intrinsically valued
Domination over nature	Harmony with nature
Market forces of most importance	Public interest of most importance
Risk and reward highly valued	Safety highly valued
Ample natural reserves	Earth's resources limited
Environment can be exploited	Nature delicately balanced
Emphasis on individual and self-help	Emphasis on collective and social action
Decision making by experts	Citizen, worker involvement in decisions

truly scientific and which are trans-scientific so that communication lines can be kept open. We must also realize that decisions about safety will cause legitimate disagreements that cannot be resolved by simple utilitarian arguments. Some opposition to our technological inventions goes beyond questions of measurable risks and economic benefits and instead focuses on social, political, and psychological risk and intangible, unmeasurable risk considerations [68]. According to Weinberg, it is the scientist's duty to inject some order into this often chaotic debate by distinguishing scientific from trans-scientific issues. An attempt is made throughout this book to do just that; that is, to provide approaches to the scientific and engineering questions while identifying those for which other types of answers must be sought.

Exercises

1. What types of changes have occurred in the risk factors associated with your field (or area of study) in the past fifty years? How much control do users or workers have over the risks of the products or workplaces in your field?
2. Which of the changing factors described in section 2.3 hold for your industry? Does your industry involve new or changing hazards?
3. What are some examples of trans-scientific questions? Explain why you think they are trans-scientific.
4. What are some specific examples of engineering and regulatory practices that might differ depending on one's social paradigm? How might each side be justified by people with conflicting value systems?
5. This chapter argues for the need to shift at least some responsibility for controlling risk from individuals and personal responsibility to organizational or public responsibility. Do you agree with this argument? Why or why not?
6. Provide an argument for why each of the following groups *should not* be responsible for safety and then why they *should* be responsible: operators and workers; managers; boards of directors; hardware engineers; software engineers; researchers; government regulatory agencies; consumers and the public; stockholders; insurance companies; and the court system. What do you think should be the role in risk management today for each of these groups? What effect does new and rapidly changing technology have on these roles?
7. Most professional societies have an ethics code. What does the ethics code for the professional society in your industry say about safety?
8. Consider the role of the courts and the legal system in safety using the following cases as examples:
   a. Johns-Mansville Corporation used to be the largest producer of asbestos in the United States. The corporate managers of Johns-Manville knew from the 1930s and 1940s onward that asbestos fibers in the lungs cause asbestosis, an incurable form of cancer. For three decades, management concealed this information from workers and the public without giving them the right of informed consent to the dangers confronting them. In 1949, Johns-Mansville's company physician defended a policy of not informing employees diagnosed with asbestosis: "As

long as the man feels well, is happy at home and at work and his physical condition remains good, nothing should be said" [50]. When Johns-Manville was finally brought to trial, company officials claimed that some 1,300 of the company's own studies of asbestos had mysteriously disappeared from its files.

One study showed that 38 percent of asbestos insulation workers die of cancer, 11 percent of them from asbestosis. It has been predicted that "among the twenty-one million living American men and women who had been occupationally exposed to asbestos between 1940 and 1980, there would be between 8,000 and 10,000 deaths from asbestos-related cancer each year for the next twenty years" [252, p. 191].

To postpone settling a large number of lawsuits, Manville filed for bankruptcy in 1982. Its assets of $2 billion made it the largest American corporation ever to do so. A court agreement reached in 1985 allowed it to continue operating while paying some $2.5 billion in lawsuits over twenty-five years. At that time, it also stopped using asbestos in its products.

b. In 1985, for the first time in history, a judge convicted three officials of a company for industrial murder [106]. Film Recovery Systems was a small corporation that recycled silver from used photographic and X-ray plates. Used plates were soaked in a cyanide solution to leach out their silver content. Other companies use this process safely by protecting workers against inhaling cyanide gas and making skin contact with the liquid. Standard safety equipment includes rubber gloves, boots, and aprons, as well as respirators and proper ventilation.

None of these precautions were used by Film Recovery Systems. Workers were not given useful paper face masks and cloth gloves. Ventilation was terrible, and respirators were not provided. Workers frequently became nauseated and had to go outside to vomit before returning to work at the cyanide vats. This continued until an autopsy on one employee, a Polish immigrant, revealed lethal cyanide poisoning. Charges were brought against the executives of Film Recovery Systems under an Illinois statute, which states that "a person who kills an individual without lawful justification commits murder if, in performing the acts that cause the death . . . he knows that such acts create a strong probability of death or great bodily harm to the individual or another" [106]. During the trial, it was proven that the company president, the plant manager, and the plant foreperson all knew of the dangers of cyanide. They also knew about the hazardous conditions at their plant. Each was sentenced to twenty-five years in jail and fined $10,000. Critics have disagreed with the conviction on the grounds that murder involves intentional and purposeful killing. At most, say the critics, the executives committed manslaughter, which is killing due to negligence or indifference (such as when drunk drivers kill).

c. Cases in the news more recently include the arrest of the *Costa Concordia* captain and Mr. Mas, who was the president of the French company that put industrial-grade silicone in human breast implants.

Questions: What role should the courts and lawyers play in ensuring public safety? Are there subcategories or distinctions you think are important?

9. Consider the question of individual responsibility in the following circumstance. Approximately 25 percent of the US population has nonlethal food sensitivity, and 1–2 percent has food allergies severe enough to cause life-threatening reactions. Although an individual could be allergic to the proteins in any food, such as fruits, vegetables, and meats, there are eight foods that account for 90 percent of all food-allergic reactions. These are: milk, eggs, peanuts, tree nuts (walnuts, cashews, and so on), fish, shellfish, soy, and wheat. For some consumers, even a trace amount of protein from one of these foods can cause life-threatening anaphylactic shock. The minimum safe level of food protein is not known, and experiments to find a safe level are ethically difficult to execute.

   Imagine you are an employee in a small food company who is intimately aware of the production practices of this firm. One of your fellow employees comes to you and indicates he heard some other employees discussing a mistake regarding allergens made on the night shift. You know the firm has not recovered from the recession, and cash is tight. A recall for allergens might be very expensive and put real pressure on the company's finances, causing the future of the firm to be in doubt. You also know that the regulatory environment is not clear with regard to managing allergens in foods.

   Questions: What do you think should or should not be done? What are some of the options? What factors could keep the "right" things from being done? How do you establish the behavior expected by employees if you are a manager?

3

Fundamental Concepts and Definitions

Defining concepts is frequently treated by scientists as an annoying necessity to be completed as quickly and thoughtlessly as possible. A consequence of this disinclination to define is often research carried out like surgery performed with dull instruments. The surgeon has to work harder, the patient to suffer more, and the chances for success are decreased.
—Russell L. Ackoff,
Towards a System of Systems Concepts [6]

Some basic concepts in safety engineering, system engineering, and Systems Theory underlying the rest of this book are introduced in this chapter. Surprisingly, even the definition of the term "safety" is controversial. Only the most basic concepts are included here; others are introduced where appropriate in later chapters.

3.1 Definitions of Safety and Risk

Even in engineering, there are different definitions of safety. A straightforward and inclusive one is simply that safety is the freedom from harm or other undesirable outcomes. In the US defense department standard for System Safety, MIL-STD-882, for all versions since the original one in July 1969, safety is defined as

Safety: freedom from conditions that can cause death, injury, occupational illness, damage to or loss of equipment or property, or damage to the environment.

This definition is similar to the standard *Oxford English Dictionary* definition of safety as "being protected from danger or harm." Other definitions may include different or additional types of losses such as a mission loss, but there is no relativity in any of these definitions, such as depending on how much it would cost to reduce losses.

Other definitions do introduce subjectivity and thus make a scientific approach more difficult. The confusion starts with the introduction of the concept of risk and confusing safety with risk.

The term *risk* is usually used as a measure of safety or the degree of safety provided in a particular system. Most often, risk is defined as a combination of the severity and likelihood of an unwanted outcome. One problem that arises from this definition is that by defining something, such as risk, as one way to measure it, alternative measurements or definitions are precluded, including nonprobabilistic ones and those that allow different approaches to assessment. Some alternative definitions are provided in later chapters of this book.

Instead of being expressed as an absolute, safety is sometimes defined in terms of being *as low as possible*, *as low as reasonably practical* (abbreviated as ALARP), or *as low as reasonably achievable* (ALARA). Safety then is operationalized in terms of weighing losses or undesirable consequences against the effort, time, and resources needed to prevent or control them. This type of definition is common in some industries or countries, but it has serious philosophical difficulties.

A major problem with these relative definitions occurs in determining how low is "possible," "practical," or "achievable." Tradeoffs are almost always necessary when multiple objectives and constraints are imposed on a system design: one might be able to reduce the number of negative outcomes, but that would require sacrificing other goals or even increasing other types of negative outcomes, and it might require more resources than available.

Who decides what is reasonably possible or practicable? In the Ford Pinto case, described in the previous chapter, the Ford decision makers might have described the car as "safe" in their definition of reasonably practical but perhaps not the victims' families. What is reasonably practical may be determined by the perch from which one is observing the system: the person paying to lower risk, the potential victim of an accident, or the government regulator. The person reaping profits from a system may have a different view of what is reasonably practical than a person who may be injured by the system.

Who decides how much money, time, or effort is reasonable to devote to making something safer? Does the definition change if one or one's loved ones are likely to be injured or killed rather than a stranger? Is it misleading to say that something is "safe" when there is still a significant potential for losses? And, of course, how is "significant" defined? To label something as safe when it could have a relatively high level of risk for a catastrophic consequence omits the possibility of simply concluding that the system should not be built or used at all.

At this point, one gets into philosophical and ethical concerns that go beyond engineering and venture in the area that Alvin Weinberg labeled *trans-scientific* as defined in chapter 2. That leaves the questions involving measurement like "how safe is safe enough?" or "what is reasonably practical" as ones that cannot be answered by scientists and engineers as they involve individual value systems.

To enhance communication and avoid the use of trans-scientific definitions in engineering, this book uses the absolute definition that is common to the System Safety and US defense industry: safety is freedom from those conditions that can cause death, injury, occupational illness, environmental damage, damage to or loss of equipment or property, or mission loss. That is, safety is freedom from losses defined as important to the system stakeholders. We can then determine how close we get to this ideal, knowing at the same time that the ideal is usually not achievable. Otherwise, everyone starts from a different definition of safety, and communication is inhibited. As an analogy, would most people say that their bank accounts were secure if there were only two break-ins a month? If there was an "acceptable" number of break-ins a month? As low as possible number of break-ins a month? In these cases, we might all decide to keep our money under our mattresses.

3.2 Hazards and Hazard Analysis

The concept of a *hazard* is basic to almost all of safety engineering. Engineers focus on eliminating, preventing, and controlling or mitigating hazards.

Hazards are, informally, states of the system that can lead to losses, not the losses themselves. Losses are usually labeled as accidents, mishaps, or, in some industries, incidents or adverse events. Here are some more formal definitions:

Accident/Mishap: an undesired and unplanned event that results in a loss.

Hazard: a system state or set of conditions that, together with a particular set of environmental conditions, will lead to an accident (loss).

The critical part of the hazard definition is that hazards are always defined as system states; that is, states *within* the system boundaries and therefore states over which the engineer, designer, manager, or regulator has control.

Why is the concept of hazard needed? Why not just deal directly with preventing losses? An accident may involve conditions over which the engineer or designer of the system has no control. Consider a chemical plant. Accidents can happen when chemicals are inadvertently released from the plant and atmospheric conditions are such that the chemicals are transferred to a place where humans are present, instead of harmlessly dissipating into the atmosphere. There must also be humans or other relevant entities in the path of the chemicals. The only part of these events that the designer of the chemical plant has control over is the inadvertent release of the chemicals. That release by itself does not involve a loss but is a system state that can *lead* to a loss under some potential worst-case environmental conditions. That state is called a *hazard*. We need a definition that will allow us to design safer systems, which means we must have design control over the conditions that can lead to losses.

The same is true when hazards are defined as features of the system environment, such as a mountain or bad weather being an aircraft hazard. We cannot eliminate the mountain or the weather, but we can design our system to prevent the aircraft from getting too close to the mountain. We can design our system to protect the aircraft from extreme weather or to avoid such weather conditions. The hazard, then, is the system state over which we have control but which, if it occurs, will lead to an accident or loss given worst-case environmental conditions.

Examples of hazards for automobiles are violation of the minimum distance from the car ahead of you or changing into a lane where there is another car. In the first case, braking distance may depend on visibility and road conditions. In the second, the other driver in the other lane may be able to avoid you, but you don't have any real control over their behavior and the state of their car. As a consequence, the focus is on preventing hazards or system states that we *can* potentially control. An example of a hazard for aircraft is flying into extreme weather conditions. An example of a hazard in health care is administering a different medication than was prescribed for that patient.

Sometimes hazards are defined as "precursors to an accident." The drawback of such a definition is that almost everything is a precursor to an accident. Our goal is to eliminate or reduce the number of hazards. A plane taking off is a precursor to a crash, starting a car is a precursor to hitting another car while moving, and treating a patient is a precursor for providing the wrong medication. We don't want to prevent planes from taking off, starting cars, or treating patients.

Once the losses are defined and the related hazards identified, then system safety engineering can begin. *Hazard analysis* involves identifying the scenarios describing the conditions under which the system will be unsafe; that is, the scenarios leading to a hazard. If a car is traveling at 100 mph, the brakes may be ineffective at preventing an accident when the road is wet even if they are effective at lower speeds and under different conditions. Or mixing up patient names can lead to a medication error. Information about the conditions in those scenarios is then used to engineer or design safety into the system by preventing the scenarios.

The primary goal during system design is to *eliminate hazards*. Hazards are eliminated either (1) by eliminating the hazardous state from system operation (that is, by creating an intrinsically safe design), or (2) by eliminating the negative consequences (losses) associated with that state. If a state cannot lead to any potential losses, it is not a hazard. Of course, philosophically, almost nothing is impossible. Theoretically, you could be hit by a meteorite while reading this book. But from a practical engineering standpoint, the occurrence of some physical conditions or events is so remote that their consideration is not reasonable.

Hazard elimination may involve substituting a nonhazardous or less hazardous material for another one. Examples include substituting nonflammable materials for combustible ones or nontoxins for toxins. There are almost always tradeoffs involved. In addition, some hazards simply cannot be eliminated without, at the same time, resulting in a system that does not satisfy its goals. Taking more risks in a spacecraft design may be more justifiable than in a commercial airliner: all losses do not have the same priority.

If hazards cannot be eliminated, then an attempt is made to reduce their occurrence. Examples of ways to achieve this goal are to incorporate interlocks in the design, to build in safety margins, or to increase reliability if component failures are a cause of a loss identified in the hazard analysis. Examples of interlocks are design features that prevent people from entering a high-voltage area without first shutting off the power or devices to disable a car's ignition unless the automatic shift is in *park*. Incorporating safety margins involves designing a component to withstand greater stresses than are anticipated to occur. Redundancy can be used to improve component reliability, although there are many disadvantages and limitations of this approach, such as common-mode and common-cause failures of redundant components. In addition, redundancy is not useful for software components or for design errors.

Because it is often impossible to ensure that the design prevents the occurrence of all hazards, an attempt is usually also made to design to control hazards before damage results if, despite efforts to eliminate or reduce them, they do occur. For example, even with the use of a relief valve to maintain pressure below a particular level, a boiler may have a defect that allows it to burst at a pressure less than the relief valve setting. To deal with this possibility, building codes may limit the steam boiler pressure that can be used in densely populated areas. In general, the types of protection designed into a system to deal with preventing or limiting damage in case the hazard occurs may include designing to limit exposure of the hazard, isolation and containment, protection systems, and fail-safe design.

Finally, the last resort is to design to reduce potential damage in the case of an accident, such as alarms and warning systems, contingency planning, providing escape routes (e.g., lifeboats, fire escapes, and community evacuation), or designs that limit damage (e.g., blowout panels, collapsible steering columns on cars, or shear pins on motor-driven equipment). Chapter 11 provides a detailed look at designing for safety.

In summary, engineers define safety in terms of hazards. Vague terms like "safe" or "unsafe" are not commonly used because a standard definition that is acceptable to everyone is not easily found, which is where this chapter started. Therefore, the focus in system safety engineering is on the goal of eliminating or controlling hazards—not achieving some general goal of the system being safe. That general goal is trans-scientific, while eliminating and controlling hazards can be defined in terms of concrete engineering processes. Those processes can be evaluated using scientific methods.

Trying to convince people that safety engineering efforts are effective and thus worth the investment when nothing bad happens can be difficult. Ironically, nothing bad happens precisely when safety efforts are effective. That is why arguments should be, and are, based on how the defined hazards have been eliminated or controlled in the system design, including the design of operations. The success of that goal can be objectively evaluated. The same is true for security and other desired system properties.

3.3 Defining Safety Requirements and Constraints

One of the first steps in creating any system is that the designers and stakeholders must agree on the goals for the system. These may be very general, such as "Produce chemicals" or "Create a rover to send to Mars." Once the goals are agreed on, they will normally be restated as more detailed "shall" statements or requirements; for example, "The plant shall produce one hundred pounds of ethylene per day," "The rover shall be able to identify the minerals on the surface of Mars," or "The rover shall be able to travel five miles in twenty-four hours." While the goals may not be achievable because they are often vague and only aspirational, the requirements by definition translate the goals into statements that are detailed enough that it is possible to determine whether the requirements have been achieved by the system design.

Along with goals and requirements, constraints on how those goals can be achieved must be identified. For example, producing the ethylene must not result in explosions, pollution of the environment outside the plant, or the death or serious injury of plant employees. The rovers must not pollute the planet surface or send back incorrect information. The goal of a nuclear power plant may be to produce power, but it must not release dangerous levels of radiation in the process. The same is true for workplace safety: the goal is to produce products or to provide services. The safety constraints are that workers (and customers in the case of services) are not injured in the process. It is possible to increase factory output—achieve the system requirements—but at the same time cause repetitive stress injuries in the employees. An example of a basic conflict for health care is provided in chapter 1.

To summarize,

Goals: the reason the system is being constructed.

Requirements: the goals specified in terms of concrete and verifiable statements.

Constraints: the acceptable ways in which the goals (and thus requirements) can be achieved.

Note that separating requirements and constraints makes it possible to identify conflicts and critical tradeoffs.

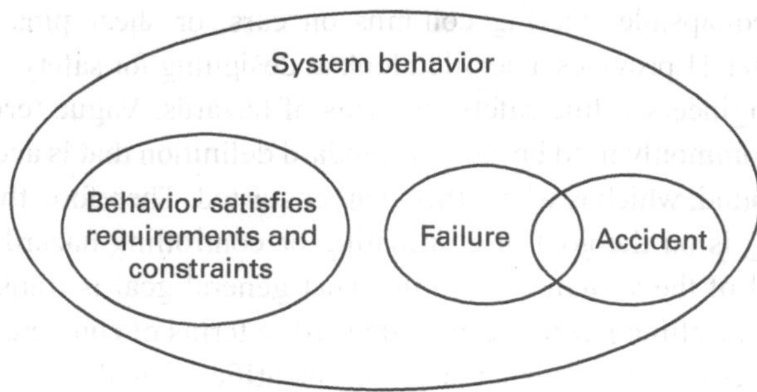

Figure 3.1
Basic engineering concepts.

Figure 3.1 shows the basic types of system behavior that can result in an engineered or designed system. *Failure* in this case is defined at the system level; that is, the system does not achieve the stated goals. An accident (loss) occurs when the constraints are violated. Note that a failure is not the same as an accident, although there may be some overlap. An accident occurs only when the safety-related constraints or requirements are violated.

Safety is usually associated with constraints rather than goals or requirements. However, that is not always true. If a goal of the system is to ensure safety, the requirements will also involve safety. As an example, one goal of an air traffic control system is to ensure that aircraft maintain a safe distance from other aircraft, obstacles, and dangerous weather conditions. When changing this goal into a measurable requirement, the definition of "safe distance" would need to be specified.

Air traffic control systems usually also have requirements unrelated to safety, for example, maximizing throughput in the controlled airspace. In addition, constraints can be related to system properties, behavior, or losses that have nothing to do with safety; for instance, the operation of the system must not damage the reputation of the company or produce litigation against the company.

The space outside the three labeled ovals represents behavior that the stakeholders, designers, and managers of the plant do not necessarily care about; for example, the plant contributes to the happiness or contentment of the employees, or it causes friction in an employee's marital relationship.

Within the large set of system behaviors, there are two subsets that can be called failures and accidents. Because anything that goes wrong could be declared a "failure," even if it has nothing to do with the requirements for the system, engineers define failure in terms of the system specification: that is, a *failure* is the nonperformance or inability of the system or component to perform its *specified* function for a *specified* time under *specified* environmental conditions.

The word "specified" is important. Sometimes, the word "intended" is used, but the problem is "intended by whom"? Who determines what type of behavior is intended? Instead, what is "intended" has to be identified in the requirements and constraints and thus be specified. Without such a specification, we would not know what to build or whether we achieved what was intended by those who commissioned the system. And when "intended" is only defined in retrospect, such as after a loss, then it becomes useless as an engineering goal because the engineers could not have possibly designed the system to prevent it without knowing about it during the design process.

Accidents and failures are overlapping but not identical. The system can fail in the sense that a requirement is not satisfied, but the safety constraints may not be affected; that is, unsafe behavior may not result. The system has failed, but no hazard has resulted.

In contrast, accidents may occur when one or more safety constraints are not satisfied, but there may not be a system failure in the sense that the nonsafety requirements are satisfied. The nuclear power plant may reliably produce energy but in the process expose humans to unsafe levels of radiation.

As another example, the automotive engineer spends much of his or her time designing the braking system to stop the car under the circumstances specified in the requirements. Because engineers know, however, that designs and the behavior of the designed systems may not always be perfect, they at the same time try to eliminate or mitigate any hazardous behavior that may occur. An example for hydraulic braking systems is a way to stop the vehicle even if hydraulic pressure is lost. The constraint here is that the system must not exhibit hazardous behavior if the hydraulic braking system does not satisfy its requirements. If the backup is successful, then the hydraulic braking component has failed but the system has not.

A critical concept in understanding safety engineering is that not all failures, noncompliance, and errors lead to accidents. Another way of saying this is that because there is only an overlap between failures and accidents—and not equivalence—a system that is completely reliable is not necessarily safe and one that is safe does not need to be reliable. Accidents can occur without any component failures. A hazard is not the same as a failure nor necessarily the result of a failure.

Understanding the difference between reliability and safety is so important in safety engineering today, and the two are so often confused that it is worth delving into the topic more deeply.

3.4 Safety versus Reliability

Confusion between safety and reliability is very common. Perhaps that is because, until recently, reliability was in many cases a reasonable proxy for safety. That is, making systems and products more reliable by making their components more reliable did make them safer because most accidents involved component failures. That is no longer true. In today's world, reliability and safety are different system properties and sometimes even conflicting.

All system components can individually operate 100 percent reliably, and accidents may still result, usually from unsafe interactions among the system components caused by system design errors. System design errors may, in turn, result from design complexity that overwhelms our ability to identify and thus prevent or handle all potential unsafe component interactions.

In addition, the larger environment beyond the system boundaries is important in determining whether something is safe or not. As a simple, real-world example, consider going out to the middle of a large, deserted area, pointing a gun away from oneself, and firing. If there is nobody or nothing in the vicinity, the gun and the activity could be considered to be both reliable and safe. Consider, however, doing the same thing in a crowded mall. The gun has not changed, the gun's reliability has not changed, and the action (pulling the trigger) has not changed. But the safety certainly has.

The safety of something, therefore, can only be determined by considering the *context* in which the thing operates. This fact is true for components as well as their encompassing systems.

Lots of accidents today occur in systems where there are no component failures. Examples are provided throughout this book. One example is an accident that occurred when an A320 aircraft was landing at Warsaw airport in a bad storm with heavy rain. The aircraft did not decelerate in time, crashed into a small mound at the end of the runway, and caught on fire. On landing, the flight crew tried to decelerate by activating the reverse thrusters, which operate by temporarily diverting an aircraft engine's thrust so that it acts against the forward travel of the aircraft. The software, however, would not allow activating the reverse thrusters because the software "thought" that the aircraft was still in the air. It is dangerous to activate reverse thrust when airborne, so protection had been built into the software to prevent such activation from occurring; that is, to protect against a flight crew error. But the engineers did not fully account for some rare environmental conditions at an airport. In this case, the usual conditions indicating the aircraft had landed did not hold for unusual reasons.

Design errors of this type are theoretically identifiable during system development. When the systems we were developing were simpler—mostly before the computer age in engineering—the designs could be exhaustively tested, and engineers could usually think through all the potential interactions during design. These two properties are no longer true. We are building systems today for which we cannot anticipate or guard against all unintended behavior. In addition, complex systems today have so many potential states that exhaustively testing the system before use is simply not feasible. The result is an increasing number of what engineers call the *unknown unknowns*; that is, unanticipated things that happen during system operation. We can no longer assume that design errors will be detected and removed before the use of a product or system.

Considering reliability at the system level, instead of at the component level, does not make reliability and safety the same. Complex systems almost always have many goals along with constraints on how those goals can be achieved. As an example, a chemical plant may very reliably produce chemicals—the goal or mission of the plant—while at the same time polluting the environment around the plant. The plant may be highly reliable in producing chemicals but not safe.

In summary, safety is a system or *emergent* property. This fact means that safety is only determinable by examining the behavior of all the components working together along with the environment in which the components are operating. System safety cannot be determined by simply looking at the behavior of individual system components, including the human operators.

Another way of saying this is to go back to where this section started; that is, that a system component failure is not equivalent to a hazard. Component failures can lead to system hazards, but a component failure is not necessary for a hazard to occur. In addition, even if a component failure occurs, it may not be able to contribute to a system hazard. Therefore, decreasing component failures may have no impact on safety whatsoever.

3.5 What Is a System?

So far, the word *system* has not been carefully defined. The basic concept of a system underlies all understanding of system engineering and the subset consisting of system safety engineering. So more careful definitions are necessary.

Here are some common but not useful definitions of a system:

- *A system is anything that consists of parts connected together.*
- *A system is a set of objects together with relationships between the objects and between their attributes.*

These definitions are adequate for many systems in nature; for example, the solar system. Physicists use such definitions as the basis for predicting the behavior of natural systems.

Engineers create new systems, and a different definition is needed:

System: a set of things, referred to as system components, that act together as a whole to achieve some *common goal, objective, or end.*

The goal is the critical part of the definition. Note that the connections are not even mentioned at all in the definition given here. My ear, my shoe, and my hair are interconnected through their connections to my body. They could be considered a system if a purpose could be conceived for considering these individual things together, but a purpose is basic to the concept. The human digestive system is an example that fits this definition: it consists of the gastrointestinal tract plus the accessory organs and hormones used in digestion. The purpose is the breakdown of food into smaller and smaller components until they can be absorbed and assimilated into the body. As another example, the goal of a transportation system is to move people from one place to another.

The system components usually are either directly or indirectly connected to each other, where indirect connections involve the system purpose only (not physical connections) and various types of nonlinear interdependencies.

The consequence of this definition is that a goal or objective for the system is fundamental. But there may be different objectives conceived by those defining a system than by those viewing that system. Consider an airport. To a traveler, the purpose of an airport may be to provide air transportation to other locales. To local or state governments, an airport may be a means to increase government revenue and economic activity in the area of the airport. To the airlines, the purpose of an airport may be to take on and discharge passengers and cargo. To the businesses at the airport, the purpose is to attract customers to whom they can sell products and services. When talking about a system, it is always necessary to specify the purpose of the system that is being considered.

The critical part of the use of the term "system" is that while the components of a system may exist, in fact

- *A system is an abstraction: that is, a model conceived by the viewer.*

Systems are not labeled as such in nature; they are a construction of the human mind. The physical components exist, but humans impose a purpose on those components that allow them to be considered together as a system.

The observer may see a different system purpose than the designer or focus on different relevant properties. Specifications that include the purpose of the system are therefore critical in system engineering. They ensure consistency of mental models among those designing, using, or viewing a system, and they enhance communication. Notice that different components of the airport may be included in particular airport systems: for example, an airline's view of an airport might contain passenger check-in counters, ramps to the planes, and taxiways, whereas a commercial view might contain shops and customers.

In summary, systems are *models* or *abstractions* laid on the actual physical world by human minds, not by nature. The components that are considered in any airport system or subsystem and the role they play in the system as a whole may be different for each concept (model) of an airport system or of airport subsystems. The basic idea here is that the purpose or goals of the system being considered must be specified and agreed on by those modeling and analyzing a particular system and that these aspects of systems are abstractions or models *imposed by the viewer* on the real-world objects.

3.5.1 Assumptions Underlying the Concept of a System

Important assumptions underly the concept of a system:

- The system goals can be defined.
- Systems are atomistic; that is, they can be separated into components, usually having interactive behavior or relationships among the components, whether the interactions are direct or indirect.

Systems themselves may be part of a larger system or be divisible into subsystems—viewed as components of the overall system. Note that the definition here is what is called *recursive* in mathematics and computer science. That is, the definition is made in terms of itself. A system is usually part of a larger system, and it can be divisible into subsystems, which are themselves systems.

Figure 3.2 shows a typical abstraction hierarchy for a system labeled A (for example the airport system) and for three subsystems, which in turn can be conceived as systems themselves (for example, security, airlines, and commercial business). Figure 3.2 is not a connection diagram. Rather it is a hierarchical abstraction that shows the different views of a system at different levels of abstraction or modeling.

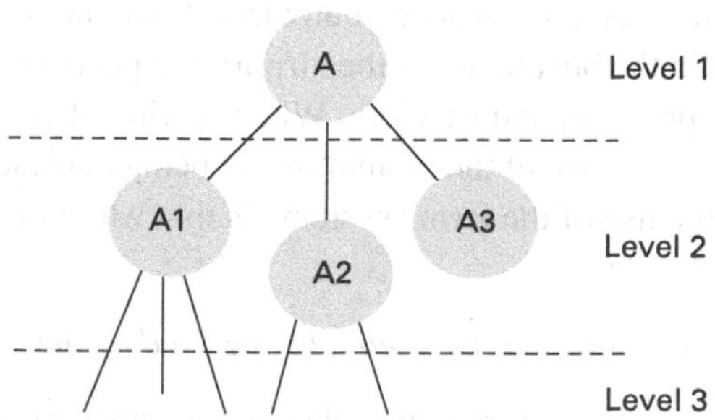

Figure 3.2
The abstraction System A may be viewed as composed of three subsystems. Each subsystem is itself a system. Subsystems may overlap in terms of their components.

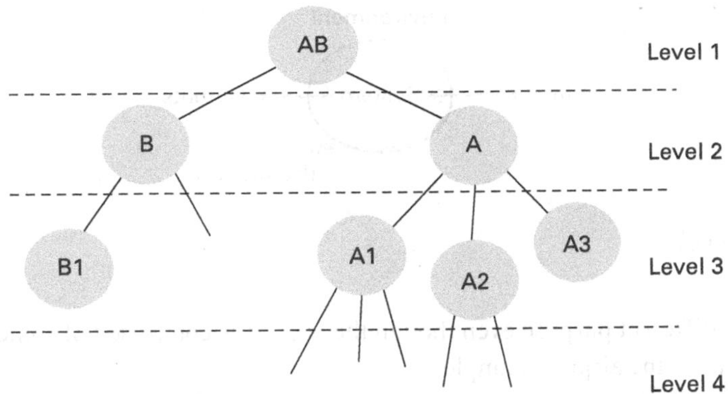

Figure 3.3
System A can be viewed as a component (subsystem) of a larger system AB.

Figure 3.3 shows an abstraction where A itself is conceived as part of a larger system AB.

The recursive nature of the definition of a system is important because many people have suggested that systems-of-systems must be treated differently than systems. In fact, the same general system engineering and analysis methods and techniques are applicable to all systems. A system-of-systems is just a system. The combining of two or more systems is not just some combination of those systems but instead is a completely new system with different system properties. The implication of this principle is that one cannot just combine the properties of individual systems and assume the combination will reflect the properties of the combined system. There will be interactions and influences that arise in the new system that did not exist in the original systems.

Systems have *states*. Informally, a state is a set of relevant *properties* describing the system at any time. Some properties of the state of an airport viewed as an air transportation system may be the number of passengers at a particular gate, where the aircraft are located, and what they are doing, such as loading passengers, taxiing, taking off, or landing. The components of the state that are relevant depend on how the boundaries are drawn between the system and its environment and, of course, on the purpose.

In safety, some subset of the total number of states of the system may be defined as *hazardous*. Examples of hazards for an airport are: the state where an aircraft is not able to decelerate when headed toward a ramp, food is served in airport restaurants that is injurious to human health, or even when large numbers of passengers are unable to reach their departing aircraft because of delays in security screening. What are considered hazardous states, once again, depends on the *losses* defined by the stakeholders.

The *environment* is usually defined as the set of components (and their properties) that are not part of the system but whose behavior can affect the system state (figure 3.4). Therefore, the system has a state at a particular time and the environment has a state. The concept of an environment implies that there is a boundary between the system and its environment. Again, this concept is an abstraction created by the viewer of the system and need not be a physical boundary. What is part of the system or part of the environment will depend on the particular system and its use at the time; that is, the definition of a system boundary may be useful for the current purpose but may be changed when

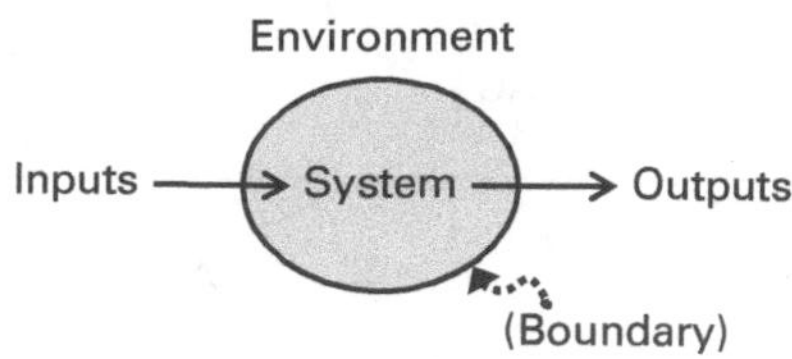

Figure 3.4
The parts of a system.

considering a different purpose even though the system is comprised of some of the same components, as in the airport example.

System *inputs* and *outputs* cross the system boundary. This model or type of system is usually called an *open* system. There is also a concept of a *closed system* (which has no inputs or outputs), but closed systems do not have much relevance for the engineered systems with which we are most concerned in system safety.

3.5.2 Sociotechnical Systems

All useful systems are sociotechnical. Ralph Miles Jr., an early practitioner of Systems Theory, wrote: "*Underlying every technology is at least one basic science, although the technology may be well developed long before the science emerges. Overlying every technical or civil system is a social system that provides purpose, goals, and decision criteria*" [261, p. 1].

No technical systems exist in a nonsocietal vacuum. However, a practice that is rampant both among engineers and social scientists is that they focus on either the social aspects or the technical aspects and not their interaction; that is, they define what is inside the system very narrowly. Engineers frequently draw the boundary of their system of interest around the technical components of the larger system and leave the social components in the environment. They then focus on the parts of the system within the boundaries they have drawn. It's not that they don't know that the social elements are important, but they leave those to social scientists.

Social scientists do use the term *sociotechnical system* more than engineers, but then they almost always focus all or most of their attention on the humans and social structures, and they don't examine the tight connections between the social and technical aspects. That is, they draw their system boundaries around the social or human aspects and put most of the technical aspects in the environment.

The problem is that when the focus is primarily only on the human operators or only on the technical components, then we have too small a keyhole to allow us to understand and solve our problems, including improving safety. The impact of the technical design and the social systems on each other (that is, the integrated sociotechnical system behavior) must be considered to deal with safety in our complex world today.

As an example, the design of cars is not simply related to the physical components of the car but is also related to the design of the road system along with other factors such as regulations. Regulations, such as fuel mileage rules, relate to social issues and not only to the technology involved in designing cars. At the same time, our social structures are related to the limits and requirements of our technology. Human error when interacting with a technological system is almost always a function of the design of the technical system with which the human is interacting.

To reduce human error, we need to design our systems by considering both the technical and social aspects. In the same way, to improve human adaptability within systems and system resilience, we need to design the context in which the human works to allow *positive* adaptability—humans can adapt in the wrong ways, and, in fact, often do. Jens Rasmussen has hypothesized that systems migrate to states of higher risk under competitive and other pressures [320]. To improve system resilience, we need to design the system *as a whole* to allow resilient behavior, not just focus on one part, such as the human operators or the social components.

In fact, the role of human operators in systems is at best staying steady or decreasing, the role of hardware is mostly decreasing, the role of designers and managers is for the most part staying the same, and the role of software is increasing rapidly. It makes no sense to concentrate only on the role of humans in trying to make systems safer unless we believe that all accidents are the fault of human operators, a topic in chapter 6. Humans operate within systems, and their actions are limited by the design of the systems in which they work. A true sociotechnical approach views all the components of the system, human and otherwise, as a whole. Safety problems cannot be tackled successfully by focusing only or primarily on the human operators or by focusing only or primarily on the technical aspects, as is too often the case.

Thus, improving safety requires focusing on the sociotechnical system as a whole and designing it to achieve our goals, including the defined safety goals. One of the reasons unintended consequences so often happen is that we focus on optimizing one part of a complex, sociotechnical system without considering how it will impact the behavior of the other parts.

While it may not be practical to train engineers as social scientists or social scientists as engineers, we can develop tools that allow them to work together on complex sociotechnical system problems.

Sometimes the concept of sociotechnical systems, which, of course, describes every system, is confused with *systems thinking.* Systems thinking has a much broader scope and is defined later in this chapter. At this point, it is enough to know that they are different.

We are left with one other important term that has been used extensively so far but not carefully defined, and that is complexity.

3.6 Defining Complexity

The term "complex system" is used frequently without ever defining what is meant by it, as is true so far in this book. The difficulty in defining complexity is that it is not static: what was complex one hundred years ago is often not considered to be complex given the knowledge and tools we have available to us today.

Another problem in defining complexity is that it is not just one property but many. For example, *interactive complexity* can be defined in terms of the interaction among system components. *Dynamic complexity* is related to changes over time. *Decompositional complexity* is introduced when the structural decomposition into system components does not match the functional decomposition. As a final example, *nonlinear complexity* arises when cause and effect are not related in a direct or obvious way. Other types of complexity could be defined.

One property of systems that makes prediction of behavior especially difficult is *coupling* between system components. In fact, one useful definition of complexity that incorporates many of those described in the previous paragraph is the degree and types of coupling or dependencies between the behavior of system components. Some of these dependencies may be direct, while others may be indirect and thus more difficult to identify. Predicting the behavior of a particular system as a whole and often even individual components within a system requires understanding all the dependencies, especially the indirect ones.

A way to connect all of these aspects of complexity into one concept is to define complexity in terms of *intellectual manageability*. The operation of some systems is so complex that it defies the understanding of all but a few experts, and sometimes even they have incomplete information about the system's potential behavior. The problem is that we are attempting to build and operate systems that are beyond our ability to intellectually manage: increased coupling of all types makes it difficult for designers to consider all the potential system states or for operators to handle all normal and abnormal situations and disturbances safely and effectively.

Complexity then must always be defined as relative to current knowledge and tools. Human minds are not changing; what is changing is the tools we have available to us for understanding and predicting how and why certain systems behave the way they do. This is not a new situation: throughout history, inventions and new technology have frequently gotten ahead of their scientific underpinnings and engineering knowledge. The result has been increased risk and accidents until science and engineering catch up. High-pressure steam engines, which powered the industrial revolution, for example, were subject to disastrous explosions until scientists and engineers amassed enough information about thermodynamics, the action of steam in the cylinder, the strength of materials in the engine, and other basic principles to determine how to prevent such tragedies.

Today, complexity of all types is similarly increasing rapidly, frequently due to the exponential escalation in the use of software in control systems. We are now in the position of having to catch up with our technological advances by greatly increasing our ability to design and manage system safety. This book suggests some promising paths forward to accomplish this goal.

3.7 Approaches to Dealing with Complexity

Scientists and engineers have devised ways to deal with the complexity of the world with which they are dealing. Three primary approaches are used: analytical decomposition, statistics, and systems thinking or Systems Theory.

3.7.1 Analytic Decomposition

The basic definition of a system presented previously in this chapter includes the assumption that systems are atomistic; that is, they can be separated into components with interactive behavior or relationships between the components, whether the interactions are direct or indirect.

Atomicity is related to the most prevalent approach to dealing with complex systems in engineering: decomposition. One of the great advances in modern science, starting around the 1600s, was this concept whereby complex things are broken down into their constituent components. The idea was that the individual components could be analyzed separately and then the analysis results combined to get a result for the whole.

An example of an analytically decomposable property for an aircraft or a car is weight. If we add up the weight of all the components, we get an acceptable value for the total weight. We can even get good estimates of a nonphysical property by combining the measures of those properties for each component into a combined measure for the system as a whole. As an example, for relatively simple systems, the reliability of the individual components and information about how they interact in the system can be used as a way to calculate the reliability of the system as a whole. As systems increase in complexity, some of these older ways of calculating system properties may need to be revised.

Clearly, engineered systems are decomposable. In fact, we create them by putting components together. Social systems are also decomposable. Remember, a system is an abstraction formed in the mind of the observer of that system. The airport example purposely was used in section 3.2 because it is easy to view it as being a sociotechnical system. In that example, the airport was decomposed into the subsystems for commercial businesses and air transportation. We might decompose the businesses into restaurants, merchandisers (e.g., books and clothes), and services (e.g., massage parlors and hotels). We could decompose the air transportation subsystem into air traffic control, aircraft, check-in counters, ramps, security gates, and so on. A different decomposition separates the airport into the airlines that operate there. Or the airport could be decomposed into commercial entities versus other services; for example, security. For various purposes, we might utilize some or all of these decompositions. As stressed, a system is an abstraction created in someone's mind; the components exist in reality, but the concept of a system is something that we impose on those components.

If we look at social systems, we may decide to focus on different parts of them, such as the people involved, the social structures, such as government agencies and political entities, or other aspects. We can decompose these and describe properties of the individual components—such as properties of the individual humans involved—or describe system properties created by the interactions among the components such as democratic or authoritarian.

Analytic decomposition is based on assumptions about the system that allow the synthesis or combination of individual component analyses to analyze system properties. The primary assumption is that the separation of the whole into separate components does not distort the phenomenon or property of interest. For this assumption to be true, the following four sub-assumptions must be true:

1. Each component or subsystem operates independently.
2. Components act the same when examined separately as when playing their part in the whole.
3. Components or events are not subject to feedback loops and nonlinear interactions.
4. Interactions can be examined pairwise; that is, via the connections between any two components.

Unfortunately, these assumptions are not usually true for complex systems today.

As an example of a system property that cannot be analyzed using analytical decomposition, consider gridlock on the roads. Gridlock cannot be predicted by simply calculating some properties of individual automobiles and then combining the results or even just looking at the cars alone. Whether gridlock occurs will depend on such system properties as the number of lanes on the highway, the time of day, the number of people who want to traverse the highway at various times of day, whether previous experiences with gridlock convince enough people to travel at nonpeak times or take alternative routes or transit, whether there is a global pandemic occurring and stay-at-home orders have been issued, and so on.

Violations of the four assumptions often arise in the form of *unknown unknowns*. These involve behavior during the operation of a system that was not anticipated. System engineering of large, complex systems usually starts with decomposition of the system into its components, specification of requirements on the individual components, and an ICD (interconnection diagram) that specifies the information that will be passed between components; that is, the interfaces between the components. A typical software interface specification includes timing, content, and other information about the data that is shared between components. An assumption is made that this process will result in a system that operates in a way that satisfies the overall system requirements. Underlying this assumption is a belief that the four required sub-assumptions specified previously are true for the system being designed or operated. They may not actually hold, however, and unexpected and often undesired behavior can result.

As an example, consider an avionics system for a modern aircraft, where the different avionics functions may be created by different manufacturers and updated individually over time [32]. The example concerns the flaps controller. The flaps on an aircraft are a hinged panel or panels mounted on the trailing edge of the wing. The position of the flaps changes the flow of air over the wing and thus the aircraft's lift and drag.

Because the position of the flaps affects the behavior of the aircraft as a whole, multiple aircraft components need to know the flaps position in order to operate correctly and safely. Information in an ICD might include whether the flaps are extended or not. The flaps control software could send this information over a bus to the components that need the information about the current state of the flaps for their own functionality. But when is this information sent? Is it sent when the flaps controller receives a command to move the flaps? Or perhaps when the flaps controller has commanded the hydraulic system to actually move the flaps? Maybe when the flaps reach the commanded position? Another possibility is to provide the information when the pilot or other aircraft component sends a command to the flaps controller to move the flaps.

What if the flaps controller sends a command to the hydraulic system to move the flaps, but there is a hydraulic system failure during movement? If the flaps controller waits to inform the other components about the change in the flaps position until it gets feedback that the flaps are in the commanded position, will that cause timing issues in the operation of other components? What if the flaps controller software is changed and operates in a way different than the other aircraft components originally assumed? What if the different software components are created by different developers or the aircraft reuses software or uses COTS (commercial-off-the-shelf) software where the internal logic is not provided and thus unknown to the users?

The problem is that the aircraft components do not truly act independently as assumed in analytic decomposition. With avionics systems now containing millions of lines of software, separate development is necessary. Decomposition is still necessary as is separate analysis and construction of the components. However, the problem is that assumptions that each developer makes about the other system components cannot all be documented and controlled in the ICD. When something unexpected occurs, it is labeled an *unknown unknown* that could not have been anticipated.

The appropriate conclusion is *not* that gridlock or other system properties cannot be predicted and analyzed or that complex engineered systems as in the avionics example cannot be developed and system properties assured without the appearance of unknown unknowns. They can be. And decomposition is still important and necessary for systems today. The point is that analyzing complex systems today requires more than analyzing the individual system components and combining them in a straightforward manner by looking only at their physical or logical interfaces with other components. Predicting and enforcing system properties requires using more sophisticated tools that do not rely on analytic decomposition. We need to change some activities in system development and operation to go beyond decomposition and to account for the interactions and coupling in the systems we are building and operating today.

3.7.2 Statistics

When systems get beyond the complexity that can be handled by analytic decomposition, an alternative is to use statistics and treat the system as a structureless mass with interchangeable parts. The *law of large numbers* is used to describe the component and system behavior in terms of averages. This law says that the larger the population, the more likely we are to observe values that are close to the predicted average values. There are some assumptions here too. The primary one is that the system components are sufficiently regular and random in their behavior that they can be studied statistically.

The problem is that the averages can be deranged by the underlying structure of the system; that is, the interactions among the components may not be completely random. In addition, there may not be enough components and interactions to use the Law of Large Numbers, and their individual behavior may not be sufficiently regular or random. Most all systems containing software do not satisfy the assumptions necessary to understand and predict their behavior using statistics.

Figure 3.5 is adapted from Weinberg [407]. Most of the systems we are engineering today are not sufficiently random to use statistics, but they are too coupled (complex) for analytic decomposition. Systems theorists label such systems as having *organized complexity*. Systems thinking, also known as Systems Theory, has been suggested as a foundational theory for such systems.

3.7.3 Systems Thinking and Systems Theory

Occasionally, systems thinking is confused with including sociotechnical considerations in system design and operations. While it is useful to consider all parts of a system in modeling and analysis, this is not the common definition of the concept of systems thinking.

Systems thinking is an approach to solving problems that is grounded in Systems Theory; that is, systems thinking is the result of applying Systems Theory in solving problems (see Ludwig von Bertalanffy [397], Peter Checkland [60], Gerald Weinberg [407], John Sterman [356], Russell Ackoff [7], Jens Rasmussen [320], and many more). It is characterized by a view of systems as teleological (purposeful), holistic, contextual, interdependent and interconnected, dynamically complex, nonlinear, and hierarchical.

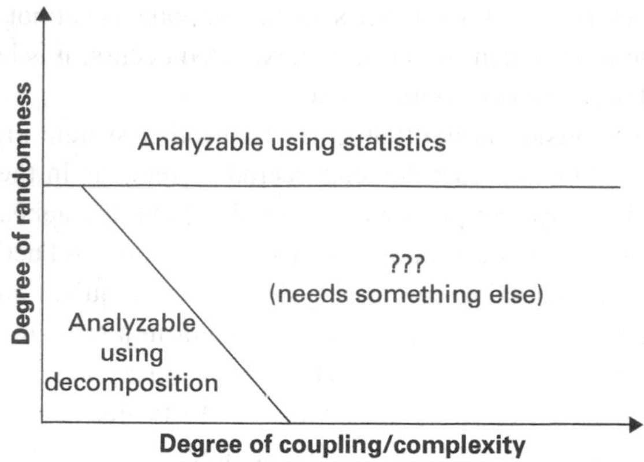

Figure 3.5
Systems with sufficient randomness can be analyzed using statistics. For those that are not sufficiently random but with a limited amount of coupling/complexity, analytic decomposition is sufficient. But most of our engineered systems today do not meet these criteria; that is, they are not sufficiently random to use statistics but are too coupled to treat the components as independent.

Teleological, purposeful Engineered (designed) systems are not just a set of connected components but are characterized by having an overall purpose or goal. We define this purpose ourselves and impose the concept of a system on the world as we see it. Systems are constructions of the human mind as emphasized earlier in this chapter.

The system purpose is achieved through the operation of the separate components, subject to constraints on their individual behavior, as well as on their interactions. System properties *emerge*—that is, are created—through the operation of the system as a whole.

Holistic Understanding and designing systems requires treating them as a whole, not by simply examining and designing the parts separately. The focus should be on the interactions and relationships between the parts. To accomplish this goal, we can use multiple perspectives on the whole, created by abstraction and viewed through different domains of knowledge.

Contextual All behavior is affected by the context in which it occurs. We cannot completely understand the behavior of something without looking at the context in which that thing is operating.

The most complex behaviors (that is, the emergent behaviors) usually arise from the interactions among the components, not from the complexity of the components themselves. In fact, attempting to simplify individual component behavior without changing overall system behavior almost always results in adding complexity to the interactions and interconnections.

Human behavior is driven not only directly by context but also indirectly by our mental models of that context. Our mental models impose a structure that allows us to deal with a "messy" world. But perception is affected by expectation, so we often interpret what we see through the lens of what we expect to see.

Interdependent and interconnected The world and the systems we impose on parts of it are interconnected and interdependent. These connections usually are not simple: we cannot look only at the behavior of individual components and understand either their behavior or the effect of their behavior in the context of the whole. This leads to the conclusion that systems must be studied and designed as a whole rather than as a conglomeration of parts.

One result of interconnectedness and interdependence is that changing one part of a complex system often creates unintended consequences and events in other parts of the system or in the system as a whole, as noted earlier. This is sometimes called the *Law of Unintended Consequences*. It results from the fact that the elements in systems are in mutual interaction. If we want to fix something or intentionally change the behavior of a complex system, we must first understand the system as a whole (holism).

Dynamically complex In dynamically complex systems, cause and effect are not related in a simple way. Dynamic complexity makes understanding and changing systems more challenging. In addition, delays between cause and effect can lead to instability. Constraints can be imposed in system design and operation to control the dynamics that prevent the system goals from being achieved.

Nonlinear To deal effectively with complex systems, views of causality must include nonlinear (nonsequential) behavior. The relationships between components and events include the potential for feedback and other types of communication leading to nonlinear causality. Goal-seeking behavior includes feedback and monitoring of information about the state of the system and the components in it.

Hierarchical We can view systems in terms of hierarchical levels and the relationships between them. Hierarchical thinking focuses on levels of organization and issues of scale as defined by an observer of the system.

These system thinking concepts have their theoretical foundation in Systems Theory.

3.7.4 Systems Theory Fundamentals

Around the middle of the last century, new technology and complexity in our engineered systems started to surpass the ability of decomposition and statistics to adequately handle them. The systems we wanted to build, such as spacecraft and automated vehicles, used heterogeneous technologies—electronic, mechanical, chemical, and digital. They included large numbers of components with complex networks of interactions and changed the role of human operators; and their behavior was influenced by financial, economic, social, and political factors.

Engineers were forced to confront the problem of how to achieve the system objectives given an enormous number of alternative designs and how to deal with unparalleled complexity. The assumptions underlying analytic decomposition and statistical analysis were no longer true in our tightly coupled, software-intensive, highly automated, and tightly interconnected engineered systems. Something new was needed.

Similar concerns arose at the same time in biology, where the complex interdependencies of the human body also created a need for more than decomposition to make significant progress. Social scientists, including anthropologists, economists, philosophers, sociologists, and others, also ran into limitations in the explanatory power of decomposition and started to think differently about systems.

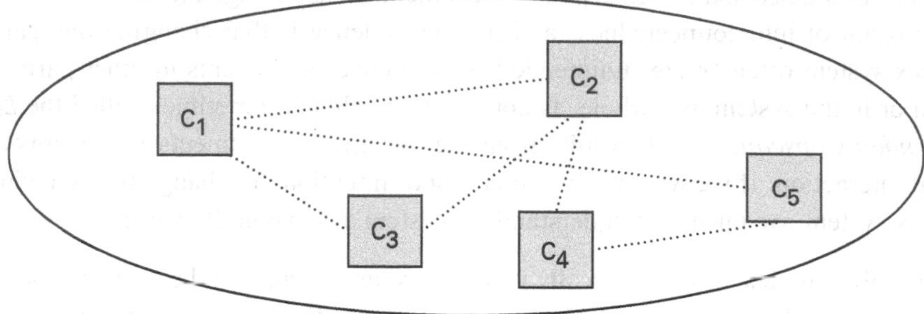

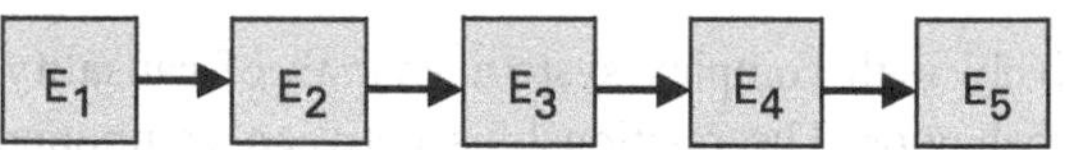

Figure 3.6
Analytic decomposition, shown at the top for system components and at the bottom for behavior over time.

Before this time, as noted previously, complexity had been handled in science and engineering for several hundred years by using analytic decomposition; that is, breaking up systems into their components, analyzing the components independently, and then combining the results to evaluate and understand the behavior of the system as a whole (figure 3.6). Events were treated as linear or sequential chains of cause–effect relationships. The properties that could not be handled this way were analyzed statistically.

The idea of decomposability and analytic decomposition still forms the basis of modern physics and much of engineering, along with many other fields. It was only after we started to stretch the limits of the power of these tools that alternatives were considered. Systems Theory was created to stretch these limits, originally in biology (von Bertalanffy [397]) and independently in engineering and math (Norbert Weiner [410] and others) where it was originally called cybernetics. Note that Systems Theory and Complexity Theory are different, although sometimes people confuse the two. Complexity Theory was an outgrowth of Systems Theory that has been used to explain natural systems.

In Systems Theory, there are some system properties that cannot be evaluated by combining the individual analysis results for the components. These are called *emergent* properties. They arise from the interactions among the components and not from the independent behavior of the components.

For these properties, separation and analysis of separate, interacting components (subsystems) distorts the results for the system as a whole because the component behaviors are coupled in nonobvious ways. Instead, the system is treated as a whole, not as a combination of its parts. A common way of expressing this concept is that "the whole is more than the sum of its parts."

The explanation of complex phenomena in terms of simpler ones no longer provided the power needed to successfully construct and analyze these new complex, tightly coupled systems. At the same time, the structure in these systems distorts the statistics and makes a statistical approach less useful. Systems Theory was the result.

In Systems Theory, the focus is on systems taken as a whole, not on the parts considered separately. Emergent system properties derive from relationships among the parts of the system, how they interact and fit together. These properties can only be treated adequately in their entirety, taking into account all social and technical aspects.

The first engineering uses of these new ideas were in the missile and early warning systems of the 1950s and 1960s. Today they provide the foundation for creating more powerful tools for dealing with emergent properties like safety and for successfully creating and operating systems that are increasing in complexity.

Understanding how Systems Theory can be used as the foundation for new approaches to system safety engineering requires understanding its basic concepts: emergence, control, hierarchy and communication, and abstraction.

Emergence For complex systems, some or all of the four assumptions required for analytic decomposition to be effective may not hold for some properties. These are the emergent properties. Most important system properties in complex systems today are emergent, and they require considering the entire sociotechnical system, not just the technical or the social parts in isolation (figure 3.7).

Safety is one of those emergent properties: An individual system component considered in isolation from the others may have some hazards associated with it, for example, sharp edges or toxic chemicals or materials. But the system operating as a whole may have other *emergent* hazards or states that result from the interactions among the components and not just from the behavior of the individual components considered separately.

For example, the hazard when driving of crashing into the car in front depends not only on the brake system components and how they interact—which given today's use of digital components and the interactions between the braking system and other auto systems is complex in itself—but also on driver reaction time, weather and visibility, driver distraction, the behavior of other drivers on the road, and so on.

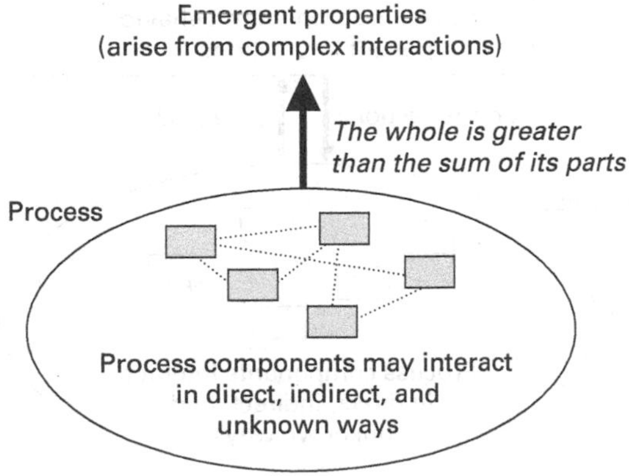

Figure 3.7
Emergent properties arise from complex interactions.

In an airport, a particular airline flight being late does not depend simply on factors related to that airline but also on complex interactions among many other systems at the airport and around the country; for example, air traffic control and congestion at the airport to which the aircraft is going or came from, crew duty cycles, weather, and so on. And it depends not only on a simple combination of these but also on how they interact with each other.

The concept of emergence does not mean that emergent properties cannot be explained or even predicted. For example, we can explain why gridlock occurs on Highway I-5 in Seattle every weekday at rush hour and why accidents occur in complex sociotechnical systems. But it does imply that they cannot be explained fully using linear causality and decomposition. Linear causality, while the basis for most existing causality models, is not the only possible type of causal explanation. New causality models can be created based on Systems Theory and emergent properties. One proposed new accident causality model based on Systems Theory is described in chapter 7.

If emergent properties arise from both individual component behavior and from the interactions among components, then it makes sense that controlling emergent properties, such as safety, security, maintainability, quality, efficiency, and operability, requires controlling both the behavior of the individual components *and* the interactions among the components.

Control The controller added in figure 3.8 provides control actions on the system and gets feedback to determine the impact of the control actions. In engineering, this is a standard feedback control loop. In other words, control actions are issued by the controller on the controlled process. *Feedback control* uses information about the current state of the process to generate corrective actions: if the process develops some undesired characteristic, it is then corrected. In this way, the system is treated as an adaptive system that is kept in a state of equilibrium over time as the system and environment change.

In contrast, *feedforward control* attempts to anticipate undesired changes in the process and issue commands to prevent them. As an example, experienced drivers increase

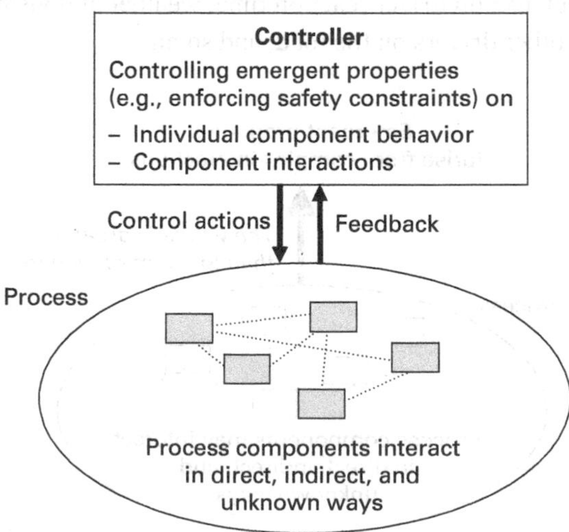

Figure 3.8
Control of emergent properties through feedback control.

pressure on the gas pedal when going uphill in order to maintain speed instead of waiting for the car to slow down and then giving it more gas.

The simplest example of a feedback control loop that almost everyone is familiar with is a thermostat. The thermostat is provided with a setpoint or goal and gets feedback about the current temperature in the controlled space. If the temperature is below the setpoint, then heat is applied until feedback (via sensors) shows that the setpoint has been reached. Then the application of heat is discontinued. The opposite is done if the measured current temperature is above the setpoint. A general model is shown in figure 5.2.

Note that variability and adaptability are not precluded in this concept; indeed, they are assumed. For the thermostat example, the temperature in the controlled space is assumed to vary over time due to a variety of influences in the system and in the environment. The thermostat is used to limit the variability to an acceptable range or to adapt when a different setpoint is entered. An acceptable range, in the case of safety, is defined as one that does not include hazardous states. In safety, the controller, which may be a human or automation, ensures that the system stays in nonhazardous states, such as maintaining a safe distance from other vehicles.

A physical—human, hardware, or software—controller is not required here. The thermostat is just one simple example of control. Control of behavior can be implemented through system design, rules and laws, procedures, processes, social pressures, incentive structures; that is, any means of controlling system behavior to satisfy particular goals and constraints.

In general, effective control requires four conditions:

1. **Goal Condition**: the controller must have a goal or goals; for example, to maintain a property or enforce a constraint on the controlled system.
2. **Action Condition**: the controller must be able to affect the system state or behavior; for instance, using actuators.
3. **Observability Condition**: the controller must be able to ascertain the state of the system; for example, through sensors and feedback.
4. **Model Condition**: the controller must be or contain a model of the system. Many accidents result from incorrect models of the state of the system; that is, the controller thinks the system is in a particular state for which a particular control action is appropriate, when in fact it is in a different state.

A more complex example of a controller and emergent property is an air traffic control system and the emergent property of *throughput* in an airspace. If each airline tried to optimize its routes independently, chaos would result as more aircraft might arrive at a popular hub airport than there are runways and gates available. Most would find that they were unable to achieve their optimum, and overall throughput in the system would suffer.

By introducing an air traffic control system, each airline might not be able to get its optimum schedule, but the system throughput is optimized, and most will hopefully do better than they would have otherwise. In fact, schedules and traffic are optimized for the US National Airspace using a central computer, and the country is divided into sectors. Real-time control commands are provided by air traffic controllers for those sectors, and aircraft are passed to other controllers as they pass between sectors.

Clearly, the air traffic controller must have an accurate model of where the aircraft are at any time, their direction and travel, speed, and so on. Inaccurate models could lead to issuing unsafe commands.

The controller (or controls) enforces constraints on the behavior of the system, both safety and other types of constraints. Example safety constraints might be that aircraft or automobiles must remain a minimum distance apart, pressure in deep water wells must be kept below a safe level, aircraft must maintain sufficient lift to remain airborne unless landing, accidental detonation of weapons must not occur, toxic substances must never be released from a plant, and dangerous viruses must not be allowed to spread unchecked through the population. Note that these constraints represent the hazards that we try to prevent in designing and operating systems.

Control is interpreted broadly here and includes everything that is currently done in safety engineering, plus more. For example, component failures and unsafe interactions may be controlled through design, such as the use of redundancy, interlocks, barriers, safety margins, and fail-safe design. Safety may also be controlled through processes, such as development processes, manufacturing processes and procedures, maintenance processes, and general system operating processes. Finally, safety may be controlled using social controls including government regulation, culture, insurance, law and the courts, or individual self-interest. Human behavior can be partially controlled through the design of the societal or organizational incentive structure and not simply by issuing rules and procedures.

Hierarchy and communication Modeling complex sociotechnical systems requires a modeling and analysis approach that includes both social and technical aspects of the problem and allows a combined analysis of both. Some techniques that are labeled as sociotechnical, as noted earlier, in fact include only analysis of the social system or the technical system in isolation and not their true integration.

Systems can be conceived as levels of control that form a hierarchy [60]. The hierarchies are characterized by control processes working at the interfaces between levels. The control actions impose constraints on the activity at a lower level of the hierarchy. The levels of the hierarchy are characterized by the emergent properties of the components at the lower levels.

Control actions impose constraints on the activity at a lower level of the hierarchy. That is, a control action at one level of the hierarchy represents constraints on the degree of freedom of the lower-level system components. Feedback must be capable of detecting potential disruptive behavior at the lower level or environmental changes and used to initiate control and remedial actions on the lower level(s).

Figure 3.9 shows an example of a hierarchical safety control structure for a typical regulated industry in the United States. Notice that the operating process, which is the focus of most hazard analysis techniques today, in the lower right of the figure makes up only a small part of the safety control structure.

There are two basic hierarchical control structures shown in this particular example—one for system development (on the left) and one for system operation (on the right)—with interactions between them. Other control structures and components, such as emergency services, could have been included. Figure 3.9 is simply one example.

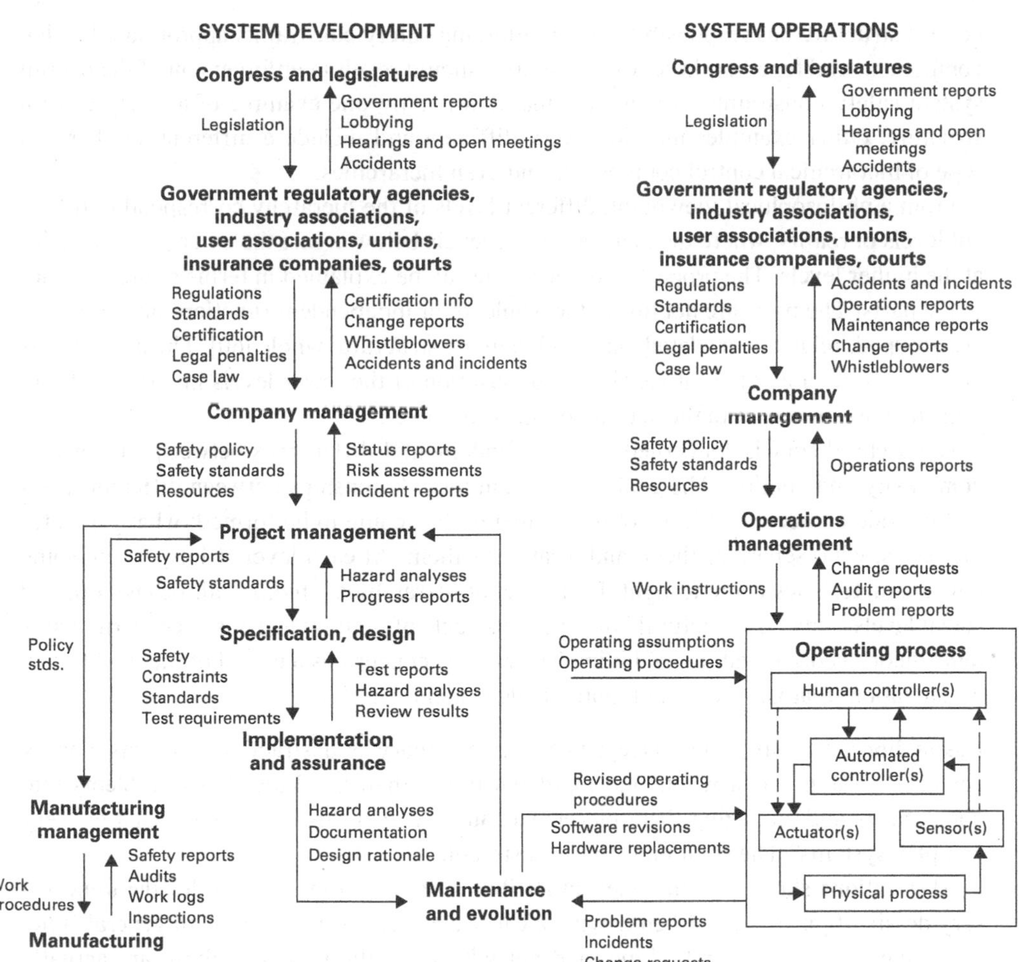

Figure 3.9
An example of a safety control structure.

Each level of the structure contains controllers with responsibility for control of the behavior of the individual components at the next lower level as well as the interactions between those components. Higher-level controllers may provide overall safety policy, standards, and procedures (downward arrows), and get feedback (upward arrows) about their effect in various types of reports, including incident and accident reports. The feedback provides the ability to learn and to improve the effectiveness of the safety controls and control actions. Communication occurs between and within levels of the hierarchy.

There are usually many interactions between the control structures; only a few are shown in the figure for readability purposes. Manufacturers must communicate to their customers the assumptions about the operational environment in which the original safety analysis was based; for example, maintenance quality and procedures, as well as information about safe operating procedures. The operational environment, in turn, provides feedback to the manufacturer and potentially others, such as governmental authorities, about the performance of the product during operations. Each component in the hierarchical safety control structure has responsibilities for enforcing safety constraints appropriate for that component, and together these responsibilities should result in enforcement of the overall system safety constraints. Again, note that this is only one example of a safety control structure. Other examples may look very different and include a different number and type of hierarchical control components and even hierarchies.

From a philosophical viewpoint, different levels of the hierarchy correspond to different levels of reality, where the behavior of the levels below create the emergent properties at the higher levels. The properties of the whole can be explained in terms of the behavior of the parts. The parts are not lost in the whole: their independent functions and activities are grouped, related, correlated, and unified in the structural whole [60]. The upper levels are the source of an alternative, simpler description of the lower levels in terms of functions that are emergent in the activity of the lower level.

Hierarchy theory is concerned with the fundamental differences between one level of complexity and another. The goal is to explain the relationships between different levels and provide an account of how observed hierarchies come to be formed: what generates the levels, what separates them, and what links them. At each level of integration, some new characteristics come to light. Each level of hierarchies—biological, behavioral, and social levels—may have natural laws and constructs of their own. Thus, a concern of systems theory is organization, and it implements this concern as a model of the world as an organization; that is, a system of mutually dependent variables.

Abstraction A final basic concept in Systems Theory is abstraction. Systems Theory considers systems as a whole and top-down rather than breaking down problems into components and analyzing them piecemeal. But how can this be accomplished for very complex systems? The solution is to use abstraction.

Abstractions show only the relevant attributes of something and "hide" the unnecessary details (figure 3.10). When driving a car, we are concerned only with general functions like accelerating or braking, and not with the details of how these are actually accomplished by the hardware and software. When something goes wrong in our car, we may shift to a different level of detail to handle it. We use abstractions or models to deal with a messy and complex world. The models filter out irrelevant information for the problem being solved at that time.

By this definition, it is clear that we will need multiple views or perspectives of the whole, changing focus and zooming in and out when necessary to accomplish our goals. All of engineering uses models, but analysis of these models using Systems Theory goes top-down rather than bottom-up. Each level is considered as a whole but at a different level of detail.

To summarize, Systems Theory combines these concepts of emergence, control, communication, hierarchy, and abstraction to view systems as interrelated components in a state of dynamic equilibrium by feedback loops of control and information.

3.8 Summary

With these definitions and theoretical concepts, you are ready to learn both traditional system safety engineering as well as the newer extensions based on Systems Theory that

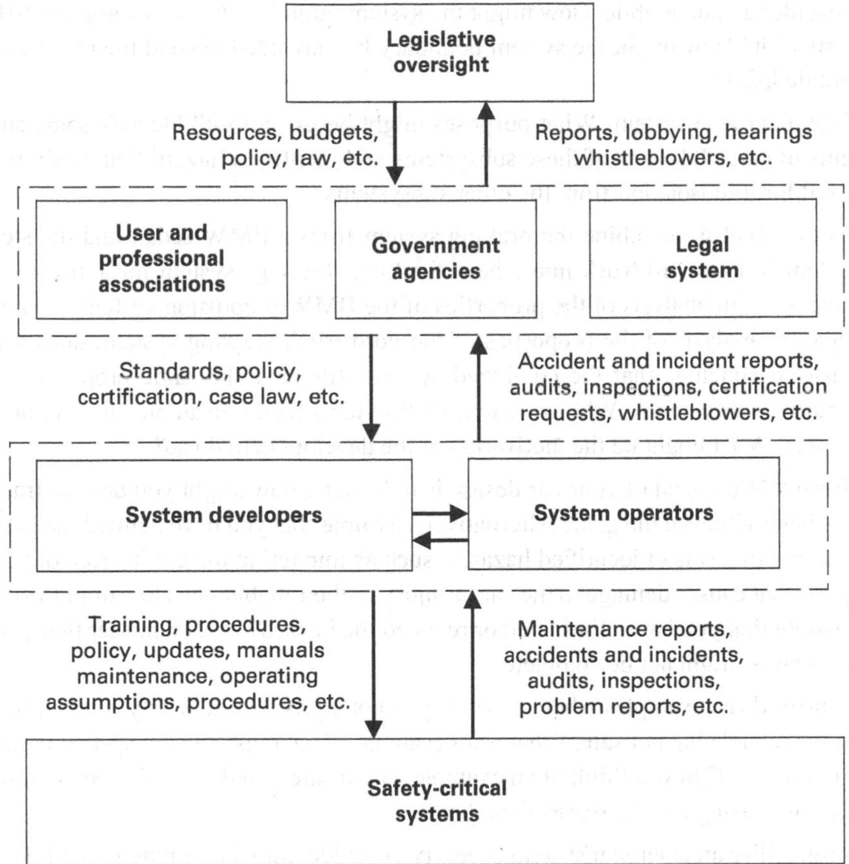

Figure 3.10
An abstraction of figure 3.9. Models or abstractions filter out information irrelevant to the property being considered.

have been created to deal with our modern, highly complex systems. The next step is to understand why accidents occur.

Exercises

1. Explain why defining something as "safe" in terms of the risk being "as low as reasonably practical" is a trans-scientific problem. Similarly, do you agree with the claim in this chapter that eliminating or reducing hazards is not trans-scientific? Why or why not? What are some difficulties in showing a system is safe? What are the limitations in showing it cannot get into hazardous states?
2. Nothing is totally safe under all conditions, no matter how extreme. That argument has been used to justify defining safety relatively rather than using an absolute definition. What do you personally think about this controversy?

3. Consider an automobile. How might the system boundary be drawn around different parts of it? How might the system boundary be expanded beyond the physical automobile itself?
4. Take a car as a system. What purposes might be given to it? Identify some subsystems of cars. Take one of these subsystems and identify a hazard that might be analyzed for it in isolation from the other subsystems.
5. Assume that we combine the braking system from a BMW sedan and the steering system from a Ford truck into a new "braking-steering" system for a Toyota. If we have done an analysis of the properties of the BMW propulsion system and another separate analysis of the properties of the Ford truck steering system, does it make sense to conclude that the combined system will have the same properties of the separate subsystems? Why or why not? Provide specific examples to support your answer. What might be the motivation of the developers to do so?
6. Given that the goal of your car design is to be safe, how might you demonstrate that you have achieved the goal? Alternatively, assume that you have defined the safety of your car in terms of identified hazards, such as impacting the car in front of you at a speed that causes damage to the car or injury to the inhabitants. How might you demonstrate that you have eliminated or reduced the hazard? Does it matter that your list of hazards might not be complete?
7. Identify three examples where reliability is not equivalent to safety; that is, the system is reliable but not safe. What is an example of the opposite: the system is safe but not reliable? Can you think of an example where safety and reliability are conflicting (i.e., increasing one decreases the other)?
8. Is reliability an emergent system property? Provide both an argument that it is emergent and an argument that it is, alternatively, analytically decomposable.
9. What are some safety constraints in your industry or field?
10. Consider the transportation system in a large city.
    a. Identify four subsystems that might be considered as part of this system.
    b. Is it reasonable to take the throughput of each of these systems and add them together to get a system throughput (i.e., the number of people transported in the city's public transportation system per day)? Why or why not? Provide specific examples.
    c. Analytic decomposition is based on four assumptions about the system (as described in the chapter). Considering your big city transportation system, do the four assumptions hold for the emergent property of "throughput" in this system? Provide an example of how each does or does not hold.
11. Does the design of the systems on which you work contain assumptions about independence of components or functions that might be compromised by indirect causal relationships?
12. Note that the use of the term "control" does not imply rigid command and control. Behavior is controlled not only by engineered systems and direct management intervention but also indirectly by policies, procedures, shared value systems, and other

aspects of the organizational culture. All behavior is influenced and at least partially "controlled" by the social and organizational context in which the behavior occurs. Engineering this social and organizational context can be an effective way to create and change a safety culture; that is, the subset of organizational or social culture that reflects the general attitude about and approaches to safety by the participants in the organization or society [348]. Provide some examples of this type of management and culture controls in a system with which you are familiar.

13. The Bad Apple Theory says that accidents are caused by incompetent people. If those people were simply excluded from the system, accidents would not occur; for example, if incompetent doctors lost their licenses to practice medicine or were not licensed in the first place, medical errors would not occur. What is wrong with the Bad Apple Theory from a systems thinking or Systems Theory perspective? Provide two specific examples where it might not be correct.

4

Why Accidents Occur

It is necessary to learn from the mistakes of others. You will not live long enough to make them all yourself.
—Adm. Hyman Rickover

Before we can successfully prevent accidents, we first need to understand why they occur. This goal is more difficult than it may appear to be. The goal of this chapter is to help you understand why it is hard and describe what is currently known about this topic. We need to start by defining causality.

4.1 The Traditional Conception of Causality

Western philosophers have debated the notion of causality for centuries. The standard models of causality used in safety are consistent with the definition by John Stuart Mill (1806–1873) that a cause is a set of *sufficient and necessary conditions.* "The cause is the sum total of the conditions positive and negative, taken together, the whole of the contingencies of every description, which being realized, the consequences invariably follow" [262].

As an example, combustion requires a flammable material, a source of ignition, and oxygen. Each of these conditions is necessary, but only together are they sufficient. The cause, then, is all three conditions, not one of them alone. The distinction between *necessary* and *sufficient* is important. An event may be caused by five conditions, but conditions 1 and 2 together may be able to produce the effect, while conditions 3, 4, and 5 may also be able to do so. Thus, there are two causes or sets of conditions sufficient for the event to occur; both of the causes have a set of necessary conditions (figure 4.1). The standard chain-of-events model used in safety (see chapter 7) is based on this definition of causality.

Note that, in this definition, simple association is not enough. A condition may precede another event without causing it. For Mill and those who use his definition, a regular association of events is causal only if it is unconditional—that is, only if its occurrence does not depend on the presence of further factors such that, given their presence, the effect would occur even if its assumed cause was not present. Thus, a condition or event may precede another event without causing it.

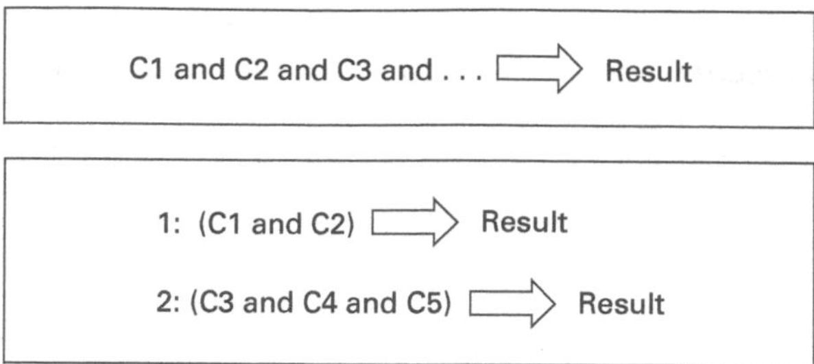

Figure 4.1
Examples of the classic definition of causality.

Furthermore, in this definition of causality, a condition is not considered to cause an event without the event happening every time the condition holds. Drunk driving, for example, is informally said to cause accidents, but being drunk while driving a car does not always lead to an accident. And a car accident does not only result from a drunk driver. The same type of relationship holds between smoking and lung cancer. Using this standard definition of causality, drunk driving does not "cause" automobile accidents and smoking does not "cause" lung cancer. The Tobacco Institute successfully made the argument that smoking does not cause lung cancer using this logic for decades. This argument does not mean, however, that in reality there is no causal connection but simply that the conditions may be indirect or depend on additional factors being true.

Both workplace safety and product/system safety have used and often still use today the traditional conception of causality. In fact, it is prevalent in almost all of Western society. Without some model of causality, we would be faced with a totally random world, with few tools to assist in engineering or in preventing accidents. But that does not mean that we are limited to the model of causality defined two hundred years ago. Simple, direct linear causality is not enough in today's world. We know there is an important relationship between drunk driving and accidents and between smoking and lung cancer, for example.

To handle accident causality in complex systems, we need to go beyond traditional definitions of causality. Accident causality today is always complex. In a well-investigated accident, there are almost always physical, technical, operational, social, managerial, and governmental contributors, not all of them conforming to the traditional narrow view of causality. In the disastrous release of MIC (methyl isocyanate) at Bhopal in 1984, there were technical factors and physical design errors in the plant as well as operator errors, management negligence, regulatory deficiencies, international agricultural policies, and more involved. The same is true for just about any major accident in the past century. Just because an accident investigation does not cite all these factors does not mean they were not present.

Why does this matter? Because taking only one of these perspectives or simplifying the explanation or cause for an accident inhibits learning and preventing similar accidents from occurring in the future. The NASA *Challenger* and *Columbia* Space Shuttle accidents had the same basic causes although the specific technical triggers were different (see appendix B). The same is true for the BP Texas City oil refinery explosion and the later BP *Deepwater Horizon* oil rig explosion and oil well blowout (see appendix C). The basic causes were never fixed after the first loss and therefore contributed to the second. Most of the same causal factors cross organizational boundaries and even industry boundaries. To prevent accidents, we need to look beyond the superficial triggering events and causes.

But before delving deeper into causes of accidents today, let's explore further the practical difficulties in understanding causality.

4.2 Subjectivity in Ascribing Causality

Descriptions of accident causes often involve subjectivity and filtering. Rarely are the causes of an accident perceived identically by corporate executives, engineers, union officials, operators, insurers, lawyers, politicians, the press, the state, and the victims. Bogard suggests that "the specification of possible causes will necessarily bear the marks of conflicting interests" [39, p. 1].

Some conditions may be considered hazardous by one group yet viewed as perfectly safe and necessary by another. In addition, each person questioned may attribute an accident to a different cause. Conflicts in causal attribution often involve normative, ethical, and political considerations. In addition, judgments about the cause of an accident may be affected by the threat of litigation. After examining the research data, Leplat concluded that causal attribution depends on some characteristics of the victim and of the analyst—hierarchical status, degree of involvement, and job satisfaction—as well as on the relationships between the victim and the analyst and on the severity of the accident [208].

Individuals who have a high position in the hierarchy tend to blame workers for accidents. Most everyone tends to cite factors in which they were not personally involved. Data on near-miss reporting suggests that causes for these events are mainly attributed to technical deviations while additional factors are cited when significant losses occur [92; 178].

Even the data collected may be systematically filtered and unreliable. While the basic events are usually not subject to interpretation, the actual events identified and how they are described may reflect bias.

Data usually is collected in the form of textual descriptions of the sequence of events preceding the accident, which tend to concentrate on obvious conditions or proximal events; that is, those closely preceding the accident in time. On the one hand, report forms that do not specifically ask for nonproximal events often do not elicit them. On the other hand, more directive report forms may limit the categories of conditions considered [179].

After the data is collected, more sources of subjectivity and bias arise. Accident causes can be described in terms of events, goals, or motives. Consider the example, provided by Leplat [209], of an accident that occurs when a person steps onto a sidewalk and trips on the pavement. This explanation (that is, stepping on the sidewalk and tripping) represents the basic events leading to the accident. An explanation based on goals would include the facts that a car was rapidly approaching and the person wanted to walk on the sidewalk. An explanation based on motive would describe the events as a car rapidly approaching and the person wanting to avoid the car.

Explanations that include goals and motives will be influenced by the analyst's mental model. Leplat illustrates this process by describing three different motives for the event "The operator sweeps the floor": (1) the floor is dusty, (2) the supervisor is present, or (3) the machine is broken, and the operator needs to find other work [209].

A final consideration is what psychology calls the Fundamental Attribution Error, also known as correspondence bias and attribution effect. This bias is the tendency to assume that someone else's actions depend on their personal characteristics rather than on the influence of social and environmental factors outside their control. In contrast, we are more likely to attribute our own mistakes to the environment in which they occur and not to our own flaws. The fundamental attribution error leads to blaming the victim or placing blame on individuals rather than on the context or system in which they are operating.

In summary, the causal factors selected for a particular accident may involve a great deal of subjectivity and need to be interpreted with care. Keep this in mind when reading the descriptions of the causal factors for the accidents in the appendices. They are one view of these accidents, although more comprehensive than usual.

4.3 Oversimplification in Determining Causality

A second trap in identifying accident causes is oversimplification. Out of a large number of necessary conditions for the accident, one is often chosen and labeled as *the* cause, even though all the factors involved were equally indispensable to the occurrence of the event. For example, a car skidding in the rain may involve many factors including a wet road, the driver's lack of knowledge about how to avoid a skid, the driver's lack of attention, and the car not being equipped with anti-skid brakes. None of these is sufficient alone to cause the skid, but one will often be cited as *the* cause.

A condition may be selected as the cause because it is the last condition to be fulfilled before the effect takes place, its contribution is the most conspicuous, or the selector has some ulterior motive for the selection. Miller [265], Lewycky [231], and others argue that although we often isolate one condition in analyzing an accident, call it the cause, and label other conditions as contributory, the basis for this distinction does not exist.

Oversimplification of the causal factors in accidents can be a hindrance in preventing future accidents. For example, in the crash of an American Airlines DC-10 at Chicago's O'Hare Airport in 1979, the US National Safety Transportation Board (NTSB) blamed only a maintenance-induced crack and not also a design error that allowed the slats to retract if the wing was punctured. Because of this omission, McDonnell Douglas was not required to change the design, leading to future accidents related to the same design error [303].

Some common types of oversimplifications include applying legalistic approaches, ascribing accidents to human error or technical failure alone, ignoring organizational factors, and looking for single causes.

4.3.1 The Legal Approach to Causality

Lawyers and insurers often oversimplify the causes of accidents and identify what they call the *proximate* (immediate or direct) cause. They recognize that many factors contribute to an accident but identify a principal factor for practical and particularly for liability reasons. The goal is to determine the parties in a dispute that have the legal liability to pay damages, which may be affected by the ability to pay or by public policy considerations.

Usually, however, there is no objective criterion for distinguishing one factor or several factors from other factors that make up the cause of an accident. Although the legal approach to causality may be useful for establishing guilt and liability, it is of little use from an engineering perspective, where the goal is to understand and prevent accidents. It may even be a hindrance because the most relevant factors—in terms of preventing future accidents—may be ignored for nontechnical reasons.

Haddon [127] argues that countermeasures to accidents should not be determined by the relative importance of the accident's causal factors, however that may be defined; instead, priority should be given to the measures that will be most effective in reducing future losses. Simplistic explanations for accidents often do not provide the information necessary to prevent them, and, except for liability reasons, spending a lot of time determining the relative contributions of causes to accidents is not productive.

4.3.2 Human Error as the Cause of Accidents

The most common oversimplification in accident reports is blaming the operator. In any system where humans are involved, a cause may always be hypothesized as the failure of the human to step in and prevent the accident: virtually any accident can be ascribed to human error in this way. Even when human error is more directly involved, considering that alone as a cause and the elimination of humans or of human error as the solution is too limiting to be useful in identifying what to change in order to increase safety most effectively.

Modern views of human behavior recognize that behavior is always affected by the context in which it occurs. Designing that context may be the most effective way to change behavior.

In general, human error is used much too loosely to describe the cause of accidents; too often, analysis stops there and ignores all the other necessary conditions that must accompany human error for the accident to occur. "*Almost any accident can be said to be due to human error and the use of the phrase discourages constructive thinking about the action needed to prevent it happening again; it is too easy to tell someone to be more careful.... Perhaps we should stop asking if human error is the "cause" of an accident and instead ask what action is required to prevent it happening again*" [190, p. 182].

Because an understanding of the role of humans in accidents is so important in reducing risk, it is discussed in depth in a chapter of its own. The role that humans play in accidents is also considered in chapter 8 on accident causal analysis.

4.3.3 Technical Failures as the Cause of Accidents

Another important oversimplification is concentrating only on technical failures and immediate physical events. This type of overly narrow focus may lead to ignoring some of the most important factors in an accident. For example, an explosion at a chemical plant in Flixborough, Great Britain, in June 1974, resulted in twenty-three deaths, fifty-three injuries, and $50 million in damages to property (including 2,450 houses) up to 5 miles from the site of the plant. Many more lives might have been lost had the explosion not occurred on a weekend when only a small shift was at the plant, the wind was light, and many of the nearby residents were away at a Saturday market in a neighboring town. The official accident investigators devoted most of their effort to determining which of two pipes was the first to rupture. The British Court of Inquiry concluded that "The disaster was caused by a coincidence of a number of unlikely errors in the design and installation of a modification," and "such a combination of errors is very unlikely ever to be repeated" [79].

Yet the pipe rupture was only a small part of the cause of this accident. A full explanation and prevention of future such accidents require an understanding, for example, of the management practices of running the Flixborough plant without a qualified engineer on site and allowing unqualified personnel to make important modifications to the equipment, of making engineering changes without properly evaluating their safety, and of storing large quantities of dangerous chemicals close to potentially hazardous areas of the plant [69].

The British Court of Inquiry investigating the accident amazingly concluded that "*There were undoubtedly certain shortcomings in the day to day operations of safety procedures, but none had the least bearing on the disaster or its consequences and we do not take time with them*" [79]. Fortunately, others did not take this overly narrow viewpoint, and Flixborough led to major changes in the way hazardous facilities were allowed to operate in Britain.

4.3.4 Organizational Factors as the Cause of Accidents

Large-scale engineered systems are more than just a collection of technological artifacts: they are a reflection of the structure, management, procedures, and culture of the engineering organizations that create them. They are also, usually, a reflection of the society in which they are created. Accidents are often blamed on operator error or equipment failure without recognition of the industrial, organizational, and managerial factors that made such errors and defects inevitable. The causes of accidents are frequently, if not almost always, rooted in the organization—its culture, management, and structure. These factors are all critical in the eventual safety of the engineered system.

Some support for this hypothesis about the cause of accidents can be found by looking at major accidents of the past. In-depth, independent accident investigations usually point to organizational and managerial deficiencies. For example, the Kemeny Commission's official report on the Three Mile Island (TMI) accident contains nineteen pages of recommendations, two of which deal with technical matters and seventeen of which concern management, training, and institutional shortcomings in the nuclear industry. The commission concluded that the nuclear industry must "dramatically change its attitudes toward safety and regulation" and recommended that the utilities and the suppliers establish appropriate safety standards; systematically gather, review, and analyze operating experiences; plan to make changes with respect to a realistic deadline; integrate management responsibilities; clearly define roles and responsibilities; attract highly qualified personnel; devote more care and attention to plant operating procedures (providing clear and concise wording, clear formats, practical procedures, and so on); and establish deadlines for resolving safety issues [173].

Similar results were found by a commission investigating the explosion on the *Bravo* oil rig in the North Sea. Seven main reasons are given for the accident, only one of which concerns a purely technical question. The others refer to organizational and training problems [361].

These are not isolated cases. In most of the major accidents of the past twenty-five years, technical information on how to prevent the accident was known and often even implemented. But in each case, the technical information and solutions were negated by organizational or managerial flaws.

4.4 Multifactorial Explanations of Accidents

In general, any particular condition in a complex technological system is unlikely to be either necessary or sufficient to cause an accident. In most systems, where some care has been taken in their design, accidents or hazards will depend on a multiplicity of causal factors and each event on a complex combination of conditions, including technical, operator, and organizational or societal conditions and actions. For example, design errors in the DC-10 caused accidents only after millions of flight hours. For such accidents to happen, the conditions involved necessarily must be very rare or must be the result of a complex interaction and coincidence of factors; preventing any of these conditions might eliminate losses.

The high frequency of accidents having complex causes probably results from the fact that competent engineering and organizational structures eliminate the simpler causes. On the positive side, the very complexity of accident processes means that there may be many opportunities to intervene or interrupt them: thorough consideration of the conditions leading to accidents will be more useful than simplistic explanations. On the negative side, the complexity of accident causes makes the process of identifying them more difficult.

Lewycky [231] has proposed a three-level model for understanding causality (figure 4.2) based on multiple hierarchical levels that includes technical, human, organizational, and regulatory perspectives.

In this general organization of causality, the lowest level describes the mechanism of the accident—the chain of events. For example, the driver hit the brake, the car skidded and hit the tree, and the driver was thrown from the car and injured. That level provides little information useful in preventing future accidents.

The second level of understanding causality includes the conditions or lack of conditions that allowed the events at the first level to occur. For example, the driver does not know how to prevent or stop the skid, the car is not equipped with antilock brakes, the driver is driving too fast, the street is wet from rain and friction is reduced, an object appears in front of the car suddenly that requires the driver to apply the brakes quickly, the driver is not wearing a seat belt, the seat belt is defective, and so on. At this level, the cause or causes may be over-specified; that is, not all conditions may have to be met before the accident will occur.

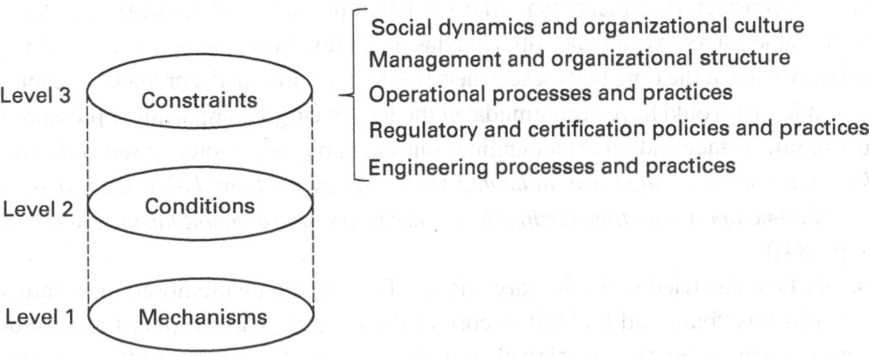

Figure 4.2
Three levels of accident causes.

The third level includes constraints or the lack of constraints that allowed the conditions at the second level to cause the events at the first level, or that allowed the conditions to exist at all. These are more commonly called *systemic factors.* This level includes causal factors related to engineering and operational processes, social dynamics and organizational culture, the management system and organizational structure, and governmental and socioeconomic policies.

This third level affects general classes of accidents, not just the accident being considered. Responses to accidents or attempts to prevent them tend to involve fixing the lower-level events and conditions while ignoring the more general or systemic factors.

Because accidents rarely repeat themselves in exactly the same way, however, patches to particular parts of the system may be ineffective in preventing future accidents. For example, the *Challenger* Space Shuttle accident was not caused simply by an O-ring problem: an O-ring failure precipitated the accident, but the systemic factors identified during the accident investigation were related to organizational deficiencies. These same organizational factors led to the *Columbia* loss, although the events involved shedding foam and not O-rings (see appendix B).

The DC-10 cargo-door saga is another example of the serious consequences of failing to eliminate systemic causes. In March 1974, one of the worst accidents in aviation history occurred when a fully loaded Turkish airline DC-10 jumbo jet crashed in the suburbs of Paris, killing 346 people. The crash was attributed to the faulty closing of a cargo door, causing rapid decompression of the cabin and collapse of the cabin floor. The collapse destroyed all the major control cables and hydraulic lines. The inherent design flaw in running the cables through the cabin floor was amply demonstrated in advance by a failed pressure test of the aircraft in 1970 that caused the cabin floor to collapse; by a near loss of a DC-10 near Windsor, Ontario, in 1972, which was avoided only by the extraordinary poise and skill of the pilot; and by the warnings of three Dutch engineers in 1968 when the first prototype was being built [345]. In addition, Convair, who was the DC-10 fuselage contractor, in a 1969 engineering analysis found that the design of the cargo door could lead to loss of the aircraft. However, the documentation required by the Federal Aviation Administration (FAA) to be submitted by Douglas before certification of an aircraft did not mention this hazard [91].

An internal memorandum written by Dan Applegate, Convair's senior engineer on the fuselage subcontract, documents that after the failed pressure test, Convair engineers discussed possible corrective actions with Douglas, including blow-out panels in the floor [91]. Applegate argued at the time that these panels would have provided a predictable cabin floor failure mode that would have accommodated the loss of cargo compartment pressure without loss of tail surface and aft center engine control. Applegate wrote, "*It seemed to us then prudent that such a change was indicated since 'Murphy's Law' being what it is, cargo doors will come open sometime during the twenty years of use ahead for the DC-10*" (cited in [91, p. 184]).

Instead, Douglas tried to fix the cargo doors. The Applegate memorandum claims that not only did this "band-aid fix" fail to correct the inherent catastrophic failure mode of cabin floor collapse, but the redesign also further degraded the safety of the original door latch system.

In June 1972, a cargo door latch system in a DC-10 failed over Detroit at 12,000 ft, leading to the collapse of the cabin floor and the disabling of most of the tail controls.

Only by chance was the plane not lost. Douglas again studied corrective actions and, according to Applegate, applied more band-aids to the cargo door latching system.

> *It might well be asked why not make the cargo door latch system really "fool-proof" and leave the cabin floor alone. Assuming it is possible to make the latch "fool-proof" this doesn't solve the fundamental deficiency in the airplane. A cargo compartment can experience explosive decompression from a number of causes such as sabotage, mid-air collision, explosion of combustibles in the compartment and perhaps others, any of which may result in damage which would not be fatal to the DC-10 were it not for the tendency of the cabin floor to collapse.* (cited in [91, p. 184])

Applegate ends his memorandum this way: "*It seems to me inevitable that, in the twenty years ahead of us, DC-10 cargo doors will come open and I would expect this to usually result in the loss of the airplane*" (cited in [91, p. 185]). The Paris accident proved that he was correct. The systemic factors related to inadequate government oversight, bad management decision-making, safety culture problems, and others allowed the design flaws to remain and the loss to occur.

When serious accidents happen, they usually occur in unforeseen ways. But the fact that the exact events leading to an accident are not foreseen does not mean the accident is not preventable. The hazard is usually known, and measures can often be taken to eliminate or reduce it. The exact events at Chernobyl, as another example, did not have to be predicted to know that the reactor design made the operators' job too difficult and would likely lead to operator errors (see appendix D).

In both the DC-10 and Chernobyl cases, decisions were made to try to eliminate the proximal events because they would require the fewest tradeoffs with other goals, such as lower cost or desired functionality. Attempting to avoid those tradeoffs, in the long run, ended up costing much more than changing the design would have.

The lesson is clear: to reduce the risk of accidents significantly, systemic factors must be identified and eliminated. Otherwise, we will find ourselves chasing symptoms, not the cause of those symptoms, and engaging in continual fire-fighting activities.

The rest of this chapter provides an overview of the primary systemic causes of accidents in the past. The categorization is, of necessity, neither complete nor the only one possible. But it does include the factors common to most, if not all, major accidents.

4.5 Systemic Causes of Accidents

The systemic causes of accidents can be divided into six general categories: (1) social dynamics and organizational culture; (2) management decision-making structure; (3) operational processes and practices; (4) regulatory, certification, and licensing policy and practices; (5) engineering processes and practices; and (6) safety information system deficiencies. To effectively eliminate or reduce accidents, an understanding of these factors is needed. The systemic factors include:

- **Social dynamics and organizational culture**
  - Overconfidence and complacency
    - Discounting risk
    - Unrealistic risk assessment

  - Ignoring high-consequence, (supposedly) low-probability events
  - Assuming risk decreases over time
  - Ignoring warning signs and other feedback
- ◦ Low priority given to safety
- ◦ Flawed resolution of conflicting goals
- ◦ Confusing safety with other system properties
  - Confusing safety with reliability
  - Equating process safety with personal safety
  - Focusing on success rather than failure

- **Management decision-making structure**
  - ◦ Ill-defined and diffused responsibility, authority, and accountability
  - ◦ Lack of independence and low-level status of safety personnel
  - ◦ Limited communication channels and poor information flow
- **Operational processes and practices**
  - ◦ Superficial, isolated, or misdirected safety efforts during operations
  - ◦ Inadequate feedback and learning from events
  - ◦ Poorly defined operating procedures
  - ◦ Inadequate training and emergency management
  - ◦ Reduced effort during operations
  - ◦ Inadequate management of change
  - ◦ Poor maintenance practices
- **Government and professional society oversight**
- **Engineering processes and practices**
  - ◦ Superficial safety efforts
  - ◦ Over-relying on redundancy and protection systems
  - ◦ Ineffective risk control
    - Not eliminating basic design flaws
    - Basing safeguards on false assumptions
    - Complexity
    - Using risk-control devices to reduce safety margins
    - Lack of defensive design
- **Safety information system**

Many of the accidents that are used in the examples for the rest of this chapter are examined in depth in the appendices. More details can be found there.

4.5.1 Social Dynamics and Organizational Culture

Social dynamics and organizational culture greatly influence individual decision-making. Culture is more carefully defined in chapter 14, but it is basically a shared set of norms and values. When applied to an organization, culture denotes the values and assumptions that underlie the decision-making of those in the organization. The organization or industry's rules, policies, and practices are based on the organizational culture, all of which can change over time.

All engineering efforts take place within a political, social, and historical context that has a major impact on decision-making. For example, consider the Space Shuttle where political and other factors contributed to the adoption of a vulnerable design during the original approval process. Unachievable promises were made with respect to performance in order to keep the manned space flight program alive after Apollo and the demise of the Cold War. While these performance goals even then seemed unrealistic, the success of the Apollo program and the *can-do* culture that arose during it contributed to the belief that these unrealistic goals could be achieved if only enough effort was expended.

The Rogers Commission study of the Space Shuttle *Challenger* accident concluded that the root cause of the accident was an accumulation of organizational problems [332]. The commission was critical of management complacency, bureaucratic interactions, disregard for safety, and flaws in the decision-making process.

Despite a sincere effort to fix these problems after the *Challenger* loss, seventeen years later almost identical management and organizational factors and cultural flaws were cited in the *Columbia* Accident Investigation Board (CAIB) report.

Social dynamics and organizational culture flaws contributed to all the accidents described in the appendices and almost all accidents in general. Preventing losses requires understanding the social dynamics and aspects of organizational culture that impact risk, including: (1) overconfidence and complacency, (2) a disregard for safety or assigning a low priority to it, and (3) confusing safety with other system properties.

4.5.1.1 Overconfidence and complacency Complacency and overconfidence are common elements in most of the major accidents of the last one hundred years. For example, the Kemeny Commission identified a major contributor to the TMI accident as a failure by the Nuclear Regulatory Commission (NRC) to believe that a serious accident could happen. One official testified that those involved had developed a belief about the infallibility of the equipment as a result of repeated assurances that the technology was safe [173]. An NRC post-TMI Lessons Learned Task Force noted that the mindset regarding serious accidents was "probably the single most important human factor with which this industry and the NRC has to contend" [9, p. 43].

According to a former Atomic Energy Commission attorney, Harold Green: "*Nobody ever thought that safety was a problem. They assumed that if you just wrote the requirement that it be done properly, it would be done properly*" [9, p. 3]. Sometimes, lessons learned from accidents, even in the same industry, do not cross international borders:

> *Eight months after the TMI accident, top Soviet government and scientific leaders told a party from the United States that they regarded nuclear safety as a "solved problem" and that the problems raised by the U.S. experience at TMI had been overdramatized. They quoted the head of the Soviet Academy of Sciences as saying that Soviet reactors would soon be so safe that they could be installed in Red Square* [201]. *Soviet authorities at the Chernobyl plant described the risk of a serious accident as extremely slight a year before it occurred, and only a month before the Chernobyl accident, the British Secretary of State for Energy repeated the often stated belief that "nuclear energy is the safest form of energy yet known to man."* [303]

A US observer in Vienna noted that even after the Chernobyl events, the Soviets were still incredulous that the accident could have happened [12].

After the Fukushima nuclear power plant meltdown in March 2011, the prime minister of Japan said that the government shared the blame for the disaster, saying that officials had been blinded by a false belief in the country's "technological infallibility," and were taken in by a "safety myth" [198]. This myth consisted of a belief that serious accidents could never happen in nuclear power plants in Japan. Risks both before and during the accident were downplayed, and complacency was widespread. This complacency resulted in an inability to consider such events as a reality, to plan for them, and to respond to the events appropriately.

Nuclear energy is not the only area where optimism prevails. After the *Challenger* accident in January 1986, the official Rogers Commission report [332] pinpointed two related causes of the loss as complacency and a belief that less safety, reliability, and quality-assurance activity was required during "routine" Shuttle operations.

The Bhopal accident is a classic case of events caused by widespread complacency. The accidental release of MIC came as a complete surprise to almost everyone, including Union Carbide's own scientists and risk assessors. They believed that such a catastrophe could not happen with such modern technology, that so many safety devices could not fail simultaneously, and that the Bhopal plant was a model facility [40]. Claims were made that overemphasized or exaggerated the safety of the production. Many employees believed that adequate precautions had been taken and that nothing more could be done to improve safety. Problematic practices were labeled "acceptable" or "necessary" risk.

Employees at Bhopal were apathetic about routine mishaps and about the value of emergency drills. Although several minor accidents and many warnings had preceded the Bhopal disaster, nobody seemed to believe that an accident of the size that occurred was possible. The Union Carbide Bhopal plant works manager, when informed of the accident, said in disbelief: *"The gas leak just can't be from my plant. The plant is shut down. Our technology just can't go wrong, we just can't have leaks"* [40]. Such disbelief was common before the accident, even though the plant had experienced so many problems with hazard detection instruments that, even when the instruments worked reliably and recorded real changes, these changes were often ignored and assumed to be faulty. One gauge in the MIC unit, right before the MIC release, recorded a fivefold rise in the storage tank's pressure, but it was ignored or disbelieved.

After the Bhopal disaster, both Union Carbide and the US Occupational Safety and Health Administration (OSHA) announced that the same type of accident could not occur at Union Carbide's plant in Institute, West Virginia (which also made MIC) because of that plant's better equipment, better personnel, and America's generally "higher level of technological culture" [303]. Yet only eight months later, a similar accident occurred at the Institute plant that led to brief hospital stays for approximately one hundred people. The consequences were less serious in this accident because of such incidental factors as the direction of the wind and the fact that the tank happened to contain a less toxic substance at the time. As at Bhopal, the warning siren at the Institute plant was delayed for some time, and the company was slow in making information available to the public [201].

A few months after the Institute accident, a leak at yet another Union Carbide plant created a toxic gas cloud that traveled to a shopping center. Several people had to be given emergency treatment, but for two days doctors and health officials did not know what the toxic chemical was or where it came from because Union Carbide denied the leak's existence [304]. OSHA fined Union Carbide $1.4 million after the Institute accident, charging "constant, willful, and overt violations" at the plant and a general atmosphere and attitude that "a few accidents here and there are the price of production" [304].

In these and most other major accidents, complacency played a major role. Examining more carefully the various aspects of complacency will help us to understand it and its contribution to accidents. These aspects include discounting risk, over-relying on redundancy, unrealistic risk assessments, ignoring high-consequence and supposedly low-probability events, assuming risk decreases over time, underestimating software-related risks, and ignoring warning signs.

Discounting risk One aspect of complacency is the basic human tendency to discount risk. Most accidents in well-designed systems involve two or more supposedly low-probability events occurring in the worst possible combination. When people attempt to predict the system risk, they explicitly or implicitly multiply events with low probability—assuming independence—and come out with impossibly small numbers, when, in fact, the events are not independent. Machol calls this phenomenon the *Titanic Coincidence* [240].

When the *Titanic* was launched in 1912, it was the largest and supposedly safest ship the world had ever known. The fact that it was unsinkable was widely touted and generally accepted. The ship was designed with a double-bottom hull having sixteen separate water-tight compartments. Calculations showed that up to four compartments could be ruptured without the ship sinking: in the history of maritime accidents, none had involved the compromise of more than four underwater compartments.

The builders had promised an unsinkable hull, and confidence in their claim was so great that one of the ship's officers assured a female passenger that "not even God himself could sink this vessel" [200]. Lloyd's of London issued the *Titanic* with a certificate of unsinkability, even though the partitions between the sixteen compartments were not high enough to shut every compartment hermetically. In case of trouble, designers thought that there would be time to intervene before the water reached the height where it could spill over into adjacent compartments.

On its maiden voyage, while the owners were trying to break the current speed record, the *Titanic* ran into an iceberg that cut a 300 ft gash in one side of the ship, flooding five adjacent compartments. The ship sank with a loss of 1,513 lives. Several telegrams had been received that day warning about the presence of icebergs, but nobody worried. When the *Titanic* hit the iceberg about 95 miles south of Newfoundland, passengers were told that there was no reason for alarm and that they should stay in their cabins. The captain finally ordered an evacuation, but "*the evacuation turned into disorder and terror: the classical evacuation exercises had not been carried out, the sailors did not know their assignments*" [200].

A number of "coincidences" contributed to the accident and the subsequent loss of life; for example, the captain was going far too fast for existing conditions; a proper watch was not kept; the ship was not carrying enough lifeboats; lifeboat drills were not held; the lifeboats were lowered properly, but arrangements for manning them were insufficient; and the radio operator on a nearby ship was asleep and so did not hear the distress call. Many of these events or conditions might be considered independent, but they appear less so when we consider that overconfidence most likely led to the excessive speed, the lack of proper watch, and the insufficient number of lifeboats and drills. That the collision occurred at night contributed to the iceberg not being easily seen, made abandoning ship more difficult than it would have been during the day, and was a factor in why the nearby ship's radio operator was asleep [252]. Multiplying together the probability of all these conditions would not have provided a good estimate of the probability of the accident occurring. In fact, given these factors, an accident was almost inevitable.

Watt defines a phenomenon he calls the *Titanic Effect* to explain the fact that major accidents are often preceded by a belief that they cannot happen. The *Titanic* effect says that the magnitude of disasters decreases to the extent that people believe that disasters are possible and plan to prevent them or to minimize their effects [402]. Taking action in advance to prevent or deal with disasters is usually worthwhile, because the costs of doing so are inconsequential when measured against the losses that may ensue if no action is taken.

Unrealistic risk assessment Unrealistic risk assessments contribute to complacency and poor decision-making. For example, instead of launching an investigation when first informed about possible overdoses by their radiation therapy machine called the Therac-25, the manufacturer responded that the probabilistic risk assessment showed that accidents were impossible, and no action was taken (appendix A). When forced to confront the fact that patients were being injured or dying, a microswitch was fixed and a claim was made that later accidents could not have been caused by their machine as the new probabilistic risk assessment showed that safety had been increased by five orders of magnitude. The microswitch had nothing to do with the design flaw creating the overdoses, however.

In the original risk assessment, software errors were excluded. Later, software was included, but the event "Computer selects wrong energy" was assigned a probability of 10^{-11} and the event "Computer selects wrong mode" was assigned four times 10^{-9}. No justification for either number was provided. After the accidents and the corrections to the machine were made, the manufacturers performed another risk analysis. This time, they included software in the fault tree, but they assigned a probability of 10^{-4} to every type of software error. This is a popular number to assign to any box in a fault tree that includes the word "software," but there is no justification for such a number. In addition, it is unlikely that all software errors would be of equal probability. In fact, the assignment of probability numbers for software doing something unsafe is not supported by science.

Software is pure design, so erroneous behavior is always a result of design errors. Even if the probability of such a design error could be assessed probabilistically, it would require so much information about the flaw that it could and should be fixed instead of ignoring it by declaring it will never or rarely occur. More generally, there is no justification for using probabilistic risk assessment on systems containing software or cognitively complex human decision-making. While it might be appropriate for human tasks involving simple motions and decisions, those tasks have mostly been automated today.

Another example of unrealistic risk assessment occurred in the Space Shuttle program after the *Challenger* accident but before the loss of the *Columbia*. William Readdy, head of the NASA Manned Space Program, for instance, wrote in 2001 that "the safety of the Space Shuttle has been dramatically improved by reducing risk by more than a factor of five" [112, p. 101]. It is difficult to imagine where this number came from as safety upgrades and improvements were deferred while, at the same time, the infrastructure continued to erode.

Risk assessment is extremely difficult for complex, technically advanced systems such as the Space Shuttle. When this engineering reality is coupled with the social and political pressures existing at the time, the emergence of a culture of denial and overoptimistic risk assessment is not surprising.

Probabilistic risk assessment might be useful in making some decisions about the reliability of particular pure hardware designs. Using it for system safety, however, is not justified. In addition, the assessments are too often misused, even for hardware, and lead to overconfidence in the small numbers obtained.

One major limitation is that probabilistic assessments measure only what they *can* measure and not necessarily what *needs* to be measured. The unmeasurable factors, such as design flaws and management errors, are ignored—even though they may have a greater influence on safety than those that are measurable. In addition, the underlying assumptions of the assessment—for example, that the plant or system is built according to the design or that certain failures are independent—are often ignored and untrue. These assumptions, which are usually easily violated, are often forgotten, and the numbers are taken at face value. The belief grows that the numbers actually have some relation to the real risk of accidents.

A US Air Force handbook description of an accident in a highly critical system provides a cogent example [13]. The system design included a relief valve opened by the operator to protect against overpressurization. A secondary valve was installed as backup in case the primary relief valve failed. The operator must know if the first valve did not open so that the second valve could be activated.

On one occasion, a position indicator light and open indicator light both illuminated; however, the primary relief valve was *not* open, and the system exploded. A post-accident examination discovered that the indicator light circuit was wired to indicate *presence of power* at the valve, but it did not indicate valve *position.* Thus, the indicator showed only that the activation button had been pushed, not that the valve had actually operated.

An extensive quantitative risk assessment of this design had included a low probability of simultaneous failure for the two relief valves but ignored the possibility of a design error in the electrical wiring; the probability of the design error was not quantifiable. No actual examination of the electrical wiring was made; instead, confidence was established on the basis of the low probability of coincident failure of the two relief valves. This same type of design error was also a factor in the TMI accident: an indicator misleadingly showed that a discharge valve had been ordered closed but not that it had actually closed. In fact, the valve was blocked in an open position.

Almost all major accidents have involved important causal factors that are not quantifiable. In fact, a case can be made that the most important causal factors in terms of accident prevention are often the unmeasurable ones. As just one example, the Bhopal accident involved such unmeasurable factors as the refrigeration being disconnected by management, an operator ignoring or not believing a recording on a gauge, operators putting off investigating the smell of MIC until after a tea break, the vent scrubbers being turned off, the insufficient design and capacity of the scrubbers and the flare tower, and not informing the community about what to do in case of emergency.

Most such "design" errors, if they are known beforehand, should be fixed rather than measured. A probabilistic risk assessment of the Bhopal facility that included only the usual failure probability of the protection system components could not have predicted this accident. In fact, each of the protection components (including the flare tower and the refrigeration unit) at the time of the accident was out of commission because of broken parts, was not designed to handle the amount of MIC released or had been turned off to save money.

Another limitation of probabilistic risk assessment is that it frequently changes the emphasis in development from making a system safer to proving that the system is safe as designed. Attention is directed away from critical assessment of the design and instead is focused on lowering the numbers in the assessment. Or the designers may work to achieve the numerical goals and proceed no further, even when additional corrective action is possible. In the worst case, such assessments start with a required probability, such as 10^{-9} per year or per use, and then a model is built to justify that number for the existing design—often an exercise in fantasy rather than engineering.

William Ruckelshaus, two-time head of the US Environmental Protection Agency (EPA) has cautioned that "risk assessment data can be like the captured spy; if you torture it long enough, it will tell you anything you want to know" [339, p. 157].

Assertions are common that decisions cannot be made without probabilistic risk assessment, but a counterexample of that belief is the US Navy's submarine safety program called SUBSAFE, one of the most successful safety programs ever devised. SUBSAFE allows decision-making about safety using only *Objective Quality Evidence* (OQE), which is defined as "any statement of fact, either quantitative or qualitative, pertaining to the quality of a product or service, based on observations, measurements, or tests *that can be verified*" [217]. Because it is not possible to verify the correctness of probabilities about what will happen in the future, such risk assessments are not allowed. No submarine in the SUBSAFE program has been lost in the fifty-nine years since SUBSAFE was created after the *Thresher* loss in 1963 [6], an astounding historical track record. Before 1963, one submarine was lost on average every three years.

The limitations of risk assessment do not mean that probabilistic analysis is not useful for some purposes, only that it should be interpreted and used with care. E. A. Ryder of the British Health and Safety Executive has written that the numbers game in risk assessment "should only be played in private between consenting adults as it is too easy to be misinterpreted" [341, p. 12].

Ignoring high-consequence, (supposedly) low-probability events Complacency often arises from a limited, unsystematic consideration of serious risks. Usually, the most likely hazards are controlled, but hazards with high severity and assumed low probability are dismissed as not worth investing resources to prevent. A common discovery after accidents is that events involved were recognized before the accident but dismissed as incredible. This was true, for example, in Japan before the Fukushima Daiichi nuclear power plant accident.

After the accident occurs, we find that independence of causative events was incorrectly assumed or that probabilistic assessments were unrealistic. Finding examples is easy—most accidents that occur in systems that have safety programs are of this type. Ayres and Rohatgi wrote of Bhopal:

> *People generally learn best by experience. People can—and do—learn to behave with reasonable caution so as to avoid accidents of kinds they have personally experienced*

or seen at first hand. This stimulus-response mechanism for learning caution depends on feedback between accidents and safety-related activities. Study after study reveals that new safety regulations in all countries are adopted largely after major accidents—not in advance of them. Shutting the barn door after the cow escapes seems to be an irremediable human trait. Obviously, this mechanism works best in avoiding repetitions of small to medium-sized accidents that are reasonably frequent, i.e., they have occurred before. Humans seem to be unwilling to be proportionally more careful to avoid larger but rarer calamities of kinds they have never personally experienced. The fact that a complex system has not (yet) failed massively is perhaps regarded subconsciously as evidence that it is fail-safe. This, in turn, leads to laxity. [27, p. 36]

A Therac-25 operator, who was involved in two of the radiation overdoses, testified that she had been told the system had so many safety devices that it was impossible for an overdose to occur (appendix A). The first accident reports were not investigated; instead, those reporting them were told that an accident was impossible on this machine. Proof of this impossibility was provided in terms of the number of safety devices on the equipment, and, later, such proof included a hardware change that increased the safety of the machine, according to the manufacturer, by five orders of magnitude. The overdoses continued to occur.

During the *Apollo 13* emergency, when the spacecraft and crew were almost lost, the ground-control engineers did not believe what they were seeing on the instruments. The reason for their reluctance was later reported by a NASA engineer: "Nobody thought the spacecraft would lose two fuel cells and two oxygen tanks. It couldn't happen." Jack Swigert, an astronaut, supported this view: "If someone had thrown that at us in the simulator, we'd have said, 'Come on, you're not being realistic'" [40, p. 232].

It is impractical to require that all hazards, no matter how remote they are judged, be eliminated or controlled. However, independence and other assumptions used in either informal or formal risk assessments should be scrutinized with great skepticism. In addition, the potential for catastrophic accidents of a type not yet experienced should not be casually dismissed.

Assuming risk decreases over time A common thread in most accidents involving complacency is the belief that a system must be safe because it has operated without an accident for many years. The Therac-25 was operated safely thousands of times before the first accident. Industrial robots operated safely around the world for several million hours before the first fatality [296]. Nitromethane was considered nonexplosive and safe to transport in rail tank wagons for eighteen years (between 1940 and 1958) until two tank wagons exploded in separate incidents [192]. Carrying out an operation in a particular way for many years does not guarantee that an accident cannot occur, yet informal risk assessments appear to decrease quickly when there are no serious accidents.

In reality, risk may decrease, remain constant, or even increase over time. Risk can increase for several reasons, including the fact that as time passes without an accident, caution wanes and attempts to build more complex and error-prone systems increase. Tradeoffs between safety and other factors start to be made in the direction of the other factors, and safety margins are cut, giving rise to a rather surprising fact: *As error rates in a system decrease and reliability increases, the risk of accidents may actually be increasing.*

Risk also may increase over time either because the system itself changes as a result of maintenance or evolution or because the environmental conditions change. Changes in software are inevitable and ubiquitous during the lifetime of any software-intensive system. In some cases of automation, the introduction of a newly automated system may actually cause the environment to change from what it was assumed to be during design. For example, human operators may change their behavior as they become more familiar with an automated control system. As operators became more familiar with the Therac-25 operation, they started to type faster and triggered a software error that had not surfaced previously. The difficulty of testing more than a small fraction of the possible software states complicates even further the problem of ensuring that hazards will not occur under any plausible environmental conditions.

The common belief that safety activities can be phased out once a system becomes operational is clearly incorrect. Many of today's systems are highly complex, the requirements are exacting, and changes are bound to occur, especially in software. After the Shuttle became operational in 1980, many of the safety-related activities were reduced. A safety committee, the Space Shuttle Program Crew Safety Panel, was disbanded, losing an important focal point for flight safety, according to the Rogers Commission report [332]. The panel had been established to ensure a minimum level of communication about safety among the engineering, project management, and astronaut offices, but NASA expected it to be functional only during the design, development, and flight test phases. After it was disbanded, according to the investigation report for the *Challenger* accident, the NASA Shuttle program had no focal point for flight safety.

One dilemma in arguing for safety programs is the impossibility of determining how many accidents a good program has avoided. The ultimate irony is that a successful safety program may lead to complacency when few or no accidents occur, and that complacency can then lead to accidents. Therefore, the more successful an organization is in eliminating accidents, the more likely that complacency will increase risk.

Ignoring warning signs and other feedback A common component of complacency is discounting the warning signs of potential accidents. Accidents are frequently preceded by public warnings or by a series of minor occurrences or other signs. Often, these precursors are ignored because individuals do not believe that a major accident is possible. In fact, there are always more incidents than there are accidents because most systems are designed to handle single failures or errors; the problems usually are corrected before they lead to hazards. Also, hazards may not lead to accidents because the specific environmental conditions necessary for serious losses may not occur while the hazardous condition exists, at least in that particular instance.

For seven years before the Flixborough explosion, the chief inspector of factories in Britain had been warning about the risks involved in the increasing use and amounts of dangerous materials in British factories [137]. He was ignored.

At least six serious accidents occurred at the Bhopal plant in the four years before the big one in 1984, including one in 1982 in which a worker was killed. People in key positions were alerted to the potential dangers by a number of sources, including a series of newspaper articles by a journalist, R. K. Keswani, who predicted the accident; however, local authorities and plant managers did nothing [27; 66]. Early warning signs were ignored.

The TMI accident was preceded by records of similar accidents and operator errors, reports predicting such an accident, and evidence of persistent and uncorrected equipment failures. All of these signals were disregarded or dismissed [302]. For example, between 1970 and the time of the accident in 1979, eleven pilot-operated relief valves (PORVs) had stuck open at other such plants—a stuck-open PORV initiated the TMI events.

A fire at London's King's Cross underground station in 1987, in which thirty-one people were killed and many more injured, was apparently caused by a lighted match dropped by a passenger on an escalator, which set fire to an accumulation of grease and dust on the escalator running track [192]. A metal cleat that should have prevented matches from falling through the space between the treads and the skirting board was missing, and the running tracks had not been cleaned regularly. No water was applied to the fire, which spread for 20 minutes and then suddenly erupted in the ticket hall above the escalator. The water spray system installed under the escalator was not actuated automatically, and the acting inspector on duty walked right past the unlabeled water valves. London Underground employees had little or no training in emergency procedures; their reactions were haphazard and uncoordinated.

Although the combination of a match, grease, and dust was an obvious causal factor in the fire, a systemic factor was the view accepted by all concerned, including the highest levels of management, that occasional fires on escalators and other equipment were inevitable and could be extinguished before they caused serious damage or injury [192]. There had been an average of twenty fires per year from 1958 to 1967—called "smoulderings" to make them seem less serious—but although some had caused damage and passengers had suffered from smoke inhalation, no one had been killed. The view grew that no fire could become serious, and fires were treated almost casually:

> *Recommendations after previous fires were not followed up. Yet escalator fires could have been prevented, or reduced in number and size, by replacing wooden escalators by metal ones, by regular cleaning, by using non-flammable grease, by replacing missing cleats, by installing smoke detectors that automatically switched on the water spray, by better training in firefighting, and by calling the Fire Brigade whenever a fire was detected, not just when it seemed to be getting out of control.* [192, p. 85]

On March 22, 1975, during plant modification at the Browns Ferry Nuclear Power Plant, a candle being used to detect air leaks ignited polyurethane foam (which was used in parts of the electrical system), causing extensive damage to the electrical power and control systems and leading to a common-mode failure. Before the 1975 fire in Browns Ferry Units 1 and 2, the record shows that "*there was extensive official fore-knowledge of safety deficiencies in Browns Ferry and that the very combination of problems responsible for the accident had been identified by Federal safety authorities but left uncorrected*" [405, p. 405].

Weil [405] details prior warnings related to Browns Ferry about the dangers of electrical cable fires arising from poor control of combustible materials, inadequate fire prevention programs, and poor separation of redundant circuitry that went back to 1969 (the Browns Ferry plant went into full operation in August 1974).

On March 23, 2005, an explosion at the BP Texas City oil refinery killed fifteen workers, injured 180 others, and severely damaged the refiner. Most of those killed were working in temporary trailers installed next to the isomerization (ISOM) unit, which is where the set of events started. Previously, in 1995, five workers had been killed in a Pennzoil refinery when two storage tanks exploded, engulfing a trailer. The conclusion was that trailers should not be located near hazardous materials. BP ignored the warnings; they believed that because the trailer where most of the deaths happened was empty most of the year, the risk was low.

The problem was not simply ignoring what was happening in other companies. In the years prior to the incident, eight serious releases of flammable material from the ISOM blowdown stack at Texas City had occurred, and most ISOM startups experienced high liquid levels in the splitter tower. These events were not investigated. Other major accidents involving fatalities and millions of dollars in damage occurred, but almost nothing was done to improve safety at Texas City in response.

Because people tend to underestimate risks and ignore warnings, one or more serious accidents are often required before action is taken. Similar small accidents tend to repeat themselves until one has such an enormous impact that it is impossible to not notice or ignore the problems.

A reasonable argument can be made that it is unfair to judge too harshly those who have ignored warnings. Too many accident warnings that prove to be unfounded may desensitize those in decision-making positions and result in real warnings being ignored. Hindsight bias, also known as the "I knew it all long" phenomenon, can be defined as the common tendency for people to perceive events as having been more predictable than they actually were. Hindsight creates warning signs from events that previously were viewed as random occurrences and not identified as suggesting an increase in risk. In hindsight, signals or precursors may appear more relevant than they seemed before the event. Nevertheless, if such events were taken more seriously and if lessons were learned from them, accidents might be avoided.

Audits, reports, and direct warnings to management about the existence of specific problems, such as deferred maintenance, are harder to ignore and misread. Most of the process safety deficiencies leading to the Texas City ISOM accident had previously been identified by internal and outside auditors, but they were never fixed. Beginning in 2002, three years before the explosion, BP and Texas City managers received numerous warnings about a possible major catastrophe at Texas City. In particular, managers were told about serious deficiencies regarding the mechanical integrity of aging equipment, process safety, and the negative safety impacts of budget cuts and production pressures. Additional audits in 2003 and 2004 identified many of the same deficiencies. A 2005 report warned that the refinery likely would "kill someone in the next 12–18 months" [372]. Other reports noted that there was an exceptional degree of fear of catastrophic incidents by the workforce.

BP had had several major accidents preceding the Texas City explosion. Following these, BP mobilized a task force to review refinery operations. One of the task force's findings was that BP was too focused on short-term cost reduction and not focused enough on longer-term investment for the future. The task force also found that health and safety was unofficially sacrificed to cost reductions. "Cost pressures inhibited staff from asking the right questions; eventually staff stopped asking" [29].

The response of BP executive management to all these written warnings was to cut budgets and increase production pressure and the focus on short-term profits. The strategy used to achieve these goals was to reduce safety meetings; training; fire drills; maintenance, engineering, supervision and inspection staff; and plant maintenance. These cuts were made despite the accidents and fatalities that were happening at the refinery and all the audit reports revealing serious safety deficiencies in the areas cut.

Two relatively recent accidents that have had far-reaching effects on legislation are the 1974 Flixborough explosion in England and the 1976 release of dioxin at Seveso near Milan, Italy. Both are described in more detail in appendix C. Neither accident was the largest of its decade, but both were accidents that people realized could happen again under existing regulations and in which the potential danger of new industrial processes to life, property, and the environment was seen to be enormous and beyond the damage caused in these particular instances. These accidents also made clear the need to plan for mitigating the effects of an industrial accident on people, property, and services beyond the plant itself and even beyond national boundaries [288].

In addition to legislative initiatives, serious accidents such as Three Mile Island, Bhopal, Fukushima, and *Challenger* have led to industry and organizational learning, as well as to advances in research. This phenomenon is not new. Boiler explosions were an impetus to learning more about the nature of steam and the causes of explosions, and they led to the first technological research grant, to the Franklin Institute in 1824, by the US government. The Morrison Inquiry into box-girder bridges, held after failures of such bridges in England, led to research resulting in improvements in design and quality control. Farmer [99] argues that there would have been no inquiry and no sense of urgency if there had been only cracks and traffic limitations on a small number of bridges.

Despite these positive results, major accidents have arguably been more expensive than the benefits of the lessons learned, especially if we consider that most of the findings on the accidents were known beforehand but not taken seriously [399]. Using this information to attempt to prevent accidents before they occur seems reasonable.

4.5.1.2 Low priority given to safety Even if complacency is not the norm and individuals in an industry or organization are concerned about safety, their efforts will most certainly be ineffective without support from top management. The entire organization must have a high level of commitment to safety in order to prevent accidents, and the lead must come from the top and permeate every organizational level. Employees must believe that they will be supported by the company if they reasonably choose safety over other goals. The informal rules—that is, the social processes as well as the formal rules of the organizational culture—must support the overall safety policy.

Exhortation or policy statements alone are not enough. In December 1988, a crowded commuter train ran into the rear of a stationary train near Clapham Junction in Great Britain. After the initial impact, the first train veered to its right and struck an oncoming train. Thirty-five people died in this accident and nearly 500 were injured, sixty-nine of them seriously. According to the official government report on the accident, there was a sincere concern for safety at all levels of management, but

> *The best of intentions regarding safe working practices were permitted to go hand in hand with the worst of inaction in ensuring that such practices were put into effect. The evidence therefore showed the sincerity of the concern for safety. Sadly, however, it also showed the reality of the failure to carry that concern through into action. It has to be said*

that a concern for safety which is sincerely held and repeatedly expressed but, nevertheless, is not carried through into action is as much protection from danger as no concern at all. [144, p. 163]

One way that management can demonstrate true commitment to safety goals is through the assignment of resources. The Rogers Commission report on the *Challenger* accident noted that the chief engineer at NASA headquarters, who had overall responsibility for safety, reliability, and quality assurance, had a staff of twenty people of which one person spent 25 percent of his time on Shuttle maintainability, reliability, and quality assurance and another spent 10 percent of his time on these issues.

While safety efforts increased after the *Challenger* accident, they slowly started to degrade again. The manned space program was operating with an increasingly constrained budget. One of the results was workforce reductions. There was little margin in the budget to deal with unexpected technical problems or make shuttle improvements. Safety standards were weakened and became nonmandatory. The priority of safety in the engineering efforts quickly fell back to the pre-*Challenger* level.

The *Columbia* accident report identified a perception that NASA had overreacted to the Rogers Commission recommendations after the *Challenger* accident, for example, believing that the many layers of safety inspections involved in preparing a Shuttle for flight had created a bloated, costly, and unnecessary safety program.

Just one indication of the atmosphere existing at that time were statements in a 1995 report, which presented the findings of a NASA review of Space Shuttle management. The report dismissed concerns about Shuttle safety by labeling those who made them as partners in an unnecessary "safety shield conspiracy" [195]. This accusation of those expressing safety concerns as being part of a "conspiracy" is a powerful demonstration of the attitude toward system safety at the time and the change from the Apollo era when dissent was encouraged and rewarded.

The report concluded that "the Shuttle is a mature and reliable system, about as safe as today's technology will provide" [195]. A recommendation in this report was that NASA should "restructure and reduce overall safety, reliability, and quality assurance elements" [195].

Staff, training, and maintenance at Bhopal had been severely reduced prior to the accident [40]. Top management justified these measures as merely reducing avoidable and wasteful expenditures without affecting overall safety.

At Three Mile Island, the maintenance force was overworked at the time of the accident and had been reduced in size to save money. There were many shutdowns, and a variety of equipment was out of order. A review of equipment history for the six months prior to the accident showed that a number of equipment items that were involved in the accident had had a poor maintenance history without adequate corrective action.

Drilling the Macondo well in the *Deepwater Horizon* accident was behind schedule and above budget. BP was spending $1 million a day at the site and almost all decision-making at the time of the accident was focused on schedule and budget ahead of safety.

In general, management's most important actions in preventing accidents involve setting and implementing organizational priorities. Many managers recognize that safety is good business over the long term; others put short-term goals ahead of safety. Government agencies and user or customer groups can force management to take safety seriously. This type of pressure will be exerted only if safety receives societal support and emphasis.

In general, user groups are much more effective than government regulatory agencies in building management support for safety activities. Regulatory agencies are often viewed as a mere nuisance to work around, while user groups that refuse to buy unsafe products have control over the continued existence of the company. Other alternatives to mandatory government standards include tort and common law, insurance, and voluntary standards-setting organizations.

4.5.1.3 Flawed resolution of conflicting goals Not only does safety need to be recognized as a high-priority goal, but procedures for resolving goal conflicts also need to be established. Desirable qualities tend to conflict with each other, and tradeoffs are necessary in any system design or development process. Attempts to design a system or a development process that satisfies all desirable goals, or to provide standards to ensure several goals without considering the potential conflicts, will result only in failure or in de facto and nonoptimal decision-making.

The *Challenger*, *Columbia*, Texas City, and *Deepwater Horizon* losses are classic cases of poorly handled conflicts between safety and schedule, but other less well-known examples abound. Nichols and Armstrong examined industrial accidents and found a large number that occurred during an interruption of production and while an operator was trying to maintain or restart production [282]. In each case, the dangerous situation was created by a desire to save time and ease operations. And in each case, the company's safety rules were violated: "*The [operators] acted as they did in order to face up to the pressure exercised by the foremen and the management who aimed at maintaining production. This pressure was continuous; the interruptions of the process were rather frequent and equally hasty methods used to deal with them were employed repeatedly*" *[282, p. 19]*.

Management often brags about policies that allow workers to stop production whenever they are worried about safety. The real question, however, is whether subtle—and sometimes not so subtle—behavior undermines these policies. "*[Workers] know where management's fundamental preoccupation lies. They see the man in charge of production burst out of his office like a cannon ball when the conveyor stops. Naturally, from time to time, management and the foremen have preached about safety; there are even some sanctions in case of an accident. But in day-to-day operations one sees clearly what is important*" [282, p. 20].

The best technology can be defeated by such management behavior. Often, accidents blamed on operator error or equipment failure can just as easily be traced to management placing higher priority on goals other than safety. Howard provides several examples of this from the chemical process industry [153]. One accident in a polymer processing plant occurred after operating management bypassed all the alarms and interlocks in order to increase production by 5 percent; their perception of upper management's priorities was maximum production over all other factors.

In another accident, a severe ethylene decomposition in a long pipeline resulted in a costly business disruption. The interlocks and alarms had failed (at a normal rate), but this was unknown because management had decided to eliminate regular maintenance checks of the safety-related instrumentation.

In a third accident, a holding tank ruptured from an exothermic decomposition of the contents. Management had for many years suspected that the material in the tank could exothermically decompose, but tests to determine the decomposition properties of the material had been postponed because they were too busy running the unit and maintaining high production records. In addition, most of the operator training had been eliminated in order to save money, so an untrained operator in charge of the tank was unaware of the significance of a gradual increase in the temperature of the tank contents over several hours.

Safety often suffers in tradeoffs between conflicting goals because of a dearth of hard data on benefits. Cost and schedule are usually the primary drivers of a project, while benefits from investments in system safety show up primarily in the long run and, even then, are observable only indirectly—as non-accidents and the avoidance of modifications or retrofits to improve safety [109]. How does one prove they have prevented an accident that does not happen? Long-term uncertainties are difficult to quantify, so short-term factors tend to be emphasized in making tradeoffs. Sacrificing a sure productivity gain in favor of a seemingly low-probability accident may not seem like a reasonable course of action. Of course, all that does is increase the likelihood of an accident.

Belief that safety conflicts with the basic benefits of technology is common. For example, Fischoff and colleagues write:

> *Controlling hazards often requires foregoing benefits. The benefits of a technology are the reason for its existence. Customarily, the benefits are as clear and tangible as the risks are ambiguous and elusive. Moreover, the sponsors of a technology, workers, and consumers all have a sizable stake in its existence. It is not surprising, therefore, that hazard management is not practiced to the fullest extent possible. For some groups, benefits may outweigh corresponding risks, and thus it often happens that the beneficiaries of technologies are at political loggerheads with hazard managers concerned with the general welfare.* [101, p. 168]

This belief is also a factor in the use of *downstream* protection; that is, adding protection features to a completed design. The alternative is *upstream* hazard elimination and control by designing safety into the system from the beginning through eliminating or reducing hazards in the original design. Fischoff and colleagues claim that "*the importance of benefits also accounts for the paucity of 'upstream' hazard management. Intervention of this type tends to conflict much more fundamentally with benefits than does more conventional 'downstream' management*" [101, p. 168].

Although it may be true in some cases that increasing safety requires foregoing some of the benefits of technology, this is not a basic law. The truth of the statement depends on what benefits are being considered and how the safety fixes are implemented.

On the surface, it does seem that the fewest compromises are required, as Fischoff and colleagues claim, if the basic control system is built without concern for safety and protection devices are added later to detect and ameliorate hazards if they occur. This downstream approach may eliminate the need for some types of design tradeoffs, but it may increase the overall cost of the system and schedule delays while also increasing risk relative to what the upstream approach might have achieved.

In contrast, the upstream approach of eliminating or controlling hazards in the early design stages and making tradeoffs early may result in lower costs during both development and the overall system lifetime, in fewer delays and less need for costly redesign, and in lower risk.

It usually costs no more to build an inherently safe system, or one in which hazards are controlled, if the system is designed correctly in the first place. In fact, an inherently safe system is often cheaper than one with a lot of overdesign in the form of added-on protection devices [235]. Moreover, designing in safety from the start is much cheaper than making retroactive design changes. And, of course, the costs may be minimal in comparison with the potential costs of a major accident.

Most of these arguments are applicable to relatively simple systems. In complex, high-technology systems, in fact, it can be enormously *more* expensive to use the downstream approach compared to designing safety into the system from the beginning. The effectiveness of downstream methods in such systems is greatly reduced. Simple redundancy and other component reliability enhancing approaches are not effective in preventing accidents in these new enormously complex systems.

A widespread belief is that increasing safety slows operations and decreases performance. In reality, experience has shown that, over a sustained period, a safer operation is generally more efficient and can be accomplished more rapidly. One reason is that stoppages and delays are eliminated [133].

Juechter [169] provides an example involving power presses. Because of a number of serious accidents, OSHA tried to prohibit the use of power presses where employees had to place one or both hands beneath the ram during the production cycle. Preliminary motion studies showed, however, that reduced production would result if all loading and unloading were done with the die out from under the ram. After vehement protests that the expense would be too great in terms of reduced productivity, the requirement was dropped. After OSHA gave up the idea, one manufacturer who used power presses decided, purely as a safety and humanitarian measure, to accept the production penalty. Instead of reducing production, however, the effect was to increase production from 5 to 15 percent, even though the machine cycle was longer.

This result should not have been a surprise. In chapter 1, a large study from over one hundred years ago was cited that found the same thing, namely that production increases as workplace safety increases. The same is true outside of workplace safety and in the operation of any safety-critical system. It is time for this fact to be recognized.

The belief that safer systems cost more or that building safety in from the beginning necessarily requires unacceptable compromises with other goals is simply not justified.

4.5.1.4 Confusing safety with other system properties The culture in an organization may not be marked by complacency and a high priority may be placed on safety, but confusion about how to achieve safety goals may inhibit the decision-making needed to be successful.

Confusing safety with reliability A detailed description of the differences between reliability and safety can be found in chapter 3. Briefly, dangerous things can be done reliably, over and over again, until an accident occurs. The Therac-25 software was highly reliable: it worked tens of thousands of times before overdosing anyone, and occurrences of erroneous behavior were few and far between. AECL (Atomic Energy of Canada,

Limited) assumed that their software was safe because it was reliable, and this led to dangerous complacency.

When systems were primarily electromechanical, nearly exhaustive testing was possible and design errors could be eliminated before operational use. What was left during operations was random wear-out failures. Safety then could be assumed to be effectively approximated by reliability.

Very few purely electromechanical systems are built today. Almost everything has software in it, and this fact has changed the nature of accidents. Software is unique in that it is pure design, namely, design abstracted from its physical realization. While the hardware on which the software is executed may fail, the design itself does not fail. The software can reliably implement its specified requirements, but almost all software-related accidents result from flawed software requirements and not from errors in the implementation of those requirements [217].

In fact, software by itself is not safe or unsafe—safety depends on the *context* in which the software is used. Much of the Therac-25 software had been used on an earlier version of this machine, called the Therac-20. The same software flaws that resulted in overdosing people with the Therac-25 did not have the same effect in the Therac-20 because of differences in the overall system that the software was controlling.

Related to this misunderstanding is the assumption that reused components, particularly reused software, will be safe because they did not contribute to accidents in earlier systems or because they have been exercised extensively in practice. The inertial reference system software on the Ariane 4 space launch system was safe, but when it was reused on the updated Ariane 5, it contributed to the loss of the launcher and its payload. A study of NASA spacecraft losses involving software over a ten-year period found that each of the losses involved reused software [215].

The safety of software cannot be evaluated by looking only at the software itself but must include the hazards of the system the software is controlling and the context in which the system and software are being used. Reusing software that was safe in one system does not mean it will be safe when used in a different system because the context may be different. Safety is a quality of the system in which the software is used; it is *not* a quality of the software itself. The assumption sometimes built into practice or even government standards that reused software is safe or safer is not justified.

Equating process safety with personal safety A second dangerous confusion results from equating process safety with personal or workplace safety. It is much easier to define and collect metrics for personal safety, such as days missed from work, than for process safety. Low personal worker injury rates at Texas City created false confidence that process safety was also high. In reality, process safety at the refinery was deteriorating. The use of personal injury rates kept BP from detecting this decrease and intervening early enough to prevent major losses.

The lack of focus on process safety can often be traced to compensation policies. For example, personnel and management compensation at BP put very little weight on safety in general, and what was included was primarily personal safety—days away from work, recordable injuries, vehicle accidents, and so on. Metrics that might have provided insight into process safety were not included.

A safety culture assessment of Texas City was performed by consultants before the 2005 accident. The assessment concluded that workers perceived the managers as "too worried about seat belts" and too little about the danger of catastrophic accidents. The Baker Panel report on the Texas City accident noted that workplace safety was more closely managed because it counted in managers' evaluations whereas the consequences of poor process safety might occur later on someone else's watch [29].

Focusing on success rather than failure A final cause of complacency is focusing on success rather than failure. Surprisingly, such a focus on "what goes right" rather than "what goes wrong" has been promoted recently as the best way to reduce accidents [148]. In fact, in engineering, almost all our learning comes from what goes wrong in terms of incidents and accidents. We learn very little when things go right.

Henry Petroski, a professor of civil engineering at Duke University, has written many thoughtful essays and books on the role of failure in engineering (for example, [306; 307]). While drawn largely from examples in civil engineering, the principles are equally applicable to any engineered system. A key theme of Petroski's writings is that a detailed understanding of the manner in which a given design fails in use allows iterative improvement of later designs. When nothing fails, it simply means that one of the necessary conditions for failure may not have occurred in this case, usually when that necessary condition is unknown—until the loss occurs. Petroski argues that nobody wants to learn by mistakes, but we cannot learn enough from success.

If we look only at the millions of times that structures do not fall down or accidents do not occur, we learn nothing. Knowledge advances when the assumptions we make about systems or their environment are violated and a loss occurs. We cannot determine that the assumptions are wrong until we encounter a situation where they do not hold, usually in odd cases that nobody anticipated. We have just not yet come across the unique and unusual conditions that lead to a hazard or loss.

We get more confidence that what we are doing is right the longer it takes to come across these conditions. If it takes a while, we may incorrectly become more convinced that our assumptions are right. In other cases, the assumptions originally were right, but the world has changed.

In fact, most major accidents result from the overconfidence that comes with success. Some of the most important lessons from the investigation of major accidents is that things appear to be going right for a long time until the exact circumstances necessary for tragedy come along. We believe that risk is decreasing or has decreased when, in fact, the risk is the same as it always was. This is the folly of trying to learn from success or lack of bad consequences. Continuing success only gives us more confidence that what we are doing is the safe thing to do. Accidents are rare, so we must learn from the failures and accidents that *do* occur. It is from failure that engineers learn how to be successful in the future.

Is the folly of relying on success true only for engineering and not for human performance? Here is a final quote from Russell Ackoff, a professor of management science and one of the great systems thinkers of our time, commenting on the human and social components of systems: "All learning ultimately derives from mistakes. When we do something right, we already know how to do it; the most we get out of it is confirmation of our rightness" [7].

4.5.2 Management Decision-Making Structure

Many accident investigations uncover a sincere concern for safety in the organization but find organizational structures in place that were ineffective in implementing that concern. Systems thinking teaches that organizational structures drive the behavior of the humans who operate in that system.

Basic management principles apply to safety as well as to any other quality goals. Nevertheless, some specific issues in organizational structure appear to be related to organizations that have many accidents: diffusion of responsibility, authority, and accountability; low-level status of safety personnel and exclusion from critical decision-making; and limited communication channels and poor information flow of safety-related information. Chapter 14 describes how to design an effective safety management system to overcome these problems.

4.5.2.1 Ill-defined and diffused responsibility, authority, and accountability In the Congressional hearings after the Macondo/*Deepwater Horizon* blowout, BP testified that the massive Gulf oil spill was caused by the failure of a safety device (the blowout preventer [BOP]) made by Transocean and was part of the Transcom-owned and operated *Deepwater Horizon* oil rig. The company argued that because BP did not own the rig, the responsibility for the safety of drilling operations belonged to Transocean. In turn, Transocean testified that BP was in charge of the oil drilling operations, that BP had leased the rig, and that BP had prepared the plan and given the go-ahead to fill the well pipe with seawater before the final cement cap was installed. Both BP and Transocean claimed that the US Minerals Management Service (MMS), the government agency overseeing offshore oil drilling in the United States) had approved the operations plan and that the MMS certified and regulated the rig and drilling operation in the Gulf. Both also said that Halliburton poured the cement to plug the well and did not do it right. Halliburton claimed that it was only following BP's drilling plan and that its work was in accordance with the requirements set by BP and followed accepted industry practices.

While such finger-pointing is common after a major accident, the real problem was the lack of coordination and leadership for overall operations. Believing that everyone is and can be responsible for safety is widespread in many industries. The problem is that where everyone is responsible for safety and for resolving conflicts between safety and other goals, then nobody is responsible for safety. Personal responsibility for doing one's job safely is, of course, important. But overall system safety requires leadership and the assignment of responsibility, authority, and accountability to individuals. Someone must ensure that safety activities are coordinated and conducted properly and that necessary communication and information flow is occurring.

The Macondo and Texas City accident reports found a lack of BP corporate oversight of both safety culture and major accident prevention programs. Management responsibilities for process safety at all levels above the refinery and oil drilling levels were unclear and mostly unspecified. There was a belief that safety decisions should be made at the lowest possible level with little or no responsibilities assigned to the higher management levels. Responsibility for coordinating decisions requires more visibility about the larger system state than is possible at lower levels of the management hierarchy. A decision that seems perfectly safe at one system level can be dangerous for the system as a whole.

When multiple companies are involved in a project, the problems become worse. In Congressional testimony on the Macondo/*Deepwater Horizon* blowout, a petroleum engineering professor testified that "*the individual contractors have different cultures and management structures, leading easily to conflicts of interest, confusion, lack of coordination, and severely slowed decision-making*" [275, p. 229]. Ultimately, while both BP and Transocean had corporate policies for risk management, neither company ensured their implementation at Macondo [275, volume 3, section 4.0–4.5].

An additional necessary form of support for safe decision-making is providing a safety engineer to provide input for such decisions. There should have been such a person assigned to and on the rig when the critical tasks were being performed. All responsibility was instead placed on managers and workers with conflicts in their responsibilities. Projects need to assign responsibility for assisting with safety-critical decisions to people who specialize in and are responsible for providing the necessary information to managers so they can make better decisions. While decision makers on the rig could have called for help from onshore experts, this type of interaction was not part of the BP culture.

But Macondo and Texas City are not the only examples of accidents where ill-defined and diffused responsibility played an important role. For *Challenger*, the responsibility for information about a factor such as safety margins and temperature limits on O-rings was several organizational levels and at least two contractual interfaces away from the people making decisions on schedules and launch.

As with BP, risks were accepted at an inappropriately low level, without the knowledge of higher levels of management [166]. In the case of BP, this lack of communication was actually encouraged, whereas at NASA the problems stemmed from cultural and other hidden communication barriers.

A factor identified by the Rogers Commission as contributing to the ineffectiveness of the *Challenger* safety program was that the safety, reliability, and quality assurance offices were under the supervision of the organizations and activities whose efforts they were to check. The report said that this lack of independence reduced their effectiveness as watchdogs.

After the *Challenger* loss, a new independent safety office was established at NASA Headquarters, as recommended in the Rogers Commission report. This group was supposed to provide broad oversight, but its authority was limited, and reporting relationships from the NASA Centers were vague. In essence, the new group was never given the authority necessary to implement their responsibilities effectively, and, most critically, nobody seems to have been assigned accountability. The later *Columbia* accident report noted in 2003 that the management of safety at NASA involved "confused lines of responsibility, authority, and accountability in a manner that almost defies explanation" [112, p. 186]. As contracting of Shuttle engineering outside NASA increased over time, safety oversight by NASA civil servants diminished and basic system safety activities were delegated to contractors.

The problems were exacerbated by the fact that the project manager also had authority over the safety standards applied on the project. NASA safety standards were not mandatory. In essence, they functioned more like guidelines than standards. Each program decided what standards were applied and could tailor them in any way they wanted. While there are advantages to being able to tailor standards, the authority to do so must be placed in the right hands and oversight provided.

At Fukushima, the utility Tokyo Electric Power Company (TEPCO), which owned as many as seventeen nuclear reactors, had never had a CEO whose expertise was rooted in nuclear technology or engineering. Only a limited number of its executives appreciated the managerial responsibilities of the nuclear power business. Past CEOs rose up through administrative departments, which were more valued in TEPCO than technical organizations such as the Nuclear Power Division.

As a result, the top management could not establish effective leadership and management for nuclear safety. They did not understand the details of the highly specialized operations of the Nuclear Power Division and, instead, entrusted matters, such as safety provisions, to underlings familiar with nuclear power. Lack of experience and expertise contributed to an inadequate sense of responsibility by top management [47; 162].

4.5.2.2 Lack of independence and low-level status of safety personnel Serious accidents are often associated with low status of the safety personnel and their lack of involvement in critical discussions and decision-making. The critical teleconference calls between the NASA Marshall Space Center and Morton Thiokol on January 27, 1986, about the *Challenger* launch decision did not include a single safety, reliability, or quality assurance engineer, and no representative of safety was on the management team that made key decisions during the countdown before the *Challenger* loss. The status of safety personnel and their involvement in decision-making is a clear sign to everyone in the organization of the real (as opposed to professed) emphasis that management places on safety.

A second problem that can occur is that the safety organization is separated from engineering, sometimes isolated in a lower-status group, such as quality assurance. The best engineers may not want to participate as a result. To have an impact on design, safety engineering must be part of and work closely with the engineering group. Otherwise, safety becomes simply an after-the-fact assurance activity that has no or little impact on the design or operation of the system.

At the same time, safety has to have independence from management decisions that are subject to conflicting pressures, such as time and budget. One of the criticisms that emerged during the official inquiry into the Flixborough explosion was that there was no engineering organization independent of production-line management responsible for assessing the overall system and ensuring that proper controls were exercised [400].

Safety engineering can be located within the engineering group while having an independent management line at higher levels of the organization with the power to ensure funding and emphasis, provide independent oversight, and provide advice about safety concerns when relevant management decision-making is occurring. Chapter 14 provides information about the design of an effective safety management system.

4.5.2.3 Limited communication channels and poor information flow An examination of organizational factors in accidents often turns up communication problems [419]. Concerns about performance and safety may be subject to many delays in transmittal up the organizational chain, and they can be edited or stopped from further transmission by some individual or group along the chain.

Communication paths and information need to be explicitly defined (figure 4.3). Two types of information flow are necessary: (1) a *control/command channel* that communicates goals

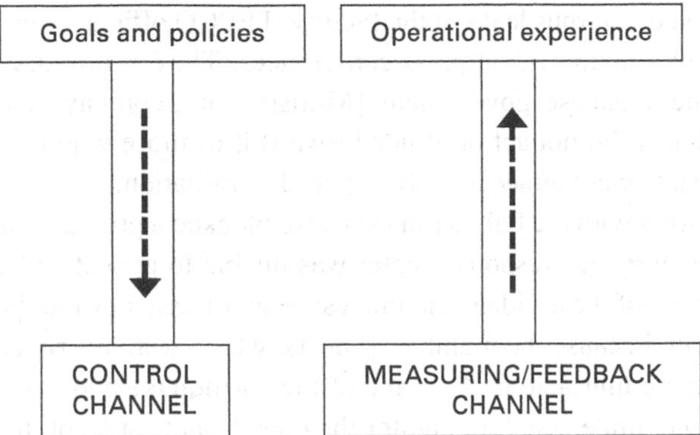

Figure 4.3
Communication paths.

and policies downward, and (2) a *measuring/feedback* channel that communicates the actual state of affairs upward.

In the control or command channel, the reasons for decisions, procedures, and choices need to be communicated downward, along with goals and policies, in order to avoid undesirable modifications by lower levels and to allow detection and correction of misunderstanding and misinterpretation. In the measuring/feedback channel, the feedback from operational experience and communication of technical uncertainties and safety issues up the chain of command is crucial for proper decision-making.

Several cultural assessments at Texas City all agreed that personnel were not encouraged to report safety problems, and some feared retaliation for doing so. For example, the 2003 audit concluded that "bad news is not encouraged." Employees reported that management did not encourage reporting of complaints about safety and sometimes were actively discouraged from reporting safety problems upward. The Baker Panel heard from the spouses of those killed that they were afraid to go to work but did not feel free to complain about unsafe working conditions [29].

BP's system for assuring communication about performance and risks used a bottom-up reporting system that originated with each business unit, such as a refinery. As information was reported up, however, data about safety and other types of risks such as time and schedule were aggregated. By the time information was formally reported at higher management levels, information about safety was effectively invisible. And, as noted, the data collected almost exclusively focused on personal safety, not process safety. As a result, the Baker Panel concluded that a substantial gulf existed between the actual state of process safety in the refineries and the management's perception of that performance [29]. The same was true for upstream oil exploration and extraction activities.

Communication before an accident is not the only problem. Emergency response also requires communication. After the events in the Fukushima Daiichi nuclear power plant accident started, communication and information flow almost totally broke down. Government agencies fell into a so-called "elite panic," in which they refused to pass on critical pieces of information for fear of inciting panic among the general public. According to Ueseke [380], leaders knew little about the measures available to them and delays occurred in releasing data on dangerous leaks at the facility. TEPCO officials were instructed not to use the phrase "core meltdown" at press conferences. The US military aircraft provided information to the Japanese government [Ministry of Economy, Trade and Industry (METI)], but officials did not act on it nor forward it to those who could act on it. As a result, some residents were unnecessarily exposed to radiation.

Miscommunication with the Fukushima site also blocked a prompt emergency response. The government emergency response center was unable to collect and share information about the progression of the accident and the response. Emergency monitoring results were impossible to obtain because the monitoring posts, which were overly concentrated along the coastline, became unusable in the wake of the earthquake and tsunami. The TEPCO head office could not understand or monitor the operational status of the reactors because the automated Safety Parameter Display System was unavailable.

Summoning and convening the NSC advisors to establish an Emergency Technical Advisory Body was delayed because group email to mobile phones was not delivered and public transportation and telecommunications were disrupted. Due to information security concerns, mobile phones could not be used in the basement of the building that contained the prime minister's office, where the relevant government officials assembled. With other communication means not available, it was difficult for these officials to gather information on the accident rapidly.

These types of communication failures were never anticipated, and no preparations were made to overcome them.

4.5.3 Operational Processes and Practices

Operational processes and practices have an important impact on system safety. Some practices are frequently associated with accidents. These include superficial, isolated, and misdirected safety efforts, inadequate feedback and learning from events, poorly defined operating procedures and tolerance of deviations from them, inadequate training and emergency management, inadequate management of change, and poor maintenance practices.

4.5.3.1 Superficial, isolated, or misdirected safety efforts during operations Sometimes a belief exists that after spending a lot of effort engineering safety into the system design during development, less effort is required after operations begin. In reality, more effort may be required to operate safely than is required during design. First, there may have been hazardous scenarios omitted during design or assumptions may have been made about the operating environment that turn out to be flawed. Designers may have decided that some scenarios would best be handled during operations. And, of course, the world changes over time and the designers may not have predicted the way the system would be operated or the changes in its environment. Many examples have already been provided in this chapter.

4.5.3.2 Inadequate feedback and learning from events Major accidents are almost never the result of one random event with no warning signs and no previous related appearance of the same causal factors. BP had many major accidents in the ten years preceding the *Deepwater Horizon* loss, with the resulting accident reports identifying the same or similar causal factors. Apparently, these factors were never fixed. Some were never even investigated.

Sometimes accidents are investigated but so poorly that nothing much can be learned. One of the techniques used in BP for root cause analysis was *5 Whys*, which suggests that there is only one root cause and one linear path to an accident. Using such approaches, as recommended in BP safety documents, can lead an investigation team to limit the investigation so much that nothing is learned about what really happened.

Halliburton and Transocean also had learning problems. Halliburton was a contractor on a well that suffered a blowout nine months before in August 2009 in the Timor Sea off Australia. The *Montara* rig caught fire, and the well leaked tens of thousands of barrels of oil over two and a half months before it was shut down. The leak occurred because a Halliburton cement seal failed, according to the government report on the accident. The report said it would not be appropriate to criticize Halliburton because the operator of the well "exercised overall control over and responsibility for cementing operations" [42, p. 63]. The inquiry concluded that "Halliburton was not required or expected to 'value add' by doing more than complying with [the operator's] instructions" [42, p. 63]. This report demonstrates the common focus on placing blame rather than learning from events and how this can interfere with learning and lead to repetition in the future.

Transocean was cited in the accident reports as having poor dissemination of lessons learned between teams. Lessons from a very similar earlier near miss were never adequately communicated to its crew.

While depending on learning from past events is not a substitute for a proper proactive attempt to prevent accidents the first time, not learning from prior events is foolish.

The MMS tried to expand data reporting requirements as part of an effort to track and analyze offshore oil incidents and to identify safety trends and leading and lagging indicators. The proposal was abandoned when industry complained about compliance cost and overlap with Coast Guard reporting requirements. The Coast Guard relaxed their requirements because they thought the MMS was handling it.

As a result, the United States has historically had no legal requirement that industry track or report instances of uncontrolled hydrocarbon releases or near misses. At the same time, the United States has the highest reported rate of fatalities in offshore oil and gas drilling among its international peers.

A few accidents have had such disastrous consequences that the government or others have stepped in to conduct a thorough investigation, including many of the examples used in this book. The results of these accident investigations are useful as they provide the only substantial information to affirm or disaffirm hypotheses and intuition and to learn the mistakes not to repeat. But the results are useful precisely because the investigations were so thorough and comprehensive. More such investigations are needed along with less oversimplification of accident reports.

In some industries, there are a large number of incidents and accidents with small consequences—far too many to investigate each thoroughly. As a result, superficial investigations are performed, and nothing is learned. While it is true that a thorough investigation takes more time and resources than a superficial one, the systemic causes are usually the same in all of these. An in-depth investigation of a few, along with useful recommendations, could reduce the overall number significantly.

The problem of inadequate investigation of events is particularly acute in health care. Killing or injuring one person at a time does not get the same attention as a chemical plant explosion or a nuclear power plant meltdown. Usually, a root cause analysis is performed in a few hours, some low-level person is blamed for the events, and things continue as usual [290; 344]. Almost never is an investigation performed to determine why the nurse, lab technician, or doctor acted in the way they did. This may be one reason why health-care errors are not being reduced over time.

Even careful investigations of the causes of accidents are subject to bias and incompleteness. Incomplete evaluations, such as those that blame everything on operator error, are worse than useless because they encourage designers to ignore other factors that contribute to accidents. Liability considerations complicate matters further because causal information can be used as a basis for lawsuits—companies may be reluctant to look too hard at an accident. Individuals may not provide complete information because of fears about job security. How to create more useful accident causal analyses is covered in chapter 8.

Besides incidents, companies can learn through internal or external audits, but audits may not provide useful information. For example, the focus of audits may be on compliance with legal requirements such as required management systems being in place and not with safe practices. Auditors may not be qualified, or they may have conflicts of interest that lead to reports that omit important feedback.

Of course, simply doing audits or investigating accidents and incidents is not enough. The results must be used to improve operations and management practices, and the lessons learned must be disseminated. Nearly all the accidents mentioned in this chapter suffered from deficiencies in these areas.

4.5.3.3 Poorly defined operating procedures After accidents, the procedures provided to the operators are often found to be flawed. At Texas City, the procedures lacked sufficient instructions to the operator to safely and successfully start up the unit. They did not explain the safety implications of deviations, did not include instructions on halting unit operations in the middle of the startup process, as happened, nor did they provide instructions for recommencing startup several hours later by a different operating crew. The hazards and safety implications of a partial startup/shutdown/re-startup were not addressed, and specific steps to initiate such activities were not provided.

In addition, management did not ensure that the startup procedure was regularly updated, even though the startup process had evolved and changed over time with modifications to the unit's equipment, design, and purpose. When procedures are not updated or do not reflect actual practice, operators and supervisors learn not to rely on procedures for accurate instructions. Other major accident investigations reveal that workers frequently develop work practices to adjust to real conditions not addressed in the formal procedures. If there have been so many process changes since the written procedures were last updated that they are no longer correct, workers will create their own unofficial procedures that may not adequately address safety issues.

The startup procedure provided to the operators for the ISOM unit on the day of the Texas City accident did not address the critical events the unit experienced during previous startups, such as dramatic swings in tower liquid level, which could severely damage equipment and delay startup. Specific instructions were incomplete. For example, tower pressure alarm set-points were frequently exceeded, yet the procedure did not address all the reasons this might happen and the steps that operators should take in response.

BP management allowed operators and supervisors to alter, edit, add, and remove procedural steps without following the required management of change policy to assess the safety impact from these changes. They were allowed to write "not applicable" (N/A) for any step and continue the startup using alternative methods. Procedural workarounds became accepted as normal.

Despite all these deficiencies, Texas City managers certified the procedures annually as up-to-date and complete.

Procedures were similarly flawed or not followed at BP's Macondo well before the blowout. Procedures changed frequently during the temporary capping of the well. Nobody saw the procedure they would be using until 11 a.m. on the day of the accident. BP provided no consistent or standardized procedures for the temporary abandonment that was being done that day. Formal written guidance was minimal. This left the Macondo engineers to determine what to do themselves.

Compounding the lack of written procedures for the test, BP did not have any policy, or at least did not enforce any policy, that would have required personnel to contact experts on shore to call for a second opinion about confusing or anomalous data.

Inadequate procedures also contributed to the Three Mile Island and Fukushima accidents. The Kemeny Commission found that at Three Mile Island, the procedures used by the operators during the emergency could cause operator confusion or incorrect action [173]. Operating procedures were not thoroughly reviewed by experts and were deficient in many ways. At Fukushima, plant workers had no clear instructions on how to respond to the events that occurred during the accident.

Even when planned operating procedures are acceptable, complacency can lead to a tolerance for serious deviations from them. BP management had long tolerated deviations from their own safety guidelines and operating procedures. Before most accidents, shortcuts may have been taken many times with a lack of consequences, convincing those involved that the deviations are safe. Procedural workarounds become standard practice. In fact, the work environment may actually encourage operations personnel to deviate from procedures. Of course, after an accident, the operators will be blamed for not following the procedures.

4.5.3.4 Inadequate training and emergency management Providing good procedures is not enough. For good decision-making, people need adequate training, information, procedures, resources, and support.

Most of the operators and others involved in the TMI accident did not fully understand the operating principles of the plant equipment. Their training was severely inadequate. The lack of depth in this understanding, even by senior reactor operators, left them unprepared to deal with something as confusing as the situation in which they found themselves. In addition, their simulator training did not prepare operators for multiple-failure accidents. The operators were only trained for an accident in the course of which only one thing went wrong. In this particular accident, three independent things went wrong.

In fact, the simulator was not programmed to reproduce the conditions that the operators faced during the accident. It was unable to simulate increasing pressurizer level at the same time that reactor coolant pressure was dropping. Nuclear power experts did not think this condition could occur.

The person responsible for training the TMI operators was a witness at the hearings. He was very proud of the last five years of his company's program. When asked what he considered his most important achievement, he replied: "*When I arrived, many courses had been given by engineers. But the engineers don't know how to talk in a way which people can understand. Consequently, the first rule I introduced was that no engineer was authorized to participate in the training of operators*" (quoted in [173, p. 65]).

The Kemeny Commission found that all theory had been taken out of the operator training program and that they were trained to be button pushers. This training was adequate for normal conditions, but the operators had not been prepared for a serious situation [174]. This training satisfied NRC requirements and had become standard practice.

Note that lack of understanding of what was happening was not just a problem for the operators: the TMI emergency and engineering personnel also had difficulty in analyzing the events, and the hydrogen bubble issues confused even outside experts. Even after supervisory personnel took charge, significant delays occurred before the full amount of core damage was recognized and stable cooling of the core was achieved. Of the people on duty at the plant when the accident started, none were nuclear engineers, none were even college graduates, and none were trained to handle complex reactor emergencies.

At Fukushima, operator training in response to a serious accident was not provided during normal operations or periodic inspections, leading to a lack of staff experience or training in activating emergency equipment. "*Plant workers had no clear instructions on how to respond to such a disaster, causing miscommunication, especially when the disaster destroyed backup generators*" [98].

At Chernobyl, operators and other staff were not trained in the technological processes in a nuclear reactor. They had also "lost any feeling for the hazards involved" [390]. In addition, as at TMI, the operators had no simulator training for the accident sequence that occurred.

The Presidential Commission report on the *Deepwater Horizon* accident noted that the rig crew had not been trained adequately to do their assigned duties. Transocean managers deliberately decided not to train their personnel in the conduct or interpretation of negative pressure tests. The rig workers were supposed to learn about these procedures through general work experience. The same was true for BP [275].

BP had been warned about this deficiency. A 2004 culture assessment at Texas City noted that the quantity and quality of training at Texas City was inadequate. Instead of fixing the problems, BP cut the budget for training and reduced staffing: the central training department staff was cut from twenty-eight to eight, and simulators were unavailable for operators to practice handling abnormal situations, including infrequent and high-hazard operations such as startups and unit upsets. BP admitted after the accident that they had not informed employees of known fire and explosion risks [374]. Not only were the workers poorly trained, but so were the managers [29].

At Macondo, a lack of contingency planning also likely contributed to the losses. The crew on the *Deepwater Horizon* rig did not seem prepared for the events that occurred during this highly dangerous temporary abandonment process. There was confusion on the rig at the same time as great personal heroism by individuals in attending to those who were injured.

Comparison with the behavior on the *Bankston* mudboat is instructive. During the temporary well-capping operation, this boat was taking on the drilling fluid (called mud) from the well. While chaos reigned on the oil platform after the explosion and fires, the *Bankston* crew performed spectacularly well. During the hearings that followed the accident, the captain of the *Bankston* explained that whenever they perform a potentially dangerous job, they plan extensively. In the morning review, they always rehearse all the potential contingency actions that may be needed during the day. When the explosion occurred and the *Deepwater Horizon* crew were panicking and jumping into the sea, the *Bankston* crew calmly went through their planned actions, including immediately releasing the hose carrying the mud to the ship in order to protect themselves and the ship and then boarding their rescue boats to save those on the *Deepwater Horizon* rig who were in the water. Without their calm and preplanned response, more lives might have been lost.

The Japanese government was unprepared to cope with the type of serious accident that occurred at Fukushima. The nuclear emergency preparedness drills conducted by the government did not anticipate severe accidents or complex disasters at all and were virtually useless as a measure to increase preparedness for nuclear accidents. [198]

The planning that was done for responding to a nuclear plant disaster in Japan was found not to function in the case of serious accidents in which radioactive substances were released into the environment at a large scale. For example, the offsite center for emergency response was located only 5 miles from the Fukushima Daiichi plant and was not equipped with air cleaning filters to insulate it from radioactive substances. Ultimately, it was forced to relocate its functions, wasting valuable time.

At the time of TMI, the existence of a state emergency or evacuation plan was not required. The Kemeny Commission suggests that the reasons for this included the agency's confidence in design reactor safeguards and their desire to avoid raising public concern about the safety of nuclear power. The report also suggests that the attitude fostered by the NRC regulatory approach and by Met Ed at the local level was that radiological accidents having off-site consequences beyond the 2-mile radius were so unlikely as not to be of serious concern. Similar reasons for the lack of emergency planning in Japan have been suggested. Emergency planning does not rise to a level of importance when accidents are not expected and a fear of public concern about nuclear safety exists.

The TMI emergency plan did not require the utility to notify state or local authorities in the event of a radiological accident, and delays occurred in doing this. Met Ed also did not notify its physicians under contract, who would have been responsible for the onsite treatment of injured and contaminated workers, and the emergency medical care training given to these physicians was inadequate. [173].

The response to the emergency was characterized by an atmosphere of almost total confusion and a lack of communication at all levels. Almost all the local communities around TMI lacked detailed emergency plans. Many key recommendations were made by people who did not have accurate information, and those who managed the accident were slow to realize the significance and implications of the events that had taken place.

At Fukushima, communication was disrupted, which led to difficulty in responding to the emergency and implementing contingency plans. Most of the communication lines and monitoring posts were washed away by the tsunami. Communication problems were exacerbated by confusion about responsibilities. To make things worse, when the nuclear accident happened, a large number of personnel both in the Fukushima Prefecture and in the municipalities were tied up with their response to the earthquake and tsunami disasters.

The emergency radiation medical system was unable to deal with accidents that involved the release of large amounts of radioactive substances over a wide area because of the inappropriate locations of primary radiation emergency hospitals: the hospitals themselves had to be evacuated and were unable to treat any patients. But even if they had been available, there was a lack of decontamination facilities and inadequate or almost nonexistent radiation training of the hospital staff. As a result, some of those who were injured at the Fukushima Daiichi Nuclear Power Plant did not have their injuries treated for 3 days. Claims have been made that the central government refused to pass on critical information for fear of inciting panic among the general public [380].

4.5.3.5 Inadequate management of change Accidents often occur after some change, either in the system itself and its procedures and policies, or the behavior of the operators, or a change in the environment. Changes are going to occur; the problem lies in not identifying and controlling unsafe changes.

At Flixborough, the process was changed, and the production capacity was tripled without a proper hazard analysis. At the time of the accident, a temporary pipe was used to replace a reactor that had been removed to repair a crack (see figure 4.4). The crack was the result of a process modification. The bypass pipe was not properly designed. The only drawing was a sketch on the workshop floor. The bypass pipe was also not properly supported; it rested on scaffolding. The pipe failed, and the resulting explosion killed twenty-eight people and destroyed the site. At Seveso, the process was changed to manufacture a different chemical (tricholophenol), which was more dangerous, without a review of the safety measures.

Some members of the chemical industry apparently did not learn the lesson of Flixborough. Many years later, a batch reactor exploded as a result of an exothermic decomposition after a temporary change. Management had decided that it was not cost-effective to connect the high-level interlock from the regular feed-tank pump to a temporary pump for only a month; instead, they instructed the operators to "be careful." An operator overfilled the feed tank, and the reactor overcharge led to the runaway decomposition. Attempts to achieve safety goals by instructing employees to be careful or not to make mistakes are not going to be effective.

In another accident, a leased crane was being used to lift a $100 million satellite into its shipping container.

> *Once the crane started lifting, it could not be stopped. When the hook reached the top of the crane boom, the cable snapped and dropped the satellite 20 feet to the floor.*

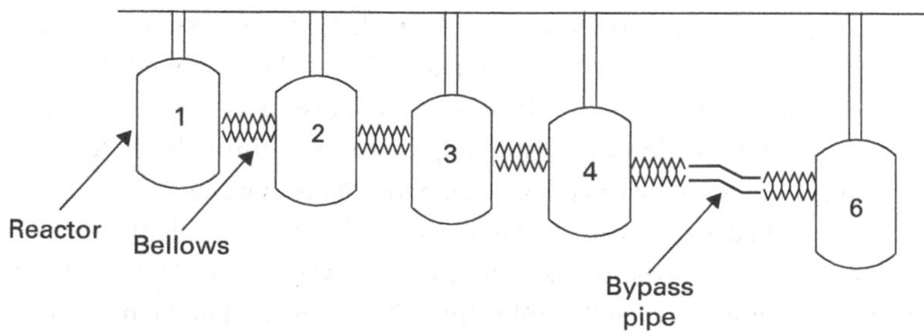

Figure 4.4
The temporary change at Flixborough.

Subsequent investigation showed that the leased crane had been modified and not tested prior to delivery to the contractor. A relay had been installed that was rated for less than 10% of the electrical load it was to carry. When electrical power was applied, the contacts fused so the power could not be shut off. [22]

Any change during operations must be evaluated to determine whether safety has been compromised. To accomplish this goal, most companies have what are called MoC (Management of Change) policies and standards. Note that changes may be planned and therefore known, or they may be unknown. The latter may result from changes in the environment in which the system is used or slow degradation of a system over time. Unplanned changes are more difficult to handle, but there are ways to detect and prevent losses that might result from them. Even when effective MoC procedures exist, they may not be followed or may be performed in a superficial manner.

One of the requirements for performing an effective change analysis is that the safety-related design decisions and the assumptions underlying them have been documented. The system engineers and the original designers of both the hardware and software parts of the system need to document the design decisions, their rationale, and assumptions underlying them so that maintainers do not accidentally violate the assumptions or eliminate the safety features. Decision makers need to know why safety-related decisions have been made so they do not inadvertently undo those decisions.

As an example of what can happen when this documentation does not exist, consider the following accident. In a test of an experimental ballistic missile launch detection satellite, the launch had to occur precisely at 11 a.m., but problems in the launch caused an estimated thirty-minute delay. One task took exactly 30 minutes to complete: spinning up the guidance system gyros and physically checking them to be sure that they were all running and functioning properly. An inquiry revealed that no problems had been experienced with these gyros in the last few launches. Against the advice of the guidance-system engineers, the gyro test was deleted to make up the 30 minutes. The missile was launched at precisely 11 a.m., but a pitch gyro did not function. The missile looped and had to be destroyed. The people who made the decision to delete the gyro test did so without knowing that it had been included to control a known hazard: the gyros were due to be replaced by a new model with the deficiency corrected [13].

In another example, a satellite was to be launched with an aged ballistic missile as a booster. During launch preparation, technicians questioned the necessity of installing a cumbersome heater hose into the engine bay. They were told that it was a carryover from the booster's use as a ballistic missile, and they decided to delete the installation. During launch, the engine gearbox coagulated because of the low temperature of an adjacent LOX (liquid oxygen) tank. The gearbox failed, and the booster fell back on the launcher. Again, appropriate documentation of the reason behind a design feature was lacking, and a procedure was deleted without an adequate analysis of its purpose [13].

In the Macondo blowout, numerous changes were made continually to the procedures that were to be used to temporarily cap the well in the two weeks before the blowout and up to the morning of the critical procedure. There is no evidence that these changes went through any sort of formal risk assessment or management of change process or review of any kind. There were no controls in place to ensure that key decisions were safe and sound from an engineering perspective.

The decision to perform any formal risk analysis was left to the team's discretion; an MoC was optional and applied mainly to decisions to deviate from well plans approved during the project creation stages and not to drilling procedures such as the temporary abandonment of the well.

At Fukushima, changes occurred that increased risk even during construction. An example is the construction of a seawall only 33 ft (10 m) high instead of the planned 98 ft (30 m) above sea level to protect the plant from tsunamis, which are common in that area. The change was made to make it easier to bring in the equipment to build the wall. No evaluation of the risk involved in that change was made.

At the Icmesa Chemical plant at Seveso in northern Italy, an accidental release of dioxin, one of the most poisonous substances known, occurred in 1976. Changes had been made without adequate review, both in the chemical being manufactured and in the production process. These changes violated local regulations and the original patented process.

At Flixborough, the process was changed and the capacity was doubled, but safety questions were not rethought. At Chernobyl, the shutdown procedures for routine maintenance were changed, without any review by experts, to accommodate doing a test that would otherwise have been delayed for another year [249]. Another complication here was that the test procedures had been developed by a station electrical engineer, and the operators thought it was an electrical test. They did not understand the implications for nuclear safety.

Management of change procedures must be enforced and used in an organization. To make this more feasible, safety-related design decisions must be documented, and this documentation must be used when making decisions, including decisions involving changes to the system design or operation. When new procedures are created, even if they are only temporary, they must be evaluated before being used.

Unplanned changes are the most difficult to manage. Rasmussen has suggested that it is common for organizations to migrate to states of higher risk resulting from pressures for greater productivity and profits in competitive industries [320]. Almost all the accidents described in this chapter displayed this phenomenon. As just one example, the working environment at Texas City had eroded over time under pressures to increase profits and decreasing budgets. This environment, coupled with unclear expectations about supervisory and management behaviors, led to rules not being followed consistently, a lack of rigor, and individuals feeling disempowered from suggesting or initiating improvements.

Dealing effectively with such migration to states of higher and higher risk requires procedures for identifying the dangerous changes as well as designing effective countermeasures to ensure that such migration stays within safe boundaries, as described in chapter 14.

4.5.3.6 Poor maintenance practices Redundancy is commonly relied on for safety, but assumed redundancy can deteriorate over time. For example, during the emergency at BP's Texas City refinery, redundant alarms did not sound because they were not functional or not properly calibrated. The same was true for other safety-critical equipment. None of the instruments that might have indicated a potential problem were working properly on the day of the accident. BP instrument technicians described the unit instrumentation as being run-down and in disrepair. Instead of preventative maintenance on safety-critical equipment, a policy of run-to-failure was used.

A variety of problems can lead to poor maintenance besides the obvious ones of not providing resources to do it correctly. Such problems include inadequate documentation and written procedures for testing and maintaining equipment.

BP management was aware or should have been aware of the maintenance deficiencies at Texas City from audit reports and previous accidents. When management was told that the maintenance budgets were not large enough to address identified risks, they responded that only the money on hand would be spent, rather than increasing the budget.

Things were not any better on the *Deepwater Horizon*. The BP and Transocean maintenance policy deviated and bypassed maintenance regulations and recommended practice for critical equipment. At the time of the accident, the *Deepwater Horizon* rig was operating with numerous maintenance issues. In September 2009, for example, about nine months before the blowout, BP conducted a safety audit on the rig before it headed to the Macondo well. The audit team identified 390 repairs that needed immediate attention and would require more than 3,500 hours of labor to fix and some downtime onshore. Those repairs were not done [29].

Safety-critical equipment should be the focus in any maintenance program. The BOP on an oil rig is supposed to be recertified every five years, but the *Deepwater Horizon* BOP had not been recertified for ten years before the accident and therefore was long overdue. The batteries, which were dead at the time of the blowout, should have been inspected and changed before it was lowered to the seabed, but they were not.

At Three Mile Island, the maintenance force was overworked at the time of the accident and had been reduced in size to save money. There were many shutdowns, and much of the equipment was out of order. A number of equipment items that were involved in the accident had had a poor maintenance history without adequate corrective action. As an example, inspection of the valves in the TMI-1 containment building after the accident showed a long-term lack of maintenance. Boron stalactites more than a foot long hung from the valves, and stalagmites had built up from the floor. At the time of the accident, Met Ed had not corrected deficiencies in radiation monitoring equipment pointed out by an NRC audit months before.

4.5.4 Government and Professional Society Oversight

Much can be learned by examining the government agencies that were tasked with oversight of industries where major accidents have occurred. Some industries and processes present so high a risk to the public that governments have stepped in to oversee them and protect their citizens and the environment. After major accidents, a common finding is that the government agencies tasked with this oversight did not operate as originally intended.

In some cases, limitations in effective oversight occurred because the same agency was responsible for both assuring safety and promoting the industry. In the United States, the conflicts in nuclear energy were resolved in 1974 by splitting the original Atomic Energy Commission into two parts: (1) a regulatory and licensing agency, the NRC, and (2) the Energy Research and Development Administration (ERDA). ERDA was assigned responsibility for promoting and developing commercial nuclear power. ERDA was later merged with another agency to become the US Department of Energy.

Oversight activities may depend on standards. Standards, however, can have the undesirable effect of limiting the safety efforts and investment of companies that feel their legal and moral responsibilities are fulfilled if they follow the standards. As the standards often represent the input of commercial companies, there are often conflicts of interest involved in producing effective standards. Changing a standard can take a decade or more, so they tend to always be behind the state of the art. The existence of a vast body of NRC regulations tended to focus industry attention narrowly on the meeting of regulations rather than on a systematic concern for safety.

Inadequate oversight by the US Nuclear Regulatory Commission was implicated in the investigation of the Three Mile Island accident. The licensing process concentrated on the equipment, assuming that the presence of operators would only improve the situation—they would not be part of the problem. Something similar is true for certification of commercial aircraft today.

The Kemeny Commission noted a persistent assumption that plants could be made sufficiently safe that they would be "people-proof." Thus, not enough attention in the licensing process was devoted to the training of operating personnel and to operator procedures [173]. More generally, there was no identifiable office within the NRC responsible for systems engineering examination of overall plant design and performance, including interaction between major systems, and also no office to examine the interface between machines and humans: "*There seems to be a persistent assumption that plant safety is assured by engineered equipment, and a concomitant neglect of the human beings who could defeat it if they do not have adequate training, operating procedures, information about plant conditions, and manageable monitors and controls. Problems with the control room contributed to the confusion during the TMI accident*" [173].

As was true originally in the United States, the Japanese government at first mixed responsibility for promoting nuclear power with protecting the populace from its potential effects. When an attempt was made to fix this problem, responsibility was spread over several ministries and offices, leading to ambiguity about who was responsible for what. Complete separation was never achieved: the Japanese Nuclear Safety Commission (NSC) avoided any regulation that appeared to be an obstacle to the promotion and utilization of nuclear power.

After the Fukushima accident, the government was accused of being too closely affiliated with the industry and with promotion of nuclear power in Japan. This closeness resulted in both limited oversight of nuclear plant operations and inadequate planning for an accident so as not to create resistance to the promotion of nuclear power in Japan. As just one concrete example, the comprehensive nuclear emergency preparedness drill conducted by the national government in cooperation with local governments was superficial in nature because it was aimed primarily at not worrying or confusing local residents. It was ineffective as a response to actual accidents.

The utility, TEPCO, that owned and operated Fukushima Daiichi as well as other nuclear power plants and which was supposed to be subject to nuclear safety regulatory supervision, strongly pressured regulatory authorities for postponement of regulations and softening of regulatory criteria.

The Nuclear and Industrial Safety Agency (NISA) was the Japanese agency responsible for regulation of all industrial safety, not just nuclear safety, but it did not have sufficient human resources to carry out this charge. It had been occupied with handling various nuclear incidents since its establishment in January 2001 and lacked sufficient personnel capable of addressing mid- to long-term nuclear safety challenges [380]. In addition, periodic personnel transfers for NISA staff members made it difficult to develop specialized technical ability along with expertise and experience with nuclear regulations.

At Bhopal, all levels of the government were lax in enforcing safety-related policies [27]. The Indian government inspection agencies were understaffed and uniformed about the safety problems related to the chemical MIC involved in the release. Ladd claims that "safety was given a low priority by all the parties involved" [199]. The government was hesitant to interfere with a plant like the one in Bhopal that was producing important chemicals and providing high-quality employment in a country with an inadequate number of such jobs.

The Federal Aviation Administration, throughout most of its history, has been highly effective in reducing aviation accidents. It benefited from the fact that the largest US manufacturer, Boeing, understood that safety was critical to ensuring the future of the company and the commercial aviation industry. With very low accident rates and increasingly limited budgets, however, the FAA reduced its oversight activities and outsourced more responsibility for oversight to both independent companies and to Boeing itself. The B737 MAX accidents have been partially attributed to reduced FAA activities.

The United States government oversight of offshore drilling provides a good example of how government agencies can contribute to a major accident. Offshore oil and gas regulation at the time of the Macondo/*Deepwater Horizon* blowout was regulated by the MMS, which was part of the US Department of the Interior. They faced conflicting pressures. Energy independence required increasing domestic production, but previous accidents, such as the Santa Barbara spill in 1969, increased mandates for environmental protection. A third pressure came from the goal of revenue generation—the expansion of offshore oil and gas production brought in billions of dollars in federal revenues from offshore leases. The incentive to promote offshore drilling conflicted with the mandate to ensure safe drilling and environmental protection. Revenue collection became the dominant objective [275].

Increasing revenues, both for the companies and for the government, meant that offshore drilling needed to move into much deeper water with the greater risks that entailed. Those increased risks were not matched by greater and more sophisticated regulatory oversight. Industry resisted such oversight, and there was not much political support to overcome that opposition.

Even if they had wanted to provide strict regulation, the MMS lacked resources, technical training, and experience in petroleum engineering. Over time, those resources fell increasingly short. Industry relied on more demanding and complex technology, but the government did not have the expertise to regulate this technology or the ability to maintain up-to-date regulations. MMS responsibilities increased at the same time budgets decreased. Offshore fires, explosions, and blowouts increased as a result.

Funding for training at the MMS was inadequate. Low salaries, compared to workers in the industry, meant that the MMS had difficulty recruiting the employees it needed to create and enforce regulations. The agency's ethical culture declined, and gifts from oil and gas companies were accepted in some offices.

The need for Coast Guard oversight also increased in the 1990s as industry drilled in deeper waters farther offshore and used more ambitious floating drilling and production systems. But, like the MMS, it faced severe budgetary constraints. As a result, it did not update its marine-safety rules to reflect the industry's new technology, and it shifted much of its responsibility for fixed platform safety to the MMS in 2002 at the same time as those in industry argued that they did not need MMS oversight because they already were being regulated by the Coast Guard.

Oversight need not be governmental but can instead consist of voluntary activity by an industrial association. The American Petroleum Institute (API) is the largest trade association for the oil and gas industry. Among other things, they produce voluntary standards for the industry. This provided a potential alternative to government regulation. API led the effort for many years to persuade the MMS not to adopt regulations about safety and environmental management systems. It instead argued for its own voluntary recommended safety practices [275].

The Presidential Commission on the *Deepwater Horizon* Accident concluded that API's ability to serve as a reliable standard-setter for drilling safety was compromised by its role as the industry's principal lobbyist and public policy advocate. Because regulations would make oil and gas industry operations potentially more costly, API regularly resisted agency rulemakings that government regulators believed would make those operations safer. API favored rulemaking that promoted industry autonomy from government oversight.

According to statements made by industry officials to the Presidential Commission that investigated the Macondo/*Deepwater Horizon* blowout, API's safety and technical standards were a major casualty of this conflicted role. As described by one representative, API-proposed safety standards had increasingly failed to reflect best industry practices and had instead expressed the lowest common denominator—in other words, a standard that almost all operators could readily achieve. Most of the safety standards were really guidelines and suggested practices but not requirements. Because, moreover, the Interior Department had relied on API in developing its own regulatory safety standards, API's shortfalls undermined the entire federal regulatory system.

With respect to the Macondo/*Deepwater Horizon* blowout, the MMS never required testing the BOP under the conditions in which it would be used. It could have required testing of the design (many of the problems were in the design) but did not. It could have performed better inspections and required specific maintenance actions, such as those for the BOP. The MMS never required testing of the deadman, the autoshear, or the underwater robots. The agency did not even require rigs to have these backup systems in place at all, although it did send out a safety alert encouraging their use.

4.5.5 Engineering Processes and Practices

A third group of systemic factors related to accidents reflects poor implementation of the specific activities necessary to achieve an acceptable level of safety. These flaws include superficial safety efforts, ineffective risk controls, and safety information system deficiencies.

4.5.5.1 Superficial safety efforts Sometimes safety organizations are caught in the same type of bureaucratic environments as other groups. Paperwork may be completed to meet milestones, irrespective of the quality of that work. This problem is most likely to arise when the system safety engineers become so enmeshed in the project development effort that they lose their objectivity and become defenders of group decisions. Alternatively, this effect can occur when they are totally separated managerially from the engineering design group, and they view their efforts as an end in itself and not as a part of creating a safer design.

Childs [63] coined the term *cosmetic system safety*, which is characterized by superficial safety efforts and perfunctory bookkeeping. Hazard logs may be meticulously kept, with a line item that amazingly supports and justifies each design decision and tradeoff made by the project management and engineers.

This type of superficial safety effort is commonly found after major accidents. All of the accidents mentioned in this chapter had inadequate system safety programs characterized by

- a checklist mentality, where personnel completed paperwork and checked off the safety policy and procedural requirements even when those requirements had not been met;
- no formal system for identifying high-level risks: even the risk assessments that were done were not very rigorous;
- lack of investigation of critical events: audits may have been performed, but corrective actions to identified problems were limited and ineffective. Lessons learned were not captured or acted on;
- compliance-based approaches to safety that emphasized completing paperwork;
- lack of follow-up: in some cases, adequate safety analyses were performed, but no follow-up was done to ensure that the recommendations were being followed or that they were effective, that the hazards were actually controlled, or that the safety devices were maintained and kept in working order.

In the process industry, it is common to design a safety system that is used to shut down or respond to hazardous events. This subsystem is designated as *safety-related* and is subjected to risk management activities, but the rest of the system is not. While this strategy may work for the relatively simple electromechanical systems of the past, it will not for the complex systems being created today. It is nearly impossible to show that nothing in the rest of the system can impact the behavior of the safety system. Safety is a system problem, and the entire system must be included in attempts to reduce losses.

At the time of TMI, strict reviews and requirements applied to the former; the latter were exempt from most requirements—even though they could have an effect on the safety of the plant. Items not labeled as safety-related did not need to be reviewed in the licensing process, were not required to meet NRC design criteria, did not need to be testable, did not require redundancy, and were ordinarily not subject to NRC inspection. At TMI, the pressure-operated relief valve (PORV), which played an important part in the events, was not considered a safety-related item because it had a block valve behind it. The block valve was not considered safety-related because it had a PORV in front of it.

A common mistake in safety practices is omitting software or human factors analysis or making unrealistic assumptions about software or humans. Operators may then be blamed for accidents in systems when the real problems were flawed engineering practices and designs.

Sometimes it is assumed that if standard software engineering or human factors practices are used, then the software and interface will be safe. One version of this belief is that "levels of safety" can be assured by using subsets of software engineering practices. Other assumptions are that easy-to-use interfaces are safe or that human-interface design can be done separately from the basic engineering design. There is no scientific evidence that these beliefs are true and little reason to argue that they are. Safety is achieved in software and human factors engineering by applying special design principles, tools, and engineering practices and by incorporating them into the basic engineering effort.

4.5.5.2 Overreliance on redundancy and protection systems Inaccurate views of risk can arise due to overreliance on the use of redundancy and diversity to increase reliability. For example, a chemical plant protection system might be diverse in that it monitors two independent parameters, such as temperature and pressure, and it might redundantly include multiple channels using majority voting. However, one factor might cause all these components to fail simultaneously, such as an electrical outage or a fire. Many examples are in this chapter and in the accident descriptions in the appendices.

A type of redundancy, common in the process industry, is the use of layers of protection or *defense-in-depth.* Defense-in-depth entails providing a succession of barriers to the propagation of a malfunction. For example, in a nuclear power plant, the layers of protection may include cladding around the fuel to prevent the release of fission products, a closed reactor coolant system to keep the fuel from becoming overheated and carry the heat away from the core, a backup perhaps in the form of a biological shield, elaborate spray and filter systems, and the final layer of protection in the form of the containment building itself.

This approach to designing for safety is subject to the *Titanic Coincidence*. More generally, the layers of defense may not be independent, as assumed. The first layers of protection usually include design features, while emergency response is used in case the design is deficient. Frequently, the system designers and operators are so sure that the design features will be effective that they begin to believe that emergency response will never be needed and put less emphasis on it. In the Fukushima nuclear accident, an assumption was made that a large tsunami would not occur in the near future. Before the accident, the Japanese government set up a study group on flooding, which concluded that no basis existed for assuming that the probability of a tsunami hitting the Fukushima Daiichi Nuclear Power Plant was extremely low. No appropriate response was taken by the government, however, in response to this new information. When the tsunami hit, nearly all preparation that had been made in advance became useless.

Consider the layers of protection provided at Bhopal (figure 4.5). The chemical was stored in underground tanks encased in concrete. The operating manual specified that the tanks were never to contain more than half their maximum volume. Standby tanks were available to which some of the chemical could be transferred in case of trouble. Regular scheduled inspection and cleaning of valves and piping were required. If staff members were doing sampling, testing, or maintenance at a time when there was the possibility of a leak or spill, they were to use protective rubber suits and airbreathing equipment. A refrigeration system was provided to keep the chemical at a nonreactive temperature. A vent gas scrubber was designed to neutralize any escaping gas with caustic soda. A flare tower was designed to burn off any escaping gas missed by the scrubber. The toxic gases would be burned high in the air, making them harmless. Small amounts of gas that were missed by the scrubber and the flare tower were to be knocked down by a water curtain that reached 40 to 50 ft above the ground. And as a final protection, a siren was installed to warn the workers and the surrounding community.

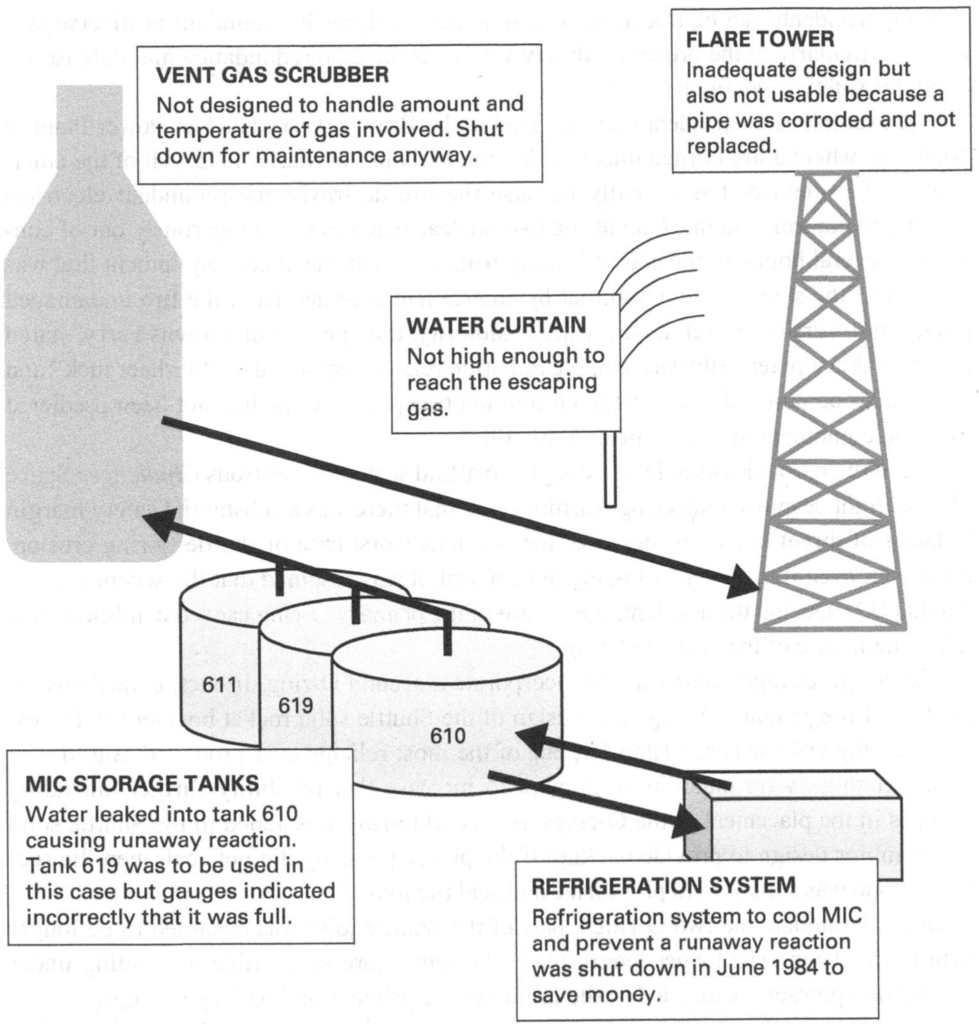
VENT GAS SCRUBBER
Not designed to handle amount and temperature of gas involved. Shut down for maintenance anyway.
FLARE TOWER
Inadequate design but also not usable because a pipe was corroded and not replaced.
WATER CURTAIN
Not high enough to reach the escaping gas.
611
619
610
MIC STORAGE TANKS
Water leaked into tank 610 causing runaway reaction. Tank 619 was to be used in this case but gauges indicated incorrectly that it was full.
REFRIGERATION SYSTEM
Refrigeration system to cool MIC and prevent a runaway reaction was shut down in June 1984 to save money.

Figure 4.5
The basic defense-in-depth protection system at Bhopal.

How could all of these layers of protection fail at the same time? It seems incredulous but, in actuality, is easily explained. *Common-cause failures* are failures that are coincident in time because of dependencies among conditions leading to the events, such as failure of a common power supply. Common-mode failures denote the failure of multiple components in the same way, such as stuck-open or silent. Redundant systems can have failures that are common-cause, common-mode, or both.

The causes of the multiple failures may appear to be different, such as at Bhopal, but are in reality related to systemic factors such as lack of investment in safety devices or inadequate maintenance practices. Note the potential for circular causality here: the use of redundancy leads to complacency which, in turn, leads to the redundancy being ineffective.

Many accidents can be traced to common-cause failures in redundant or diverse systems, particularly in the process industry where reliance on redundancy and defense-in-depth for safety is common.

One example is an accident that occurred at the Browns Ferry Nuclear Power Plant in Alabama, where a fire burned uncontrolled for 7.5 hours in March 1975. All of the emergency safety devices failed totally because the fire destroyed the redundant electrical power and control systems. One of the two nuclear reactors was dangerously out of control for several hours. In the end, it was controlled by some available equipment that was not part of the safety system and that by chance had emerged from the fire undamaged [405]. Engineers of the Tennessee Valley Authority, the operator of Browns Ferry, stated privately that a potentially catastrophic radiation release was avoided "by sheer luck" and by human operators who jury-rigged a solution to a problem that had not been predicted by using equipment in an unexpected way [9].

One of the rationales used in deciding to go ahead with the disastrous *Challenger* Space Shuttle flight despite engineering warnings was that there was a substantial safety margin (a factor of three) in the O-rings over the previous worst case of shuttle O-ring erosion. Moreover, even if the primary O-ring did not seal, it was assumed that the secondary one would [332]. During the accident, the failure of the primary O-ring caused conditions that led to the failure of the backup O-ring.

The design changes necessary to incorporate a second O-ring, in fact, contributed to the loss of the primary O-ring. The design of the Shuttle solid rocket booster (SRB) was based on the US Air Force Titan III, one of the most reliable ever produced. Significant design changes were made in an attempt to increase that reliability further, including changes in the placement of the O-rings. A second O-ring was added to the Shuttle solid rocket motor design to provide backup: if the primary O-ring did not seal, then the secondary one was supposed to pressurize and seal the joint.

To accommodate the two O-rings, part of the Shuttle joint was designed to be longer than in the Titan. The longer length made the joint more susceptible to bending under combustion pressure, which led to the failure of the primary and backup O-rings.

In this case, and in a large number of other cases [213], the use of redundancy requires design choices that in fact defeat the redundancy at the same time that the redundancy is creating unjustified confidence and complacency.

In the Fukushima nuclear accident, most of the protection systems in the plant were unavailable because they depended on a common power source, which failed. An unjustifiable amount of confidence was placed on redundancy and in defense-in-depth that did not actually exist due to lack of independence and single point failure modes. Planned control and feedback channels did not operate as designed or expected.

In November 1961, all of the US Strategic Air Command (SAC) special communication lines with early warning radars and with early warning command headquarters (called NORAD [North American Aerospace Defense Command]) suddenly went dead. Commercial phone circuits also were not working. Because the outage could have been caused by enemy action, an emergency alert was sounded at SAC bases across the United States. Later, it was discovered that although redundant communication links had been established between SAC and NORAD, they had been run through a single relay station, despite assurances to the contrary. An overheated motor at that station had cut all communications [342].

Accident investigations often find that the safeguards or barriers intended to prevent such an event were not properly constructed, tested, or maintained. In some cases, they had been removed. Consider the *Deepwater Horizon* oil rig accident.

The immediate cause of the Macondo/*Deepwater Horizon* blowout was a failure to contain hydrocarbon pressures in the well. Three physical barriers could have contained those pressures: cement at the bottom of the well, drilling fluids (called "mud") in the pipes connecting the oil rig with the well, and a device called a blowout preventer. Each of these barriers in turn included various types of redundancy. But mistakes and misunderstanding about and sometimes disregard for risk compromised each of those potential barriers, depriving the crew of safeguards until the blowout became inevitable and, in the end, uncontrollable. The design was filled with common-cause failure modes, such as inadequate maintenance and common power sources, but these were either never identified or never eliminated. Poor decision-making, mostly related to saving time and money, invalidated the assumption of multiple lines of defense.

Juechter warns that a poorly designed safety device is worse than no safety device at all because its presence creates a sense of security that allays natural fears and caution [169]. As noted earlier in this section, providing redundancy may lead to the complacency that defeats the redundancy. Chapter 11 discusses the general flaws in the use of redundancy for safety and the alternatives that are available.

4.5.5.3 Ineffective risk control The majority of accidents are not the result of a lack of knowledge about how to prevent them but of failure to use that knowledge [185] or to use it effectively. As suggested earlier, the particular causal factors of an accident often are identified, but nothing is done about them because of complacency or underestimating risks. Even with good intentions, the ways to prevent particular hazards, though well understood, may not be known to the people concerned. This education problem is particularly acute when using computers in safety-critical systems, where those actually building the software may know little about basic safety engineering practices and safeguards.

In some accidents, the hazards are identified and efforts are made to control them, but that control is inadequate. One of the frustrations of those who attempt to reduce risk is that risk reduction often results merely in risk displacement: a risk that appears to be eliminated reappears in a different guise. Workers in the scrap metal industry are exposed to toxic materials originally used to protect metal in bridges or in steel frameworks used for large buildings [245]. X-ray examinations may help diagnose illness, but they may also contribute to future disease. Chemicals introduced to reduce the incidence of particular risks, for example, the use of nitrites and nitrates to counter botulism, are suspected causes of cancer. A flame-retardant material (TRIS or dibormopropyl phosphate) used in children's pajamas in the 1960s and early 1970s was found to be a carcinogen. We often create a new problem while trying to solve an old one.

In fact, technological safety fixes themselves sometimes create accidents. A partial meltdown at the Detroit Fermi nuclear reactor, for example, was caused by the breaking loose of a flow-deflecting zirconium plate that had been installed to *reduce* the likelihood of a core meltdown [110]. The triangular piece of zirconium became detached and blocked the flow of sodium coolant.

In another accident, seventy-two out of 141 French weather balloons were inadvertently destroyed by a software control program feature originally intended to protect the public from out-of-control balloons [21]. The computer in the French meteorological satellite was supposed to issue a *READ* instruction to the high-altitude weather balloons but instead ordered an *EMERGENCY SELF-DESTRUCT*. It is interesting to ponder why the other sixty-nine balloons did not self-destruct. In a chemical plant, an error in a watchdog card, installed to protect against errors or failures in another device, opened some valves at the wrong time, and several tons of hot liquid were spilled [189].

To make our technological fixes more effective, we need to understand why they sometimes do not work. Five major factors appear to explain why past efforts to reduce risk have been unsuccessful: (1) patches are used to eliminate the specific causes of past accidents but not the basic underlying design flaws; (2) the design of the safety devices is based on false assumptions; (3) the safety fixes increase complexity and cause more accidents than they prevent; (4) the safety devices are effective in achieving their original goals, but then are used to justify reducing safety margins; and (5) a lack of defensive design. Each of these needs further explanation.

Not eliminating basic design flaws In some cases, technological fixes (safety devices) do not work because they are added to compensate for poor organization or for poor system design, but they do not offset the effect of the organizational or design problems. The overuse of and reliance on checklists of hazards is related to this problem. Such checklists are often constructed using information from previous accidents. Although these lists may be useful in learning from the past and making sure that obvious things are not forgotten, blindly using checklists alone can also limit the factors that are considered. As Johnson writes, one finds triple locks on those doors from which horses have been stolen, while other doors are left wide open [166]. The rapid introduction of new technology into systems today makes such lists even less useful.

Basing safeguards on false assumptions A second reason that risk-reduction measures may not be effective is that they are predicated on false assumptions. Kletz [185] relates an example of an attempt to increase safety that actually increased loss and damage. A tank was filled once a day with sufficient raw material to last for 24 hours. An operator was tasked to watch the level and then switch off the filling pump and close the inlet valve when the tank was 90 percent full. The operator performed this task without incident for five years, until one day he allowed his attention to wander, and the tank was overfilled. The spill was small, as the operator quickly noticed it and switched off the pump. To prevent another such spill, a high-level trip was fitted to the tank to switch off the pump automatically if the level exceeded 90 percent.

Everyone was surprised when the tank overflowed again after about a year. The designer's intention was that the operator would continue to watch the level and that the trip would take over on the rare occasion when the operator forgot to do so. The designer had assumed that the chance of the operator and the trip failing at the same time was negligible. Unfortunately, this assumption was unrealistic. The operator left the control of the tank level to the trip and turned his attention to other duties, relying on the trip to turn off the pump. Both the manager and the foreman decided that the operator's time was better spent elsewhere. The trip failed, and the tank overfilled. This type of trip is known to have a mean time between failure of a little over a year, so the tank was likely to overflow around that time, but the spill was much greater than before because the operator was doing something else and not paying attention to the level of the tank [188].

This story is not unusual. Often, operators are blamed for an accident when equal blame should be put on the designers who failed to understand the problems in using humans as monitors (see chapter 6) and who made unrealistic assumptions about the operation of the plant.

Predicating safety devices on assumptions that may not be true—especially assumptions about human behavior—is a serious mistake.

Complexity A third reason why safety devices may not reduce risk is that they may increase system complexity so much that they cause more accidents than they prevent. This increased complexity results more often when safety devices are added as an afterthought than when safety is considered in the original design. If hazards can be identified early in the design process, removing the hazards through system design may be possible, rather than attempting to control them by adding protective equipment. Risk may be reduced without increasing complexity or cost, and the risk reduction measures may be more effective.

Protection systems are those features added to a standard design in order to restore the system from an unsafe to a safe state or to warn that an unsafe state has been reached. Such systems are usually expensive and are effective only if they exhibit ultrahigh reliability. Techniques to achieve ultrahigh reliability often involve redundancy and diversity in the protection-system design, adding to the complexity of the system and perhaps causing failures themselves. A NASA study of an experimental aircraft found that all of the software problems occurring during flight testing resulted from errors in the redundancy management system and not in the basic aircraft control software, which was very simple and worked perfectly [241]. The redundancy management system, used to protect against errors in the aircraft control software, was much more complex than the original flight software.

The seeming ability to add software functions easily may unwittingly encourage the addition of functions to computer-based protection systems, such as different types of *stop* functions, additional inputs reporting the state of the system, or additional status checks of the protection system itself. Complexity may be increased to the point where the increased risk of software errors is greater than the added protection. Adding functions to software is easy; determining whether you *should* do so is much more difficult.

More information and examples can be found in chapter 11 on "Design for Safety."

Using risk control devices to reduce safety margins In some cases, the technological fix does reduce risk, but it then is used to justify the reduction of safety margins—for example, performance is increased by running the system faster or in worse weather or with larger explosives. The net effect may not be reduced risk and may even be greater risk than before the safety device was introduced. Perrow notes for aviation: "*As the technology improves, the increased safety potential is not fully realized because the demand for speed, altitude, maneuverability, and all-weather operations increases*" [303].

As radar and collision avoidance systems are introduced, minimum separation distances between planes may be reduced. The responsibility lies with managers who use the safety devices as an excuse to reduce operating expenses, to increase productivity and profits, to gain time, and to reduce staffing.

Lack of defensive design The Therac-25 software did not contain self-checks or other error-detection and error-handling features that would have detected the inconsistencies and coding errors. There were no independent checks that the machine was operating correctly. Patient reactions were the only real indications of the seriousness of the problems with the Therac-25. There seemed to be so much confidence in the software that standard hardware safety and defensive design features (described in chapter 11) were omitted.

Such verification cannot be assigned to human operators without providing them with some means of detecting errors: The Therac-25 software "lied" to the operators, and the machine itself was not capable of detecting that a massive overdose had occurred. Engineers need to design for the worst case. We will make little progress if human operators are blamed for accidents when the problems are really in the machine design.

In the Therac-25, audit trails were limited because of a lack of memory in computers at that time. However, today larger memories are available and audit trails and other design techniques must be given high priority in making tradeoff decisions. The problem today is usually the opposite one: so much data can be and therefore is collected, for instance during aircraft operations, that it is difficult to detect problems until after a serious event has occurred. The problem today is not in collecting data, but identifying trends and behaviors of the hardware, software, and operators that are increasing risk *before* an accident occurs. Simply collecting data alone will not help here. We need sophisticated modeling and analysis tools to analyze that data. Data, whether "big" or not, is not the same as information.

Of course, care must be taken that adding defensive design does not increase complexity to the point where the added design features actually cause the problems.

4.5.6 Safety Information System Deficiencies

The quality of an organization's safety information system has been found to be the second most important factor distinguishing organizations that have high accident rates [179]. The most important is a sincere commitment and priority assigned to safety by management.

Good decision-making about risk is dependent on having appropriate information. Without it, decisions are often made on the basis of past success and unrealistic risk assessment, as was the case for the Shuttle. Lots of data was collected and stored in multiple databases, but there was no convenient way to integrate and use the data for management, engineering, or safety decisions [10].

Creating and sustaining a successful safety information system requires a culture that values the sharing of knowledge learned from experience. Necessary data may not be collected, and what is collected is often filtered and inaccurate or tucked away in multiple databases without a convenient way to integrate the information to assist in management, engineering, and safety decisions. Methods may be lacking for the analysis and summarization of causal data, and information may not be provided to decision makers in a way that is meaningful and useful to them. In lieu of such a comprehensive information system, past success and unrealistic risk assessment are often used as the basis for decision-making.

As a specific example in the *Challenger* accident, the ineffectiveness of the added O-ring was actually known. An engineer at NASA concluded after tests in 1977 and 1978 that the second O-ring was ineffective as a backup seal. Nevertheless, at the time of the *Challenger* accident, the SRB joint design was still classified as redundant in the Marshall Space Center information system.

The difficulty of systematizing knowledge is an important factor in explaining why similar accidents occur. The large volume of accident case histories in most fields is uncollated and often nonexistent in a usable form. Although accident investigations can provide very useful information for improving the design of new systems and for avoiding repetition of similar accidents, information that is detailed enough to be useful is difficult to acquire.

Software errors or computer hardware failures have caused potentially serious problems for the US military, but learning from them has at times been diminished because of a failure to disseminate this information. In October 1962, in the midst of the Cuban Missile Crisis, the radar operators at an early warning site in Moorestown, New Jersey, reported to the national command center (NORAD) in Colorado that a missile had just been launched from Cuba and was about to detonate 18 miles west of Tampa, Florida [342]. NORAD alerted the rest of the command centers and asked the radar operators in New Jersey to recheck their data. The reply confirmed that a missile was coming. The NORAD officers immediately passed the warning information to the Strategic Air Command (SAC) and checked with the bomb alert system (a network of nuclear detection devices placed on telephone poles around cities and military bases in the United States) but were told that no detonation had been reported.

Afterward, they discovered that a software test tape simulating a missile launch from Cuba had been mistakenly loaded into the radar operators' computer. At the same time, an object in space, most likely a satellite, came over the horizon right in the location that a missile launched from Cuba would be. The early warning system included overlapping redundant radars to provide more reliability, but the redundant radars were not turned on when the incident occurred. In addition, the facility that was supposed to provide advance information about satellites passing overhead had been taken off that function, ironically, to help provide early warning of missiles during the Cuban Missile Crisis. None of this information was reported in any of the classified after-action reports—it appeared only on the NORAD command post log. Sagan concludes that this reporting failure had very unfortunate consequences: "*It ensured that higher authorities would fail to learn from the incident, increasing the likelihood of a later repetition of this particular failure mode in nuclear command and control systems*" [342, p. 130].

Indeed, seventeen years later it happened again. In November 1979, a realistic display of a Soviet nuclear attack appeared on NORAD and SAC computer screens: a large number of Soviet missiles appeared to have been launched in a full-scale attack on the United States. The air defense interceptor force was launched along with the president's special doomsday plane (the National Emergency Airborne Command Post), but without the president. FAA air traffic controllers were instructed at some locations to order commercial aircraft in their area to prepare to land immediately. "For a few frightening moments, the U.S. military got ready for nuclear war" [342, p. 229]. After 6 minutes, the alert was canceled when direct contact with the warning sensors (satellites and radars) indicated that no attack was underway and that the information on the computer displays was false.

Officials later reported to the press and Congress that the incident had been caused by an operator error—which, in fact, was untrue—when an "exercise or training tape" simulating an attack had been mistakenly inserted into the warning system at Cheyenne Mountain. As a result of this incident, stringent test procedures and rules were instituted and an off-site computer testing facility was constructed near Cheyenne Mountain, reducing the need for software test tapes to be stored near or used on the live computer again [342, p. 231].

Other false warning incidents have resulted from computer problems, including one (described in chapter 11 on "Design for Safety") involving a computer chip fault. Not repeating such mistakes requires recording such problems and learning from them.

An important use for operational experience is for feedback on the hazard analysis and assessment process. Were the failures or errors not included in the accepted hazards, were the controls ineffective, or were the hazards overlooked? When numerical risk assessment techniques are used, operational experience should be used to provide insight into the accuracy of the models and probabilities used. Often, this type of reevaluation is not done.

As one example, a large explosion occurred at a Shell chemical plant at Moerdijk in the Netherlands in June 2014. In 1977, Shell scientists had shown that a particular catalyst was inert in the presence of a chemical called ethylbenzene. This assumption was never reassessed even though the composition of the catalyst had changed over time and two incidents had occurred in Shell reactors, where the catalyst exploded in the presence of ethylbenzene. Those incidents should have prompted a reexamination of the original assumption. The likelihood of an explosion cannot be zero if it has happened twice before. When the Shell Moerdijk plant was designed in 1997, however, the risk analysis used zero as the likelihood that an explosion could occur under the same conditions that had previously occurred in two of Shell's own plants.

In various studies of the DC-10 by McDonnell Douglas, the chance of engine power loss with resulting slat damage during takeoff was estimated to be less than one in a billion. However, this highly improbable event occurred four times in DC-10s before changes were made. Even one event should have warned someone that the models used might be incorrect. A device was finally installed in 1982 to prevent slat retraction in such emergencies [303].

A study of the safety information systems of various companies by Kjellan [179] showed that, in general, these systems were inadequate to meet the requirements for systematic accident control. For example, collected data was improperly filtered and thus inaccurate, methods were lacking for the analysis and summarization of causal data, and information was not presented to decision makers in a way that was meaningful to them. As a consequence, learning from previous experience was delayed and fragmentary.

An effective system, either formal or informal, for reporting incidents requires delegation of responsibility for reporting; a government, industry, or organizational policy that includes protection and incentives for informants; a system for handling incident reports; procedures for analyzing incidents and identifying causal factors; and procedures for using reports and generating corrective actions [399]. More can be found about the design of safety information systems in chapter 14.

4.6 Summary

This chapter has described the most common systemic factors leading to accidents. The rest of this book will show you how to eliminate or reduce these factors.

Exercises

1. Read the transcript (from a video) at http://sunnyday.mit.edu/citichem-transcript.doc. Make a list of the causes that you see in these events. Which ones are listed as systemic causes in this chapter? Or, alternatively, look at the systemic causes listed and see if you can find them in the citichem video. This video was created to train accident investigators and is not a real accident.
2. Take a major accident in the last ten years for which a nonsuperficial accident investigation was performed. Examine the accident report to identify any of the factors mentioned in this chapter with respect to whether they were cited in the report as being involved. You will need to look at the details of the investigation results and not just the conclusion section that summarizes the causes. Which factors not mentioned in the report (but noted in this chapter) might have been appropriate to investigate? What questions might have been asked about them during the investigation? In the final conclusions of the accident report about the cause of this accident, which of the factors were identified as root causes or contributory causes? Is there some reason you can identify that might account for leaving other potential causes out?
3. Take a few major accidents in your industry, examine the accident reports, and identify systemic factors described in this chapter or additional factors that seem to be common among them.
4. Change is inevitable in all engineered (designed) systems.
   a. Why do you think most accidents follow some kind of change, or, in other words, what role does change play in accidents? What implications does this have for designers? For managers?
   b. The changes preceding an accident may be planned—and thus can be handled with management of change (MoC) procedures—or unplanned and perhaps known. Given that a management of change procedure exists for the organization, what are some examples of why accidents might still occur?
   c. Describe two ways unplanned changes might be handled to prevent accidents?

5

The Role of Software in Safety

And they looked upon the software, and saw that it was good. But they just had to add this one other feature . . .
Software temptations are virtually irresistible. The apparent ease of creating arbitrary behavior makes us arrogant. We become sorceror's apprentices, foolishly believing that we can control any amount of complexity. Our systems will dance for us in ever more complicated ways. We don't know when to stop. . . . We would be better off if we learned how and when to say no.
—G. Frank McCormick,
When Reach Exceeds Grasp

Most of the traditional (and current) methods used to engineer safety into systems were created in the 1960s or before. At that time, system design was very different than today, most notably the use of computers. If they were used at all in complex system control, they were used for calculations. Safety-critical systems were made up primarily of hardware and humans. As a result, safety engineering focused on those two types of components.

In the 1970s, however, computers began to be introduced into the control and design of safety-critical systems. Safety engineering tried to force fit software and computers into the traditional safety engineering methods or extend those methods in some way to accommodate software, but with little real success. Understanding why this is true is important in making progress. This chapter describes the essential difficulties that software presents in safety engineering and why traditional methods no longer work adequately for our new, software-intensive systems.

5.1 The Use of Software in Systems Today

The advantages of computers have led to an explosive increase in their use, including their introduction into potentially dangerous systems. Few systems today are built without computers to provide control functions, to support design, and sometimes to do both. Computers now control most safety-critical devices, and they often replace traditional hardware safety interlocks and protection systems—sometimes with devastating results. Even if the hardware protection devices are kept, software is often used to control them.

This new use of software began in the late 1970s. Frola and Miller in 1984 described the new problem of computer-related hazards in military aircraft at that time:

> *A relatively new breed of hazards and associated problems has appeared. They appear primarily in flight control systems, armament control systems, navigation systems, and cockpit displays. They add new dimensions to the human-error problem. Some of the hazards are passive until just the right combination of circumstances arrives. . . . Some result from the crew's multitude of choices in aircraft system management, often during prioritization of tasks. Conversely, computer-based systems are supposed to relieve pilot workload, but perhaps too much so in some instances, with resultant complacency and/or lack of situation awareness.* [109]

Human factors experts have defined levels of automation with ten levels of increasing control by automation over a task with the decision-making versus oversight varying between the computer and the human. A five-level model is popular in the automotive industry. The problem is that these models assume a relatively simple allocation of control between the human and automation over a single task. In fact, humans and automation are now working as partners or even independently with shared or interacting decision-making.

Figure 5.1 shows, on the left, the relatively simple control allocation where the human controls software that in turn controls the actual physical device or process. In this simple case, the decision-making can shift in increments between the human and the computer. The right shows an example of the more complex problem facing designers today where decision-making is distributed and coordination between human and automated controllers is required.

Basically, humans and computers are teaming to achieve overall system goals. Poorly coordinated parallel decision-making can lead to hazards whether the decision makers are

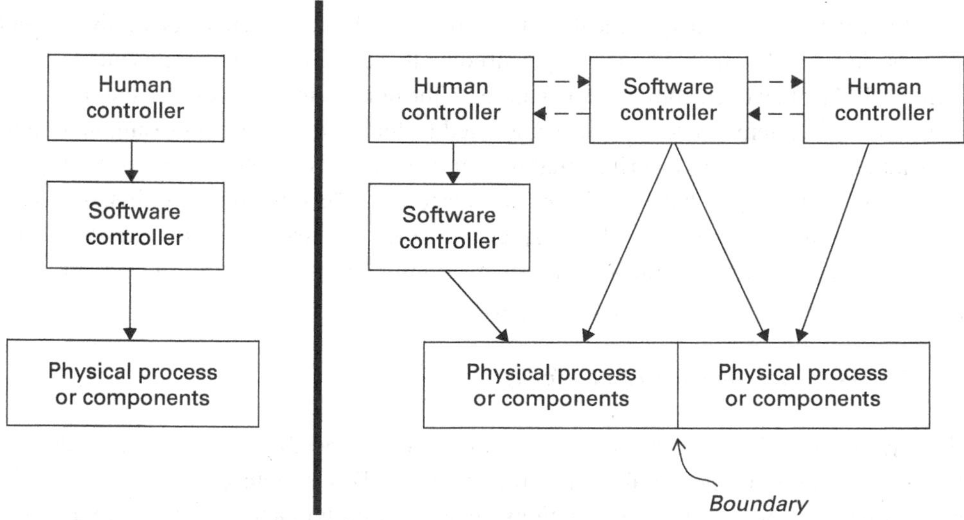

Figure 5.1
On the left is the common design today of a human controlling software, which in turn controls a physical process or physical components. The figure on the right illustrates examples of more complex control schemes, including shared control by computers and humans over the same or different processes with related or at least nonconflicting goals. The humans and computers may have to communicate information and coordinate actions to achieve the system goals.

automated or human. Examples of general problems are conflicting commands being given to two different processes or the common boundary case where each controller thinks that the other is providing control in the boundary between processes, but in reality, neither is.

An example of the boundary problem occurred after the Persian Gulf War when the US enforced a *no-fly zone* over northern Iraq to protect Kurdish refugees. Air traffic over the zone was controlled by two human controllers: one was in charge of the air traffic outside the no-fly zone and the other controlled traffic within the zone. A handoff occurred between the two controllers when aircraft outside the zone crossed into it. The problem occurred because there was a US Army headquarters just inside the boundary line. Helicopters were entering the zone frequently, only to land at the headquarters. Over time, the overhead of constantly switching control led to the handovers being informally abandoned and the helicopters continuing to be controlled by the outside controllers when they crossed the boundary. One day, a helicopter did not remain at the headquarters but traveled far inside the no-fly zone. The US Air Force fighters protecting against enemy incursions did not know it was a friendly aircraft and shot it down. There were, of course, many more details of this accident not included here. For a full description, see Leveson [217].

Potential safety implications of computers exercising direct control over dangerous processes are obvious. Perhaps less obvious are the dangers when:

1. Software-generated *data* is used to make safety-critical decisions (such as air traffic control and medical blood analyzers).
2. Software is used in design analysis.
3. Safety-critical data is stored in computer databases, which can be compromised.

Examples abound of the problems that can result. For example, the FDA has received reports of software errors in medical instruments that led to mixing up patient names and data as well as reports of incorrect outputs from laboratory and diagnostic instruments, such as patient monitors, electrocardiogram analyzers, and imaging devices [33]. In 1979, the discovery of an error in the software used in the design of nuclear reactors and their supporting cooling systems resulted in the Nuclear Regulatory Commission's temporary shutdown of five nuclear power plants that did not satisfy earthquake standards. There exists a serious danger of overreliance on the accuracy of computer outputs and databases.

In some cases, arguments have been advanced that software that generates data but does not make decisions is not safety critical or is less so than direct-control software because the human controller makes the ultimate decision, not the computer. As an example of such an argument: "*If diagnostic devices produce incorrect results, the errors may be readily noticed or may be inconsistent with other clinical signs. Thus, the risk to the patient is less than in the case of software-driven devices that directly affect patients*" [167, p. 6].

While risk *may* be reduced by the use of a human intermediary, this reduction is by no means assured. In fact, information from a computer or technological device may be more readily believed than conflicting information that is directly observed. Even more frequently, systems are built that require the operator to rely on computer-generated information that the human has no independent way to check. In almost all cases, risk from incorrect computer operation is not eliminated by having a human in the loop, and it may not even be reduced. An argument could be made, in fact, that it may increase risk due to complacency by the human operator when the computer rarely produces a wrong answer. Poorly designed software can cause or contribute to human errors.

Arguments that the human operator makes the ultimate decision and therefore is always responsible may be appropriate in a courtroom when affixing blame and determining financial liability, but not when the goal is to build safer systems. Engineers need to consider all the contributors to accidents if their efforts to reduce risk are to be effective. Often the argument that software providing information or advice to humans, who make the ultimate decisions, is not safety critical is used to downgrade the criticality of the software and to avoid or reduce the difficult task of ensuring that the system is safe. If system safety truly is to be increased, then all the components whose operation can directly or *indirectly* affect safety must be considered, and the related hazards must be eliminated or controlled.

5.2 Understanding the Problem

The first step in creating safer software-controlled systems is recognizing that software is an abstraction, and abstractions are not unsafe: software cannot explode, catch on fire, or release toxins and it does not have sharp edges. It can only have an impact on safety by providing commands or not providing commands to hardware or by providing incorrect information to humans. An important implication of this fact is that: *Any attempt to determine whether software is safe or not without including the context in which it is to be used will not work.*

The implication of this fact is that the attempts to define and use a SIL (*Safety Integrity Level*) are useless. A belief that a safety level can be assigned to software and that the level has any relevance in a different system is not technically justified. Software can only be shown to be safe within a given context. In any other context, conclusions about safety cannot be made without examining the use of the software in that specific context.

This conclusion makes sense when one considers that software can only be unsafe when controlling a potentially unsafe physical process. When the hardware is different than originally considered in the software design, the control commands provided by the software may lead to system hazards. The bottom line is that *software itself is not safe or unsafe*. Safety related to software is a relevant concept only with respect to the overall encompassing system in which the software is operating.

Figure 5.2 represents this general idea. The hazards are in the physical (controlled) process that may be controlled or partially controlled by the software or humans or both. In most systems today, the software itself is controlled by humans.

Considering figure 5.2, software and human controllers themselves cannot get into a hazardous state, as a hazard is defined in system safety. The software does not have sharp edges, for example, and cannot explode or catch on fire. Software and human controllers are not themselves unsafe, in such a system, but they can produce unsafe control signals that can contribute to a hazardous state in the controlled process.

People sometimes informally talk about "safe" or "unsafe" software. But these phrases only have meaning if one interprets them as *the software can produce unsafe control signals that will result in the controlled process getting into a hazardous state within certain contexts.*

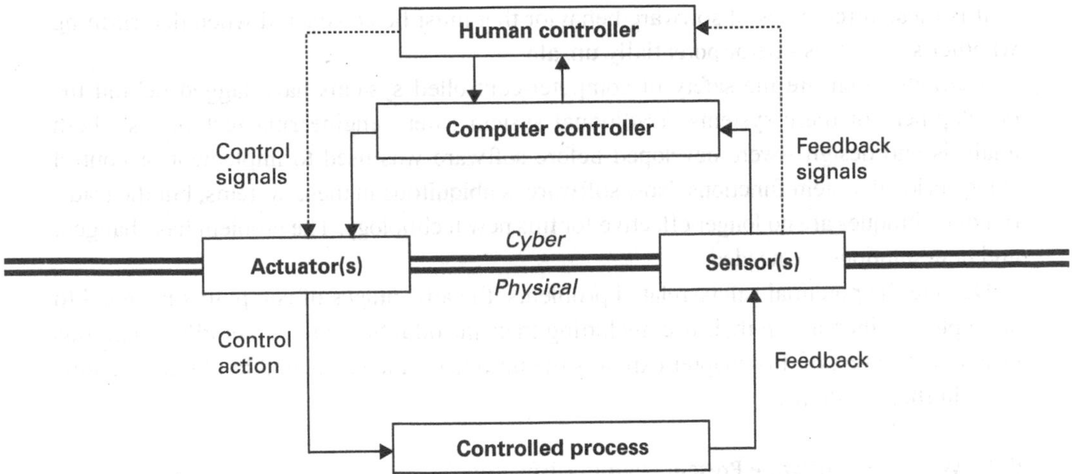

Figure 5.2
A generic control loop.

Related to this misconception is a belief that software implementation errors are important in accidents. While this belief could theoretically be true, in practice almost all accidents related to software can be traced to flaws in the software requirements and not to errors in the implementation of those requirements in some programming language [217]. Potential approaches to assuring the safety of the software behavior that involve applying sophisticated methods to verify the correct implementation of the requirements or using a higher LOR (*Level of Rigor*) in the software development will have little to no impact on the safety of the software. The focus needs to be placed on the identification of the safety-related software constraints or requirements related to the hazards of the particular process that the software is controlling.

Software can be unreliable and incorrect and lead to the nonsatisfaction of the system goals and constraints that are not related to safety, but that software will be safe if the control signals satisfy the safety-related requirements. Such software may be incorrect and unreliable, but it will be safe.

At the same time, the software may satisfy system requirements and constraints unrelated to safety but not the safety-related ones. Such software is potentially unsafe in operation but may be useful in terms of satisfying non-safety-related system goals. Software, like any system component, can be reliable but unsafe or safe but unreliable.

To understand these statements, consider that the only way software can impact controlled process safety is to provide unsafe control actions; that is, control signals that contribute to a controlled process reaching a hazardous state. There are three ways this can happen:

1. The software provides a control action that contributes to a hazard in the controlled process.
2. The software does not provide a control action that is needed to prevent a hazard in the controlled process.
3. The software provides a potentially safe control action, but does so too late or too early, resulting in a controlled process hazard.

It is these three types of software behavior that must be considered when determining whether software is safe or potentially unsafe.

Methods to ensure the safety of computer-controlled systems have lagged behind the development of these systems. Traditional system safety engineering techniques—both analysis and design—were developed before software was used to implement or control safety-critical system functions. Now software is ubiquitous in these systems, but the traditional techniques are no longer effective for this new technology. The problem has changed, and new solutions are needed.

Despite the potential safety-related problems, the advantages of computers have led to an explosive increase in their use, including their introduction into potentially dangerous systems. The rest of this chapter explores the unique problems created by the use of software in these systems.

5.3 Why Does Software Present Unique Difficulties?

Just as James Watt's invention of the first practical steam engine fueled the Industrial Revolution, the invention of the first practical computers about fifty years ago has drastically altered our society. The uniqueness and power of the digital computer over other machines stem from the fact that, for the first time, we have a general-purpose machine (figure 5.3). We no longer need to build a mechanical or analog autopilot from scratch, for example, but simply to write down the "design" of an autopilot in the form of instructions or steps to accomplish the desired piloting goals.

These steps are then loaded into the computer, which, while executing the instructions, in effect *becomes* the special-purpose machine (the autopilot). If changes are needed, the instructions can be changed instead of building a different physical machine from scratch.

Machines that previously were physically impossible or impractical to build become feasible to create. In addition, the design of the machine can be changed quickly without going through an entire retooling and manufacturing process. In essence, the manufacturing phase is eliminated from the life cycle for these machines: the physical parts of the machine (namely the computer hardware) can be reused, leaving only the software design and verification phases. Although duplication of software might be considered to be manufacturing, it is usually a relatively trivial process. The design phase also has changed: emphasis is placed only on the steps to be achieved without having to worry about how those steps will be realized physically.

Because software and hardware engineering developed separately, are usually taught in siloed classes, and have partitioned development efforts and practitioners, the engineering of systems with extensive software components presents unique difficulties.

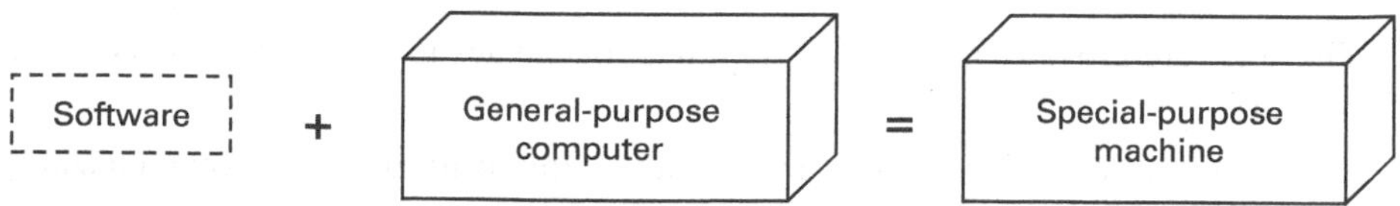

Figure 5.3
Software plus a general-purpose computer creates a special-purpose machine.

Communication problems, in particular, are making efforts to create safe systems more difficult. Hardware and software engineers have few common models or tools. Even their vocabulary is different. Computer science, by developing separately from engineering, has created its own technical vocabulary—sometimes the two groups use different names for the same things, or they give the same name to different things. These problems must be solved before we can make progress in creating safer software-intensive systems.

Let's start our examination of the special difficulty of dealing with software in safety engineering—in fact, in engineering in general—with some common misunderstandings about software among non-software professionals.

5.3.1 Software Myths

If there are difficulties in using software in engineering, why are computers being used so widely? The basic reason is that computers provide a potential level of speed, control, power, and flexibility not otherwise possible; they are also relatively light and small. Many other supposed advantages of using computers are myths, however, usually stemming from looking only at the short term. Understanding these myths is important if we are to make good decisions about using computers to control safety-critical processes.

Myth 1: The cost of computers is lower than that of equivalent analog or electromechanical devices.
Reality: This myth, like most myths, has some superficial truth. Microcomputer hardware is cheap relative to other electromechanical devices. However, the cost of writing and certifying highly reliable and safe software to make that microprocessor useful, together with the cost of maintaining that software without compromising correctness and safety, can be enormous.

The onboard Space Shuttle software, for example, while relatively simple and small (about 400,000 words) compared to more recent control systems, cost NASA approximately $100 million a year to maintain [212].

Designing an electromechanical system is usually much easier and cheaper, especially when standard designs can be used. Of course, software *can* be built cheaply, but then lifetime costs—including the cost of accidents and required changes when errors are found—increase and may become exorbitant.

Myth 2: Software is easy to change.
Reality: Again, this myth is superficially true. Changes to software *are* easy to make. Unfortunately, making changes without introducing errors is extremely difficult. And, just as for hardware, the software must be completely reverified and recertified every time a change is made, at what may be an enormous cost. In addition, software quickly becomes more "brittle" as changes are made—the difficulty of making a change without introducing errors may increase over the lifetime of the software as changes accumulate.

Myth 3: Computers provide greater reliability than the devices they replace.
Reality: Although true in theory—software does not "fail" in the sense this term usually implies in engineering—there is little evidence to show that erroneous behavior by software is not a significant problem in practice.

When systems were made only from electromechanical and human components, engineers always had to worry about mechanical failure and operator or maintenance error. Techniques were developed to reduce greatly, but not eliminate, random hardware failures and to mitigate their consequences. Those techniques are largely very effective in making them highly reliable.

Now that computers are being introduced into safety-critical, real-time systems, however, a new, completely abstract factor—software—has been added. Because software is pure design, there is no need to worry about the random wear-out failures found in physical analog and electronic devices, but system behavior now can be significantly affected by software design errors and by maintenance and upgrade actions on the software that can introduce new errors. Human errors can also be impacted and indeed induced by software design, as described elsewhere in this book.

Little hard data is available on the reliability of operational software, especially data that compares software reliability to the reliability of equivalent systems that do not use computers. What is available mixes too many types of software (such as game, business, and control software) to be useful. And the complexity of the systems is usually not equivalent. A little useful data does exist, however.

One example is a study by the British Royal Signals and Radar Establishment that used commercially available tools to examine the number of errors in software written for some highly safety-critical systems [72]. Up to 10 percent of the program modules or individual functions were shown to deviate from the original specification in one or more modes of operation. Discrepancies were found even in software that had undergone extensive checking using sophisticated test platforms. Many of the detected anomalies were too minor to have any perceptible effects—for example, a discrepancy of 1 part in 32,000 in a computation using 16-bit arithmetic. However, about 1 in 20 of the functions found to be faulty; that is, about 1 in 200 of all new modules, contained errors with direct and observable effects on the performance of the system being controlled. For example, potential overflows in integer arithmetic were detected that caused a change in the direction of deflection of an actuator—such as "turn left" when the correct action was "turn right."

On the surface, because wear-out over time is not applicable to software, it seems that the solution to the software reliability problem is simply to get the software right in the first place. Although human error is a factor, because software is designed by humans, ample time *appears* to be available to apply sophisticated techniques to eliminate any design errors before the software is used. However, accomplishing this goal has turned out to be harder than expected. Very few, if any, sophisticated software systems have been fielded that have not contained a significant number of errors.

These are not just "teething" problems that go away after the software is used for a while. Software-related errors usually occur over the entire system lifetime, sometimes after tens or hundreds of thousands of hours of use. The Therac-25 worked correctly thousands of times before the first known overdose, and the accidents were spread out over two and a half years [229].

The Space Shuttle software was used for about thirty years, and NASA invested an enormous amount of effort and resources in verifying and maintaining this software. Despite this effort, during the Shuttle operation, sixteen severity level 1 software errors, defined as errors that could result in the loss of the Shuttle or its crew, were discovered in released software.

Eight of those remained in code that was used in flights, but none was encountered during flight. An additional twelve errors of lower severity were triggered during flight: none threatened the crew, three threatened the achievement of the mission, and nine were worked around. These problems occurred despite NASA, at the time, having one of the most thorough and sophisticated software development and verification processes in existence.

Although hardware reliability is easily augmented by the use of redundancy to prevent or reduce the impact of random hardware failures, techniques that provide the equivalent reliability enhancement for software errors, which again are only design errors, have not been found. For various reasons, described in the next section, highly effective techniques are unlikely to be found. Vendors sometime tout tools or approaches that will lead to "zero-defect" software, but these claims are more sales than science.

Even if the techniques did exist to produce perfect software, there is usually not enough time to accomplish this goal: the ideal conditions—including unlimited funds and time—seldom if ever exist. Instead, there are competing needs to reduce software costs, which already exceed hardware cost in many large systems. Time pressures also become severe because the time required to develop software often controls the pace of the overall project.

But even if this myth were true—that computers are more reliable than other devices—this fact would not necessarily mean that they are *safer* than the devices they replace.

Myth 4: Increasing software reliability will increase safety.
Reality: Software reliability can be increased by removing software errors that are unrelated to system safety, thus increasing reliability while not increasing safety at all.

In addition, software reliability is defined as compliance with the requirements specification, while most safety-critical software errors can be traced to errors in the requirements—that is, to misunderstandings about what the software should do. Most software-related accidents have occurred while the software was doing exactly what it was intended to do—the software satisfied its specification and did not fail. Software can be correct and 100 percent reliable and can still be responsible for serious accidents. Safety is a system property, not a software property.

Myth 5: Testing software or "proving" (using formal verification techniques) software correct can remove all the errors.
Reality: The limitations of software testing are well known. Basically, the large number of states of most realistic software makes exhaustive testing impossible; only a relatively small part of the state space can be covered during testing. Although research has resulted in improved testing techniques, no great breakthroughs are on the horizon, and mathematical arguments have been advanced for their impossibility [130].

The use of mathematical approaches to verify the consistency between the software instructions and the specifications is another way to gain assurance. Although not currently practical, mathematical verification of software may be so in the future. Unfortunately, such verification will not solve all our problems. The process requires that the correct behavior of the software first be specified in a formal, mathematical language. This task is not easy and is turning out to be as difficult and error-prone as writing the code.

In addition, although the basic computations and algorithms are easily specified and verified, the most important errors may not lie in these aspects of the code. For example, many software-related accidents have involved overload. An instance occurred in England when a computer that was dispatching emergency ambulance services stopped working because it was unable to handle the number of calls it received [294]. Sophisticated timing and overload problems are much more difficult, and perhaps impossible, to verify formally because they involve more than just the application software itself but the design of the system as a whole.

Most important, as stated earlier, practical experience and empirical studies, for example Lutz [239], have shown that most safety-related software errors can be traced to the requirements and not to coding errors, which tend to have less serious consequences in practice. Writing adequate software requirements is a difficult and unsolved problem. Some techniques have been proposed that may help, but the problem is far from solved and may remain unsolved for quite some time. Simply proving that the software implements the requirements (as specified) will not solve the safety problem when the problem is unsafe requirements.

Myth 6: Reusing software increases safety.
Reality: Although reuse of proven software components can increase reliability, reuse has little or no effect on safety. In fact, reuse may actually *decrease* safety because of the complacency it engenders and because the specific hazards of the new system were not considered when the software was originally designed and constructed. The Therac-25 and Ariane 5 examples are described in other chapters. There are many others, for example:

- Software used successfully for air traffic control for many years in the United States was reused in Great Britain with less success. The American developers had not worried about handling zero degrees longitude (because that was not relevant in the United States); as a result, the software basically folded England along the Greenwich Meridian, plopping Manchester down on top of Warwick [416].
- Aviation software written for use in the northern hemisphere often creates problems when used in the southern hemisphere. In addition, software written for American F-16s caused accidents when reused in Israeli aircraft flown over the Dead Sea, where the altitude is less than sea level.

Safety is not a property of the software itself, but rather a combination of the software design and the environment in which the software is used: safety is application-, environment-, and system-specific. Therefore, software that is safe in one system and environment may be unsafe in another. Reuse is *not* a solution to the safety problem.

Myth 7: Computers reduce risk over mechanical systems.
Reality: Computers have the *potential* to decrease risk, but not all uses of computers achieve this potential. Computers can automate tedious and potentially hazardous jobs such as spray painting and electric arc-welding, thus reducing the risk to workers in these particular jobs. However, other arguments that computers can reduce risk are debatable:

1. *Argument*: Computers allow finer control in that they can check parameters more often, perform complicated computations in real time, and take action quickly.

   Counterargument: Computers *do* provide finer control computations in real time, and they *can* take action quickly. But finer control allows the process to be operated closer to its optimum, and the safety margins may then be cut in response. The resulting systems have economic benefits because they will, theoretically, shut down less often, and productivity may be increased by allowing more optimal control. However, any potential safety benefits of the finer control may be negated by the decrease in safety margins—perhaps without the concomitant attainment of the high software reliability on which the arguments for smaller safety margins were based. There is no way to know before extensive use outside the test environment, namely during operations, whether high software reliability and safety have been achieved. And that environment will change over time, perhaps even because of the introduction of the software.

2. *Argument*: Automated systems allow operators to work farther away from hazardous areas.

   Counterargument: Because of lack of familiarity with the hazards, more accidents may occur when operators *do* have to enter hazardous areas. Assumptions that plants controlled by robots will not require operators to intervene physically are usually wrong and can lead to accidents. For example, a computer-controlled robot killed a worker in a plant that the designers had assumed would require minimum intervention—they did not include walkways for humans or standard safety devices such as audible warnings that the robot was in motion [111]. After the plant was operational, the operators found that they needed to enter the hazardous areas fifteen to twenty times a day to bail out the robots and maintain adequate productivity. The original assumption that all robots (and thus the plant) would be shut down before humans entered hazardous areas became impractical. The designers had overestimated the ability of the plant to work adequately without human intervention and had not foreseen changes that would be required to the planned operating procedures to meet productivity goals.

3. *Argument*: By eliminating operators, human errors are eliminated.

   Counterargument: Operator errors are replaced by software design and maintenance errors, which require humans. Humans are not removed from the system; they are merely shifted to different jobs. It should not be too much of a surprise that human software designers have been found to make the same types of errors as human operators (see chapter 6).

   Moreover, as noted in chapter 6, when humans are removed from direct contact with the system, they lose information that is necessary for correct decision-making. Physically removing operators from the processes that they supervise may simply lead to new types of human errors and hazards.

4. *Argument*: Computers have the potential to provide better information to operators and thus to improve human decision-making.

   Counterargument: While theoretically true, in reality, this potential is very difficult to achieve. The subject is complex, and a detailed discussion is deferred until later. Briefly, computers make it easy to provide too much information to operators and to provide it in a form that is less usable or error-prone for some purposes than traditional instrumentation.

5. *Argument*: Software does not fail.

   Counterargument: This common belief is true only for a very narrow definition of "failure" as discussed earlier in this chapter. The important point here is that computers can exhibit incorrect and hazardous behavior, whether we call this behavior failure or not.

One of the results of substituting computers for mechanical devices is a reduced ability to predict failure modes. Most mechanical systems have a limited number of failure modes, and often they can be designed to fail in a safe way. For example, a valve can be designed to fail closed or a relay can be designed to fail with its contacts open. In comparison to software, the limited number of physical failure modes also simplifies (1) the analysis of a system for potentially unsafe behavior, (2) the process of assuring that the design is adequately safe, and (3) the elimination or control of hazards to make the system safer. The unpredictability of software behavior and the potentially large number of incorrect behaviors often preclude the same type of failure-mode analysis and fail-safe design.

In summary, computers have the potential to increase safety, and surely this potential will be realized in the future. But we cannot assume that we know enough now to accomplish this goal. In addition, any increased potential may not be realized if those building the systems use it to justify taking more risks.

Computers will not go away: their use and importance in complex systems is only going to increase. Software engineers, often with little training or experience in safety engineering, are building software for safety-critical systems. At the same time, safety engineers are finding themselves faced with ensuring that computer-dominated control systems are safe. To achieve and ensure safety in these systems, software must be included in the system safety activities, and the software must be specifically developed to be safe using the results of system hazard analysis.

5.3.2 Why Software Engineering Is Difficult

Why do we have so much trouble engineering software when, for the most part, the software is performing the same functions as the electromechanical devices it is replacing? Shouldn't the same engineering approaches apply when the same type of design errors can be made in both? Shouldn't they be equally hard or easy to construct?

Parnas [297] and Shore [351] have written excellent descriptions of the unique engineering problems in constructing complex software. Much of the following discussion comes from these two sources.

5.3.2.1 Analog versus discrete state systems In control systems, the computer is usually simulating the behavior of an analog controller. Although the software may be implementing the same functions previously performed by the analog device, the translation of the function from analog to digital form may introduce inaccuracies and complications. Continuous functions can be difficult to translate to discrete functions, and the discrete functions may be much more complex to specify.

In addition, the mathematics of continuous functions is well understood; mathematical analysis often can be used to predict the behavior of physical systems. The same type of analysis does not apply to discrete (software) systems. Software engineering has tried to use mathematical logic to replace continuous functions, but the large number of states and lack of regularity of most software result in extremely complex logical expressions. Moreover, factors such as time, finite-precision arithmetic, and concurrency are difficult to handle. There is progress, but it is very slow, and we are far from being able to handle even relatively simple software. Mathematical specifications or proofs of software properties may be the same size as the program, more difficult to construct, and often harder to understand than the program. They are therefore as prone to error as the code itself [78; 113].

Physical continuity in analog systems also makes them easier to test than software. Physical systems usually work over fixed ranges, and they bend before they break. A small change in circumstances results in a small change in behavior: a few tests can be performed at discrete points in the data space, and continuity can be used to fill in the gaps. This approach does not work for software, which can behave in bizarre ways anywhere in the state space of inputs; the incorrect behavior need not be related in any way to normal behavior [130].

5.3.2.2 The curse of flexibility A computer's behavior can be easily changed by changing its software. In principle, this feature is good—major changes can be made quickly and at seemingly low cost. In reality, the apparent low cost is deceptive, as discussed earlier, and the ease of change encourages major and frequent change, which often increases complexity rapidly and introduces errors.

Flexibility also encourages the redefinition of tasks late in the development process to overcome deficiencies found in other parts of the system. During development of the C-17, for example—a project that ran into great difficulties largely because of software problems—the software was changed to cope with structural design errors in the aircraft wings that were discovered during wind tunnel tests rather than changing the physical design of the wings. This case is typical. As Shore [351] writes, "Software is the resting place of afterthoughts."

With physical machinery, major design modifications are much more difficult to make than minor ones. The properties of the physical materials in which the design is embedded provide natural constraints on modification. The design of a computer application, on the other hand, is stored in electronic bits and presents no physical barriers to manipulation. Thus, while natural constraints enforce discipline on the design, construction, and modification of a physical machine, these constraints do not exist for software.

Shore explains this difference by comparing software with aircraft construction, where feasible designs are governed by mechanical limitations of the design materials and by the laws of aerodynamics. In this way, nature imposes discipline on the design process, which helps to control complexity. In contrast, software has no corresponding physical limitations or natural laws, which makes it too easy to build enormously complex designs. The structure of the typical software system can make a Rube Goldberg design look elegant in comparison (see figure 5.4). In reality, software is just as brittle as hardware, but the fact that software is logically brittle rather than physically brittle makes it more difficult to see how easily it can be broken and how little flexibility actually exists.

The myth of software flexibility also encourages premature construction before we fully understand what we need to do. The software medium is so forgiving that it encourages us to begin working with it too soon. Although we often intend to go back and start again after the details are worked out, this iteration process rarely happens in practice, and poorly considered design decisions made in prototypes and early design efforts

Pencil Sharpener

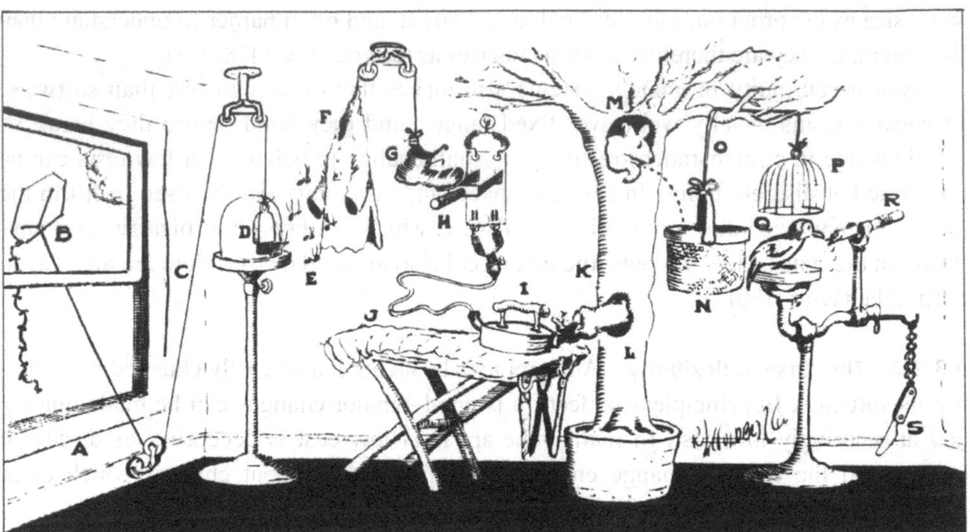

The Professor gets his think-tank working and evolves the simplified pencil sharpener.

Open window (**A**) and fly kite (**B**). String (**C**) lifts small door (**D**), allowing moths (**E**) to escape and eat red flannel shirt (**F**). As weight of shirt becomes less, shoe (**G**) steps on switch (**H**) which heats electric iron (**I**) and burns hole in pants (**J**). Smoke (**K**) enters hole in tree (**L**), smoking out opossum (**M**) which jumps into basket (**N**), pulling rope (**O**) and lifting cage (**P**), allowing woodpecker (**Q**) to chew wood from pencil (**R**), exposing lead. Emergency knife (**S**) is always handy in case opossum or the woodpecker gets sick and can't work.

Figure 5.4
Rube Goldberg's simplified pencil sharpener.

usually remain unchanged. Few engineers would start to build an airplane before the designers had finished the detailed plans.

Another trap of software flexibility is the ease with which partial success is attained, often at the expense of unmanaged complexity. The untrained can achieve results that appear to be successful but are really only partially successful: the software works correctly most of the time, but not all the time. Attempting to get a poorly designed, but partially successful, program to work all the time is usually futile; once a program's complexity has become unmanageable, each change is as likely to hurt as to help. Each new feature may interfere with several old features, and each attempt to fix an error may create several more. Thus, although it is extremely difficult to build large software that works correctly under all required conditions, it is easy to build software that works 90 percent of the time. Shore notes that it is difficult to build reliable aircraft too, but it is not particularly easy to build planes that fly 90 percent of the time [351].

Few people would dare to design an airplane without training or after having built only model airplanes, but there seem to be few such qualms about attempting to build complex software without appropriate knowledge and experience. Shore explains: "*Like airplane complexity, software complexity can be controlled by an appropriate design discipline. But to reap this benefit, people have to impose that discipline; nature won't do it. As the name implies, computer software exploits a "soft" medium, with intrinsic flexibility that is both its strength and its weakness. Offering so much freedom and so few constraints, computer software has all the advantages of free verse over sonnets; and all the disadvantages"* [351].

A final type of discipline is also necessary: limiting the functionality of the software. This discipline may be the most difficult of all to impose. Theoretically, a large number of tasks can be accomplished with software, and distinguishing between what *can* be done and what *should* be done is very difficult. Software projects often run into trouble because they try to do too much and end up accomplishing nothing. When we are limited to physical materials, the difficulty or even impossibility of building anything we might think about building limits what we attempt. The flexibility of software, however, encourages us to build much more complex systems than we have the ability to engineer correctly. A common lament on projects that are in trouble is, "If we had just stopped with doing x and not tried to do more." McCormick notes that *"A project's specification rapidly becomes a wish list. Additions to the list encounter little or no resistance. We can always justify just one more feature, one more mode, one more gee-whiz capability. And don't worry, it'll be easy—after all, it's just software. We can do anything. In one stroke we are free of nature's constraints. This freedom is software's main attraction, but unbounded freedom lies at the heart of all software difficulty" [254].*

5.3.2.3 Complexity and invisible interfaces One way to deal with complexity is to break the complex object into pieces or modules. For very large programs, separating the program into modules can reduce individual component complexity. However, the large number of interfaces in software introduces uncontrollable complexity into the overall design: the greater the number of small components, the more complex the interface becomes. Errors occur because the human mind is unable to fully comprehend the many conditions that can arise through the interactions of these components [297].

An interface between two software modules comprises all the assumptions that the modules make about each other. Shore notes that such dependencies can be subtle and almost impossible to detect by studying the modules involved. For example, one module might work properly only if another module can be relied on to finish its job in a specific amount of time. When changes are made, the entire structure collapses.

Finding good software structures has proven to be surprisingly difficult [297]. In the design of physical systems, like nuclear power plants or cars, the physical separation of the system functions provides a useful guide for effective decomposition into modules. Equally effective decompositions for software are harder to find, and some new software design approaches meant to solve the problem actually exacerbate it for control software.

In addition, the relatively high cost of the connections between physical modules helps to keep interfaces simple. As Shore points out

> *Physical machines such as cars and airplanes are built by dividing the design problems into parts and building a separate unit for each part. The spatial separation of the resulting parts has several advantages: It limits their interactions, it makes their interactions relatively easy to trace, and it makes new interactions difficult to introduce. If I want to modify a car so that the loudness of its horn depends on the car's speed, it can be done, at least in principle. And if I want the car's air conditioner to adjust automatically according*

to the amount of weight present in the back seat, that too can be done—again in principle. But in practice such changes are hard to make, so they require careful design and detailed planning. The interfaces in hardware systems, from airplanes to computer circuits, tend to be simpler than those in software systems because physical constraints discourage complicated interfaces. The costs are immediate and obvious. [351]

In contrast, software has no physical connections, and logical connections are cheap and easy to introduce. Without physical constraints, complex interfaces are as easy to construct as simple ones, perhaps easier. Moreover, the interfaces between software components are often "invisible" or not obvious: it is easy to make anything depend on anything else. Again, discipline and training arc required to control these problems, but when the software reaches a certain size, which is often found in control systems today, the complexity can overwhelm even the few tools we have to control it. McCormick suggests: "*The underlying premise is suspect, namely that we really can build any system, no matter how complicated. The right tool, the right process will let us do anything, or so the salesmen assure us" [254].*

Those waiting for tools to solve our problems are likely to be disappointed:

Regrettably, humans can cope with very little complexity. Better tools and methods can help us with many of the rote aspects of system development; the tools and methods we use are valuable, even indispensable. But consultants and tool vendors often perpetuate a delusion, the delusion that we can cope with endless complexity, if only we would use a better tool or a different method.

Tools can only be an aid to judgment. Tools cannot substitute for the physical constraints encountered naturally in other disciplines. Without a harsh and uncaring nature forcing us to make hard choices, we tend to rationalize the complexity we see growing before us on each new project. . . . Despite the best intentions of highly skilled people, each new increment of complexity seems entirely plausible on its own. We are willingly seduced.

I submit that the grand failures of big, software-intensive systems have been due primarily to this willing seduction. Post-mortem analysis of such projects routinely reveals a specification that grew in complexity until project cancellation. Natural constraints simply do not apply to software, and nobody knew when to say no. After all, it was only software. [254]

5.3.2.4 Lack of historical usage information A final difficulty with software not found in hardware systems is that no historical usage information is available to allow measurement, evaluation, and improvement on standard designs. Software is almost always specially constructed, whereas physical systems benefit from the experience gained by the use of standard designs over long periods of time and in varied environments. Consider the difficulty that would ensue if every part of an airplane or car was completely changed for each new model or version and the entire design process started anew. That basically describes the situation for software. Software reuse is not the answer; the problems that occur in reuse were described earlier in the chapter.

To complicate matters further, the features most likely to change from one complex system design to another are exactly those that are most likely to be controlled by or embedded within software.

5.3.3 The Reality We Face

When systems were composed only of electromechanical and human components, engineers knew that random wear-out failures and human errors could be reduced and mitigated but never completely eliminated. They accepted the fact that they had to devise ways to build systems that were robust and safe despite random failures. Design errors, on the other hand, could be handled fairly well through testing and reuse of proven designs.

Because software has only design errors, the primary approach used to deal with reliability and safety problems has been simply to get the software correct. Theoretically, the possibility does exist for finding a set of techniques or methodology that will allow us to build perfect software. Much energy has been invested in looking for this methodology and less in finding ways to build software and systems that are robust and safe for the hardware and the systems that the software controls.

In reality, the time to create perfect software is never there, and perhaps it never can be. We may be seeking an impossible goal: software that is free of requirements and implementation flaws and that will always do what is required under all circumstances, no matter what changes occur to it or to the environment in which it operates. We need to set realistic goals if we want to create real progress.

5.4 The Way Forward

An important factor in dealing with software in safety engineering is that the failure modes are very different, although it is not even clear that it makes sense to talk about software as "failing." This, of course, depends on the definition of failure, but software certainly does not misbehave for the same reasons that hardware does. Software does not fail randomly or wear out over time. Software, in fact, is pure design; that is, it is design abstracted from the physical components used to implement the design. As was noted at the beginning of this chapter, abstractions are not by themselves unsafe—they cannot catch on fire, explode, produce energy, or impact the physical world in any way except through the control signals they provide to a physical entity or a social process.

This statement is not true, of course, for hazards not related to physical entities, such as losing money or losing customer goodwill. But the conclusion in these cases is the same; that is, the impact of the software behavior is indirect and only relevant when that behavior impacts a social process (versus a physical process).

Because of the relatively recent introduction of computers to control potentially dangerous systems, few safety engineering techniques have been developed to cope with the new problems introduced by computers. For the most part, standard software engineering techniques and processes are being used to develop safety-critical software without any consideration of the special factors and unique requirements for enhancing safety. Alternatively, new approaches are being used that are not likely to have a positive impact on safety while being sold as doing so. The same is true for the traditional safety engineering methods: they do not work adequately for software without changes, but they are being used anyway.

Our goal should be to integrate software into the overall system safety engineering process, recognizing its differences and special properties and the challenges it presents. Simply using standard safety engineering and software engineering techniques is not going to be enough. Software is different and requires new approaches in safety engineering.

Exercises

1. Identify some accidents where software was implicated. What role did the software play in the events?
2. Consider the definition of failure when applied to hardware (look it up if necessary). Does it make sense to you to apply that definition to a pure abstraction like software? If so, why? If not, why?
3. Why is it more difficult to test and find errors in software design than in hardware designs?
4. Provide an example of (1) software that satisfies its requirements but leads to an accident, and (2) software that does not satisfy the requirements but is safe.
5. Provide specific examples of the four ways, described in section 5.2, that software can impact safety using air traffic control (or some system with which you are familiar) as an example.
6. What is an example of a software-intensive system or software-implemented feature in a system that would be very difficult to implement only with hardware?
7. Why is it possible to use fewer test cases to get confidence in hardware behavior than is necessary for equivalent confidence in software? Is exhaustive testing possible for software?
8. Read about mode confusion, perhaps in *Engineering a Safer World.* Consider the Bangalore A320 landing accident (either in *Engineering a Safer World* or on the web). What role did mode confusion play in this accident? What aspects of the software design contributed to the mode confusion in this accident?

6

The Role of Humans in Safety

For a long time, people were saying that most accidents are due to human error and this is true in a sense but it's not very helpful. It's a bit like saying that falls are due to gravity.
—Trevor Kletz,
An Engineer's View of Human Error, 1990

Human control tasks are increasingly being taken over by machines. There are several reasons for this change, one of which is the widespread belief that operator errors account for the majority of accidents. Ironically, when designers attempt to eliminate humans from systems or to take over some of their tasks, they almost inevitably make the operators' jobs more complex and, perhaps, more error-prone.

The automation of control systems and attempts to eliminate human operators from them has led to increased interest in human factors and human–machine interaction. To successfully engineer for safety, the role of humans in accidents needs to be understood.

Automation usually does not eliminate humans but instead raises their tasks to new levels of complexity. The easy parts of tasks may be eliminated, leaving the difficult parts or increasing the difficulty of the remaining tasks. Operators may be assigned monitoring tasks for which they are not suited or responsibility for intervening at the extremes. Often, automation merely assigns humans to new functions related to maintenance and moves them to higher levels of supervisory control and decision-making. The total number of humans in the system may stay constant or even increase.

Centralization and advanced automation also typically increase not only the number of humans involved but also the consequences of human decisions and actions. At the same time, the basis for decision-making becomes more obscure because of increased system complexity [314].

Finally, human error will still be a factor in accidents, even if operators are eliminated completely, because automation is designed by humans. The bottom line is that safety engineers need to have a good grounding in human factors and learn to work with human factors experts.

This chapter examines several important questions about humans and risk in safety engineering. Is replacing humans with automation always the right thing to do with respect to safety? Are humans really responsible for most accidents as seems to be the common belief? If humans are necessary in systems, what role should they play?

We start by considering the current push to replace humans with automation.

6.1 Why Replace Humans with Machines?

Technological advances, especially computers, have made the automation of many manual operations feasible. Some potential advantages of automation are increased capacity and productivity, reduction of manual workload and fatigue, relief from and more precise handling of routine operations, and elimination of individual human differences [409].

Economics plays an important role in automation. For example, automation can create potentially enormous savings in energy costs through economical utilization of machinery. In the aviation industry, for instance, fuel consumption can be reduced if flight time is reduced and more fuel-efficient climb and descent patterns are implemented [409].

However, all of these savings are potential and not necessarily completely realizable. *How* one automates will determine the actual savings. In addition, (1) automated equipment is not cheap, and additional large maintenance and training costs are usually incurred; (2) overall personnel requirements may not be reduced but the types of human tasks simply changed, perhaps requiring greater skill and expertise; and (3) the redundancy needed to achieve the required reliability levels in automation may add to the cost. From the operational point of view, it is unclear whether overall workload and total operational costs are reduced or increased by automation. In addition, the long-range effects on operators and other personnel are still largely unknown, but they are starting to come into focus.

Aside from economic considerations, there are technical reasons to automate some manual operations. As control requirements of systems grow more complex, they may exceed human capabilities. In some systems, automated controls are required because humans cannot react fast enough to control the process adequately and safely. The control of the aerodynamic surfaces on unstable aircraft and spacecraft reentry maneuvers are two examples of control functions that require automation. Whether such advanced systems can and should be built is in part a trans-scientific question, as described earlier, which cannot be answered completely by scientific means. But if they are built, at least partial automation is a necessity.

A third possible reason for automation is to increase safety. Many accidents are attributable to human error. Automation, for example, can eliminate small errors that result from human boredom, inattention, or fatigue. Increasing safety through automation is not usually a simple matter, however. The introduction of automation may create a risk–risk situation—there may be risk in both the use and nonuse of automated equipment. Aircraft collision avoidance systems, for example, warn pilots about potential collisions but also may reduce pilot alertness and visual scanning. Ground terrain warning systems reduce terrain-strike accidents, but they have also been criticized for generating frequent false alarms that are annoying and potentially dangerous and may lead to reduced pilot altitude monitoring.

The long-term effects of automation, including complacency resulting from overconfidence in automated equipment, are not well understood. Whether automation increases safety and the role of human errors in accidents are important questions for system designers to consider. The answers have implications for the design of safer systems.

6.2 Do Human Operators Cause Most Accidents?

A common assumption, supported by a great deal of data, is that human operators cause the majority of accidents. Certainly, operators are actively involved in many industrial accidents, but this fact is not surprising because human operators play a fundamental role in the operation of industrial production systems. The real question is whether they cause the accidents or are merely swept up in the events and blamed for them.

A commonly cited statistic is that 85 percent of work accidents are due to unsafe acts by humans rather than unsafe conditions [166]. Heinrich [141] claimed that 88 percent of all accidents are caused primarily by dangerous acts of individual workers. Studies in the steel and rolling mills, construction, and railroad industries conclude that 60 to 80 percent of accidents are the direct result of the operator's loss of control of energies in the system [1987]. Pilots have been blamed for 70 to 90 percent of aircraft accidents. Other studies show similar numbers.

This data, however, should not be merely taken at face value but needs to be examined more closely. Are there other explanations that do not cast so much blame on the operator?

- **The data may be biased and incomplete.**

  Reports of human error as the cause of accidents and incidents have been found to be based on highly subjective judgments by classifiers using questionable definitions. Often these reports are written by supervisors, whose motives may be self-serving. Johnson claims that the less that is known about the accident, the more likely it will be attributed to human error: thorough investigations of serious accidents almost invariably find unsafe conditions and nonoperator factors [166].

  The study cited in the previous paragraph, which found that 60 to 80 percent of the accidents in various industries were the result of the operators' loss of control, also determined that in 75 percent of these cases, various malfunctions of production and safety control systems *preceded* the operator actions. Perrow claims that even in the best of industries, there is rampant attribution of accidents to operator error, to the neglect of errors by designers or management [303]. He cites a US Air Force study of aviation accidents that concludes that the designation of human error, or pilot error, is a convenient classification for mishaps whose real cause is uncertain, complex, or embarrassing to the organization.

  After the 1979 crash of a DC-10 into Mount Erebus in Antarctica during a sightseeing flight, the initial report of the New Zealand inquiry board found pilot error as the cause [244]. The conclusions of the report were contested by the pilot's widow and the pilots' union, and a subsequent, more thorough investigation was conducted [1982]. The later investigation found that some autopilot headings—used during briefing and simulator training by the crew—had been altered by employees of the airline right before the flight without informing the crew. The controversy over this accident continued with the airline claiming that the airplane was flying too low. The second investigative report noted, however, that the airline advertised the fact that low flights were made to improve sightseeing, and, therefore, the captain was expected to fly low. Similar controversy over pilot culpability occurred over the cause of the crash of an A320 during an air show in Habsheim, Germany, in June 1988.

After a false warning of a Soviet nuclear attack appeared on NORAD displays at the Cheyenne Mountain command and control center (mentioned in chapter 4), a government spokesman informed the press and Congress that the November 1979 incident was the result of a simple operator error—mistakenly inserting an "exercise or training tape" that simulated an attack into the operational computer warning system. The NORAD commander, General James Hartinger, announced that the false warning was "100-percent personnel error" [342, p. 230].

In fact, Sagan reports, recently declassified internal NORAD documents and congressional investigations confirm that the test data was purposely being run on the computer and that the false warning was not just a matter of an operator erroneously loading the "wrong tape" [342]. At the time of the incident, NORAD was in the process of deploying an upgraded computer system to improve the reliability of the early warning system. During this deployment, some of the software development and testing was being done on the online NORAD missile warning network because no other computer system was available for this purpose.

NORAD was unable to determine why the test information appeared on the operational warning displays and was sent to the Strategic Air Command, the Pentagon, and Fort Richie. The command's internal investigation of the incident (as reported in a private letter from one general to another) stated that they were unable to replicate the mode of failure [342]. Blaming the incident on a human operator loading an incorrect tape was apparently more politically acceptable than suggesting that there might be technical problems in the early warning system, that it had been a poor decision to run tests on the computers responsible for warning about nuclear attacks, and that they could not determine the exact cause of the incident.

In commenting on these events, Sagan warns: "*The common tendency to assign blame for accidents on operator errors, and thereby protect the interests of those who designed the system and the leaders of the organization, was also found to increase the likelihood of repeated mistakes . . . the November 1979 incident did not produce a thorough search for potential failure modes that remained in the system*" [342, p. 246].

- **Positive actions are usually not recorded.**

  One reason why human error reports may be misleading is that human actions are generally reported only when they have a negative effect on safety. A human action with a positive effect, such as restoring operation after a technical failure, is considered to be normal operational performance. A US Air Force study of 681 in-flight emergencies showed 659 crew recoveries from equipment and maintenance deficiencies, with only ten pilot errors. The proportion of pilot errors was about 1.5 percent in these incidents, while the proportion of equipment failures was 91 percent.

- **Blame may be based on the premise that operators can overcome every emergency.**

  Given the role of human operators in highly automated plants as monitors and supervisors, the major risk often is not that they will cause accidents but that they may not succeed in preventing them [314]. Aviation accident boards have predicated the finding of pilot error on the premise that pilots should be able to and must overcome any emergency. In many instances, pilots were overwhelmed by failures due to causes beyond their control; many of these failures could have been prevented by the incorporation of suitable precautions in the design stage [131].

Almost every accident can be traced to human error of some kind. It appears, however, that the operator who does not prevent accidents caused by design deficiencies or lack of properly designed controls is more likely to be blamed than the designer.

- **Operators often have to intervene at the limits.**

  Not only are operators expected to overcome any possible emergency, but they often are required to intervene at the limits of system behavior when the consequences of not succeeding are likely to be serious. The emergency may involve a situation that the designer never anticipated and that was not covered by the operators' training. Operators not only have to diagnose and respond quickly to the situation, but they must do so using creativity and ingenuity under conditions of extreme stress and perhaps with limited information about the state of the system. We should be more amazed at how well humans often do in these circumstances than surprised at the times they fail.

- **Hindsight is always 20/20.**

  Hindsight often allows us to identify a better decision, but detecting and correcting potential errors before they have been made obvious by an accident is far more difficult [399].

  In an emergency situation, operators are required to detect a problem, diagnose its cause, and determine an appropriate remedial action—all of which are much easier to do after the fact. Before the accident, the operator cannot always be expected to know what is going on and what should be done when confronted by unexpected failures and interactions among components and when working with often limited information. The situation may lead the operators to construct erroneous mental models of the system state, which are obviously incorrect only when after-the-fact knowledge of the real state is considered; they were just as likely to be correct as the real model at the time they were constructed.

  The official report of the Clapham Junction railroad accident concluded that "*there is almost no human action or decision that cannot be made to look more flawed and less sensible in the misleading light of hindsight. It is essential that the critic should keep himself constantly aware of that fact*" [144].

  The Three Mile Island (TMI) accident has been widely judged to be the result of operator error. However, examining the factors involved suggests that this judgment is a classic example of the misattribution of an accident to operators and the use of hindsight to label operators' actions as erroneous. The accident sequence was initiated and compounded by equipment failure. The major errors of the operators could only have been seen after the fact; at the time, there was not enough information about what was going on in the plant to make better decisions.

  The major operator "error" at TMI was throttling back on two high-pressure injection pumps to decrease water pressure, thus allowing the core to become uncovered and overheat [302]. However, unless the operator knows that an accident involves a loss of coolant to the core (called a *loss of coolant accident* or LOCA), the recommended procedure is to throttle back in order to avoid other kinds of damage. Theodore Taylor, a theoretical physicist from Princeton University, argued at the hearings of the Kemeny Commission on the accident [173] that there was no way for the operators to know what kind of an accident they were experiencing when they cut back on the high-pressure injection: the decision to cut back had to be made *before* anyone could know that it was the wrong thing to do.

There was no direct way at TMI to determine that the core was being uncovered and superheated. A Babcock and Wilcox official testified at the Kemeny hearings that such indicators would be difficult to provide, would be too expensive, and would create other complications. In fact, the readings of the core that are normally used to determine the amount of coolant present indicated that there *was* enough coolant. Although there were several indirect measures that might have shown the true state of the plant, each was faulty or ambiguous. A drainpipe pressure indicator would have suggested a LOCA, but it was located on the back side of the 7 ft control panel; unaware that they were in a LOCA, the operators had no reason to look at it. The temperatures on a drainpipe would similarly have indicated the problem, but the operators had been discounting these readings prior to the accident because the drainpipe had leaky valves; they assumed that decay heat had caused a particularly high reading.

Knowing that the pressure had dropped in the core would have warned the operators that throttling back on the high-pressure injection was the wrong thing to do. The core pressure indicator was next to the indicator showing a rise in pressure in the pressurizer. These two indicators are supposed to move together, but one dropped as the other rose. The operators believed that the indicator measuring pressure in the pressurizer was more accurate than the indicator measuring pressure in the core because various other factors pointed to that conclusion. They were used to receiving faulty readings—there had been several during the accident—so they relied on those that made sense and discounted or explained away those that did not [302].

The control room quickly filled with managers and engineers, none of whom knew that the problem was a LOCA. It was several hours later, and much too late to matter, before even the outside experts realized what had happened:

> *Consider the situation: 110 alarms were sounding; key indicators were inaccessible; repair-order tags covered the warning lights of nearby controls; the data printout on the computer was running behind (eventually by an hour and a half); key indicators malfunctioned; the room was filling with experts; and several pieces of equipment were out of service or suddenly inoperative. In view of these facts, a conclusion of "severe deficiency in training" seems overselective and averts our gaze from the inevitability of an accident even if training were appropriate.* [302, p. 178]

According to Richard Hornick, past president of the Human Factors Society, "*The chain of events at TMI was not the result of immediate human errors in the usual sense. Rather, it reflected the prior failure of owners, managers, and engineers to appropriately design interactions between humans and machines*" [151].

Victor Gilinsky, a Nuclear Regulatory Commission official, argued that "*in emphasizing the human failures, and thereby vindicating the equipment, the report does not stress enough that the equipment could have been designed to avoid this kind of trouble*" [250].

Finally, Malcolm Brookes, a member of the Kemeny Commission investigating the accident, concluded: "*There were no operator errors as such: the events that occurred seem 'inevitable,' given existing instrumentation (which is typical to the industry). The events were a direct function of the electro-mechanical system design and detail, e.g., computer update rates, alarming, wrong placement of controls and displays, wrong instrumentation giving the wrong sort of information. The fundamental errors were in systems design"* (quoted in [415, p. 220]).

Operators often are placed in situations where they must make choices between several actions, any of which could turn out to be wrong after the fact. In the unsuccessful attempt to rescue the American hostages in Tehran in 1979, when the pilots inappropriately grounded the helicopters in the desert, changes in atmospheric pressure had triggered a false alarm indicating failure of the rotor blades. This mechanical misdiagnosis presented the pilots with two possibly erroneous actions: (1) they could trust their own instincts and ignore the alarm, thus risking a real emergency, or (2) they could ground the helicopters in perfect running condition. The correct decision could only be known in hindsight [146].

- **Separating operator error from design error is difficult and perhaps impossible.**

All human activity is affected by the environment in which it takes place. In complex, automated plants, the human operator is often at the mercy of the system design and operational procedures.

Nuclear power plants at the time of the TMI incident have been reported to display the following examples of poor design [146; 151; 313]:

1. Dials measuring the same quantities are calibrated in different scales.
2. Normal ranges are not uniformly marked.
3. Irregular scale divisions are used on seemingly identical dials.
4. The location of critical decimal points is unclear.
5. Recorders are cluttered with excess information.
6. Labels and colors are inconsistent and confusing.
7. The left-hand pair of displays is driven by the right-hand pair of controls.
8. Panel meters cannot be read from more than several feet away, but controls for these panel meters are located 30 ft away.
9. Critical displays are located on the back side of a panel, while unimportant displays occupy central, front-panel space.
10. Two identical (unmarked) scales are placed side by side, one of which the operator must remember differs from the other by a factor of ten.
11. Labels on alarm enunciators differ from the supposedly corresponding labels in the written procedures.
12. Refresher training in emergency procedures uses a control board on which displays and controls are laid out differently from those on the control board the operators will actually use.

In some plants, operators have jury-rigged their own marking systems in order to eliminate some of the confusion, using colored tape, homemade control knobs, and homemade supplemental equipment to highlight the logic of the system [302].

While many of these problems have been eliminated in newer digital displays, the new graphical displays simply have different limitations and potential design flaws. We often replace the problems in analog displays with digital displays that have different problems. These new problems are discussed in chapter 12.

Figure 6.1
The labeling on seven pumps.

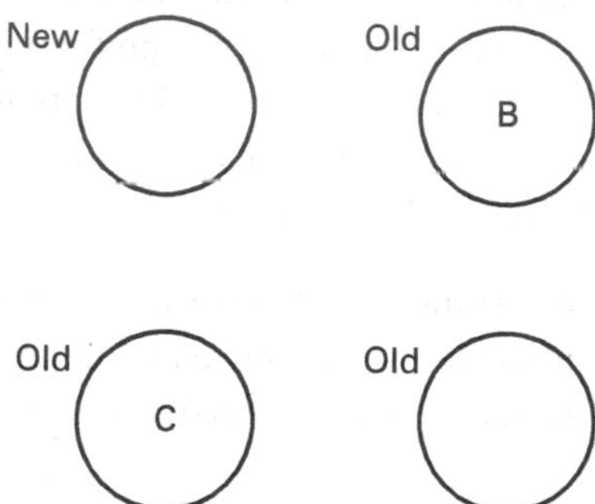

Figure 6.2
Labeling on crystallizers.

The nuclear industry is not the only one with these types of design errors. Kletz describes two accidents in the chemical process industry involving similar human–machine interface design deficiencies. In the first, a row of pumps was labeled as shown in figure 6.1. Not unreasonably, the operator assumed that number 7 was the last one, and so he did not check the numbers. Hot oil came out of the pump when he dismantled it [188].

In a second accident, a plant had four crystallizers—three old ones and one just installed (figure 6.2). A worker was asked to repair crystallizer A. When he went onto the structure, he found that two were labeled B and C, but the other two were not labeled. He assumed that A was the old unlabeled crystallizer and started work on it. Actually, A was the new crystallizer. The original three were called B, C, and D; A had been reserved for a possible future addition. Again, introducing graphical digital displays does not solve the problems and usually adds more.

In addition to simple labeling and layout problems, designers may fail to understand fully the system characteristics or to anticipate all of the environmental conditions under which the system must operate. The operator can intervene effectively only if the system has been designed to allow the operator to build a complete and accurate mental model of its operational status, including providing the information necessary for the operator to understand the system state and providing it in a form that is understandable under stressful conditions.

Williams examined failures in US nuclear power reactors over a two-year period [129]. He found that in boiling water reactors, slightly over half of the reported occurrences were caused by design deficiencies and one-third by human error, whereas in pressurized water reactors, the exact opposite proportions were found—over half were caused by human error or oversight and one-third by design errors. Because the classification was done by one person, misclassification does not explain the differences: if human error were independent of design, the percentage of accidents attributed to human error would be constant over the different reactor designs.

Aerospace studies show that about 80 percent of pilot-error accidents are due to poor training or neglect of human engineering in controls and instruments, not to stupidity or panic [358]. Claims have been made that the same ratio is true for the nuclear industry. Perrow claims that the most serious human error may be in not being able to overcome the inadequacies and complexities of the equipment humans must use [302].

By simply blaming accidents on human error and not looking at the design features that might have contributed to those errors, future accidents cannot be reduced. In June 1987, the pilot of a Delta Boeing 767 taking off from LAX accidentally turned off both engines while reaching for a fuel control switch [342]. The plane, originally at 1,700 ft, dropped 1,100 ft before the pilots got the engines started and the plane leveled off at 600 ft above the ocean. Instead of simply blaming the incident on pilot error, the Federal Aviation Administration (FAA) ordered that a safety guard be placed over the two engine shutdown switches in order to reduce the likelihood that a future operator error would inadvertently turn off both engines.

In summary, human actions both prevent and contribute to accidents in high-technology systems. Accidents and incidents occur because of human error, machine failure, poor design, inadequate procedures, social or managerial inadequacies, and the interaction of all these factors. Assignment of blame to operator error alone ignores the multiple causes of accidents and the relationship between human actions and the context in which they occur. It also suggests that little can be done about reducing accidents aside from removing humans from systems. Recognition of other causal factors in the work environment and system design presents more possibilities for improvement.

This conclusion does not imply that humans do not make mistakes, only that ascriptions of blame to humans for accidents in complex systems is often simplistic and may get in the way of reducing risk through improved engineering and management. Chapters 8 and 12 look more carefully at human error in accident analysis and system design and how the interaction between humans and machines can lead to accidents.

But even if humans have been blamed incorrectly for accidents, they certainly have contributed to many of them. The question then becomes, why should we not just eliminate humans from systems and replace them with computers?

6.3 The Need for Humans in Automated Systems

Computers and other automated devices are best at trivial, straightforward tasks. An a priori response must be determined for every situation: an algorithm provides predetermined rules and procedures to deal only with the set of conditions that have been foreseen. Not all conditions are foreseeable, however, especially those that arise not from a single event but from a combination of events. Even those conditions that can be predicted are programmed by error-prone humans. Artificial Intelligence (AI) and Machine Learning do not solve this problem and perhaps make things worse through limitations in the training data.

As system design becomes more complex and requires coordination and integration of various types of expertise, system behavior and performance become increasingly difficult to understand and anticipate. For example, the designers of the TMI reactor had anticipated a crisis of high pressure in the concrete containment building; in the actual accident, however, the pressure was not high enough to trigger the automatic feedback mechanism. Later in the accident sequence, the pressure was too high to permit activation of an emergency heat removal device. The designers of the system had not anticipated and correctly handled the conditions that actually occurred. In addition, because they had assumed that the system was self-regulating, they did not design the system (or train operators) for effective human intervention [146].

It is difficult, and in some cases impossible, to anticipate all conditions that might occur in the environment of a system or all undesired interactions in the system components. Although designers may work under conditions of less stress than operators, designers also make mistakes. Many of the same limitations of human operators that contribute to accidents are also characteristic of designers. Designers have been found to have (1) difficulty in assessing the likelihood of rare events, (2) a bias against considering side effects, (3) a tendency to overlook contingencies, (4) a propensity to control complexity by concentrating only on a few aspects of the system, and (5) a limited capacity to comprehend complex relationships [180]. AI and computers will not save us here. They are much worse than both human operators and designers at these tasks.

Designers may fail to understand fully the system characteristics or to anticipate all the environmental conditions under which the system must operate. There will remain events, usually of low probability, against which the automated system does not provide protection because they were unforeseen, they were handled improperly, or their likelihood was estimated as below the designer's cutoff level [206].

Human operators are included in complex systems because, unlike computers, they are adaptable and flexible. In a computer, the response to each situation is preprogrammed. Humans are able to look at tasks as a whole and to adapt both the goals and the methods used to achieve them. Thus, humans evolve and develop skills and performance patterns that fit the peculiarities of a system very effectively, and they are able to use problem-solving and creativity to cope with unusual and unforeseen situations. For example, the pilot of a Boeing 767 made use of his experience as an amateur glider pilot to land his aircraft safely after a series of equipment failures and maintenance errors caused the plane to run out of fuel while in flight over Canada [230]. Humans can exercise judgment and are unsurpassed in recognizing patterns, making associative leaps, and operating in ill-structured, ambiguous situations [337].

Human error is the inevitable side effect of this flexibility and adaptability. The behavior involved when operators take undesirable shortcuts in prescribed procedures is the same behavior that gives them the ability to diagnose problems and find unique paths to goals when unforeseen events occur that were not covered by their training.

Examples abound of operators ignoring prescribed procedures in order to solve a problem, but those instances when the decision was incorrect are more likely to get attention. For instance, a US NRC regulation laid down after the TMI accident required operators to leave the emergency cooling pumps on for at least 20 minutes after a reactor scram. In one incident in 1979, the operators at the North Anna nuclear power plant could see that obeying this regulation would lead to a dangerous temperature shutdown profile. They decided to disobey the regulation and turned off one of the two pumps for 4 minutes without knowing precisely what reduction in total output could be expected, and what the net result on pressure and level might be, but it was considered a step in the right direction [85]. In this case, it turned out to be the correct thing to do.

Other examples of imaginative jury-rigging or deviation from rules occurred in the TMI and Browns Ferry accidents. At TMI, two pumps were put into service to keep the coolant circulating, even though neither was designed for core cooling. Something more complex, but similar, happened at Browns Ferry.

An incident at the Crystal River nuclear power plant would have been much worse if the operator had not stepped in and overridden erroneous computer commands. The trouble began when, for some unknown reason, a short circuit occurred in a section of the control room not related to the reactor controls. The utility believes that the short circuit was caused either by a bent connector pin in the control panel or by some maintenance work being done on an adjacent panel. In any event, the short did not knock out the whole system, but it did distort readings flowing through all the controls.

The coolant in this type of reactor must remain within a relatively narrow temperature band. The computer, which was responsible for controlling the temperature, "thought" that the coolant was growing too cold. It therefore began to accelerate the nuclear reaction in the core by withdrawing control rods. At the same time, the computer reduced the flow of coolant. The reactor overheated, drove the pressure up to the danger level, and then automatically shut down. The computer then erroneously ordered the pressure relief valve to open and stay open, supposedly because of a design defect in the electrical system [398]. The emergency core cooling system began pumping water into the reactor. An alert operator noticed the computer error in opening the relief valve and, several minutes later, plugged the leak by shutting a block valve. Water filled the reactor to the top and then poured out through two safety valves. One of these did not at first reseat properly. Before the incident was over, 43,000 gal of radioactive water were dumped on the floor.

These and other incidents demonstrate the danger of trying to prescribe regulations, procedures, or algorithms. Human error is often defined as behavior that is inconsistent with normal, programmed patterns and that differs from prescribed procedures [133]. Sometimes, however, deviating from the prescribed procedures is exactly the *right* thing to do to avoid an accident:

> *Accidents have resulted precisely because the operators did follow the predetermined instructions that were given to them for a particular situation. Automating those instructions will not solve the problem or make it any easier to design the instructions. Note the chaos that results when workers "work to rule" during a limited job action. [In fact] working in accordance with the written rules has become an effective replacement for formal strikes. Deviation from the rules is the hallmark of experienced people, but it is bound to lead occasionally to human error and related blame after the fact.* [318]

6.4 Human Error as Human–Task Mismatch

It appears that the term *human error* provides limited assistance in terms of understanding accidents and preventing them. Rasmussen argues that *human error* is not a useful term and should be replaced by considering such events to be *human–task mismatches*. The term human error implies that something can be done to humans in order to improve the state of

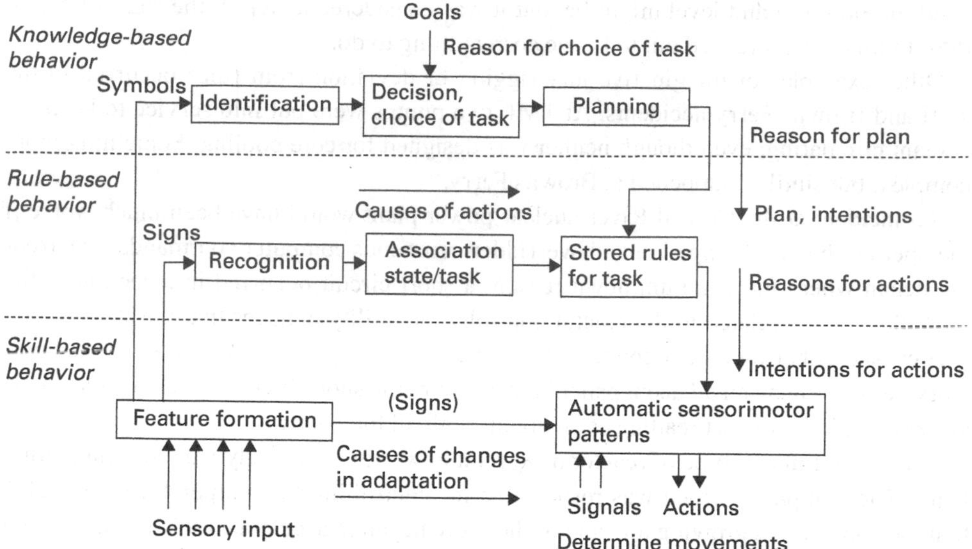

Figure 6.3
Rasmussen's model of cognitive control.

affairs; however, the "erroneous" behavior is inextricably connected to the same behavior required for successful completion of the task, as argued in the previous section. The tasks in modern automated systems involve problem-solving and decision-making that, in turn, require adaptation, experimentation, and optimization of procedures.

Rasmussen explains the relationship between these three behaviors and human error using a three-level model based partly on Newell and Simon's general theory of problem-solving (GPS) [281]. After analyzing error reports from nuclear power plants, Rasmussen identified three levels of cognitive control (figure 6.3): *skill-based behavior*, which is characterized by patterns of movement generated from a dynamic, internal model of the world; *rule-based behavior*, which depends on implicit models of the environment embedded in procedural rules; and *knowledge-based behavior*, which depends on structural models of the environment [315]. Each of the three levels of behavior can be mapped to different types of mental representations of the system and to different activities.

6.4.1 Skill-Based Behavior

Skill-based behavior involves smooth, automated, and highly integrated sensorimotor behavior arising from a statement of intention. It is characterized by almost unconscious performance of routine tasks, such as driving a car on a familiar road. Such behavior requires a very flexible and efficient world model and the use of sensorimotor inputs, or *signals*.

Efficiency results from fine-tuning human sensorimotor performance to the features of the task environment. Frequently, however, a mismatch is required before behavior is updated. If the criteria being used to optimize skill are speed and smoothness, operators can only discover an appropriate speed–accuracy tradeoff through the experience gained by occasionally crossing the limits, and the proper limits can only be determined if surpassed occasionally. Such errors, therefore, have a function in maintaining a skill at its proper level, and they neither can nor should be eliminated if skill is to be maintained.

6.4.2 Rule-Based Behavior

The second level, *rule-based behavior*, involves more conscious solving of familiar problems. At this level, sequences of actions in familiar work situations are controlled by stored rules or procedures (heuristics). The rules may be derived empirically from previous experience, communicated or taught in the form of a set of instructions or a recipe, or prepared by conscious problem-solving and planning. Performance is goal-oriented but structured by feedforward control using a stored rule. Often, the goal is not explicitly formulated but is found implicitly in the situation releasing the rules.

Selection of appropriate rules is controlled by inferences about the current state and events. When interacting with complex systems, operators often do not have direct interaction with the components and thus are controlling more or less invisible processes. They have to infer the state of the world and select the proper action on the basis of a set of physical measurements that are seldom presented in a way that allows perceptual identification of the state.

To avoid the mental effort necessary to determine the state and events from scratch each time the physical measurements change, operators generally use indications that are typical of normal events and states as convenient *signs*. The signs, which are associated with appropriate rules by training or experience, allow the operator to interact efficiently with a complex system using a large number of rules along with know-how for linking and updating those rules. This interaction is susceptible to mistakes, however, if the environment changes in a way that does not affect the signs but makes the related behavior inappropriate.

When the environment changes, the rules need to change. Often, the signs or inputs are underspecified and do not discriminate between alternatives for action in the new situation [320]. The operator cannot tell that the same or similar signs should not be relied on to select appropriate behavior until receiving feedback that they have not worked as expected.

Basic experimentation is necessary to develop and adjust efficient rules and to identify the conditions (signs) under which the rules should be applied. Initially, the rules may be based on rational planning (using knowledge-based behavior) or on a set of procedures supplied by an instructor. In both cases, the process of adapting rules and learning to apply them successfully leads to experiments, some of which are bound to end up as human errors under some environmental conditions.

6.4.3 Knowledge-Based Behavior

During unfamiliar and unanticipated situations, when the controller is faced with an environment for which no know-how or rules for control are available from previous experience, control of performance moves to a higher conceptual level where performance is goal-controlled and *knowledge-based.* In this situation, the goal is explicitly formulated, based on an analysis of the environment and the overall aims, and a plan is constructed. The plan may be formulated (1) by selection, where different plans are considered and their effect is tested against the goal, (2) by physical trial and error, or (3) by a conceptual understanding of the functional properties of the environment and prediction of the effects of the plan being considered.

When operators must adapt their behavior to the requirements of a system in a unique and unfamiliar state and switch to knowledge-based reasoning, they interpret information about the system as *symbols* to be used in symbolic reasoning rather than as signs that trigger rules. Rasmussen suggests that this switch is difficult because the use of signs basically means that information from the system is not actually observed, but is obtained by asking questions that are heavily biased by expectations based on a set of well-known situations. Reasoning at the knowledge-based level requires overcoming those biases.

One of the tools humans use to deal with unusual situations and solve problems at the knowledge-based level of control is experimentation accompanied by hypothesis formation and evaluation. Robert [329] studied human exploratory behavior in learning about a personal computer system. He found that humans try to understand the facts they observe and find relations between them: "*They draw analogies, form hypotheses, run experiments for testing and refining their knowledge, put up explanations, make diagnoses when problems occur, and so on. They are curious, motivated, full of initiative for extracting the secrets of the machine, persevering at finding a solution, sensitive to their environment*" [329].

At the same time, humans have the ability to monitor their own performance and devise strategies to recover from errors. In fact, some types of human error could be defined as a lack of recovery from the unacceptable effects of exploratory behavior [318]. Eliminating these types of human error requires eliminating the very same problem-solving ability, creativity, and ingenuity needed to cope with emergencies and the unexpected situations for which automated systems have not been preprogrammed or for which the preprogrammed recovery algorithms are not effective. Emergencies are likely to be caused by subtle combinations of malfunctions and to require diagnostic and problem-solving behavior, not just the skill- and rule-based behavior used for routine tasks. Thus, again, human errors are the inevitable side effect of human adaptability and creativity and are very different from faults in technical components [230].

6.4.4 The Relationship between Experimentation and Error

At the knowledge-based level, mental models of unfamiliar situations often are not complete enough to test a set of hypotheses entirely conceptually—that is, by "thought" experiments; the results of applying unsupported mental reasoning to a complex causal network may be unreliable [319]. Reliable conclusions may require testing the hypotheses, judging their accuracy by comparing predicted responses from the environment with actual responses, and making corrections when the hypotheses are in error. To accomplish the same task, designers are often supplied with tools such as experimental rigs, simulation programs, and computational aids. In contrast, operators usually have only their heads and the plant itself. Operator actions that appear quite rational and important during a search for information and a test of a hypothesis may be judged as erroneous in hindsight: If the test is successful, the operator is often judged to be clever; if it is not successful, the operator is blamed for making a mistake.

The designer relies most on knowledge-based behavior, but the operator at times will employ all three levels of behavior. During training in a particular task, control moves from the knowledge- or rule-based levels toward skill-based control, as familiarity with the work situation develops. In an unfamiliar situation, knowledge-based behavior may be used first to develop rules or heuristics to accomplish explicit goals. After a while, explicit knowledge and rules will no longer be needed for behavioral control of normal scenarios, and they may eventually deteriorate.

Knowledge needs to be maintained, however, to enable error detection, even though high skill has been developed. Because goals do not explicitly control activity at the skill- and rule-based levels, errors may not be observable until a very late stage—an error in following a recipe may not be evident until someone tastes the finished dish. Early detection of problems depends on an ability to monitor the process using an understanding of the underlying processes.

As control naturally moves downward toward skill-based behavior during training and adaptation, the only information the operator has to judge the proper limits of adaptation and to detect mistakes is occasional mismatches between the behavior and the environment. In this way, conscious as well as subconscious experiments providing feedback are integral to adaptation at all three levels of cognitive control.

In summary, human errors can be considered to be unsuccessful experiments performed in an unkind environment, where an unkind work environment is one in which it is not possible for the human to correct the effects of inappropriate variations in performance before they lead to unacceptable consequences. Rasmussen concludes that human performance is a balance between a desire to optimize skills and a willingness to accept the risk of exploratory acts. Attempts to eliminate such behavior by admonishing humans to take more care or by enforcing predefined procedures will have only short-term effects.

Instead, Rasmussen suggests that the ability of the operator to explore should be supported by the system design and that a means for recovery from the effects of errors should be provided. This goal can be achieved by the design of work conditions in which errors are immediately observable and reversible—an approach he calls *design for error tolerance* [316; 317; 319]. Some ways to accomplish this goal, along with other approaches to enhancing safety through the design of human–machine interaction, are described in chapter 12.

A large number of models to explain human error have been proposed. Rasmussen's model of human error as human–task or human–system mismatch is included here because it has had great influence in the safety engineering community. The interested reader is directed to the large amount of literature in this area for others.

6.5 The Role of Mental Models in Safety

Before we can identify the appropriate role of human operators in safety-critical systems, we need to consider the role of mental models in safety. Mental models are the way that humans understand the world: they shape not only what we think but how we act. Humans use mental models to simplify the complexity in our world and to determine the most relevant parts of our environment and how to react to them. Mental models help us reason about the world around us.

In general, a mental model is simply a representation of how something works. We cannot keep all the details of the world in our brains, so we use models to simplify the world into understandable and organizable chunks. Note that software also needs a model of the state of the world with which it is interacting in order to provide effective outputs and control, but that model is usually much simpler than human mental models.

In dealing with complexity, abstraction is one of our most powerful tools. Abstraction allows us to concentrate on the relevant aspects of a problem while ignoring the irrelevant. Without abstraction, humans would be unable to cope with complex natural and man-made systems.

Abstraction involves forming mental models that provide predictive and explanatory power. Because these models contain the most relevant aspects of an individual's interaction

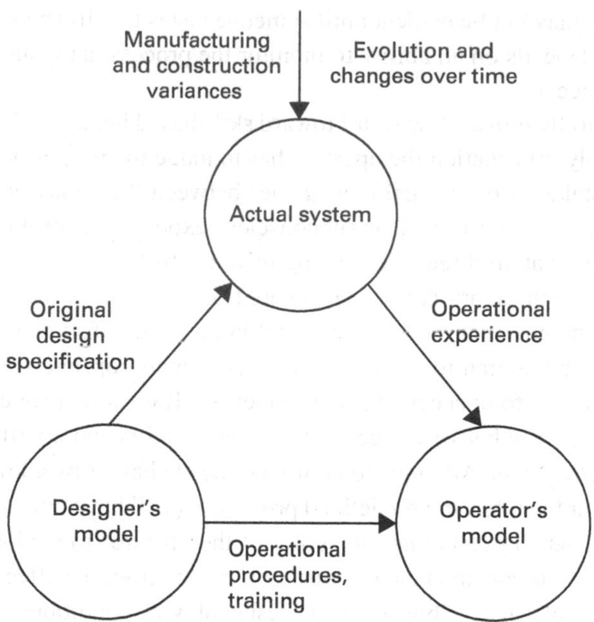

Figure 6.4
The relationship between mental models.

with a system, different people may form different models of the same system, depending on their goals, experience, and potential use for the model. Some mental models may be very simple while others may be very elaborate. Carroll suggests that multiple mental models may exist within the same person [58]. DeKeyser (as noted by Lucas) has found that even having two contradictory models of the same process does not seem to constitute a problem for people [238].

Because mental models reflect individual goals and experience, the designer's model of a system may differ greatly from the operator's model (figure 6.4). The designer's vision of the plant is often based on engineering or mathematical models of a control loop and is appropriate for situations where important decisions need not be made quickly. According to Duncan, the designer's model of a control loop involves taking an output signal to some comparator and maintaining a set point by feeding signals to an actuator. At a more detailed level, the designer's model may represent the quality of sensor information, the power and other response characteristics of the controlled device, and the comparator functions that ensure adequate tracking of the set point to avoid hunting when perturbations occur [86].

The operator most likely has some type of model of the physical locations of the points of control in the plant, along with system states, values, and limits, together with a model of the consequences of changes in those states [52].

It is quite natural for the designer's and operator's models to differ (see figure 6.4). In addition, both the designers' and operators' mental models may have significant differences from the actual plant as it exists. During design, the designer evolves a model of the plant to the point where it can be built. The designer's model is an idealization formed before the plant is constructed.

Significant differences may exist between this ideal model and the actual constructed system. Besides construction problems, the designer always deals with ideals or averages, not with the actual components themselves. Thus, a designer may have a model of a valve with an average closure time, while real valves have closure times that fall somewhere along a continuum of behavior that reflects manufacturing and material differences.

The operator's model will be based partly on formal training and partly on experience with the system. The operator must cope with the system as it is constructed and not as it may have been envisioned by the designer. Also, physical systems will change over time, and the operator's model must change accordingly.

Humans process data about a system in terms of their model of how that system works. Although an operator may be taught the designer's model, models are adjusted on the basis of experience. When operators receive inputs about the system being controlled, they first try to fit that information into their model and find reasons to exclude information that does not fit. Because operators are continually testing their mental models against reality, the longer a model has been held, the more resistant it will be to change due to conflicting information. Thus, experienced operators may act differently than novices.

Physical environments and systems change, however, and operators do change their models in the face of continued conflicts between the evidence and their perception of what is going on. This ability to change mental models is what makes the human operator so valuable. Based on current inputs, the operators' actual behavior may differ from the prescribed procedures. When the deviation is correct—the designers' models are less accurate than the operators' models at that particular instant in time—then the operators are considered to be doing their job. When the operators' models are incorrect, they are often blamed for any unfortunate results, even though their incorrect mental models may have been reasonable given the information they had at the time.

6.6 What Is the Appropriate Role for Humans in Complex Systems?

Given that human operators are unlikely to be eliminated and, in fact, play an important role in safety, the question arises about how humans should be integrated into system design and the role they should play in the operation of systems.

Rasmussen, Duncan, and Leplat 1987 [320] have stressed the ethical component of system design: people should not be caught by the irreversible consequences of typical everyday human behavior; instead, systems should be designed that tolerate human error. Volvo, for example, in the past has expressed this approach: "*Our philosophy is that every operator or maintenance mistake that can be made will be made sooner or later. Therefore, we have taken all safety measures against human error. This approach is difficult to develop. It is much more convenient to stick to the idea of negligence*" [176, p. 206]. To take this approach, designers need to have a good knowledge of human behavior. Human factors is the study of the capabilities and limitations of humans relative to the systems they operate.

The importance of human factors first gained recognition during World War I; the early work combined research in psychology, physiology, engineering, and education. Traditionally, human factors research has been concerned with anthropomorphics, which is the study of body measurements and how humans fit the layout of a system—for example, whether the operator can reach the controls and see the console. While these questions are still important, cognitive (versus physical) issues have taken the forefront.

Human factors engineering plays an important role in commercial and military aircraft design, but it is often an afterthought in other fields. The inclusion of human factors in engineering design may have been inhibited by a cultural clash between human factors researchers, who typically have a background in psychology, and design engineers, who typically have little educational background in this area [352].

Nevertheless, interest in human cognition and error mechanisms has increased rapidly in the past few decades as a result of two factors: (1) accidents, some receiving a great deal of attention, where humans played an important role either in preventing or precipitating the accident, and (2) risk analysts' need for data on human error to use in probabilistic risk assessments, particularly for licensing purposes.

The accident at Three Mile Island brought a great deal of attention to human factors in engineering design. Before the TMI accident, human factors issues were largely ignored in the design of nuclear power plants. A pre-TMI survey of human engineering aspects of reactor control panels found serious deficiencies. The following comment from a designer was cited in the results of the survey as typical: "*I have no pride of authorship in the layout of these boards. The client has to live with them. Nobody here cares that much. The NRC is only interested in knowing whether or not there is a certain function covered on the boards—either in front or in back*" (quoted in [358, p. 63]).

A participant at a symposium sponsored by the Electric Power Research Institute (EPRI) commented: "*There is little gut-level appreciation of the fact that plants are indeed man–machine systems. Insufficient attention is given to the human side of such systems, since most designers are hardware-oriented*" [358 p. 63].

It is now widely recognized that the TMI accident was not the result of operator errors as usually defined, but instead reflected a lack of appropriate design of the interactions between humans and machines. This conclusion implies the need for greater integration of human factors into system safety and general engineering efforts.

If we are to avoid the trap of oversimplifying human error and do not attempt simply to replace humans with automation, then we need to determine their proper role in automated systems. Three roles are possible: as monitors of the automated equipment, as backups to automated equipment, and as partners with machines in the control of the process. Each of these roles presents difficult design problems.

6.6.1 The Human as Monitor

With the automation of control systems, the operator's role changes from active control to monitoring: the human becomes responsible for detecting problems and providing a repair capability. A reasonable argument can be made that this change is an improvement: malfunctions theoretically can be detected rapidly because the operators are not burdened by mundane and attention-consuming control tasks [97]; they can devote all their attention to the monitoring task.

Unfortunately, experience shows that humans make very poor monitors of automated systems. There are several posited explanations for their poor performance.

- **The task may be impossible.**

  Bainbridge points out the irony that automatic control systems are put in because they can do the job better than a human operator, but a human is then asked to monitor that the automated system is doing the job effectively [28]. Two problems arise:

  1. The human monitor needs to know what the correct behavior of the process should be; however, in complex modes of operation—for example, where the variables in the process have to follow a particular trajectory over time—evaluating whether the automated control system is performing correctly requires special displays and information that may be available only from the automated system being monitored.
  2. If the decisions can be specified fully, then a computer can make them more quickly and more accurately than a human can. Whitfield and Ord found that air traffic controllers' appreciation of the traffic situation was reduced at the high traffic levels made feasible by using computers [411].

  There is therefore usually no way for a human to check in real time whether a computer is operating correctly or not. As a result, humans must monitor the automated control system at some metalevel, deciding whether the computer's decisions are acceptable rather than completely correct. If the computer is being used to make decisions because human judgment and intuition are not satisfactory, then which one should be trusted as the final arbiter?

  The same argument applies when a computer is monitoring another computer. Unless there is a different way to make the same decisions that is just as good, the better decision-making algorithm will be employed in the primary system rather than in the monitor. In most cases, identically accurate and efficient algorithms are not available.

- **The operator is dependent on the information provided; it is easy to provide too much or too little information or to display it poorly.**

  Computers provide the means to overload the operators with massive amounts of information. Humans respond more quickly, however, to the minimum amount of information needed to make a correct decision than to a lot of relevant and irrelevant data supplied in an uncoordinated fashion. Kletz [189] writes about an accident in which a computer printed a long list of alarms after a power failure in the plant. The operator did not know what had caused the upset and did nothing. After a few minutes, an explosion occurred. The designer admitted that he had overloaded the operator with too much information, but had presumed that the operator would assume the worst and trip the plant. When people are overloaded with information, however, they tend to do nothing until they have figured out what is going on.

  Information often needs to be condensed into a manageable form [46]. At TMI, so many alarms had to be printed that the printer fell behind by as much as 2 hours in recording them [173].

  A current trend is to provide "smart" alarm and warning systems that, among other things, prevent obvious false alarms and assign priorities to alarms [409]. However, the necessarily complex logic of these systems may be too complex for operators to perform validity checks on it and thus may lead to the operator over-relying on them. It may be difficult to preset appropriate alarm priorities for all situations, and operators may not have the information to recognize when the priorities are incorrect.

Although the automated system may be designed to provide only a certain amount of information to an operator, the operator may try to get more information by manipulating the system to figure out how it works. Such manipulation may lead to serious problems, but it may also be the only way that operators have to get the information they need to form the mental model necessary to operate the system correctly [46].

Miller [263] tells of a test pilot who, having some free time on the flight line, randomly pushed buttons on the computer to see what would happen. We can criticize, but, in fact, the information obtained from this exploratory behavior may be exactly what the pilot will need to handle situations that the designer has not anticipated. Actions that in hindsight are judged as mistakes may merely be reasonable attempts to gain information about the actual state of the system or to learn about how the system works. The designer provides the information necessary to handle a range of anticipated decisions, but the operator is usually given the task of handling the unforeseen decisions, perhaps without the information needed to perform this task.

- **The information is more indirect with automated systems, which may make it harder for the operator to obtain a clear picture of the state of the system.**

Technology changes not only the amount but also the basic character of the information available to operators. When operators control systems directly, their mental models of the current system state and how the system works develop from direct experience. Central control rooms separate operators from the system—the equipment cannot be seen, heard, or touched. Information about the operation of the system or about abnormal events must be obtained indirectly via remote sensors and instrumentation. Any particular set of indications may not be unique to a particular system state or to the cause of an abnormal event [95].

When automated control systems are placed between the operators and the system operation, the operators' relationship with the system is changed along with the process of developing adequate mental models to guide decision-making.

Brehmer claims that modern technology makes the connection between actions and outcomes opaque. The relationship between the operator and the system being controlled becomes more indirect and abstract: "*Both the information he receives and the outcomes he creates are mediated by complex processes that are hidden from direct view. Work thus becomes mental and abstract, rather than physical and concrete. The problem, then, is how people are able to form those mental models that can help them to make decisions under conditions where they have little insight into the process they want to control*" [46, p. 113].

Thus, to create and update mental models that will help in decision-making about unexpected events, operators may need to maintain manual involvement at some level.

In addition, a designer's lack of understanding of the mental models used by operators may limit the operators' monitoring and decision-making ability. In a highly automated system, the information operators receive about the system state through the instrumentation is not reality but merely a representation of reality [46]. That representation is a model that contains the information designers think operators need for decision-making based on the designers' mental models of the system. The operators, however, may require a different type of information to make rational and correct decisions based on their own mental models of the system state. Thus, much of the decision-making power rests with the designers, who, by giving only certain information to the operators, limit the range of possible decisions the operators can make.

At TMI, the instrumentation gave the operators information about the system state that caused them to make the situation worse. For example, the instruments reported that a particular valve had closed when it had not; it took the operators 2 hours to discover this error. Turbulence and voids in the reactor vessel caused the instrumentation to indicate incorrectly that the vessel was full when, in fact, the control rods were nearly uncovered. The reactor coolant-system water-temperature sensors registered values only between 520°F and 620°F. If the actual values fell outside these limits, the devices showed only the limit value; the operators did not know how much above or below these limits the temperature had moved. In addition, temperatures were averaged across the hot and cold poles of the system, which misled the operators; it appeared that the reactor temperature had stabilized at 570°F for 11 hours (which was within limits), but the actual cold temperature of the reactor had far exceeded the cold limits, resulting in damaging temperature differentials [52]. Instruments can be designed to prevent such ambiguous information, but such a design requires extra sensors and increases construction costs. More important, it requires that the designers recognize the need for the information.

Once an operator, because of ambiguous information, interprets a complex situation incorrectly, their flawed mental model will be difficult to change. An observer's pattern matching determines the category into which an input is placed. Expected evidence is much more readily accepted as conclusive than is unexpected evidence [269]. A warning that does not fit our mental model of what is going on can be swamped by the multitude of signals that fit our expectations and thus may be discounted as noise in the system [303].

Human observers are often reluctant to accept evidence that requires them to change their hypotheses. Computers complicate this problem by encouraging trust, which leads operators to ignore external cues that suggest a computer error has occurred. In the Air New Zealand accident at Mount Erebus in Antarctica mentioned earlier, Green [121] hypothesizes that the crew trusted the inertial navigation computers and were "probably seduced into interpreting external visual information in a way that conformed with the world model generated for them by the aircraft." Green suggests that because the pilots cannot possibly understand the technology involved in the generation of the display, they are compelled to use the display itself as their mental model of the world instead of creating their own model from the raw data.

The decisions made in these situations are perfectly rational, but the decision makers have the wrong mental models:

> Selecting a context ("this can happen only with a small pipe break in the secondary system") is a pre-decision act, made without reflection, almost effortlessly, as a part of a stream of experience and mental processing. We start thinking or making decisions based upon conscious, rational effort only after the context has become defined. And defining the context is a subtle, self-steering process, influenced by long experience with trials and errors. If a situation is ambiguous, without thinking

> about it or deciding upon it, we sometimes pick what seems to be the most familiar context, and only then do we begin to consciously reason. [318, p. 303]

Thus, when a control system's behavior conflicts with the operator's experience or training, the operator may ignore information or bypass the control system, assuming that it must be malfunctioning. On the other hand, it *may* be malfunctioning, and in that case, we *do* want the operator to ignore it. To resolve this dilemma, the designer must provide ways for the operator to check the functioning of the instrumentation.

- **Failures may be silent or masked.**

Automated control systems are designed to cope with the immediate effects of a deviation in the process—they are feedback loops that attempt to maintain a constant system state, and as such, they mask the occurrence of a problem in its early stages. An operator will be aware of such problems only if adequate information to detect them is provided. That such information is often *not* provided may be the result of the different mental models of the designers and experienced operators, or it may merely reflect financial pressures on designers due to the cost of providing operators with independent information [85].

In some cases, the required information may be impossible to provide. Norman [286] points out that building an automated system with a self-monitoring capability that would allow it to recognize that conditions are changing would require a higher level of awareness—that is, a monitoring of its own monitoring abilities. Although automated systems that monitor themselves at a metalevel for a few specific conditions could be constructed, the general case is currently unachievable.

As an example, autopilots may fail so gracefully that a decoupling may not be noticed by the crew until the plane is badly out of limits. One example is a PAA B-707 that experienced a graceful autopilot disconnect while cruising at 36,000 ft above the Atlantic. The aircraft went into a steep descending spiral before the crew took action; it lost 30,000 ft before the crew recovered [409]. In another instance, an Eastern Air Lines L-1011 slowly flew into the Florida Everglades after an autopilot became disengaged, and the crew (and air traffic controllers) failed to notice.

Problems with masked or delayed feedback in the monitoring of automated systems may be exacerbated by designs that cause referred symptoms. The place where the signs of trouble first appear may not be the place where the failures have occurred. For example, petrochemical plant design has responded to social and financial pressures, especially in the 1970s, to conserve energy and improve thermal economy. A designer may use heat exchangers to transfer recovered heat to another plant unit that is driven by heat. This design makes sense, but it introduces diagnostic difficulty because the symptoms of a problem may first appear in an unrelated part of the plant. Duncan asks:

> *Do operators always develop a process model that incorporates the 'referred symptoms' complication? If they do, then their model is again different from the designers, which I suspect is very much a heat extraction and transport model. So we have another diagnostic problem for the operator which, as it follows from other sensible plant design considerations, was probably not envisaged by anyone: the problem of symptoms referred forward, or referred back by major loops in the overall process. This problem is a design error, by definition, since we can be*

reasonably sure that the diagnostic complications were not intended. Moreover, it is a design error resulting rather directly from the designer's model of the plant, which is limited in a way that an experienced operator's model is not. [85, p. 268]

- **Tasks that require little active operator behavior may result in lowered alertness and vigilance and can lead to complacency and overreliance on automated systems.** Vigilance studies show that it is impossible for even a highly motivated person to maintain effective visual attention for more than about a half hour to a source of information on which very little happens [28]. This limitation makes it virtually impossible for humans to monitor for unlikely abnormalities, since the operator cannot monitor an automated system effectively if it has been operating acceptably for a long time. Substituting an automated alarm system only raises the question of who monitors the alarm system to determine when the alarm is not working properly. In chapter 4, an accident was described in which a high-level trip was fitted onto a tank to determine when it was full. The operator stopped watching, and the resulting accident was worse than the accidental spill that led to the automated trip being installed in the first place.

  The higher the reliability of the automated system, the more likely the operator is to become complacent and less vigilant. An intermediate level of reliability may create an impression of high reliability, and the operator may not be able to handle a failure when it occurs. Ironically, if the equipment is very unreliable, the operator will expect malfunctions and will be adept at handling them.

  Danaher [73] has examined human error in air traffic control incidents and also voices concern about human reliance on automated systems leading to accidents. Experiments have shown that the likelihood of an operator taking over successfully when the automated system fails increases as the operator's subjective assessment of the probability of equipment failure increases [1980].

  Finally, there are the well-known problems with false alarms. On the one hand, false alarms can lead to operators taking corrective action when, in fact, nothing is wrong with the system (other than the spurious alarm). Weiner and Curry tell of such an error that occurred during the takeoff of a Texas International DC-9 from Denver, when a stall warning spuriously activated [1980]. Believing that a stall was imminent, in spite of normal airspeed and pitch attitude indications, the crew aborted the takeoff, resulting in a runway overrun, severe damage to the aircraft, and nonfatal injuries to some passengers. The pilots had both experienced spurious stall warnings on takeoff before, but they probably had little choice but to regard this warning as real. The crew had to choose between aborting the takeoff, with an almost inevitable but perhaps noncatastrophic accident, and continuing the takeoff with a plane that might not be flyable, which could have resulted in a much worse accident [409].

  Too many false alarms, on the other hand, can lead the operator to ignore an alarm. Indications of a problem at Bhopal, for example, were not taken seriously because so many previous alarms had been spurious. The problem of spurious alarms is well known but very difficult to solve.

  As discussed later in chapter 12 in the section on designing to combat lack of alertness, accidents are ironically more likely on less challenging tasks. Challenging tasks usually are performed extremely well and, in the vast majority of cases, without accidents. Therefore, awareness or lack of awareness of the elevated risk of a procedure may provide a better measure of accident potential than does an evaluation of the risk level itself. If a distraction (such as an unplanned event) occurs during the more difficult operations—those that require planning and preparation and have high visibility with management—accident potential is relatively low. When the normal or routine elements of a job are performed with standard levels of visibility and interest and an identical unplanned event occurs, causing loss of the same amount of awareness, accident potential is high.

A classic method for enforcing operator attention is to require entries in a log; unfortunately, people can write down numbers without noticing what they are [28]. Various studies have attempted to determine appropriate operator workload levels. Weiner has suggested that "the burning question of the near future will not be how much work a man can do safely, but how little" [408].

6.6.2 The Human as Backup

A second role an operator may perform in an automated system is backup in the event of an emergency. Again, poorly designed automation may make this role more difficult.

- **A poorly designed human–machine interface may leave operators with lowered proficiency and increased reluctance to intervene.**

  Operators need both manual and cognitive skills, but both decline in the absence of practice. An experienced operator performs the minimum number of actions while controlling a process, with the process moving smoothly and quickly to the desired level of operation. With an inexperienced operator, the process is likely to oscillate around the target value. Physical skills deteriorate when not used, which means that the skills of an experienced operator who has been relegated to monitoring may degrade to the level of an inexperienced operator:

  > *If he takes over he may set the process into oscillation. He may have to wait until feedback, rather than controlling by open-loop, and it will be difficult for him to interpret whether the feedback shows that there is something wrong with the system or more simply that he has misjudged his control action. He will need to take action to counteract his ineffective control, which will add to his workload. When manual takeover is needed, there is likely to be something wrong with the process, so that unusual actions will be needed to control it, and one can argue that the operator needs to be more rather than less skilled, and less rather than more loaded, than average.* [28, p. 272]

  Cognitive skills also decline. The efficient retrieval of knowledge from long-term memory depends on frequency of use and on feedback about the effectiveness of that knowledge [28]. If operators assume the role of problem solver and decision maker only when things go wrong, they will have little experience to go on when they *do* need to take over. And, again, the mental models necessary for correct decision-making will be inaccurate or incomplete. Therefore, theoretical instruction must be accompanied by practical exercises and associated with retrieval strategies that are integrated with the other parts of the task. Simulation provides a partial solution to these problems, but it is not always obvious how to provide effective simulation facilities.

A common but mistaken belief is that automated systems require less skilled users. A counterexample can be found in computer-aided design systems, which tend to work well with the original builders and users of the system, who are often very skilled designers. As time passes, less skilled users may be substituted or skills may decline from nonuse. As a result, flawed outputs sometimes are not detected by the less skilled staff until product flaws show up much later [398].

Lack of control experience or declining skills may lead to reluctance to intervene in an emergency. The investigation of a productivity problem in a strip-rolling steel mill in the Netherlands found that the operators did not always know when to step in and take over for the computer [398]. The operators became so unsure of themselves that they sometimes left the control panels unattended. They also had difficulty in observing the process. The designers had enclosed the steel strips being rolled, which seriously lessened the ability of the staff to determine whether the computer was controlling the operation effectively.

The investigation also found that the operators did not fully understand the control theory used in the control software, and this lack of understanding reinforced their reluctance to intervene except when things were very clearly going awry. By intervening late, the operators let productivity drop below that of plants using traditional (not computer-based) control systems. In this case, automation had led both to lower productivity and to operator alienation.

Proficiency seems to be important to operators. In one automated plant, management had to be present during the night shift or the operators would switch the process to manual [28]. Many pilots regularly turn off the autopilot or other automated systems in order to retain their manual flying skills. Proficiency allows operators to be confident that they can take over if required to do so. Bainbridge notes that the worst type of job is one that is boring yet requires assuming great responsibility without the opportunity to acquire or maintain the skills necessary to handle that responsibility [28].

- **Fault-intolerant systems may lead to even larger errors.**

  Sometimes designers are so sure that their systems are self-regulating that they do not include appropriate means for human intervention. For example, in aircraft inertial navigation systems, the latitude and longitude of the initial position of the aircraft and a series of checkpoints (waypoints) defining the desired track across the earth are loaded into the computer by keyboard before the flight. During one initial setup, the crew loaded their position with a Northern latitude rather than the Southern latitude of their actual position. This error was detected neither by the computer nor by the crew until after takeoff. The aircraft had to return to the departure point because the navigation computer could not be reset in flight [409].

  Hirschhorn argues that automated systems do not eliminate failures—they merely raise them to a new level of complexity [146]. The system may fail in ways that the designer did not anticipate, and unless the designer provided appropriate means for the operator to intervene, serious problems can result.

- **The design of the automated system may make the system harder to manage during a crisis.**

  Systems may respond reasonably in a calm atmosphere, but they may add to frustration and errors during a crisis. The semiautomatic system may require the operator to make too many decisions in a short time. Some nuclear reactors, for example, use computers to make split-second judgments and adjustments because a human operator cannot be expected to respond quickly enough when the reactor's equilibrium is disturbed. The operator may have a grace period of only 30 to 60 seconds after a loss of coolant before the reactor blows dry. In contrast, other reactors typically allow 30 to 60 minutes for the same thing. This unforgiving quality of the reactor design greatly magnifies the demand placed on human capabilities when something unexpected happens, as was the case at TMI, or when the computer makes an error, as was the case at Crystal River [251].

Computer-based systems may provide the operator with many more choices during a crisis, but, at the same time, may provide inadequate information to decide among the choices. In addition, the human operator, who has to take over quickly, may be disoriented and may need a significant warm-up period to change effectively from passive monitor to active controller; in an emergency, that time may not be available. Automated systems have made some operator workload (such as that for pilots) nonuniform, with long periods of inactivity and short bursts of intense activity [303]. Long periods of passive monitoring make operators unprepared to act in emergencies.

Control decisions are made on the basis of the operator's knowledge of the current state of the process. This knowledge is accumulated by making predictions and decisions about the process, which are validated by feedback. Obviously, this knowledge takes time to build up. Manual operators often arrive at the control room 15 to 30 minutes before they are due to take control so they can get a feel for what the process is doing [28]. In aircraft, this feel is often referred to as *situational awareness*. Without this awareness, it may be very difficult for the operator to make decisions quickly with minimum information.

Norman notes that the crew in modern automated aircraft is isolated both physically and mentally from the moment-to-moment activities of the aircraft and of the controls because the automated equipment provides the crew with little or no trace of its operations [286]. When problems arise and the operators need to take over, they may not be sufficiently in touch with the current state of the system to diagnose the problems in a reasonable amount of time. He argues that the culprit is not automation but rather the lack of continual feedback and interaction.

The situation may be further complicated by disorientation resulting from the sudden appearance of many alarms. In an incident from CHIRP, a British confidential incident-reporting service, a pilot wrote:

> I was flying in a Jetstream at night when my peaceful reverie was shattered by the stall audio warning, the stick shaker, and several warning lights. The effect was exactly what was *not* intended; I was frightened numb for several seconds and drawn off instruments trying to work out how to cancel the audio/visual assault rather than taking what should be instinctive actions. The combined assault is so loud and bright that it is impossible to talk to the other crew member, and action is invariably taken to cancel the cacophony before getting on with the actual problem (quoted in [299, p. 37]).

In summary, automation can change the role of a human operator from an active controller of the process to a passive monitor and occasional backup controller during emergencies. In addition, computers allow systems to be built that are naturally unstable and more difficult to control, requiring more complex and faster control maneuvers. Thus, at the same time that the opportunities for learning and practicing skills are decreasing, the demands for those skills—on the rare occasion when they are needed—may be growing because of the steadily increasing complexity of the processes being controlled.

6.6.3 The Human as Partner

In a third type of human–machine interaction, the human and the automated system may both be assigned control tasks. However, unless this partnership is carefully planned, the operator may simply end up with the tasks that the designer cannot figure out how to automate. The number of tasks that the operator must perform is reduced, but, surprisingly, the error potential may be increased. There are several possible explanations for this increase:

1. The operator may be left with an arbitrary collection of tasks for which little thought was given to providing support.
2. The remaining tasks may be significantly more complex, and new tasks may be added, such as maintenance and monitoring. Partial automation may not eliminate or even reduce the operator workload but may merely change the type of demands on the operator.
3. By taking away the easy parts of the operator's job, automation may make the more difficult parts even harder [28]. The importance of the operator's mental model in controlling a system has been described. Eliminating some tasks may make it more difficult or impossible for the operator to receive the feedback necessary to maintain an accurate model of the system.

An interesting experiment by Allnutt and colleagues [16] on the effects of sleep deprivation in operators compared performance on a battery of cognitive tests and performance on a simulator. The simulator tasks were stimulating and varied, provided good feedback, and required the active involvement of the subjects in most operations. Relatively few of the simulator activities involved either passive or repetitive action—the designers deliberately resisted pressures to remove human error by automation except to help overcome operator overload. The results of the experiment showed the usual deterioration from sleep deprivation on the battery of cognitive tests, but performance held up remarkably well on the simulator.

Some of the issues in the automation of functions currently performed by humans are exemplified by the automation of air traffic control. The FAA is investigating the possibility of eliminating human control over minute-to-minute traffic decisions. The human role would change from controller of every aircraft to a manager who handles exceptions while the computer takes care of routine commands to aircraft.

Doubts about such a goal were raised some time ago in a Rand Corporation report commissioned by the FAA, which characterizes the program as potentially jeopardizing the safety of the system [210]. The Rand report cites two principal drawbacks to such a system design:

1. The design would not really allow a human backup in case the computer failed to handle a given situation correctly. The volume of traffic handled by each control

station would be about double that handled today, so the controller probably would not have time to check the computer's assessment of possible conflicts.

2. Even if controllers had the time to handle conflicts, their passive role in the system would over time tend to make them unreliable monitors. They would lose touch and skill in dealing with traffic.

The Rand report also questioned whether any software could handle the job because of unexpected situations such as military aircraft flying outside prescribed routes, pilots making errors in their flight paths, contingencies, and so on. The FAA responded that the system would have two independent backups: (1) central computers would have a separate checking algorithm that would monitor all flight paths for conflicts independently of the main algorithm, and (2) TCAS (an airborne collision avoidance system) would alert both the pilots and controllers if the other two systems were wrong.

The Rand report proposed an alternative concept, called shared control, in which primary responsibility for air traffic control would rest with human controllers, but the automated system would assist them by continually checking and monitoring their work and proposing alternative plans. In high-traffic periods, controllers could turn increasing portions of the planning over to the automated system. Thus, they could keep their own workloads relatively constant, neither overtaxed in high-traffic periods nor underused in low-traffic periods. The most routine functions requiring the least intellectual ability, such as monitoring plans for deviations from agreed flight paths, would be the only functions fully automated.

In a shared control system, controllers could use any module of the automated system to assist them. The automated system could perform the automated monitoring, trajectory projection, and conflict resolution functions of the FAA proposal, but in this design, these functions would be requested by the controller instead of being preassigned to the computer.

FAA officials argued that the Rand shared-control concept would be unprofitable, since only marginal productivity gains could be realized until the responsibility for maintaining aircraft separation was passed from the human controller to the machine. Only at that point could productivity gains of as much as 100 percent be achieved [210]. The European approach to the same problem, which appears to have a philosophy similar to that expressed in the Rand report, is a partnership between computer and human that is superior in effect to either of them working alone [398].

How any such partnership should be defined is an open question that highlights important technical problems. For example, task allocation can be dynamic, with either the computer or the human in charge; tasks can be partitioned statically (preassigned) between the two partners; or one partner can assist the other in performing a task. The best way to make this allocation is still unknown.

In addition, a partnership of human and computer requires some means of communication between the two: "*But what should be the nature of the communication? If communication is explicit, there is less uncertainty as to what is being communicated, but the human must invest resources in receiving and transmitting information. This resource demand may be less if communication is implicit, but there may be less certainty as to what is communicated. There may also be a need for the human to invest resources into determining what the computer is doing*" [272, p. 67].

Designing a partnership ventures into the trans-scientific domain. Margulies argues that economic as well as social, ethical, and humanitarian reasons dictate that man should be dominant in this partnership, using most modern machines as supporting tools. Robots would take over the heavy, hazardous work, and other machines would do the boring and monotonous work [248]. Humans would not be left to fill in the gaps left by automation or to be laid off. Instead, system designs would allow operators to apply their own discretion, skills, and creativity while the machine handles the unintelligent, repetitive, and routine parts of a job.

> *This crucial relationship must not stop at making man and machine equally important parts of the system or at trying to find an optimal order of precedence between man and machine; it will rather have to clearly subordinate the machine to the interests, requirements, and strivings of man. The human–machine system of the future will have to offer to its users an improvement of working and living conditions, greater individual freedom for action and decision-making, and a wider scope to apply their skills and creativity.* [248, p. 11]

To summarize, automation allows almost unlimited freedom in the design of human–machine interaction—designers can either enhance the difficulty of the operator's task by a poor human–machine interaction design or simplify the operator's task by matching tasks and data to human abilities and preferences. Human–machine interaction becomes especially critical in an emergency, when the operator is under stress or is tired. Much more information about the design of human–machine interaction is needed if we are to reduce accidents in automated systems.

6.7 Conclusions

Human skill needs may be changing in modern systems. With automation increasingly taking over routine tasks, operators will use less skill-based and rule-based behavior and more knowledge-based behavior. Operator training will need to focus less on building skills and rules for action and more on general ability to understand how the system functions and to think flexibly when solving problems. Because effective action may require teams, each worker may need some familiarity with the tasks and skills of other workers. Research on work teams suggests that workers with broader knowledge can function much more effectively than workers trained in a single skill [146]. In the near future, humans will be teamed with automated systems, which magnifies the potential design problems.

Hirschhorn suggests that managers and engineers in traditional industries are highly reluctant to introduce operators to questions of system design or to train them to think conceptually beyond a limited list of specified responses to anticipated problems: "*Engineers have not learned to design a system that effectively integrates worker intelligence with mechanical processes. They seldom understand that workers even in automated settings must nevertheless make decisions; rather they tend to regard workers as extensions of the machines*" [146, p. 44].

This chapter has laid out some of the reasons why this approach may be exactly the wrong one. TMI and other accidents are exemplars of what can happen in complex systems when operators do not understand how equipment works and what they are required to do in an emergency, and when the system is not designed to provide them with the information necessary to intervene effectively.

The term *human error* is generally used in two different ways: (1) the human did something that he or she should have known was wrong, and (2) the human did something that could not have been known at the time to be wrong but in retrospect was. The distinction is useful only if we are trying to assign blame or guilt rather than reduce accidents.

Several authors have suggested that the whole concept of human error as a cause of accidents is outdated and that the term should be eliminated from use. For one thing, the individual operator in a complex, automated plant will be, to a large extent, at the mercy of the system design.

In addition, removing dependence on an operator by installing an automated device to take over the operator's functions only shifts that dependence onto the humans who design, install, test, and maintain the automated equipment—who also make mistakes.

Thus, almost any accident can be attributed to human error, and doing so does not provide much help in determining how to prevent it. Indeed, Kletz suggests that the phrase actually discourages constructive thinking about how to prevent future accidents [192]. He argues that outdated scientific concepts such as ether, phlogiston, and protoplasm have been determined to be unnecessary. They do not explain anything that cannot be explained without them or explained better differently—that is, they serve no useful purpose:

> *Perhaps the time has come when the concept of human error ought to go the way of phlogiston, the ether, and protoplasm. Perhaps we should let the term "human error" fade from our vocabularies, stop asking if it is the "cause" of an accident, and instead ask what action is required to prevent it happening again. Perhaps "cause" as well as "human error" ought to go the way of phlogiston, ether, and protoplasm.* [192]
>
> *According to Dewey, "Intellectual progress usually occurs through sheer abandonment of questions together with both of the alternatives they assume—an abandonment that results from their decreasing vitality. . . . We do not solve them: we get over them. Old questions are solved by disappearing, evaporating, while new questions corresponding to the changed attitude of endeavour and preference take their place."* [192, p. 181]

The new question may be how to prevent human–machine mismatches and how to design an interaction between human and machine that allows a symbiotic relationship enhancing the natural abilities and advantages of both.

Exercises

1. Consider the argument at the beginning of the chapter about the role of human error in accidents. What do you personally think is the role of human error?
2. Consider Rasmussen's three-level model of cognitive control: skill-based, rule-based, and knowledge-based. Identify an example from your own life for each of the skill-based, rule-based, and knowledge-based behaviors. What type of mistakes might you make for each of your examples, and how might Rasmussen's model explain your behavior in doing this? How might this model assist in the design of interfaces and in operator training for complex systems?

3. A statement was made in the chapter that software also needs a representation of the system it is controlling in order to provide effective control, but that representation is simpler or perhaps different than that required by a human controller.
   a. Consider a human driver versus an automated highway driving system, consisting of steering, braking, and passing. What does the software need to know about the car design and its environment in order to provide safe performance in this case? How might it get that information? Now consider a human driver who is supervising this automated system. What is needed in the mental model of the human to perform this task? Is it the same information as the software? How does the human get this information?
   b. How docs your answer to the first part of the question explain some of the difficulty in designing automated systems?
4. Describe three types of situations in which you use experimentation to learn how to interact with an unfamiliar system or to detect if changes have occurred.
5. Because operators are continually testing their mental models against reality, the longer a model has been held, the more resistant it will be to change due to conflicting information. Thus, experienced operators may act differently than novices. What are the implications of this for training? For designing systems that novices use rather than experts?
6. Do you agree or disagree with the following statement, and why or why not? Requirements for the mental models of the human operators will depend on what role they are playing in the system. Provide examples.
7. The use of automation in cars today (for example, the Tesla Autopilot) is predicated on the assumption that the drivers will monitor the automated systems and respond appropriately if there is a problem. Discuss whether this is reasonable given the information presented in the section on humans acting as monitors and as backups. What potential solutions can you think of?
8. Select a report of a well-investigated accident, perhaps by the NTSB (National Transportation Safety Board)? What role did the report say the human operators played? Were there extenuating circumstances that might explain the human operators' behavior?
9. What additional reasons can you think of for the widely cited belief that human operators are responsible for most accidents beyond those provided in this chapter?
10. The chapter argues that designers' mental models of systems often differ from those of the operators of the system.
    a. Why might they differ? What implications does this have for designers? For imposing requirements that operators always follow procedures rigorously (which is very common)? What are the implications for management of operations?
    b. The "following procedures dilemma" has been described as operators being blamed when they follow procedures and it turns out to be the wrong thing to do but also being blamed when they deviate from procedures and it turns out they should not have done so in that case. Is imposing requirements for always following procedures a good way to ensure safety? If managers discover that operators are not rigorously following procedures, what is the best way for them to respond?

7

Accident Causality Models

Reasoning about causes and effects is a very difficult thing . . . to trace sometimes endless chains of causes and effects seems to me as foolish as trying to build a tower that will touch the sky.

—Umberto Eco,
The Name of the Rose

Man is so intelligent that he feels impelled to invent theories to account for what happens in the world. Unfortunately, he is not quite intelligent enough, in most cases, to find correct explanations. So that when he acts on his theories, he behaves very often like a lunatic.

—Aldous Huxley

Given the informal presentation of why accidents occur in the previous chapters, let's shift to a more theoretical treatment of the topic. Our theoretical or formal models or beliefs about why and how accidents are caused will determine how we try to engineer and operate safety-critical systems to prevent them. To be effective, they must obviously represent the knowledge obtained from experience, as described in the first six chapters.

Our theoretical models of accident causality underlie everything we do to engineer for safety. The way we explain why accidents occur—that is, our model of causality—determines the way we go about preventing them and learning from those that do occur. Most people are not aware that they are using an accident causality model, but they are. Basically, models impose patterns on the events we see and represent our assumptions about how the world operates. For example, if an underlying assumption is that accidents are caused by human operator error, then the analysis will focus on what the operators did to contribute to the loss. Such assumptions about the causes of accidents always underlie engineering for safety, but those doing the analysis may be unaware of any subconscious assumptions they are making.

Four different types of causality models have been proposed: (1) the energy model; (2) the linear chain-of-failure events model and some subsets such as the domino model, the Swiss cheese model, and the recent functional resonance model; (3) the epidemiological model; and (4) systems-theoretic models, of which the most widely used is STAMP (Systems Theoretic Accident Model and Processes). Most other models are simply different manifestations of one of these.

7.1 Energy Models

The oldest and most pervasive model views accidents as the result of an uncontrolled and undesired release of energy (figure 7.1). The energy can take a variety of forms—chemical, electrical, radiant, mechanical, thermal, nuclear—and has an even larger number of potential carriers, but the energy constitutes the direct cause of injuries in this model. The type of energy is an important variable in predicting or explaining expected hazard or accident consequences and in designing safety measures.

If accidents are the result of uncontrolled energy flows, then the obvious way to reduce them is to use barriers or other flow control mechanisms. In this way, accidents are prevented by altering or controlling the path on which the energy flows between the energy source and the at-risk object or the potential victim. In turn, controlling the energy flow can be achieved by controlling the source of the energy or the carrier through which the energy reaches the object, such as bullets, boiling water, or sharp knives.

The basic energy model as described is simple, but it does not account for many types of accidents, such as those involving suffocation. Extensions have been proposed to handle more types of accidents. For example, MacFarland extended the basic energy model by describing accidents as resulting either (1) from the application of specific forms of energy in amounts exceeding the resistance of the structures on which they impinge, or (2) from interference in the normal exchange of energy between an organism and its environment including lack of oxygen and exposure to the elements [166].

Another variant divides accidents into two types: energy transformation and energy deficiency [142].

An *energy transformation accident* occurs when a stable or controlled form of energy is transformed in a way that damages property or injures people. For example, a fire might result when the chemical energy of gasoline is transformed into thermal energy through some form of combustion. An energy transformation accident requires both an energy source and an associated energy transformation mechanism. The mechanisms for transmitting or altering the energy may be either passive or active. Prevention of such accidents requires controlling the sources or the mechanisms of energy transformation or both.

An *energy deficiency accident* occurs when the energy needed to perform a vital function, such as powering the engines in an airplane, is not available and results in damage to

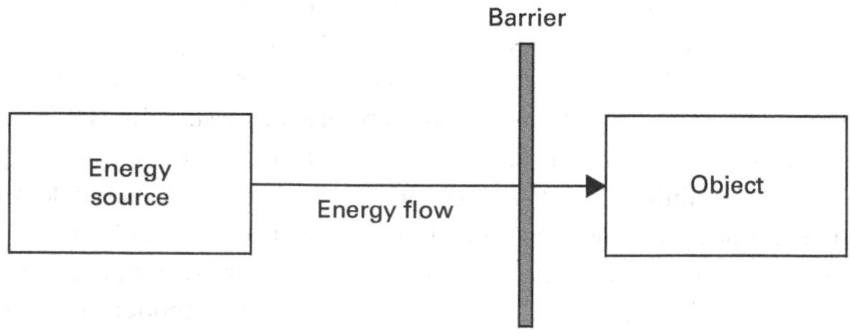

Figure 7.1
Simple energy model of accident causation.

property or injury to humans. The energy deficiency may be either a direct or indirect cause of the accident.

A final energy model variant divides systems into two types: those that produce energy (action systems) and those that constrain energy (nonaction systems, such as pressure vessels, buildings, and support structures) [13]. A complex system will usually consist of both action and nonaction components. Action components contribute directly to the switching and modification of energy or the operation of the system; nonaction systems do not contribute to system operation but only to the support or containment of energy. The action components may include safety devices, such as relief valves or fuses, that control the buildup of energy that could exceed minimum design allowances and thus affect the nonaction components.

Hazards in action systems are controlled by imposing limitations on the operation of the system. Hazards in nonaction systems are controlled by the application of a fixed standard, design allowance, or minimum safety factor.

A drawback to the energy model and all of its variants is that their scope is limited to energy processes and flows, and so may not include accidents involving nonphysical losses caused by logical errors during operations. If accidents include things like loss of mission, then energy models will not be sufficient.

The biggest limitation is that the model does not contain information about why an uncontrolled energy flow or lack of necessary energy occurs in the first place. For example, why energy needed to power the engines on an aircraft is not available. Including causal factors expands the possibilities for preventive action, beyond simply controlling the energy (or lack of energy) directly, to dealing with the conditions and events contributing to the loss of energy control.

It also does not go beyond physical causes of accidents: social and human aspects are omitted. Because of these omissions, the influence of energy models in safety engineering has waned.

7.2 Linear Chain-of-Failure Events Models

The almost universal model of causality underlying engineering for safety today posits that they result from a linear chain-of-failure-events. Note that the mathematical definition of linearity and nonlinearity is not being used here. In mathematics and science, a *nonlinear system* is a system in which the change of the output is not proportional to the change of the input. Nonlinear problems are of interest because most natural systems are inherently nonlinear.

In contrast, *linear causality* (versus mathematical linearity) is a conception of causation as following a single, linear direction; that is, event A causes effect B, where event B has no demonstrable effect on event A. A synonym for linear causality here might be "sequential causality." The latter would be a more accurate term, but it is difficult to change terminology after decades of use.

Linear causality models are built on the assumption that accidents are caused by sequential chains of failure events, each failure being the direct consequence of the one before. For example, someone enters the lane in front of your car, you slam on the brakes but are too late in applying them, and you hit the car in front of you.

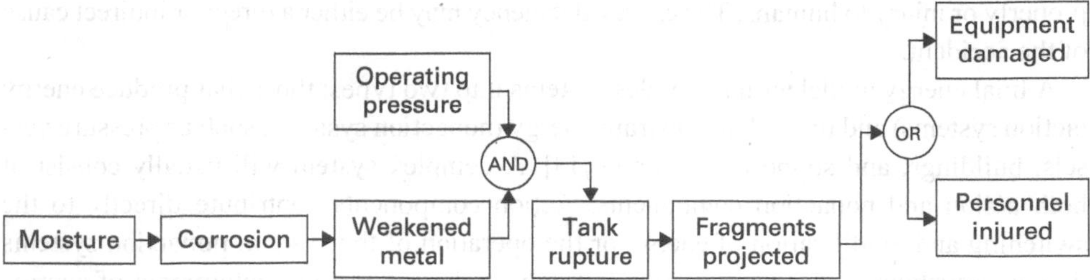

Figure 7.2
Chain-of-events model for a tank explosion.

Figure 7.2 shows an example of applying a simple chain-of-events model for a tank explosion. Note that the chain can have logical ANDs and ORs in it—it is still considered a linear chain. An OR provides a shorthand for condensing two different chains into one chain. The AND provides more power to this simple model in terms of allowing the possible requirement for multiple events to precede the next one in the chain. The OR provides no additional power but is merely a convenience in specifying the chain more succinctly.

In the accident described in figure 7.2, moisture gets into the tank, which leads to corrosion, which in turn causes weakened metal. The weakened metal along with too high an operating pressure for the condition of the tank leads to a tank rupture, which causes fragments to be projected. The fragments lead to damaged equipment or personnel injury or both. The AND in the chain means that both weakened metal and an operating pressure above a specific value are required for the tank to rupture. If neither event occurs, then the rupture will not occur, according to this example model.

Using a chain-of-events model of accident causation, it appears that the simplest way to prevent such an accident is to eliminate one of the events in the chain. An alternative is to put barriers between the events so that the consequences of one event do not lead to the next event in the chain. An example of a barrier in this case is to put a screen around the tank so that in the event of a rupture, the fragments cannot be projected outside a protected area.

The use of barriers is common practice in the nuclear power and the process industries where *defense in depth* is commonly used to prevent accidents. In such approaches, multiple barriers are provided, where each barrier is used to back up the previous one. In the Bhopal chemical plant, the hazard was release of methyl isocyanate (MIC) into the air. The barriers preventing the hazard included valves and other devices to prevent water from coming into contact with the MIC, cement-encased tanks, a refrigeration system, a vent gas scrubber, a flare tower, and a water curtain.

Figure 7.3 shows an annotated model of the tank explosion accident chain shown in figure 7.2 where possible protection or control activities are added. A wide selection of preventive activities is included in the example, including both activities or design features that prevent the event and those that prevent propagation (barriers between the events). For example, moisture might be kept out of the tank by using a desiccant or the tank might be coated with stainless steel to prevent corrosion.

There are a few things to note here. The first is that direct causality is assumed; that is, each event leads to the next event in the chain. Also, the preceding event is necessary for the following event to occur; that is, if moisture does not get into the tank, then corrosion

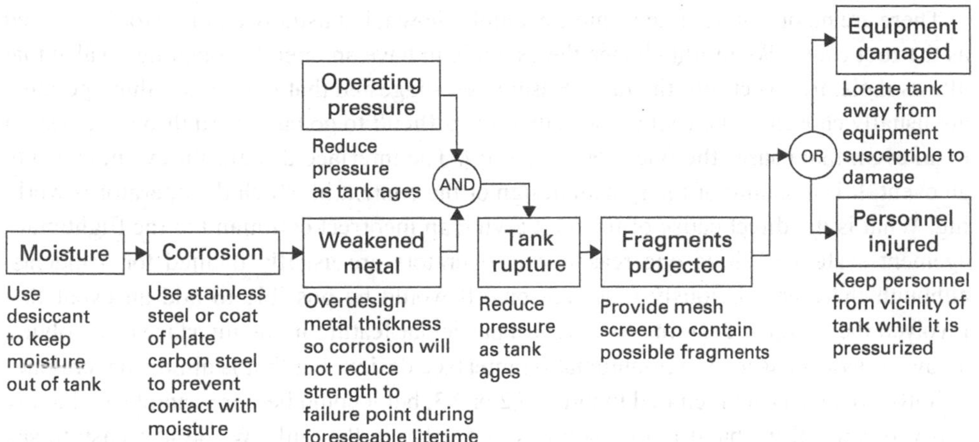

Figure 7.3
Tank explosion example shown with added protections.

does not occur. The previous event (or events in the case of an AND) in the chain is both necessary and sufficient to produce the following event, as John Stuart Mill suggested in his definition of causality two hundred years ago (see chapter 4) [262].

Using this model to explain how a loss might occur, the analysis works backward from the loss event to identify the causal chain. The initial event is often called the *root cause*. While the event labeled the root cause does not have to be the first one documented in the chain, it usually is. Note that the stopping point is almost always arbitrary, and often the root cause is identified as a human operator. In figure 7.2 or 7.3, more previous events could be added, which would then be the root cause. In practice, the search works backward until something is found that is easy to prevent or the search cannot easily go backward any farther. That event is labeled the root cause. Sometimes politics or liability become involved in the selection of a root cause.

As an example of how events could be added, consider the first event in figure 7.2 or 7.3, which is moisture entering the tank. The moisture must be introduced somehow, and there probably were design features used to prevent water and moisture reaching the tank. An example might include valves on pipes leading to the tank to allow it to be isolated. Water getting into the tank and the provision of protection devices to prevent this event could be added to the beginning of the chain. Is the failure of those valves the root cause? Again, the event that is chosen as the root cause is arbitrary because any of the events in the chain could theoretically be labeled as such or the chain could be followed back farther.

While the events in this example all are physical events, there is nothing to preclude including flawed management decision-making or other types of social events. These might include such events as management requiring unsafe operational procedures, policies, or training, or requiring the use of a flawed development process. Other examples could involve social or governmental activities such as the regulatory agency approving a flawed design, regulatory procedures not being created that prohibit the specific design flaws, and so on. These are all events in the chain of events. Including social and managerial factors does not in itself mean that the result is not a chain-of-events model.

There are no operators in the simple example shown, but usually an operator is selected as the root cause. We might change the example to have an operator opening a valve that allows moisture to get into the tank. Rasmussen suggested that one reason that operators are usually chosen as the root cause is that it is difficult to go backward through a person to an event that causes the operator error [319]. The interface design, for example, is not an event. It is a feature of the system design or the context in which the operator is working. What is the direct cause of the pilot giving an incorrect command to the flight management system? This is one reason why operators are usually blamed for accidents, although there are obviously other reasons. It would be possible to add an event that involved the designer deciding to include a particular feature in the interface or involving a flawed review of the human-automation interface design, but that is almost never done.

Software also is not included in figure 7.2 or 7.3, but it could be, for example, by having software control the burst diaphragm or relief valve in the tank. While it is easy to say "Software does not open the relief valve" (perhaps in a box preceding "Tank Rupture"), it is more difficult to think of a way to protect against this behavior. Software is an abstraction (set of instructions) that cannot fail—it does exactly what it was told to do, so the problem must involve a system design or requirements error on the part of the engineers. There usually are no simple ways to protect against design errors.

Note that the other boxes in the chain might also contain design errors, such as the design of the tank, but those causes are omitted from the chain-of-events model because they are not events. As seen in the previous chapters, the most common root cause selected is a human error. In the accident report of the American Airlines B757 crash while landing at Cali, Colombia, for example, there were four causes cited, all of them involving something the flight crew did or not do. The strangest was the fourth one: "*Failure of the flight crew to revert to basic radio navigation at the time when the FMS-assisted navigation became confusing and demanded an excessive workload in a critical phase of flight.*"

Note that the cause was that the flight crew did the wrong thing rather than the cause being that the automation (the FMS or flight management system) was confusing and led them to do the wrong thing.

The problem is not that the chain-of-events model cannot be used to describe the events leading to any accident. It can. The important question (discussed later) is whether it provides the power to prevent accidents in today's complex systems or whether important causal factors are left out of this model.

While almost all engineering safety techniques are based on this general type of linear causality model, a few special cases of this general model have been proposed and used widely, including the domino model, the Swiss cheese model, and the functional resonance model.

7.2.1 The Domino Model

Heinrich, who created the domino model, worked for an insurance company and was primarily interested in workplace safety. He was convinced that workers were the cause of most accidents and, in 1931, created his domino model to explain this phenomenon (figure 7.4) [141].

Note that this is simply a special or limited case of the general chain-of-events model where the events are depicted as five dominos, with each domino falling and causing the

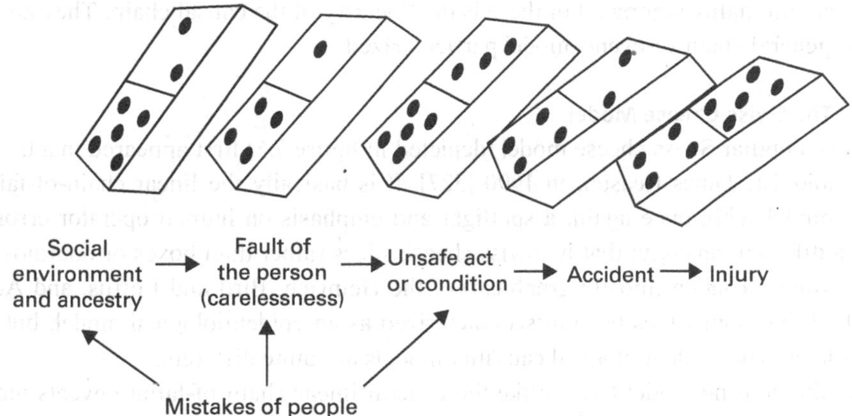

Figure 7.4
Heinrich's domino model of accident causation.

next domino to fall instead of using an arrow between square boxes. While unsafe conditions are included in the third domino, they are assumed (as shown) to be the result of "mistakes of people."

Removing any of the dominoes will break the sequence and prevent the injury, but Heinrich argued that the easiest and more effective domino to remove was the third one, representing an unsafe act or condition. Unfortunately, Heinrich was very successful in promoting worker error as the primary cause of accidents in the workplace, and his causality model persists today, particularly in workplace safety.

Heinrich's model was extended by others to include more factors. For example, Bird and Loftus in 1976 extended it to include management decisions as a factor in accidents [28]. The modified chain or sequence of events was defined as:

1. lack of control by management, permitting
2. basic causes (personal and job factors) that lead to
3. immediate causes (substandard practices/conditions/errors), which are the proximate cause of
4. an accident or incident, which results in
5. a loss.

In this model, the four major elements of a business operation—people, equipment, materials, and environment—individually or in combination are the factors involved in a particular accident.

Adams, also in 1976, suggested a modified and more general management model [2] that included:

1. management structures (objectives, organization, and operations)
2. operational errors (management or supervisor behavior)
3. tactical errors (caused by employee behavior and work conditions)
4. accident or incident
5. injury or damage to persons or property.

The commonality among all of these is the linearity of the causal chain. They are simply the general chain-of-events model particularized.

7.2.2 The Swiss Cheese Model

The very popular Swiss cheese model (depicted in figure 7.5) first appeared in a book by a psychologist, James Reason, in 1990 [327]. It is basically the linear chain-of-failure-events model with, once again, a spotlight and emphasis on human operator error and using a different analogy; that is, Swiss cheese slices rather than boxes or dominos. It is just a minor variation (mostly graphical) of the Heinrich, Bird and Loftus, and Adams models. It has sometimes been mischaracterized as an epidemiological model, but as is shown later, true epidemiological causality models are quite different.

Like the domino model (but unlike the general linear chain-of-failure events model), the Swiss cheese model concentrates on human error, here called "unsafe acts"—unsafe conditions appear to be omitted. In this depiction of the chain-of-events model, the failure events are the holes in the cheese; an accident occurs if the holes or failure events line up. Note that there is still a chain of events and direct causality only (see figure 7.5). There are ORs in the Swiss cheese representation (represented by multiple holes in one slice of cheese), but it is not clear how or whether an AND (a combination of nonsequential failure events) could be represented. It does not appear that it can be, but this is a problem with models that are simply drawings without a formal, mathematical underpinning.

Whether the prevention measures or barriers are drawn next to the box as in figure 7.3 or the prevention measures or barriers are depicted as slices of cheese as in figure 7.5 with the holes being the failure events, the basic underlying causality model is identical: failure events—which include failures of protection devices—occur in a linear (sequential) order where the preceding event is necessary and sufficient for the following event to occur. Note that in the Swiss cheese depiction of this abstract model, the holes must line up and no hole

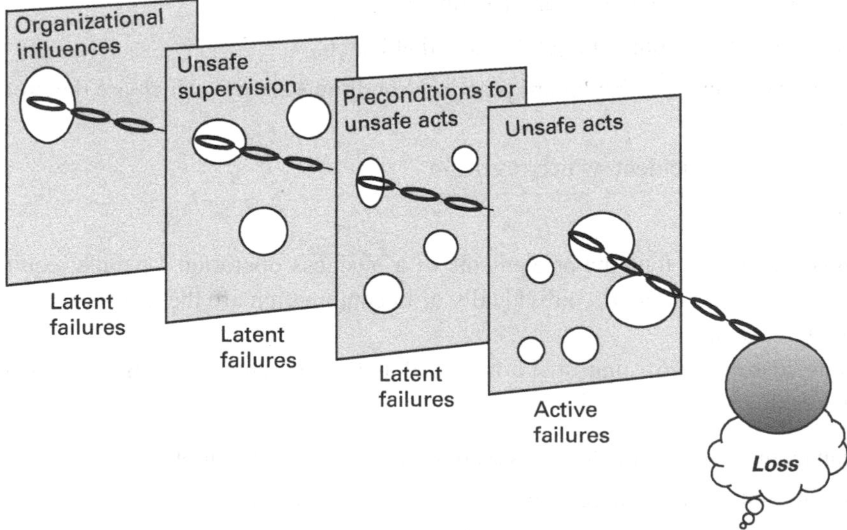

Figure 7.5
Reason's Swiss cheese model.

(or slice) can be skipped. The depiction of the chains of events as dominoes falling, holes in slices of Swiss cheese, or using other analogies are only graphical differences. It is the same identical causality model.

It is not clear what Reason means by a "latent failure." In engineering, failures cause erroneous states, which in turn can lead to later failures. The term here seems to mean an erroneous state. Unless all the failures involved are coincidental in time—in which case there is no chain but simply one box labeled with all the coincident failures or only one slice of cheese with many holes—causal events always precede in time the effect of the event, which is an erroneous state. Therefore, the term "latent" appears to imply simply a forward chain over time where the last failure in the chain is termed an "active failure." Any chain of events depicts events over time.

There are some differences between figures 7.4 and 7.5 and the more general chain-of-events model depicted in figures 7.2 and 7.3. The domino and Swiss cheese depictions emphasize human error as the cause of events, particularly direct operational human error, over everything else. For example, if the wings fell off of a plane or a freak weather event occurs leading to a crash, it would appear, in the domino model, as somehow the result of a pilot's action along with the pilot's social ancestry or, in the Swiss cheese model, the pilot's actions would be assigned major responsibility although the pilot's actions here might result from previous management "failures." Technical system design errors or physical failures do not seem to be included.

7.2.3 The Functional Resonance Model

Resonance in physics is defined as the reinforcement or prolongation of sound by reflection from a surface or by the synchronous vibration of a neighboring object. Hollnagel has created a related concept that he calls *functional resonance*, which he defines here:

> *Functional resonance is defined as the detectable signal that emerges from the unintended interaction of the everyday variability of multiple signals. The signals are usually subliminal, both the 'target' signal and the combination of the remaining signals that constitutes the noise. But the variability of the signals is subject to certain regularities that are characteristic for different types of functions, hence not random or stochastic. Since the resonance effects are a consequence of the ways in which the system functions, the phenomenon is called functional resonance rather than stochastic resonance.* [149]

He claims that functional resonance is an alternative to linear causality along with (or including) a lack of resilience on the part of human operators and human performance variability. In reality, his principle of functional resonance still assumes linear causality in the nonmathematical (sequential) sense. The different outcomes are the events.

Consider the linear causality chain in figures 7.2 and 7.3 where moisture leads to corrosion, which leads to weakened metal. The *mechanism* for the causation is not specified but is represented by the arrow. That is where what he calls functional resonance or approximate adjustments would lie; that is, in the arrow but not in the events caused by them. The difference appears to be in the way the linear or sequential chain progresses, in this case, by increasing variability. But there is still a linear chain of events leading to the loss.

Perhaps the closest to Hollnagel's resonance model, as noted earlier, is the model of accident causation underlying the HAZOP analysis method. This model still involves a linear chain of events, but instead of the events only being caused by general failures of components, they are caused by deviations in the operating parameters. The deviation is in the arrow or mechanism that creates the next event. Of course, the difference depends on the definition of "failure" that is being used. In HAZOP, the parameter deviations lead to failures. In Hollnagel's Functional Resonance theory, it appears that the mechanism for transitioning to another event is based on approximate adjustments.

No matter what the mechanisms in generating the outcome(s) from the preceding event(s)—such as component failures, parameter deviations, or approximate adjustments—the sequence of events, outcomes, or effects is linear, and the result is a chain of events that leads to the outcome.

While functional resonance might be used to explain a few accidents, it is very far from a general accident causality theory. Neither of his recent books on what he calls Safety-II [148, 149] provides any examples of how it can be used to explain real accidents. It would be helpful to see how this model and concept could be used to explain and prevent accidents beyond assigning operator error as the cause.

7.2.4 Limitations of Linear Chain-of-Events Models

Most of safety engineering technology has been built on the linear chain-of-events causal model of how and why accidents occur, although not the more limited domino, Swiss cheese, and functional resonance models that focus primarily on human error.

The example accident in figures 7.2 and 7.3 is quite simple. Real systems today may have hundreds and probably thousands of such chains of failure events that could lead to losses. One Fault Tree Analysis (which generates linear causal chains, just as do all traditional hazard analysis techniques, see chapter 10) for one aircraft's Integrated Modular Avionics system required more than 2,000 pages to document the results. And this was only for one part of the aircraft.

During Space Shuttle development, an FMEA (Failure Modes and Effects Analysis) identified 40,000 critical items. It's not clear what to do with the information that the failure of 40,000 individual items could lead to a serious loss, but only a government project like the Space Shuttle could have the resources to identify all of these, let alone provide protection against them. And, of course, Space Shuttle design errors and poor management decision-making are omitted from this analysis. This omission includes the causes attributed to the two actual Space Shuttle losses.

Abstractions or models are used by humans to understand complex phenomena. By definition, they leave out factors—otherwise they would be the thing itself. For abstractions or models to be useful, they need to include the important factors in understanding the phenomenon and leave out the unimportant. Unfortunately, the simple linear chain-of-events causality model leaves out too many types of important factors to be useful in understanding and preventing accidents in complex, sociotechnical systems. In short, the second and third levels of the model in figure 4.2 are omitted.

There are other inherent limitations of this traditional and almost universal chain-of-failure-events model. First, there is an assumption that the events and barriers fail *independently*; that is, there is nothing that will reduce or eliminate the effectiveness of all of them at the same time. In the Swiss cheese model (where the barriers are depicted as Swiss cheese slices), the slices or barriers are assumed to be independent. Given this assumption, the risk of an accident, if all defenses are implemented correctly, is theoretically low. However, the independence assumption is almost always untrue in real systems. For example, accidents commonly occur because budget cuts, demands for increased productivity, inadequate maintenance procedures, or competitive pressures; that is, level 3 systemic factors, make all the barrier or protections ineffective at the same time. Many examples were provided in chapter 4. A poor organizational or safety culture, such as management pressures to ignore safety rules and procedures, can also undermine the effectiveness of all the safety controls and the applicability of the model. The systemic factors do not appear in linear chain-of-events models, which depict only the level 1 causes. For the most part, systemic factors have to be ignored to perform quantitative or even qualitative risk assessment based on linear causality.

Another critical omission from the linear chain-of-failure events model are accidents that involve nonfailures, where all the components may operate as designed, but their interactions lead to a hazardous condition in the system as a whole. Accidents resulting from the unsafe interaction of non-failed components (that is, that operate according to their specification) may stem from complexity in the overall system design and the incorporation of software controls and autonomy. What system components failed in the Warsaw A320 reverse thruster accident described in chapter 3? Certainly not the flight crew or the software, both of which did exactly what they were instructed to do.

Most accidents today result from these types of system design errors, although they are often incorrectly blamed on the pilots or, more generally, human operators. The problem lies in trying to identify them using the traditional hazard analysis techniques that assume that accidents are caused by chains of component failures. More powerful causality models and hazard analysis techniques built on such models are needed to identify these more complex accident causes in order to design and operate systems to prevent them.

Consider the Bhopal MIC accident, which is considered the worst industrial accident of all time (see chapter 4 and appendix C). There was no lack of protection devices and barriers to prevent this accident. Because of the dangers associated with the production of MIC, primarily related to contact of the chemical with water, many physical barriers were used to prevent such contact, including slip discs in valves, concrete around storage tanks, and operational procedures. Those are the first level of barriers in the defense-in-depth approach taken. In addition, if the water did get through the barriers, which could cause an enormous increase in pressure and heat, there were relief valves, procedures specifying that the tanks were never to be more than half their maximum capacity along with spare tanks to bleed the contents of one tank into an empty one, a refrigeration unit to limit the reactivity of the MIC, and a high temperature alarm along with lots of other gauges and instrumentation to keep the operators informed. If, despite all these protections, a release of MIC occurred, there was a vent scrubber to neutralize any escaping gas with caustic soda, a flare tower to burn off any escaping gas missed by the scrubber, and a water curtain to knock down any gas missed by both the scrubber and the flare tower. There was, of course, also warning sirens, protective equipment, and frequent testing of the alarms and practice of emergency procedures.

Despite all these preparations and an extensive defense-in-depth design, tens of thousands of people were killed or injured in an inadvertent release of MIC. How could the vent scrubber, flare tower, waterspouts, refrigeration unit, alarms and monitoring instruments, and so on, all fail simultaneously? In fact, a probabilistic risk assessment of this plant would have combined the probability of the "failure" of all these devices and come up with an extremely low—basically impossible—likelihood.

The answer is that a defense-in-depth strategy does not protect against systemic factors that impact all the barriers and protection devices at once; that is, all the Swiss cheese slices or design features associated with the boxes in the event chain in figure 7.3. These are the factors described in chapter 4. At Bhopal, these systemic causal factors included design errors in most of the protection devices and severe pressures on the company to cut costs due to a sharp decrease in demand for MIC. The cost-cutting pressures led to cutting maintenance and operating personnel in half and reducing maintenance procedures—the scrubber and flare tower were out of operation at the time of the accident because of a lack of maintenance—turning off the refrigeration unit to save money, unskilled workers replacing skilled ones, minimal training of workers in how to handle emergencies, and reductions in educational standards and staffing levels, among other things.

There had even been warnings in the form of a less serious accident the year before, several serious incidents involving MIC in the previous three years, and an audit report two years before the tragic events that noted all the deficiencies in the plant that led to the loss. The deficiencies in the audit report were never corrected. The more one learns about this accident—and, indeed, most accidents—the less important the role of the operators appears to be in causing or not preventing the accident. In this case, no matter how adaptable or resilient the operators had been, they could not have prevented the tragedy. Using too simplistic an accident causality model can lead to tragedies.

Past assumptions about the role of humans in systems also do not fit systems today, where the humans are mostly managing complex automation rather than directly controlling physical devices or computer-automated functions. The future will see even more changes to the human role away from active control and toward being a manager or monitor of computers and even partnering with automation to achieve common goals as responsibilities are divided between machines and humans. Autonomy does not usually mean that humans are totally eliminated from systems, except for the simplest functions, but only that the roles played by humans are changed. None of these new roles and human factors considerations are included in the traditional model of accidents caused by chains-of-failure events. They cannot be represented using a simple linear failure model.

Again, note that descriptions of chains of events as dominoes falling, holes in Swiss cheese, similarities to men's formal attire such as bow ties, or approximate adjustments are only graphical or superficial differences. The chains of events may be drawn differently, use different notations, or apply different analogies, but they all are describing the same underlying formal chain-of-events model. They are not different causal models but simply different names or notations for the same model.

Safety engineering has been built on this limited causality model of how and why accidents occur. And despite these limitations, no alternative to this traditional linear accident causality model has been suggested until relatively recently except for epidemiological models.

7.3 Epidemiological Models

Hollnagel has labeled the Swiss cheese model as an epidemiological model, but the Swiss cheese model has little relationship to the basic science underlying the field of epidemiology.

Historically, a true epidemiological model of accident causality was proposed. In the 1940s, John Gordon, a professor of epidemiology at Harvard University, stressed the multifactorial nature of accidents [118]. He and others suggested that accidents should be viewed as a public health problem that can be handled using an epidemiological approach.

In Gordon's epidemiological model of causality, accidents are conceptualized in terms of an agent (physical energy), the environment, and the host (victim). Accidents are considered to result from complex and random interactions among these three things and cannot be explained by considering only one of the three factors or by simple linear interactions between events.

Two types of epidemiology are relevant [376]:

- **Descriptive epidemiology**: The general distribution of injuries in the population is described by determining the incidence, prevalence, and mortality rates for accidents in large population groups according to characteristics such as age, sex, and geographical area.
- **Investigative epidemiology**: The specific data on the causes of injuries is collected to devise feasible countermeasures.

The epidemiological approach to accident causation assumes that some common factors are present in accidents and that these can be determined by statistical evaluation of accident data, in much the same way that epidemiology handles disease. Because specific relationships between factors are not assumed, previously unrecognized relationships can be discovered. A claim is made that determinant as opposed to chance relationships can be distinguished [36; 247].

This model has not been widely used. The validity of the conclusions from such epidemiological studies of accidents is dependent on the quality of the database used and the statistical significance of the anomalies found in the sample [150]. In practice, the data reported by accident investigators may be limited or filtered (see chapter 14) and almost never of high enough quality to use to determine causality relationships. Also, the sequencing and timing relationships between events and conditions is not captured by a purely statistical approach. This omission might be considered a limitation or a feature, but sequencing and timing relationships can provide important information when considering causality.

7.4 More Sophisticated Models of Causality

The basic chain-of-events and energy models are limited in the causal factors and types of losses that they can model and explain. While linear conceptions of causality may be adequate abstractions for the simpler world of the past, they no longer provide the explanatory power to prevent undesired events in complex systems today. Using too simple a causality model leads to unexplained and unexpected behavior and to unknown unknowns. More complex relationships between the components of the system are necessary to explain the reason why accidents occur than a simple one-way sequential chain.

Figure 7.6 shows four explanations of accident causality (labeled a, b, c, and d) in complex systems. In addition to linear chains shown in figure 7.6(a), multiple independent causal factors can produce an effect or event as depicted in figure 7.11(b); conditions and

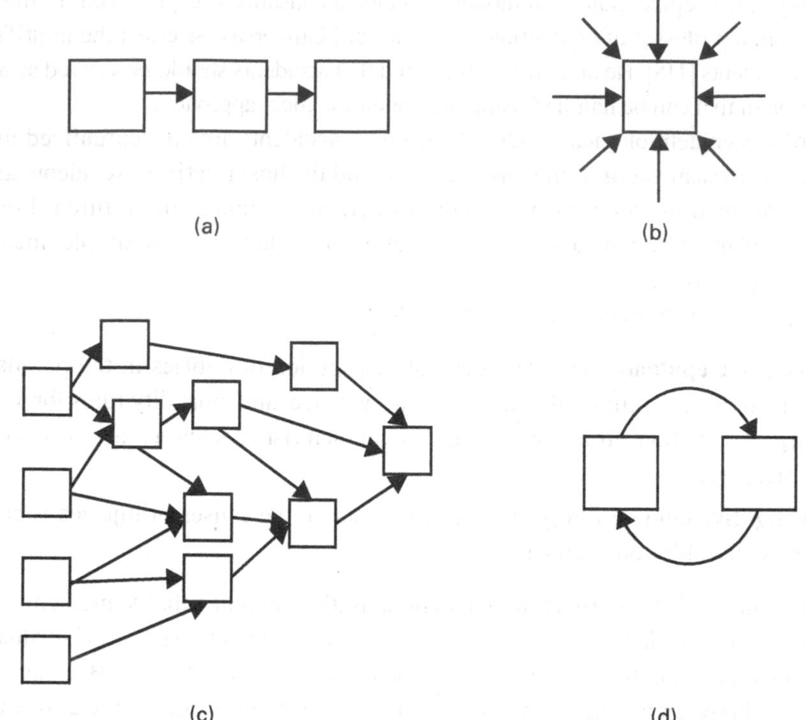

Figure 7.6
Four types of causality included in STAMP.

systemic factors can impact multiple events shown in figure 7.6(c); and causality can even take the form of circular loops as shown in figure 7.6(d). As Senge says, "Reality is made up of circles but we see straight lines" [346].

In figure 7.6(c), there may be multiple conditions or systemic factors that affect all the events. For example, some systemic factors, such as budget cuts or efforts to increase productivity, can defeat all the barriers at once, undermining the assumptions of linear chains of events. Figure 8.5 (in the next chapter) shows a model of the causality in a ferry capsizing accident where several hundred people died. Causality in this case involved multiple different groups in the system all making independent decisions that interacted in unexpected ways with decisions made by people working in other parts of the system. The decisions made by port designers, ship designers, ship operators, traffic schedulers, and so on, which were made independently and at different times, all factor into understanding why the accident occurred. A simple chain-of-events model of the accident leads to ignoring the most important factors and simply placing all responsibility on the captain and crew and leaving open the possibility of future accidents related to the unidentified systemic factors. In this form of causality, multiple causes can lead to a loss where there is no simple linear relationship between the events and causes.

Causes can also be thought of in terms of circular feedback or feedforward relationships between events leading to an accident as shown in figure 7.6(d). For example, chapter 4 includes the description of several major accidents involving circular causality where the

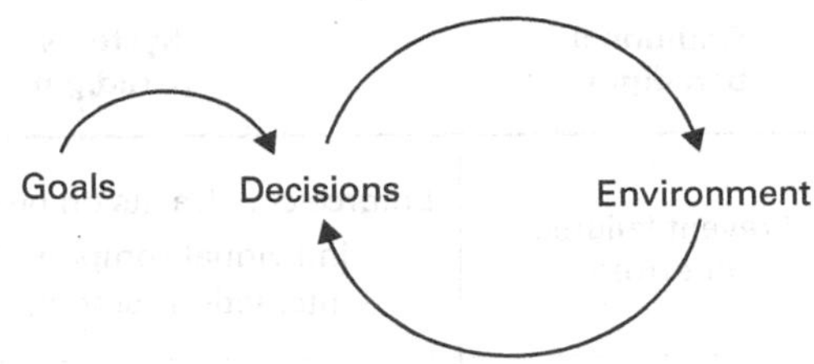

Figure 7.7a
Decisions may change the environment, which leads to new decisions.

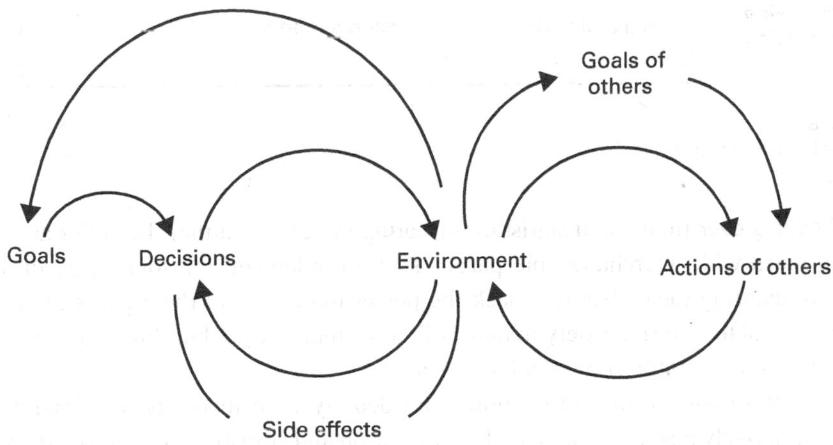

Figure 7.7b
Changes in the environment may also change our goals, impact the goals of others, have side effects that lead to new decisions, or trigger interventions by others.

use of redundancy led to complacency, which, in turn, led to behavior that defeated the redundancy.

Sterman [356] provides a simple example, shown in figures 7.7a and 7.7b. Our goals affect our decisions, which affect our environment, which can in turn affect our decisions (figure 7.7a). The world is even more complicated than that, however, because decisions trigger side effects, delayed reactions, changes in goals, and interventions by others, as depicted in figure 7.7b.

In chapter 3, systems thinking is described as viewing systems as dynamically complex, interconnected, interdependent, and nonlinear. In our complex systems today, we cannot ignore these factors, including the role played by feedback, context, and mental models in influencing our behavior and the behavior of the systems we design and operate. An argument can be made that we need more powerful accident models than simple chains-of-failure events to understand why accidents occur and how to prevent them.

7.5 The STAMP Model of Causality

Chapter 3 describes three ways to deal with complexity: analytic decomposition, statistics, and systems theory. Analytic decomposition—considering behavior as chains of directly

	Traditional paradigm	Systems paradigm
Goal	Prevent failures or errors	Enforce constraints on behavior of – individual components – interactions among components
Approach	Treat safety as a *reliability* problem	Treat safety as a *control* problem
Underlying foundation	Reliability theory	Systems theory

Figure 7.8
The paradigm change in STAMP.

related events over time—and statistics—treating systems and their behavior as a structureless mass with interchangeable parts—still are adequate for some uses in system safety engineering today. But they lack the power to handle all the aspects of causality that are critical to ensuring safety in many of the systems we are building today, including those aspects of causality shown in figure 7.6.

For complex systems, something more is needed. Systems theory is described in chapter 3 as a relatively new alternative to decomposition and statistics. As described, systems theory is based on the concepts of emergence, control (including feedback), hierarchy, communication, and abstraction. It is being used today in many different disciplines. STAMP is a new causality model that was created to apply these system-theoretic concepts in system safety engineering.

In STAMP, safety and other emergent system properties, such as security, are treated as a dynamic control problem rather than a failure prevention problem (figure 7.8). The goal is to enforce constraints on system behavior in order to prevent losses. That process may include enforcing constraints on individual system component behavior, including preventing component failures that can lead to losses, so it includes the traditional model of accident causality. But it also includes other types of unsafe component behavior that do not involve failure. In addition, it includes controlling the interactions *between* system components; that is, preventing unsafe interactions.

STAMP uses hierarchical control loops to model dynamic system behavior and the role of hardware, software, human operators, and managers in systems.

Figure 7.9 shows a basic component from which the structures are built. There is a controller, which may be software or human (or hardware, but few hardware-only controllers are used today). The controller has responsibilities or requirements including goals and constraints, for example, maintaining a minimum distance between aircraft. It also contains a decision-making component—ranging from an algorithm if the controller is software to a complex decision-making process for human controllers. Note that the usual approaches to describing management functions—authority, responsibility, and accountability—are included so that managers can be included in the model.

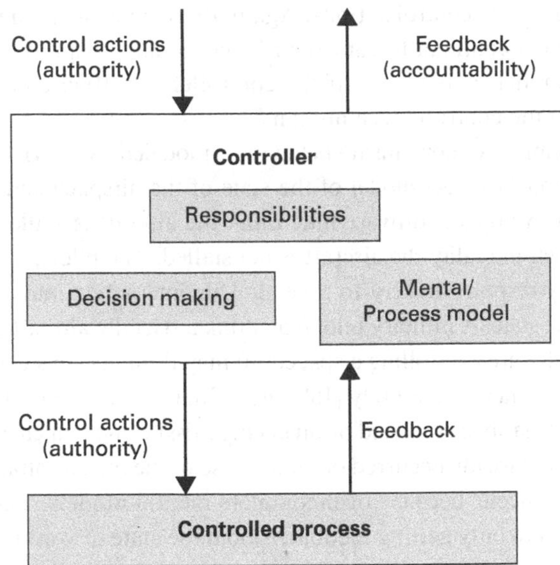

Figure 7.9
The basic component of STAMP hierarchical control models.

The controller also contains a model of the process being controlled, which may also include information about the environment in which the process is operating and other things, dependent on the process being modeled. This model is usually called a *mental model* for humans, but it is much simpler for automation and may consist of only a few variables for software. The model in automation is usually just a model of the current state of the process being controlled.

The controller issues control actions on the controlled process and gets feedback about the current state. The feedback can be used by the controller to update the current state of the process in order to determine the effect the previous control actions had on the process or to identify changes in the process that occur independent of the controller. For example, the air traffic controller issues *advisories* to pilots who then apply control commands to the aircraft. The controller gets feedback about the current location of the aircraft.

Every controller (human or software) needs to contain a model of the controlled process as well as other relevant aspects of the system in order to provide effective control. Humans make decisions based on their mental model, that is their understanding of the state of the system components they are controlling, the other system components not directly under their control, and the environment. Human mental models also need to include system goals and constraints and perhaps other information depending on their responsibilities. Humans adapt their behavior with respect to their view of reality and of the systems they are controlling.

These models are used by the controller to decide what types of control actions are needed. A pilot, for example, needs a model of the state of the aircraft, the state of any software that is controlling the aircraft such as an autopilot, other parts of the system not under the pilot's direct control, the environment, and so on.

Accidents in complex systems often result from inconsistencies between the model of the process used by the controller and the actual controlled process state, resulting in the controller providing unsafe control actions. Again, these controllers may be automated or human. The controllers continually vary their behavior and adapt to the current circumstances either in a predetermined way if the controller is software or in a more flexible and adaptable way if the controller is a human.

Here are some examples of how unsafe behavior is modeled using STAMP. The air traffic controller may have an incorrect model of the state of the airspace and provide an unsafe advisory to the pilot. Autopilot software may think the aircraft is stalled and issue a command to descend when in reality the aircraft is not stalled. The pilots, surprised by this aircraft movement, may respond and try to ascend. This interaction may end up moving the aircraft into an unsafe state. A military pilot may think a friendly aircraft is hostile and shoot a missile at it. The software controlling a spacecraft may think the spacecraft has landed and turn off the descent engines prematurely [168]. As a final example, an early warning system may think the country has been targeted by an enemy missile and launch a counterattack. All of these examples have already occurred or, in the case of the missile attack, come close.

Accidents can also occur because of incomplete mental models. The managers at the Texas City refinery were only getting feedback about the state of workplace safety and not process safety. As a result, because the feedback they received indicated few worker injuries, they cut budgets for process safety efforts and maintenance. The NASA Shuttle management were under pressure to increase flights and decrease budgets and used their flawed models about the current state of Shuttle safety to make decisions that led to the loss of the *Challenger* and *Columbia* space shuttles.

Part of the challenge in designing an effective safety control structure is providing the feedback and inputs necessary to keep the controller's model of the controlled process consistent with the actual state of the controlled process so that unsafe control actions are not issued. Mental models must include the risk involved in the controllers' decisions, particularly at the higher levels of control where controllers may have limited visibility of the lowest levels of the hierarchy. Most accidents involve some type of complacency at the management levels, leading to bad management decisions.

With the changing role of operators in systems today from following procedures to monitoring and complex decision-making, the way we treat humans in our models of causality needs to change. When viewed as "unreliable" system components, humans have to be either eliminated altogether and replaced by computers and automation, or they have to be made as reliable as possible by making them rigidly follow procedures. Alternatively, when humans are viewed as flexible and adaptable system components in a dynamic and complex system, they can be seen as providing the ability to learn and avoid losses when assumptions made during development do not hold during operations, perhaps because of changes in the system or the environment.

The role of management in accidents can be included using STAMP. Formal modeling and analysis of safety must include social and organizational factors. Controllers may be managers or government agencies. System safety engineering cannot be very effective if it focuses only on the technical aspects of the system. We have learned from major accidents (see chapter 4) that managerial and organizational factors and often governmental controls (or lack of them) are as important as technical factors in accident causation and prevention. The technical, human, and social/organizational parts of a complex system are included in the STAMP control models and the analysis tools based on STAMP.

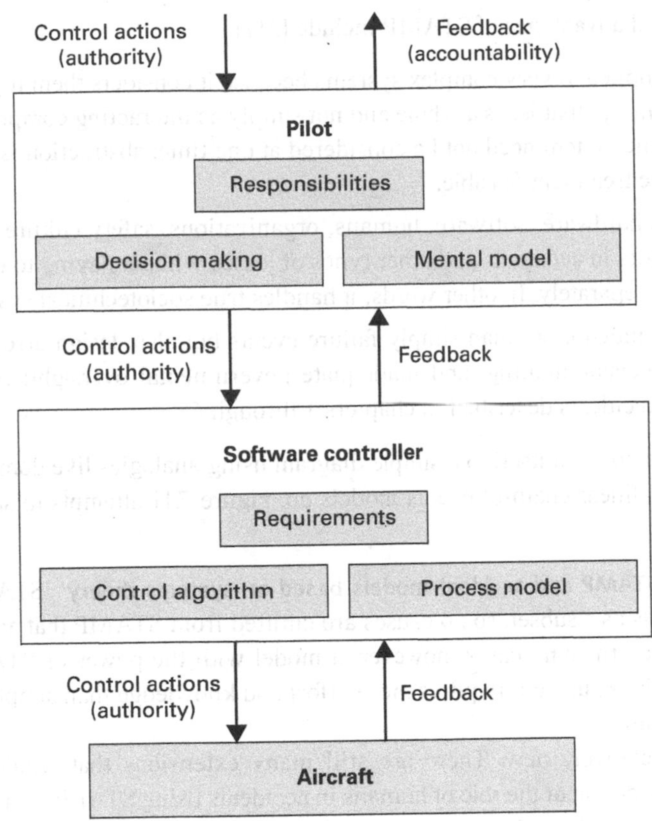

Figure 7.10
An example control structure showing a pilot controlling software that controls the actual aircraft hardware.

Complex control structures are created by putting the basic components together into a hierarchical control structure. Figure 7.10 shows an example. In this example, a pilot controls software, which controls the physical components of the aircraft. There are several controllers above the software that are not shown. For example, air traffic control and the airline dispatcher. The use of STAMP and its control structures in hazard analysis is described in chapter 10 with the other hazard analysis methods.

No causes are omitted from the STAMP model that are included in the older causality models, but more can be represented and the emphasis changes from increasing safety by preventing failures to enforcing constraints on system behavior. Note that the use of the term "control" does not imply rigid command and control. Behavior is controlled not only by engineered systems and direct management intervention but also indirectly by policies, procedures, shared value systems, and other aspects of the organizational culture. All behavior is influenced and at least partially controlled by the social and organizational context in which the behavior occurs. "Engineering" this social and organizational context can be an effective way to create and change a safety culture; that is, the subset of organizational or social culture that reflects the general attitude about and approaches to safety by the participants in the organization or society [348]. Chapter 14 includes more information about safety culture and how to improve it.

Some claimed advantages of STAMP include [217]:

- It can be applied to very complex systems because it considers them top-down rather than bottom-up; that is, as a whole and not simply as interacting components. All the details of the system need not be considered at one time: abstraction is used to make this holistic treatment feasible.
- It includes hardware, software, humans, organizations, safety culture, and so on as causal factors in accidents and other types of losses without having to treat them differently or separately. In other words, it handles true sociotechnical systems.
- It includes much more than simply failure events but also design errors, poor management decision-making, and inadequate governmental oversight; that is, all the factors in accidents described in chapters 1 through 6.

STAMP does not lend itself to a simple diagram using analogies like dominos or Swiss cheese slices as linear chain-of-events models do. Figure 7.11 attempts to show the basic concept.

Limitations of STAMP and accident models based on systems theory STAMP includes the linear models as a subset, so no causes are omitted from STAMP that are in the older causality models. In some cases, however, a model with the power of STAMP may be overkill. In addition, using it requires more effort and knowledge than simply looking for component "failures."

STAMP is relatively new. There are still many extensions that would be helpful. Although the modeling of the role of humans in accidents using STAMP is more complete than simply "human failure" or "human error" as in other models, there is more that can

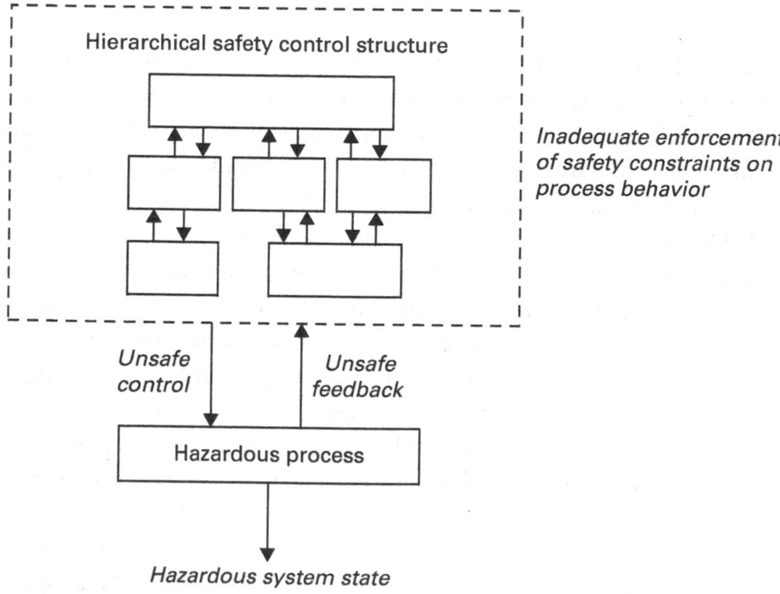

Figure 7.11
A representation of the STAMP model of accident causality.

be added to it to include sophisticated human factors decisions. The same is true for other types of complex accident causes.

One way to compare causality models is to compare the analysis methods based on those models. Chapter 10 describes the most widely used hazard analysis methods, including one based on STAMP, and provides examples and some comparisons of the results from using them.

7.6 Looking Ahead

The following chapters in this book describe how the different underlying accident causality models are used in modeling, analysis, and design in order to prevent accidents. This description starts with after-the-fact accident causality analysis. The major emphasis should always be on preventing accidents that have not yet occurred. But accidents and especially incidents (close calls) will almost surely still occur due to limitations in our tools or resources, and they will need to be investigated to identify the causes and determine how to prevent repetitions.

Exercises

1. Consider a rear-end car collision. What are some necessary and sufficient conditions for this accident? Put the conditions you identified into a causal chain for the accident. Are there other conditions that are not so easily put into the causal chain?
2. What is an example of indirect causation other than the ones mentioned in the chapter?
3. Take an accident and describe its cause as a linear event chain. Then use STAMP to model the same accident. What differences did you find in the causes that can be specified?
4. What are some examples of systems or conditions in which the epidemiological approach might be particularly useful?
5. Reason has written about what he calls *organizational accidents*. In the systems theoretic view, however, all accidents have organizational factors along with technical and other social factors. What types of accidents might be classified as "organizational accidents"?
6. A statement is made in the chapter that the Swiss cheese model is identical to the chain-of-events model with a different analogy (cheese slices). Make an argument either for or against this statement. In other words, is the Swiss cheese model a subset of, equivalent to, or a superset of the classic chain-of-events model? Justify your answer.
7. In figure 7.3, the design features are shown as annotations on the event boxes. Using circular loop causality, suggest a couple of ways that the protective design features might become ineffective over time, leading to an accident. What might you add to the design or operation of the system to prevent such accidents?
8. Consider the generic control structure component in figure 7.9. Where might the complacency aspects of the *Columbia* loss arise in this type of control structure model? Rasmussen has suggested that systems migrate toward states of high risk. In general, where would this migration appear in such a control structure model? How might we explain complacency (as defined in chapter 4) in the STAMP model?

8

Accident Analysis and Learning from Events

A learning experience is one of those things that says, "You know that thing you just did? Don't do that."
—Douglas Adams,
The Salmon of Doubt

You have now read a lot about why accidents occur, what has caused accidents in the past, and ways to think about the cause of accidents. It's now time for you to learn how to analyze the causes yourself.

Many industries are not making much progress in reducing accidents: major accidents keep occurring that seem preventable and that have similar systemic causes. One contributor to this lack of progress is that we often fail to learn from the past and make inadequate changes in response to accidents. While depending on learning from accidents and incidents is a poor long-term or overall strategy, it is also clear that we should learn everything we can from the events that do happen: It would be irresponsible if we did not learn from tragedies. The goals for such learning may include improving specific systems, learning in general about causal factors in accidents, and improving our processes and methods by identifying incorrect assumptions or missing causal factors.

8.1 Why Are We Not Learning Enough from Accidents?

The overriding goal of accident analysis should be to increase the effectiveness of the controls used to prevent accidents. Investing in this learning process provides an enormous return on investment. In contrast, superficial analysis of why accidents are occurring in the organization or industry will primarily be a waste of resources and have little impact on future events.

Many reasons could be posited for why we sometimes fail to learn from accidents, but the most important are oversimplification of causality in accident reports, hindsight bias, narrow views of human error, and a focus on blame. Most of these topics have been touched on in previous chapters, but some unique aspects related to accident analysis are emphasized here.

8.1.1 Oversimplification and Root Cause Seduction

Humans appear to have a psychological need to find a straightforward and single cause for a loss, or at least a very limited number of causes. John Carroll [59] calls this *root cause seduction.* We want simple answers to complex problems. Not only does that make it easier to devise a response to a loss, but it provides a sense of control. If we can identify one cause or even a few that are easy to fix, then we can "solve the problem" and shift our attention to other concerns.

Searching for a root cause and claiming success once something plausible is identified results in the real problems remaining unfixed and further accidents occurring. We end up in continual fire-fighting mode: fixing the symptoms of problems but not tackling the systemic causes and processes that allowed those symptoms to occur. Our intentions may be sincere, but too often we play a sophisticated *whack-a-mole* game—removing one supposed root cause after another and never finding out why the losses continue to occur. An enormous amount of resources may be expended with few lasting results.

Here are some examples of oversimplification of causal analysis leading to unnecessary accidents. Chapter 4 describes two of these. The first was the crash of an American Airlines DC-10 at Chicago O'Hare Airport in 1979, where only a maintenance problem was identified as the cause but not a design error in the wing slats that also contributed to the loss. Because of this omission, McDonnell Douglas was not required to change the design, leading to future accidents related to the same design error.

The explosion of a chemical plant in Flixborough, Great Britain, in June 1974 is also described in chapter 4. In this accident, a temporary pipe was used to replace a reactor that had been removed to repair a crack. The crack itself was the result of a poorly considered process modification. The British Court of Inquiry concluded that "The disaster was caused by a coincidence of a number of unlikely errors in the design and installation of a modification" and that "such a combination of errors is very unlikely ever to be repeated" [245]. Wisely, the British government ignored the limited insight provided by the official accident report and made important changes to the way chemical plants are operated and regulated in that country.

In many cases, the whack-a-mole approach leads to so many incidents occurring that they cannot all be investigated in depth, and only superficial analysis of a few are attempted. If instead, a few were investigated in depth and the systemic factors fixed, the number of incidents would decrease by orders of magnitude.

In some industries, when accidents keep happening despite their attempts to remove the identified root causes, the conclusion is reached that accidents are inevitable and that providing resources to prevent them is not a good investment. Like Sisyphus, they feel like they are rolling a large boulder up a hill with it inevitably crashing down to the bottom again until they finally give up, decide that their industry is just more dangerous than other industries with better accident statistics, and conclude that accidents are the price of productivity. As with being caught in any vicious circle, the solution lies in breaking the cycle, in this case by eliminating oversimplification of causal explanations and expanding the search for answers beyond looking for a few root causes.

Accidents are always complex and multifactorial. Almost always there is some physical failure or physical equipment that had flaws in its design, operators who did not prevent the loss or whose behavior may have contributed to the hazardous state, flawed management decision-making, inadequate engineering development processes, safety culture problems, regulatory deficiencies, and so on. Jerome Lederer, considered the Father of Aviation Safety, wrote:

> *Systems safety covers the total spectrum of risk management. It goes beyond the hardware and associated procedures of systems safety engineering. It involves: attitudes and motivation of designers and production people, employee/management rapport, the relation of industrial associations among themselves and with government, human factors in supervision and quality control, documentation on the interfaces of industrial and public safety with design and operations, the interest and attitude of top management, the effects of the legal system on accident investigations and exchange of information, the certification of critical workers, political considerations, resources, public sentiment and many other non-technical but vital influences on the attainment of an acceptable level of risk control. These non-technical aspects of system safety cannot be ignored.* [205, p. 8]

Our accident investigations need to include all of these factors and more.

8.1.2 Hindsight Bias

A lot has been written about the concept of *hindsight bias*. At the risk of oversimplifying, hindsight bias means that after we know that an accident occurred and have some idea of why it occurred, it is psychologically extremely difficult, if not impossible, for people to understand how someone might not have predicted the events beforehand. After the fact, humans understand the causal connections, and everything seems obvious. We have great difficulty in placing ourselves in the minds of those involved who have not had the benefit of seeing the consequences of their actions (see figure 8.1).

Hindsight bias is usually found throughout accident reports. A glaring clue that hindsight bias is involved is when you see the words "he/she should have . . . ," "he/she could have . . . ," or "if only he/she would have. . . ."

Here is an example from a real accident report: after an accident involving the overflow of SO^2 (sulfur dioxide) from a tank in a chemical plant, the investigation report concluded that "The Board Operator *should have* noticed the rising fluid levels in the tank" [emphasis added]. Sounds bad, right? Let's examine that conclusion.

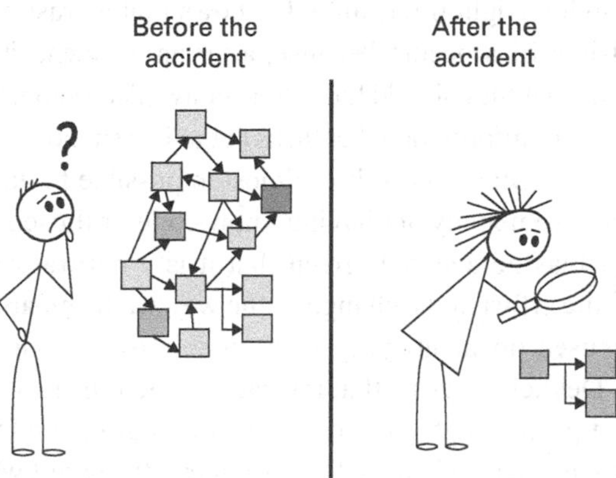

Figure 8.1
Hindsight bias makes things clear after the fact.

The operator had turned off the control valve, allowing fluid to flow into the tank, and a light came on saying it was closed. All the other clues that the operator had in the control room showed that the valve had closed, including the flow meter, which showed that no fluid was flowing into the tank. The high-level alarm in the tank did not sound because it had been broken for eighteen months and was never fixed. There was no indication in the report about whether the operators knew that the alarm was not operational. Another alarm that was supposed to detect the presence of SO^2 in the air also did not sound until later.

One alarm did sound, but the operators did not trust it as it had been going off spuriously about once a month and had never in the past signaled anything that was actually a problem. They thought the alarm resulted simply from the liquid in the tank tickling the sensor. While the operators could have used a special tool in the process control system to investigate fluid levels over time (and thus determine that they were rising), it would have required a special effort to go to a page in the automated system to use the tool. There was no reason to do so—it was not standard practice—and there were, at the time, no clues that there was a problem. At the same time, an alarm that was potentially very serious went off in another part of the plant, which the operators investigated instead. Despite all this evidence explaining why the operators could not have known what was happening, they were identified in the accident report as the primary cause of the SO^2 release because they did not stop it in time.

It is interesting that the report writers could not, even after careful study after the incident, explain why the valve did not close and why the flow meter showed no flow, in other words, why the tank was filling when it should not have been and why no clue was provided to the operators that this was happening. But the operators were expected to have known this without any visible clues at the time and with competing demands for their attention. This is a classic example of the investigators succumbing to hindsight bias. The report writers knew, after the fact, that SO^2 had been released and assumed that the operators should have somehow known too.

The words "should have" or their equivalent may not appear in the report, but hindsight bias may still be at work. As an example, one of the four probable causes cited in the accident report of the American Airlines B757 crash while approaching Cali, Columbia in 1995 was "Failure of the flight crew to discontinue the approach into Cali, despite numerous cues alerting them of the inadvisability of continuing the approach." In fact, the so-called "cues" are only cues in hindsight if one already knows that a crash resulted.

In summary, hindsight bias occurs because, after an accident, it is easy to see where people went wrong and what they should have done or avoided doing. It is also easy to judge others for missing a piece of information that turns out to be critical only after the causal connections are clarified by ensuing events. It is almost impossible to go back and understand how the world looked to somebody not having knowledge of those later events.

How can hindsight bias be avoided, given that it is a natural result of after-the-fact reasoning? It takes some effort and a change in the way we think about causality. Instead of spending time focused on identifying what people did wrong when analyzing the cause of an accident, Dekker suggests that instead we need to start from the premise that the operators were not purposely trying to cause a loss but instead were trying to do the right thing. Learning can occur when we focus on identifying not what people did wrong but *why it made sense to them at the time to do what they did* [76]. By answering this type of question, more useful ways to prevent such behavior in the future can be identified.

As an example, in the tank explosion described, the goal should not be to identify what the operators did wrong in hindsight. That is obvious, and it leads only to a way to assign blame to them for the loss. The question to ask instead is why it made sense to them to do what they did. Answering this question assists in identifying how to change the system to improve responses to the same or similar events in the future.

8.1.3 Misunderstanding the Role of Humans in Accidents

Another reason for lack of learning from events results from limited understanding of the role of human operators in accidents. As discussed in chapter 6, much research has been published that concludes that operator error is the cause of most accidents. The problem is that most of this research involves examining accident reports. Does the belief arise from the fact that operators actually are the primary cause of accidents or simply that they are usually blamed in the accident reports? Most likely, the latter is true. At best, such beliefs cannot be justified by simply looking at the conclusions of accident reports.

A type of self-fulfilling prophesy is at play here: An assumption is made at the beginning of an accident investigation that the operators must be the cause. That leads to a focus on what role the operators played in the events. Unsurprisingly, focusing primarily on what the operators did (and not *why* they acted the way they did) leads to them being identified as the cause.

The Bad Apple Theory The emphasis on human error as the cause of accidents is partly due to the Bad Apple Theory, which arose a hundred years ago and was thoroughly discredited scientifically about seventy years ago. The concept goes back to 1925 when both German and British psychologists were convinced that they could solve the safety problem by identifying and getting rid of the bad apples in an organization. They used statistical analysis over fifty years to determine that there was a cohort of accident-prone workers. These were people with personal characteristics that, they claimed, predisposed them to making errors and thus precipitating accidents. The data seemed to imply that a small percentage of people are responsible for a large percentage of accidents. If those people were removed, the system would become much safer.

The Bad Apple Theory existed until World War II, when the complexity of the systems we were creating and that humans had to work within started to increase dramatically. Human error in these systems was not so simple to explain in terms of the personal characteristics of the humans involved. The theory was finally put to rest scientifically in 1951 by two statisticians, Arbous and Kerrich [25]. Unfortunately, this theory is still believed by many today.

The theory is not true because of a major statistical flaw in the argument. For the accident-prone or Bad Apple Theory to work, the risk of error and accidents must be equal across every system. But, of course, it is not. Newer views of accident causation conclude that context and system design have more explanatory power for why accidents occur than personal characteristics:

> *When faced with a human error problem you may be tempted to ask "Why didn't they watch out better? How could they not have noticed?" You think you can solve your human error problem by telling people to be more careful, by reprimanding the miscreants, by issuing a new rule or procedure. They are all expressions of the "Bad Apple Theory" where you believe your system is basically safe if it were not for those few unreliable people in it. This old view of human error is increasingly outdated and will lead you nowhere.* [76]

The Bad Apple Theory is especially prevalent in medicine, but it permeates almost every industry in the form of arguments that most accidents are caused by human operators. If, for example, one posits that 5 percent of bad doctors cause most accidents, then simply identifying and getting rid of the doctors who get the most complaints and are involved in adverse events should drastically reduce medical error. Unfortunately, it does not. It may simply get rid of a group of doctors who do the really difficult, tricky work, such as some oncological cases with a negative prognosis, or those who work under the most difficult circumstances, perhaps with inadequate resources and support.

In fact, if there are system features that are likely to induce human mistakes under some circumstances, such as poor facilities with limited resources and financial and other stresses, then swapping the humans involved, but not changing the system features that are creating the erroneous behavior, will have little impact on the accident rate. We know now that the design of systems has the potential to create various classes of human errors or, conversely, that design can reduce them or eliminate them.

As an example, most of our systems today require humans to work with computers and other complex designs that can more easily induce an incorrect mental model on the part of the human about the state of the system or the state of the automation. Blaming the human, then, for erroneous behavior without looking at why that behavior occurred does nothing to reduce accidents due to flawed system design. Operators today are doing the wrong thing (in hindsight) in certain situations because of factors like inconsistent behavior of the automation, misunderstanding about how the automation is designed, confusion about the current mode of the system being controlled, and so on. Identifying these error-inducing design features during accident causal analysis, rather than simply stopping with the conclusion of operator error, will allow us to improve safety in both current systems and in our future designs. In addition, of course, we need to use hazard analysis during the original system design effort to identify system design aspects that cause human errors before losses occur, but that is the topic of later chapters.

Why focusing on human error is a problem Focusing on operators as the cause of accidents results in recommendations that do not reduce future accidents. The common solution to a finding of human error is to fire or retrain the operators involved. Sometimes, in addition, something may be done about operators in general. For example, their work may be more constrained by creating more rules and procedures—which may be impossible or unrealistic to expect them to always follow or which may themselves lead to an accident. Or the response may be to marginalize the operators by adding more automation. Adding more automation may introduce more types of errors by moving operators farther from the process they are controlling. Most important, by focusing on the operators, the accident investigation may ignore or downplay the systemic factors that led to the operators' behavior and the accident.

As just one simple example, many accident investigations find that operators had prior knowledge of similar previous occurrences of the events but never reported them in the incident reporting system. In many of these cases, the operators did report them to the engineers who they thought would fix the problem, but the operators did not use the official reporting system. A frequent conclusion of such an accident report then is to attribute the cause of the accident to the operators not using the reporting system. This conclusion then leads to a recommendation to make new rules to require that operators always use it and perhaps recommend providing additional training in its use.

In most of these cases, however, there is no investigation of *why* the operators did not use the official reporting system. Often their behavior results from the system being hard to use, including requiring the operators to find a seldom-used and hard-to-locate website with a clunky interface. Reporting events in this way may take a lot of time. When the operators never see any results or hear anything back, they assume the reports are going into a black hole. It is not surprising then that they instead report the problem to people who they think can and will do something about it. In the end, fixing the problems with the design of the reporting system will be much more effective than simply emphasizing to operators that they have to use it.

Human error defined as noncompliance with procedures Violating safety rules or procedures is commonly considered *prima facie* evidence that the operator caused the accident. The investigation rarely goes into why the rules were violated. There is usually a very good reason, often related to achieving other management goals or pressures such as productivity and efficiency.

Compliance with rules and procedures puts operators and workers into an untenable situation when they are faced with the "following procedures" dilemma. Chapter 6 explains that there are always at least two mental models of the actual system (see figure 6.4): the designer's model and the operator's model. Management may have a third model. The actual system state may differ from what was originally assumed by the designer and implied in the training provided to operators. The only way operators can check that their mental model of the system is currently correct is to use feedback and operational experience. Rasmussen [319] suggests that operators will test their own mental models of the actual system state by performing "experiments" or informal tests. They are continually testing their own models of the system behavior and current state against reality.

The procedures provided to the operators by the system designers may not apply when the system behaves differently than the operators and designers expected. For example, the operators at Three Mile Island recognized that the plant was not behaving the way they expected it to behave. They could either continue to follow the utility-provided procedures or strike out on their own. They chose to follow the procedures, which, after the fact, were found to be wrong. The operators received much of the blame for the incident because they followed those procedures.

In general, operators must choose between

1. sticking to procedures rigidly when cues suggest they should instead be adapted or modified; or
2. adapting or altering procedures in the face of unanticipated conditions.

The first choice, following the procedures they were trained to follow, may lead to unsafe outcomes if the trained procedures are wrong for the situation at hand. The operators will then be blamed for their inflexibility and for applying rules without understanding the current state of the system and conditions that may not have been anticipated by the procedure designers.

If, instead, the operators make the second choice to adapt or alter procedures, they may take actions that lead to accidents or incidents when they do not have complete knowledge of the current circumstances (system state) and what might happen if they violate the written procedures. They will then be blamed for deviations and rule violations.

Following rules and procedures does not necessarily guarantee safety. Instead, Dekker suggests that safety comes from people being skilled in judging when and how to apply them [76]. The traditional approach to safety contends that safety improvements come from organizations telling people to follow procedures, enforcing them, and assigning blame to people when accidents occur and there was a rule or procedure violation.

In contrast, a systems approach to safety assumes that safety improvements come from organizations *monitoring and understanding the gap* between written procedures and practice (behavior) and then updating procedures and rules accordingly. Asking why the operators are violating the specified procedures provides the information necessary to redesign the system to improve both safety and productivity. There usually is a good reason why procedures are not being followed: changing the system design can eliminate those reasons. Exhortations about following procedures or punishment for not following them will be much less effective.

After accidents or incidents where written procedures or rules are violated, the focus should be placed on determining why the operators felt the need to disregard the procedures or rules. Simply concluding that the operators were at fault because they did not follow the written rules provides no useful information in preventing related accidents in the future.

In general, the role of operators in modern systems is changing. Rather than directly controlling the system, humans are increasingly supervising automation, which is actually implementing most of the detailed control tasks. At the same time, software is allowing enormously complex systems to be created, which are stretching the ability of people to understand them and leading to human behavior that under some conditions could be unsafe. In addition, systems are sometimes designed without using good human-centered and human-factors design principles. The result is that we are designing systems in which operator error is inevitable and then blaming accidents on operator error rather than designer error.

The systems view of human error A systems view of human error starts from the assumption that all behavior is affected by the context (system) in which it occurs. Therefore, the best way to change human behavior is to change the system in which it occurs. That involves examining the design of the equipment that the operator is using, carefully analyzing the usefulness and appropriateness of the procedures that operators are given to follow, identifying any goal conflicts and undue production pressures, evaluating the impact of the safety culture in the organization on the behavior, and so on.

Human error is always a symptom, not a cause. An example is the previously noted accident report of the American Airlines Cali crash where the operators were blamed for being confused by the automation design and continuing to use it when the automated system demanded an excessive workload in a critical phase of flight. When accident reports state causes in this way, learning is inhibited and attention is directed to the wrong aspects of the causal scenario, in this case, deflecting from the fact that the automation was confusing and needs to be changed.

A systems approach starts from the premise that *human error is a symptom of a system that needs to be redesigned.* Accident analysis should identify the design flaws and recommend ways they can be fixed, not blame the operators for the consequences of those design flaws.

8.1.4 Focusing on Blame: Blame Is the Enemy of Safety

Focusing on blame seriously hinders what we learn from accidents. Blame is a legal or moral concept, not an engineering one. The goal of courts is to establish blame and liability. The goal of engineering, in contrast, is to understand *why* accidents occurred so they can be prevented, not to establish blame and decide who was responsible.

A focus on blame in accident analysis has many unfortunate results that reduce learning from accidents and impede preventing future ones. One result is that important information is often hidden: those involved resort to pointing fingers at everyone else and searching for someone else to blame. An example of this behavior was provided in chapter 4 for the Macondo/*Deepwater Horizon* accident. When such finger-pointing occurs, the search for causes usually devolves to identifying the immediate actors in the loss, usually the human operators or low-level managers who participated in the events and have no way to deflect attention onto others. The spotlight then is placed on the aspects of the loss that are least likely to provide important information about preventing future accidents.

The goal in accident analysis should not be to assign blame but to determine why well-meaning people acted in ways that contributed to the loss and to identify the impact of system design on the loss.

An exercise in understanding blame C. O. Miller and Gerry Bruggink [265] provide a useful exercise in distinguishing between an accusatory approach to causal analysis and an explanatory one. Consider the following conclusion from a real accident report.

Exercise Part 1. Description of the cause of the accident.

WHO

Accident Board **A** determined the *probable cause* of this accident was:

1. the flight crew's failure to use engine anti-icing during ground operations and takeoff
2. their decision to take off with snow/ice on the airfoil surfaces of the aircraft
3. the captain's failure to reject the takeoff during the early stage when his attention was called to anomalous engine instrument readings

WHY

Contributing factors:

1. the prolonged ground delay between de-icing and receipt of air traffic control (ATC) clearance during which the airplane was exposed to continual precipitation
2. the known inherent pitch-up characteristics of the B-737 aircraft when the leading edge is contaminated with even small amounts of snow or ice
3. the limited experience of the flight crew in jet transport winter operations

Questions:

Using this description of the cause of the accident (and before looking at the discussion that follows), where would you assign responsibility for this loss? What recommendations would you provide as a result? Are there additional questions you might ask during the investigation?

Discussion of your answer:
Did you assign responsibility to the flight crew? Why? What was the focus of the recommendations that you generated? Did they involve flight crew training and procedures? What other causal factors and recommendations did you generate? You might have mentioned something about the ground delay and avoiding that, but most likely all or most of your causes and recommendations involved the flight crew. That would not be surprising as their actions are identified in the analysis as the probable cause of the accident.

Exercise Part 2: Now consider a second description of the cause for the same accident.

WHAT

Based on the available evidence, Accident Board **B** concluded that a thrust deficiency in both engines, in combination with contaminated wings, critically reduced the aircraft's takeoff performance, resulting in a collision with obstacles in the flight path shortly after liftoff.

WHY

Reasons for the thrust deficiency:

1. Engine anti-icing was not used during takeoff and was not required to be used based on the criteria for "wet snow" in the aircraft's operations manual.
2. The engine inlet probes became clogged with ice, resulting in false high-thrust readings.
3. One crew member became aware of anomalies in cockpit indications but did not associate these with engine inlet probe icing.
4. Despite previous incidents involving false thrust readings during winter operations, the regulator and the industry had not effectively addressed the consequences of blocked engine inlet probes.

Reasons for the wing contamination:

1. deicing/anti-icing procedures
2. the crew's use of techniques that were contrary to flight manual guidance and therefore aggravated the contamination of the wings
3. ATC procedures that resulted in a 49-minute delay between departure from the gate and takeoff clearance

Discussion:
Now, to whom or what would you assign responsibility for this accident? Was your list larger and different than the list you generated in part 1 of the exercise? Are there additional questions that you might like to investigate? Do you now think the cause was more than just "flight crew failures"? Did you identify other factors? What recommendations would you create from this analysis? How do the recommendations you generated differ from those for Exercise Part 1? Why? Did you generate design recommendations or recommendations involving groups or people other than the flight crew?

The first description of the accident (in part 1) was *accusatory*. It focuses on the operators and their role in the loss; that is, on who was responsible. The second description (in part 2) was *explanatory*. It does not focus on *who* but instead on *what* and *why*. Very different recommendations will result from each of these, although they are describing the same accident. Learning is enhanced by using an explanatory approach rather than an accusatory one; that is, putting the focus on what and why and not on who (figure 8.2).

Another thing to notice in these two exercises is the use of the word "failure" in the first but not in the second. Failure is a pejorative word; that is, using it implies judgement and assignment of blame. Consider the following two statements, differing only in the use of the word "failure":

1. The captain failed to reject the takeoff during the early stage when his attention was called to anomalous engine instrument readings.
2. The captain did not reject the takeoff during the early stage when his attention was diverted to anomalous engine instrument readings.

The first suggests a conclusion that the captain did something wrong. Further exploration is not encouraged because it is stated as a conclusion. A judgment has been made about the captain's actions. A cause has been determined, and that cause was a failure on the part of the captain. The second is a simple statement of what the captain did without any judgment in it. It encourages further exploration about why the captain's attention might have been drawn to the anomalous engine instrument readings and, even more important, why anomalous readings were produced.

Is this really a problem in accident reports? In fact, the conclusion of "failure" pervades most accident reports. Here is an example of the conclusions from an aircraft CFIT (Controlled Flight into Terrain) accident at Birmingham Shuttlesworth International Airport [278], which is typical. The NTSB concluded that t*he probable cause of this accident was the* flight crew's *continuation of an unstabilized approach and their* failure *to monitor the aircraft's altitude during that approach, which led to an inadvertent descent below the minimum approach altitude and subsequently into terrain.*

The report also concludes that contributing to the accident were [emphasis added]:

(1) the *flight crew's failure* to properly configure and verify the flight management computer for the profile approach
(2) the *captain's failure* to communicate his intentions to the first officer once it became apparent the vertical profile was not captured
(3) the *flight crew's expectation* that they would break out of the clouds at 1,000 ft above ground level due to incomplete weather information
(4) the *first officer's failure* to make the required minimums callouts

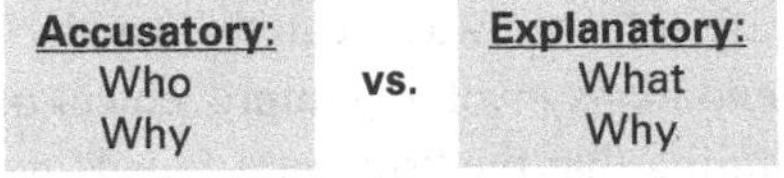

Accusatory: Who Why	vs.	**Explanatory:** What Why

Figure 8.2
Two opposing views of accident explanation.

(5) the *captain's performance deficiencies* likely due to factors including, but not limited to, fatigue, distraction, or confusion, consistent with performance deficiencies exhibited during training
(6) the *first officer's fatigue* due to acute sleep loss resulting from her ineffective off-duty time management and circadian factors

Notice that the conclusions about the probable cause and contributing factors for this accident identify only flight crew behavior and the events that reflect flight crew "failures." *Why* the flight crew behaved the way they did is not included (except for fatigue, which does not fully explain it). One contributory factor does mention a reason; that is, (3) the flight crew's expectation that they would break out of the clouds at 1,000 ft above ground level due to incomplete weather information. But the emphasis here is on the flight crew's behavior and expectations and not why incomplete weather information was provided to the crew. The system design flaws that are related to providing weather information and that need to be fixed are not mentioned in the causal summary contained in the accident report.

A more complete accident analysis would not only look at what the pilots did that contributed to the accident but, more important, *why* they believed it was the right thing to do at that time. The official conclusions of the Birmingham accident report omit most of the information that would be useful in preventing such accidents in the future. We don't learn why the flight crew behaved the way they did because the explanation stops with the conclusion that they "failed." What recommendations might result from the conclusions provided above?

In addition, the entire system for preventing CFIT, which has been a focus of accident prevention efforts for many years, needs to be examined in the accident report—not just the pilot's behavior. Some of the many contributors to the Birmingham accident were the design of the automation; landing on a runway that did not have an instrument landing system (which was not operating due to scheduled maintenance); lack of an air traffic control alert to indicate that the aircraft was too low; maintenance practices at the airport; the ground proximity warning system not providing an alert until too late, which was due in part to the airline not upgrading the warning system with the new software that had been provided; and a dozen other factors that are more likely to lead to useful recommendations than simply identifying the "failures" of the flight crew [246].

The use of the word "failure" pervades the causal description of most accident reports. It is erroneously applied not only to humans but to software, operators, and management decision-making. Failure is well defined for hardware but not for all the other things that people apply it to. By overusing the word "failure" for almost everything, useful information about the actual behavior and its causes are omitted and lost.

Software does not fail; it simply executes the logic that was written. There needs to be an examination of why unsafe software was created—usually it can be traced to requirements flaws—and recommendations involving improvement of the process that produced the unsafe software. Concluding that the "software failed" not only makes no technical sense, but it provides no useful information. Did the software not provide any outputs? Did it provide incorrect outputs? Did it provide correct outputs but too late?

Humans also do not fail (unless their heart stops). They simply react to the situations in which they find themselves. In hindsight, what they did may turn out to have been the wrong thing to do. But why it seemed to them to be the right thing at the time needs to be examined in order to make useful recommendations. Again, simply recounting what people did wrong provides no useful information beyond creating a convenient scapegoat for the accident. Understanding why they behaved the way they did will focus the accident analysis in a useful direction.

Finally, companies do not fail unless they go out of business. A typical example found in accident reports is a statement that "Company X failed to learn from prior events." Companies are pieces of paper describing legal entities. Documents do not learn or, for that matter, fail. The company may be made up of hundreds, thousands, or even tens of thousands of employees. It would be more useful to determine why learning by the appropriate company employees did not take place and to ask questions such as: Is there a safety information system that captures prior adverse events so that learning can occur? If so, were the prior events recorded in the safety information system? If not, then why not? Were they just missed accidentally, was there no process to include them, or was the process not followed? If the latter, why was the process not followed? If the previous events *were* recorded, then why were they not used? Were they hard to find, or was enough information not included to prevent repetition? Were there procedures in place to require using recorded information? And so on.

Answering these questions will provide information useful for creating recommendations that will improve future learning from events and prevent repetition of the losses. Simply concluding that the company failed to learn from events only assigns blame to an abstract entity "company" without providing the information necessary to improve the situation in the future.

Eliminating the use of "failure" will greatly enhance learning from accidents. By not short-circuiting the search for understanding through concluding that there was a failure, we could immeasurably improve accident investigation and learning and, hopefully, safety.

More generally, to increase learning from losses, we need to eliminate the focus on assigning blame in accident reports. This may be hard to do as there seems to be a human need to find someone responsible for a tragedy. But a choice has to be made between personal satisfaction in identifying a villain or seizing the opportunity to prevent more such accidents in the future. Engineers should leave blame to the legal system. And management should not be allowed to use the official accident investigations to shift liability away from themselves and onto someone else. A report that simply describes what happened, including all the contributions to the events and why, in a blame-free manner will shift the effort to assign or avoid liability to the legal process where it belongs.

8.2 Goals for Improved Accident Analysis

Using the information in the last section about why we do not learn enough from accident investigation and analysis, we can generate goals for improvement. Identifying what is labeled a "root cause," biasing our findings by hindsight, oversimplifying human error, and focusing on blame unnecessarily limits what is learned. Too many factors are missed and never fixed. Blame also limits the scope and effectiveness of the investigation because, after a loss, everyone will try to point at someone else as the cause.

Instead, the goal of the investigation should be to learn as much as possible from a loss in order to prevent future ones—not to find someone or even something to blame. Many of the people on whom blame is assigned, usually low-level operators, have simply been caught up in a system where a loss was inevitable, and that person was unlucky enough to be the one that triggered or was involved in the events for an accident that was "waiting to happen."

An accident analysis technique that maximizes learning from accidents and incidents has the following characteristics:

1. It includes all causes; that is, does not focus on a few so-called "root" or "probable" causes.
2. It reduces hindsight bias.
3. It takes a systems view of human behavior: What role did the overall system design play in the events and the human behavior? Why were the hazards not effectively prevented or controlled?
4. It provides a blame-free explanation of why the loss occurred: focuses on *what* and *why* rather than *who*. Why did people behave the way they did?

Accident investigation necessarily starts with identifying the events that occurred and the role played by individuals in those events. Unfortunately, most accident analysis methods that are used or have been proposed stop there and focus on the events and not an explanation of why the events occurred. To learn enough to be able to prevent future accidents, we need to know not just what events occurred and what bad decisions were made, but why they were made; that is, the systemic factors that shaped and influenced the behavior of all the system components and the system as a whole. Understanding the "why" should be the goal of the investigation. Once that is understood, creating recommendations to prevent future accidents is usually straightforward.

8.3 Example: The Zeebrugge Ferry Accident

To help understand how the goals outlined in the previous section can be achieved, an example analysis is provided here for a real accident. Note that all of the information used was in the original accident report. The only difference is in how the information was interpreted and used to explain the cause of the accident. Any missing information might have easily been acquired by asking a few additional questions that arose from trying to understand why the events occurred. The problem usually is not that the information is unknown, but that it is not used to provide a full causal analysis. Collecting information after an accident is not covered in this book; the focus is instead on how the questions to be asked are generated and how the information is analyzed once it has been collected.

The example starts, as suggested in the previous section, by examining the events and how individuals contributed to those events. It then goes on to (1) explain why the individuals acted the way they did, including flawed mental models and contextual factors, and (2) identify the systemic factors that created the flawed mental models and the behavior-influencing contextual factors. The process used in the example is an informal version of a real accident analysis method called CAST (Causal Analysis based on System Theory) [217; 222].

In the process of explaining why the individuals behaved the way they did, questions will be raised that need to be answered to fully understand the accident. These questions should be recorded and used to inform the rest of the investigation and causal analysis. At the end of the investigation, all questions should have been answered. Some examples of questions are included below, but all potential questions are not included for space reasons. As you read the causal analysis, try identifying additional questions that you think are important in understanding the events.

The following analysis is not totally complete, but it is complete enough to be instructive. Let's start with the basic events involved. A car ferry named the *Herald of Free Enterprise* in 1987 was working the route between the Belgium town of Zeebrugge and Dover in the United Kingdom. When it left port, the bow doors were not closed, water entered the ship, and it capsized; 193 passengers and crew died.

There is always a set of events that can be strung together in a way that makes them appear to have caused the accident. In this case, the assistant bosun overslept and did not close the car deck doors before the ferry left the dock in Zeebrugge. The first mate assumed that the assistant bosun would close the doors, did not check that he had, and returned to the bridge to assist the captain. The chief officer and master did not ask whether the doors had been closed and left port for the trip to Dover. A bow wave filled the car deck with water, and the ferry capsized. Simple, right?

These events can be illustrated by the linear chain of events shown in figure 8.3. Usually, such chains focus on human operators or physical events (failures) because they are easy to identify in hindsight. Indeed, many of the individuals involved in the events were prosecuted for manslaughter, as was the operating company. The official Court of Inquiry into the accident identified three main causal factors:

1. the assistant bosun's failure to close the bow doors
2. the chief officer's failure to make sure the bow doors were closed
3. the captain leaving port without knowing whether the bow doors were closed

Note the description of the human behavior as "failures."

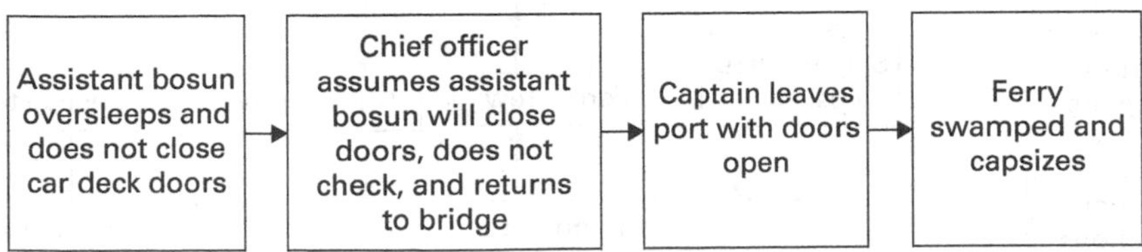

Figure 8.3
Chain of events for the *Herald of Free Enterprise* accident.

The court concluded that the three men exhibited serious negligence in the discharge of their duties and suspended the licenses of the chief officer and captain. While punishing these operators and putting them at the center of blame for the accident, the official accident report did note contributory factors such as "poor workplace communication," a stand-off relationship between the ship operators and shore-based managers," and a "disease of sloppiness at every level of the corporation's hierarchy."

If only the linear chain of events is considered, the three men directly involved do seem to be primarily responsible for the loss. If true, it appears justified to take their licenses away in order to prevent such behavior by them in the future. The problem is that the chain of events does not include important information about *why* they behaved the way they did. Without that information, individuals can be identified to punish for the accident, but information about how to prevent such behavior by others in the future is not provided, nor is information about what systemic causal factors need to be changed both within this company and in companies generally to prevent the observed behavior. The bottom line is that learning from the loss is minimal.

Simply identifying more components to blame, such as the shipping company or its managers, does not help but merely spreads the blame to more individuals. Instead, determining why all the components of the system behaved the way they did allows learning how to change the system design and how to design future systems to avoid such losses. In almost all accidents, when this information is elicited, the causal factors are usually widespread and not limited to the particular company or even the industry. Figure 8.4 shows the basic control structure (see figure 3.9 and section 10.3) involved in the accident.

Let's start by examining the ship and port design more fully; that is, the physical system involved in the accident. The Dover–Calais crossing of the Channel is the shortest route between England and France, and in 1987, it was the quickest route. To remain competitive with other ferry operators on the route, Townsend Thoresen (the owner and operator of the *Herald of Free Enterprise*) needed ships that were designed to allow fast loading and unloading and quick acceleration.

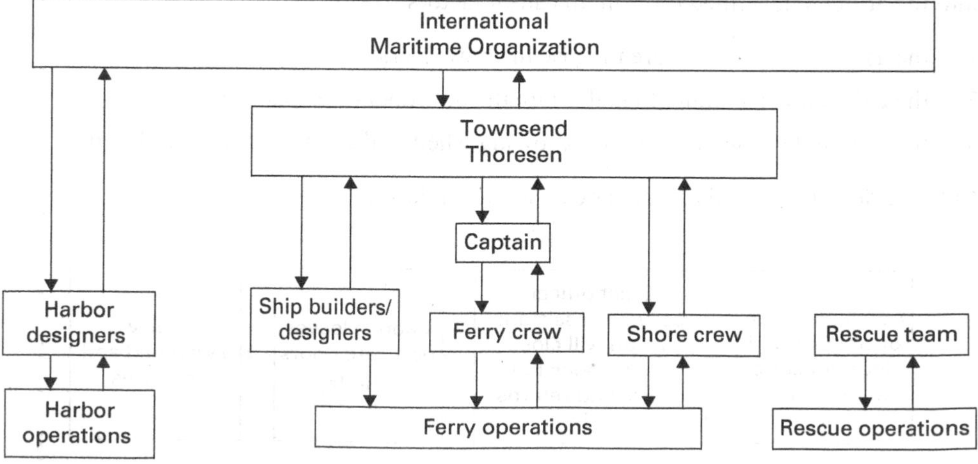

Figure 8.4
The basic control structure involved in the loss.

The ships had eight decks, named A to G from top to bottom:

A deck: crew accommodation and radio room

B deck: crew accommodation and galley

C deck: passenger areas and galley

D deck: suspended vehicle deck within E deck

E deck: upper vehicle deck

F deck: mezzanine level

G deck: main vehicle deck

H deck: engine rooms, stores and passenger accommodation

The loading of vehicles onto G deck was through watertight doors at the bow and stern. Both sets of doors were hinged about a vertical axis, which meant that the status of the bow doors could not be seen from the wheelhouse (the enclosed area where the ship is piloted by the captain and crew). The loading of vehicles onto E deck and F deck was through a weathertight door at the bow and an open portal at the stern. Vehicles could be loaded and unloaded onto E and G deck simultaneously using the double deck linkspans (car loading ramps) in use at Dover and Calais.

Causal factors related to the ship design and the rationale behind the design On the day the ferry capsized, the *Herald of Free Enterprise* was working the route between Dover and the Belgium port of Bruges-Zeebrugge. This was not her normal route. Note that there was a change here. As emphasized repeatedly in this book, most accidents follow a change in the system design, operations, or the environment in which the system is operating.

The linkspan at Zeebrugge had been designed for loading onto the bulkhead deck of single deck ferries. The *Herald* had two car decks. The linkspan could be raised or lowered, but decks E and G on the *Herald* could not be loaded simultaneously.

In addition, the ramp could not be raised high enough to meet the level of deck E due to the high spring tides being encountered at that time. In order to load the upper deck of the *Herald* at Zeebrugge, the forward ballast tanks were filled to lower the ferry's bow in the water. The ship's natural trim was not restored immediately after loading as the process took a while and the ferry would not have been able to meet its tight schedule. The *Herald* was due to be modified during its refit later in 1987 to overcome this problem. As a result of these conditions, not only were the bow doors open but they were also 3 ft lower in the water than during usual passage.

The *Herald* was to make the voyage in the C1 condition. In that condition, the ship was capable of accepting damage to any one compartment in the ship without either losing stability or submerging the margin line at any time during flooding. (The margin line is an imaginary line 76 mm below the bulkhead deck.) The ship's displacement on sailing consisted of her basic design together with the sum of all consumables on board (fuel oil, diesel oil, fresh water and stores, and so on), the weight of her crew and their effects, and the weight of the passengers, cars, luggage, commercial vehicles and coaches. As will be seen, the *Herald* was significantly overloaded at sailing. The investigation found that such overloading of the *Herald* was a regular occurrence.

This type of ferry was designed for rapid acceleration. It was built with very powerful engines in order to make the crossing at high speed. The intention was to unload passengers and vehicles rapidly and then, without any delay, load passengers and vehicles for the return voyage.

Another physical factor that contributed to the capsizing was the *squat effect*. When a vessel is under way, the movement under it creates low pressure, which has the effect of increasing the vessel's draft. Informally, the draft on a ship is the depth of the ship below the surface of the water. Draft determines the minimum depth of water a ship or boat can safely navigate. The more heavily a vessel is loaded, the deeper it sinks into the water, and the greater is its draft.

In deep water, the squat effect is small, but in shallow water it is greater because, as the water passes underneath, it moves faster and causes the draft to increase. This reduced the *Herald*'s clearance between the bow doors and water line to between 1.5 m (4.9 ft) and 1.9 m (6.2 ft). After extensive tests, the investigators found that when the ship traveled at a speed of 18 knots (33 km/h), the wave was enough to engulf the bow doors. The combination of acceleration and squat led to an increase in the bow wave and caused water to flow into G deck.

The accident report cited other ship design factors related to the accident such as the absence of an indicator in the bridge that the bow door was open, the top-heavy design of the vessel, and a lack of bulkheads to restrict water ingress.

Life-saving equipment The capsizing happened very rapidly and thus precluded the deployment and use of any of the life-saving equipment except the lifejackets. The lifejackets were stowed principally in lockers adjacent to the muster stations on C Deck, with a port and starboard locker at each station. The lockers were locked. The keys were in small glass crash boxes adjacent to the doors. When the ship capsized, the doors on the starboard side would have opened downward, and those on the port side were submerged. All survivors reported masses of lifejackets floating in the water. It is not clear whether the doors on the starboard side burst open or were opened by unknown persons. The latter would have been extremely difficult physically. It is unlikely that the jackets on the port side were released. They were probably forced out by their own buoyancy.

The mass of floating jackets impeded some swimmers and prevented others from floating to the surface. People having access to these lifejackets complained of difficulty in donning the jackets, untangling the tapes from other jackets, and then discovering how to manipulate them. As hypothermia set in, fingers became too numb to tie the tapes. The standard lifejacket is intended to be donned under supervision in an orderly manner while waiting embarkation into lifeboats. Note that the design of the life jackets is not in the set of events, and neither are most of the physical factors noted so far. But there is much more.

None of the lifeboats was launched. Those on the starboard side, however, were the source of much useful equipment that was used by the crew in the rescue operation.

Zeebrugge port design Beyond the physical design of the ship, the design of the port at Zeebrugge had an influence. The berth design at Calais and Dover differed significantly from that at Zeebrugge, as described previously.

Analysis of the role played by human controllers At this point in the analysis, let's switch from examining the physical system to examining individual roles in the loss. For each individual, the following are identified:

1. their responsibility as related to this accident
2. the role they played in the loss
3. *why* they behaved the way they did
   a. flaws in their mental models or beliefs that contributed to their behavior
   b. contextual factors that affected their behavior

Notice as you read the analysis below that if you stop after reading 1 and 2 (the individual's responsibilities and the role they played in the loss), the individual appears very much a guilty party. After reading 3 (why they behaved the way they did), the picture starts to look very different. In fact, at the end of the entire analysis, it usually seems like nobody was to blame—that each had a good reason for acting the way they did. That will be the result in most thorough accident causal analyses. The cause of the accident is in the overall system design and the features of the system in which each individual had to work and make decisions at the time of the loss. The solution to preventing accidents is to change this system design. Otherwise, other people in the future are likely to behave in the same unsafe ways.

Most people are trying their best to avoid accidents, and there is no reason to believe that was not the case here. In fact, those directly involved in the events were cited in the accident report as being instrumental in saving lives after the ship capsized, risking their own lives to do so. There needs to be some reason, therefore, why they made what turned out to be dangerous decisions or acted in what, in hindsight, appears to be a dangerous and irresponsible manner. The contextual factors will provide clues as to what systemic factors need to be considered. As more is learned about what happened before the loss, questions will be raised that should be answered during the investigation. A few examples of such questions are included below.

The system components considered are the ship designers, the harbor designers, the assistant boatswain, the boatswain, the first officer, the second officer, the master (captain), the senior master, shore management (management of Townsend Thoresen, the owner of the ferry), the board of directors of Townsend Thoresen, the government regulators, and those involved in rescue operations.

Ship Designers

Responsibility: Design seaworthy, safe vessels.

Role in loss: Designed a vessel with:

- no mechanism for preventing anyone from reopening the doors until an order had been given to do so
- no indicator light in the wheelhouse about the state of the bow doors
- no high-capacity ballast pump
- no draft indicator in the wheelhouse
- no communication channel for the captain to communicate with the deck crew
- no bulkheads to restrict water ingress as well as a ship that was top-heavy

Why?

Beliefs contributing to the behavior: The designers did not believe they were designing an unsafe ship.

Contextual factors:

- The ship designers wanted to satisfy their customers. The customer in this case, Townsend Thoreson Ferries, wanted to save money and did not think the omitted features were necessary for their ships. They needed ships that were designed to permit fast loading and unloading and quick acceleration in order to remain competitive in the ferry business.
- The designers followed the minimum ship design requirements at the time.

Harbor Designers

The car loading ramp (linkspan) at Zeebrugge was designed for loading onto the bulkhead deck of single deck ferries. To load the upper deck of a multi-deck ferry like the *Herald*, the ferry needed to trim ballast tanks (put water in the tanks) so that the ferry sat lower in the water. The accident report recommended changes to berth design so that ships can close their doors before leaving the dock. But there was no explanation provided for the harbor design decisions. Several can be hypothesized. Asking this question at the time of the accident would have been useful.

Assistant Boatswain (Mark Stanley)

What happened: Mark Stanley had opened the bow doors on the *Herald*'s arrival at Zeebrugge. He was supervising members of the crew in maintenance and cleaning the ship until he was released from work by the boatswain. He then went to his cabin, where he fell asleep and was not awakened by the call "harbor stations," which was given over the loudspeakers. He remained asleep on his bunk until he was thrown out of it when the *Herald of Free Enterprise* began to capsize.

Responsibilities related to accident: Close doors before the moorings are dropped.

Role in loss: Did not close the bow doors.

Why?

Beliefs contributing to the behavior:

- He assumed he would wake up in time after the call to stations to be able to close the doors before leaving the berth.

Contextual factors:

- He had returned to his cabin for a short break after cleaning the car deck upon arrival. He had been released from work at that time by the boatswain (Terence Ayling).
- He was tired from his shift, overslept, and did not hear the call to stations. He remained asleep on his bunk until the *Herald* began to capsize. This trip was the return leg of a second round trip during a 24-hour shift. Many of the crew members were tired.

  Question raised: *Were the crew rest periods adequate?*
- The "harbor stations" call is the alert to close the doors and for all crew members to report to their stations. Frequently the order to report to harbor stations was given before loading was complete. The order was given as soon as the

loading officer decided that by the time the crew arrived at their stations, everything would be ready for the ship to proceed to sea.

Question raised: *Why was the absence of Mark Stanley from his harbor station not noticed?*

- This sailing was not the first time in which the closing of bow doors was not performed. In October 1983, the assistant bosun of the *Pride* had fallen asleep and had not heard "harbor stations" being called. The result was that he neglected to close both the bow and stern doors on the sailing of the vessel from Dover. The *Pride* made it safely to her destination with no problems. It was therefore generally believed that leaving the bow doors open alone should not cause the ship to capsize.

Potential recommendations: At this early point in the analysis, it appears that, at the least, recommendations about clarifying roles are needed in the instructions about closing the bow doors, that backup assignments are needed, that some way of determining that the bow doors have actually been closed is needed before leaving port, and that an investigation about fatigue and recommended work and rest periods is justified.

Boatswain (Terence Ayling)

What happened: The bosun, Terence Ayling, told the investigation that he thought he was the last man to leave G deck, where he had been working in the vicinity of the bow doors and that, so far as he knew, there was no one there to close the doors. He had put the chain across after the last car was loaded. There was no reason why the bow doors could not have been closed as soon as the chain was in position.

Responsibilities related to accident: It was not his assigned responsibility to close the bow doors.

Role in loss: He was the last person on G deck before the *Herald* left the dock, but he did not close the doors or make sure someone did.

Why?

Beliefs contributing to the behavior: It was not his assigned duty to close or check on the doors.

Contextual factors:

- When asked during the investigation whether there was any reason why he did not shut the doors, he replied "It has never been part of my duties to close the doors or make sure anybody is there to close the doors." He also said: "At that stage it was 'Harbor Stations' so everybody was going to their stations."
- A general instruction had been issued in July 1984 stating that it was the duty of the officer loading the main vehicle deck (G deck) to ensure that the bow doors were "secure when leaving port." The officer loading G deck in this case was the first officer, Leslie Sabel, not the boatswain.

  Questions raised: How were duties assigned? Who did this? How were they enforced?

First Officer (Leslie Sabel)

What happened: He returned to the wheelhouse after "harbor stations" was called but before ensuring that the doors were closed. The wheelhouse (bridge) was his harbor station.

Responsibilities related to accident: Stay on deck to make sure the doors are closed and the ferry is ready to sail.

Role in loss: He returned to the wheelhouse before checking if the doors were closed. The official accident report concluded that he "failed to check the doors were closed upon departure." The report labeled this as negligence and recommended that his license be suspended. Again, note the use of the word "failed."

Why?

Beliefs contributing to the behavior:

- He assumed the assistant boatswain would close the doors.
- He thought he saw Mark Stanley (the assistant boatswain) approaching.
- When he got to the bridge, he thought the bow doors were shut and the *Herald* was ready to sail.

Contextual factors:

- The call to harbor stations had occurred. His harbor station was on the bridge.
- He knew the captain wanted to stay on schedule and was under pressure to do so.
- He left G deck with the expectation that the assistant boatswain would arrive soon.
- He said he saw a man in orange overalls who he assumed was the assistant boatswain while doing his rounds and assumed the assistant boatswain was on deck.
- A crucial question is: Why did he not remain on G deck until the doors were closed before going to his harbor station on the bridge? That operation could be completed in less than 3 minutes. But the officers always felt under pressure to leave the berth immediately after the completion of loading. The usual practice was for the officer on the car deck to call the bridge and tell the quartermaster to give the order "harbor stations." In fact, frequently, the order "harbor stations" was given before loading was complete. The order was given as soon as the loading officer decided that by the time the crew arrived at their stations everything would be ready for the ship to proceed to sea. Captain Lewry testified that, on the Zeebrugge run, it would have been necessary to delay the order "harbor stations" until the bow doors had been closed if the chief officer was required to remain on G deck until this had been done.
- Management rules specified that the first officer must be on bridge 15 minutes before sailing time and remain on bridge thereafter. Another captain had complained previously to shore management that it was impossible for a person responsible for loading the ship (the first officer) to be on the bridge 15 minutes before sailing time. This complaint was ignored by management.
- The *Herald* sailed 5 minutes late. This may have contributed to the chief officer's decision to leave G deck before the arrival of the assistant boatswain, which he anticipated would happen shortly. The ship's captains and officers were under pressure to sail at the earliest possible moment. An earlier shore management internal company memorandum read:

  There seems to be a general tendency of satisfaction if the ship has sailed two or three minutes early. Where, a full load is present, then every effort has to be made to sail the ship 15 minutes earlier. . . . I expect to read from now onwards,

. . . , that the ship left 15 minutes early. . . . put pressure on the first officer if you don't think he is moving fast enough. Have your load ready when the vessel is in and marshall your staff and machines to work efficiently. Let's put the record straight, sailing late out of Zeebrugge isn't on. It's 15 minutes early for us.

- There were three crews and five sets of officers that manned the *Herald* during different crossings. There was no uniformity in the duties of each set of officers and of the members of the crew. The Zeebrugge run carried only two deck officers in addition to the captain (master) versus three on the Calais run. The officers did not always have the same crew.
- A few years earlier, one of the *Herald*'s sister ships sailed from Dover to Zeebrugge with the bow doors open, but she made it to the destination without incident. It was therefore believed that leaving the bow doors open should not alone have caused the ship to capsize.
- A general instruction had been issued in July 1984 stating that it was the duty of the officer loading the main vehicle deck (G deck) to ensure that the bow doors were "secure when leaving port." That instruction had been regularly flouted. It was interpreted as meaning that it was the duty of the loading officer merely to see that someone was at the controls and ready to close the doors. While that was not the intention of the instruction, the instruction is not clearly worded. Whatever its precise meaning, it was not enforced.

Questions raised: How were duties assigned? Who did this? How were they enforced? Lots of questions arise about pressures to be on time and shore management lack of response to feedback from the operational staff. What information was collected and stored in the Safety Information System? Was there one?

Second Officer (Paul Morter)

What happened: The second officer went to G deck during loading to relieve the chief officer. Despite his arrival, the chief officer remained on G deck for a time. In due course, the chief officer left Mr. Mortar in charge of loading. There was then confusion about who was in charge. As is common after a serious accident, the testimony was conflicting as the participants tried to minimize their role in the loss.

Why?

Beliefs contributing to the behavior:

He assumed that the first officer was responsible and in charge at the time.

Contextual factors:

Mr. Morter testified that cargo duties were shared between himself and the chief officer and were not set down precisely. About 10 or 15 minutes before the ship was due to sail, the chief officer, who had overheard difficulties between Mr. Morter and the shore staff, returned. He testified that he suggested that the second officer should go aft and stand by for harbor stations while he completed the loading. That statement did not accord with the statement by Mr. Morter. Mr. Morter testified that he did not expect the chief officer to return before departure. When there were still twenty to twenty-five cars to load, Mr. Morter overheard on his radio the chief officer giving orders. The two officers did not meet face to face. Mr. Morter assumed that once the chief officer had arrived and started issuing orders that he (Mr. Morter) was no longer responsible for exercising the responsibilities of loading officer.

Master/Captain (David Lewry)

What happened: He left port with the bow doors open, ballast in the tanks that caused the bow to lie low in the water (not enough draft) and accelerated, causing a bow wave that swamped the car deck.

Responsibilities related to accident:

- Add ballast to allow upper car deck to be loaded.
- Remove ballast before the ship gets into an unsafe state.
- Do not sail until ship is in a safe state.

Role in Loss:

- He left the dock before the doors were closed and the ballast was removed.
- He accelerated after leaving the dock.
- He did not ascertain the draft of the ship before leaving port. This fact is important because a legal restriction in number of passengers was required if the draft exceeded 5.5 m. This rule was even more important at Zeebrugge because of the necessity to add ballast (lower the bow of the ship) in order to load vehicles onto E deck. Fictitious figures were regularly entered in the Official Log that did not take account of the trimming water ballast.
- He regularly allowed the ship to be overloaded.

The report accused him of negligence because he failed to check the doors were closed upon departure and recommended his license be suspended. With just the information provided above, this result seems justified.

Why?

Beliefs contributing to the behavior:

He thought the doors were closed.

Contextual factors:

- This was a competitive route, and he was under pressure to stay on schedule.
- It was not possible to see whether the doors were closed from the wheelhouse.
- There was no indicator in the wheelhouse of whether the doors were closed. The only way to determine if the doors were open or not would be a visual check of the doors themselves.
- There are many ways that leaving the dock with the doors open could have been prevented, including mechanical interlocks that gave a visual signal that the bow doors were shut in the control room, interlocks that prevented the engines from engaging if the doors were open, and TV cameras that showed the bow door's status. Most of these protective devices had been recommended, but shore management did not think such interlocks and safety devices were necessary.
- The absence of a communication channel with deck crew meant that the captain had to make assumptions about the status of the bow door. This deficiency had previously been raised to shore management as an issue by a captain of a similar vessel, which had also gone to sea with bow doors open but the report was ignored by the shore-based managers of the operating company, Townsend Thoresen.

- In lieu of any of these protection devices, the captain relied on the first officer's report that all was well. There was a standing order: "Heads of Departments are to report to the Master immediately they are aware of any deficiency which is likely to cause their departments to be unready for sea in any respect at the due sailing time." It appears that that is not the way in which the order was interpreted by deck officers. Masters came to rely on the absence of any report at the time of sailing as satisfactory evidence that their ship was ready for sea in all respects. And, of course, the first officer thought the doors were closed so had no reason to report that the ship was not ready to depart. In addition, the standing orders issued by the company made no mention about the opening and closing of the bow and stern doors.
- On the day of the accident, Captain Lewry saw the chief officer come to the bridge. Captain Lewry did not ask him if the ship was all secure, and the chief officer did not make a report. The captain followed the system that was used by all masters of the *Herald* and was approved by the senior master, Captain Kirby.
- Before this disaster, there had been at least five occasions when one of the company's ships had proceeded to sea with the bow or stern doors open. Some of those incidents were known to the management, who had not drawn them to the attention of the other masters. Captain Lewry told the court that if he had been made aware of any of those incidents, he would have instituted a new system under which he would have required a report that the doors were closed. But even those masters who were aware of the occasions when ships proceeded to sea with bow or stern doors open did not change their orders.
- Ship's officers always felt under great pressure from management to be on time and to leave the berth immediately after loading was completed. They were already 15 minutes behind schedule, and the captain was trying to make up time. That also partly explains why he accelerated after leaving the dock.
- It is difficult for the master to ascertain the draft on the ship so it was usually "guesstimated." A senior master had suggested that automated draft recorders be installed with readout in the wheelhouse, but such requests were ignored by shore management.
- It was believed by nearly everyone that leaving the bow doors open should not alone have caused the ship to capsize.

Senior Master (Captain Kirby)

The ship had multiple masters and officers. Captain Kirby was the senior master.

Responsibilities: His responsibility was to act as coordinator between all masters and officers of the ship in order to achieve uniformity in the practices operated by the different crews. This responsibility was important because there were three different crews serving with five different sets of officers for this ship. Also, frequent changes occurred among the officers. When the *Herald* started the Zeebrugge route, deck officers were reduced from fifteen to ten, and the extra five were distributed around the fleet along with those from ships that were being refitted.

Role in loss: Did not issue and enforce clear orders to all the masters to check that the doors were closed.

Why?

Beliefs contributing to the behavior:

- He thought the orders were clear and complete. His set of general instructions included "The officer loading the main vehicle deck, G deck, to ensure that the watertight bow and stern doors are secured when leaving port." He thought that there had been sufficient compliance with that instruction if the loading officer ensured that the assistant boatswain was actually at the control position and going to operate the doors before the officer left to go to his own harbor station.
- It was believed by nearly everyone that leaving the bow doors open should not alone have caused the ship to capsize.

Contextual factors:

- Before the accident, nobody had questioned the clarity of the order.
- He warned shore management before the accident that the continual changing of crews and officers had led to degradation of shipboard maintenance, safety gear checks, crew training, and "overall smooth running of the vessel." Noting that the vessel had had an unprecedented seven changes in sailing schedule in the six previous months, he argued that the vessel badly needed a permanent complement of good deck officers. Nothing was changed as a result.

Shore Management (Townsend Thoresen)

The specific responsibilities of most of the management positions are not stated in the accident report so it is not possible to go to the same level of detail as has been done so far. It really does not matter, however, as assigning blame to individuals is not the goal here. In addition, it appears from the accident report that the role of the different managers in the company was not defined well anyway. That turned out to be one of the major factors in the loss.

The accident report mentions the following roles in shore management: head of department, maintenance master, chief officer, operations director, and technical director. The board of directors is also mentioned, but they are treated separately here.

Responsibilities related to the accident:

- Buy, design, and maintain safe vessels.
- Operate vessels safely.
- Manage cargo.
- Manage passenger load.
- Manage traffic.
- Provide a clear definition of ship operators and shore-based management roles related to safety.

Role in the accident: Shore management seemed to have little understanding of their safety-related responsibilities.

- Many of the managers were not qualified to deal with nautical matters and did not listen to their masters, who were well qualified.
- The company did not have an express instruction that doors should be closed in the Bridge and Navigation Procedures Guide, did not have a positive reporting system, and did not ensure the closure of the doors was properly checked, did not

introduce a monitoring or checking system. When asked who was responsible to ensure that company orders were properly drafted, the answer was "nobody." Shore management thought it was preferable not to define the roles associated with the various shore management positions but to let them "evolve."

- The standing orders issued by the company for the ship made no mention about the opening and closing of the bow and stern doors. This omission led Captain Lewry to assume that his ship was ready for sea in all respects merely because he had had no report to the contrary.
- Management did not draw attention to the masters warning about at least five occasions when one of the company's ships had proceeded to sea with bow or stern doors open.
- There was no management of change (MoC) policy with respect to the new Zeebrugge run. At Zeebrugge the turnaround was different than that at Calais in four ways: (1) only two deck officers were available, (2) only one deck could be loaded at a time, (3) it was frequently necessary to trim the ship using ballast, and (4) the two doors could be closed at the berth. These differences required changes in the duties of the deck officers at Zeebrugge.
- Shore management pressured masters to maintain schedules over all else.
- Shore management allowed overloaded ships to sail, which was not only illegal but dangerous. The rule was that the number of persons on board be restricted to 630 when the draft exceeded 5.5 m or to ensure that the draft did not exceed 5.5 m if more than 550 passengers were on board. Shore management relied on the ship's officers to ensure the ship was not overloaded, although it was not possible for the officers to read the draft marks, and therefore they did not know what the current draft was.
- Before the accident, a senior master sent a memorandum to shore management saying that his ship had been grossly overloaded. This was not an unusual case. He asked for assurance that there would be a proper review of the system for counting heads. Nothing was done to fix this problem. The shore staff relied on the ship's officers to ensure that the ship was not overloaded. The shore staff continued to send traffic to the vessel until he was told not to do so by the ship's officer. The ship's officers did not know the draft of their ship and had no way to do so as they could not read the draft marks. The actual number of passengers and the load was frequently undercounted. As just one factor, trucks often reported lower weights, for a variety of reasons, than actually existed.
- The company purchased and operated ships with inadequate safety features with respect to the hazard of a ship going to sea with the bow doors open. Potential physical controls were feasible that could have prevented the events including mechanical interlocks and visual signals on the bridge.
- The lack of a communication channel between the captain and the deck crew had been raised as an issue by a captain of a similar vessel that had also gone to sea with bow doors open. This warning was ignored by the shore-based managers of Townsend Thoresen.
- Shore management never checked the ship's stability book, which showed that the ship was operating outside her specified safety conditions and therefore not complying with the conditions under which the Passenger Ship Certificate was issued.

- It had been suggested that high-capacity ballast pumps be installed on the company's vessels. The cost to install a high-capacity ballast pump for trimming for the Zeebrugge berth was regarded as prohibitive. Pumping with the installed pumps required close to 2 hours using one pump to fill or empty the two tanks. Using two pumps, the discharge time was reduced to 1.5 hours. The pumping time amounted to approximately half the normal passage time. Management concluded that it was not a safety issue and the cost was too high.
- In 1981, another passenger ferry owned by the same company had capsized after a collision off Harwich. Following that casualty, the company investigated passenger safety on their ships and the resulting report stated:

  The Company and ships' masters could be considered negligent on the following points, particularly when some are the direct result of "commercial interests."

  (a) The ship's draft is not read before sailing, and the draft entered into the official log book is completely erroneous.

  (b) It is not standard practice to inform the master of his passenger figure before sailing.

  (c) The tonnage of cargo is not declared to the master before sailing.

  (d) Full speed is maintained in dense fog. (quoted in [347])

  None of these problems seem to have been corrected.

Why?

Beliefs contributing to the behavior:

Because of safe passage previously with the bow doors open, they did not believe closing the bow doors was an important safety issue.

Contextual factors: The accident report focused on listing what management did wrong and not why they made their flawed decisions. The following, therefore, is likely incomplete. But the following factors were noted:

- Competition on this ferry route across the channel was intense. The company needed ships that were designed for fast loading and unloading and quick acceleration to be competitive.
- There was ambiguity in the responsibilities of the shore management roles of head of department, maintenance master, and chief officer. Shore management thought it was preferable not to define these roles but to let them evolve and allow them to operate as a team.
- A lack of clear, complete, and uniform standing orders for all ships of one class was critical because there were three crews and five sets of officers that manned the *Herald* during different crossings. There was no uniformity in the duties of each set of officers and of the members of the crew.
- They most likely had concluded that the closing of the doors was in practice irrelevant to or of low importance to safety. There is no indication in the accident report about how they determined the relative importance of factors that could contribute to accidents. But it is common for people to discount risk in the face of no accidents for the same behavior in the past. In addition, this was not the first occasion on which this hazard had occurred. In October 1983 the

assistant bosun of the *Pride* had fallen asleep and had not heard "harbor stations" being called, with the result that he neglected to close both the bow and stern doors on the sailing of the vessel from Dover.

- There was a poor relationship between ship operators and shore-based management resulting from and compounded by ambiguity regarding definition of each other's duties and responsibilities.
- Shore management, who included qualified naval architects, ignored previous requests by at least one senior master to install automated draft recorders with readout in the wheelhouse on the company's ships. The complaints stated that the masters had to estimate the draft because there was no way to ascertain it from the bridge. They have no way of knowing how much cargo they are carrying as loading is controlled from shore. Shore management during the accident inquiry stated that they believed the complaint was an exaggeration and that "if he had been unhappy with the problem he would have come in and banged my desk."
- At the time of the accident, there was no space for entry of the draft in the logbooks provided to ships of this class by the company
- Many of the managers were not qualified to deal with nautical matters but did not listen to their masters, who were well qualified.

If more information had been provided in the accident report that explained the context that management was working under, their behavior might have been more understandable.

Board of Directors

Responsibility: oversee company operations to ensure that large losses do not occur.

Role in the loss: did not include safety in their oversight activities.

Why?

Beliefs: the board of directors did not think their job included responsibility for the safe management of the company's ships.

Context: The members of the board of directors are often experts in management but not necessarily experts in engineering or, in this case, naval engineering. This lack of understanding of responsibility of the board of directors for safety is very common to find after accidents in many industries. For example, it was true of the Boeing board of directors when the B737 MAX was being developed and sold to airlines.

Government Regulators

After the accident, regulations for ships at sea were changed, as were designs. The United Kingdom asked the IMO (International Maritime Organization, an agency of the UN responsible for regulating shipping) to require:

- clear and concise orders
- strict discipline
- attention at all times to all matters affecting the safety of the ship and those onboard; there must be no "cutting of corners"
- the maintenance of proper channels of communication between ship and shore for the receipt and dissemination of information
- a clear and firm management and command structure

The United Kingdom decided after this accident that it needed an independent, unbiased investigative body. The formation of the Marine Accident Investigation Branch (MAIB) occurred as a result in 1989.

Context: unknown, but often it takes a major accident to get regulators to act.

Rescue Operations

What happened: Water quickly reached the electrical systems, leaving the ship in darkness. Most of the victims were trapped inside the ship and died of hypothermia due to the frigid water. During rescue operations, the tide started to rise, and the rescue team was forced to stop all efforts until morning.

Contextual factors:

The accident report gave high marks to the rescue operations. It did suggest that more thought be given to the need to communicate with the rescuers and rescued alike. The noise of the helicopters and other rescue craft made communication very difficult, if not impossible. The helicopters' down-draft made it difficult to stand on the side of the ship. Lights that were used to try to improve the rescue efforts in fact blinded both the rescuers and victims.

Factors related to the overall system design Fully understanding why an accident occurred and making changes to prevent future ones is not complete without considering the systemic factors that influenced the physical design and the human behavior involved in the loss. For example, internal and external economic conditions such as declining profits or external market competition may lead to management reducing safety margins and ignoring established safety practices, practices that lead to fatigue on the part of workers and poor morale, and pressures to increase productivity beyond a reasonable level.

Many of the systemic causal factors involved in this accident and accidents in general cannot be linked by direct arrows to the events and to the behavior of individual components in the control structure. Instead, they result from the overall control structure design and factors that impact the behavior of many or all of the individual system components and their interactions. That does not make them less important than those that can be put into a box and arrow structure. Direct causal factors may be easier to identify than indirect ones, but many times the indirect, systemic factors may be more critical to fix in order to prevent a broad range of future accidents, not just the specific circumstances that happened this time.

Only a few examples of the systemic factors involved are provided here: company cultural issues, communication and coordination problems, confusion about responsibilities related to safety, inadequate management of change, and poor risk management in regard to tradeoffs between safety and profits. These were primarily identified during the contextual analysis of the individual system components. Chapter 4 has more types of common systemic factors in accidents. Chapter 14 explains how systemic factors in accidents can be prevented through the design of the safety management system.

Cultural issues There was clearly nobody taking responsibility for safety at the management level in this company. Instead, there was pressure to remain on schedule at all costs and nobody providing safety leadership. When complaints or suggestions were made by those working on the ships about improving safety, they were ignored. Shore management totally ignored feedback and suggestions by the masters and labeled them as exaggerations or too expensive. Even at the ship level, nobody felt the responsibility to tell the captain that the doors were open or to shut the doors themselves. More generally, a senior shore manager, when asked who was responsible for the safety of the company's ships, replied that it was the responsibility of the senior masters. The masters would probably have pointed to the shore managers.

As an example, consider the lack of an indicator light for the state of the doors in the bridge of the vessels. When informed previously of the need to install such a light by several of the ship's masters, the response of a deputy superintendent in shore management was typical: "Do they need an indicator to tell them whether the deck storekeeper is awake and sober? My goodness!!" Another manager commented: "Don't we already pay someone to do this?" Management concluded that the indicator light was unnecessary and not the real answer to the problem. Instead, according to the accident report, they suggested that if the doors were left open, then the person responsible for them should be punished.

The accident report makes it clear that by autumn of 1986 the shore staff of the company was well aware of the possibility that one of their ships would sail with her stern or bow doors open and knew that it had happened several times before. They rejected the suggestion by their masters to install indicator lights in their ships, suggesting that it was the job of the person who shuts the doors to tell the bridge if there is a problem. The fact that this was a simple fix is illustrated by the fact that within a matter of days after the disaster, indicator lights were installed in the remaining Spirit class ships and other ships of the fleet.

The accident investigation report concluded that the shore management took very little notice of what they were told by their masters. The masters met only intermittently. There was one period of two and a half years during which there was no formal meeting between management and senior masters. Even when they did meet, the Marine Department did not listen to the complaints or suggestions of their masters. The report notes four areas in which the voices of the masters fell on deaf ears ashore:

1. Complaints that ships proceeded to sea carrying passengers in excess of the permitted number.
2. The request to have lights fitted on the bridge to indicate whether the bow and stern doors were open or closed.
3. The information that the draft marks could not be read. Ships were not provided with instruments for reading drafts. At times ships were required to arrive and sail from Zeebrugge with ballast tanks full but without any relevant stability information.
4. The request to add a high-capacity ballast pump to deal with the Zeebrugge trimming ballast.

Responsibility for safety seemed to be pushed down to the ship level and not regarded as the responsibility of shore management. Ship workers and operators but not company management were regarded as having responsibility for safety. In addition, safety seems to have been given less importance than profits by management.

Communication and coordination problems Communication and coordination were flawed throughout this operation. There were five masters with three sets of crews, with officers and crew changing frequently. This led to no uniform practices. Standing orders were incomplete, and nobody seemed to be responsible for making sure they were communicated and followed. Coordination between shore and ship procedures were deficient in controlling passenger loads or even reporting how many passengers were loaded and when maximum loads were violated. Feedback to identify potential hazards was either missing or ignored.

Management of change The ship was built and operated for the Dover-Calais run. There appears to have been little to no planning for the changes necessary when the run between Dover and Zeebrugge was initiated.

For example, at Zeebrugge the turn-around was different from the turn-around at Calais in four main respects. At Zeebrugge, (1) only two deck officers were available, (2) only one deck could be loaded at a time, (3) it was frequently necessary to trim the ship by the head, and (4) the bow doors could be closed at the berth. Because of these differences, the duties of the deck officers at Zeebrugge needed to be organized differently from their duties at Calais. No such thought was given to the matter.

Previous ships were built such that the master could see whether the door is open or closed. When the Spirit class ships were built with clam doors, it appears not to have crossed the mind of any manager to include in the standing orders an order that the closure of those doors must be reported to the bridge and recorded in the logbook.

The *Herald* had had an unprecedented seven changes in sailing schedule in the six previous months with little analysis of whether these changes might create hazards.

Confusion about responsibilities/inadequate safety management system It appears that almost everyone working for this company was confused about their safety-related responsibilities. There was no clear attempt to specify and communicate them rigorously. The standing orders were unclear and inadequate.

In general, the safety management system needs to be clearly defined and responsibilities assigned. That does not seem to be the case in this company. In fact, some of the statements from managers in the accident report appear to argue that it was better *not* to assign responsibilities for safety or other important system properties.

Summary and conclusions about the accident The analysis of this accident provides a good illustration of the inadequacy of linear chain-of-events causality models and linear causality in general. Looking only at the events and the linear causality relationship between them makes it appear that the officers on the ship were primarily responsible for the accident and that the accident report's recommendation that the licenses of the *Herald*'s chief officer and master be suspended seems justified.

The accident report, as is usually the case, was almost totally focused on making arguments that particular individuals (the ship's operators) or classes of people (management) were to blame for the accident and not the real problem, which was unsafe design of the system as a whole and the dynamics of the system. Focusing on blame detracts from learning.

Let's review the accident analysis process used and the causal factors identified in this chapter. The first step in the process is to understand the physical factors involved in the loss, including

- The limitation of the physical system design: for example, the *Herald of Freedom* had a loading ramp that was too low to reach the upper car deck at high tide

- The failures and unsafe interactions among the physical system components: for example, the bow doors not being closed together with the squat effect and acceleration of the ship on leaving port
- The environmental factors: for example, the high spring tides, that interacted with the physical system design

Most accident analyses include this information, although they may omit unsafe interactions and look only for component failures.

Understanding the physical factors leading to the loss is only the first step, however, in understanding why the accident occurred. The next step is to understand how operating practices for the ship along with the engineering design practices contributed to the accident and how they could be changed to prevent such an accident in the future. Why was the hazard—capsizing as a result of flooding—not adequately controlled in the ship's design? Some controls were installed to prevent this hazard, for example, the doors themselves and the assistant bosun's assignment to close them, but some controls were inadequate or missing such as a lack of watertight compartments and indicators in wheel house to show that the doors had closed. Important questions to consider are what parts of the design and analysis process along with design of operations for the ship allowed this flawed design to be accepted and operated in an unsafe manner? What changes in that process—for example, better hazard analysis, design, review, or oversight processes—could be used to ensure that designs have adequate hazard controls in the future?

Many of the reasons underlying poor design and operational practices stem from management and oversight inadequacies due to conflicting requirements and pressures. Identifying causal factors beyond the physical design starts with identifying the safety-related responsibilities (requirements) assigned to each system component along with the role they played in the accident. For example, a responsibility of the first officer on the *Herald of Freedom* was to ensure that the doors were closed before the ferry leaves the dock, management had the responsibility to ensure their ferries have a safe design and are operated safely, and the International Maritime Organization (IMO) had the responsibility to provide regulations and oversight to ensure that unsafe ships are not used for passenger transportation [157]. Using these safety-related responsibilities, the unsafe control provided by each of the system components can be identified. In most major accidents, there is unsafe control over hazards exhibited throughout the system and not just a few components.

To understand *why* people behaved the way they did, we must examine their mental models and the environmental factors or context affecting their decision-making. All human decision-making is based on the person's mental model of the state and operation of the system being controlled. For example, the first officer assumed that the assistant bosun had closed the doors, the assistant bosun was asleep after limited rest time after the last crossing, other people on the deck may have thought that someone else would notice that the doors were open and close them, and the captain thought the doors had been closed. He was under pressure to meet schedules, and the departure was already late.

Preventing unsafe behavior in the future requires not only identifying the flaws in the controllers' mental models—including those of the management and government system components—but also *why* these flaws existed. For example, the captain's flawed mental model (thinking the doors were closed) was not corrected in time to prevent the accident due to lack of feedback about the state of the doors. All of them thought that leaving the doors open would not cause a loss of the ferry because of the lack of an accident a year before when the same thing happened with one of the *Herald*'s sister ships; that is, they had incomplete knowledge about the potential ferry hazards.

The impact of the operating environment, including environmental conditions, cultural values, and so on must also be identified. For example, the problematic ferry design features were influenced by the competitive ferry environment in which the ferry was to operate.

The accident report blamed a "disease of sloppiness and negligence at every level of the corporation's hierarchy" [347]. But this superficial level of analysis (that is, management sloppiness and negligence) is not useful in preventing future accidents—it simply provides someone to blame and to prosecute. It does not eliminate the underlying pressures that led to the poor decision-making, the inadequate design of the safety management system, or the poor overall safety culture within the company. Without changes that respond to these causal factors, similarly flawed and risky decision-making is likely again in the future, although the actual accident details may be very different. The complex environmental, social, and economic factors contributing to poor decision-making must be identified in accident analysis in order to provide policy and other changes to improve risk-related decision-making in the future.

A critical factor in this accident, as well as most accidents, is that those making decisions about vessel design, harbor design, cargo management, passenger management, traffic scheduling, and vessel operation were unaware of the impact of their decisions on the others and the overall impact on the process that led to the ferry accident. The type of bottom-up decentralized decision-making (promoted in the so-called High-Reliability Organization [HRO] theory [404]) that was in operation in the overall ferry system can easily lead to major accidents in complex socio-technical systems. Each local decision may be "correct"—and "reliable," whatever that might mean in the context of decisions—within the limited context within which it was made. But those independent decisions can lead to an accident when decisions and organizational behaviors interact in unsafe ways.

Figure 8.5 illustrates this principle in the context of the *Herald* accident. Flawed decisions may not only result from limitations in the boundaries of the system view available to individuals, but the boundaries relevant to a particular decision maker may depend on the activities of several other decision makers found within the total system [320]. In this case, there were unknown and unexpected interactions among those making independent decisions about vessel design, cargo management, passenger management, harbor design, vessel operation, and traffic scheduling. Accidents may then result from the interaction of the potential side effects of the performance of the decision makers during their normal work. It is difficult if not impossible for any individual to judge the safety of their decisions when it is dependent on the decisions made by other people in other departments and organizations.

As the interactive complexity grows in the systems we are building, accidents caused by unsafe interactions among system components become more likely. As has been emphasized repeatedly in this book, high-reliability system components do not imply safety. Safety is a system property, not a component property, and must be controlled at the system level, not the component level.

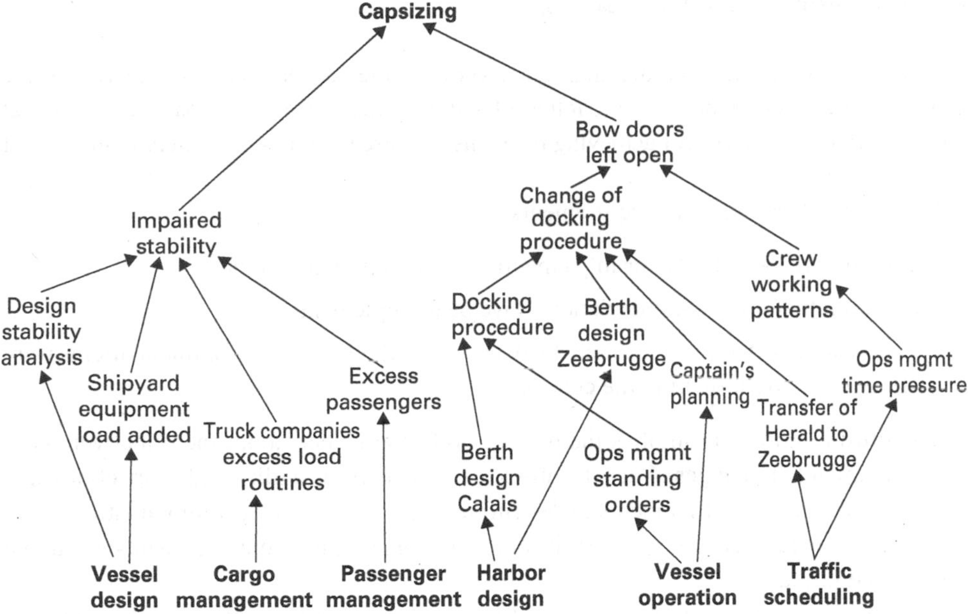

Figure 8.5
The complex interactions in the *Herald of Free Enterprise* ferry accident. (Adapted from Rasmussen [320].)

8.4 Generating Recommendations

Once the other parts of the analysis are completed, generating recommendations should be straightforward. The biggest complaint about a thorough accident analysis is that it generates too many recommendations. This complaint is only justified if the goal of the accident investigation is to make as few recommendations as possible. Accident investigation has had the goal of identifying root causes or probable causes and focusing on the role of the human operators for so long that it may be a cultural shock to start looking at more causal factors that generate more recommendations but, in turn, are more effective at preventing future accidents.

One of the objections raised to including a large number of recommendations is that responding to them is overwhelming. This is simply a logistical problem and not one that should be solved by learning less from each accident. There is no reason that recommendations cannot be prioritized according to specified criteria. There is also no implication that all the recommendations must be implemented immediately. Some recommendations will be relatively straightforward to implement immediately while others may take longer. Some may require such extensive changes that implementing them will take a great deal of effort and resources. Examples of the latter include establishing a new oversight agency or changing standards and laws. Difficulty of implementation is not an excuse to omit a recommendation from the accident report, but it may be a good reason to categorize it as a longer-term goal rather than an immediate fix.

8.5 Implementing Long-Term Learning

Sometimes recommendations are made but never implemented. Not only must there be some way to ensure recommendations are followed, but there must also be feedback to ensure that they are effective in terms of achieving the goals and strengthening the safety management system.

Essentially there are three requirements:

1. assigning responsibility for implementing the recommendations
2. checking that the recommendations have been implemented
3. establishing a feedback system to determine whether the recommendations were effective in strengthening the controls

The third requirement implies the need to collect evidence about the effectiveness of the recommended changes. Such feedback can come from audits and inspections and from the analysis of later incidents to determine whether previous recommendations were successful. Such an activity is a critical component of any safety management system, but it is often omitted.

Subsequent accidents or losses, particularly if they are analyzed thoroughly, provide a rich source of information to understand why previous recommendations were not effective and what else is needed. Was the original causal analysis flawed? Were assumptions about the potential effectiveness of particular improvements incorrect? Did other changes occur that thwarted the attempt to strengthen the safety control structure? Did the planned changes result in unforeseen consequences?

The goal here is to ensure that the organization is continually learning and improving its risk management efforts so that the potential for losses is reduced over time.

8.6 The Cost of Thorough Accident Investigation

One question remains regarding costs. Does it cost more to use more sophisticated and expanded models of causality and to do the things described in this chapter? Surprisingly, the answer to this question is usually no. The application of the principles in this chapter requires only a different interpretation of the information usually collected in a thorough accident or incident investigation. The cost of collecting the information remains the same.

In addition, the alternative to better accident analysis is to continue to have preventable accidents, the cost of which usually dwarfs the cost of more sophisticated accident analysis. Every major accident has had precursors that might have been used to prevent the major loss, as argued in chapter 4.

Another argument sometimes raised is that it is too expensive to analyze all the accidents and incidents that occur in some industries and companies. How does one select the incidents requiring in-depth analysis? That question is the wrong one to ask. Given the number of incidents and accidents that have identical systemic causes, simply investigating one or two in depth can potentially eliminate dozens of incidents. For example, a superficial accident analysis might blame the occurrence of a loss on the flawed design of a relief valve, leading to replacement of all the relief valves with the same design throughout the plant or company. The result is prevention of incidents due to that particular flawed valve design. Understanding why the flawed design was used, however, such as inadequate design, hazard analysis, review, testing, or other development or management practices, and improving those practices could potentially have a much greater impact on reducing future losses due to design inadequacies of all kinds.

The cost and time required to do a complete causal analysis will usually pale in contrast to the cost of a future accident due to the same factors that were never fixed. Doing a series of superficial investigations that omit the most important factors that could contribute to future accidents cannot be considered cost effective.

8.7 Summary

This chapter has focused on causal analysis to identify why a particular incident or accident has occurred. We, of course, need to do this type of analysis so that we don't repeat our mistakes. But our primary goal in safety engineering is to prevent accidents from occurring in the first place. The next two chapters focus on hazard analysis, which is the process for proactively identifying how accidents might occur for a particular system so the causal scenarios can be prevented in the first place. How to use that information in the design of safer systems is described in chapters 11 and 12.

Exercises

1. Do you think that the label "failure" make sense for system components beyond hardware; that is, humans and organizations? Why or why not?
2. The following is a quote from a forest service helicopter report: "*Contributing to the accident was the failure of the flight crewmembers to address the fact that the helicopter had approached its maximum performance capability on their two prior departures from the accident site because they were accustomed to operating at the limit of the helicopter's performance" [278, p. 14].*

   How might this conclusion be rewritten in an explanatory versus accusatory fashion? What impact might such a rewording have in terms of raising questions that might be examined in the analysis?
3. Consider the tank overflow accident report provided in the supplementary materials for this textbook at httl://sunnyday.mit.edu/supplementary-materials. Can you find instances of hindsight bias in the report? Who or what is cited as the cause of the accident? Are there instances of blame in the report? What is left out of the report that you think would be useful to include in understanding why the accident occurred and how such events could be prevented in the future (in other words, what additional questions would you have asked)? Are there additional recommendations you think should be included?
4. Examine a selection of accident reports in your field. How many of them conclude that the root cause is an operator error? Why do you think operators are so often selected as the basic or root cause of accidents? Do you think it stems from the truth about accident causation or some other reasons?

5. Take an accident that interests you and for which an accident report (and ideally additional information) exists.
   a. Do you see any instances of hindsight bias in the report? If so, who or what was blamed, and what type of bias might have been involved?
   b. Was blame assigned in the conclusions (i.e., who versus what)? If so, was it focused on the human operators or something else? Management? What role do you think assigning blame played in the investigation and causal analysis? Is the word failure used to describe the behavior of the system components involved? Were the causal factors described in an accusatory (blame-laden) way or in an explanatory fashion?
   c. Is there a chain of events described? If so, what was it? Do you see anything missing from it that might explain why the accident occurred?
   d. Create the chain of events for the accident if not explicitly included in the report. Considering the events alone, what do the causal factors appear to be?
   e. Was there a root cause identified? If so, does it explain why the accident occurred and provide enough information to prevent it in the future?
   f. What conditions (versus events) contributed to the accident? Does the cause of the accident appear to be different when the conditions/context are added to the events? If so, in what way?
   g. Were any systemic factors mentioned in the report as a whole? Or in the final conclusions about the cause of the accident? Are there any other potential systemic factors that you might have investigated if you were part of the investigation team?
   h. How does the report treat operator error?
6. Take the same accident you used in exercise 3 and analyze it using the approach shown in section 8.3 (the Zeebrugge ferry accident). Compare your results, including potential recommendations, with the official accident report. Feel free to use information from online resources or newspaper articles to supplement the information in the report if you want. Be sure to cite any resources you use. You may not have access to information that was not included in the official accident report. What questions, however, were raised in your analysis that might have been appropriate to investigate further? Did you find any causal factors in your analysis that are not described as causal factors in the official accident report? If so, what were they? Did you identify any of the systemic factors described in chapter 4 in your analysis?

   Hints:
   a. Start by describing the hazard. Make sure your hazard describes the overall state of the system and not specific behavior of an individual component such as software, engines, pilots, and so on. See chapter 3.
   b. Create a control structure describing the system as it existed at the time of the accident. Your control structure should include the responsibilities for each controller, the control actions involved in the loss, and any feedback arrows. Examples are in section 10.3 and in [222; 228]. Label each arrow with the particular

actions/information they represent. Add controllers, if necessary, to capture any additional causes you find you need to answer the questions generated as you do the analysis. Include, using dotted lines, the things that you find missing as you analyze the accident causes.

c. Analyze the physical process/system first. Identify:
   - safety responsibilities of physical system (what it must do)
   - physical controls (what design features were meant to fulfill the safety responsibilities)
   - failures and inadequate controls (what went wrong)
   - contextual factors (why those things went wrong)

   This should not include anything above the physical components in the control structure; that is, any of the controllers of the physical process.

What are some of the important questions raised by looking at the physical process/system?

d. Analyze the controllers: remember that a controller can be automated software, a human operator, a human manager, or an organization such as the FAA or congress). Identify:
   - accident-related responsibilities (what goals they should achieve)
   - decisions and control actions contributing to the loss (what they did that was unsafe)
   - process model flaws (what did they believe about the system that explains their actions)
   - context (what other factors—pressures, incentives, history, and so on—explain their actions)
   - additional questions raised that could not be answered in information contained in the accident report

e. Analyze any factors related to the control structure as a whole.

7. Examine the safety information system in your field or your company (or look at a government one, such as the ASRS [Aviation Safety Reporting System]). with respect to information about past accidents, incidents, and so on. Is the most useful information included? How is it gathered? Is the information complete? Are there any biases in the collection mechanism? Is the information used by the people who could benefit from it? Is it usable? Can the most important information be found when needed?
8. Why do you think most accidents follow some kind of change?
9. The changes preceding an accident may be planned (and thus can be handled with management of change procedures or unplanned and perhaps unknown. Given that an MoC policy may exist, why might accidents still occur? How might unplanned changes be handled to prevent accidents?
10. Why do you think it is so hard for people to let go of the concept of blame in accident investigation and analysis?

9

Hazard Analysis: Basic Concepts

The argument that the same risk was flown before without failure is often accepted as an argument for the safety of accepting it again. Because of this, obvious weaknesses are accepted again and again, sometimes without a sufficiently serious attempt to remedy them, or to delay a flight because of their continued presence.

—Richard Feynman,
Personal Observations on Reliability of Shuttle

Unfortunately, everyone had forgotten why the branch came off the top of the main and nobody realized that this was important.

—Trevor Kletz,
What Went Wrong?

Hazard analysis is at the heart of any effective safety program. The term hazard, which is fundamental in safety engineering, is defined in chapter 3. This chapter covers basic concepts in hazard analysis, mainly what it is. The next chapter examines specific hazard analysis techniques: that is, how to do it.

9.1 What Is Hazard Analysis?

Hazard analysis is the process used to identify system hazards and their potential causes. While hazard analysis is not the only activity in system safety, it is difficult to see how safety can be effectively handled in a complex technological system without a powerful hazard analysis process at its core. Hazard analysis techniques provide formalisms for systematizing knowledge, drawing attention to gaps in knowledge, helping prevent important considerations from being missed, and aiding in reasoning about systems and determining where improvements are most likely to be effective.

Although hazard analysis alone cannot ensure safety, it is a necessary first step before hazards can be eliminated or controlled through design or operational procedures. Simply knowing that a hazard exists may provide sufficient information to eliminate or control it, even without in-depth analyses of its causes. Often, general safeguards can be provided even if little is known about the hazard's precursors. A larger number of options for

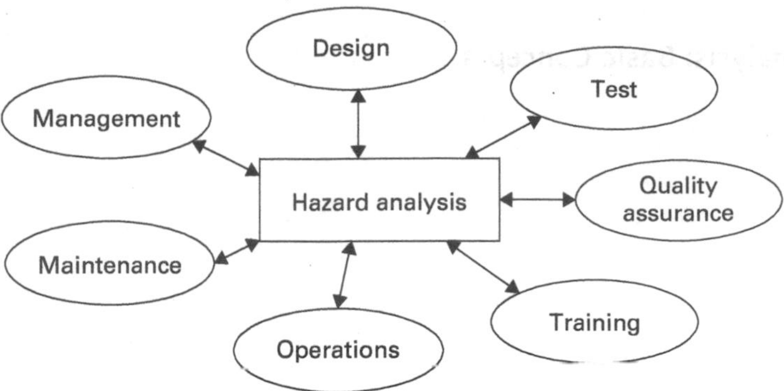

Figure 9.1
Hazard analysis is at the heart of any system safety program.

elimination and control usually exist, however, if more is known about the hazard and the conditions and events leading to it. In addition, condition-specific or event-specific safeguards are frequently more effective and less costly in terms of the tradeoffs required.

Figure 9.1 shows some of the uses of hazard analysis. The results provide guidance for all aspects of development and operations. For example, information from hazard analysis can assist system designers in eliminating or mitigating hazards and in performing trade studies. Hazard analysis results also form the basis for identifying test requirements and can be used in designing production facilities. Finally, hazard analysis is crucial in operating the system safely throughout its life, such as identifying required inspections and audits, making safe upgrades and changes, identifying when risk is increasing over time because of both planned and unplanned changes, and investigating incidents and accidents. In essence, hazard analysis establishes the basic goals for the system safety program.

Hazard analysis should not just be performed at the beginning of a project or at fixed stages but should be continuous throughout the life of the system, with increasing depth and extent as more information is obtained about the system design. As the project progresses, the uses of hazard analysis will change, such as identifying hazards, testing basic assumptions and various scenarios about the system operation, specifying operational and maintenance tasks, planning training programs, evaluating potential changes, and evaluating the assumptions and foundations of the models as the system is used and feedback is obtained. In an operational system, analyses and their assumptions act as preconditions for safe operation and as constraints on management and operational procedures.

For some well-understood hazards and simple systems, an analysis may consist merely of comparing the design with various standards and codes that have been developed over time to deal with known hazards. Standards and codes provide a means of encoding historical information derived from accidents and near accidents so that safety reviews and analyses need not start from scratch each time. They help us encapsulate and learn from experience.

Nevertheless, as new technology is developed and systems are scaled up in size, new hazards arise and the possibility of introducing hazards increases. Systems with new and complex designs usually require sophisticated, formal, documented analytical procedures.

In some countries or industries, the particular hazards to be analyzed and the depth and types of analysis may be established by regulatory authorities or by legislation. In others, these decisions must be made by the engineers early in project development.

Because hazard analysis results are so widely used in system development and operations and for the entire lifetime of the system, documentation of the results and assumptions is critical for successful safety programs.

9.2 The Hazard Analysis Process

The hazard analysis process, especially the comprehensiveness, depends on the use for which the analysis is intended. The process described in this section is a general process appropriate for complex systems. For some relatively simple systems, only parts of this process may be necessary.

As suggested in the previous section, hazard analysis is not only useful during system development, but should be a continual and iterative process that is performed throughout the entire system lifetime, including decommissioning and disposal. Starting early in the system design process is imperative if safety considerations are to be incorporated into trade studies and important early design decisions, when hazards can be most effectively and economically handled. Planning for hazard elimination and control should begin as soon as hazards are uncovered—preferably before unsafe features become firmly embedded in the design. Even the hazards of decommissioning and disposal are important to consider early in some systems, such as nuclear power and space satellites.

The forms of analysis will be different as the system matures, but all are part of a single analysis process (figure 9.2). Each stage of analysis acts as a baseline on which later steps build. The stages reflect the quality of information available, with analysis depth and breadth increasing as more information is obtained and design decisions are made. A potential gap exists between development and operations, when a handoff usually occurs to a totally different organization or company. It is critical that this handoff includes all the information that operations management and operators need to use the system safely.

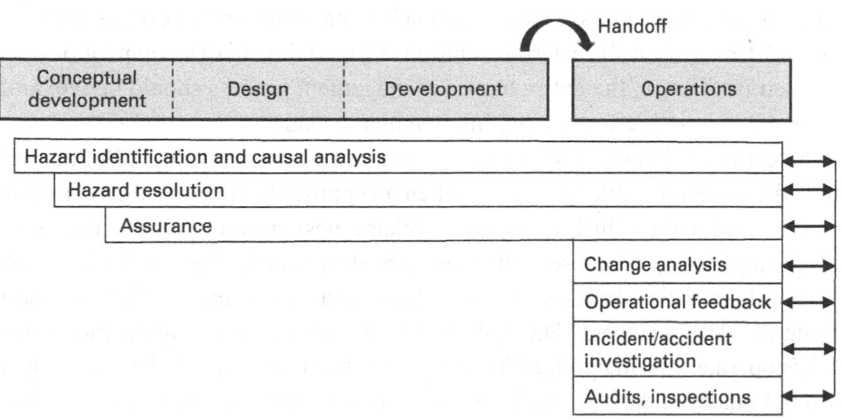

Figure 9.2
Hazard management is needed throughout the entire lifetime.

9.2.1 The Overall Process

Hazard identification and causal analysis should begin at the beginning of conceptual development. As the project progresses and the design is elaborated, more detailed analyses and tests may uncover new hazards or eliminate old ones from consideration. New hazards may be identified throughout the entire lifetime of the system as more experience is gained and, particularly, as the system use and environment change. New causes will also be identified either due to limitations of the original causal analysis process or changes to the system and environment.

The need for continual updating of hazard analyses makes important demands on both the types of analysis performed and how the results are documented. Both the analysis itself and assumptions underlying it need to be documented in a usable way. Hazard analysis results often are not used during operations because the information needed is too difficult to locate.

In the early project phases, *hazard resolution* may involve simply getting more information about the hazard or generating alternative design solutions. If a hazard cannot be resolved at a particular stage, follow-up evaluation and review may be necessary. Because hazards can be identified during operations, resolution processes will be needed throughout the system lifecycle.

Similarly, *assurance* is a continual process. The earlier assurance processes begin, the more cost-effective they will be. When assurance starts late in the development process, fewer options exist to resolve the problems identified, and the cost of resolving them usually soars. The best assurance methods will be usable at the time design and other decisions are made so that few safety-related decisions need to be changed later on.

Planning for operations and operational assurance, including the design of feedback channels, also needs to start early. Organizational controls, such as audit trails and tracking systems, must be installed to ensure that this follow-up occurs. Because hazards can be identified during operations, these types of controls are necessary throughout the system lifetime.

The establishment of processes to determine whether a hazard was actually eliminated or controlled effectively is a critical component of the assurance process. Ideally this process can be done before a serious loss occurs. The occurrence of minor incidents and anomalies should be used as well as feedback from users and operators through some type of reporting system. If hazard resolution is found during development or operations to have been ineffective, the entire hazard management process should be reexamined to determine why it was ineffective and improvements made.

System test is an obvious place for safety assurance to be embedded. But that will only happen if the testers are able to identify when a supposedly eliminated or controlled hazard has occurred and not just a non-safety-related design flaw. This requirement means that safety engineering processes will need to be deeply embedded in the test process.

The operational safety achieved depends on the accuracy of the assumptions and models underlying the design process. The system must be monitored during operations to ensure that (1) it is operated and maintained in the manner assumed by the designers; (2) the models and assumptions used during initial decision-making and design were correct; and (3) the models and assumptions are not violated by changes in the system, such as workarounds or unauthorized changes in procedures, or by changes in its environment. Operational feedback on trends, incidents, and accidents should trigger reanalysis. This reanalysis requires that the original analysis was documented in a way that it can be easily examined and that changes do not require starting from the beginning again; starting again in most cases will be prohibitively expensive.

If a change is proposed to a baseline or completed design, or to operational procedures, or if an unplanned change is detected during operation, the change must be analyzed for its potential effect on safety. This process involves reviewing previously generated analyses to identify the impact of the proposed change, updating the analyses and documentation to reflect the changes, and identifying new hazards and hazard causes. The reanalysis must start at the highest level of system design at which the change becomes visible and must demonstrate that the change does not create a new hazard, affect a hazard that has already been resolved, or increase the severity level of a currently existing hazard.

During operations—and even sometimes during development—assumptions used in the analysis may change, such as how the system will be operated, as well as important factors in the environment. Continual updating of the hazard analysis will be necessary to ensure that the system does not experience increasing risk over time. Such updating during operations may be triggered by incidents or accidents or may be scheduled periodically or after specific predicted changes occur. It may also be identified during audits or inspections. The assumptions underlying the hazard analysis and control processes should be a focus of audits. This requires, of course, that the assumptions have been recorded.

Hazard identification and causal analysis is the focus of the rest of this chapter, and techniques to perform it are discussed in the next chapter. Hazard resolution and assurance through design is described in the three chapters that follow. Managing the entire process as shown in figure 9.2 is the topic of chapter 14.

9.2.2 Detailed Steps

In general, a hazard analysis program consists of the following steps:

1. *Definition of the objectives and losses of relevance to the stakeholders.*

   Identification of stakeholder expectations, including identifying the losses of importance in the system and, optionally, prioritizing the losses by the stakeholders to assist in decision-making.

2. *Definition of the scope.*

   Definition and description of the system, system boundaries, and information to be used in the analysis as well as the creation of a system model if it is to be used in the analysis. This model may start at a very high level of abstraction and be elaborated as design and other decisions are made. Depending on the system, this step may involve collecting data important for the analysis, such as historical data and related standards and codes of practice.

3. *Identification of system hazards.*

   Identification of the hazardous states that can lead to a loss and linking them to the identified potential losses.

4. *Identification of safety constraints.*

   Restatement of the hazards as constraints on the development and operation processes. These constraints define the safety-related goals for developers and operators.

5. *Ranking hazards by risk based on potential effects and perhaps their likelihood.*

   This common step is controversial. If the hazards have been traced to their potential losses (step 1), then theoretically identifying severity is trivial. However, the list of hazards is often poorly created and long, with causes and hazards confused. In addition, there is no way to determine likelihood in any scientific way as the system has not been designed yet, so the results of such a ranking are often fanciful. How one designs the system will determine the risk associated with the identified hazards. Without design information, likelihood can only be based on prior experience with past systems, and that experience may not be applicable to future systems and new technology. We create new systems because the old ones are no longer satisfactory.

   Few evaluations have been done to determine the validity of such risk assessments, but the few that have been done have not been encouraging. One evaluation of the assessed risk on a real defense system found that hazards labeled as "marginal" actually could lead to catastrophic results because the ranking ignored everything but failures of standard components and ignored causes such as design flaws, operator errors, and software errors [5].

6. *Identification of the causal scenarios leading to the hazardous state.*

   The hazard causes are identified so they can be eliminated or controlled in the system design or operations. The causes may involve complex scenarios and not just single events.

7. *Identification of preventive or corrective measures and general design criteria/constraints and controls.*

   This step might include some type of evaluation of cost and relative rankings of preventive measures as well as tradeoffs with conflicting goals. For hazards that cannot be eliminated or deemed to be controlled adequately, operating limitations or procedures should be specified. In practice, this step will be integrated into the mainline system design process and performed by a team consisting of engineering, human factors, and system safety experts.

8. *Verification that the controls have been implemented correctly and are effective.*
9. *Certification or licensing, if applicable.*
10. *Providing hazard analyses and information about safe operation to system operators, including unresolved hazards.*

    Completely eliminating or controlling hazards in the system design may not be possible or practical due to political factors, lack of time, potential cost, tradeoffs with important system goals, or the magnitude or nature of the hazard. Information must be provided to operators to assist them in designing operational processes, training, and so on, to cope with any residual hazards. This information may include requirements for operational procedures and other inputs to the operator's safety management system.

11. *Hazard analysis during operations to identify new hazards, changed hazards due to changes in the system or its operation, assumptions made about hazards during development that turn out to be untrue or later change, and information to assist in an accident investigation.*

    This step can be triggered by periodic audits and inspections as well as investigation of incidents and accidents. At various times in its operational life, a system will be exposed to different environments, processes, conditions, and loads, and the effects will differ according to when the stress or condition occurs. These conditions and environments cannot always be anticipated and handled during system development.

12. *Feedback and evaluation of operational experience to developers and any regulators.* This feedback is important in creating system upgrades and modification, in designing future systems, and in regulation to determine if additional operational controls or oversight is needed.

Each step requires documentation of the results and of any underlying assumptions and models. In general, the purpose of the hazard analysis process is to use the results as a reference for judgment in design, maintenance, and management decisions [316]. Accomplishing this goal requires an explicit formulation of the models, premises, and assumptions underlying the safety analysis and the design features used to eliminate or control hazards. Documenting design rationale is particularly important: critical design decisions can be undone unintentionally if the reason for the decision is forgotten. Not only is such documentation itself critical, but the necessary information must be able to be located easily when required and exist in a form that is usable for decision-making.

Not all of the steps may need to be performed for every system and for every hazard. For standard designs with well-established risk mitigation features (perhaps included in standards and codes), only some of the steps are needed. For new designs and complex systems, most of the steps are required. The last step, feedback and evaluation of operational experience, is critical for all systems.

The output of the hazard analysis process is used in developing system safety requirements and constraints, preparing performance and design specifications, test planning, preparing operational instructions and training, and management planning. The results serve as a framework or baseline for later analyses.

Hazard analysis itself can be divided into five basic functions: (1) identifying losses, (2) defining the system boundaries or scope of the analysis, (3) identifying hazards, (4) specifying safety constraints, and (5) identifying hazard causal factors. Although names such as *Preliminary Hazard Analysis* (PHA) and *Subsystem Hazard Analysis* (SSHA) have been applied to these functions, these are merely phases, as noted earlier, in a continuing and iterative process rather than separate sets or types of analyses. Following hazard analysis, the results should be used in a comprehensive safety engineering program.

Identifying losses Identifying the losses that must be prevented is necessarily the job of the system stakeholders. These stakeholders may be management, users, customers, operators, the public (perhaps represented by a user group or government regulatory agency), and so on. They must identify the goals of the system and also the constraints on how those goals can be achieved. The result will be a list of losses.

The hazard analysis process starts from and is focused on this list of losses. For example, the losses associated with a vehicle might include such things as:

1. human death or injury
2. damage to the vehicle
3. damage to objects outside the vehicle
4. loss of mission (e.g., transportation, surveillance, science, or defense)

5. inadequate customer satisfaction
6. environmental damage

While the list need not be prioritized, such priorities will assist designers later when and if tradeoffs are required in the design process.

Defining the scope or boundaries of the system Given the definition of a hazard as within the system boundaries, it follows that those boundaries must be defined before hazards are identified. Basically, the goal is to define what is within the design space of the developer and what is outside. Are the operators included? What about social and management structures?

Identifying hazards Identifying hazards can be the trickiest part of performing a hazard analysis. If a good list is not created from the beginning, the analysis that follows can be incomplete and not very useful. A common mistake is to use brainstorming to generate a long and unstructured list that cannot be determined to be complete, that includes duplicates, or that includes the causes of the hazards rather than the hazards themselves. Here are some basic tips for identifying hazards.

- **Start from losses.**

  While there is no formal way to identify hazards, there are structured steps one can take that will improve the results. Clearly, one should start from the identified losses. Then system-level hazards are listed that can lead to those losses given worst-case environment conditions. Linking the hazards to the losses will assist in keeping the process on track and will be needed later to provide traceability in the hazard analysis results. It will also assist in prioritization of hazard resolution as the hazards usually will inherit the priority of the losses they can cause.

- **Use abstraction and refinement.**

  By using abstraction in the process, the results can be greatly improved. The first, highest-level list should have no more than five to ten hazards included. If you have more than that, you are probably starting at too detailed a level or including causes of the hazards and not just the hazards themselves. Causes should be identified later. Make sure the high-level hazards you generate are disjoint. That is easier if you limit them to a small number. Ensuring completeness is assisted by using high-level abstraction and then refinement to generate a more detailed list.

  As an example, the hazards associated with an aircraft may be identified as loss of lift and control over the flight angle, violating separation requirements with terrain or other aircraft, coming too close to other objects on the ground, and loss of structural integrity. Notice that these all refer to the aircraft as a whole, not the components such as braking failure.

  In the next steps of the analysis, these high-level system hazards are broken down into more detailed hazardous states. If too many hazards are being generated in one step, simply go up to a higher level of abstraction. For example, one aircraft hazard might be stated as:

  H1: Aircraft comes too close to other objects on the ground.

  Coming too close to other objects is related to the aircraft deceleration, acceleration, and steering functions:

H1–1 Deceleration:

H1–1.1 Deceleration is insufficient on landing, rejected takeoff, or during taxiing.

H1–1.2 Asymmetric deceleration maneuvers aircraft toward other objects.

H1–1.3 Deceleration occurs after V1 point during takeoff (the V1 point is the time where it is safer to continue the takeoff than to try to abort it).

H1–2 Acceleration:

H1–2.1 Excessive acceleration is provided while taxiing.

H1–2.2 Asymmetric acceleration maneuvers aircraft toward other objects.

H1–2.3 Acceleration is insufficient during takeoff.

H1–2.4 Acceleration is provided during landing or when parked.

H1–2.5 Acceleration continues to be applied during rejected takeoff.

H1–3 Steering:

H1–3.1 Insufficient steering to turn along taxiway, runway, or apron path.

H1–3.2 Steering maneuvers aircraft off the taxiway, runway, or apron path.

Essentially a tree structure is being created as shown in figure 9.3. The decomposition structure provides an organized way to structure the process of generating hazards and to identify gaps and redundancy. Decomposition, in fact, is a standard way that humans cope with complexity, as described in chapter 7. It will help both with generation of hazards and with review by others. It should be easier, for example, for reviewers to identify missing hazards related to deceleration in the above list than if all of the aircraft hazards are mixed up in one long list.

- **Do not include causes.**

A common mistake is to include failures and other causes of hazards in the list of hazards themselves. Causes will be identified in a later step. A particularly egregious example is the common inclusion of "human error" as a system hazard. Not only is it not a hazardous state (at best it is an event that leads to a hazard), but it provides no useful information to the designer or even operator of the system. It leads to the useless requirement/constraint: *System operation must not include human errors.* As another example, "Brake failure" is a cause of the system hazard "Deceleration is inadequate upon landing, rejected takeoff, or during taxiing." There are many potential causes of inadequate deceleration, including such things as not applying the brakes or failure of another type of aircraft braking mechanism such as reverse thrusters. Starting with one or even a few causes will short circuit the process of identifying as many causes as possible. Trying to list all the causes by brainstorming at the beginning of the

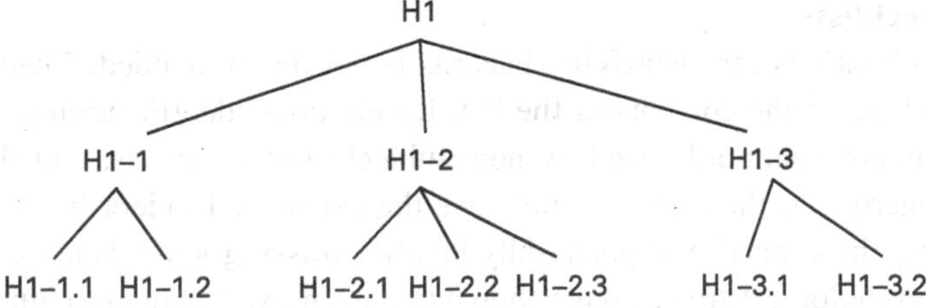

Figure 9.3
A decomposition structure for generating hazards.

causal analysis is likely to lead to an incomplete analysis. Most hazard analysis techniques, as described in the next chapter, provide an organized process for identifying causes from the hazards.

- **Include only system-level hazards.**

  The highest-level hazards in the tree should be at the system level; system components should not be mentioned. For example, "Engine provides inadequate acceleration" is a cause of the system hazard "Inadequate propulsion to maintain lift in the aircraft." The components involved will be generated when the causes of the hazards are identified later. Hazard identification should start at the beginning of system concept development, before detailed design decisions, such as the type and number of system components, have been made. Later in development, the system-level hazards can be traced to the hazards associated with the system components and their interactions. Even better, the system components to be used may be determined by the identified system hazards.

  Attempting to include component-level hazards and causes of hazards at the beginning of the process is very likely to result in a disorganized and incomplete analysis, including omission of hazard causes related to interactions among the system components. The overall goal should be to structure the hazard identification process to get the best results.

- **Separating hazards by phases can lead to incompleteness.**

  The cause and especially the result of a hazard may differ depending on the phase of operation in which the hazard occurs. As an example, loss of control of a missile immediately after liftoff might result in more damage (since the missile is filled with propellant) than loss of control far down range [131]. Similarly, failure of a missile launch while the missile is sitting on the pad might simply result in a lost opportunity for launch while failure immediately after launch could result in a fallback and total destruction.

  To make these distinctions, sometimes hazards are separated into operational phases (for example, takeoff, cruise, landing), but this separation does not work for hazards that span phases and may cause them to be missed. Instead, the phase can simply be included in the context in which the hazard occurs, such as deceleration during takeoff or after the V1 point in the above list.

  A hazard always involves a state of the system and perhaps information about the context in which that state might occur. If the phase or phases in which the hazard occurs is important, it is best to simply document all the conditions under which the hazard can occur rather than to separate the analysis into phases.

- **Avoid checklists.**

  Using checklists for identifying hazards is not recommended. Checklists tend to limit thinking to the contents of the list. Rarely does identification go beyond that. Checklists are commonly used by non–subject matter experts, and this use is the most dangerous as the user does not have the expertise to identify what is missing. But having the expertise to potentially identify missing items is not enough to prevent it. Psychological studies have shown that even experts tend to ignore things not on the checklist or in a specification.

As one example, Fischoff, Slavin, and Lichtenstein conducted an experiment in which information was left out of fault trees. Both novices and experts failed to use the omitted information in their arguments, even though the experts could be expected to be aware of this information [102]. They attributed the results to an "out of sight, out of mind" phenomenon. In related experiments, an incomplete problem representation actually impaired performance because the subjects tended to rely on it as a comprehensive and truthful representation—they failed to consider important factors omitted from the specification. Thus, being provided with an incomplete problem representation can actually lead to worse performance than having no representation at all [396].

Another problem is that checklists commonly include causes of hazards; for example, oxidation or high pressure, rather than the actual hazard state. Including causes in the list of hazards, as stated, results in generating long and incomplete lists that are very difficult to review.

For systems that are very similar to those that have been designed before, the hazards may be similar or the same so reuse of the early parts of hazard analysis can be justified. But steps must be taken to ensure that any additional hazards not in the previous system are considered.

- **Special cases.**

In a few special cases, hazards are considered so serious and unacceptable that the systems are regulated by government agencies, and the hazards are set by them. For example, the US Department of Defense identifies four hazards that must be prevented when constructing nuclear weapon systems:

1. nuclear weapons producing a nuclear yield when involved in accidents or incidents, or jettisoned weapons producing a nuclear yield
2. deliberate pre-arming, arming, launching, firing, or releasing nuclear weapons except on execution of emergency war orders or when directed by competent authority
3. inadvertent pre-arming, arming, launching, firing, or releasing of nuclear weapons
4. inadequate security of nuclear weapons

Special analysis processes may also be mandated and required for such programs.

Identifying design constraints and criteria As the hazard identification process proceeds, the identified hazards should be translated into design constraints and criteria to be used by the engineers who are designing the system. This provides the information and goals the designers need to prevent accidents. Figure 9.4 shows an example of hazards and their translation into design constraints for the doors of a rapid transit system. Note that some of the design constraints are potentially conflicting, for example, the third and sixth. Identifying potential conflicts early in the design process will allow better resolution of such conflicts.

An example of a design constraint or criterion for an aircraft collision avoidance system is that maneuvers must be avoided that require the objects to cross paths. A typical design constraint for a pressure system is that there is a way to reduce pressure for all pressurized components when the pressure exceeds a specific amount above normal operating pressure. Like the hazards from which they are derived, the design criteria are very general and are refined later into specific requirements on the system components.

HAZARD	DESIGN CONSTRAINT
Train starts with door open	Train must not be capable of moving with any door open
Door opens while train is in motion	Doors must remain closed while train is in motion
Door opens while improperly aligned with station platform	Doors must be capable of opening only after train is stopped and properly aligned with platform unless emergent exists (see below)
Door closes while someone is in the doorway	Door areas must be clear before door closing begins
Door that closes on an obstruction does not reopen or reopened door does not reclose	An obstructed door must reopen to permit removal of obstruction and then automatically reclose
Doors cannot be opened for an emergency evacuation	Means must be provided to open doors anywhere when the train is stopped for an emergency evacuation

Figure 9.4
Design constraints for the doors of a rapid transit train.

Ideally, design criteria or constraints should state *what* has to be achieved, leaving the designer free to use ingenuity in deciding *how* the goal may best be achieved. In the pressure example above, it is best not to require use of a pressure relief valve but to instead leave the actual design decision for later in the development process when more information is available. Chapters 11 and 12 describe how to design to satisfy the design criteria and constraints.

Identifying causes of hazards After the hazards are identified, the next step is to determine the causal scenarios associated with each hazard. The causal scenarios can be high-level at first and later elaborated in greater detail as design decisions are made. Ideally, the causal scenarios will be used to assist in *making* the system design decisions. This approach intertwines hazard analysis and design in order to better direct and optimize the design process. Hazard analysis results obtained after a design is completed can, at best, result only in designs having to be changed after they are created, which is very costly. Sometimes finding problems late in the design process precludes the ability to provide the best solutions.

Each hazard will probably have several causal scenarios that can lead to it. The causal factors considered and the process of identifying them will depend on the underlying accident causality model used (see chapter 7), either consciously or subconsciously by the analyst. In complex systems, using accident models and their associated analysis techniques that only consider system component failures will lead to missing many important scenarios. Other limitations of some analysis techniques include the omission of humans, software requirements flaws, and accidents related to system design errors.

Causal analysis is sometimes divided into system (whole) and subsystem (unit or part) analyses. These two types of analysis are merely different aspects of a total process and different ways of looking at the system. Both are part of any complete hazard analysis, and separation can, again, lead to missing scenarios. Separating the analysis of various parts of the system, such as doing "software hazard analysis" or "human error analysis" is also a mistake: hazards arise from the system components working together, not separately. The next chapter describes and evaluates the most commonly used hazard analysis techniques.

Sometimes *Operational System Hazard Analysis* (OSHA) is considered as a separate analysis, and considerations of how the system will be operated are left until the design or development is complete or at an advanced stage. This separation is another mistake. How the system will be operated should be considered from the beginning and should be an integral part of the total system hazard analysis and design process.

9.3 Types of System Models

Every hazard analysis requires some type of model of the system, which may range from a fuzzy idea in the analyst's mind to a complex and carefully specified mathematical model. It may also vary in level of detail. Nevertheless, information about the system must exist in some form, and that information constitutes the system model on which the analysis is performed. The analysis may start with a high-level abstraction and continue with more detailed lower-level models.

A model is a representation of a system that can be manipulated in order to obtain information about the system itself. Different types of models can be used, depending on the type of system being considered and the type of information needed for the analysis.

The information can be stored or displayed in various ways. Simple representations, such as tables or drawings, can provide a way of storing or organizing information. but is difficult to use for formal manipulation or analysis methods. More formal structures may allow formal manipulation or analysis, such as trees or graphs, state machines (which show the states of the system and the events that cause state changes), or full unsteady-state models comprised of algebraic and differential equations. Some of these structures can be combined: models used in traditional hazard analysis techniques, for example, often employ tree structures to show logical relationships along with tables to describe additional information.

Selection of the model will determine what information can be specified and what can be derived from the model through analysis. For example, a particular model may not include any mechanism for showing changes over time, or it may have no way of representing dependencies between the system components being modeled. The better a representation of the original system and the better the match between the properties of the model and the properties of interest in the analysis, the more useful the model will be in providing useful information about system behavior.

When using any model, it is important to specify the boundaries of the system being modeled and what is not included, assumptions about the system (such as independence of components), and what assumptions are most likely to be incorrect or invalidated by changes and thus need to be checked periodically.

9.4 General Types of Analysis

Different types of models allow for various types of analysis or manipulation of the model to learn more about a system or focus the analysis on various system aspects. For example, a model and its associated analysis methods may focus on logical and functional structure of a technical system to evaluate the propagation of events and conditions, or it may include a detailed description of work or tasks to evaluate the effect of various human or software operations on the system state. The models and analysis techniques will influence the hazards and causes that can be identified and considered.

There is often a relationship between the difficulty of building and analyzing the model and the quality of information that can be derived from it. To make modeling and analysis practical, simplification of complex system behavior may be required. For example, process variables are intrinsically continuous, but for many models and analysis techniques, they can be treated as having a discrete and small number of values. Thus, flow may be characterized as *normal*, *high*, or *low*, or the state of a valve may be limited to *normal movement*, *stuck open*, or *stuck closed* if these are the states that are of interest.

No one model or analysis technique is useful for all purposes, and more than one type may be required on a system engineering project. A relatively simple system with only a few well-understood hazards may require only simple comparisons with previous systems or codes of practice. On the other hand, when what could happen or has happened involves complex sequences or combinations of conditions or events, then formal, documented analysis procedures will most likely be necessary.

Analysis techniques can be differentiated by their goals, whether they are quantitative, the phase in the life cycle when they are used (such as pre- or post-design), the depth of analysis, the domain on which they are defined (such as the structure and function of a technical system, the description of work or tasks, or the structure and function of an organization [360]).

Analysis techniques usually involve searching. The search strategy will depend on the type of structure being searched, including the basic elements of the underlying model (such as physical or logical components, events, conditions, or tasks) and the relationship between those elements. Typical relationships are temporal (time or sequence related) and structural (whole–part). For example, if the relationship is temporal, the search may identify prior or succeeding events or conditions that may or may not be causally related to the original event or condition. If the relationship is structural, the search may involve refining the event or condition into its constituents.

Search techniques of particular interest in hazard analysis can be classified as (1) forward and backward, (2) top-down and bottom-up, or (3) combinations of these two.

9.4.1 Forward and Backward Searches

Forward (sometimes called *inductive*) and *backward* (also called *deductive*) searches are useful when the underlying structure is temporal and the elements are events, conditions, or tasks (figure 9.5).

A forward search takes an initiating event and traces it forward in time. The result is a set of states, where a state is a set of conditions. An example of such a search is determining how the loss of a particular control surface will affect the flight of an aircraft.

The purpose of a forward search is to look at the effect on the system state of both (1) an initiating event, and (2) later events that are not necessarily caused by the initiating event. In fact, causal independence is often assumed.

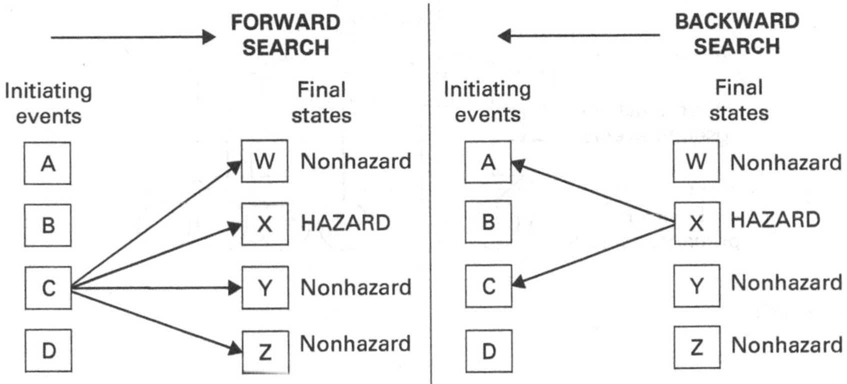

Figure 9.5
The states found in a forward search and in a backward search will probably not be the same.

Tracing an event forward can generate a large number of states, and the problem of identifying all reachable states from an initial state may be impossible using a reasonable set of resources. For this reason, forward analysis is often limited to only a small set of temporally ordered events, usually system component failures.

In a backward or deductive search, the analyst starts with a final event or state and determines the preceding events or states. This type of search can be likened to Sherlock Holmes reconstructing the events that led up to a crime.

The results of forward and backward searches are not necessarily the same. Tracing an initiating event forward will most likely result in multiple final states, not all of which represent hazards or accidents: there is one initiating state and multiple final states (figure 9.5).

Tracing backward from a particular hazard or accident to its succeeding states or events, in contrast, may uncover multiple initiating events. Forward searches could, of course, consider multiple events, but combinatorial explosion usually makes this goal impractical, and so the number of initiating events that can be considered is usually limited.

It is easy to see that if the goal is to explore the precursors of a specific hazard or accident, the most efficient method is a backward search procedure. On the other hand, if the goal is to determine the effects of a specific failure, a forward search is most efficient. Backward searches are usually used in safety analysis while forward searches are most appropriate for reliability analyses.

Backward and forward search methods fit well with traditional chain-of-event accident models, where the goal is to determine the paths (sets of states or events in temporal ordering) that can lead to a particular hazard or accident. Energy, epidemiological, and system-theoretic accident models may involve different types of searches.

9.4.2 Top-Down and Bottom-Up Searches

A second categorization of search methods is top-down (figure 9.6) or bottom-up (figure 9.7). Here the relationship being investigated is structural (whole–part): higher-level abstractions are refined or broken down into their constituent parts. A basic event, set, task, or system may be broken up into more basic events, conditions, tasks, or subsystems in a top-down

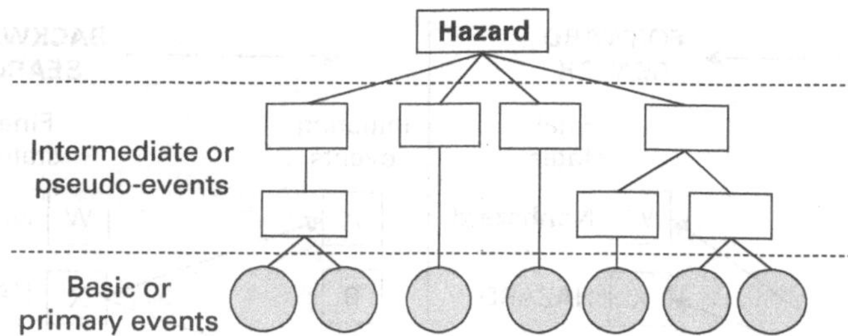

Figure 9.6
Top-down search.

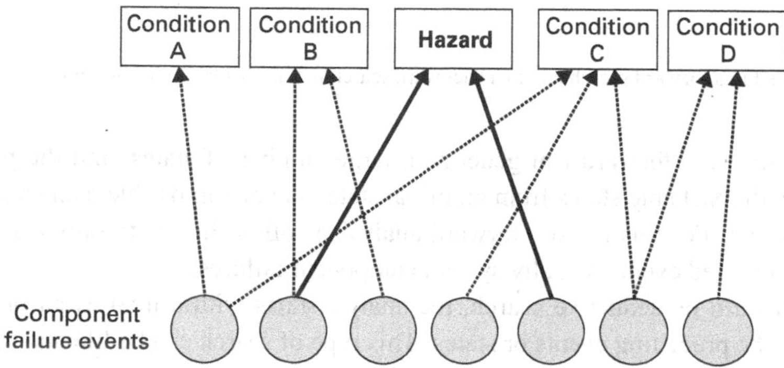

Figure 9.7
Bottom-up search.

search. When the search is bottom-up, subcomponents are put together in different ways to determine the result. An example of a top-down search is identification of all the ways that power can be lost. A bottom-up search might examine the effect of an individual battery failure on the system as a whole.

As with forward and backward searches, the results of top-down and bottom-up searches are not the same. For example, examining only the effects of individual component failures on the overall behavior or state of the system (a bottom-up search) misses hazardous system behavior that results from combinations of subsystem failures or from combinations of non-failure (correct) behavior of several subsystems. As in forward searches, considering the effects at the system level of all possible combinations of component behavior is not practical. Top-down searches that start from a hazardous system behavior will be more practical in this case.

On the other hand, determining the cause of a particular component failure is, theoretically, most efficiently accomplished using a bottom-up search. For complex systems, such a bottom-up search to determine the effect at the system level is usually very difficult or impossible.

9.4.3 Combined Searches

Some search strategies do not fit into one of these categories. Instead, the search starts with some event or deviation and goes forward and backward or top-down and bottom-up to find paths between hazards (or accidents) and their causes or effects. The search may start with deviations, failures, changes, unsafe control actions, and so on.

9.5 Who Should Do Hazard Analysis?

Successful hazard analysis requires an in-depth understanding of the system under consideration. That means that those doing the hazard analysis necessarily must be subject matter experts. In many instances, hazard analysis is separated from design engineering and performed by people who specialize in doing hazard analysis but who are neither subject matter experts nor experts on the specific system being designed. The results of such disconnected analyses are usually of limited usefulness.

Even if they have the requisite subject matter expertise, safety engineering specialists may be isolated in completely separate organizations, such as quality assurance. Interaction with design activities is limited or nonexistent until it is too late to make the changes needed to use the results of the hazard analysis and, at best, compromises in the design must be made.

In the worst case, the safety engineer's job consists of filling out lots of forms, documenting known failure rates for standard parts and, at the end of the project, providing the paperwork necessary to argue that the system is safe to operate as it is [63]. The results of the analysis then have almost no significant impact on the design and operation of the system. Such approaches have been cited as the cause of major accidents, such as the loss of UK Royal Air Force Hawker Siddeley Nimrod, which suffered an in-flight fire and crashed in Kandahar, Afghanistan in 2006, killing all sixteen crew members on board [128].

As repeatedly stressed in this book, hazard analysis should provide design engineers with tradeoff information as design decisions are made, not after they are made. Either design engineers themselves should be doing hazard analysis or experts on the analysis techniques should be deeply embedded in the design organization and not part of some separate organization. In addition, simply analyzing component failure rates may provide reliability information but not most of the information needed to make a system safe. Remember, safety and reliability are not the same thing.

Successful engineering, including safety engineering, requires interdisciplinary communication and a team of people with a wide variety of knowledge and skills. System safety engineers provide expertise in hazard analysis, while other team members bring expertise in specific engineering disciplines along with alternative viewpoints. The approaches to solving a problem suggested by system engineers, subsystem engineers, reliability engineers, safety engineers, human factors engineers, application experts, operators, and management may be entirely different, but all contribute significantly to finding a satisfactory design [13]. When humans and computers play important roles in system operation, operators, human factors experts, and software engineers should also be involved in the analysis and in the design of hazard reduction measures.

9.6 Limitations and Criticisms of Hazard Analysis

Hazard analysis serves as the basis of judgment for many aspects of system safety. Therefore, understanding its limitations and common problems is important. Because qualitative analysis always precedes quantitative analysis, all limitations of the former apply to the latter. Quantitative techniques have additional limitations, however.

Hazard analyses, along with their underlying accident and system models, may not match reality. They often are based on unrealistic assumptions, such as assuming that (1) the system is designed, operated, and maintained according to good engineering standards; (2) quality control procedures will ensure that all equipment conforms to the design specifications and is inspected, calibrated, maintained, repaired, and tested at suitable intervals; (3) testing is perfect and repair time is negligible; (4) operators and users are experienced and trained; (5) operational procedures are clearly defined; (6) the system operates perfectly from the beginning; and (7) key events are independent and random.

Phenomena unknown to the analysts obviously cannot be covered in the analysis, and discrepancies between the written documentation and the real system mean that important accident contributors may not be considered [364]. Sometimes, the boundaries of the analysis are drawn incorrectly, and relevant subsystems, activities, or hazards are excluded.

Even if all the assumptions are right to begin with, conditions change, and the models and analyses may not accurately reflect the current system. In general, there is no way to assure completeness; that is, that all factors have been considered.

Harvey claims that, in practice, the majority of errors in hazard analyses result from faults in the model or failure to foresee hazards and not from errors in the data:

> *The mistake made by many hazard analysts is to quantify (with ever greater accuracy) the particular hazards that they have thought of and fail to foresee that there are other hazards of much more importance. This means that so-called confidence limits have a very restricted meaning. They are telling you the error that can arise because of the sample size of the data, but they do not tell you that errors can arise because the analyst did not realize that there was some other way in which the hazard can occur.* [137]

Hope and colleagues [150] suggest that the general nature of some of the methodologies can easily lead to misinterpretation and misuse of the results.

The limitations discussed so far relate to the particular models constructed. A second group of limitations relates to simplifications in the modeling and analysis techniques themselves. Examples include requiring that continuous variables be specified as discrete variables when continuity is important in the analysis, not allowing consideration of certain timing factors such as time delays or the ordering of events, and assuming independence so that common-cause and common-mode failures are not handled.

Simplifications may stem from the inability to represent particular aspects of the system or to evaluate them in the analysis. Policy and principles of management are rarely included, for example, nor are the safety culture of the organization, organizational structure, training factors, and the safety engineering process employed. Models that require assigning numbers to everything may omit important factors because they are not easily quantified. In addition, there may be limitations in the search patterns, the system models, or the underlying accident models that restrict the factors that can be considered.

Some oversimplifications stem from limitations in knowledge. All the data used in a quantitative analysis must be relevant to the situation being analyzed, but in practice data is scarce, incomplete, and often not directly applicable [150]. Sometimes assumptions about infrequent events have to be based on extremely limited data with necessarily low confidence levels. It is difficult to collect statistically sufficient data on component behavior that includes different operating conditions, component types, failure modes and distributions, and so on.

Not only are there inaccuracies in data, but the information required for consequence modeling and human error data is often unknown or inaccurate, as discussed earlier.

Other limitations of hazard analysis stem not from the techniques themselves but from the fact that they must be used by humans. To varying degrees, the analysis represents the analyst's interpretation of the system; the analyst may inadvertently introduce bias, especially when the system being analyzed is complex.

In addition, many of the techniques are complicated and demand appreciable specialist manpower and time that may not be available. Automation of the analysis does not completely solve the human limitation problems—it merely shifts them to the computer programmer, who, in effect, becomes the analyst. Automated analyses also usually require a manually constructed system model, with all its inherent limitations. The system model used may omit aspects of the system that are not easily incorporated in that particular model in order to allow automated generation or analysis. The result is usually very detailed analyses of very limited aspects of the system.

All of this does not mean that hazard analysis should not be done, only that the limitations of the results must be considered when using them and when selecting a particular analysis approach. It also implies that hazard analysis will be most useful if performed by people having expertise and knowledge about the system being designed.

9.7 Analysis versus Assessment

Analysis can be defined as a detailed examination of anything complex in order to understand its nature or to determine its essential features. Less formally, it is a thorough study of something. There is no implication that quantification must be done; analysis can be purely qualitative and commonly is. Assessment usually implies some type of quantitative or comparative analysis.

Qualitative analysis is more useful in improving designs than quantitative information. For example, knowing the potential scenarios—events and conditions—leading to a hazard provides engineers and designers with the information necessary to eliminate or reduce the occurrence of the hazard in the design or in operations.

Quantitative assessment, for example, determining the probability of a specific hazard occurring is 10^{-7} per hour of operation, is most useful in determining how much effort and resources to place on dealing with the hazard, if any, but not what to do about it. Quantitative information about all the hazards plus the likelihood of the environmental conditions that will lead to a loss is useful in certification, licensing, insurance, and other decisions after the system development is completed, but does not provide the design engineer with much useful information at the beginning of or during the development process.

This chapter has described the general hazard analysis process. The next chapter examines specific hazard analysis techniques. The related but different topics of assurance and risk assessment are covered in chapter 13.

Exercises

1. Select a type of common system in your field or in a safety-critical industry.
   a. Identify the system hazards and losses associated with it. Make sure your losses and associated hazards are very high level so that you have no more than ten of each (and you may have fewer). What do you think will happen if you identify a large number of more detailed hazards (say 100 to 200)? The words used to describe each of your hazards should not include any mention of system components; for example, for an automobile a system hazard would be "the car is unable to stop" rather than "the brakes fail." Why do you think this is important? Keep your list at a high enough level of abstraction that it contains ten or fewer things.
   b. Take one of the high-level system hazards and refine it into more detailed hazards, again with a limit of ten to twelve. Do you think this process of successive refinement will minimize duplications and omissions (as claimed in the text)? Why or why not?
2. Identify some stakeholders in your field and what their concerns might be in terms of identifying losses important to them. Will the different stakeholders have different concerns?
3. In a forward search, almost always failures are examined to find the future states. What other alternative is there to using failures? What are the implications of using the alternative(s) you identified?
4. What type of information might be useful for an airline and pilots to get from the aircraft manufacturers to operate the aircraft safely? Some products are mostly sold to individuals rather than to large companies, for example, cars. What types of safety information should be provided to individual car owners? The standard method used to pass information to car owners is through a written owner's manual. What are the drawbacks to this type of information passing and can you think of a better way or ways to do it?
5. Give a specific example of a forward search and a backward search that provide different results.
6. Consider figure 9.1. How might the results of hazard analysis be used by each of the components in the figure?

10

Hazard Analysis Techniques

Accidents on the whole are becoming less and less attributable to a single cause, more to a number of contributory factors. This is the result of the skill of the designers in anticipating trouble, but it means that when trouble does occur, it is inevitably complicated.
—DeHavilland and Walker
(after reviewing failures of the *Comet* aircraft, quoted in [381])

There is no such thing as an accident. What we call by that name is the effect of some cause which we do not see.
—Voltaire

A whole book could be written just on hazard analysis techniques, and many have been. The goal of this chapter is not to teach the techniques. Instead, the intent is to provide an overview of the most commonly used hazard analysis techniques and their relative usefulness and power in order to assist in selecting an appropriate one for a particular use. Details about how to perform them can be found elsewhere.

The techniques employed in each industry tend to differ partly because the industries have unique hazards, such as chemical exposure versus flying into a mountain, and partly for historical reasons. Although many hazard analysis techniques have been proposed, very few are actually used extensively in practice.

One thing to keep in mind is that almost no validation of most of these techniques has ever been done, even those that are widely used. That does not mean that the techniques are not useful, only that they must be used carefully and combined with a large dose of engineering judgment and expertise.

Any hazard analysis method reflects the underlying causality model. Chapter 7 described the four main types of causality model: (1) energy models, (2) linear chain-of-failure-events models, (3) epidemiological models, and (4) systems-theoretic models. Because no special hazard analysis methods have been defined for the epidemiological model, only the other three are covered in this chapter.

10.1 Energy Model Techniques: Hazard Indices

The energy model is so simple that most techniques support it. The only widely used approach created directly for the energy model is Hazard Indices.

Description Hazard Indices measure loss potential due to fire, explosion, and chemical reactivity hazards in the process industries. They were originally developed primarily for insurance purposes and to aid in the selection of fire protection methods, but they can be useful in general hazard identification, in assessing hazard level for certain well-understood hazards, in the selection of hazard reduction design features for the hazards reflected in the index, and in auditing an existing plant.

The oldest and most widely used index was developed by the Dow Chemical Company. The *Dow Chemical Company Fire and Explosion Index Hazard Classification Guide* (usually abbreviated as the *Dow Index*) was first published in 1964 and was originally the basis for calculating a fire and explosion index. Later, it was expanded to calculate the maximum probable property damage and the maximum probable days of outage. Any operation where a flammable, combustible, or reactive material is stored, handled, or processed can be evaluated with the *Dow Index*. Auxiliary plant components, such as power generation equipment, office buildings, control rooms, or water systems, are not covered.

The *Dow Index* first requires dividing the plant into units, a unit being a part of a plant that can be readily and locally characterized as a separate entity. Generally, a unit consists of a segment of the overall process: in some cases, it may be a part that is separated by distance or by walls; in others, it may be an area in which a particular hazard exists [150; 236]

The fire and explosion index indicates the fire and explosion hazard level of a particular unit. The calculation of this index uses a measure, called the *Material Factor* (MF), of the energy potential of the most hazardous material or materials in the unit in sufficient quantity to present a hazard. This measure is a number from one to forty and is calculated on the basis of flammability and reactivity. For some properties, the MF can be found in a table; for others, it must be calculated [206]. General and special hazards, including factors such as properties of the materials, quantities involved, the type of process and whether it is difficult to control, process conditions, and construction materials, are treated as penalties applied against the MF. A toxicity index can also be calculated to evaluate the exposure level of toxicity hazards.

Basically, these calculations combine a number of empirical hazard factors that reflect the properties of the materials being processed, the nature of the process, the spacing of equipment, and the judgment of the analyst about them [150]. The index is then used to determine the fire protection required. Basic fire protection design features, including minimum separation distances, are recommended in the *Dow Index*.

Attempts have been made to improve on this index or to come up with alternative indices, but most alternatives have not found widespread acceptance outside the organizations in which they were developed [292]. One that has been used in the chemical industry, called the *Mond Index*, was proposed in 1979. It expands the *Dow Index* to include additional factors related to the effects of toxic materials and layout features, such as spacing, access, height, and drainage, on the hazard level.

Evaluation Hazard Indices provide a quantitative indication of the potential for hazards associated with a given design. They work well in the process industry, where designs and equipment have historically been standard and changed little, but they are less useful for systems where designs are unique and technology changes rapidly. In fact, most industries, including the process industries, now are introducing new technology such as software and digital systems.

Lowe and Solomon [236] claim that the *Dow Index* and others are particularly useful in the early stages of hazard assessment because they require a minimum of process and design data and can graphically demonstrate which areas within the plant require more attention. The indices can also help to identify which of several competing process designs contain the fewest inherent hazards, and they provide information useful for site selection and plant layout [293].

The indices only consider a limited set of hazards, and even for these, they determine only hazard level. No attempt is made to define specific causal factors, which are necessary to develop hazard elimination or reduction measures beyond the standard equipment information provided in tables. Thus, the indices do not provide a complete picture and are useful primarily to supplement other hazard analysis methods.

10.2 Techniques Based on the Chain-of-Failure-Events Causality Model

The analysis techniques that in the past have been used the most are based on the linear chain-of-events causality model, with different techniques used in different industries. The most commonly used are FMECA (Failure Modes and Effects Criticality Analysis), FTA (Fault Tree Analysis), ETA (Event Tree Analysis), and HAZOP (HAZard and Operability Analysis).

10.2.1 Failure Modes and Effects Criticality Analysis

Description FMECA is a derivative of FMEA (Failure Modes and Effects Analysis). FMEA was originally developed in the 1940s by reliability engineers to allow them to predict equipment reliability. As such, it is a form of reliability analysis that emphasizes successful functioning rather than hazards and risk. The goal is to establish the overall probability that the product will operate without a failure for a specific length of time, or, alternatively, that the product will operate a certain length of time between failures. A FMECA adds criticality information about potential failures to the FMEA and thus can provide safety-related information, but still considers only component or functional failures as causes of accidents.

FMECAs use forward search based on an underlying chain-of-failure-events model, where the initiating events are failures of individual components. The first step in a FMECA is to identify and list all components and their failure modes, considering all possible operating modes. For each failure mode, the effects on the overall system are determined. Then the probabilities and seriousness of the results of each failure mode are calculated. Figure 10.1 shows a simple example. Other columns with additional information may be included.

For standard components, failure probabilities can be predicted from generic rates that have been developed from experience and are often published. Manufacturers usually have this data for their own products. Care must be taken that the environment in which

FAILURE MODES AND EFFECTS CRITICALITY ANALYSIS						
Subsystem______		Prepared by______			Date____	
ITEM	FAILURE MODES	CAUSE OF FAILURE	POSSIBLE EFFECTS	PROB.	LEVEL	POSSIBLE ACTION TO REDUCE FAILURE RATE OR EFFECTS
Motor case	Rupture	a. Poor workmanship b. Defective materials c. Damage during transportation d. Damage during handling e. Overpressurization	Destruction of missile	0.0006	Critical	Close control of manufacturing processes to ensure that workmanship meets prescribed standards. Rigid quality control of basic materials to eliminate defectives. Inspection and pressure testing of completed cases. Provision of suitable packaging to protect motor during transportation.

Figure 10.1
A sample FMECA.

the component will be working is identical to the one for which the statistics were collected. Probabilities are based on averages collected over large samples, but individual components may differ greatly from the average, perhaps because of substandard manufacturing or extreme environments. Confidence levels and error bounds are often omitted from FMECAs, but they should be included.

Problems in performing FMECA arise when nonstandard hardware components are used without extensive information about their failure modes, failure rates, or failure probability. Even more serious problems occur when the system includes software; the ways that software can do the wrong thing are almost boundless, and the probabilities of these behaviors are unknowable. Including humans in the FMECA creates similar problems, so they often are simply ignored or their behavior drastically oversimplified.

The FMECA results are documented in a table with column headings such as failure modes, cause of failure, possible effects, probability, severity level, and possible actions to reduce the effects. Variations in what column headings are included in the table are common.

Standard probability may be used, if it is known, or frequencies may be used, such as the number of failures of a specific type expected during each one million operations performed in a critical mode. A description may be included of the preventive and corrective measures that should be taken and the safeguards to be incorporated.

Sometimes a Critical Items List (CIL) is generated from the results of the FMEA or FMECA. This list might include the item name, a list of possible failure modes for each item, failure probability (for each operational mode), effect on the mission (such as abort, degradations of performance, or damage), and criticality ranking within the subsystem (perhaps using a numerical scale).

More recently, functional FMEAs have become popular where the unit of concern is not a physical unit but a function, where the function is defined as a series of steps. The analysis considers the failure of the individual steps in the function. FMEA has also been similarly used to analyze processes, services, and manufacturing. The definition of what is meant by a "failure" in these instances needs to be specified. Obtaining probability data may be even more difficult in these usages.

Life-cycle phase Traditional FMEAs are appropriate when a design has progressed to the point where hardware items may be easily identified on engineering drawings and functional diagrams. The analyst needs a detailed design that includes schematics, functional diagrams, and information about the interrelationships between component assemblies. At that point, however, little can be done to improve safety except in very simple systems.

While functional FMEAs have been used at earlier life-cycle phases, it is not clear how a likelihood can be determined without a detailed design of the system. Different designs of particular functions will have different likelihoods of failure.

Evaluation Traditional FMECA is effective for analyzing single units or single failures to enhance individual component reliability or to resolve tradeoffs between differing physical component designs. It has been used to identify redundancy and fail-safe design requirements, single-point failure modes, and inspection points and spare part requirements. It is also useful in determining how often the components must be serviced and how components and designs must be improved in order to extend the operational life of a product.

One strength of the technique is its completeness and relative simplicity, at least with respect to standard hardware components, but completeness means it is also very time-consuming and can become tedious and costly if applied to all parts of a complex design. An aircraft today, for example, can have millions of parts.

All the significant failure modes must be known in advance, so FMECA is most appropriate for standard parts with few and well-known failure modes. The technique itself does not provide any systematic approach for identifying failure modes or for determining their effects and no real means for discriminating between alternate courses of improvement or mitigation. In fact, for systems that exhibit any degree of complexity, identifying all possible component hardware failure modes—both singly and in combination—is impossible [394].

FMECA does not normally consider effects of multiple failures; each failure is treated as an independent occurrence with no relation to other failures in the system except for the subsequent effects it might produce. By limiting the analysis to single units and not considering multiple- or common-cause failures, the technique becomes simple to apply and the examination is very orderly, but the results may be of limited use if time sequences and the interrelationships among the elements of a system are not considered. Even before the extensive use of software in products, failures more often resulted from connector problems than failures in the components themselves [133].

Hammer points out that, as usually applied, FMECAs pay little attention to human errors in operating procedures, to software, to hazardous characteristics of the equipment, or to adverse environments [131]. Although environmental conditions are considered in identifying the stresses that could cause hardware to fail, the probabilities of occurrence of such environmental stresses are rarely used. Instead, a usage factor is incorporated for the type of system application, such as shipboard, aircraft, or missile use, and another factor is applied for reduction of theoretical reliability that could result from substandard manufacture or assembly. This latter factor is extremely rough, even over a large sample. Hammer writes: "*Oddly enough, in spite of all those factors affecting a system but whose probability of occurrence can only be estimated imprecisely, reliability engineers carry out their calculations to six or seven significant figures*" [131].

Because they include the end effects of failures, FMECAs are sometimes used in safety analyses. If the limitations are understood, there is no problem with this. Not all failures result in accidents, however, so analyzing all parts, the ways each part can fail, and the resultant effects is generally a time-consuming and inefficient way to obtain only part of the needed safety-related information. One limitation of the information obtained is that the probability of damage determined by a FMECA is related to individual failures only; it rarely involves investigating damage or injury that could arise if multiple components fail or if the components operate successfully but the problems arise in their interactions. And, of course, there is no way to obtain the historical reliability numbers for nonstandard or new components and for software.

Given the changes in engineering in the eighty years since FMEA was invented, its expense, and its limited applicability to most systems today, it is surprising that it is still used as much as it is. One reason may be that the analysts do not need to be subject matter experts or to have much knowledge of the design involved beyond a list of the standard parts included. For the same reason, FMECA is not very useful for safety.

Functional FMEAs, which examine the failure modes of the process steps, have many of the same limitations. The most important limitations are that only failures of individual process steps can be included and not their interactions. Processes today usually involve many interactions among parts of the processes and the encompassing systems. And, of course, Functional FMEA is still a reliability analysis and not a hazard analysis. Hazards, beyond those related to individual component failures, are not considered.

10.2.2 Fault Hazard Analysis

Fault Hazard Analysis (FHA) was used on the Minuteman missile system and thus dates from at least the 1950s [394]. It is basically a FMECA with both a broader and more limited scope, and therefore essentially a reliability analysis technique. But while FMECA considers all failures and their effects, FHA considers only those effects that are safety related. FHA is often performed on functions rather than hardware components.

The scope is broadened by considering human error, procedural deficiencies, environmental conditions, and other events that might result in a hazard caused by normal operations at an undesired time [75]. At the same time, its scope is more restricted than that of a FMECA because supposedly only failures that could result in accidents are considered, although it is difficult to understand how a forward analysis of this type can be done without all failures being considered.

Two new pieces of information are added about upstream and downstream effects: (1) upstream components that could command or initiate the fault in question, and (2) factors that could lead to secondary failures. The effects on the system are briefly stated in terms of associated damage or malfunction. Once again, the results are documented in a table. The column headings may include component or function, failure probability, failure modes, percent failure by mode, effect of failure (traced to some relevant interface), upstream components that could command or initiate the failure or fault, and factors that could cause secondary failures.

Table 10.1 shows an example FHA from a standard used in the certification of commercial aircraft. Note that hazards are limited to failure conditions and do not include other types. Consideration of human behavior is limited to the assumed consequence of the

Table 10.1
Aircraft FHA (partial only) [from SAE ARP 4761, p. 178]

Function	Failure Condition (Hazard Description)	Phase	Effect of Failure Condition on Aircraft/Crew	Classification	Reference to Supporting Material	Verification
Decelerate Aircraft on the Ground	Loss of Deceleration Capability	Landing/ RTO/Taxi	See below			
	a. Unannunciated loss of deceleration capability	Landing/ RTO	Crew is unable to decelerate the aircraft resulting in a high-speed overrun	Catastrophic		S18 Aircraft Fault Tree
	b. Annunciated loss of deceleration capability	Landing	Crew selects a more suitable airport, notifies emergency ground support and prepares occupants for landing overrun.	Hazardous	Emergency landing procedures in case of loss of stopping capability	S18 Aircraft Fault Tree
	c. Unannunciated loss of deceleration capability	Taxi	Crew is unable to stop the aircraft on the taxi way or gate resulting In low speed contact with terminal, aircraft, or vehicles	Major		
	d. Annunciated loss of deceleration capability	Taxi	Crew steers the aircraft clear of any obstacles and calls for a tug or portable stairs	No Safety Effect		
	Inadvertent Deceleration after V1 (Takeoff/RTO decision speed)	Takeoff	Crew is unable to takeoff due to application of brakes at the same time as high thrust settings resulting in a high-speed overrun	Catastrophic		S18 Aircraft Fault Tree

functional hazard/failure on their behavior. Of course, different uses of FHA could include different types of information in the columns and humans could be included in a more substantial way.

Note that in the aircraft FHA example shown in table 10.1, best-case behavior and environment is assumed to evaluate the effect of the failure on the aircraft and crew, not necessarily worst-case. For example, consider case "d" where the loss of deceleration capability (perhaps a brake failure) is considered and the crew is informed; that is, the loss of braking ability is annunciated to the crew. An assumption is made that the crew will be able to safely stop the aircraft without any danger, and therefore this case is considered to have no safety effect. It is, however, easy to think of conditions where this assumption is not true and an accident could result.

Evaluation Like FMEAs and FMECAs, FHA primarily provides guidance on what information to obtain, but it provides no help in actually getting that information. And, again, FHA in use tends to concentrate primarily on single events or failures. As with FMECA, it can also be very expensive if done thoroughly.

In addition, probability of failure leading to hardware-related hazards ignores latent defects introduced through substandard manufacturing processes. Therefore, some hazards may be missed [388].

Finally, emphasis may be placed on obtaining numbers and not in obtaining the information necessary to eliminate hazards. One of the pitfalls in FHA, and in other techniques, is over precision in mathematical analysis. Analysts may try to obtain exact numbers from inexact data and spend too much time improving the preciseness of the analysis rather than eliminating the hazards.

Once again, software and human errors may be omitted or treated superficially. Because human error analysis is so important, it is treated separately at the end of this chapter.

10.2.3 Fault Tree Analysis

FTA (Fault Tree Analysis) is widely used in the aerospace, electronics, and nuclear industries. It was originally developed in 1961 by H. A. Watson at Bell Telephone Laboratories to evaluate the Minuteman Launch Control System for an unauthorized or inadvertent missile launch. Boolean logic methods had been used at Bell Labs for communications equipment, and these were adapted to create FTA. Engineers and mathematicians at the Boeing Company developed the procedure further and became its foremost proponents.

Description FTA is primarily a means for analyzing causes of hazards, not identifying hazards. The top event, the hazard, in the tree must have been foreseen and thus identified first in other ways. FTA uses Boolean logic to describe the combinations of individual faults that can constitute a hazardous event. Each level in the tree lists the more basic events that are necessary and sufficient to cause the problem shown in the level above it.

FTA is a top-down search method. Backward or forward search techniques are chronological orderings of events over time, but each level of the fault tree merely shows the same thing in more detail. The intermediate events (events between the top event and the leaf nodes in the tree) are pseudo-events; that is, abstractions of real events—they are simply combinations or sets of the basic or primary events and are usually removed during the later formal analysis of the tree (figure 10.2).

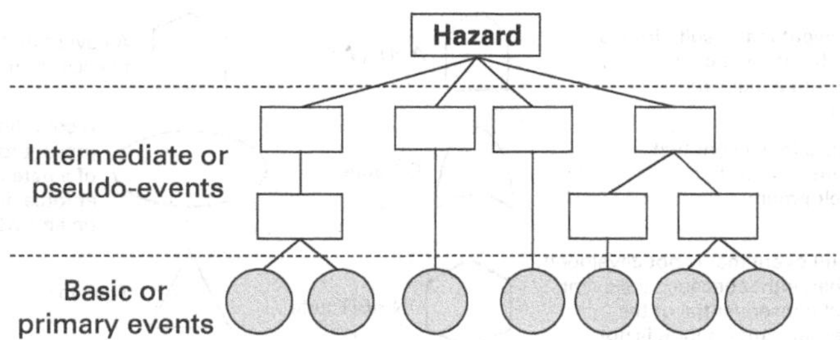

Figure 10.2
The leaf nodes of a fault tree represent the basic or primary events.

Once the tree is constructed, it can be written as a Boolean expression and simplified to show the specific combinations of identified basic events sufficient to cause the undesired top event. If a quantitative analysis is desired and feasible (that is, the individual probabilities for all the basic events are known), the frequency of the top event can be calculated.

Fault Tree Analysis has four basic steps: (1) system definition, (2) fault tree construction, (3) qualitative analysis, and (4) quantitative analysis.

Step 1: System definition

This step is often the most difficult part of the FTA task; it requires determining the top event, initial conditions, existing events, and impermissible events. The selection of top events is crucial because the assessment of hazards in the system will not be comprehensive unless fault trees are drawn for all significant top events.

A thorough understanding and definition of the system and its interrelationships is essential for this step and all other steps in FTA. The analyst may use system functional diagrams, flow diagrams, logic diagrams, or other design representations, or may rely on his or her knowledge of the system. The physical system boundaries must be carefully defined. But the analysis itself is performed in the analyst's head, not on a model of the system. The fault tree simply shows the results of the analysis.

For any component that has more than one possible state, the analyst must decide on the system state (initial state) to be analyzed for the occurrence of the top event. If the top event is an inadvertent weapon release from an aircraft, for example, the events in the tree will be very different depending on whether the aircraft is on the ground, in flight and cruising to target, or over the target but not in proper position for the release [333]. Similarly, the fault tree for the collision of two automobiles will depend on traffic speed and density.

Step 2: Fault tree construction

Once the system has been defined, the next step is fault tree construction. Briefly, the analyst first assumes a particular system state and a top event and then writes down the causal events related to the top event and the logical relations between them, using logic symbols to describe the relationships. Figure 10.3 shows the symbols used for fault trees, of which the most frequently used connectors are AND gates and OR gates. The output of an AND gate exists only if all the inputs exist: it represents combinations of events. In contrast, the output of an OR gate exists provided at least one of the inputs exists: it shows single-input events that can cause the output event. The input events to an OR gate do not

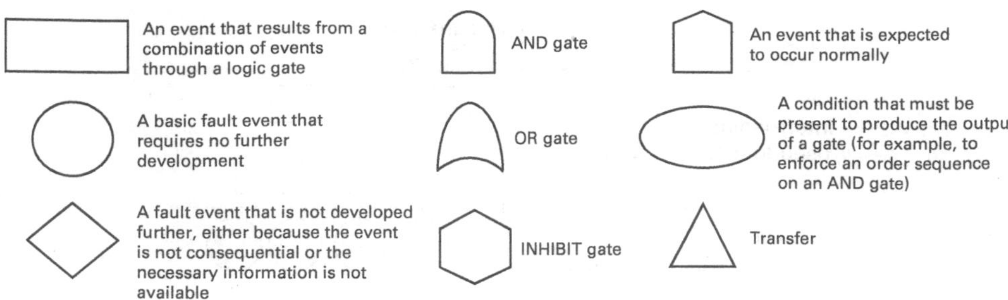

Figure 10.3
Fault tree symbols.

cause the event above the gate but are simply re-expressions of the output event. In contrast, the events attached to the AND gate are the causes of the event above the gate [394]. This causal relationship is what differentiates an AND gate from an OR gate.

INHIBIT (NOT) gates can also be used but are less common. They may be needed in a situation where there are two flows X and Y, and the top event occurs if there is either no X flow or no Y flow. The simple OR gate is inclusive—it states that the top event occurs if there is a failure of X flow, Y flow, or both. If the goal is to specify that the top event does not occur if there is a failure of *both* X and Y flows (exclusive OR), then the simple OR gate will not suffice and an INHIBIT or other type of gate is needed.

The relationships between the events shown in the fault tree are just standard logical relations and therefore can be expressed using any of the alternative forms of Boolean algebra or truth tables. The tree format, however, seems to have advantages in terms of readability.

The process continues, with each level of the tree considered in turn until basic or primary events are reached. These are completely arbitrary, and the analyst must determine the stopping rule for the analysis or, in other words, the resolution limit of the analysis. The events considered to be basic in the analysis will depend on its purpose, scope (a first estimate or a fully detailed analysis), and intended users; the available knowledge about the causes of events; and the availability of statistical data if a quantitative analysis is desired. Figure 10.4 is an example of a fault tree where the top event is identified by an FHA.

Step 3: Qualitative analysis

After the tree is constructed, qualitative analysis can begin. The purpose is to reduce the tree to a logically equivalent form showing the specific combinations of basic events sufficient to cause the top event. In essence, the intermediate pseudo-events are removed, and only relationships between the top event and the primary events are described. These are called *cut sets*. The goal of the analysis is to find the *minimal cut sets*, which represent the linear sequence of basic events that will cause the top event and which cannot be reduced in number—that is, a cut set that does not contain another cut set. Cut sets are defined such that if even one event in the cut set does not occur, the top event will not take place. Figure 10.5 shows an example.

The minimal cut set representation as a tree corresponds to one OR gate with all the minimal cut sets as descendants. The same primary events usually will occur in more than one of the minimal cut sets; thus, the minimal cut sets are generally not independent

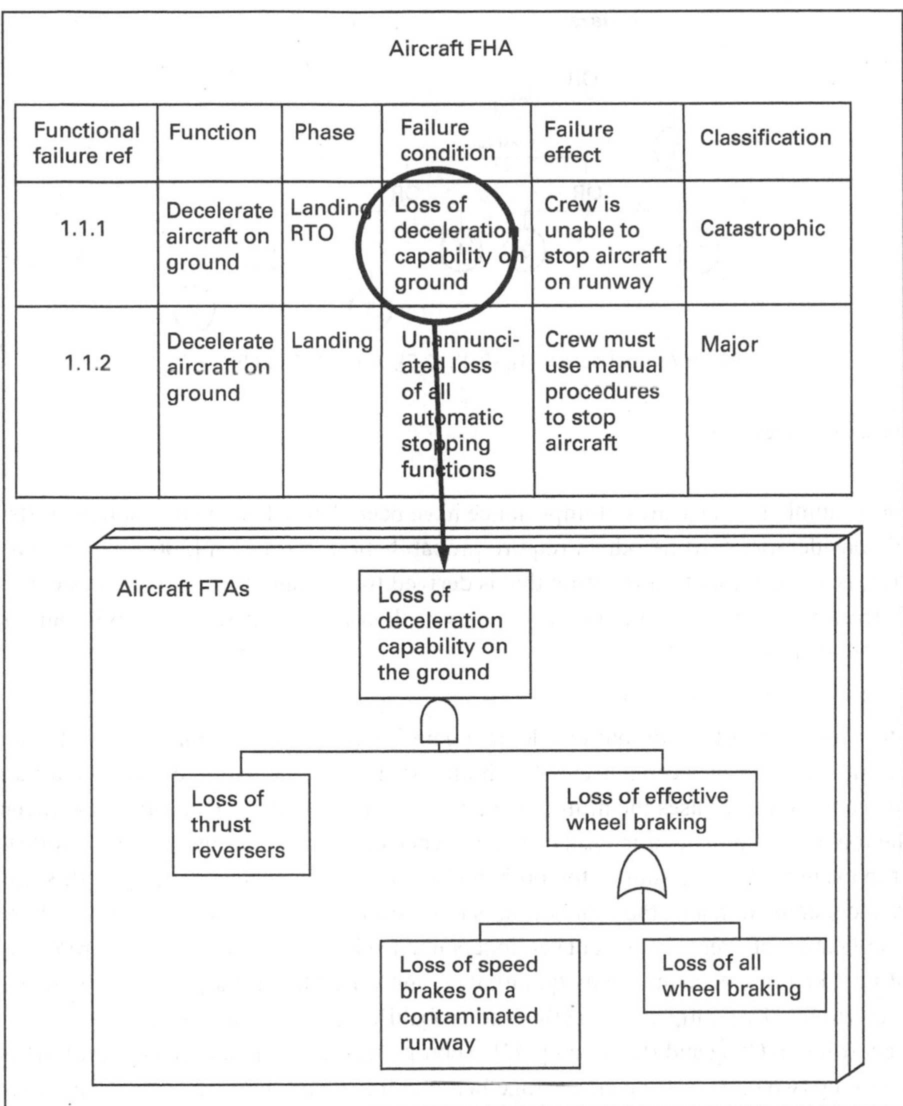

Figure 10.4
Portion of a fault tree with the initial event identified by an FHA [from SAE ARP 4761].

of each other. A medium-sized fault tree can have millions of minimal cut sets, so computer programs have been developed to calculate them.

The procedures for reducing the tree to a logically equivalent form are beyond the scope of this book; the interested reader is referred to one of the many books on this subject. In general, the procedures employ Boolean algebra or numerical techniques, such as using the logical structure of the tree as a model for trial-and-error testing of the effects of selected combinations of primary events.

Minimal cut sets provide information that helps identify weaknesses in the system. For example, they determine the importance or ranking of each event with respect to the top

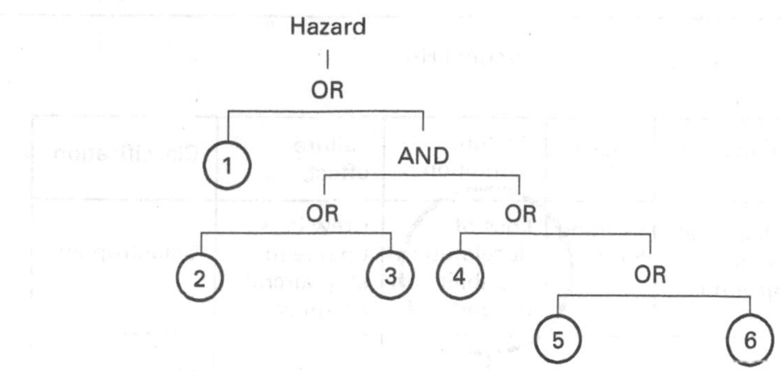

Figure 10.5
Minimum cut sets example.

event. A number of measures of importance have been defined; some rely solely on structural considerations, while others require probabilistic information [206]. A problem, of course, is the amount of information that is derived from a fault tree on a real system. The fault trees themselves can be hundreds or even thousands of pages long, with an enormous number of minimal cut sets.

Step 4: Quantitative analysis

The probability of the output of a logical gate is equal to the probability of the corresponding function of the input events—both characterize the same event. Quantitative analysis of fault trees uses the minimal cut sets to calculate the probability of occurrence of the top event from the probability of occurrence of the basic events. The probability of the top event will be the sum of the probabilities of all the cut sets if they are all statistically independent; that is, the same event is not present in two or more cut sets. If there is any replication of events in any cut set, independence is compromised, and the replication must be taken into account in any quantitative analysis. The probabilities of each cut set are determined by multiplying together the probability of the basic events.

According to Ozog and Bendixen [293], a common mistake in quantifying fault trees is multiplying two or more frequencies together, yielding meaningless results. To help avoid this mistake, they suggest changing the tree symbols to clarify which events are frequencies and which are probabilities.

If the probabilities of the basic events are given by a probability density function—the range of probabilities over which the event can occur—rather than by a point probability value, then the probability of the top event also must be expressed as a density function. Monte Carlo simulation can be used to determine these functions [150].

Figure 10.6 shows an example of a quantified fault tree for aircraft. The probabilities are combined bottom-up from the probabilities of the basic events. The question remains, of course, about how to get these probabilities, especially if they include software, humans, or even hardware based on new technology.

Automatic synthesis Several procedures for automatic synthesis of fault trees have been proposed, but these work only for systems consisting purely of hardware elements [15;

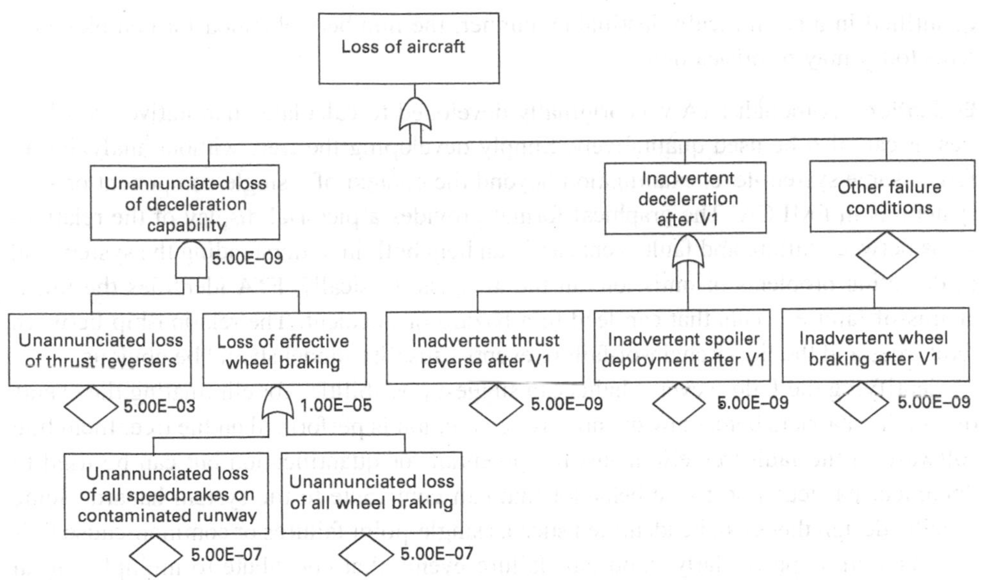

Figure 10.6
Part of the quantified fault tree for loss of aircraft including just a deceleration hazard [source: SAE ARP 4761]. (V1 is the point in the takeoff procedure where it is safer to take off than to try to brake.)

20; 206]. Basically, a model of the hardware, such as a circuit diagram, is used to generate the tree.

Taylor's technique, which is typical, takes the components of the hardware model and describes them as transfer statements [371]. Each statement describes how an output event from the component can result from the combination of an internal change in the component and an input event. Such statements can also describe how the component changes state in response to input events. In general, the transfer statement will be conditional on the previous component state. Together, the transfer statements form the transfer function for the component.

Both the normal and failure properties of the component are described, and each transfer statement is represented as a small fragment of a fault tree or mini-fault tree. The synthesis process consists of building the fault tree by matching the inputs and outputs of the mini-fault trees. A serious limitation, of course, is that only hardware failures are included.

Life-cycle phase Although generic fault trees can be constructed before the details of design and construction are known, they are of limited usefulness. To be most effective, FTA requires a completed system design and a thorough understanding of the system and its behavior in all operating modes. Information is usually too incomplete to perform detailed Fault Tree Analysis at the preliminary design stage [13], although alternative designs may be compared. Fault trees can also be used early in the design process to identify where interlocks are required and will be most effective, but they are not the most efficient model for identifying this information [54].

FTA is usually applied to completed or existing systems to demonstrate a particular failure probability. However, as software and human errors are not included or cannot be quantified in a scientifically justifiable manner, the numbers obtained for complex systems today may be misleading.

Evaluation Although FTA was originally developed to calculate quantitative probabilities, it can also be used qualitatively. Simply developing the tree, without analyzing it, forces some system-level examination beyond the context of a single component or subsystem as in FMECA. The graphical format provides a pictorial display of the relationships between failure and fault events and can help both in understanding the system and in detecting problems or omissions in the analysis. Basically, FTA identifies the linear chains-of-failure events that can lead to a hazard or accident. The relationship between fault trees and the linear chain-of-failure events causality model should be obvious.

Identifying the failure event chains can suggest possibilities for eliminating the events or creating barriers before any quantitative evaluation is performed on the tree. Including software in the fault tree eliminates the potential for quantification but can be used to document particular software behavior that can contribute to the system hazard. Some specific design flaws can be identified such as single-point failures or common-cause failures as well as particularly important failure events that contribute to multiple causal chains. Potential common-cause failures resulting from fault propagation (domino effects) cannot be handled in fault trees [230].

Another positive and important aspect of Fault Tree Analysis is that problems may also be found because the analyst has to think about the system in great detail during tree construction.

Fault Tree Analysis has serious limitations, however. The most useful fault trees can be constructed only after the product has been designed; they require detailed knowledge of the design, construction, and operation of the system. A good safety program, in contrast, requires concentrating on the early stages of the system life cycle. Waiting until designs are complete enough to be analyzed using fault trees either limits the ability to prevent the faults or requires often unrealistic and costly backtracking during development. Generic fault trees can be built early, but they may provide only information that is well known and already part of the project standards and design criteria. Hammer suggests that it may be better to spend time ensuring that the design criteria have been incorporated than building fault trees [133]. Childs [63] notes that sometimes Fault Tree Analysis finds only what is intuitively obvious.

Fault Tree Analysis shows cause-and-effect relationships but little more. Additional analysis and information are usually required for an effective safety program. Moreover, reliability analysts usually concentrate only on failure events in fault trees, whereas hazard analysis (as opposed to reliability analysis using fault trees) requires a broader scope. Thus, the use of fault trees by reliability analysts should differ from their use by system safety analysts, but confusion between these two different system properties is widespread. Applications of Fault Tree Analysis that focus primarily on failures are essentially just reliability analyses.

A fault tree, like any other model, is a simplified representation of a generally very complex process, but its relative simplicity can be deceptive [206]. Much of the research on FTA is concerned with correcting the oversimplifications, but the problems might be better overcome by using different types of models and analyses to handle these factors directly rather than trying to force fit everything into one—perhaps inappropriate—analysis framework.

For example, the technique is particularly suited to discrete events, such as a valve opening or closing, but time- and rate-dependent events, such as changes in critical process variables, degrees of failure (partial failure), and dynamic behavior are not so easily represented [150]. And, of course, system design errors that do not involve component failures or faults cannot be identified with fault trees.

Consider the following real explosion caused by overpressurization that was mentioned in chapter 4. The fault tree created is shown in figure 10.7. Two relief valves are used in this design although only one is needed to respond to overpressurization; the second is included for redundancy in case the first relief valve fails shut. An operator is responsible for making sure that the primary relief valve opens. If it does not, the operator must manually open the backup relief valve.

This system had a design error that was not—and would not have been—detected by the use of a fault tree. In the actual loss scenario, overpressurization occurred, and the open position indicator light and open indicator light both illuminated. However, the primary valve was *not* in fact open. Because the light indicated that the primary valve was open, the operator did not open the secondary (backup) valve and the system exploded.

Post-accident examination discovered that the indicator light circuit was wired to indicate presence of power at the valve, but it did not indicate valve position. Therefore, the indicator showed only that the "open valve" button had been pushed, but not that the valve had actually opened. An extensive quantitative safety analysis of this design had assumed a low probability of simultaneous failure for the two relief valves but ignored the possibility of design error in the electrical wiring. The probability of design errors was not

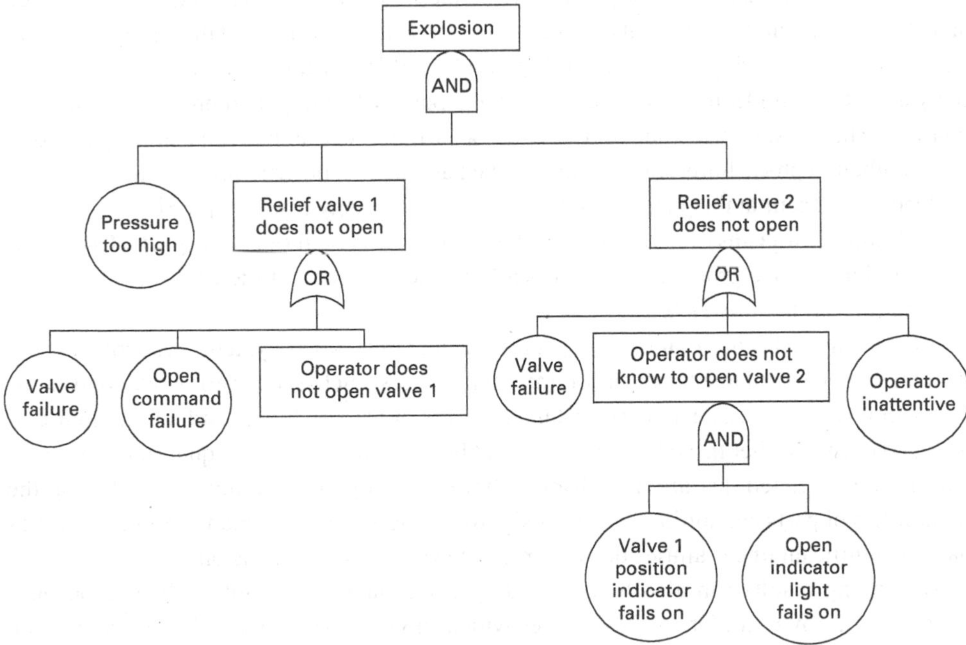

Figure 10.7
Example of a fault tree for an overpressurization event.

quantifiable. No hazard analysis of the electrical wiring was made; instead, confidence was established on the basis of the low probability of coincident failure of the two relief valves. This same type of design error has been involved in many different accidents.

Another limitation is that the use of simple AND and OR gates does not convey any notion of time ordering or time delay; the fault tree is a snapshot of the state of the system at one point in time. In some cases, time spans or chronological ordering of events may need to be specified. Other types of gates, such as DELAY and INHIBIT, allow some treatment of time in fault trees and in the tree reduction process [310]. They complicate the evaluation of the tree, however, and somewhat negate one important advantage of FTA—the ease with which the trees can be read and understood and thus reviewed by experts and used by designers.

Of course, this reviewability advantage only holds for relatively small fault trees. Very large fault trees may not be very reviewable at all. If chronology is important, using a model and analysis technique that involves backward or forward search may be more appropriate than force fitting the analysis into this type of top-down modeling technique.

Transitions between states are not represented in fault trees, which deal best with binary states: partial failures and multiple failures can cause difficulties [74]. Because system states rather than sequences of states are shown, fault trees are used less often in studies of batch systems and plants (where sequence is important) than in continuous systems. Nonaction or static systems, such as pressure vessels, are also difficult to handle, because their state depends primarily on environmental events or event combinations rather than on the component state itself [13].

Problems also occur in the analysis of *phased-mission* systems, which pass through more than one phase of operation [206]. Typically, in these systems, the same equipment is used at different times and in different configurations for different tasks, and thus a separate fault tree is needed for each phase. While it is possible to think of this type of system as essentially an OR gate under the top event, where the inputs to the OR gate represent the different phases of the mission, the standard OR will not suffice because the inputs are separated in time. Although phased-mission systems can be handled by constructing several fault trees, problems can occur at the phase boundaries that are not easily resolved [206].

Additional criticisms of Fault Tree Analysis relate to its quantitative aspects. Common-cause failures cause problems and can lead to orders-of-magnitude errors in the calculated failure probability [1981].

As with any technique that tries to quantify factors in complex systems probabilistically, data may not be available for the most important factors, such as operator work conditions, management decisions, design errors, human errors of various kinds, and nonrandom failures and events. Either these factors are left out because they cannot be quantified, or probabilities are assigned that are unrealistic or have very large uncertainties. Combining the reliabilities of parts containing five or six significant figures with human error probabilities having significant uncertainties does not produce very useful conclusions.

Misleading results can also be obtained by using data that is not applicable because conditions are not similar to those under which the data was obtained or by averaging widely different data—one can drown in a lake with an average depth of 6 inches [187].

Actually, most errors in hazard analysis are not due to errors in the data but by not foreseeing all the ways in which the hazard could occur. According to Kletz, "Time is usually better spent looking for all the sources of hazard than in quantifying with ever greater precision those we have already found" [187].

10.2.4 Event Tree Analysis

Description

FTA is the most widely used method for the quantification of system failures, but it becomes very difficult to apply in complex systems. WASH-1400 was a probabilistic risk assessment of nuclear power plants performed in the early seventies. The study team first attempted to draw a fault tree starting with the top event "accidental release of radioactivity," but they gave up when this led to a hopelessly complicated fault tree [325]. Instead, they adapted the general decision tree formalism, widely used for business and economic analysis, to break up the problem into smaller parts to which FTA could be applied.

The decision tree formalism, called *Event Tree Analysis* when used in this way, uses forward search to identify the various possible outcomes of a given initiating event, such as the rupture of a pipe, by determining all sequences of events that could follow it. The initiating event might be a failure of a system component or some event external to the system.

The problem in any forward search, of course, is in knowing where to start. In nuclear power plants, which is the principal application for this technique, an accident is defined as any failure of the operating system that might result in the release of radioactivity. Thus, the starting point for listing the initiating events to be considered is the potential failures previously identified and defined by the many years of safety analysis and by the licensing process for commercial nuclear power plants.

The states in the forward search are determined by the success or failure of other components or pieces of equipment. In nuclear power or other applications where the principal design approach for safety is to use a protection system, all the layers of defense that can be used after the accident are first defined and then structured as headings for the event tree. The layers of defense are listed left to right in chronological order after the initiating event. The ordering of the headings on the event tree is critical.

As shown in figure 10.8, the event tree is then drawn from left to right, with branches under each heading corresponding to two alternatives: (1) successful performance of the defense (the upper branch), and (2) failure of the defense (the lower branch). After the tree is drawn, paths through it can be traced by choosing a branch under each successive heading, where each path corresponds to an accident sequence.

Figure 10.8 shows an example from the WASH-1400 report. Here, the headings are pipe break (the initiating event), electric power, emergency core cooling system (ECCS), fission product removal, and containment integrity failure. Each of these system components or functions is assumed either to succeed or fail in performing its required function. The expression at the right of each path is the probability for that path. Because the probability of failure is assumed to be very small, the probability of success is always close to 1. Therefore, the probability associated with the upper (success) branches of the tree is assumed to be 1, as in the example shown.

Event trees tend to get quite large. They are reduced by eliminating sequences whose functional and operational relationships are illogical or meaningless. Because the system states on a given branch of the tree are conditional on the previous states having already

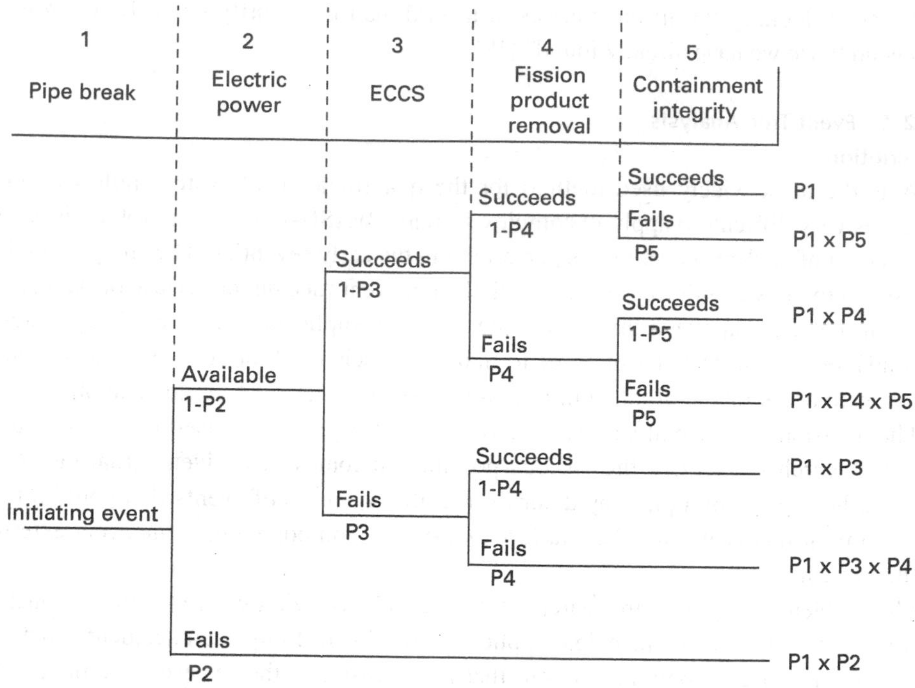

Figure 10.8
A reduced event tree for a loss of coolant accident. (Source: *Reactor Safety Study*. US Nuclear Regulatory Commission, WASH-1400, NUREG 75/014, October 1974.)

occurred, another way to prune an event tree is to eliminate all branches that have a zero conditional probability for at least one event.

A path's total probability is found by multiplying together the probabilities at the various branches of the path. The total risk of an accident is then found by combining the path probabilities for all paths leading to an accident. The initial event is expressed as a frequency (events per year), while the other, secondary events, are probabilities (failures per demand). The probabilities for the secondary events are often determined using fault trees.

Event Tree Analysis is usually applied using a binary state system, as explained earlier, where each branch of the tree has one failure state and one success state. If a greater number of discrete states are defined for each branch (for example, if partial failures are included), then a branch must be included for each state. A problem is the possible explosion in the number of paths—for a sequence of N events, there will be 2^N branches of a binary tree. The number can be reduced by eliminating impossible branches, as described earlier, but a large number of paths can still result.

Usually, a finite number of branches is defined at each node, but there is no conceptual problem with introducing a continuous random variable in an event tree [310]. Graphically, the spectrum of possible values of the continuous variable is represented by a fan originating at the event node. The analysis, in this case, uses a continuous conditional probability density function and provides continuous joint distributions. In practice, a discrete variable may be more convenient, but in theory a continuous variable could be used [310].

Timing issues can cause problems in event tree construction. In some cases, failure logic changes depending on when the events take place. This happens, for example, in the operation of emergency core cooling systems in nuclear power plants [1981]. As with fault trees, phased-mission analysis techniques are then needed to model the system changes during the accident sequence, even though the protection system does not change.

Another consideration is possible dependencies between the various probabilities arising from common-cause failures. In the nuclear reactor example shown in figure 10.8, the value of the probability of ECCS function failure may depend in some way on the conditions created by the pipe break itself. Such dependencies must be identified and assessed in the analysis, or the results can be distorted [324].

Life-cycle stage Like Fault Tree Analysis, ETA is appropriate only after most of the design is complete. Thus, it has been used primarily to evaluate existing plants or designs. Note that, by definition, a decision is made in advance that the solution to the problem of safety will be to use protection systems or layers of defense. ETA does not require such headings on the columns, but it is difficult to determine which events to use for the headings otherwise. A general forward analysis of this type that did not drastically limit the events to be considered would be potentially enormous or incomplete.

Evaluation Fault trees lay out relationships between events: they are snapshots of the system state. Event trees, in contrast, display relationships between juxtaposed events (sequences of events) linked by conditional probabilities. As a result, at least in theory, event trees are better at handling notions of continuity—logical, temporal, and physical—while fault trees are more powerful in identifying and simplifying event scenarios.

Event trees allow the direct introduction of time factors and continuous random variables. Combinations of events can be more concisely represented in fault trees using logical functions. Figure 10.9 shows the same event represented by a fault tree and an event tree.

Notice that a top-down search model like a fault tree loses the information about the ordering of relief valve operation shown in figure 10.9, although it could be added by adding more complex types of tree structures, while the forward-search event tree model does not include detailed evaluation of the individual events.

The fault tree example is the same as that shown in figure 10.7, but a computer is substituted for the operator's responsibility to open the primary relief valve to show how computers might be used in fault trees. Note that the accident is not eliminated by the introduction of a computer. Also, little information is gained about the hazards related to the software beyond the obvious. Finally, the ability to obtain a probability is lost when software is included because software has only design errors.

Event trees are useful within the scope for which they were devised—probabilistically evaluating the effects of hardware protection system functioning and failure in an accident sequence, particularly when events can be ordered in time. They are practical when the chronology of events is stable and the events are independent of each other [207].

Event trees can be helpful in (1) identifying the protection system features that contribute most to the probability of an accident so that steps can be taken to reduce their failure probability, (2) identifying top events for subsequent Fault Tree Analysis, and (3) displaying various accident scenarios that may result from a single initiating event.

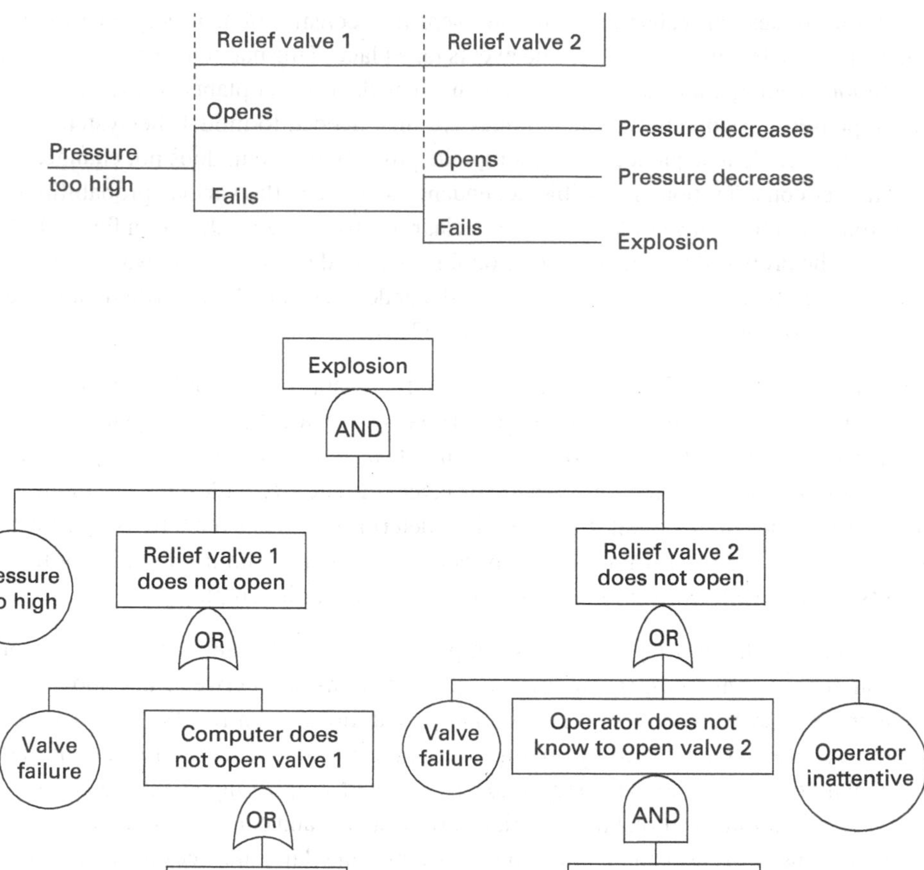

Figure 10.9
A fault tree and event tree comparison.

Like all the analysis techniques discussed in this chapter that are based on the linear chain-of-failure-events model, event trees have many limitations. For one, they can become exceedingly complex, especially when a number of time-ordered system interactions are involved [65]. A complete risk analysis of a complex plant, using a combination of event trees and fault trees, will require many person-years of effort along with a number of simplifying assumptions.

A question arises about how the probabilities for the event trees are determined. One way that has been suggested, and was used in Wash-1400, is to use FTA for the events that appear in the event tree. The use of FTA to determine the probabilities for many of the event tree branches, however, may make it more difficult to identify common causes of failures [293]. A separate tree is required for each initiating event, making it difficult to represent interactions between event states in the separate trees or to consider the effects of multiple initiating events.

In addition, while the event tree enumerates all possible combinations of component states related to an initiating event, it offers no help in determining whether the component failure combinations (paths) lead to system failure. Either the system is simple enough and the mapping can be done for each failure scenario without more formal analysis, or the system is more complex and fault trees have to be used to identify the failure modes [310].

The usefulness of event trees depends on being able to define the set of initiating events that will produce all the important accident sequences. For nuclear power plants, where all the risk is associated with one hazard—serious overheating of the fuel—and designs are fairly standard, defining this set of hazards may be easier than for other systems. Whether it can be done completely, even for nuclear power plants, is unknown. And, as discussed previously, design errors are omitted.

Similarly, defining the functions across the top of the event tree and their order is difficult. Again, in nuclear power plants, where responsibility for safety is vested in a specific set of layers of defense, the events to use are more obvious than in other systems, although the problem of ordering is still there. Order is important when the performance of one system affects the performance of another. To solve the ordering problem, the analyst needs a detailed understanding of all plant systems, how they operate, and how they interact with one another [319]. As in most of these analysis techniques, building the model requires the interaction of analysts with different areas of expertise.

Finally, as with fault trees, continuous, nonaction systems such as dams are not appropriate for Event Tree Analysis.

10.2.5 Combinations of Analysis Techniques

Combinations of these techniques are possible, particularly as they are all based on the same underlying causality model; that is, a linear chain-of-failure-events leading to the loss.

Some example combinations have been shown in the previous sections, such as the use of FHA to identify failure conditions that are then evaluated using FTA (see figure 10.4). The previous section described how event trees and fault trees have been used together such that the probability on the event tree branch is generated using a fault tree.

Fault trees and event trees have also been "hooked together" in other ways. Cause–Consequence Analysis (CCA) is a technique developed by Nielson in the 1970s that combines fault trees and event trees and thus top-down search with forward search. The cause–consequence diagram thus shows both time dependency and causal relationships among events.

The procedure starts with the selection of a critical event, which is followed by a search for factors that constitute the critical event and by a propagation of the potential effects of the event. Finally, the interrelationships of the factors are described by a graphical model. Figure 10.10 shows an example. Note that an "upside down" fault tree is attached to the basic event chain by decision boxes. For example, whether relief valve 1 opens or not is dependent on the fault tree showing two OR conditions: valve failure or computer does not open the valve.

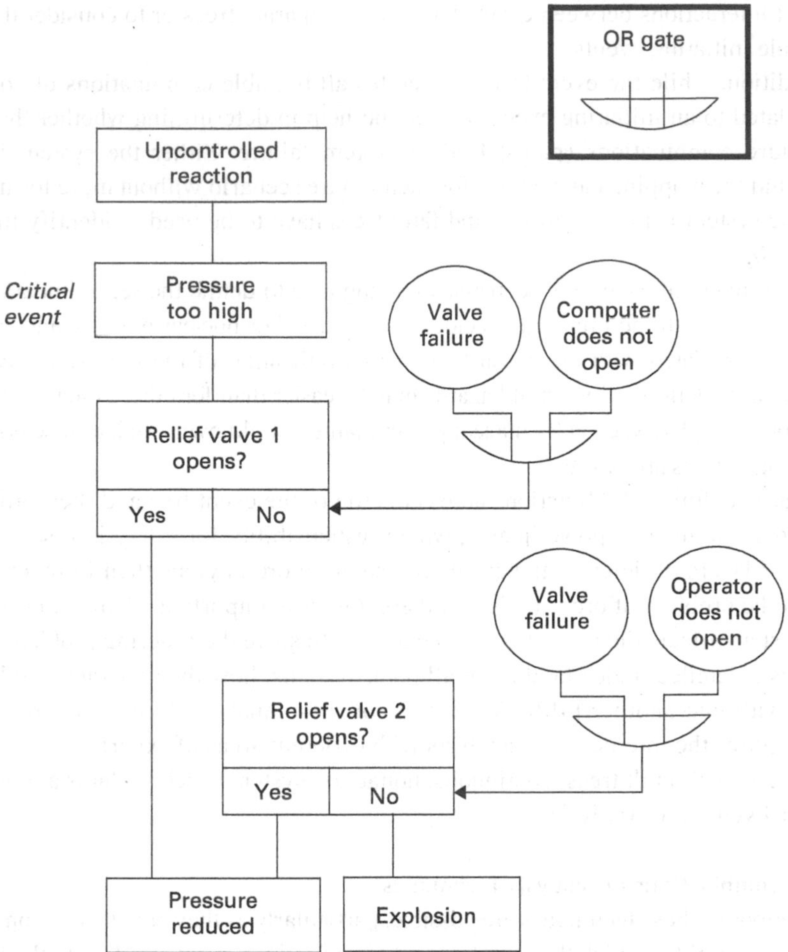

Figure 10.10
An example of a Cause–Consequence Analysis for an overpressurization hazard.

Several cause charts may be attached to a consequence chart. The cause charts describe the alternative prior event sequences that can lead to the critical event and the conditions under which these sequences can occur. According to Nielson, the initiating events should be traced back to spontaneous events covered by statistical data [284]. Other cause charts attached to the consequence chart may be conventional fault trees, which show the combination of conditions under which a certain event sequence in the consequence chart can take place. Fault trees are used in the diagram not only for the critical event but also for abnormal conditions [206]. Taylor [370] has shown how cause–consequence diagrams can be formalized to provide a semiautomated analysis method.

Cause–Consequence Analysis has the same advantages of combining event trees and fault trees described previously. But they also provide a formal notation for describing the results of the analysis. On the negative side, they inherit the same disadvantages; that is, the diagrams can become unwieldy, separate diagrams are required for each initiating event, and outcomes are related only to the cause being analyzed, although they could be

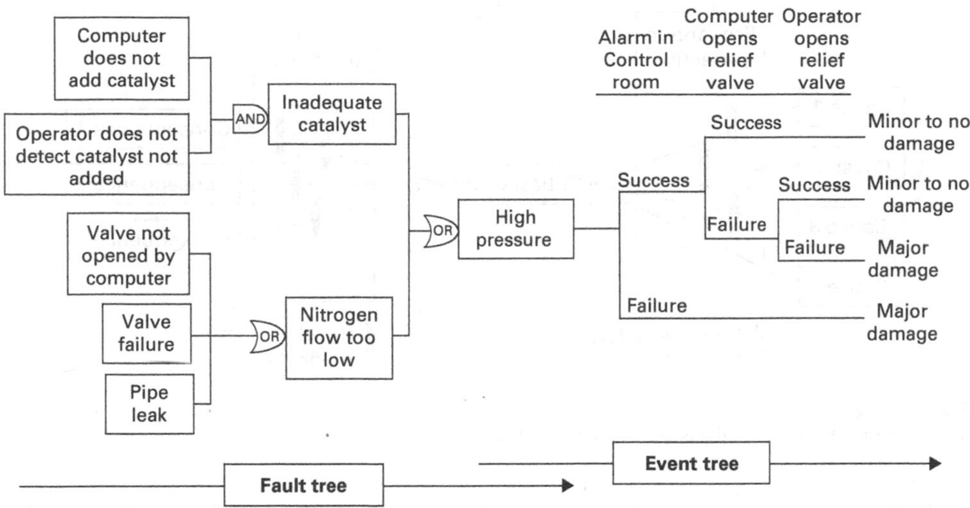

Figure 10.11
An example of Bow Tie Analysis with the fault tree on the left and the event tree on the right.

caused by other initiating events [74]. Cause–Consequence Analysis was used some in the 1970s in Europe, but never got wide use. Other attempts to combine fault trees and events trees were suggested but never caught on.

In the early 1970s, Bow Tie Analysis put fault trees and event trees together horizontally. An example is shown in figure 10.11. It was called Bow Tie Analysis because of its resemblance to the shape of a man's bow tie. Around the 1990s, the model was greatly simplified by allowing only one box in the fault tree (oversimplifying causality greatly), but protection and mitigation activities were added to the model. An example is shown in figure 10.12. Bow Tie Analysis is currently popular in the oil industry.

Cause–Consequence/Bow-Tie analysis was later simplified in terms of what could be included. Figure 10.12 shows the simplified version created in the 1990s and used today. In the simplified form, the causal event chain is limited to one event although there may be multiple single-event causes. The chains of events after the critical event are also simplified. The Bow Tie model does show mitigation measures, which are equivalent to the annotations on the events in the basic event chain model shown in chapter 7 on accident causality models.

The simplified accident causation depicted in bow tie diagrams can potentially provide management with a distorted view of the operational risk which, in itself, could lead to accidents [220]. In bow tie analysis, virtually all accidents in today's systems, such as air transportation, railroads, hospitals, chemical plants, and so on, are effectively characterized as having a chain of events comprised of a single independent failure event. Therefore, only a small fraction of the potential accident causes will be identified and controlled, leaving companies open to very high levels of risk and liability unless more powerful and complete methods are used.

None of these combined models overcomes the problems of handling software and human error nor do they go beyond the assumptions of the linear failure-event-chain causality model that these techniques are based on. In addition, putting together two techniques that

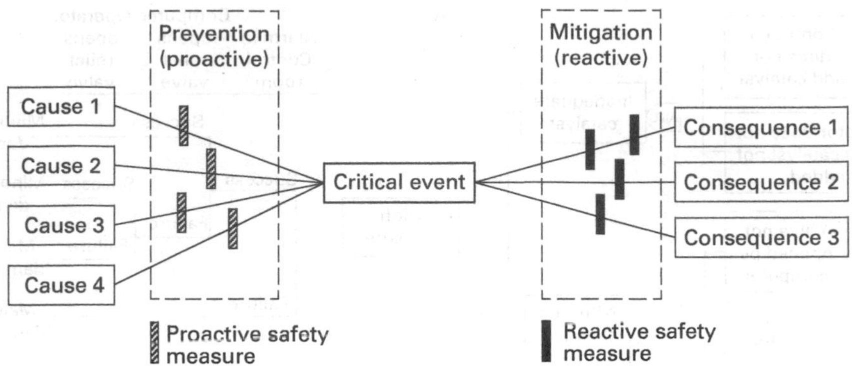

Figure 10.12
The simplified Bow Tie Analysis created in the 1990s.

each alone generate masses of information simply exacerbates that problem for complex systems. Bow tie has solved this problem by greatly reducing the amount of information provided, but that can lead to omitting critical information necessary to prevent accidents.

Before looking at analysis techniques not based on the linear chain causality assumptions, there is one other technique, called HAZOP, that is widely used in the chemical and process industries.

10.2.6 Hazards and Operability Analysis (HAZOP)

HAZOP was developed by Imperial Chemical Industries (ICI) in England in the early 1960s and later improved on and published by the Chemical Industries Association in London. A large percentage of the chemical process industry uses HAZOP for all new facilities [292].

As the name suggests, the technique focuses not only on safety but also on efficient operations. Although it is usually applied to fixed plants, Kletz describes an application to tank trucks, in which several previously undetected hazards were identified and eliminated or controlled [192].

While HAZOP is based on the same linear chain-of-events causality model of accidents, it generalizes the definition of failure by defining the cause of accidents as deviations from the design of operating intentions, for example, no flow or backward flow when there should be a forward flow. The events caused by the deviations are still assumed to occur in a linear sequence. The analysis may start in the middle of the chain and continue in either or both directions to identify previous and subsequent events related to the deviation.

Basically, the technique encourages creative thinking about more possible ways in which hazards or operating problems might arise than simple random failure of components. To reduce the chance that anything is forgotten, HAZOP is performed systematically, considering each process unit in the plant (such as pipelines, tanks, and reactors) and each hazard in turn. Questions are generated about the design by a small team of experts. Although prompted by a list of guidewords, the questions arise creatively out of the interaction of the team members [192]. The analysis, unlike those described so far, is performed on a physical model of the plant. Figure 10.13 shows a simple example of such a physical model. A complete model of a plant would be much more detailed and comprehensive.

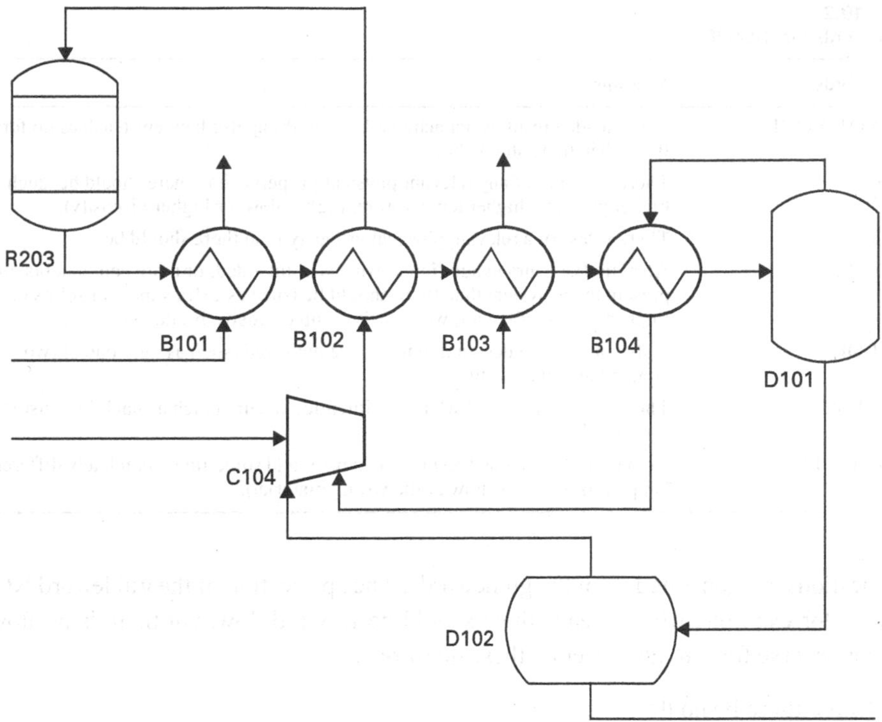

Figure 10.13
A model of part of a chemical plant.

HAZOP is a qualitative technique whose purpose is to identify all possible deviations from the design's expected operation and all hazards associated with these deviations. In comparison with hazard identification techniques like checklists, HAZOP is able to elicit hazards in new designs and hazards that have not been considered previously.

Using a model of the proposed process plant, such as that shown in figure 10.13, a HAZOP team, which is composed of experts on different aspects of the system along with an independent team leader who is an expert on the technique itself, will consider

- the design intention of the plant;
- potential deviations from the design intention;
- the causes of these deviations from the design intention;
- the consequences of such deviations.

The guidewords used in the HAZOP process are shown in table 10.2. They are applied to any variables of interest such as flow, temperature, pressure, level of composition, and time. Each line in a line drawing of the plant is examined in turn, and the guidewords are applied. As each process deviation is generated, the members of the team consider every potential cause, such as a valve closed in error or a filter blocked, and its effect on the system as a whole, such as a pump overheating, a runaway reaction, or a loss of output.

Table 10.2
Guidewords for HAZOP

Guidewords	Meaning
NO, NOT, NONE	The intended result is not achieved, but nothing else happens (such as no forward flow when there should be).
MORE	There are more of any relevant physical property than there should be (such as higher pressure, higher temperature, higher flow, or higher viscosity).
LESS	There is less of a relevant physical property than there should be.
AS WELL AS	An activity occurs in addition to what was intended, or more components are present in the system than there should be (such as extra vapors or solids or impurities, including air, water, acids, and corrosive products).
PART OF	Only some of the design intentions are achieved (such as only one of two components in a mixture).
REVERSE	The logical opposite of what was intended occurs (such as backflow instead of forward flow).
OTHER THAN	No part of the intended result is achieved, and something completely different happens (such as the flow of the wrong material).

Questions are generated from the guidewords. The application of the guideword NONE to flow, for example, which means there should be forward flow, but there is no flow or there is reverse flow, might generate these questions:

- Could there be no flow?
- If so, how could it arise?
- How will the operators know that there is no flow?
- Are the consequences hazardous, or do they prevent efficient operations?
- If so, can we prevent no flow (or protect against the consequences) by changing the design or method of operation?
- If so, does the size of the hazard or problem justify the extra expense?

Figure 10.14 shows a detailed flow chart of the HAZOP process [187] while table 10.3 shows a typical entry in the table that might result.

Note that table 10.3 simply details a linear causal chain where, for example, a pump failure leads to no flow which leads to overheating in the heat exchanger. The guidewords, however, provide more assistance than simply looking for failures. Moreover, performing the analysis on a model of the plant also provides assistance.

The procedure differs for continuous and batch processes. In the HAZOP for a continuous plant, the process is as described previously. In addition to normal processing, the study should include operability and safety during commissioning of the plant and during regular startup and shutdown.

For a batch plant, not only the flow diagrams but also the operating procedures are examined. The guidewords are applied to the instructions (whether written for operators or executed by a computer) as well as to the pipelines. Time is important in batch operations. In applying the guidewords to time, such factors as duration, frequency, absolute time, and sequence may be relevant.

Life-cycle phase HAZOP uses process descriptions; flowsheets; control logic diagrams; piping and instrumentation diagrams; a plant layout; draft operating, maintenance, and

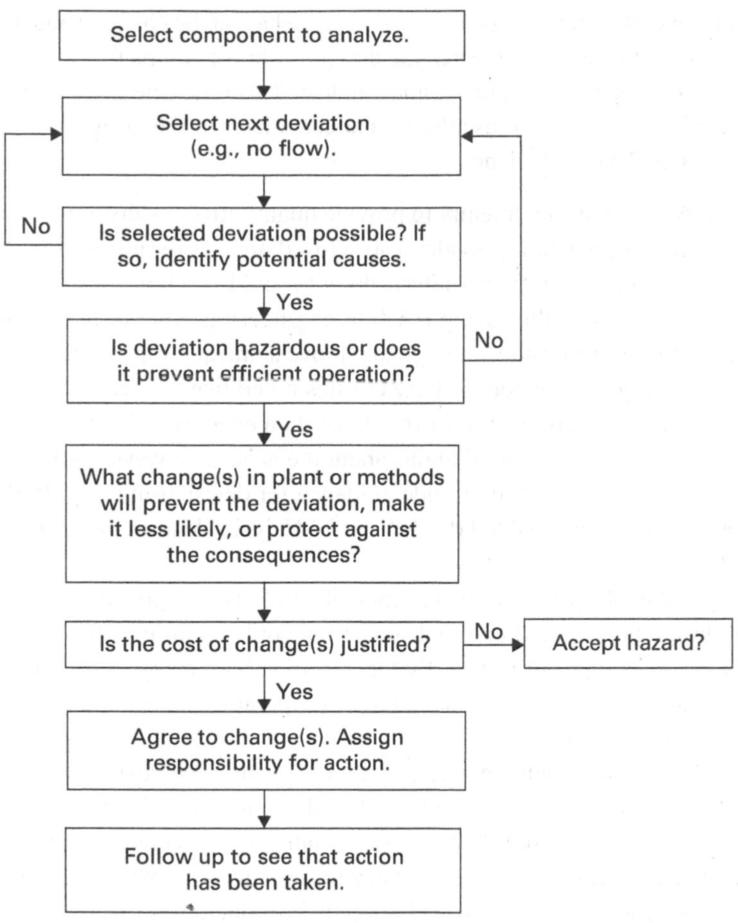

Figure 10.14
Flowchart of the HAZOP process.

Table 10.3
Example entry in a HAZOP report

Guide Word	Deviation	Possible Causes	Possible Consequences
NONE	No flow	1. Pump failure 2. Pump suction filter blocked 3. Pump isolation valve closed	1. Overheating in heat exchanger 2. Loss of lead to reactor

emergency procedures; safety and training manuals; and data on the chemical, physical, and toxicological properties of all materials, intermediates, and product. By the time this much information is available, it is usually too late to make major changes in the design if uncontrolled hazards are identified. Therefore, hazards are usually controlled by the addition of protection devices rather than removed by changes in design [186].

For this reason, many companies conduct preliminary HAZOPs on conceptual flowcharts and preliminary layout diagrams—noting only safety aspects, not operability problems. At this stage, for example, it is possible to replace a flammable piece of equipment with a nonflammable one. But there are still limitations on what can be changed and how early the process can start. At a later stage, when the design is almost complete, it may only be possible to reduce the risk by adding fire insulation, leak detectors, emergency isolation valves, and so on [186]. A full HAZOP usually is conducted later in the design process even if a preliminary HAZOP has been done.

Evaluation HAZOP does not attempt to provide quantitative results, but instead systematizes a qualitative approach. It provides a structured process for generating the chains of events. Fault Tree and Event Tree Analysis do not provide such a process. In most situations, once a hazard is identified using HAZOP, engineering experience or a code of practice is adequate to determine how far to go to remove it and quantification is not necessary. Kletz, one of the original creators of HAZOP, has noted that: "There is no need, and we do not have the resources, to quantify every hazard on every plant" [192].

In situations where uncertainty remains about the hazard, however, numerical analysis may help to clarify priorities and provide guidance for decision-making. In the chemical process industry, the term HAZAN (for HAZard ANalysis) denotes the related quantification method.

The strength of the HAZOP lies in its simplicity and ease of application and in the identification of design problems. It does not concentrate only on failures but has the potential to find more complex types of causes. Reductions of at least an order of magnitude in the number of hazards and problems encountered in operation have been claimed to result from the use of this technique [206].

Although HAZOP is closely connected with the chemical industry, the basic idea can be adapted to other industries. HAZOP has the advantage over checklists of being applicable to new designs and design features and of not limiting consideration to previously identified hazards. Complex, potentially dangerous plants with which there is as yet relatively little experience and procedures that occur infrequently, such as commissioning a new plant, are especially good subjects for this type of study [391].

In addition to its open-ended approach to identifying potential problems, a fundamental strength of HAZOP is the encouragement of cross-fertilization of ideas among members of the study team. People from different disciplines working together often find problems that are overlooked by functional groups working in isolation [186]. HAZOP's success, however, depends on the degree of cooperation between individuals, their experience and competence, and the commitment of the team as a whole. Except for the team leader, who is an expert on HAZOP, the members of the team must be experts on the process being analyzed: the HAZOP procedures allow their knowledge and experience to be applied systematically.

The drawbacks of the technique are the time and effort required—it is very labor-intensive—and the limitations imposed by the search pattern. HAZOP relies very heavily on the judgment of the engineers performing the assessment [375]. For example, the extent to which the guideword AS WELL AS is applied will restrict the number of simultaneous faults that can be considered, and evaluation is done by human reasoning alone. Again, all the methods described so far have these same limitations.

Each hazard analysis method has its own search pattern, limiting the factors that will be considered. HAZOP covers hazards caused by process deviations, which is certainly more comprehensive and inclusive than considering failures only, but it still leaves out other types of causal factors that occur in complex systems.

Examples of causes covered well are failures of the main operating equipment, such as pumps, compressors, heat exchangers, critical valves, and instrumentation, and human errors in manual operations that involve the main process equipment and its functions, such as opening or closing valves and starting or stopping pumps.

Suokas says that it is unusual for HAZOP to consider deviations or determining factors related to organizational factors such as the information or management systems [364]. An argument can be made that causes related to these factors will be reflected in the process units as a change from the normal state or from acceptable values of the operating parameters, and that tracing back to the causes can reveal some limited management factors. The problem may be more that this type of causal analysis is not encouraged by the technique; the process stops when more proximal factors such as a pump failure are uncovered without necessarily tracing the failure back to a maintenance error and perhaps back from that to a management problem. And again, a linear chain of events is assumed.

Another limitation of the technique is that it uses a physical, connection model of the plant, which may not be adequate for today's complex, software-intensive systems. Techniques that use other types of models for search can find different types of causes and can be used before the physical design is complete.

In the oil industry, bow tie models are sometimes used as a filter to determine when a full HAZOP should be performed. The problem, of course, with this scheme is that bow ties dangerously oversimplify and omit important causes and thus can filter out the most important events for HAZOP to consider.

10.2.7 Miscellaneous Techniques

The techniques described so far are the most widely used. Other less structured approaches have been proposed and used, but they either provide little guidance or limit the search and thus provide dangerously misleading results. Most have come out of management schools rather than engineering. These techniques include brainstorming, checklists, 5 Whys, and fishbone diagrams.

Brainstorming While brainstorming is an appropriate technique to identify creative, out-of-the-box solutions to problems, it is very poor at analysis of existing or potential systems. Without a structure to the process of identifying hazardous scenarios, critical factors are very likely to be omitted. Chapter 13 describes the problem of heuristic bias. Humans have been found to be very poor at assessing risk and at identifying causes without having a structured method to perform these tasks and to ensure a more than superficial analysis.

Checklists The problems with using checklists to identify hazards (discussed in chapter 9) are equivalent and a subset of the problems that arise when using checklists for identifying hazard causes. Beyond being necessarily incomplete, they tend to limit what the analysts identify or examine. For complex systems, checklists are nearly as dangerous as simple brainstorming.

Checklists are most useful in the design of well-understood and relatively simple hardware systems, for which standard design features and knowledge have been developed over time. On the negative side, checklists often contribute to false confidence—a belief that if everything is checked off, the system is safe.

Another problem arises when the lists are used without giving careful thought to the specific situation being considered. Ozog and Bendixen [293] provide an example from the process industry: a checklist might reasonably require flame arrestors in vents from flammable liquid storage tanks, but if the vapors are susceptible to polymerization, venting directly to the atmosphere might be safer. In this case, relying on the checklist without considering special circumstances might create a more hazardous situation.

While checklists may be useful in very limited circumstances, more sophisticated analyses for all but the simplest systems are essential for an effective safety program.

5 Whys The *5 Whys* technique is surprisingly popular given its obvious limitations and simplifications. Perhaps the simplicity is exactly why it is so popular. It was originally developed and used by the Toyota Motor Corporation for problem-solving training during the evolution of its manufacturing methodologies. It basically involves asking the question "why" five times when trying to understand a problem. The solution to the problem then supposedly becomes clear. Figure 10.15 shows a standard example.

Note that the argument in figure 10.15 is equivalent to a five-box event chain with no ANDs or ORs; that is, simply one backward chain with five boxes. While this technique was developed for management problems, it is surprisingly fairly widely used and touted in the petrochemical industry and others for hazard analysis. Perhaps those in management are unaware of the other, much more powerful, techniques that have existed for decades in engineering to do these types of analyses, even for management problems.

5 Whys Example

Problem: The Washington Monument is disintegrating.

Why is it disintegrating?

Because we use harsh chemicals.

Why do we use harsh chemicals?

To clean pigeon droppings off the monument.

Why are there so many pigeons?

They eat spiders and there are a lot of spiders at the monument.

Why are there so many spiders?

They eat gnats and there are a lot of gnats at the monument.

Why so many gnats?

They are attracted to the lights at dusk.

Solution: Turn on the lights at a later time.

Figure 10.15
Standard 5 Whys example.

The 5 Whys technique assumes that there is only one potential cause for anything, although that is almost surely incorrect for anything but the simplest of systems. For example, in figure 10.15, there could be other potential causes for the disintegration of the Washington Monument beyond the use of harsh chemicals for cleaning, such as air pollution, water seepage, acid rain, or a combination of these and other things.

In fact, the memorial damage is a real problem and not just a made-up example. In a 1989 report prepared for the National Park Service, Messersmith found that the act of cleaning the memorials was only one contributing factor among many to the damage [260]. In addition, cleaning chemicals were not causing most of the deterioration. Instead, most were a result of the large amount of water used in the cleaning process. Reducing the amount of water used dramatically reduced the damage.

The technique ignores all but one potential reason for the problem as well as other potential solutions. In 1990, the National Park Service did try turning the lights on later, but tourists complained about their inability to get iconic photographs of the lighted memorials, and the decision was reversed. More serious was the increased potential for injured tourists resulting from the low lighting such as tripping on unseen obstacles or uneven surfaces. Using this simplistic analysis technique, other potential solutions with perhaps less serious side effects will be ignored, such as using less water, finding less harsh chemicals, or treating the surface in some way to protect it.

In summary, this technique can identify only a single causal scenario starting from a single cause. The analysis is so superficial that it is difficult to imagine when it might be useful. The ubiquity of the 5 Whys method shows the attraction of simplistic solutions to difficult problems.

Fishbone (Ishikawa) diagrams Fishbone diagrams, which are claimed to identify the root causes of problems, were invented by Kaoru Ishikawa, a Japanese quality control statistician. They are slightly more powerful than the 5 Whys technique but much less powerful than almost any of the engineering analysis techniques They are equivalent to a very simple and limited fault tree diagram.

Figure 10.16 shows the format of a fishbone diagram. The problem to be solved is written on the right. The "spines" represent categories of potential causes. In the example, these categories include people, process, equipment, materials, environment, and management. Other categories can be used. The potential causes are listed on the branches off the spines. Note that the causal analysis that results is fairly superficial. Fishbone diagrams are created using brainstorming. Figure 10.17 shows an example.

In comparison to 5 Whys, at least there are some ORs in the scenarios or causal factors, but no ANDs and almost no real technical information about the cause of the accident. In addition, the different categories are treated separately so that interactions between the design of the equipment and the behavior of people, for example, or of process interactions with people and the technical design cannot be included.

While there may be some justification for using a fishbone diagram for simple management problems, it is difficult to justify its use for engineered systems and hazard analysis or even for more in-depth analysis of complex management systems.

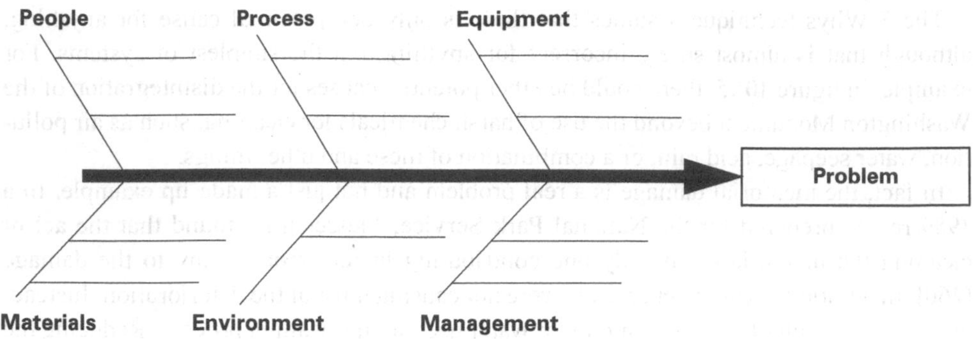

Figure 10.16
The form of a fishbone diagram.

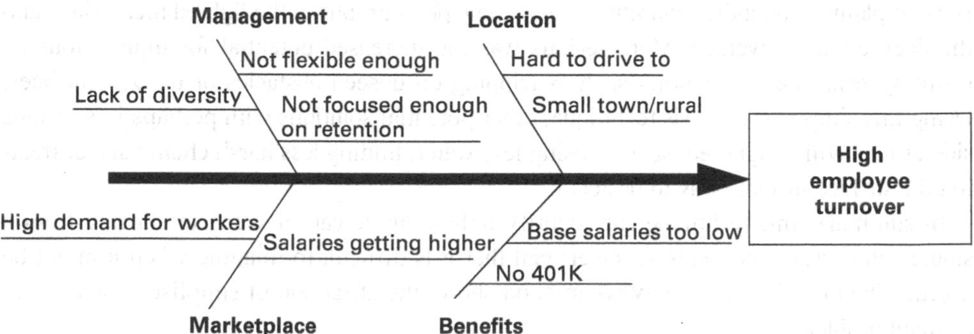

Figure 10.17
An example of a fishbone analysis.

10.3 STPA: A Technique Based on STAMP

In contrast to the techniques described so far, which all date from the 1960s or before, STPA (System Theoretic Process Analysis) is a new hazard analysis method that claims to handle recent advances and changes in engineering practice and technology [217]. It is based on STAMP, the system-theoretic accident causality model described in chapter 7.

Because STAMP considers accidents to be caused by unsafe control of system behavior rather than component failure, the causes identified will include those generated by the older techniques (that is, component failures and faults) plus additional ones involving unsafe control of the system as a whole. These additional causal scenarios include unsafe interactions *among* the components of the sociotechnical system, none of which may have failed. The entire sociotechnical system—that is, the hardware, software, humans, and social entities (including management and regulatory agencies)—can be included in the analysis.

STPA starts the analysis with hazards rather than the component failures. As such, it is a backward analysis technique (see chapter 9) that identifies only the hazardous scenarios and not all failure-related scenarios and can identify complex interactions among the causal factors.

A further difference with the standard hazard analysis techniques is that STPA is embedded within the system engineering process and not standalone.

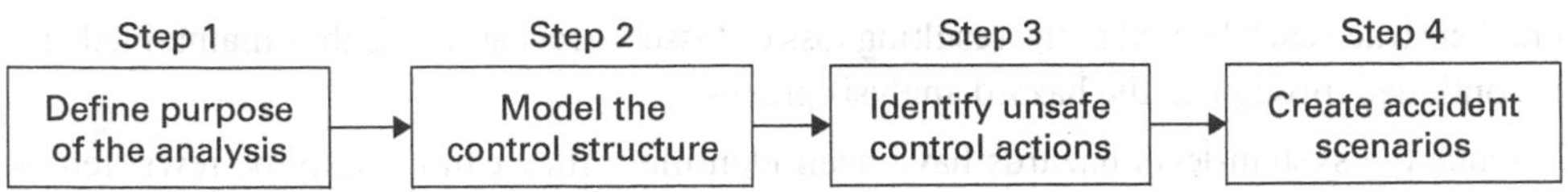

Figure 10.18
The four steps in STPA.

Description Figure 10.18 shows the STPA process. It consists of four steps.

Step 1: Define the scope and purpose of the analysis.

STPA starts by specifying the losses to be considered. STPA can be used to target any loss of interest to stakeholders, including those not traditionally considered in hazard analysis. It also can be used to analyze the system for any emergent system property, such as security. The system to be analyzed and the system boundary must also be identified.

Examples of losses are:

L-1: Loss of life or injury to people

L-2: Loss of or damage to vehicle

L-3: Loss of or damage to objects outside the vehicle

L-4: Loss of mission (e.g., transportation mission, surveillance mission, scientific mission, defense mission, and so on)

L-5: Loss of customer satisfaction

L-6: Loss of sensitive information

L-7: Environmental loss

L-8: Loss of power generation

Next, the high-level system hazards associated with the losses are identified. The level of abstraction of the hazards to start the analysis is very high, and thus very few are involved at the start of the process. Examples of high-level hazards are:

H-1: Aircraft violates minimum separation standards in flight [L-1, L-2, L-4, L-5]

H-2: Aircraft airframe integrity is lost [L-1, L-2, L-4, L-5]

H-3: Aircraft leaves designated taxiway, runway, or apron on ground [L-1, L-2, L-5]

H-4: Aircraft comes too close to other objects on the ground [L-1, L-2, L-5]

H-5: Satellite is unable to collect scientific data [L-4]

H-6: Vehicle does not maintain safe distance from terrain and other obstacles [L-1, L-2, L-3, L-4]

H-7: UAV (Unmanned Air Vehicle) does not complete surveillance mission [L-4]

H-8: Nuclear power plant releases dangerous materials [L-1, L-4, L-7, L-8]

Iterative refinement of these system-level hazards can be done at this time if desired, as explained in chapter 9. But they must all be at the system level. The goal is to use a process that will ensure that hazards are not missed by employing a structured, top-down way to identify them starting with a few at the top level, as described in chapter 9. The numbers in brackets trace each hazard to the resulting loss or losses. Traceability is thus maintained and is continued throughout the hazard analysis process.

Once the system-level hazards have been identified, they can trivially be rewritten as system-level constraints or requirements. Later steps identify detailed hazards associated with the system components. As an example of a constraint derived from H-1 above:

> SC-1: Aircraft must satisfy minimum separation standards from other aircraft and objects [H-1]

Constraints can also define how the system must minimize losses in case the hazards do occur. For example, if aircraft do violate minimum separation, then the violation must be detected, and measures must be taken to keep the aircraft from colliding. If a chemical plant does release toxic chemicals, then the toxic environment must be detected and appropriate measures taken. For example:

> SC-3: If aircraft violate minimum separation, then the violation must be detected and measures taken to prevent collision

Note that the constraints generated are simply the set of safety-related requirements and constraints that should be included in any system requirements specification. They are also at a very high level. Later steps will identify detailed hazards associated with each system component and the component constraints and the requirements necessary to maintain the system-level safety constraints.

The syntax of the system hazards and safety constraints is formally defined to allow automated generation and processing:

<Hazard> = <System> & <Unsafe Condition> & <Links to Losses>

<Safety Constraint> = <System> & <Condition to Enforce> & <Link to Hazards>

<Safety Constraint> = If <hazard> occurs, then <what needs to be done to prevent or minimize a loss>

Step 2: Model the control structure.

Like HAZOP, but unlike the other traditional hazard analysis methods, STPA is performed on a model. HAZOP uses a physical model of the plant, while STPA uses a functional model in the form of a hierarchical safety control structure.

Hierarchical safety control structures were defined in chapter 7 when describing the STAMP accident causality model. An aircraft example is shown in figure 10.19. The hazard analysis is performed on this model. Higher levels of the control structure might be included, such as the Federal Aviation Administration, which runs air traffic control in the United States, or other levels of government or nongovernment groups.

During the process of refining this high-level model, component responsibilities, control actions, process models, and perhaps feedback information are identified. Because this detailed design information is not needed to start STPA, the hazard analysis can start during high-level conceptual design development; safety can then be built into the system design as it is created. Feedback requirements are generated in later steps of the STPA analysis, so one need not start by identifying them. Figure 10.20 shows a refinement of

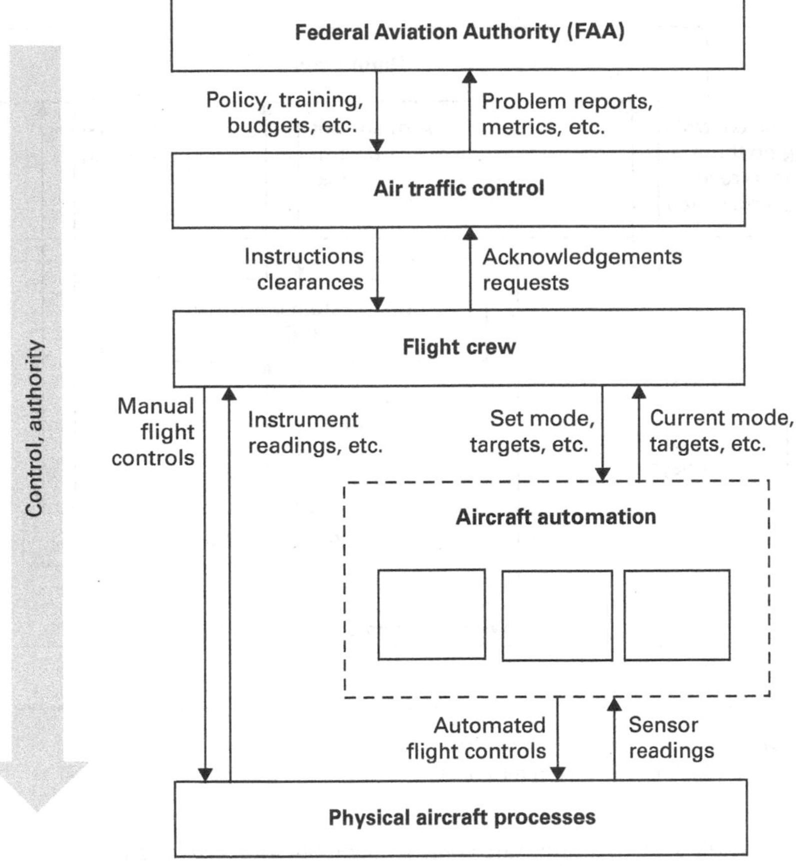

Figure 10.19
An example high-level safety control structure for an aircraft.

this control structure involving the Wheel Braking System on the aircraft that might be generated later as the system design and STPA analysis evolves. The process models for each of the controllers are omitted to simplify the figures.

Step 3: Identify unsafe control actions.

After the initial control structure is created, the analysis can start to be performed on it. The first part of the analysis identifies unsafe control actions.

Definition: An *Unsafe Control Action* (UCA) is a control action that, in a particular context and worst-case environment, will lead to a hazard.

There are four ways a control action can be unsafe:

1. Not providing the control action could lead to a hazard (e.g., the air traffic controller does not provide a control action to avoid a collision when two aircraft violate minimum separation requirements).
2. Providing the control action could lead to a hazard (e.g., the air traffic controller provides a control action that leads to a minimum separation violation).

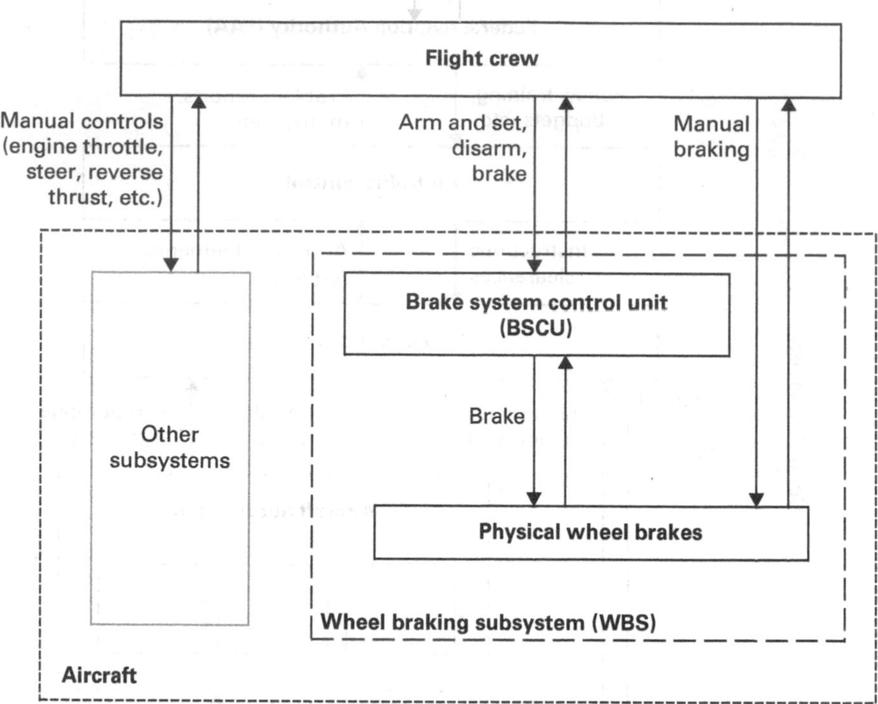

Figure 10.20
An aircraft Wheel Braking System control structure.

3. Providing a potentially safe control action but too early, too late, or in the wrong order could lead to a hazard (e.g., the air traffic controller provides the appropriate instruction but too late to avoid a collision).
4. The control action lasting too long or stopping too soon could lead to a hazard (for continuous control actions, e.g., the air traffic controller tells the pilot to ascend to 25,000 ft but the pilot levels off at 20,000 ft, which does not result in avoiding the collision).

Hazardous states may also be reached by hardware failures, which are the focus of the traditional hazard analysis techniques. To identify such hazard causes, the STPA analysis also includes the examination of how hazardous states might be reached even if the control actions provided are safe. The primary way this can happen is if the commands provided by the controller are not executed by the controlled process. Thus, failures are included, but only those that can lead to hazards are considered, saving a lot of the effort that goes into a failure analysis technique like FMECA.

The unsafe control actions can be documented in a tabular form as shown in table 10.4 or in any format convenient to the analyst.

Each of the control actions in the model is analyzed to identify all contexts in which it will be unsafe. Again, traceability is maintained. The system controllers can be automated (usually using software), or they may be human operators, so both software and humans

Table 10.4
Examples of unsafe control actions for the BSCU (Brake System Control Unit) (partial example only)

Control Action	Not Providing Causes Hazard	Providing Causes Hazard	Too Early, Too Late, out of Order	Stopped Too Soon, Applied Too Long
Brake	UCA-1: BSCU Autobrake does not provide the Brake control action during landing roll when the BSCU is armed [H-4.1]	UCA-2: BSCU Autobrake provides Brake control action during a normal takeoff [H-4.3, H-4.6] UCA-5: BSCU Autobrake provides Brake control action with an insufficient level of braking during landing roll [H-4.1] UCA-6: BSCU Autobrake provides Brake control action with directional or asymmetrical braking during landing roll [H-4.1, H-4.2]	UCA-3: BSCU Autobrake provides the Brake control action too late (>TBD seconds) after touchdown [H-4.1]	UCA-4: BSCU Autobrake stops providing the Brake control action too early (before TBD taxi speed attained) when aircraft lands [H-4.1]

are included directly in the analysis. Controls may also be in the form of physical devices, rules and procedures, standards, and so on.

The control structure is simplified for this example, but if higher levels of the organization were included, then the controllers might be managers or regulators. In this way, the entire sociotechnical system can be included in the hazard analysis. Hardware failures that can lead to hazards, the primary or only focus of the traditional techniques, are also identified as noted earlier, but this very straightforward process is omitted for the example included here.

Using the model, the unsafe control actions are generated and documented. An example is shown in table 10.4 for the control structure shown in figure 10.20. This example involves the potential unsafe control actions that can be generated by the Brake System Control Unit (BSCU), a sophisticated software component of the aircraft ground braking system. The BSCU includes an autobraking function that can be activated by the pilot to automatically brake on landing. Notice how these are simply refinements of the initial system-level hazards identified in Step 1 of STPA but associated with the system components, here the braking system. The analysis process assists in the generation of more specific hazards from the initial short list of high-level hazards.

Table 10.4 shows examples of the unsafe control commands for a software-based system. The unsafe control commands provided by a human controller/operator are generated and documented in the exact same way. An example of a human controller/operator UCA for this system is that: *The crew provides a BSCU Power Off Command when anti-skid functionality is needed and the Wheel Braking System is functioning normally.*

In general, every UCA contains five parts:

<Source> <Type> <Control Action> <Context> <Link to Hazards>

For example,

UCA-2: BSCU Autobrake provides brake command during normal takeoff [H-4.3]

The first part (source) is the controller that can provide the control action. The second part is one of the four types of unsafe control action—provided, not provided, too early or

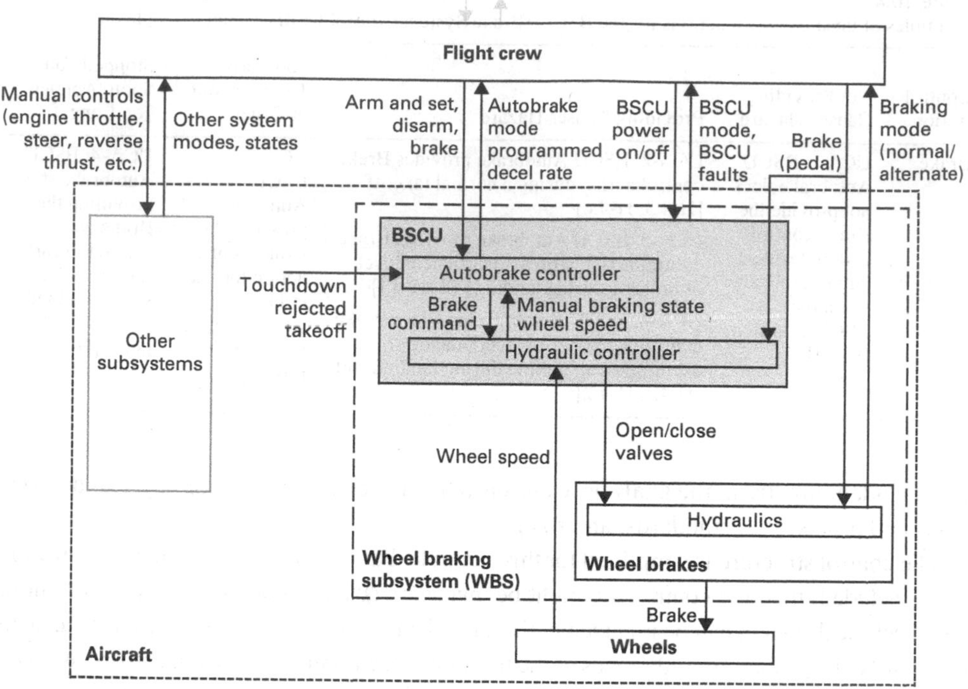

Figure 10.21
Refinement of the control structure as design progresses.

too late, and stopped too soon or applied too long. The third part is the control action or command itself, which is included in the control structure. The fourth part is the relevant context in which the braking command is given when the command is hazardous, and the last part is the link to hazards (or subhazards). UCAs are often written with each part in the same order shown, but, in some cases, it may be clearer or more natural to use a different ordering. The ordering is not critical. Algorithms and tools have been created to generate the UCAs automatically [373].

At the end of this step, additional system safety requirements and constraints can be generated from the identified UCAs if desired. For example: *The BSCU Autobraking system must not provide a brake command during normal takeoff* (UCA-2).

The identified hazards, constraints, and causal scenarios (identified in Step 4) can be used to generate more detailed versions of the system design. In this way, the design is refined as more is learned about the hazard causes. In essence, the hazard analysis is used to design safety into the system as the design unfolds.

Figure 10.21 shows a refinement of the control structure in figure 10.20. Here the BSCU is modeled in more detail as being composed of an autobrake controller, which in turn controls the controller of the physical hydraulic braking system. Using this more detailed control structure, the hazard analysis performed so far can be refined into more detailed hazards and requirements.

Step 4: Identify loss scenarios.

The final analysis step is to identify the loss scenarios that can lead to a UCA. The results are used to eliminate or mitigate the identified scenarios in design or operations.

Definition: A loss scenario describes the causal factors that can lead to the unsafe control actions and to hazards.

Two types of loss scenarios must be considered:

a) Why would unsafe control actions occur?

b) Why would potentially safe control actions be improperly executed or not executed, leading to hazards?

The second type represents physical hardware failures, again, and the types of scenarios that would be identified by the traditional hazard analysis techniques, as noted previously. An example is that the pilot provides a braking command on landing, but the braking hardware fails. Because the standard scenarios involving failures can be identified using STPA, there is no need to perform the traditional failure-oriented techniques, such as FMECA or FTA, in addition to STPA.

In addition to physical failures, STPA can be used to identify more sophisticated causal scenarios such as flawed software control algorithms, missing or incorrect feedback, or flaws in human mental models and errors related to human factors design. How the scenarios are generated is beyond this simple explanation of STPA. Two example causal scenarios for UCA-2 are:

UCA-2: *BSCU Autobrake does not provide the Brake control action during landing roll when the BSCU is armed* [H-4.1]

Scenario 1:

The autobrake (controller) believes the aircraft has already stopped (incorrect process model).

How could this happen? The BSCU is armed, and the aircraft begins the landing roll. The BSCU does not provide the Brake control action [UCA-2] because the BSCU incorrectly believes the aircraft has already come to a stop. This flawed process model can occur if the received feedback momentarily indicates zero speed during landing roll. The received feedback may momentarily indicate zero speed during antiskid operation, even though the aircraft is not stopped.

Scenario 2:

The autobrake believes the aircraft is in flight (incorrect process model).

How could this happen? The BSCU is armed, and the aircraft begins landing roll. The BSCU does not provide the Brake control action [UCA-2] because the BSCU incorrectly believes the aircraft is in the air and has not touched down. This flawed process model will occur if the touchdown indication is not received on touchdown. <Why? The rest of this scenario is omitted for brevity reasons.>

Heuristics and clues have been created to assist in generating the scenarios. However, during extensive use of STPA in practice, subject matter experts, such as the engineers designing the system, have been found to be quite adept at coming up with causal scenarios. Safety specialists can perform the rest of the analysis to this point and work with the engineers to generate application-specific causal scenarios.

The scenarios, once generated, can be used to:

- drive the system architecture development
- create system or component requirements
- identify design recommendations to eliminate or mitigate the causal scenarios for further design refinement or identify changes to an existing design
- define test cases and create test plans
- drive new design decisions (if STPA is used during development)
- evaluate existing design decisions and identify gaps and changes needed (if STPA is used after the design is finished)
- develop leading indicators of risk
- design more effective safety management systems
- analyze data collected during operations

Extensions to STPA have been created to allow sophisticated human factors analysis [105] and cybersecurity analysis [417], but these extensions are beyond the scope of the simple introduction provided here.

Life-cycle phase STPA can be used in every phase of the system life cycle, including operations. Unlike the traditional hazard analysis methods, it does not need a detailed system design to be performed—just a high-level control structure—and thus can be used during early concept formation to derive the system safety and security requirements and constraints that guide the development of the system and component architectures and detailed designs.

Evaluation Like any hazard analysis method, STPA generates causal scenarios. Some important differences exist both in the process to identify the scenarios and in the types of scenarios that result, however. STPA is a backward analysis technique (see chapter 9) and starts from hazards.

Forward hazard analysis techniques start from failures and go forward to see if failures result. Only single failures can usually be considered for complex systems, and, even then, the amount of work involved may be enormous. Software and humans are usually only considered superficially, if at all, in terms of causes and thus often mitigations. Much of the work may involve following paths that do not turn out to have safety implications.

In contrast, the backward analysis used in STPA generates only hazardous scenarios so no unnecessary analysis is performed. In fact, empirical studies in industry have found that STPA requires orders of magnitude fewer resources than the traditional hazard analysis methods. At the same time, more causal factors and scenarios can be identified, including interactions among multiple causal factors, software flaws, and human errors. Software and human errors are included directly in the system analysis and are analyzed along with hardware failures.

Fault Tree Analysis also starts with hazards, but the branches only contain faults and failures. Hazards cannot be identified that result from system design errors where no components may have failed but their interaction leads to the hazardous state. This type of causal scenario is common in complex systems today.

To show the difference between STPA and Fault Tree Analysis, the control structure for the example in figure 10.10 is shown in figure 10.22. Assumptions need to be made about the design of the system as the system design is not included in the fault tree, only the

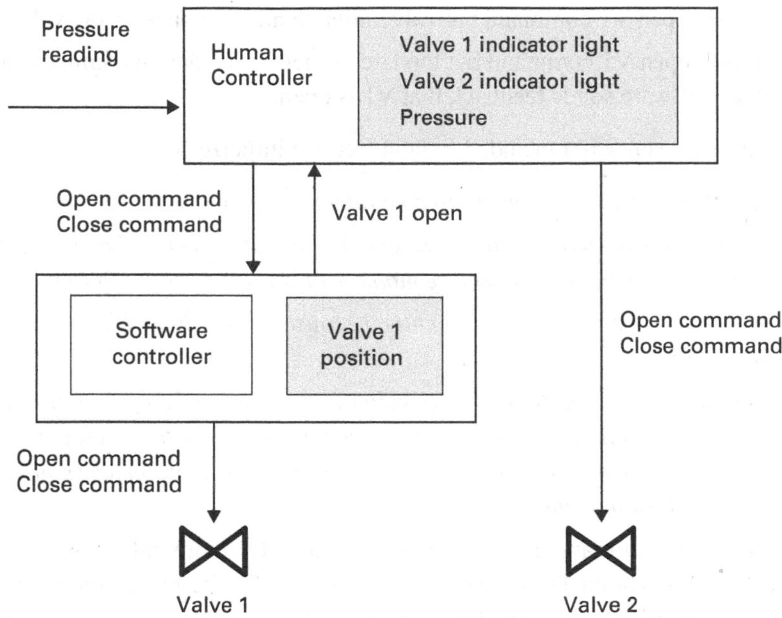

Figure 10.22
The control structure for the system analyzed in figure 10.10. The process model contents are shown in the shaded boxes. It is assumed that the pressure reading only goes to the human and not to the software valve controller. Performing STPA on different designs can provide information to the designers to use in making design tradeoff decisions.

failure events that occurred in the event chain. A different design would provide slightly different results in the STPA, but the differences demonstrated by this example would still remain.

The potential system losses for this example are human injury or death and equipment damage. The hazard is overpressurization. In identifying the unsafe control actions, an assumption is made that opening a relief valve when pressure is not too high is only an efficiency and loss of work issue, which is not included as a loss (although it could have been). Unsafe control actions for the human controller are:

OP-UCA-1: Operator does not send command to software to open V1 (and also does not open V-2) when pressure is too high.

OP-UCA-2: Operator sends command to software to open V1 when pressure too high, but software does not open V1 and operator does not open V2.

OP-UCA-3: Operator sends command to software to open V1 but too late after pressure too high condition occurs.

OP-UCA-4: Operator sends open V2 command too late.

OP-UCA-5: Operator thinks pressure is no longer too high and closes relief valve2 when pressure is still too high.

Unsafe control actions for the software controller are:

S-UCA-1: Software does not issue open V1 command when operator commands it. It sends feedback that indicates valve 1 is open.

S-UCA-2: Issues open V1 command but valve fails. Sends feedback to say valve is open.

S-UCA-3: Issues open V1 command but too late or there is a delay in command getting to the valve. Software sends feedback that V1 is open.

Some scenarios (the ones not included in fault tree are italicized):

1. *Operator does not send command to computer to open valve 1.*
   a. *Operator distracted by other emergencies or duties, related or not to the pressure, and does not read pressure monitor or does not read it in time.*
   b. *Operator misreads pressure monitor because of human factors issue in control panel design.*
   c. *Pressure monitor failure, wrong value provided (e.g., calibration issue somewhere in pressure sensing or some other failure in plant) or delay in displaying of value. Operator does not know to open relief valves (process model incorrect). Explosion occurs.*
2. Operator sends command to software to open Valve 1. Command to software does not arrive (communication failure) or arrives too late and therefore computer does not know to open valve. *Operator does not notice that indicator 1 light has not gone on until too late* or indicator light fails on. *Therefore, operator process model incorrect (thinks valve 1 open) and does not open Valve 2. Process model flaw could be related to operator being distracted or other human factors issues perhaps related to control panel design. The potential human factors issues should be identified and handled in design]. Explosion.*
3. *Operator sends command to software to open Valve. Software sends open valve command, but valve fails or command is not received or not received in time because of failure or delay in communication line. Software does not know valve has not opened. Process model updated when sends command to valve to open. Process model says valve has opened so computer sends message back to operator that valve 1 has opened. Operator therefore does not open Valve 2. Explosion.*
4. *Same as 3 but computer sends command too late because receives it too late from operator (communication delay). Software does not know valve has failed and sends message back to operator that valve 1 has opened. Operator therefore does not open Valve 2. Explosion.*

There are many more. Note that everything *could* be included in a fault tree because it is always possible to write anything in a box or circle. But the *why* will be missing most of the time, and that is what is needed to eliminate or mitigate the problem through system design. In addition, it will not be possible to get probabilities for such circles (root causes).

A unique feature of STPA is that it is performed on a model of the process. HAZOP uses a model, but the difference is in the type of model: HAZOP models are physical system models that allow identifying physical system failures. In contrast, STPA is performed on functional models and allows a functional hazard analysis. Another similarity is that HAZOP also has a structured process for identifying causal factors. The HAZOP process is less complete than that used in STPA, but the HAZOP guidewords do provide assistance in the analysis. The traditional hazard analysis techniques described in this chapter, except for HAZOP, do not have any process for generating the results but simply a format for displaying them once they are generated. A fault tree simply shows the results of the analysis that has been done in the analyst's head.

Another important difference between STPA and HAZOP is that HAZOP is based on the chain-of-events accident causality model and has the limitations associated with that model, including limitations in identifying potential scenarios.

An important advantage of STPA is that it can be done early to generate system requirements and evaluate proposed designs. The analysis techniques based on component failure or deviation analysis require that a detailed design already exists. Such a detailed design is only available late in the development process, when changing the design if serious problems are found is difficult or impossible and backing up in the design process is extremely expensive. Options for preventing or controlling hazardous scenarios may be limited at that time.

Some limitations of STPA can be identified, of course. It does not provide quantitative results as the old techniques do. The hardware failure–based techniques can use known failure probabilities for standard hardware components. Although STPA finds causes that include hardware failures, and thus those scenarios could theoretically be used to produce a probability, the STPA causal scenarios also include events and system behavior—such as software- and human-related behavior—for which no probabilities exist or could exist as they are not stochastic. Risk analysis and failure probabilities that include only those factors that can be quantified will not be very useful in terms of accuracy of the results for complex systems, but managers and regulators often want a number for decision-making, even if that number is not accurate.

A second criticism has been made that doing STPA requires subject matter expertise. Failure probabilities and historical information is available for standard components, and therefore hazard analyses, which are really failure and reliability analyses, can be generated without great subject matter knowledge The results of the failure-focused methods may not be terribly useful in ensuring safety, but they often can be used to satisfy regulators and certification standards that require probabilistic risk assessments.

A large number of evaluations of STPA have been done on real systems. The results almost uniformly show that it finds more scenarios, including the scenarios found by the older techniques, with much lower costs and resource investment.

10.4 Task and Human Error Analysis Techniques

The hazard analysis techniques described so far, except for STPA, include human error in only a limited way, if at all, and instead concentrate on equipment failures. As a result, separate human task and error analysis have been developed. These traditional approaches to performing hazard analysis on human operators can be divided into qualitative and quantitative techniques.

10.4.1 Qualitative Techniques

Most of these techniques involve an examination of the procedures either specified or assumed to be used by the operator. The simplest methods identify procedural steps and analyze them in some way.

A procedure is an ordered set of instructions or actions to accomplish a task. *Procedure* or *Task Analysis* [131] involves reviewing procedures to verify that they are effective and safe within the context of the mission tasks, the equipment that must be operated, and the environment in which the personnel must work. Such analyses involve determination of the required tasks, exposures to hazards, criticality of each task and procedural step, equipment characteristics, and mental and physical demands.

As with FMEAs, the results of the analysis are entered on a form with columns labeled Task, Danger, Effects, Causes, Corrective Measures, and so on. Possible results include recommendations for corrective or preventive measures to minimize the possibilities that an error will result in a hazard, changes or improvements in hardware or procedures, warning and caution notes, special training, and special equipment, including protective clothing.

Operator Task Analysis [206] is similar to Procedure Analysis. The operator's task is broken down into separate operations, and the analysis looks for difficulties in executing either the individual operations or the overall plan. Neither of these first two analyses (Procedure Analysis and Operator Task Analysis) seems to have a specific procedure associated with it, and they may simply be generic terms for the goals involved.

Action Error Analysis (AEA) [360; 362] uses a forward search strategy to identify potential deviations in human performance. The analysis consists of a systematic description of the operation, task, and maintenance procedures along with an investigation of the potential for performance deviations—such as forgetting a step, wrong ordering of steps, and taking too long for a step. Internal phases of data processing associated with an operator's tasks are usually excluded; instead, only the external outcomes of the error modes in different steps are studied. Some information about physical malfunctions may result from the analysis because it includes the effects of human errors on the physical equipment. This method is very similar to FMEA, but it is applied to the steps in human procedures rather than to hardware components or parts. The results are entered in a table, this time with columns labeled Work Step, Action Error, Primary Consequences, Secondary Consequences, Detection, and Measures.

Work Safety Analysis (WSA) [360, 392] was developed by Suokas and Rouhiainen in Finland in the early 1980s. It is similar to HAZOP, but the search process is applied to work steps. The goal is to identify hazards and their causes. The search starts, as in the other methods, by breaking a task down into a sequence of steps. Each of the steps is examined with respect to a list of general hazards and examples of their causes—both deviations and determining factors. All types of system functions and states, including normal states, are considered. The analyst examines the consequences of (1) forgetting a work step, (2) performing a step too early or too late or too long, and (3) unavailability of the usual equipment. Because of the nature of the search pattern, certain types of hazards will not be identified, such as those related to management procedures or those related indirectly to the operator's task but not to the task being analyzed, for example, contact with chemicals or an explosion in the proximity of the operator [362].

A more recent and somewhat more structured search method, called Hierarchical Task Analysis (HTA), uses hierarchical decomposition to identify the task steps. It provides a description of tasks in a hierarchical structure showing the goals, subgoals, operations, and plans. Tasks are broken down into progressively smaller units. Figure 10.23 shows part of the results of an HTA for a pilot.

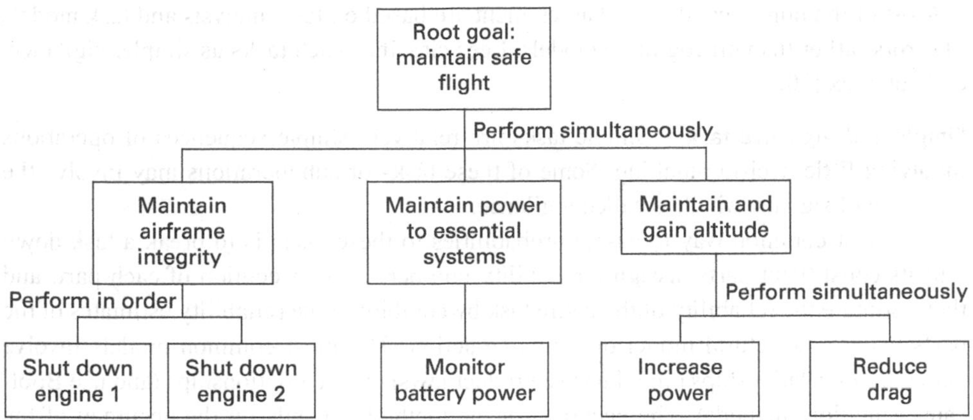

Figure 10.23
An example of HTA.

With the introduction of more automation into systems, operator task analysis requires more sophisticated cognitive analysis, and many assumptions in these simple task analysis techniques described so far do not hold. Various sophisticated cognitive task analyses to identify decision-making tasks and processes have been proposed based on relatively new psychological concepts such as naturalistic decision-making and ecological psychology. These approaches try to understand tasks requiring decision-making, problem-solving, memory, attention, and judgment. A collection of methods has been proposed for conducting a cognitive task or work analysis, but none are widely used in practice.

10.4.2 Quantitative Techniques

All the human error analysis methods described so far focus on the operators' tasks. The goal is to obtain the information necessary to design a human–machine interface that reduces human behavior leading to accidents and improves the operators' ability to intervene successfully to prevent accidents. Human error is not considered inevitable, but a result of human–task mismatches and poor interface or operating procedures design. When the focus is design, qualitative or semi-quantitative results are usually adequate to achieve the goals.

Probabilistic assessment of human error, on the other hand, necessarily accepts the inevitability of human error. Despite its limited usefulness in improving the human–machine interface, the application of reliability engineering, which focuses on numerical assessment, to process control systems (especially nuclear power plants) has led to a demand for assessing the reliability of the process operator in order to assess risk for the system as a whole. The assignment of probabilities to human error is especially important in system risk assessment because of the large proportion of accidents that are attributed to human error. Chapter 13 provides an analysis of the accuracy and usefulness of such data.

Simply having a need is not enough to guarantee that the need can be satisfied. Probabilistic assessment of human error is not very advanced. The rest of this section describes the current state of the art; readers can determine for themselves how much confidence they want to place on the resulting numbers.

Most of the numerical data and assessment are based on task analysis and task models of errors rather than on cognitive models. Lees classifies such tasks as simple, vigilance, and complex [206].

Simple and vigilance tasks Simple tasks are relatively simple sequences of operations involving little decision-making. Some of these tasks or suboperations may involve the detection of signals, which is called vigilance.

The most common way to assign probabilities to these tasks is to break a task down into its constituent parts, assign a reliability number to the execution of each part, and then estimate the reliability of the entire task by combining the reliability estimates of the parts using a structural model of their interaction. The most common models involve either series relationships (and thus use product laws) or tree relationships (and use Boolean evaluation methods). The accuracy of the method depends on the accuracy of the individual part reliabilities and the appropriateness of the structural model. Notice that an assumption is made here that the tasks are completely independent.

The sophistication of the quantitative reliability estimates varies greatly [206]. The simplest approaches often use an average task error rate of 0.01. This number is based on the assumption that the average error rate of the constituent task components is 0.001 and that there are, on average, ten components per task.

A second approach to assigning human error rates uses human experts to rank tasks in order of their error likeliness and then uses ranking techniques to obtain error rates. Sophisticated statistical methods, such as paired comparisons, can be used to produce a ranking [154]. The problem is that these sophisticated statistical methods all rest on the accuracy of the largely subjective ranking done by human experts.

The techniques described so far rely on human judgment to assign error rates to tasks, or they make simplifying assumptions. Other approaches collect and use empirical and experimental data evaluated with respect to performance-shaping factors. *Data Store*, as an example, was developed by the American Institute for Research in 1962 to predict operator performance in the use of controls and displays [133]. The data indicates the probability of successful performance of a task, the time required to operate particular instruments, and the features that degrade performance. To analyze a task using Data Store, the task components are identified and assigned probabilities using tables for standardized tasks. The reliabilities are then multiplied to determine the task reliability.

Data Store and similar techniques assume that the discrete task components are independent. THERP (Technique for Human Error Rate Prediction), developed by Swain at Sandia National Laboratories, relaxes this assumption. Later, Bell and Swain described a methodology for HRA (Human Reliability Analysis) that combines both task analysis and THERP [305]. THERP and HRA have been used primarily for risk assessments of nuclear power plants.

Most of the errors identified and analyzed in HRA involve not following written, oral, or standard procedures. Only occasionally are actions that are outside the scope of the specified operations, such as extraneous acts, considered.

The first part of HRA (and of most similar methods) involves task analysis, where a task is defined by Bell and Swain as a quantity of activity or performance that the operator views as a unit, either because of its performance characteristics or because the activity is required as a whole to accomplish some part of a system goal. The correct procedure for accomplishing an operation is identified and then broken down into individual units of physical or mental performance. For example, the tasks involved in pressurizing a tank to a prescribed level from a high-pressure source [131] include:

1. opening the shutoff valve to the tank
2. opening the high-pressure regulator from the source
3. observing the pressure gauge downstream from the regulator until the prescribed level is reached in the tank
4. shutting off the high-pressure regulator
5. shutting the valve to the tank

Note that most of these types of straightforward tasks have been automated in today's control rooms. Humans are, instead, mostly monitoring the automation's performance of the tasks or providing high-level supervisory control actions.

Next, specific potential errors (human actions or their absence) are identified for each unit of behavior in the task analysis. Acts of commission and omission are considered errors if they have the potential for reducing the probability of some desired system event or condition. In the previous example, the operator could forget to open the high-pressure regulator from the source (step 2), open the wrong valve (step 1), or execute the actions out of proper sequence. The actions actually considered are limited. For example, if the error being examined is the manipulation of a wrong switch, perhaps because of the control panel layout, the analysis does not usually try to predict which other switch will be chosen, nor does it deal with the system effects of the operator selecting a specific incorrect switch. It also assumes there is only one way to accomplish some goal.

The next step in HRA is to determine the likelihood of specific event sequences using event trees. Each error defined in the task analysis is entered on the tree as a binary event. If order matters, then the events need to be ordered chronologically. Care must be taken to consider all alternatives, including "no action taken." Other logical models, such as fault trees, can also be used.

Probabilities are assigned to each of the events in the tree, using handbooks or tables of human error probabilities. If an exact match of errors is not possible, similar tasks are used, and extrapolations are made. Table 10.5 is a small example of this type of table [206]. The accuracy and usefulness of such tables is highly debatable. Systems thinking and systems theory would suggest that human behavior is not independent of, and, in fact,

Table 10.5
Typical human error data

Probability	Activity
10^{-2}	Not returning a manually operated test valve to the proper position after maintenance when no display in control room of state of item omitted.
3×10^{-3}	Omission of items embedded in a procedure (rather than at the end).
3×10^{-2}	General human error of commission, such as selecting the wrong switch.
3×10^{-2}	Simple arithmetic error with self-checking.

is greatly influenced by the specific context in which it occurs and thus that such general tables are not useful or valid.

The data in the THERP handbook is based on a set of assumptions that limit the applicability of the data [305]:

- The operator's stress level is optimal.
- No protective clothing is worn.
- The level of administrative control is average for the industry.
- The personnel are qualified and experienced.
- The environment in the control room is not adverse.
- All personnel act in a manner they believe to be in the best interests of the plant (malevolent action is not considered).

Because these assumptions may not hold and because of natural variability in human performance, environmental factors, and task aspects, the THERP handbook gives a best estimate along with uncertainty bounds. The uncertainty bounds represent the middle 90 percent range of behavior expected under all possible scenarios for a particular action; they are based on subjective judgment rather than empirical data. The analyst is expected to modify the probabilities used in HRA to reflect the actual situation. Examples of performance-shaping factors that can affect error rates are:

- level of presumed psychological stress
- quality of human engineering of controls and displays
- quality of training and practice
- presence and quality of written instructions and methods of use
- coupling of human actions
- personnel redundancy (such as the use of inspectors)

Bell and Swain [35] suggest that if, for example, the labeling scheme at a particular plant is very poor compared to labeling at other plants, the probabilities should be increased toward the upper uncertainty bound; if the tagging is particularly good, the probabilities for certain errors might be decreased. These performance-shaping factors either affect the whole task or affect certain types of errors regardless of the types of tasks in which they occur. Other factors may have an overriding influence on the probability of occurrence of all types of errors under all conditions.

Consider, however, that if the labeling example is known to be very poor, system safety engineering would focus on improving the labeling or using human factors experts to design better labels in the first place. The ultimate goal is to create a safe system not to create a measurement.

Dependencies or coupling may exist between pairs of tasks or between the performance of two or more operators. The dependencies in the specific situation need to be assessed in Bell and Swain's approach and estimations made of the conditional probabilities of success and failure.

Once all these steps have been accomplished, the end point of each path through what is essentially an event tree can be labeled a success or a failure, and the probability of each

Table 10.6
Typical error rates used for emergency situations

Probability	Activity
0.2–0.3	The general error rate given very high stress levels and dangerous activities are occurring rapidly.
1.0	Operator does not act correctly in first 60 seconds after the start of an extremely high-stress condition.
9×10^{-1}	Operator does not act correctly in the first 5 minutes after the start of an extremely high-stress condition.
10^{-1}	Operator does not act correctly in the first 30 minutes of an extreme stress condition.
10^{-2}	Operator does not act correctly in the first several hours of a high-stress condition.

path can be computed by multiplying the probabilities associated with each path segment. Then the success and failure probabilities of all the paths are combined to determine the total system success and failure probabilities. The results of HRA are often used as input to fault trees and other system hazard analyses, although care must be taken that the limitations and assumptions are not violated.

Humans make errors, but they also often detect their errors and correct them before they have a negative effect on the system state. If it is possible to recover from an error in this way, the actual error rate for the task may be reduced by orders of magnitude from the computed rate [206]. The probability of recovery depends greatly on the cues available to the operator from the displays and controls and from the plant in general. Bell and Swain suggest that the effects of recovery factors in a sequence of actions not be considered until after the total system success and failure probabilities are determined [35]. They say that these may be sufficiently low, without considering the effects of recovery, so that the sequence does not represent a dominant failure mode. Sensitivity analyses (manipulating a particular parameter to determine how changes to its value affect the final value) have been suggested to identify errors that have a very large or very small effect on system reliability.

Most of these probabilities do not apply to tasks under emergency conditions, where stress is likely to be high. Analyses usually assume that the probability of ineffective behavior during emergencies is much greater than during normal processing. In general, error probability goes down with greater response time. For short response times, very little credit is normally given for operator action in an emergency. Table 10.6 shows some typical error rates used for emergency situations [206].

One other factor needs to be considered when computing or using these numbers, and that is sabotage or deliberate damaging actions by the operator, including suicide. Most of the available data on human behavior assumes that the operator is not acting malevolently. Instead, according to Swain, an assumption is made that any intentional deviation from standard operating procedures is made because employees believe their method of operation to be safer, more economical, or more efficient, or because they believe the procedure is unnecessary [305]. Ablitt, in a UK Atomic Energy Authority publication, discusses the possibility of suicide by destruction of a nuclear power plant:

> *The probability per annum that a responsible officer will deliberately attempt to drop a fuel element into the reactor is taken as 10^{-3} since in about 1000 reactor operator years, there have been two known cases of suicide by reactor operators and at least*

one case in which suicide by reactor explosion was a suspected possibility. The typical suicide rate for the public in general is about 10^{-4} per year although it does vary somewhat between countries. (Quoted in [206, p. 41])

Another widely touted method for human error analysis is called SHERPA (Systematic Human Error Reduction and Prediction Approach). It was specifically designed for situations where a set of clearly defined task steps for achieving a goal exists [96]. SHERPA starts with HTA (Hierarchical Task Analysis) as described in the previous section, breaking down each step additionally into the types of errors possible, what they would entail, the consequences, what is involved in recovering from the error, design recommendations, and criticality. A taxonomy of errors is used based on five types: action, retrieval, check, selection, and information communication errors. Failure modes are assigned to the task steps identified in HTA, and human reliability assessment techniques are used to identify the total task reliability.

Other human reliability estimation techniques have been proposed, although THERP and SHERPA are probably the most widely used. A weakness of all these techniques, as noted, is that they do not apply to emergency situations. In fact, very little data on human errors in emergencies is available. If one accepts Rasmussen's Skill–Rule–Knowledge model (see chapter 6), the error mechanisms embedded in a familiar, frequent task and in an infrequent task will differ because the person's internal control of the task will be different [314]. Therefore, error rates obtained from general error reports will not apply for infrequent responses.

Another weakness is that the techniques cannot cope with human decisions and tasks that involve technical judgment. Factors other than immediate task and environmental factors are also ignored.

Many of these human reliability assessment techniques were proposed and the data collected before plants became highly automated. We are automating exactly those tasks that can be measured and leaving operators with the tasks that cannot. Therefore, measurement of this type is bound to be of diminishing importance.

Another limitation is that all these methods focus on reliability, not safety. Complying with written procedures or designed tasks does not necessarily imply that hazards will not occur. Many accidents have occurred when the human operators were doing exactly what they were told to do. They were, in fact, acting completely reliably, but the specified procedures were unsafe. The difficulty, of course, is determining how to define "human mistake" without using hindsight to determine what should have been done in a specific unplanned and unanticipated situation.

Another problem arises if a systems approach is taken and safety is assumed to be an emergent property, as described in chapter 7. Looking only at human error does not begin to address the safety any more than focusing only on hardware failures. Losses occur when the interactions among all the components lead to a hazardous system state. Using a system's view of safety, looking at human behavior without considering the specific system design in which that behavior occurs will neither provide a useful assessment of safety nor clues to how to design the system to be safer.

Complex control tasks The measurement approaches described in the previous section consider human performance as a concatenation of standard actions and routines for which error characteristics can be specified and frequencies determined by observing similar activities in other settings. In such analyses, the task is modeled rather than the person. Rasmussen and others argue that such an approach may succeed when the rate of technological change is slow, but it is inadequate under the current conditions of rapid technological change [317].

Computers and other modern technology are removing repetitive tasks from humans, leaving them with supervisory, diagnostic, and backup roles. Tasks can no longer be broken down into simple actions; humans are more often engaged in decision-making and complex problem-solving for which several different paths may lead to the same result. Only the goal serves as a reference point when judging the quality of performance—task sequence is flexible and very situation and person specific. Analysis, therefore, needs to be performed in terms of the cognitive information processing activities related to diagnosis, goal evaluation, priority setting, and planning—that is, in the knowledge-based domain, not the skill- or rule-based domains.

From this viewpoint, performance on a task can no longer be assumed to be at a relatively stable level of training. Learning and adaptation during performance will have a significant impact on human behavior. If the models of behavior used simply consider external characteristics of the task but have a significant cognitive component, then measurement—and, of course, design—needs to be related to internal psychological mechanisms in terms of capabilities and limitations [316]. If, as Rasmussen recommends, the concept of human error is replaced by human–task mismatch, then task actions cannot be separated from their context. Measurements for these human-task mismatch errors do not exist, and deriving them will be difficult, however, as the cognitive activities involved in complex and emergency situations cannot easily be identified in incident reports. In addition, of course, they will be system-specific and not general quantities that can be applied to all system designs.

In addition, human integration with system design is increasing in importance along with new types of human errors, such as mode confusion and errors related to situation awareness, that are related to specific system designs. It seems that the focus, in these cases, should be more on designing systems—including software, hardware, human tasks, and so on—to eliminate or control human error than on trying to measure it.

10.5 Conclusions

Once created during development, the hazard analysis should be used as an ongoing operational and training tool. There are all too many situations where a comprehensive hazard analysis is created and then it sits on a shelf, gathering dust, with no one paying attention to it. In such situations, there is a danger that operations personnel may take the attitude that "we know we are safe because we have a hazard analysis." For continuing use to be practical, of course, the documentation of the analysis has to be usable: it needs to be feasible for operations safety personnel and others to easily access and use the information they need to operate the system safely and to determine when significant factors have changed over time that could decrease safety.

The hazard analysis is not an end in itself; the results need to be used in system design, operation, and management. The next two chapters describe how the results of hazard analysis can be used to design safer systems.

Exercises

1. HAZOP and STPA are both performed on a system model and provide a defined set of flaws to investigate. What is the difference between the two techniques? Will they identify the same set of causal factors/scenarios? If the causal factors they identify are different, then what will be the differences and why?
2. (a) Take the 5 Whys example for the Washington Monument and try rewriting the analysis in the form of a chain of events. Try writing this as a fault tree. Describe the general result and compare the power of each. (b) Take the fishbone diagram example and rewrite it as a fault tree. Describe the general result and compare the power of each. (c) Compare the power of all three techniques using the results you got from the first two parts of the exercise.
3. How might Event Tree Analysis be applied to a complex system such as air traffic control? Where might any difficulties lie, if there are any?
4. Describe the general types of causal scenarios that will be identified by FMEA, Fault Tree Analysis, HAZOP, and STPA. Would there be any advantage to using more than one of these for a particular project? Disadvantages?
5. (Large Project) Do the steps in this example in the order provided. Do not do them out of order. You will be provided with additional material to learn more about these analysis techniques than provided in this chapter.
   a. Select a system to be analyzed. Have there been accidents in the past with it (if it is not new)? What are the technical system goals and requirements? Make sure you select a system that has hardware, software, and humans in it.
   b. Identify the accidents and hazards of importance to the stakeholders. Select one or two to analyze.
   c. Perform FMEA or FMECA on the system. Did you experience any problems in doing this analysis? If so, describe them.
   d. Compare the results you got from parts b and c.
   e. Create a fault tree for one of the hazards. Compare the results you got from FMEA and FTA.
   f. Perform STPA for the same hazard that you used in parts b and c. Be careful not to let the results from parts b and c influence your STPA analysis, but do determine if your STPA analysis could find the same causal scenarios. Do not let the experience of doing the FMEA and Fault Tree Analysis influence the type of causal scenarios you find with STPA; that is, do not just look for failures as the cause of the scenarios. Follow the detailed STPA steps.
      i. Create a model of the hierarchical safety control structure for your system that either exists or needs to be created (if the system is new).
      ii. Identify the potential UCAs for at least one human operator and one automated (software) part of your control structure. Rewrite the results as requirements.

iii. Identify at least two scenarios for each of the UCAs you identified in step ii. They cannot both be just physical failures. Then create a couple of scenarios that involve cyber security.

iv. Generate the requirements for the system design to eliminate or control those scenarios.

v. What information should be passed to operations that you have generated in your analysis? Create a plan for operators to use that information.

vi. What recommendations do you have for the overall safety control structure for the organization that will use this system?

vii. Is there an existing hazard analysis for this system? If so, compare the results with what you got using STPA.

g. Compare the results from your FMEA, STPA, and STPA analyses, and describe the differences in general terms between the types of results you found.

11

Design for Safety

Engineers should recognize that reducing risk is not an impossible task, even under financial and time constraints. All it takes in many cases is a different perspective on the design problem.
—Mike Martin and Roland Schinzinger,
Ethics in Engineering

Before a wise man ventures into a pit, he lowers a ladder so he can climb out.
Rabbi Samuel Ha-Levi Ben Joseph,
Ben Mishle

Safety must be designed into a system. Identifying and assessing hazards is not enough to make a system safe; the information obtained in the hazard analysis needs to be *used* in design, testing, and operations. Most accidents are not the result of lack of knowledge about hazards and their causes but the lack of effective use of that knowledge by the organization.

Safeguards may be designed into the product, or they may be designed into the procedures that operators are given for specific situations. Often, the most complex and tricky problems are left to the operators, who are then blamed when they are unsuccessful and accidents occur. When operators are replaced by computer software that is supposed to carry out the same procedures, the onus is—or should be—shifted to the software.

There are a few general principles that should be followed in designing for safety. The first is that simple design features often can improve safety without increasing complexity or cost, but this improvement requires considering safety early in the design process. One old example is the use of railroad semaphores, which were gravity- and weight-operated devices that lowered and automatically assumed the STOP position if the cable broke. Another example that is still used today is a deadman switch, which ensures a system is powered only so long as pressure is exerted on a handle or foot pedal. Other examples are provided later in the chapter. Many of these simple and safe physical devices are now being replaced by computers that may not provide an equivalent level of safety, but, of course, have other advantages.

The second general principle is that poorly designed risk reduction measures can actually increase risk and cause accidents. Such designs, for example, can increase system complexity. Adding unnecessary complexity is common when safety is not considered early in design but is instead added on at the end, often in the form of redundancy or protection systems. In addition, predicating designs on false assumptions about human or software behavior or independence between components may defeat attempts to reduce risk and encourage the reduction of safety margins on the basis that risk has been supposedly reduced.

In some cases, safety devices are an attempt to compensate for a poor basic system design. Operators may come to rely on them and take fewer precautions; when the safety devices fail, serious accidents may result. System design should make it possible to work in the vicinity of machines without disturbing production unnecessarily. A poorly designed safety device that slows down production or makes operations more difficult will encourage operators to bypass or trick it. "*It is a much better strategy to design a practical safety system than one that cannot be tricked*" [365].

Both software and system design are affected by the introduction of computers into safety-critical systems. Software now controls dangerous systems, and system safety protection previously provided by hardware is now commonly being implemented in software. Software engineers need to understand the basic principles behind safe *system* design so they can implement protection that was previously provided by physical protection mechanisms in the software functions used to replace those mechanisms. It is not enough to simply check the software requirements for consistency with system safety goals and then implement those requirements. Not everything can be written down in requirements specifications, and software developers must understand enough about safety in system design, both physical hazards and unsafe interaction with humans, that the software does not inadvertently contribute to hazards.

On the one hand, the use of computers introduces new possibilities for improving system safety in terms of increased functionality and more powerful protection mechanisms. In a chemical plant, for example, a rapid rise in the temperature in a reactor vessel can indicate a runaway reaction long before the temperature actually reaches a dangerous level; a computer can potentially monitor the temperature and the rate of increase and provide warning early enough to avoid a hazard [189].

At the same time, such software may introduce new hazards such as reduced attention or increasing complacency on the part of the human operators. Often, new safety mechanisms implemented by software are introduced while assuming that their introduction will not change human behavior, which is usually not a valid assumption.

In addition, the system must be protected against the software behaving in an unsafe way. Many hardware backups and safety devices are now being replaced or controlled by software, making safety almost totally dependent on the software not doing anything unsafe. Unfortunately, building physical protection against the software behaving unsafely may be more difficult than protecting against hardware failures.

Failure modes of electromechanical systems are well understood, and components can often be built to fail in a particular way. For example, a mechanical relay can be designed to fail with its contacts open, or a pneumatic control valve can be designed to fail closed or open. Those components can then be used to design the system to fail into a safe state, such as shutting down a dangerous machine. It is difficult, however, to plan for unsafe software behavior because it is not predictable. Many computer hardware failures and some software unsafe behavior can be detected and handled, but doing so requires a great deal of planning and effort. Simply leaving it to the human operators to detect and protect against unsafe software is unrealistic.

These issues along with others are discussed in this chapter and the next one. Standard system design approaches to prevent or mitigate hazards are described, but the introduction of software into almost every system design creates new problems that standard designs do not solve. New problems also are arising in the design of human–computer interaction. Discussion of the design of the safety aspects of human–automation interaction is so important that it is covered separately in chapter 12.

11.1 The Design Process

There are two basic approaches to safe design: (1) applying standards and codes of practice that reflect lessons learned from previous accidents, and (2) guiding design using the results of hazard analysis.

11.1.1 Standards and Codes of Practice

For standard hardware, general safety design principles for relatively simple systems have been incorporated into standards and codes of practice to pass on lessons learned from common accidents. For example, the proper use of pressure relief valves is specified in standards for pressure vessels in order to avoid explosions, while the use of electrical standards and codes reduces the probability of fires.

There are no equivalent standards for safe software. Many of the software design features suggested in various standards are aimed at reliability, maintainability, readability, and so on—although little scientific, empirical evaluation of the efficacy of these features has ever been done. When these other qualities coincide with safety requirements in a particular system, they may make the software safer; when they conflict with safety or have little to do with specific system hazards, they may have little effect on safety and may even increase risk. Unless the software design practices protect against specific identified hazards for the system being designed, simply using "good" software practices will have no significant impact on safety.

Design checklists are another way to systematize and pass on engineering experience and knowledge. Safe design checklists identify design features or criteria found to be useful for specific hazards. Table 11.1 gives examples of partial checklists for mechanical hazards and pressure systems.

Although a large number of design errors are possible in hardware, the checklists usually focus on those that lead to known hazards. Checklists are often also used in software design reviews, although they are much less well developed than for hardware and are oriented toward coding errors in general and not toward safety in particular. Because software is not by itself hazardous—software behavior can only be identified as potentially hazardous when it is used within a particular context or system design—there is no generic software hazardous behavior to consider in design checklists. What is hazardous software behavior in one context that needs to be avoided may be exactly the right thing to do in another. More sophisticated design processes will be necessary in complex systems today.

With the introduction of computers into safety-critical systems, many of the lessons learned and incorporated into hardware standards and codes are being lost—either because of a lack of knowledge on the part of the software engineers or because the principles are not translated into the language of the new medium or into the different and sometimes more

Table 11.1
Checklist examples

Mechanical hazards checklist (incomplete)

1. How are pinch points, rotating components, or other moving parts guarded?
2. Have sharp points, sharp edges, and ragged surfaces not required for the function of the product been eliminated?
3. How have bumpers, shock absorbers, springs, or other devices been used to lessen the effect of impacts?
4. Are openings small enough to keep people from inserting fingers into dangerous places?
5. Do slide assemblies for drawers in cabinets have limit stops to prevent them from being pulled out too far?
6. If a product of assembly must be in a particular position, how is this guaranteed? Is it marked with a warning and a directional arrow?
7. How are hinged covers or access panels secured in their open positions against accidental closure?
8. How are the rated load capacities enforced? Is the equipment at least posed with rated load capacities?

Pressure checklist (incomplete)

1. How have connectors, hoses, and fittings been secured to prevent whipping if there is a failure?
2. Is there any way to accidentally connect the system to a source of pressure higher than that for which the system or any of its components was designed?
3. Is there a relief valve, vent, or burst diaphragm?
4. How will the exhaust from the relieving device be conducted away safely for disposal?
5. How can the system be depressurized without endangering the person who will work on it?
6. Do any components or assemblies have to be installed in a specific way? If so, what means are used to prevent a reversed installation or connection?

complex designs made possible by computers. Some design principles may not hold when computers replace electromechanical systems. If they do still hold, they must be incorporated into system and software designs to avoid needless repetition of past accidents.

11.1.2 Design Guided by Hazard Analysis

Design checklists and standards are not going to be adequate in complex systems today, especially for new types of functionality: The design must eliminate or control the specific causal scenarios identified in the hazard analysis for that system.

Sometimes standard design processes without special safety-related activities are followed for safety-critical systems, and focus is placed instead on after-the-fact assurance of the design or of the other activities used in the safety program. It is not possible, however, to assure something that is not there. The primary emphasis should be on designing safety into the system from the beginning. Trying to eliminate hazards from a complex system design after the fact is nearly impossible, but it is often possible to eliminate them if thought is given to that goal from the beginning of development.

Consideration does, of course, also need to be given to how it will be demonstrated that the hazards have been eliminated or controlled in the design. That will require, at least for software, unique designs to facilitate the process.

How *should* safety be handled in the design process?

Use the results of hazard analysis An effective hazard analysis process will identify the specific safety-related requirements and constraints on the behavior of the individual system components and on their interactions. That was the topic of the previous two chapters. The components then need to be designed to satisfy those requirements and constraints. In addition, the system design as a whole must enforce the identified constraints on component interactions and eliminate the identified causal scenarios.

Focus on off-nominal cases Too often, careful consideration in design and testing is focused on the normal or nominal operation of a system (that is, the use cases), while much less attention is paid to erroneous or unexpected (off-nominal) states. The common focus on *use cases* in design, particularly software design, almost guarantees that safety will not be handled well. Accidents occur most commonly under conditions that were not carefully considered in design and during off-nominal operations. The system as a whole and the software in particular must be designed to be robust against unexpected inputs and environments.

Allow designer flexibility As is true for all requirements, safety requirements should state what is to be achieved rather than how to achieve it so that the designer has the freedom to decide how the goals can best be accomplished and the constraints enforced. Some example software requirements might be that the software must ensure the controlled process moves into a safe state if events A, B, or C occur; that collision avoidance software must not generate avoidance maneuvers that cause unnecessary crossing of paths; or that robot control software must not issue instructions for the robot to move without receiving proper operator inputs.

Identify safety-related requirements While all requirements need to be designed into a system, the safety-related ones must be identified as such to developers to guide their design and tradeoff decisions. Special analysis, design, and verification efforts can then be focused on those functions. Requirements on human controllers should be used to design human–automation interaction and in operator training, development of procedures and checklists, audits, and so on.

Design for assurance From the start, testability and analyzability of the design—which often demand simplicity—should receive serious consideration in decision-making. Certification and verification of safety are extremely costly procedures and may be impossible or impractical for some large systems unless the design is specifically tailored to be testable and certifiable. The design should leverage the certification effort by minimizing the verification required and simplifying the certification procedures. Systems today can have millions of physical parts and hundreds of millions of lines of code. Assurance is impossible in these systems unless the design has been created to allow practical assurance processes.

Ensure traceability A basic component of assurance, maintainability, and so on is traceability from the safety requirements and constraints into the physical and logical system components that satisfy them and that control safety-critical functions. That traceability may have implications for the design of hardware and particularly software that do not exist for non-safety-critical systems or functions and may require the use of different software processes than are currently popular. As is true for any requirements, a traceability matrix and tracking procedures within the configuration control system need to be established to ensure traceability of safety requirements and their flow through the documentation. Isolating and modularizing critical software functions may be the only way to make assurance practical or even possible. Isolation will also assist in dealing with inevitable changes to the software.

Put more effort into early development New types of early system development processes may also be needed, such as the development of conceptual architectures [221]. A conceptual architecture can be used to make tradeoffs and architectural design decisions before detailed architectures are created. Fundamental changes to the basic system architecture often become impossible or impractical later in development.

Record safety-related decisions and their rationale Conditions change, of course, and decisions need to be reviewed periodically. The system design and software will change over time as a result. The design specification should include a record of safety-related design decisions including both general design principles and criteria and detailed design decisions, the assumptions underlying these decisions, and why the decisions were made. Without such documentation, critical design decisions can easily be undone accidentally when a change occurs. Finally, incidents that occur during the life of the system can be used to determine whether the design decisions were well founded and allow for learning from experience.

Consider alternative ways to control hazards Ideally, hazards can be eliminated or controlled through the system design, but that may not always be possible or practical. Hazards may also be controlled through development and operating processes, such as manufacturing processes and procedures, maintenance processes, and operational procedures. Social controls are also possible such as government or regulatory processes, culture, insurance, law and the courts, and of course individual self-interest, for example through the design of the incentive structure. This chapter considers only controlling hazards in the technical system design. The design of other types of controls—management, culture, and social structures—are outlined in chapter 14 and also described in Leveson [217].

To summarize, the design process must include at least:

- Developing system-specific safety design criteria and requirements, testing requirements, and computer–human interface requirements based on the safety constraints and requirements and the causal scenarios identified in the system hazard analysis.
- Tracing safety requirements and constraints to the detailed design of the physical components and the software code, including identifying those parts of the software that control or implement safety-critical functions and the paths that lead to their execution. The design and development steps must then enforce the safety-related constraints and requirements in the system design. Isolation of critical functions may also have benefits. Recording the design rationale is critical.

11.2 Types of Design Techniques and Precedence

The idea of designing safety into a product is not new. As described in chapter 1, it was advocated by John Cooper and Carl Hansen over one hundred years ago. The later development of system safety engineering increased the emphasis on preventing accidents through design. The basic safety design goal is to eliminate identified hazards or, if that is not possible, to control or mitigate those hazards.

While risk is often defined in terms of severity and likelihood, it can alternatively be defined as a function of the strength of the controls to protect against hazards and losses. Hazards can in general be controlled by eliminating them from the system design,

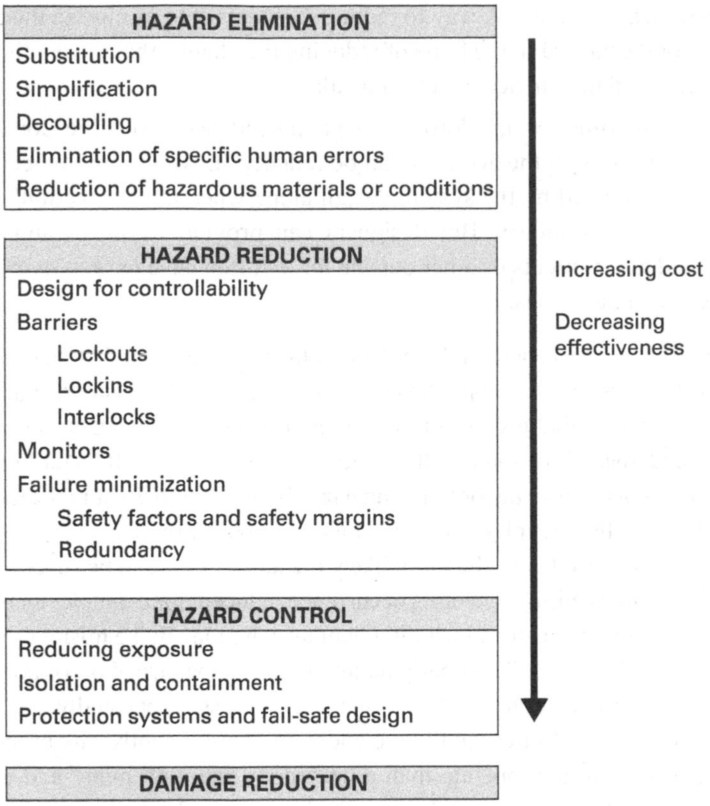

Figure 11.1
The safety design precedence and examples of design techniques.

reducing the ways a hazardous system state can occur, controlling a hazard if it does occur, and, as a last resort, reducing the severity of the consequences. Figure 11.1 shows the basic safety design precedence.

Specifically, risk can be reduced by

1. **Hazard elimination**: Designs are made intrinsically safe by eliminating hazards. Hazards can be eliminated either (a) by eliminating the hazardous state itself from system operation, or (b) by eliminating the negative consequences (losses) associated with that state, and thus eliminating the hazard by definition: if a state does not lead to any potential losses, it is not a hazard.
2. **Hazard reduction**: The occurrence of hazards is reduced by strengthening the controls to prevent hazards. Accidents are prevented if the hazards that precede and contribute to them do not occur. For example, if two aircraft do not violate minimum separation standards, which is a standard hazard in air traffic control, they will not collide.
3. **Hazard control**: If a hazard does occur, actions can be taken to prevent it from leading to an accident. One type of hazard control is to detect the hazard and transfer to a safe state as soon as possible. Accidents do not necessarily follow from a hazard; usually, other conditions must be present in the environment of the system that, together with

the hazard, lead to losses. A way to reduce accidents is to minimize the duration and exposure of the hazard in the hope of reducing the chance that those other conditions will develop and that an accident will result.

4. **Damage reduction**: At the lowest design precedence level, the consequences or losses associated with the accident can be reduced. Losses from an accident usually cannot be eliminated by the system design alone when the loss actually occurs outside the system boundary. But designers can provide warnings and contingency actions, and governments or other outside forces often have options available to them to reduce potential losses.

A clear precedence exists here in the order that the techniques are listed, both in terms of the general categories and the approaches within each category. The design precedence, however, does not imply that just one of these approaches should be taken. If the hazard cannot be eliminated, then all or some will be necessary because not all hazard causes will be foreseen, the costs of eliminating or reducing hazards may be too great in terms of money or required tradeoffs with other objectives, and mistakes will be made.

The higher in the precedence, the more likely the measures are to be successful in avoiding losses; if a hazard is eliminated or its occurrence reduced, for example, there needs to be less reliance on control measures, including humans who may make mistakes or protection devices that may fail, may not be properly maintained, may be turned off or not functioning, or may be ignored in an emergency. Also, as has been stressed repeatedly in this book, it is easier and cheaper to build safety in than to add it on. An inherently safe process will usually be cheaper to build and operate than a hazardous one with many add-on protection devices. So not only does efficacy go down in the lower precedence levels but cost also goes up at the lower levels.

The specific design features chosen to prevent hazards will often depend on the accident causality model used and thus on the types and causes of hazards that are hypothesized (see chapter 7). The models including more causal factors will produce more types of design protection possibilities.

For example, if accidents are defined as resulting from a loss of control of energy, then basic accident prevention strategies will often include the use of controls on the energy and the use of barriers or physical separations between it and humans or property. Energy model hazard control measures include such design goals as limiting the energy used in the process, safe energy release in the event of containment failure, automated control devices to maintain control over energy sources, barriers, strengthening targets, and manual backups to maintain safe energy flow if there are control system failures. Barriers, such as containment vessels and safety zones, depend only on reliable operation of the barriers rather than on any particular hypothesized chain of events or causal factors. Accordingly, safe design in energy containment systems often involves the use of design allowances and safety factors rather than hazard analysis.

A chain-of-failure events model may focus design protections on eliminating the identified failures leading to the hazard or providing barriers to their propagation.

A system model that focuses on component interactions and systems theory will suggest design features that eliminate or control hazardous states, including protecting against unsafe interactions and component failures that can lead to hazardous states.

The specific design features applied will therefore be affected by the accident model selected. The rest of the chapter describes some of these design features. The next chapter presents approaches to design of human–machine interaction.

11.3 Hazard Elimination

The most effective way to deal with a hazard is to eliminate it or to eliminate all possibility that it will lead to an accident, which, by definition, eliminates the hazard. "*If the meat of lions was good to eat, farmers would find ways of farming lions. Cages and other protective equipment would be required to keep them under control and only occasionally, as at Flixborough, would the lions break loose. But why keep lions when lambs will do instead?*" [185, p. 66].

In the energy model of accidents, an intrinsically safe design is one that is incapable of generating or releasing sufficient energy or causing harmful exposures, under normal or abnormal conditions (including outside forces and environmental failures), to cause a hazard, given that the equipment and personnel are in their most vulnerable condition. In other models, an intrinsically safe design is one in which hazardous states or conditions cannot be reached by chains-of-failure events or by unsafe interaction or component failures. Of course, philosophically speaking, nothing is impossible. I could potentially be attacked by a lion in downtown Cambridge that had escaped from a zoo. But from a practical engineering standpoint, the occurrence of some physical conditions or events is so remote that using resources to provide protection against them is not reasonable.

Sometimes a hazard can be eliminated by a design change. Figure 11.2 shows an example where exchanging the position of a battery and a motor in a switch eliminates the potential for a short to discharge the battery and render it useless.

Several techniques can be used to create an intrinsically safe design: substitution, simplification, decoupling, elimination of the potential for human errors, and reduction of hazardous materials or conditions.

11.3.1 Substitution

One way to eliminate hazards is to substitute safe or safer conditions or materials for them, such as substituting nonflammable materials for combustible materials or nontoxins for toxins. As Kletz states with respect to the design of chemical plants, "What you don't have, can't leak" [184].

Of course, substitution may introduce other hazards, but the goal is for these new hazards to be less serious. For example, using pneumatic or hydraulic systems instead of electrical systems may eliminate the possibility of fatal injuries from an electrical hazard, but not the more minor hazards associated with compressed air. Examples of substitution include:

- In the chemical industry, water or oils with high boiling points have been substituted for flammable oils as heat transfer agents; silicious materials have been eliminated from scouring powders; flammable refrigerants have been replaced by fluorinated hydrocarbons [184]; and hydraulic instead of pneumatic systems have been used to avoid violent ruptures of pressure vessels that could generate shock waves [133]. Similarly, pressure vessels are generally tested with water or other liquids and not with gas because the rupture of a vessel containing pressurized gas can generate a shock

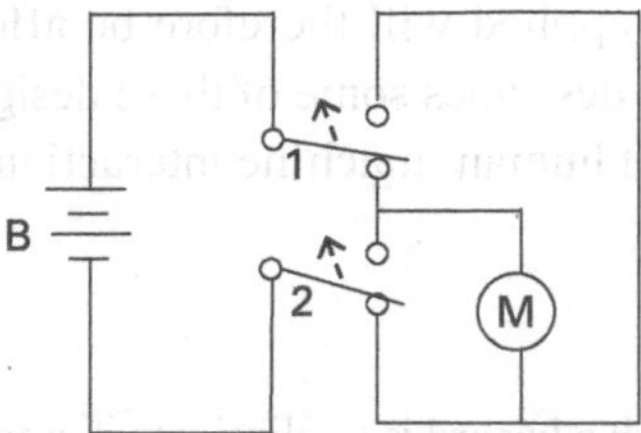

(a) Arms 1 and 2 of the switch are both raised by a solenoid (not shown). If either one does not move (for example, a contact sticks) while the other does, there is a short across the battery. The battery will discharge and be useless even after the trouble is detected.

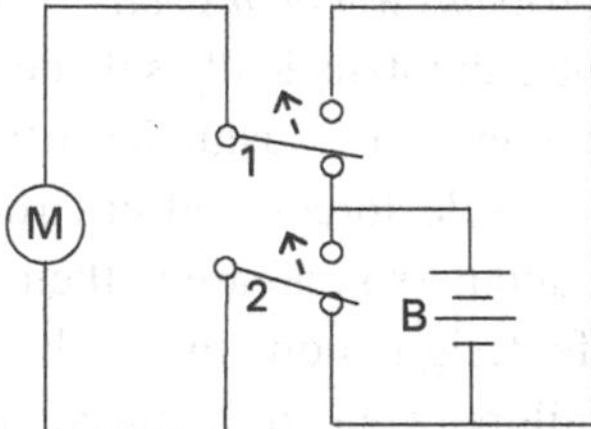

(b) By exchanging the position of the battery and motor, a stuck switch will cause no harm to the battery (the motor can be shorted without harm).

Figure 11.2
Eliminating a potential failure mode through design change. (From Mike W. Martin and Roland Schinzinger. *Ethics in Engineering*. New York: McGraw-Hill Book Company, 1989).

wave and damage similar to that caused by a high explosive. A liquid will not expand the way a gas will when pressure is released, and therefore no shock waves will be created after rupture [132].

- Some missiles have used hybrid propulsion systems, containing both a solid fuel and liquid oxidizer, which eliminate the possibility of combustion and explosion as long as the two are separated. They also eliminate the possibility of uncontrolled combustion due to cracks, voids, and other separations in the solid propellant [131].
- Kletz [185] tells of a plant where a chlorine blower was to be made from titanium, a material suitable for use with wet chlorine but which burns in dry chlorine. A decision was made to have the chlorine pass through a water scrubber before reaching the blower, and an elaborate trip system was designed to make sure that the chance of dry gas reaching the blower would be small. Following a study of the design, the complex trip system was scrapped and a rubber-covered blower installed instead. Although this blower needed to be replaced more often than the titanium one, it eliminated the hazard resulting from the blower coming into contact with dry chlorine and was less costly in the long run.
- After the Apollo 13 near accident, the potential danger of using pure oxygen was eliminated or reduced by design changes to the command and service modules, for example, replacing all Teflon insulation in the oxygen tanks with stainless steel [41].
- In the nuclear industry, pressurized water reactors depend on engineered cooling systems. If the normal cooling system fails, emergency systems are needed to prevent

overheating. Newer designs include more inherent safety features that, for example, cannot overheat even if all the cooling systems fail completely, or they include more fail-safe design features.

- This approach works for electronic and digital systems as well as for mechanical systems. An example is shown in figure 11.2. The design in figure 11.2(a) potentially allows sticky contacts to short out battery B, making B unavailable for further use, even after the contacts are loosened. A rearrangement of wires, as shown in figure 11.2(b), removes the problem altogether.

Even if hazards cannot be eliminated completely, they may be controlled in the design, such as reducing the amount of hazardous material used or using it at a lower temperature or pressure. For example, chemical plant designers may be able to avoid the need for relief valves and associated flare systems by using stronger vessels.

During a hazard analysis for a computer-based nuclear power plant shutdown system, engineers found that simple coding changes, such as rearranging the order of statements, eliminated certain hazardous failure modes [43].

In some extremely critical cases, using inherently safe hardware protection mechanisms may be safer than introducing computer control and the necessarily greater complexity inherent in implementing an analog function on a digital computer. An example is running the wiring providing power to high-voltage equipment through a door that blocks access to the equipment. When the door or panel is opened, the circuit is broken. This design is much safer than sophisticated electronic devices that detect a human entering an area and send the information to a computer, which then must send a command to an actuator to shut down the equipment. Any of these components and the communication channels between them are subject to not working correctly. There is no technological imperative that says we *must* use computers to control hazardous functions.

Eliminating or reducing hazards often results in a simpler design, which may, in itself, reduce risk. The alternative is to add protective equipment to control hazards, which usually adds complexity. The Refrigerator Safety Act was passed because children were being trapped and suffocated while playing in unused refrigerators. Manufacturers had insisted that they could not afford to design safer latches, but when forced to do so, they introduced magnetic latches that permit the door to be opened from the inside without major effort. The new latches not only eliminate the hazard, but also happen to be cheaper than the older mechanical latches. No child has died in a refrigerator that satisfies the design requirements of the act [133; 252].

Adding a protection system approach to the refrigerator problem might have retained the original latch but inserted sensors into the refrigerator to detect a human inside and then sounded an alarm or piped oxygen into the closed space. This admittedly far-fetched solution would have been extremely expensive and less effective than redesigning the latch, but it is often the approach used in engineered systems.

11.3.2 Simplification

One of the most important aspects of safe design is simplicity. A simple design tends to minimize the number of parts, functional modes, and interfaces, and it has a small number of unknowns in terms of its operational interactions [309].

William Pickering, a director of the Jet Propulsion Laboratory, credits the success of early US lunar and planetary spacecraft to simplicity and a conservative approach:

> *The most conservative designs capable of fulfilling the mission requirements must be considered. Conservative design involves, wherever possible, the use of flight-proven hardware and, for new designs, the application of state-of-the-art technology, thereby minimizing the numbers of unknowns present in the design. New designs and new technologies are utilized, but only when already existing flight-proven designs cannot satisfy the mission requirements, and only when the new designs have been extensively tested on the ground.* [309, p. 136]

New technology and new applications of existing technology often introduce *unknown unknowns*. After the accident investigation, discussions of these events are generally prefaced by "Who would have thought. . . ."

Simpler systems provide fewer opportunities for error and failure. The existence of many parts is usually no great problem for designers or operators if their interactions are expected and obvious. But when the interactions reach a level of complexity where they cannot be thoroughly planned, understood, anticipated, and guarded against, accidents occur.

Interfaces are a particular problem in safety: design errors are often found in the transfer of functions across interfaces. Simple interfaces help to minimize such errors and make the designs more testable.

It is easier to design and build complex interfaces with software than with physical devices. Normally, increasing the complexity of physical interfaces or decreasing them greatly increases or decreases the difficulty of designing and analyzing the safety of those interfaces. The same is not true for software, which can be used relatively easily to implement extremely complex interfaces. Analyzing hazards associated with those easily created interfaces may be very difficult or even impossible.

Reducing and simplifying interfaces can reduce hazards. Interface problems often lie in the control systems; thus, a basic design principle is that control systems not be split into pieces [394]. In complex systems, the control functions may be separated into pieces and executed on multiple microprocessors, thus increasing the number of interfaces within the control system itself. Where obvious and natural interfaces exist, this separation is reasonable. But sometimes more interfaces are created than necessary, leading to accidents. And, occasionally, eliminating even natural interfaces between components where communication is safety-critical, can eliminate many hazard causes.

As an example, in one modern military aircraft, the weapons management system was originally implemented on one microprocessor, which both launched the weapon and issued a weapon release message to the pilot. Pilots quickly learned the timing relationships between messages and weapon release, and they timed their maneuvers accordingly. For some reason, the two functions were later divided up and put on separate computers, changing this timing relationship. An accident resulted when a pilot, after seeing the weapon release message, dove and the plane was hit with its own missile.

Kletz has written extensively about simplifying chemical plant designs. Most of the serious accidents that occur in the oil and chemical industries result from a leak of hazardous material [185]. Leaks can be eliminated or reduced by designs with fewer leakage points, such as substituting continuous, one-piece lines for lines with connectors [131]. If equipment does leak, design features can ensure that it does so at a low rate that is easy to stop or control.

Major chemical plant items, such as pressure vessels, do not often fail unless they are used well outside their design limits or are poorly constructed. Instead, most failures occur in subsidiary equipment—pumps, valves, pipe flanges, and so on. Designs can be changed to eliminate as many of these subsidiary devices as possible. For example, using stronger vessels may avoid the need for relief valves and the associated flare system [328]. As another example, adipic acid used to be made in a reactor fitted with external coolers. Now it is made in an internally cooled reactor, which eliminates pump, cooler, and pipelines; mixing is achieved by using the gas produced as a byproduct.

According to Kletz [185], some of the reasons for complexity in system design are

- **The need to add on complicated equipment to control hazards**: If the design is made intrinsically safe by eliminating or reducing hazards, less added-on equipment will be necessary. If hazards are not identified early, when it is still possible to change the design to avoid them, the only alternative is to add complex equipment to control them. This basic design principle once again implies that safety engineering must begin early in the conceptual design process.
- **A desire for flexibility**: Multistream plants with numerous crossovers and valves, so that any item can be used on any stream, are flexible but have numerous leakage points, and mistakes in valve settings are easy to make.
- **Using redundancy to increase reliability**: The use of some types of redundancy to increase reliability may at the same time increase complexity and decrease safety.

Adding computers to the control of systems often results in increased complexity. Adding new functions to a system using computers is relatively easy: engineers are discovering that they can add functions that before were impossible or impractical, and they are finding it difficult to practice restraint without the experience of long years of failures in these attempts—although we are quickly building up that experience. Even when there are failures, they are often attributed to factors other than the inherent complexity of the projects and designs attempted.

The seeming ease with which complexity can be added both to a system through software and to the software itself is seductive and misleading. As in any system, complexity in software design leads to errors. The complexity of the design in the Therac-25 software (see appendix A) added myriad possibilities for unplanned interactions and was an important factor in the accidents.

In contrast, the Honeywell JA37B autopilot, the first full-authority fly-by-wire system ever deployed, flew for more than fifteen years without an in-flight anomaly. Boebert [39], one of its designers, attributes its success to the purposeful simplification of the design. Using what they called a *rate structure*, the design rules allowed no interrupts and no back branches in the code; the control flow was "unwound" into one loop that was executed at a fixed rate. This design is an example of simplifying control flow at the expense of data flow.

Many software designs are unnecessarily complex. The nondeterminism in many popular software design techniques is inherently unsafe because of the impossibility of completely testing or understanding all of the interactions and states the software can get into.

Nondeterminism also makes diagnosing problems more difficult because it makes software errors look like transient hardware faults: if the software is executed again after a failure, it is likely to work because internal timings will have changed. By eliminating some forms of nondeterminism and multitasking, many of the problems associated with synchronization and possible race conditions are eliminated.

In many real-time process control applications, the users, tasks, and communication are known in advance. All processing and communication among system components can be determined at design time, which creates the opportunity for significant reductions in operating system complexity. It can also be argued that effective and safe human–automation interfaces require a predictable, repeatable system response and thus software that is repeatable in operation and predictable in performance. Various ways to achieve deterministic software behavior in control systems exist but are beyond the scope of this book.

The problems may arise in education. Simplicity is not often emphasized or taught in computer science curriculums. Paradoxically, a complex software design is usually easier to build than a simple one, and the materials, being abstractions, contain almost unlimited flexibility. Constructing a simple software design for a nontrivial problem usually requires discipline, creativity, restraint, and time.

McCormick is another voice arguing for reducing complexity in software:

What I am urging is simply an institutional recognition that software differs from other engineering pursuits: software complexity can grow without bounds, and nature provides few clues for guidance. It is that -absence of nature that leads to grief.

Engineers specifying software-intensive systems need mechanisms that will rein in their wilder urges. Ideally, such mechanisms would act in the same role as do the physical limitations of the natural world. Where software is concerned, these mechanisms are by definition unnatural and must be created artificially. Technical management must be obliged to find the right incentives. [254]

Software engineers do not yet agree on what features a simple design should have, however. Defining the criteria such features should satisfy is easier:

- The design should be testable, which means that the number of states are limited, implying the use of determinism over nondeterminism, single-tasking over multitasking, and polling over interrupts.
- The design should be easily understood and readable; the sequence of events during execution, for example, should be readily observable by reading the listing or design document.
- Interactions between components should be limited and straightforward.
- Worst-case timing should be determinable by looking at the code.
- The code should include only the minimum features and capabilities required by the system and should not contain unnecessary or undocumented features or unused executable code.

The software design should also eliminate hazardous effects of common hardware failures on the software. For example, critical decisions (such as the decision to launch a missile) should not be made on the values often taken by failed components (such as all ones or all zeros). As suggested by Brown,

> *Safety-critical functions shall not employ a logic 1 or 0 to denote the safe and armed (potentially hazardous) states. The armed and safe states shall be represented by at least a four bit unique pattern. The safe state shall be a pattern that cannot, as a result of a one or two bit error, represent the armed pattern. If a pattern other than these two unique codes is detected, the software shall flag the errors, revert to a safe state, and notify the operator.* [54, p. 11]

Software can be designed in ways that eliminate the possibility of computer hardware failures having hazardous consequences. In June 1980, warnings were received at US command and control headquarters that a major nuclear missile attack had been launched against the United States [342]. The military commands prepared for retaliation, but the officers at Cheyenne Mountain were able to ascertain from direct contact with the warning sensors that no incoming missiles had been detected and the alert was canceled. Three days later, the same thing happened again.

The false alerts were caused by the failure of a computer chip in a multiplexor system that formats messages sent out continuously to command posts indicating that communication circuits are operating properly. This message was designed to report that there were 000 ICBMs and 000 SLBMs detected; instead, the integrated circuit failure caused some of the zeros to be replaced with twos. After the problem was diagnosed, the message formats were changed to report only the status of the communication system and nothing about detecting ballistic missiles, thus eliminating the hazard.

The design of software to control a turbine provides another example of how it is possible to eliminate many potentially dangerous software design features with careful thought and planning [145]. The safety requirements for the generator were that (1) the governor must always be able to close the steam valves within a few hundred milliseconds (ms) if overstressing or even catastrophic destruction of the turbine is to be avoided, and (2) under no circumstances can the steam valves open spuriously, whatever the nature of the internal or external fault.

The software to control the turbine was designed such that the safety-critical software was separated from the much larger software responsible for the less important control functions. Erroneous behavior of the noncritical software could not endanger the turbine and did not cause it to shut down. The noncritical software used conventional hardware and software and resided on a processor separate from the safety-critical software.

The critical software was written so that it could detect significant failures of the hardware that surrounded it. It included self-checks to decide whether incoming signals were sensible and whether the processor itself was functioning correctly. A failure of a self-check caused reversion of the output to a safe state through the action of fail-safe hardware. Numerous design techniques were used to make this software easier to test and assure. For example, no interrupts were allowed in this code other than one nonmaskable interrupt used to stop the processor in case of a fatal store fault. The avoidance of interrupts meant that the timing and sequencing of processor operation could be defined for any particular state at any time, allowing more rigorous and exhaustive testing. It also meant that polling must be used. A simple design in which all messages were unidirectional and in which no contention or recovery protocols were required helped ensure a higher level of predictability in the operation of the base-level software. Other types of design features were included to make development, assurance and change simpler and less error-prone.

Carefully designed systems, such as the one described here, are less common now that engineers are more comfortable with the replacement of hardware by software. As a result, more accidents and losses are occurring that involve unsafe software behavior.

The increase in the use of AI (artificial intelligence) techniques for safety-critical functions conflicts with most design approaches for designing simpler and safer systems. The benefits of using AI and other relatively complex programming practices are usually not worth the extra work required to ensure safety, if that is possible at all. Often the same or at least adequate functionality can be provided by non-AI techniques. If not, difficult decisions need to be made about whether the functionality and perhaps the system itself is worth the extra risk. Risk may also be reduced by changes to the system outside the software involved. New approaches are being developed to try to provide greater assurance for software that uses AI techniques, but it is still necessary to make difficult decisions about whether the functionality provided is worth the risk and, if so, how to reduce that risk.

11.3.3 Decoupling

A tightly coupled system is highly interdependent: each part is linked to many other parts, so that a failure or unplanned behavior in one can rapidly affect the status of others. A malfunctioning part in a tightly coupled system cannot be easily isolated, either because there is insufficient time to close it off or because its behavior affects too many other parts, even if the unsafe behavior does not happen quickly.

Accidents in tightly coupled systems are a result of unplanned interactions. These interactions can cause domino effects that eventually lead to a hazardous system state. Coupling exacerbates these problems because of the increased number of interfaces and potential interactions: small deviations in behavior propagate unexpectedly.

Two simple examples of the use of physical decoupling to eliminate hazards are (1) firebreaks to restrict the spread of fire, and (2) overpasses and underpasses at highway intersections and railway crossings to avoid collisions.

Why not just decouple all systems? Complex and tightly coupled systems are more efficient in terms of production than loosely coupled ones. In addition, transformation systems require many nonlinear interactions. Decoupling in software takes careful planning and extra effort.

Computers tend to increase coupling in systems because they usually control multiple system components; in fact, they become the coupling agent unless steps are taken in the system design to avoid it.

The principle of decoupling can be applied to software as well as to system design. Modularization can be used to control complexity, but how the software is split up is crucial in determining the effects and may depend on the particular quality that the designer is trying to maximize. In general, the goal of modularization is to minimize, order, and make explicit the connections between modules. The basic principle of information hiding is that every module encapsulates design decisions that it hides from all other modules; communication is allowed only through explicit function calls and parameters. Besides basic information hiding, some of the principles of software coupling and cohesion [1979] also can be used to decouple modules.

When the highest design goal is safety, modularization may involve grouping together the safety-critical functions and reducing the number of such modules (and thus the number of interfaces) as much as possible. An additional advantage of isolating the safety-critical parts of the code is that it greatly simplifies the assurance process.

After safety-critical functions are separated from noncritical functions, the designer needs to ensure that errors in the noncritical modules cannot impede the operation of the safety-critical functions. Adequate protection of the safety-critical functions will need to be verified.

The software requirements and constraints identified by the hazard analysis can be used to identify safety-critical functions that need special treatment, including being protected by firewalls. Firewalls may be physical, such as in the turbine design described earlier where the critical code is executed on a separate computer, or they may be logical. In a logical firewall, a virtual computer is created for each program by making the computer act as if a set of programs or files is the only set of objects in the system, even though other objects may be present.

Even when the computer is dedicated to one program, the application code still needs to be protected against the operating system: usually, barriers between the operating system and the application programs are designed to protect the operating system from the application, but not vice versa.

Logical separation is enforced by providing barriers between programs or modules. To implement a firewall, the design must somehow prevent unauthorized or inadvertent access to or modification of the code or data. This form of protection obviously includes preventing self-modifying code.

Logical separation must also include reducing coupling through computer hardware features, such as not using the same input–output registers and ports for both safety-critical and non-safety-critical functions. The difficulty of providing logical separation may make physical separation a much better solution.

In some systems, critical code or data can be protected physically from unintentional change by being placed in permanent (read-only) or semipermanent (restricted write) memories.

11.3.4 Elimination of Specific Human Errors

Human error is often implicated in accidents. One way to eliminate operator and maintenance error is to design so that there are few opportunities for error in the operation and support of the system. For example, the design can make incorrect assembly impossible or difficult. In the aircraft industry, it is common to use different sizes and types of electrical connectors on similar or adjacent lines where misconnection could lead to a hazard. The same approach has been adopted by the personal computer industry. If incorrect assembly cannot be made impossible, then the design should make it immediately clear that the component has been assembled or installed incorrectly, perhaps by color coding.

Many other human factors issues apply here, such as making instruments readable or making the status of a component clear—whether a valve is open or closed, for example.

This topic, including the implications for the design of the software–operator interaction, is covered in more depth in the next chapter.

An important way to reduce potential human developer error is to write specifications and programs that are easily understood. Many languages produce or encourage specifications and programs that are not very readable and thus are subject to misinterpretation or misunderstanding and increase difficulty for independent review.

11.3.5 Reduction of Hazardous Materials or Conditions

Even if completely eliminating a dangerous material or substituting a safer one is not possible, it may still be possible to reduce the amount of the material to a level where the system operates properly but the hazard is eliminated or significantly reduced. Thus, a plant can be made safer by reducing inventories of hazardous materials in process or in storage.

The chemical industry started paying attention to this principle after the Flixborough accident in England in 1974 (appendix C), in which the large scale of the losses were due to the presence of large amounts of flammable liquid in the reactors at high pressure and temperature. The methyl isocyanate (MIC) in the Bhopal accident was also an intermediate that was convenient but not essential to store.

The success of this approach in the chemical industry has been striking. For example, inventories have been reduced by factors of up to 1,000 or more by redesigning separation equipment and heat exchangers, improving mixing, and replacing batch reactors with continuous ones.

A reduction in inventory is sometimes achievable only as a result of *intensification*; that is, by an increase in the pressure or temperature of the reaction. Thus, some of the advantage of less material in the process is offset by the additional energy available to expel the contents of the vessel [328]. Essentially, as in most attempts to eliminate hazards, there is simply the substitution of one hazard for a different one. The result may be better or worse than the original design in terms of safety. The advantage of intensification, and one reason for its rapid acceptance, is that it reduces cost; smaller vessels, pipes, valves, and so on can be used, which may reduce other hazards.

Another way to eliminate or reduce hazards is to change the conditions under which the hazardous material is handled—that is, to use hazardous materials under the least hazardous conditions possible while achieving the system goals. Equipment is often oversized to allow for future increases in throughput, but the tradeoff may be greater risk. For example, potential leak rates can be reduced by reducing the size of pipes—a severed 3 in pipe will produce more than twice the release rate of a severed 2 in pipe.

Other hazards can be eliminated or reduced by processing the hazardous materials at lower temperatures or pressures. In the chemical industry, this approach, which is the opposite of intensification, is called *attenuation*. As an example, the manufacture of phenol traditionally has been carried out close to the temperature at which a runaway reaction can occur. Automated equipment has to be (or should be) provided to dump the contents of the reactor into a large tank of water if the temperature gets too high. In newer designs, the operating temperature is lower, and the dump facilities may not be necessary [185]. Again, significant cost savings can be achieved at the same time that hazards are reduced.

The principle of reduction of hazardous materials or conditions can also be applied to software. For example, software and/or critical-software modules should contain only the code that is absolutely necessary to achieve the desired functionality: operational software should not contain unused executable code. Unused code should be removed and the program recompiled without it.

Eliminating unused code has important implications for the use of commercial off-the-shelf (COTS) software in safety-critical systems. Usually, code that is written to be reused or that is general enough to be reused contains features that are not necessary for every system. Although there is a tradeoff here, the assumed higher reliability of COTS and reused software may not have any effect on safety and it may even increase risk, as discussed in chapter 4. At the same time, the extra functionality and code may lead to hazards and may make software safety assurance processes more difficult and perhaps impractical.

11.4 Hazard Occurrence Reduction

Even if hazards cannot be eliminated or significantly reduced in terms of their danger, in many cases their occurrence can be reduced or alternatively, the controls to prevent hazards can be strengthened. Once again, the approach used will reflect the underlying accident causality model being assumed.

Most of the general approaches described in the previous section apply. For example, even if a perfectly safe material cannot be substituted, there may be one available with a much lower likelihood—but still within the realm of reason—of leading to the hazard.

Various types of safeguards can be used to limit hazards. The primary categorization is *passive* versus *active*.

Passive safety devices either

1. maintain safety by their presence, for example, shields and barriers such as containment vessels, safety harnesses, hardhats, passive restraint systems in vehicles, and fences; or
2. fail into safe states, such as weight-operated railroad semaphores that drop into a STOP position if a cable breaks, relays or valves designed to fail open or shut, and retractable landing gear for aircraft in which the wheels drop and lock in the landing position if the pressure system that raises and lowers them fails.

Passive protection does not require any special action or cooperation to be effective and therefore is preferable to active protection, which requires that a hazard or condition be detected and corrected. Passive safety devices are not perfect, however. For example, snow and ice may jam weighted railroad semaphores so that they do not fail into a safe state.

Active safeguards require some actions to provide protection: detecting a condition (monitoring), measuring some variable(s), interpreting the measurement (diagnosis), and responding. Thus, they require a control system of some sort. More and more often, these control systems involve a computer.

There are tradeoffs between these two approaches. Whereas passive devices rely on physical principles such as gravity, active devices depend on less reliable detection and recovery mechanisms. Passive devices, however, tend to restrict human activity and design freedom more and are not always feasible to implement.

Given this basic introduction, the general hazard reduction measures discussed in this section include designing for controllability, using barriers, providing monitoring, and attempting to minimize failures.

11.4.1 Design for Controllability

One way to reduce hazards is to make the system easier to control, for both humans and computers. Some processes are inherently more "operable" than others [206]. For example, some processes have an extreme reaction to changes while others change more gradually and take more time. Time pressures increase stress, which in turn increases the likelihood of making a mistake.

Nuclear reactors provide an example of designs that can differ greatly in their ease of control and their dependence on added control and trip systems. Compared to other designs, gas-cooled nuclear reactors give the operator more time in which to react to problems and thus more time to reflect on the consequences of an action before needing to intervene. At the other extreme, Chernobyl-style boiling-water reactors, which have a positive power coefficient at low output rates rather than the negative power coefficients of other commercial designs, are more difficult to control. As Kletz says, "*It is easier to keep a marble on a concave-up saucer than on a convex saucer. Chernobyl was a marble on a convex surface*" [189].

Incremental control One important aspect of controllability is incremental control; that is, allowing critical actions to be performed incrementally rather than as a single step [116]. With incremental control, the controller can (1) use feedback from the behavior of the controlled process to test the validity of the models on which the decisions were made, and (2) take corrective action before significant damage is done.

Intermediate states Ease of control also results from a design that gives the operator more options than just continuing to run under the given conditions or shutting down the process completely. Various types of fallback or intermediate states can often be designed into the process.

As an example, some systems can be continued safely, although at the expense of efficiency, with a small subset of essential functions. Levels of control may be designed to allow full functionality, reduced capability, and emergency mode. The emergency mode is used to protect against massive system failures. In emergency mode, only a small set of critical functions may be implemented to maintain minimal system safety. Emergency mode may be entered when the essential functions have degraded to a performance threshold below which safety would be compromised. Of course, great care needs to be taken in designing for fallback states, given the large number of problems and accidents that occur when changing modes.

Decision aids Computers can also be used to provide assistance to operators in controlling the plant, including alarm analysis, disturbance analysis, and valve sequencing. Normally, one of the primary functions of a process control computer is checking if process measurements have exceeded their limits and an alarm needs to be raised. In large, complex plants or other systems such as hospitals, a great many alarms may be raised at the same time, and operators may have difficulty sorting the alarms and diagnosing faults. This problem is discussed further in chapter 12 on human factors in designing for safety.

A second type of aid is used for analyzing process disturbances. Instrumentation collects information, which is preprocessed to check limits, validate the data, filter and derive process variables from measured variables, and separate noise from true deviations. To detect disturbances, the decision aid uses models that represent the anticipated flow of events. These models are then compared with the actual plant data; consequences are predicted, and, if possible and feasible, corrective actions and primary causes are suggested to the operator [34]. This type of analysis is quite difficult to perform reliably and is fairly controversial from a safety standpoint. The drawbacks are again discussed more fully in chapter 12. Briefly, the potential drawbacks include reduction in vigilance, increased complacency, and dependence on the automated decision aid accompanied by reduction in operator skills.

Valve sequencing programs are a third type of computer aid used in process control. Incorrect sequencing of valve operations is a common cause of accidents in process plants. Two types of computer assistance in the safe sequencing of valves have been proposed: (1) analyzing a valve sequence proposed by an operator to determine whether it is hazardous, and (2) synthesizing safe valve sequences [206]. The latter is much more difficult and potentially error prone. Physical interlocks to protect against unsafe sequences of valve operation, described in the next section, have advantages over automated valve-sequencing checks.

11.4.2 Barriers

One way to prevent the system from getting into a hazardous state is to erect barriers, either physical or logical, between physical objects, incompatible materials, system states, or events. Three general types of barriers can be used:

1. A *lockout* makes access to a dangerous process or state impossible or difficult.
2. A *lock-in* makes it difficult or impossible to leave a safe state or location.
3. An *interlock* enforces a sequence of actions or events.

Barriers may be applied redundantly—in series or in parallel—and may have passages, such as gates or channels, between them whose use is controlled.

Lockouts A lockout prevents a dangerous event from occurring or prevents someone or something from entering a dangerous area or state. The simplest type of lockout is a wall or fence, or some type of physical barrier used to block access to a dangerous condition, such as sharp blades, heated surfaces, or high-voltage equipment.

Lockouts are useful when electrical or magnetic signals can interfere with programmable devices. This phenomenon is called *electromagnetic interference* (EMI). Examples of EMI include radio signals, electrostatic discharges, or electromagnetic particles, such as alpha or gamma rays.

EMI can be especially difficult to diagnose because of its transient nature. In one case, a programmable device in a ship's crane was intermittently behaving strangely. It turned out that the radio officer was stringing an aerial between the jibs of the cranes in order to increase the range of the ship's transmitter. The crane's cables became receiving antennas [416].

EMI is a major problem for sophisticated military aircraft. In the UH-60 Black Hawk helicopter, for example, radio waves caused complete hydraulic failure, effectively generating false electronic commands. Twenty-two people were killed in five Black Hawk crashes before shielding was added to the electronic controls. After the problem was discovered, the Black Hawk was not permitted to fly near approximately one hundred transmitters throughout the world [280].

Densities in computer chip technologies can be subject to cosmic ray interference at the electron level. As microminiaturization increases, so does the probability of this interference [279]. Electronic components, including computers, need to be protected in some way against electromagnetic radiation, electrostatic interference, power interrupts and surges, stray voltages, and gradual depletion of power supplies.

Electrical interference can be eliminated or minimized in three ways [416]: (1) it can be reduced at its source, for example, suppressing arcing at switch contacts with capacitors; (2) the source and the electronic device can be separated as much as possible, for example, providing an independent electrical supply to the system; or (3) a barrier can be erected around the programmable device, for example, installing shielding or an interference filter.

Authority limiting is a type of lockout that prevents actions that could cause the system to enter a hazardous state. As an example, the control surfaces on an aircraft or the mechanisms that drive them may be designed so that an autopilot hard-over command causes a worst-case maneuver that is still within the aircraft maneuvering envelope; no matter what the autopilot does, the aircraft structure cannot be compromised. Such authority limitations have to be carefully analyzed to make sure they do not prohibit maneuvers that may be needed in extreme situations.

Lockouts in software include design techniques to control access to and modification of safety-critical code and variables. Safety-critical software often has a few modules or data items that must be carefully protected because their execution or, in the case of data, their destruction or change, at the wrong time can be catastrophic: an insulin pump administering insulin when the blood sugar is low and a missile launch routine activated inadvertently are two examples.

Techniques involving authority limitation can be used to protect safety-critical routines and data. For example, the ability of the software to arm and detonate a weapon might be severely limited and carefully controlled by requiring multiple confirmations. Again, there may be a conflict between reliability and safety: to maximize reliability, errors should be unable to disrupt the operation of a weapon, while for safety, errors should often lead to nonoperation. In other words, reliability requires designs with multipoint failure modes, while safety may, in some cases, be enhanced by a design with a single-point failure mode.

Authority limitation with regard to inadvertent activation is usually implemented by retaining a human controller in the loop and requiring a positive input by that controller before execution of hazardous commands. The human will obviously require some independent source of information on which to base the decision besides the information provided by the computer.

Lock-ins Lock-ins maintain a condition. They may be used

- To keep humans within an enclosure, where leaving under certain conditions would involve proximity with dangerous objects or not allow them to continue to control the system—for example, seat belts and shoulder harnesses in vehicles, safety bars in Ferris wheels and roller coasters, and doors on elevators.
- To contain harmful products or byproducts, such as electromagnetic radiation, pressure, noise, toxins, or ionizing radiation and radioactive materials.
- To contain potentially harmful objects—for example, cages around an industrial robot to protect anyone in the vicinity in case the robot throws something.
- To maintain a controlled environment—for example, buildings, spacecraft, space suits, and diving suits.

- To constrain a particular sequence of states or events—for example, using speed governors on moving objects, such as on industrial robots or other mobile machinery, to eliminate damage in case of collision or to allow people to get out of the way if the objects move unexpectedly. Slowing robot operation speed when a human is approaching allows workers to know that their presence has been detected. Safety valves, relief valves, and other devices maintain pressure below dangerous levels.

Interlocks Often, the sequence of events is critical. Interlocks are commonly used to enforce correct sequencing or to isolate two events in time. An interlock ensures that

1. Event A does not occur inadvertently. An example is requiring two separate commands such as pushing two different buttons before an event or state change is allowed.
2. Event A does not occur while condition C exists. An example is putting an access door over high-voltage equipment so that if the door is opened, the circuit is broken and the power shuts off.
3. Event A occurs before event D. An example is ensuring that a tank will fill only if a vent valve is opened first.

The first two types of interlocks are called *inhibits*; the third is a *sequencer*. Examples of interlocks include

- A pressure-sensitive mat or light curtain that shuts off an industrial robot if someone comes within reach.
- A deadman switch that must be held to permit some device to operate—when released, the power is cut off and the device stops.
- Guard gates and signals at railroad crossings to ensure that cars and trains are not in the intersection at the same time. Traffic signals are a similar example.
- A device on machinery that ensures that all prestart conditions are met before startup is allowed, that the correct startup sequence is followed, and that the process conditions for transition from stage to stage are met.
- Pressure relief valves equipped with interlocks to prevent all the valves from being shut off simultaneously.
- A device to prevent the disarming of a trip system or a protection system unless certain conditions are met first and to prevent the system from being left in a disabled state after testing or maintenance.
- Devices to disable a car's ignition unless the automatic shift is in PARK.
- The freeze plug in an automobile engine cooling system whose expansion will force the plug out rather than crack the cylinder if the water in the block freezes. Similarly, to protect against excessive heat, a fusible plug in a boiler becomes exposed when the water level drops below a predetermined level and the heat is not conducted away from the plug, which then melts. The opening permits the steam to escape, reduces the pressure in the boiler, and eliminates the possibility of an explosion.

The system should be designed so that hazardous functions will stop if the interlocks fail. In addition, if an interlock brings something to a halt, adequate status and alarm information must be provided to indicate which interlock was responsible [206].

People have been killed or endangered when the equipment they had de-energized to repair was inadvertently activated by other personnel. One way to avoid such accidents is to install an interlock that only the person making the repairs can operate. The possibility still exists, however, that the interlock can be inadvertently bypassed. In one incident, the doors to a weapons bay on an aircraft were held open by compressed air. An airman working on the system accidentally released the pressure by loosening a fitting while standing between the doors; the doors caught and crushed him [131]. Physically blocking open the doors so that motion is locked out is an alternative in this case.

When computers are introduced, complexity may be increased, and physical interlocks may be defeated or omitted. In one example, a workman initiated a mechanical interlock while working on a weapons bay door so that the door would not close. Some other workers in the cockpit pushed a "close weapons bay door" switch on the control panel during a test. The command was not executed because of the mechanical inhibit, but it remained active. Several hours later, when the maintenance was completed and the inhibit removed, the door unexpectedly closed.

Engineers are now often removing physical interlocks and safety features in systems and replacing them with software. That was a disastrous mistake in the Therac-25 design, but they are not alone in making it (see appendix A). Most weapon systems now have either replaced hardware interlocks and safety features with software or use software to control them. Other types of systems are quickly following suit. Software control or implementation of interlocks introduces a more complex design involving active control and many additional paths to hazards.

Hardware interlocks may be important in systems with computer control in order to protect the system against unsafe software behavior. Examples of circuitry or other hardware independent of the computer and software include hardwired deadman switches to permit termination of computer-controlled X-ray exposures, electrical interlocks for collision avoidance when motions are computer-controlled, and hardwired electrical sensors to assess the status of critical software-controlled system elements.

If facilities must be provided to override interlocks during maintenance or testing, the design must preclude any possibility of inadvertent interlock overrides or of the interlocks being left in an inoperative state once the system becomes operational again. If software is used to monitor hardware interlocks, it should verify before resuming normal operation that the interlocks have been restored after completion of any tests that remove, disable, or bypass them. While the interlocks are being overridden, information about their state should be made available to any person who might be endangered.

In general, if humans are interacting with dangerous equipment, the software controller or physical interlocks or both should ensure that no inadvertent machine movement is possible.

The software also may need to ensure that proper sequences are followed and that proper authority has been given to initiate hazardous actions. For example, before firing a weapon, the software may be required to receive separate arm and fire commands to avoid inadvertent firing. Similarly, after an emergency stop of some kind, the operator or the software should be required to go through a restart sequence to ensure that the machine is in the assumed proper state before it is activated; the equipment should not simply go to the next operation. As an example, the software controlling an amusement park ride was taken offline to be upgraded. Control was switched to a human operator. When control switched back to the operator, the software assumed that the cars were in the same position as they were when the software went offline and ran several cars into each other.

Computers provide the ability to create unnecessarily complex designs. The more complex the design, the more likely that errors will be introduced by the safety design features themselves. In general, the simplest design possible should be used.

Example: nuclear detonation systems The approach to safety in nuclear weapon systems in the United States illustrates the use of several types of barriers. The nuclear detonation safety problem is somewhat unique in that safety, in this case, depends on the system *not* working. The goals in a nuclear system are (1) to provide detonation when authorized, and (2) to preclude inadvertent or unauthorized detonation under normal and abnormal conditions. The concern here is with unintended operation.

Three basic techniques, called *positive measures*, are employed [353]:

1. *Isolation*: separating critical elements whose association could lead to an undesired result
2. *Incompatibility*: using unique signals
3. *Inoperability*: keeping the system in a state that is incapable of detonation

Figure 11.3 shows a general view of these systems. The nuclear device itself is kept in an inoperable state, perhaps with the ignition device removed or without an arming pin: positive action has to be taken to make the device operable. The device is also protected by various types of barriers (isolation).

Nuclear detonation requires an unambiguous indication of human intent to be communicated to the weapon. Trying to physically protect the entire communication system from all credible abnormal environments (including sabotage) is not practical. Instead, nuclear systems use a signal pattern of sufficient information complexity that it is unlikely to be generated by an abnormal environment. Not needing to protect the communication (unique signal) lines minimizes or eliminates many design, analysis, testing, and software/computer

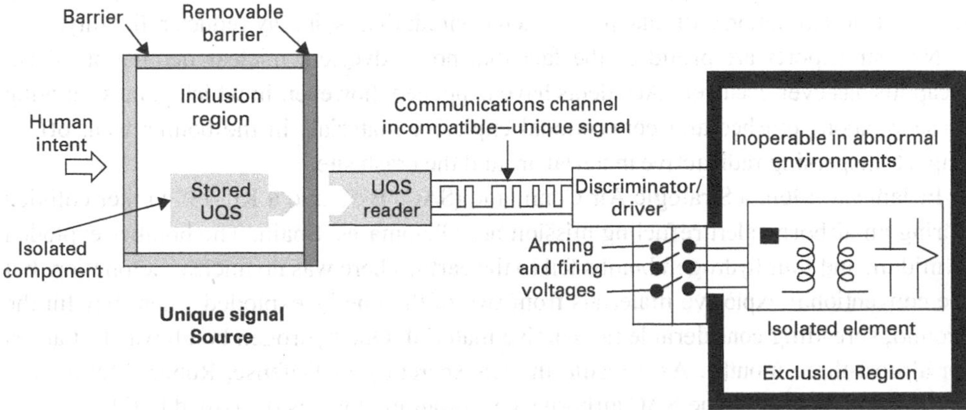

Figure 11.3
Subsystem using a unique signal, barriers, and inoperability for nuclear detonation safety from [353].

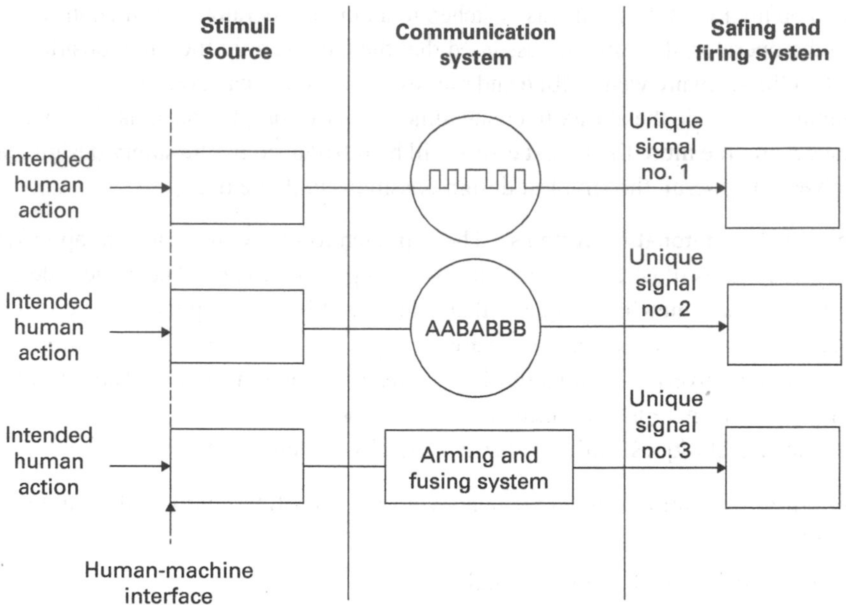

Figure 11.4
The use of multiple safety subsystems requiring unique signals (double direct intent with arming) to ensure proper intent [353].

vulnerability problems. However, the unique signal discriminators (1) must accept the proper unique signal while rejecting spurious inputs, (2) must have rejection logic that is highly immune to abnormal environments, (3) must provide predictably safe response to abnormal environments, and (4) must be analyzable and testable.

The unique signal sources are protected by barriers, and a removable barrier is placed between these sources and the communication channels. Multiple unique signals may be required from different individuals along various communication channels, using different types of signals (energy and information), to ensure proper intent to detonate the weapon (figure 11.4). While this approach enhances safety, it most likely reduces the probability that nuclear detonation will take place when desired; that is, it may reduce reliability.

Nuclear experts are proud of the fact that no inadvertent nuclear detonation of US weapons has ever occurred. Accidents have happened, however, in which planes carrying these weapons crashed and conventional explosive materials in the bombs went off on impact, dispersing radioactive material around the crash site.

In January 1966, a Strategic Air Command (SAC) B-52 and a KC-135 tanker collided during an airborne alert refueling mission near Palomares, Spain. The bomber exploded in midair, and four hydrogen bombs fell to the earth. There was no nuclear detonation, but the conventional explosive materials from two of the bombs exploded when they hit the ground, spreading considerable radioactive material. One hydrogen bomb was lost at sea for almost three months. As a result, the US Secretary of Defense, Robert McNamara, argued for eliminating the SAC airborne alert program but was overruled [342].

In January 1968, a similar accident occurred when a Strategic Air Command B-52 bomber was on an airborne alert mission over Thule, Greenland. The co-pilot turned the cabin heater to its maximum heat to combat the cold, and a few minutes later a crew member detected the smell of burning rubber. A search found a small fire in the rear of the lower cabin. The flames grew out of control, the flight instruments became unreadable because of the smoke, and all electrical power was lost. Six of the seven crew members ejected successfully and landed safely in the snow. The plane crashed with a speed at impact of 500 mph, and the jet fuel exploded. Once again, no nuclear detonation occurred; as in Palomares, however, the conventional high explosives in the thermonuclear bombs on board went off, dispersing radioactive debris over a wide expanse of ice.

The international protests against American nuclear weapons policy after these incidents eventually resulted in the termination of nuclear-armed airborne alert flights [342]. Sagan and others have expressed concern about whether organizational and management factors, as described in chapter 4, might override the technical safeguards.

Perhaps of more concern now is that our nuclear weapon systems are being redesigned using newer technology. Hopefully, the use of simple, non-computer-based protection devices will continue in these new systems, but that is probably unlikely.

11.4.3 Monitoring

Detecting a problem so that something can be done about it before a hazardous state occurs requires some form of monitoring, which involves both (1) checking conditions that are assumed to indicate a potential problem in the process, and (2) validating or refuting assumptions made during design and analysis. As an example of validating assumptions, a simulation of the controlled process might be executed in parallel with the control software during operations and compared with the actual process measurements versus those assumed during development. This idea seems to have been reinvented recently with a different name and without understanding why it did not work well in the past.

Monitoring can be used to indicate

- whether a specific condition exists;
- whether a device is ready for operation or is operating satisfactorily;
- whether required input is being provided;
- whether a desired or undesired output is being generated;
- whether a specified limit is being exceeded; or
- whether a measured parameter is abnormal.

In general, there are two ways to detect equipment malfunction: (1) by monitoring equipment performance, and (2) by monitoring equipment condition [206]. Condition monitoring usually requires the operator to check the equipment physically; performance checks can be made from the control room using instrument displays. Performance checks compare redundant information, where the redundant information may be provided by expected values, prior signals, duplicate identical instruments, other types of instruments, and so on.

Monitors, in general, should (1) detect problems as soon as possible after they arise and at a level low enough to ensure that effective action can be taken before hazardous states are reached; (2) be independent from the devices they are monitoring; (3) add as little complexity to the system as possible; and (4) be easy to maintain, check, and calibrate.

The independence of monitors is always limited [19]. One reason is that checks require access to the information to be checked. In most cases, providing access introduces the possibility of corrupting the information. In addition, monitoring depends on assumptions about the structure of the system and about the types of faults and errors that may—or may not—occur. These assumptions may be invalid under certain circumstances.

Common (and incorrect) assumptions may be reflected both in the design of the monitor and in the devices being monitored. In fact, the success of monitoring depends on how good these assumptions are—that is, how well they reflect the assumptions of the monitored device or program.

A monitoring system provides feedback to an automatic device or the operators (or both) so that they can take remedial action. Measurement should, as much as possible, be made directly on the critical variables or on closely related functions. The monitor must be capable of detecting critical parameters in the presence of environmental stresses that may degrade performance, such as vibration, temperature variations, moisture, and pressure changes. The feedback must be timely, easily recognizable, and easily interpreted as to whether a normal or unusual condition exists.

For example, a simple way to provide feedback is to mark a display, such as a dial, with a predetermined limit where an indicator points to the existing level. An automobile gauge may monitor engine oil pressure and sound an alarm when an abnormality is detected. A less effective type of feedback is a light that goes on to warn the driver when the oil pressure is less than a preset level—by the time the driver notices it, they may be in trouble [133].

Monitoring is especially important when performing functions known to be particularly hazardous, such as startup and shutdown or any non-normal operating mode. The monitor should ensure that the system powers up in a safe state and that safety-critical circuits and components are operating correctly. Similar tests should be performed in the event of power loss or intermittent power failures. Periodic tests should then be run to ensure that the system is operating safely.

Monitoring should be capable not only of detecting out-of-limit parameters (limit-level sensing control) in the process but also of detecting problems in the instrumentation system itself. Sometimes, distinguishing exactly where the problem arose—in the instrumentation or in the process—is difficult. Instrumentation error should not be assumed automatically. The detection of the ozone hole over the Antarctic was delayed because a computer was programmed to assume extreme deviations were sensor faults and to ignore them.

In extremely critical applications, monitors must be designed to indicate any failures of their own circuits or, in the case of computer monitors, any errors in execution. It may be possible, if the software is designed carefully, to trace identified hazardous behavior (that is, the violation of the safety-related requirements or constraints) into the software in the form of self-checks or assertions on the internal software state.

Detecting circuit failures is much easier (in general) than detecting computer errors; for this reason (and others), hardware monitors may be safer than software monitors in extremely critical situations where an indication of a monitor failure or error is critical.

A final problem in creating a monitor is that it is likely to be based on the same incorrect assumptions underlying the thing being monitored. Serious accidents occur when things happen that we did not expect and that violate our assumptions about the system. Monitors are very likely to have the same underlying assumptions and unknown unknowns.

11.4.4 Failure Minimization

Although many hazards are not the result of random individual component failures, some hazards *are*, and reducing the failure rate will reduce the probability of those hazards. Many reliability-enhancing techniques have been devised. The three most applicable to complex systems are safety margins, redundancy, and error recovery.

Safety factors and margins Engineered devices and systems have many uncertainties associated with them: the materials from which they are made; the skill that goes into designing and manufacturing them; their behavior in extreme environmental conditions such as very low or high temperature; and incomplete knowledge about the actual operating conditions, including unexpected stresses, to which they are exposed. Engineering handbooks contain failure rates for standard components, but these rates are subject to implied limits under different conditions and are statistical averages only: failure rates of individual components may vary considerably from the mean [252].

To cope with these uncertainties, engineers use safety factors or safety margins, which involve designing a component to withstand greater stresses than are anticipated to occur (see figure 11.5). A safety factor is expressed as the ratio of nominal or expected strength to nominal stress (load). A part with a safety factor of two, for example, is theoretically able to stand twice the expected stress.

The problem with this concept is that the strength of a specific material will vary because of differences in its composition, manufacturing, assembly, handling, environment, or usage. Therefore, a calculated safety factor of two for a component *in general* may be much less for a *particular* component: averages imply a range of values over which a particular characteristic may vary.

The problem is alleviated somewhat by the use of measures other than expected value or mean in the calculations—such as comparing minimum probable strength and maximum probable stress (called the safety *margin*) or computing the ratio at specified standard deviations from the mean—but the problem is not eliminated.

Most solutions involve increased cost for individual components: (1) increase the nominal strength, (2) decrease the nominal stress that will be imposed, or (3) reduce the variations in strength or stress. Even then, computing the margin of safety is difficult for ordinary, stable stresses and even more difficult when continually changing stresses must be considered.

Redundancy Redundancy involves deliberate duplication to improve reliability. Note that the general goal of redundancy is to improve reliability, which will not necessarily improve safety. Many serious accidents are described in chapter 4 that relied on redundancy and diversity to prevent them. They are not unique. There is a reason why redundancy is so low on the design precedence scale.

Redundancy may be achieved (1) through standby spares, which involves switching in a spare device when a failure is detected in the one currently being used, or (2) by concurrent use of multiple devices to duplicate a function, voting on the results, and then using the majority vote. If only two devices are used in parallel, then fault detection—but not fault correction—is possible.

Use of this technique requires that there be some highly reliable failure detection mechanism. For *fail stop* hardware, it is easy to detect failures, and a simple comparison of outputs may be all that is needed to vote on multiple hardware outputs. Complex failure

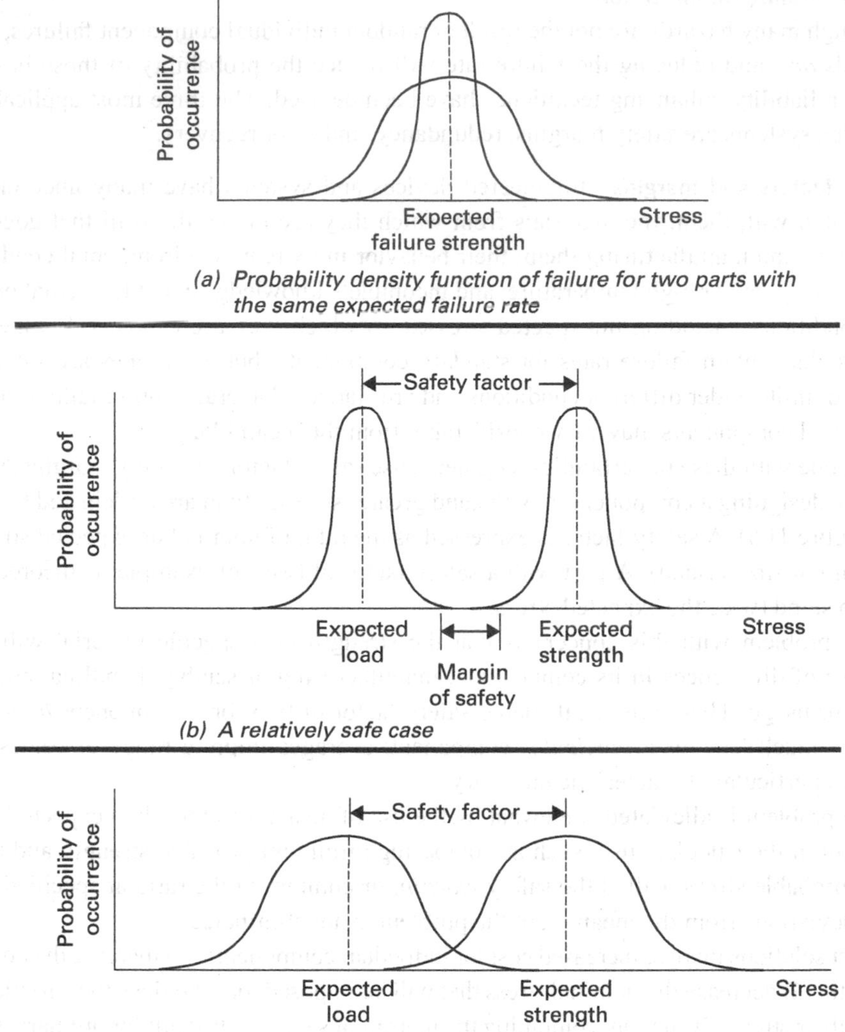

Figure 11.5
A problem in using safety margins and factors (adapted from [252]).

detection and comparison voting schemes, however, may be required in some situations. In addition, reconfiguration may be required to switch out failed parts and switch in spares. All of this additional functionality to implement the redundancy scheme, of course, is also subject to failure and design errors.

The use of redundancy in nuclear power plants has resulted in a large number of spurious scrams [403]. To avoid this problem, the redundancy may instead involve independent channels, all carrying the same kind of information and connected so that no protection action will be taken unless a certain number of these channels trip simultaneously. This approach results in some reduction in system reliability compared with alternative redundant designs, but it does reduce spurious shutdowns.

The more reliable a component, the more likely it is to operate spuriously [403]. In some cases, spurious operation may be as or more hazardous than the failure of the system to function at all. In the Ranger 6 spacecraft, redundancy caused spurious activation of a backup system that drained the power from the spacecraft and ruined the mission. In describing this incident, Weaver concludes that redundancy is not always the best design option to use.

Functional redundancy duplicates function but may do it with different designs. One of the redundant components should be able to achieve the functional goals regardless of the operational state of the other components. Functional redundancy may be accomplished through identical designs, called *design redundancy*, or through intentionally different designs, called *design diversity*. Diversity is used to try to avoid common-cause and common-mode failures, but providing complete diversity is difficult. In fact, diversity must be carefully planned and applied.

Finding and eliminating all potential dependencies in redundant or diverse systems can be extremely difficult. Examples include the following:

- A military aircraft was lost when supposedly diverse components, made from titanium, all failed at the same vibration level [41].
- A fire in the cable-spreading room of the Browns Ferry nuclear power plant (described in chapter 4) disabled many electrical and control circuits, which resulted in the loss of the redundant protection systems. Before the accident, common-cause failure of all the protection systems had been deemed not "credible." Harry Green, the superintendent at Browns Ferry, said after the fire: "We had lost redundant components that we didn't think you could lose" [405].
- The Turkish Airlines DC-10 crash outside Paris resulted from the cargo door of the baggage hold, which was underneath the passenger compartment, opening at altitude. This event caused the baggage hold to depressurize, which in turn caused the collapse of the cabin floor. The triplicated control lines were all under the floor, so when it collapsed all control of the aircraft was lost.
- The common failure of the auxiliary feedwater valves was instrumental in initiating the loss-of-coolant accident at Three Mile Island. The common-cause failure of the high-pressure injection system resulted in the uncovering of the core. Weaver believes that additional diversity in the feedwater system probably would not have prevented the accident [403].
- Just when *Challenger*'s primary O-ring gasket failed, allowing hot gases to escape, a second adjacent O-ring, designed originally for redundancy, was unseated from its groove by the movement of the rocket casing under pressure [332].
- None of the five very different ways of shutting off the Macondo well and preventing a blowout was effective in the *Deepwater Horizon* accident case (see appendix C).
- All of the electronic systems necessary to respond to the Fukushima Daiichi nuclear plant meltdown were disabled during the emergency because of unknown common-point failure modes in their design (see appendix D).

Dependencies may be introduced between redundant or diverse components not only through design but during routine maintenance, testing, and repair. If maintainers perform a task incorrectly on one piece of equipment, they are likely to do it incorrectly on all pieces of equipment [230].

In addition, functional redundancy tends to instill false confidence, which leads to the relaxing of test regimes and schedules and the care with which independent checks are made by inspectors or maintenance personnel.

A vicious circle begins to appear as redundant components introduce more complexity, which adds to the problem of common-cause failures, which leads to more equipment being installed:

> *The defense most often advocated for protection against common-mode/common-cause failures has been diversity. However, while diversity in instrumentation has been used for a long time, failures have still occurred. Conditions develop that cannot be anticipated by the designer, with the result that the improvement gained through diversity is limited. Then, too, diversity defeats attempts at standardization and may even result in increased random failures as well as increased plant costs. With functionally designed and periodically tested diverse and redundant systems, the real concerns are those caused by common external influences and inadvertent human responses.*
>
> *The pattern is recognizable. Systematic common-mode/common-cause failures are the result of adding complexity to system design. They are the product of a philosophy that has become circular. To date, all proposed "fixes" are for more of the same—more components and more complexity in system design.* [129, p. 191]

Redundancy appears to be most effective against random failures and less effective against design errors. Attempts have been made to apply it to software—which, of course, only has design errors and does not fail randomly—in an attempt to make the software fault tolerant.

Software can contain two types of redundancy: data and control. In data redundancy, data structures or messages used in one program or exchanged between computers include extra data for detecting errors, such as parity bits and other error-detecting and error-correcting codes, checksums, cyclic redundancy check characters, message sequence numbers, sender and receiver addresses, and duplicate pointers or other structural information. There is nothing very controversial about this type of redundancy.

Control or algorithmic redundancy has also been proposed for software. The problem is that there must be a way of detecting that an error has occurred. Two error-detection means have been suggested: (1) built-in reasonableness checks on the computations of the computer and the execution of alternative routines if the test is not passed, or (2) writing multiple versions of the algorithms and voting on the result.

The problem with the first approach, reasonableness checks, is the difficulty in writing them. For some limited types of mathematical computations, such as matrix inversion, there are reverse operations that, when applied to the results of a computation, should produce the inputs. In general, this type of reverse operation does not exist. Instead, the outputs or intermediate results are checked to see if they are reasonable given the type of operation being performed. Reasonableness checks are difficult to formulate in general, and writing them may be as error prone as writing the original algorithm [223].

An alternative that has been proposed is to write multiple versions of the software and to vote on the results during operation. If multiple algorithms for a particular computation are known to have singularities in different parts of their input space, then this approach might be useful. Here, the multiple algorithms used can be carefully planned, just as Weaver suggests is necessary for the effective use of diversity in hardware design.

The planned use of known multiple algorithms is also not controversial, although it is useful in very few cases. An alternative, which is to write multiple versions of the software using different teams, is more controversial. An assumption is made that different people are likely to make different mistakes and design very different algorithms. This assumption has not been supported by any carefully designed scientific experiments. In fact, every experiment with this approach that has checked for dependencies between software failures has found that independently written software routines do not fail in a statistically independent way [48; 49; 89; 193; 194; 223]. This result is not surprising: people tend to make mistakes in the harder parts of the problem and in handling nonstandard and boundary cases in the input space—they do not make mistakes randomly.

The problem of common-cause failures between independently developed software routines is not easily solved. Any shared specifications can lead to common-cause failures. But without shared requirements specifications, for example, how do we ensure that the multiple versions of the software will solve the same problem? In order to check the results against each other, they must be solving the exact same problem and the format of the results must be carefully specified. The same limitation exists in developing test data to check the software—the testers may omit the same off-nominal or unusual cases that the developers overlooked.

Claims have been made that ultrahigh software reliability will be achieved by writing multiple versions of the software and voting on the results [61]. These claims are not supported by any data, and, in fact, all the experimental tests of this assumption have been unsuccessful.

In fact, the added complexity of providing fault tolerance in this fashion may itself cause runtime failures, just as it can in hardware redundancy. Examples include the synchronization problems arising from software backup redundancy on the first Space Shuttle flight and the NASA experiences (described in chapter 4) where all the digital control system failures during flight testing of an experimental aircraft were traced to errors in the redundancy management system. In addition, mathematical models have shown that there are limits in the potential software reliability increases possible using this approach [90].

The cost of multi-version programming is not only at least *n* times the cost of producing one version—where *n* is the number of versions to be produced—but also *n* times the cost of maintenance, which is already high for software. Although arguments have been advanced that the increase in cost will be less than *n*, these arguments rest on the assumption that some aspects of the software development process will not have to be duplicated. Anything not duplicated, however, can potentially contribute to common-cause errors.

Furthermore, in experiments with this technique, Knight and Leveson found that in order to get the versions to vote correctly, the specifications had to be much more complete than usually necessary. In other words, many aspects of the processing and outputs, including those nobody really cared about, had to be specified in greater detail than usual to make the results comparable. In the end, the specification phase took more time and effort than would normally have been required. Most important, the overspecification essentially eliminated the potential for diversity in the alternative designs, which was the original goal. Thus, the safety of the system will depend on the existence of a quality that has been inadvertently eliminated by the development process.

There is no way to examine software designs and determine how different they are in their erroneous or unsafe behavior—which is all that counts in this case. Even when very different algorithms are used, the differences may not help because the problem may not be in the algorithm but in the handling of difficult input cases. In the experiments on multi-version programming, the dependencies almost all arose from the difficulty of the common problem being solved, not from dependencies in the solution techniques. In one experimental evaluation of this technique, the algorithms used in most of the versions were very different, as were the programming errors made, yet the programs failed on the same inputs [49; 193].

Given the poor results and the costs, there seems to be no good argument for spending the limited resources of any project on producing multiple versions of the software. And spending more on producing multiple versions of the software usually means that costs must be cut somewhere else. Some people have suggested saving costs by simply testing the multiple versions against each other. This type of testing allows large numbers of test cases to be executed, but it is dangerous because it ensures that the errors will not be tolerated during operational use; that is, the errors that cause identical incorrect results will not be found during testing.

Despite forty years of careful scientific experimentation and analysis showing that this approach does not work for software, it is surprising that some companies are still using it to try to assure safety.

In fact, the most important problem with attempts to tolerate software errors using redundancy is that they may not be directed to where the safety problem lies. Multiple versions of the software written from the same requirements specification can potentially only protect against coding errors, and sometimes only a limited set of these. However, as stated earlier in this book, empirical evidence suggests that most safety problems stem from errors in the software requirements, especially misunderstandings about the required operation of the software. Any redundancy, then, will simply duplicate the misunderstandings.

Any solutions that depend on assumptions about the randomness of behavior will not be effective for systems that do not behave randomly.

Recovery If errors are detected by the monitoring and checking procedures described earlier, then failures can potentially be reduced if successful recovery from the error occurs before the component or system fails. Recovery can be performed by humans, or it can be automated. Comparisons between these two approaches are complex and are left for the next chapter on human factors in designing for safety.

Recovery from software errors is sometimes possible. In general, software error recovery can be forward or backward. In *backward recovery*, the computer returns to a previous state—hopefully one that preceded the creation of the erroneous state—and continues computation using an alternate piece of code. No attempt is made to diagnose the particular software error that caused the erroneous state or to assess the extent of any other damage that may have been caused.

In *forward recovery*, the erroneous part of the state is repaired, and processing continues without rolling back the state of the machine.

Backward recovery procedures assume that the alternate code will work better than the original code. There is, of course, a possibility that the alternate code will work no better than the original code, particularly if the error originated from flawed specifications and misunderstandings about the required operation of the software. In fact, this creates the same problem discussed in the previous section on using redundant software.

Backward recovery may be adequate if it can be guaranteed that an erroneous computer state will be detected and fixed before any other part of the system is affected. Unfortunately, this property usually cannot be guaranteed. An error may not be readily or immediately apparent: a small error may require hours to build up to a value that exceeds a prescribed safety tolerance limit. Forward recovery relies, on the other hand, on being able to locate and fix the erroneous state, which can be difficult.

In practice, forward and backward recovery are not necessarily alternatives; the need for forward recovery is not precluded by the use of backward recovery. For example, containment of any possible radiation or chemical leaks may be necessary at the same time software recovery is being attempted. In such instances, forward recovery to repair any physical system damage or minimize hazards will be required [211].

Forward recovery is needed when

- backward recovery procedures fail;
- redoing the computation means that the output cannot be produced in time;
- the software control actions depend on the incremental state of the system (such as torquing a gyro or using a stepping motor) and cannot be recovered by a simple software checkpoint and rollback [334];
- the software error is not immediately apparent and incorrect outputs have already occurred.

Not only is it difficult to roll back the state of mechanical devices that have been affected by undetected erroneous outputs, but an erroneous software module may have passed information to other modules, which then must also be rolled back. Procedures to avoid domino effects in backward recovery are complex and thus error prone, or they require performance penalties such as limiting the amount of concurrency that is possible. In distributed systems, erroneous information may propagate to other nodes and processors before the error is detected.

Forward recovery techniques attempt to repair the erroneous state, which may simply be an internal computer state or the state of the controlled process. Examples of forward recovery techniques include using robust data structures, dynamically altering the flow of control, and ignoring single-cycle errors.

Robust data structures use redundancy in the structure such as extra pointers or data such as extra stored information about the structure to allow reconstruction if the data structure is corrupted [368]. Linked lists with backward as well as forward pointers, for example, allow the list to be reconstructed if only one pointer is lost or incorrectly changed.

Reconfiguration or dynamic alteration of the control flow is a form of partial shutdown that allows critical tasks to be continued while noncritical functions are delayed or temporarily eliminated. Such reconfiguration may be required because of temporary overload, perhaps caused by peak system usage or by internal conditions, such as excessive attempts to perform backward recovery.

Real-time control systems usually have tasks that are iterated many times per second. In general, this type of software is insensitive to single-cycle errors, which are corrected on the next iteration. Single-cycle errors may originate from bad input data, which is fixed in the next sensor reading. Alternatively, such errors may result for specific inputs, but the algorithm or code will produce correct results for slightly different input data. The rate at which new data is received may make it possible to ignore single-cycle errors and simply "coast," that is, repeat the last output or produce a safe output, until new data is received.

Again, the problem with both forward and backward recovery procedures is that they usually depend on assumptions about the state of the system and the safe operation of the software that may be flawed.

Many mechanisms have been proposed for implementing these procedures in software. The real problem is in detecting errors and figuring out how to recover from them, not in devising programming language mechanisms to implement the detection procedures. Some programming languages contain special exception-handling mechanisms that reduce the implementation effort. Many of these language features for error and exception handling, however, are so complex that they may cause the introduction of errors in the error-handling routines and create more problems than they solve. In general, error-handling mechanisms, like everything else, should be as simple as possible.

11.5 Hazard Control

So far, this chapter has examined ways to prevent hazards from occurring. But even if hazards do occur, accidents can sometimes still be prevented by detecting the hazard and controlling it before any damage occurs. Note, however, that these design features are at the lowest level of precedence and, therefore, the least effective and most costly.

By the definition of a hazard, there must exist other conditions in the environment for a loss to occur. Even though toxic chemicals may be released from a plant, any resulting losses will depend on weather conditions, such as which way the wind is blowing or if the chemicals just dissipate into the atmosphere, and on the location of humans in relation to the release.

Reducing the level or the duration of the hazard may increase the likelihood that the hazardous condition is reversed before all the necessary preconditions for an accident occur. For example, keeping hazardous materials under lower pressure or transporting them in smaller amounts or through smaller pipes will reduce the rate at which they are ejected or lost from the system: the basic design can help make the hazards controllable.

Resources, both physical and information, such as diagnostics and status information, may be needed to control hazards in an emergency. These resources therefore need to be managed so that an adequate amount will be available when an emergency arises.

As discussed in chapter 6, too many alarms or too much information may hinder hazard control. Warning signals should not be present for long periods or be too frequent, as people quickly become insensitive to constant stimuli. An operator was killed in an automated factory, for example, when a 2,500 lb robot came up behind him suddenly. The robots had red rotating warning lights to show they were armed, but the lights shone continually and indicated only that the robots were *capable* of starting up—no real warning of movement was provided [111]. The reason that the designers had not included an audible warning when the robot was about to move may have been an incorrect assumption that humans would never have to enter the production area while the robots were operational.

Hazard control measures include limiting exposure, isolation and containment, protection systems, and fail-safe design.

11.5.1 Limiting Exposure

In some systems, it is impossible to stay only in safe states. In fact, increased risk states may be required for the system to accomplish its functions. A general design goal for safety is to stay in a safe state as long and as much as possible. For example, nitroglycerine used to be manufactured in a large batch reactor. Now it is made in a small continuous reactor, and the residence time has been reduced from 2 hours to 2 minutes [184].

Another way to reduce exposure is to start out in a safe state and require a change to a higher risk state. The command to arm a missile, for example, might not be issued until the missile is near its target. In the automated shutdown system at the Darlington Nuclear Power Generating Station in Canada, the software contained variables that were used to determine, on the basis of sensor inputs, whether to shut down the plant. Each time through the code, the software initialized these internal variables to the *tripped* value. If a software control flow or other error occurred that resulted in the omission of some or all of the checks on the sensor inputs, the plant would be tripped. Basically, the safe state for the software in this instance was for the variables to contain a value that would result in plant shutdown when the shutdown decision had to be made; therefore, the variables were assigned this value at all times except right after a check had been made that determined that the condition of the plant was safe.

In general, critical flags and conditions in software should be set or checked as close as possible to the code that they protect. In addition, critical conditions should not be complementary. The absence of an arm condition, for example, should not be used to indicate that the system is unarmed.

11.5.2 Isolation and Containment

Protection may take the form of barriers between the system and the environment, such as containment vessels and shields, which isolate hazardous materials, operations, or equipment away from humans or the conditions that can lead to an accident.

The proximity of a hazard to an unprotected population will influence the severity of its consequences. The explosion of a chemical plant at Flixborough, which was relatively isolated from an urban population, caused twenty-eight deaths, while the explosion at Bhopal, which was located in the midst of a crowded residential area, involved at least 10,000 deaths and 200,000 injuries. Even if plants are located in an isolated area, the transport of dangerous materials can bring them into contact with large populations: the explosion of a road tanker in San Carlos, Spain, in 1978 killed more than 200 people.

The problem with isolation is that over time, for many good reasons, the isolation may diminish or even disappear. Hazardous facilities are usually initially placed far from population centers, but the population shifts after the facility is created. People want to live near where they work and do not like long commutes. Land and housing may be cheaper near smelly, polluting plants. In developing countries, utilities, such as power and water, and transportation may be more readily available near heavy industrial plants, as was the case at Bhopal. The lure of providing jobs and economic development may encourage government officials to downplay risks and not rigorously enforce zoning requirements.

Over time, the land available for buildings and structures may only be available near dangerous parts of the plant, as happened at the Texas City refinery where trailers were built to house employee offices next to the isomerization (ISOM) tower that exploded in 2005. With increasing population, local emergency facilities, such as firefighting and medical resources, may lag behind the increasing requirements due to constraints on resources and competing priorities.

11.5.3 Protection Systems and Fail-Safe Design

Hazard control may also take the form of moving the system from a hazardous state to a safe or safer state. First, some definitions are provided as different industries use slightly different terms. In the process industry, particularly nuclear power, the term *protection system* typically refers only to the electronic systems that detect conditions necessitating some type of safeguarding action but not the equipment that performs the action. The latter are called *safety systems* or *engineered safety features*. In this section, protection system includes both.

The feasibility of building effective fail-safe or protection systems depends on the existence of a safe state to which the system can be moved and the availability of early warning. These, in turn, require a suitable delay in the course of events between the warning and an accident.

A system may not have a single safe state—what is safe may depend on the conditions in the process and the current system operating mode. A general design rule is that hazardous states should be difficult to get into, while the procedures for switching to safe states should be simple.

Typical protective equipment includes gas detectors, emergency isolation valves, trips and alarms, relief valves and flare stacks, steam and water curtains, flame traps, nitrogen blanketing, fire protection equipment such as insulation and water sprays, and firefighting equipment.

A common example is a *panic button* that stops a device quickly, perhaps by cutting off power. This feature might be useful when an operator has to enter an unsafe area containing equipment that moves. One of the problems with a panic button is making sure it is within reach when needed. In the past, a rope was sometimes strung around an industrial robot work area, and a pull anywhere on the rope would operate the panic button. Today, more high-tech solutions are commonly used. Operators need to be trained to exhibit the correct panic reaction in response to an unexpected event.

Once again, protection against one hazard may simply create a worse one. Hammer provides an example involving a non-propulsive attachment (NPA) on the US Navy's Sidewinder missile [133]. The NPA was made to fit on the end of the missile's rocket motor: it looked like an automobile piston with four equidistant holes bored around its side. If the motor accidentally ignited while the missile was in storage, being transported, or being installed on an aircraft, the NPA would direct the exhaust gases out at right angles rather than straight back, making the missile unmovable. The NPA had almost perfect reliability; it never failed to work. Unfortunately, sometimes the ordnance personnel forgot to remove the NPAs after they had hung the missiles under the wing of the launch aircraft. When the pilots tried to launch the missiles in flight, the hot gas discharge hit the wings and caused damage so severe that the planes had to be scrapped [133]. After the third aircraft was lost this way, the Navy eliminated the use of the NPA.

As stated earlier, passive devices are safer than active devices. Some equipment can be designed to fail into a safe state: pneumatic control valves, for example, can be designed to open or close in the event of a failure. Another example is a mechanical relay that can be designed to fail with its contacts open, shutting down a dangerous machine. Occasionally, programmable electronic systems can be designed to fail into a safe state.

If a passive design is not possible, then designers must add control components—an operator or an automated system—that will detect a hazardous state and provide an independent way of moving the equipment into a safe state. Any shutdowns by a computer, including shutting itself down, must ensure that no potentially unsafe states are created in the process.

A watchdog timer is an example of a simple protection device to detect failures of critical system components. This device has a timer that the system must keep restarting. If the system fails to restart the timer within a given period, the watchdog initiates some type of protection action. Care must be taken to eliminate the possibility of common-cause failures of the watchdog and the thing being monitored. For example, if a watchdog timer is used to check software, the software should not be responsible for initializing the watchdog, and protection should be provided against the software incorrectly resetting it. An infinite loop in the software routine that resets the watchdog, for example, could destroy the watchdog's ability to detect the infinite loop and lack of software function.

Wray relates a case of a common-cause failure of a watchdog and the computer it was monitoring:

> *It happened recently to a system that was operating a network of relays, known as "output contactors," which controlled the power supply of some electrical equipment. The user had, unwittingly, fitted the system with a mains transformer that was too small for the job. There were no problems during the first 18 months of operation because no one had used more than two of the contactors at the same time. One day, however, the system's microcomputer called for all of the contactors to operate simultaneously. This overloaded the transformer, whose output voltage dropped and reduced the electrical supply to the microcomputer. The microcomputer stopped working—and it did so before switching off the contactors and closing down the machine. The result was some expensive damage as the machine continued working longer than it was supposed to. Although there was a watchdog on the system, it too was affected by the low supply voltage and failed to cut out the primary contactor.* [416]

In general, a device such as a watchdog at the interface between a computer and the process it is controlling can use timeouts to detect a total computer failure and bring the system to a safe state. If the computer fails to send the interface device a signal at the end of a time interval, such as every 100 ms, the device assumes that the computer has failed and initiates fail-safe action. The interface device might be another computer: a multiprocessing system, for example, might require regular transmissions between the computers. Such transmissions can simply be interrupts because they need not convey any other information. These checks have been given various names: *keep-alive signals*, *health checks*, and *sanity checks*.

Sanity checks can also be performed by the computer on other devices to determine whether input data from that device or information about the status of the device is self-consistent and reasonable. Earlier in this chapter, an incident was described where a hardware multiplexor failure caused the sanity checking operation to falsely signal incoming ballistic missiles.

The protection system itself should provide information about its status and control actions to the operator or bystander. For example, a light might flash or a warning might sound if a person enters a hazardous zone to indicate that the protection system is working and has noticed the intrusion. The status of various sensors and actuators involved in the control system might also be displayed for the operator. Whenever a system is powered up, a signal or warning should be provided to operators and bystanders. Conversely, if the software or other controller has to shut down the system or revert to a safe state, the operator should be informed about the anomaly detected, any action taken, and the current system configuration and status. Before shutting down, the controller may need to recover or undo some damage.

A common design goal is to control the hazard while causing the least damage or interruption to the system. Achieving this goal requires making tradeoffs between safety and interrupting production or damaging equipment. Stressing equipment beyond normal loads—and thus increasing equipment damage or required maintenance actions—may be necessary to reduce the risk of human injuries.

Besides shutting down, some action may be necessary to avoid harm, such as blowing up an errant rocket. At the same time, such safety devices may themselves cause harm, as in the case of the emergency destruct facility that accidentally blew up seventy-two French weather balloons described in chapter 4.

The designer also has to consider how to return the system to an operational state from a fail-safe state. The easier and faster it is to do this, the less likely it is that the safety system will be purposely bypassed or turned off.

Because of these requirements and because some systems must continue to operate at a minimum level in order to maintain safety, various types of fallback states may be designed into a system. For example, traffic lights are safer if they fail into a blinking red or yellow state rather than fail completely. The X-29 is an experimental, unstable aircraft that cannot be flown safely by human control alone. If its digital computers fail, control is switched to an analog backup device that provides less functionality than the computers but at least allows the plane to land safely. Such a feature is sometimes called "limp home mode."

Fall-back states might include

- *Partial shutdown*: the system has partial or degraded functionality.
- *Hold*: no functionality is provided, but steps are taken to maintain safety or limit the amount of damage.
- *Emergency shutdown*: the system is shut down completely.
- *Manually or externally controlled*: the system continues to function, but control is switched to a source external to the computer; the computer may be responsible for a smooth transition, which can be problematic if the fallback is due to computer malfunction.
- *Restart*: the system is in a transitional state from non-normal to normal.

The conditions under which each of the control modes should be invoked must be determined, along with how the transitions between states will be implemented and controlled.

There may also be requirements for multiple types of shutdown procedures, for example:

- *Normal emergency stop*: cut the power from all the circuits.
- *Production stop*: stop as soon as the current task is completed. This facility is useful if shutting down under certain conditions, such as mid-cycle, could cause damage or problems in restarting.
- *Protection stop*: shut down the machine immediately, but not necessarily by cutting the power from all the circuits, which could result in damage in some cases. The protection stop command may be monitored to make sure it is obeyed, and there may be a provision to implement an emergency stop if it is not.

If the system cannot be designed to fail into a safe state or to passively change to a safe state in the event of a failure, then the hazard detectors must be of ultra-high reliability, be designed to fail safe themselves, or be designed so that a failure can be detected. For example, equipment can be added to test the detection subsystem periodically by simulating the condition that a sensor is supposed to detect [296]. If the sensor fails to respond to a challenge or if it responds when no challenge is present, then a warning signal is generated. Park [296] provides an example of such a design for an industrial robot:

> *For example, an appropriate challenge to a light barrier used as an intrusion detector would be a small motor-driven vane which repeatedly passes through the light curtain. If the sensor fails to respond when the vane is supposed to be in the path of the light beam, then either the sensor or the barrier have [sic] failed, or the motion of the vane has been interfered with. If the sensor shows that an object is present in the sensing area when the vane is not supposed to be, then either a real intrusion has occurred, or the vane is stuck, or the sensor has failed. Only if the signal from the sensor changes from "safe" to "unsafe" in step with the motion of the vane can we be certain that no obstruction is present and that the safety device itself is operating properly.*

Park sees three design criteria as important in such safety devices. First, the challenge must not obscure a real hazard. In the light curtain example, the vane must pass through the light beam many times per second because a real object intruding into the protected space might be undetected for as long as one entire challenge interval. Second, the sensor and challenge subsystems must be highly reliable—Park suggests using redundancy. Third, the sensor and challenge subsystems must be as independent as possible from the monitor subsystem.

In general, a hazard detection system may consist of three subsystems: (1) a sensor to detect the hazardous condition, (2) a challenge subsystem to exercise and test the sensor, and (3) a monitor subsystem to watch for any interruption of the challenge-and-response sequence [296].

The astute reader will notice that the complexity level is creeping up and that these protection systems are starting to resemble the Rube Goldberg design of a pencil sharpener shown in chapter 5, thus decreasing the likelihood that the protection systems will work when needed. This violates the simplicity principle much higher up in the design precedence.

When possible, a passive protection system will be both safer and cheaper. In the industrial robot example described previously, simple interlocks, such as using a pressure-sensitive mat that cuts power when a human steps on it, are commonly used.

11.6 Damage Reduction

Designing to reduce damage in the event of an accident is almost always required because it may not be possible to eliminate, prevent, or control all hazards. In addition, the analysts and designers may not even foresee all hazards. In particular, they may not recognize the consequences of modifications: changes to plants and methods of operation often have unforeseen side effects.

Furthermore, humans will on occasion do the wrong thing, and our dependence on them is not eliminated by installing automatic devices. Finally, resources are usually limited: designers need to determine which hazards should be dealt with immediately and which can be left, at least for the time being.

In an emergency, there probably will be no time to assess the situation, diagnose what is wrong, determine the correct action, and then carry out that action [13]. Therefore, emergency procedures need to be prepared and practiced so that crises can be handled effectively. In the Macondo well blowout, as described in chapter 4, the *Bankston* mudboat crew had reviewed and practiced contingency actions the morning before the explosion and fire. While the *Deepwater Horizon* crew were jumping into the water and chaos reigned at the lifeboat stations, the mudboat crew calmly did what was necessary first to save themselves, which they had practiced, and then proceeded to save the people in the water.

Contingency planning usually involves determining a "point of no return," when recovery is no longer possible or likely and damage minimization measures should be started. Without predetermining this point, people involved in the emergency may become so wrapped up in attempts to save the system that they wait too long to abandon the recovery efforts or may abandon them prematurely.

Warning systems, like any alarm system, should not be on continuously or frequently because people quickly become insensitive to constant stimuli, as noted earlier. A distinction should be made between warnings used for drills and those used for real emergencies. At Bhopal, the practice alert, which went off twenty times a week, was the same as the alert indicating a real emergency.

Damage minimization techniques include providing escape routes (such as lifeboats, fire escapes, and community evacuation), safe abandonment of products and materials (such as hazardous waste disposal), and devices for limiting physical damage to equipment or people. Some examples of the latter are

- Providing oil and gas furnaces with blowout panels that give way if overpressurization results from delayed ignition or accumulations of fuel vapors and gases. This feature prevents or reduces damage to furnace walls, boiler tubes, and other critical parts of the equipment and structure. Blowout panels or frangible walls are also used in explosives-processing plants, where an explosion could destroy a structure completely [206].
- Collapsible steering columns on automobiles or signposts on highways: if an accident occurs, the steering column or signpost collapses and the possibility of injury is reduced.

- Shear pins in motor-driven equipment: if there is an overload, the torque causes shearing of the pin, thus preventing damage to the drive shaft.

11.7 Design Modification and Maintenance

Designs must be maintained, just as physical devices are. Change may be necessary because of changes in the environment or workplace, changes in procedures, changes in requirements and needs, the introduction of new technology, accumulated experience that shows that the design does not satisfy the requirements adequately or that assumptions on which the design was analyzed or implemented do not hold, or the occurrence of accidents or incidents.

As repeatedly emphasized in this book, many accidents can be attributed to the fact that the system did not operate as intended because of changes that were not fully coordinated or fully analyzed to determine their effect on the system [308]. Flixborough and the Hyatt-Regency walkway collapse are classic examples. In 1981, a walkway at the Kansas City Hyatt-Regency Hotel collapsed, killing 114 people and injuring 200. Investigation showed that the design was changed during construction without an appropriate structural analysis of the new design. After a partial roof collapse during construction, the owner requested an analysis of the redesign, but it was never done [151].

Reanalysis of the safety features of the design must occur periodically and must always be performed when some known change occurs or when new information is obtained that brings the safety of the design into doubt.

To make design changes safely, the design rationale—why particular design features were included—is needed, as again is emphasized repeatedly in this book. Changes can inadvertently eliminate important safety features or diminish the effectiveness of hazard controls. The design rationale documentation must be updated and compared to accident and incident reports to ensure that the underlying assumptions are correct and have not been invalidated by changes in the system or the environment.

Exercises

1. The chapter suggests that design for testability is important. What design techniques or features might be used to make a complex system, perhaps software, more testable for safety?
2. How does the underlying causality model affect what design features are used to increase safety? Provide several specific examples for the energy model, the chain-of-events model, and a systems-theoretic model.
3. Take your STPA hazard analysis you did for the class and at least one causal scenario that contains hardware, software, and humans (need not be one scenario but could be three). What type of design techniques might you include to control these scenarios? What are some considerations that you might need to take into account when making your design decisions?
4. Consider the safety design precedence shown in the four levels of figure 11.1. Justify the ordering of these levels in terms of effectiveness and cost. In other words, why is each level placed where it is in the precedence?

5. Redundancy, one of the most popular types of design protection techniques, is pretty far down in the precedence. Why do you think that is? Protection systems are even farther down. Why? Why do you think that protection systems and redundancy are so popular?
6. The chapter states that the general goal of redundancy is to improve reliability, which will not necessarily improve safety. Explain why this is true; that is, explain why designs that use redundancy may not improve safety.
7. What are some examples of interlocks to protect against hazards that are used in cars? What hazard is each of these protecting against?
8. Do the following for a typical design in your field or industry. If you don't have one, pick a field you know something about. What are the primary hazards?

   Identify an example of how each of the following are used or could be used to handle the hazards:

   a. Hazard elimination
      By substitution?
      By simplification?
      By decoupling?

   a. Hazard reduction
      Design for controllability?
      Barriers?
      Safety factors?
      Redundancy?

   b. Hazard control
      Reducing exposure?
      Isolation and containment?
      Protection systems and fail-safe design?
      Damage reduction

   Is anything else commonly used in your field that does not fit in one of these categories?

12
Human Factors in System Design

[The designers] had no intention of ignoring the human factor. . . . But the mechanical and technological questions became so overwhelming that they commanded the most attention.

—John Fuller,
Death by Robot [104]

In childhood, falling down may be part of growing up. But do we need the falls of hundreds of . . . Tacoma Narrows bridges and Hyatt walkways? Do we need DC-10s to plummet, . . . , nuclear plants to release radioactive fallout, and other such falls to wake up to the critical importance of human factors in design, analysis, production, installation, maintenance, training and operation of our products?

—Richard Hornick,
Dreams: Design and Destiny [151]

Chapter 11 examines safe system design as a whole. This chapter focuses on topics related to the design of systems from the human operator standpoint. This division does not imply that these two design aspects should be separated and applied in isolation. Allocation of functionality, communication among components, and system properties such as safety and security need to be optimized for the system as a whole, including any hardware, software, and humans. In fact, many of the current problems arising when humans control complex automation, such as mode confusion or situation awareness problems, can only be solved by such an integrated approach. Solving these problems will require more communication and joint problem-solving by human factors experts, hardware engineers, and software engineers.

Human factors in design, even when concentrating on safety, is a huge topic. Only a basic introduction to a few of the most important concepts can be included here. Given the goal of ensuring that the system as a whole provides safe and effective system operations, three important topics involve the design of the automation to optimize the role of human controllers, the design of the interface between them to promote communication, and the training provided to human controllers to interact with the automation.

The *interface*—that is, the controls and displays provided to the user to operate the system—acts as the communication channel between the human and the automation. Computers have replaced most of the mechanical and traditional instrumentation used in the past:

many systems have only digital displays. Controls too have changed, with the interface controls usually only indirectly connected to the mechanical system components through digital intermediaries, which is usually called *x-by-wire*, such as drive-by-wire.

The change in connections allows more flexibility in design but introduces more potential scenarios to consider in hazard analysis. Safety problems often arise because of flaws in the interface design or, in other words, in the communication channel between the human and the automation.

Hazards may also result from automation design that induces erroneous or dangerous operator behavior. Sometimes interface changes can alleviate these human errors, but often interface design fixes alone are not enough. Human–machine *interactions* are greatly affected by the design of both the software and hardware in concert with design of the activities and functions provided by the operator. Changing the software, hardware, and human activities is the most direct and effective way to eliminate interaction problems. In this approach, the design or redesign of the functionality of the software and hardware and of the activities assigned to the operator and to the automation is involved rather than just the design of the displays and controls.

The unfortunate use of the same acronym, HMI, for both human–machine *interaction* and human–machine *interface* complicates communication and sometimes understanding. The acronym is not used in this chapter. Interaction will refer to the larger topic, which includes the behavior of the operator, the design of the automation, *and* the design of the interface used for communication between humans and automation. The term *interface* will refer only to the communication channel, the displays, and the controls.

12.1 Determining What Should Be Automated

It is common to classify systems in terms of levels of automation, such as human-controlled, human-controlled with computer monitoring, computer-controlled with human monitoring, and so on. This classification, however, is too simplistic as discussed in chapter 5. In the most effective system designs, humans will have some responsibilities, automation may have others, and a large number may only be achieved by various types of cooperation between the two. How best to assign these responsibilities in order to optimize safety or other properties is still largely unknown. Some of what is known is described in this chapter. Much more research is needed.

The simplest approach to design is just to automate as much as possible, but the simplest approach is not necessarily the best one. As an example, a negative effect was achieved when a major airline, known for having one of the best maintenance programs in the industry, introduced an expert system to assist their maintenance staff. The quality of maintenance fell: the staff began to depend on the computer decision-making and stopped taking responsibility and making their own decisions. When the software was changed to provide only information and only when requested, quality again rose.

Simply labeling automation as intelligent does not make it so. Systems today, particularly where safety is involved, need the adaptability and flexibility that only humans can provide. How to make decisions about what to automate and what to leave to humans in order to optimize safety is largely unknown and perhaps less important than determining how to design the sharing of the tasks required between humans and computers. We know more about what not to do than what should be done instead. Some of the difficulties were discussed in chapter 6, such as the fact that humans perform poorly in monitoring and backup roles. Unique roles that go beyond the obvious may provide the best solutions.

As an example, Hirschhorn [146] describes the approach used a while ago in a Canadian chemical plant where computers were used to improve human decision-making rather than to replace it. In chemical plants, the production process often needs to be continually adjusted to fit varied specifications for different products. As a result, the plants often operate below their potential capacity.

In this case, a new plant was built to manufacture alcohol for use in carpets, soaps, containers, and so on. The necessary customizing process could have been programmed into a computer and the production automated. Instead, the computer was used to help workers become technologically sophisticated and inventive enough to solve the problems created by the need to customize. Rather than having the computer control production, it was instead used to assist and teach workers by providing them with technical and economic data and allowing them to test their own production decisions. The experience they got could then be used to develop new ways of controlling production: "*Because workers make experimental decisions rather than routine ones, the production process is being continually upgraded, even as workers become more knowledgeable. For this to happen, of course, entails a fundamentally different conception of the interaction of worker and machine*" [146, p. 46].

The designer needs to remember that different is not necessarily better—many changes to traditional work practices do not necessarily represent progress. The goal of replacing humans with computers will not necessarily lead to improved results with respect to safety. At the same time, humans are going to need assistance in controlling advanced system designs where complexity and coupling are increasing rapidly. Disturbances often have a much greater effect because they no longer occur in isolation: the effects of a disturbance can spread more rapidly over a larger number of processes than in the older, more decoupled systems [366]. As a result of this evolving system design, hazards have often changed in both type and character. These changes are significantly impacting the operator's job as well as the difficulty in overall system design for safety.

We need to create different types of human–computer interaction such as the novel approach described in the Canadian chemical plant. Success requires understanding more about how to augment human capabilities through the design of automation and the design of the overall system in which both the humans and the automation are operating.

Perhaps the best path for now is to use human-centered design methods and what is known about how to reduce human errors through design, to be realistic about the safety of automating functions where serious hazards are involved, and to involve a wide range of expertise in the design process.

12.2 The Need for Wide Participation in Design Activities

Successful system design can only be achieved by engineers, human factors experts, and application experts working together. Obstacles to this type of collaboration stem from limitations in training and education, the lack of common languages and models among different specialties, and too narrow a view of one's responsibilities. These obstacles need to be overcome to successfully build safer systems.

Too often hardware and software engineers design systems, and human factors experts are relegated to designing the human interfaces to the hardware and software and to somehow making it all usable by humans. The problems arising from this separation have resulted in a call by psychologists and human factors experts for *human-centered design*, which suggests that human factors should reign supreme or at least equal in the design effort.

In the same way, safety experts are often isolated from the early design efforts and relegated to after-the-fact assurance activities. As a result, safety is also not handled adequately in system design. Identifying design flaws and fixing them late in development is almost never cost-effective.

Application experts and system users may also be omitted from the early design process and only introduced during testing. The result can be systems that are usable by humans in general and effective at providing hardware and software functionality but are not usable or are difficult to use in the particular application for which they are being designed, such as air traffic control or health care. Including users in the design process can result in a better design and safer operations. Experienced operators of complex systems, such as pilots, nuclear reactor control room operators, and air traffic controllers, can provide invaluable information on which features would be helpful and which would be distracting or error-prone for their particular applications.

The obvious solution is for a systems approach to design and the use of integrated teams. As is often the case, however, the solution is clear but difficult to implement: where to go is obvious, but how to get there is not. At the least, a solution will require major changes in both education and practice as well as some advances in technology.

12.3 Safety versus Usability and Other Common Goals

The use of "good" human factors design practices does not necessarily imply that the resulting design will be safe. Differences in basic system designs and use, along with differences in the social and educational backgrounds of operators and users, may invalidate for one system what is appropriate for another. Most important with respect to design for safety, there are often conflicts between safety and other desirable system properties.

As an example, enhancing ease of use suggests that users be required to enter data only once and that the automation then access that information any time it is needed. An argument for this principle is that requiring duplicate entry of data creates extra work for users and increases the possibility of entry errors.

A counterargument is that duplicate entry allows the automation to monitor for entry errors. While extreme entry errors may be detected by reasonableness checks, small slips usually cannot. Starting with an emphasis on safety and a hazard analysis may lead to a design where multiple entry is required for identified critical data but is not required for noncritical data.

In general, usability and safety often conflict. For example, a design that involves displaying data or instructions on a screen for an operator to check and verify by pressing the *enter* button will, over time and after few errors, get the operator in the habit of pressing the *enter* key multiple times in rapid succession. In the Therac-25 console display, although the operators originally were required to enter the treatment parameters at the treatment site as well as on the computer console, they complained about the duplication (see appendix A). As a result, the interface was changed so that the parameters were displayed and the operator only needed to press the *return* key if they were correct. Operators quickly became accustomed to pushing the *return* key quickly the required number of times without checking the parameters carefully because, in the vast majority of cases, the displayed parameters were correct.

As has been stressed repeatedly in this book, conflicts between various design qualities often exist. The only way that lists of general interface design principles can be justified is if all system qualities are complementary and are achieved in exactly the same way. This is rarely the case. More often qualities are conflicting and require tradeoffs and decisions about priorities. An interface that is easy to use may not necessarily be safe, and vice versa. A focus on safety implies that the first step in the process is to identify the hazards to be avoided, just as in the more general system design principles described in the previous chapter. The hazards and their causal scenarios can then be used in the design of the human–automation interface and interactions.

In the standard interface development process, human factors experts perform a task analysis and then often use simulation and other methods to evaluate and refine the human–computer interface design. Care has to be taken, however, when applying information obtained by laboratory or simulator studies to normal operations. Humans may act very differently in these two situations.

In safety-critical system development, a different process is needed; that is, one that starts from identification of hazards and then uses the hazard analysis results during design. Figure 12.1 shows the general process. It starts with a general hazard analysis, which is refined as the detailed design features are created. Later, performance auditing and feedback can be used to further evaluate and improve the design during operations. This process will only be successful if operators and operator behavior are included in the hazard analysis technique used.

Feedback may come from incident and accident reports, but earlier design improvements are possible and accidents avoided if specific information sources and feedback loops during operations are established to validate or refute the assumptions made during design about operator behavior. Today, systems can be and are instrumented to provide a great deal of information about operations, but the amount of such information that is collected can be difficult to process. Using the hazard analysis to identify the critical assumptions about operator behavior and providing ways to check these assumptions can result in problems being identified sooner.

12.4 Reducing Safety-Critical Human Errors through System Design

The overriding assumption in this chapter is the systems theory principle that human behavior is impacted by the design of the system in which it occurs. If we want to change operator behavior, we have to change the design of the system in which the operator is working. If the design of the system is confusing the operators, for example, we can (1) try to train the operators not to be confused, which will be of limited usefulness; (2) try to fix the problem by providing more or better information through the interface; or (3) redesign the system to be less confusing. The third approach will be the most effective.

Simply telling operators to follow detailed procedures that may turn out to be wrong in special circumstances or relying on training to ensure they always do what, in

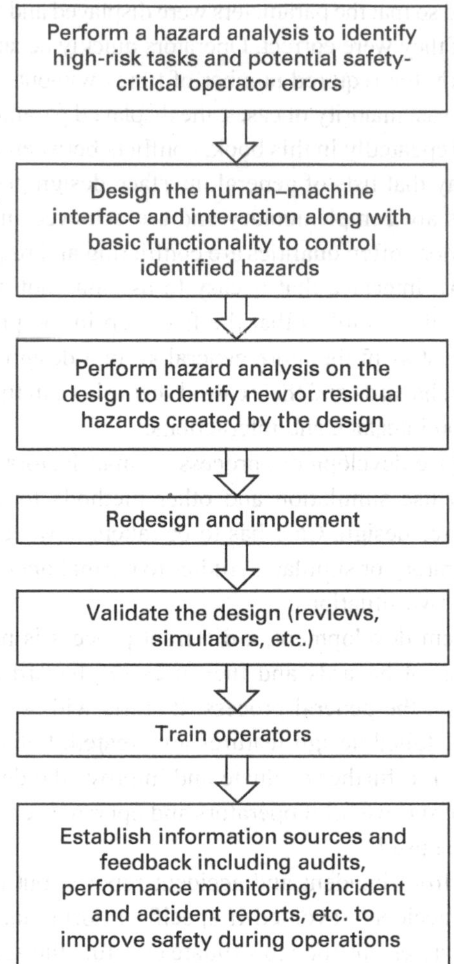

Figure 12.1
Human-interaction design process.

hindsight, is the right thing simply guarantees that unnecessary accidents will occur. The alternative is to ask how we can design to reduce operator errors or, conversely, identify what design features induce human error. In other words, how do we design to support the operator.

Much is known or has been hypothesized about how to design the human–automation interface for safety. Much less is known about how to design human–automation *interaction* to enhance safety. The rest of this chapter describes some of what is known about how to accomplish the latter goal.

Human–machine interaction is a communication process—a social interaction between the operator and the machine's designer [311]. As such, it is prone to the same misunderstandings and communication problems of any social interaction.

The communication has the additional problem, however, of being indirect: the nonverbal cues and body language that assist human–human communication are not present. These same issues often arise in electronic mail communication, but misunderstandings are mitigated by the possibility of feedback and follow-up queries, both of which are less feasible and more complicated with an electronic interface. The entire interaction and all possible miscommunications and requests for additional information must be anticipated before the interaction occurs.

Even when carefully designed, the interface may have the effect of dictating the communication process—the operator must conform to the specific structure incorporated in the interface by the designer. Therefore, the designer must try to anticipate the form of communication that the operator will require and want.

Figure 12.2 shows the basic components of communication in systems today. There are three primary components: the operator, the automation, and the controlled process.

At the top of the figure is the operator, who uses controls to provide commands to automation as well as directly to the controlled process. The operator gets feedback about the state of the controlled process from the automation through electronic displays and directly from the controlled process. Even with highly automated systems, human operators often get feedback in addition to that provided by the displays from sound, vibration, and so on that cannot easily be communicated through an electronic interface.

Figure 12.2 shows two components within the operator box. A human mental model contains the information the operator uses to make control or monitoring decisions. The

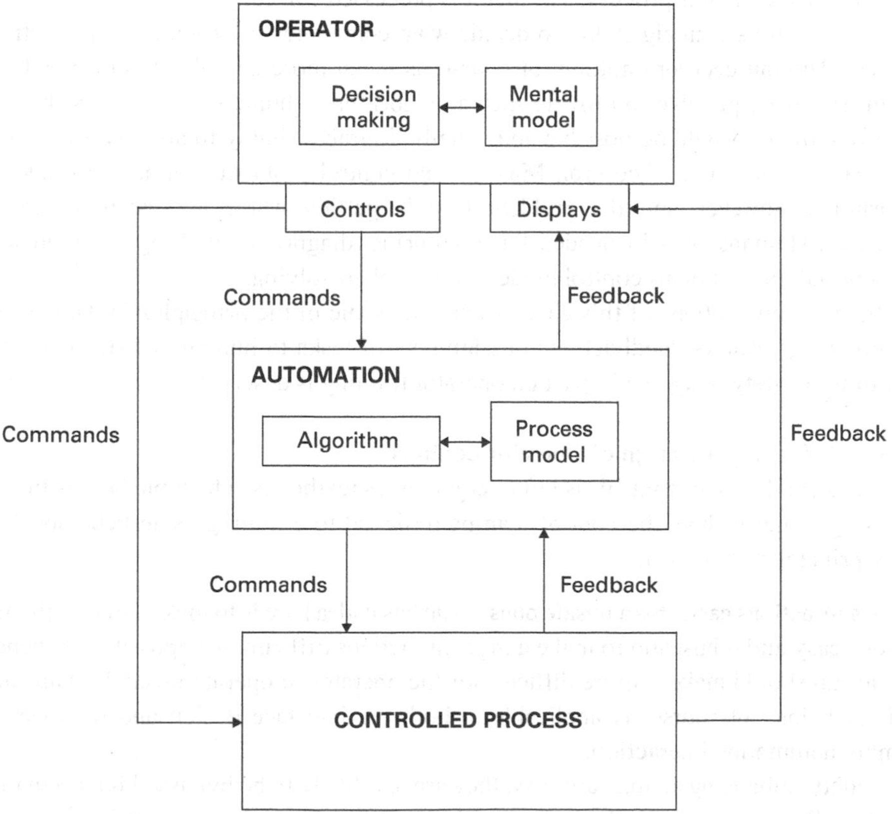

Figure 12.2
Model of the components of human–machine communication.

mental model contains what the operator *thinks* is the current state of the controlled process, the automation, and the environment. The mental model is updated by various means but primarily from feedback. Other information operators may use to update this mental model include beliefs they have about how the process can change and inferences about the effect of previous commands the operator issued to the automation—which the operator may assume were executed correctly.

Note that computer automation also has a model of the state of the process. This model is usually much simpler than human mental models and may simply be represented as a few variables in the memory of the computer.

Accidents often happen when the operator's mental model or the automation's process model becomes inconsistent with the real state of the controlled process and the environment. For example, the driver or the car automation thinks that the lane to the left is clear when it is not, and moves into that lane. Another example is that the human operator or the automation thinks that the helicopter state is fine when, in fact, some equipment is overheating and a control action is required to prevent an accident.

Human mental and software process models are updated through feedback. The operator and the automation need to receive the feedback necessary to keep the two models of the controlled process consistent with both the actual controlled process state and with each other. Note, however, that simply receiving the feedback, particularly for humans, does not mean that it is processed or that it is processed correctly.

Automation uses an algorithm to decide what commands to provide to the controlled process. Human decision-making, of course, is much more complex than a fixed algorithm. If it were possible to provide such an algorithm to humans that always should be followed, then it would be possible and usually desirable simply to automate everything and leave humans out of control. Most system control problems cannot be reduced to algorithms, however, and the problem is only getting worse as system complexity increases. Humans are still needed for monitoring, diagnosis, handling emergencies or non-normal (off-nominal) control modes, and problem-solving.

The next few sections of this chapter examine some of the principles in the design of the controls, displays, feedback, and assignment of tasks to humans versus automation that impact safety. Then the impact on operator training is described.

12.4.1 Safety in the Design of Operator Controls

The physical design of controls is left to ergonomics textbooks. The topic here is the logical design; that is, how the controls can be designed to encourage safe behavior. Some basic principles are known.

Make safe actions easier than unsafe ones The basic idea here is to make safety-enhancing actions easy and robust and to make dangerous actions difficult or impossible. In general, the design should make it more difficult for the operator to operate unsafely than safely. This principle, of course, is applicable to both the interface design and the design of human–automation interaction.

If safety-enhancing actions are easy, they are less likely to be bypassed intentionally or accidentally. If the safe method of operation is difficult or time-consuming, an unsafe method is likely to be substituted. Shortcuts taken while entering the takeoff longitude of the Korean Air Lines plane shot down by the Russians in 1983, for example, have been hypothesized as the cause of the plane being off course [22].

If protection systems slow down work or make it more difficult, humans will try to bypass or disconnect them. Restarting the process after a stop command from the protection system should be easy and fast, as discussed in chapter 11.

Stopping an unsafe action or leaving an unsafe state should be possible with a single keystroke that moves the system into a safe state. The design should make fail-safe actions easy and natural, and difficult to avoid, omit, or do wrong. Actions required to maintain safety must also be robust. Multiple physical devices and logical paths should be provided so that a single hardware failure or system error cannot prevent the operator from taking action to maintain a safe system state and avoid hazards.

At the same time, two or more unique operator actions should be required to initiate any potentially hazardous function or sequence of functions. In general, potentially hazardous actions that move the system from a safe to a hazardous state should require fairly complicated procedures so that if any of a sequence of steps is violated, the system will remain in the safe state.

Minimize inadvertent activation of critical functions A second related principle is that hazardous actions should be designed to minimize the potential for inadvertent activation of a function. For example, they should not be initiated by pushing a single button. In one cyclotron facility, designers found that a touch computer screen was being accidentally activated by operators when they leaned against the panel while chatting in the control room of a nuclear facility. A button was subsequently installed that had to be pushed at the same time the screen was touched to activate dangerous commands. Here again is an example of safety conflicting with ease of use. Informing operators of the reasons behind such design features, in this case an interlock, can help reduce complaints or attempts to bypass them.

Another example of this design principle is that initiating a launch sequence should require multiple keystrokes or actions, while stopping a launch should require only one. In general, the design goal should be to preserve the ability to intervene positively, while deliberately making harmful intervention as difficult as possible.

Design safeguards into the operations themselves Safety may be enhanced either (1) by using procedural safeguards where the operator is instructed to take or avoid specific actions, or (2) by designing safeguards into the system design and operations themselves. The second is much more effective. If the potential error involves leaving out a critical action, for example, either the operator can be instructed to always take that action, or it can be made an integral part of the process.

Consider the following: A typical error during maintenance is not to return equipment to the operational mode. The accident sequence at Three Mile Island (TMI) was initiated by such an error. An action that is isolated and has no immediate relation to the *gestalt* of the repair or testing task is easily forgotten. Instead of stressing the need to be careful, which is the usual approach, change the system by integrating the act physically into the task or make detection a physical consequence of the tool design.

In other words, change the system design in order to change human behavior rather than trying to change the human by training or exhortation [316]. An example of a way to make decisions easier is to provide references for making judgments, such as marking meters with safe or unsafe limits.

Separate critical functions In some situations, it is important to separate critical functions, such as *arm* and *fire* commands for weapon operation. One reason is to make the system more controllable, as described in chapter 11. If the operator must be able to halt a sequence when new information is received, conditions change, or monitoring reveals that an error has been made, then the operational steps need to be separated and made incremental.

Follow human stereotypes Because humans often revert to stereotype, stereotypes should be followed in design. This criterion includes making computer displays look similar to the analog displays they are replacing. Keeping designs simple, natural, and similar to what has been done before (that is, not making gratuitous design changes) is a good way to avoid errors when humans are working under stress, are distracted, or are performing tasks while thinking about something else.

Embed proper sequencing in the design A design feature to enhance safety is to place controls in the sequence in which they are to be used; similarity, proximity, interference, or awkward location of critical controls should be avoided. Where operators have to perform different classes or types of action, sequences should be made as dissimilar as possible.

Make errors obvious or not physically possible One of the most effective design techniques for reducing human error is to design so that the error is not physically possible or so that errors are obvious. For example, valves can be designed so that they cannot be interchanged because the connections are different sizes, or connections can be made asymmetric or male and female so that they can be assembled in only one way. Connection errors can be made obvious by color coding. Equivalent techniques can be used in the design of the operator–computer interface.

Use interlocks to preclude actions that cause hazards Physical interlocks can be used to preclude human error. For example, if the operator is not to proceed with startup until a particular valve has been opened, an interlock can be installed to prevent this sequence. Most cars, for example, require that the gear be in park before ignition is allowed. Another example is an interlock that prevents the operation of switch B when switch A is set to the *on* position.

12.4.2 Designing Feedback for Safety

The goal of feedback is to assist operators in maintaining accurate mental models. The system must be designed to provide both the appropriate type and amount of information. The information that needs to be provided will depend on the nature of the task.

Hazard analysis tools should have the ability to provide the designers with the information they need to design appropriate feedback. The scenarios identified that can lead to hazardous states theoretically should provide that information, but they need to include the human operators in the scenarios. Most of the traditional hazard analysis methods based on the chain-of-failure events model cannot do this.

12.4.2.1 The role of feedback and independent information Feedback is essential for both humans and automation performing control tasks. Norman discusses the difficulties in keeping pilots "in the loop" in modern automated aircraft:

> *Although the human operators are indeed no longer in the loop, the culprit is not automation, it is the lack of continual feedback and interaction. . . . The informal chatter that normally accompanies an experienced, socialized crew tends to keep everyone informed of the complete state of the system, allowing for the early detection of anomalies. Hutchins has shown how this continual verbal interaction in a system with highly socialized crews serves to keep everyone attentive and informed, helps the continual training of new members of the crew, and serves as natural monitors for error.* [286, p. 140]

System design must make up for the changes that are occurring in the role played by such interaction. In a position statement of the Airline Pilots Association, Hoagland says that pilots do not object to the monitor role; what concerns them is the inadequacy of the means provided for them to fulfill that role. He contends that there is no difference between the information required to perform a task manually and the information necessary to determine that an automated system is satisfactorily performing that task:

> *The pilot must at all times be able to assess airplane performance. If the airplane is being controlled automatically, he must be able to resume manual control at any time, and he must at all times be able to recover from any situation the automatics may get him into. This means that for flight-critical systems, the pilot requires timely and adequate information for the control of those systems. The fact that a system or process is being controlled automatically does not mean that the pilot's information needs are any different than if being controlled manually. The pilot needs the same quality of information for assessing the output of an automatic system as he does for assessing the output of manual operation. Automation is not a substitute for information.* [147]

More generally, Lucas [238] notes that automated control systems impoverish the cognitive coupling between the operator and the process. There is no longer direct sensory feedback between the operator and the task—the computer intervenes and mediates the interaction. Making up for this deficiency is difficult.

There is an added issue involving the independence of the information provided. If the instruments and computers being monitored provide the only information about the system state, the human monitor is providing little, if any, additional assistance. The operator must have access to independent sources of information in order to monitor performance, except in the case of a few extreme failure modes, such as total inactivity.

In the command-and-control warning system computer errors described in chapter 11, NORAD officers at Cheyenne Mountain were able to determine, through direct contact with the warning sensors, including satellites and radars, that the computer displays indicating that the United States was under nuclear attack were incorrect. The direct contact showed that the sensors were operating and had received no evidence of incoming missiles [342]. This error detection would not have been possible if the humans could only get information about the sensor outputs from the computer—which had incorrect information.

The system needs to provide feedback of two types:

1. the effect of the operator's actions
2. the state of the system, in order to
   a. update the operator's mental model so that correct actions can be taken when required
   b. provide the information necessary to detect faults in the automated system

Feedback about the effects of actions Without feedback, operators do not know whether their requests or instructions were received and performed properly or whether these instructions were effective in achieving the operators' goals.

Feedback is essential to allow the operators to monitor the effects of their actions, to allow for the detection and correction of errors, and to maintain alertness. The interface should also provide confirmation of valid data entry and the acceptance and processing of operator-entered commands.

Another important use of feedback is for operators to learn about the system and how it will respond to a variety of situations [286]. Care must be taken, however, not to overwhelm operators with a large amount of marginally relevant or irrelevant information [366].

Feedback to update mental models *Situation awareness* is a common term used to describe the large number of problems involving human operators becoming unaware or misinformed about the current state of the system. In general, such problems occur when the mental model of the operator becomes inconsistent with the actual state of controlled process, the automation, other parts of the system, or the environment. Situation awareness problems have increased as system complexity has increased.

Solutions to such problems involve identifying what information is critical for safety and ensuring both that feedback is provided when the state changes and that the operator receives this feedback and processes it correctly. For example, an avionics system during an automatic approach may display an enlarged runway while obliterating the rest of the scene. In case something goes wrong, the image is dislocated and the pilot takes over manual control. The assumption here is that the pilot will take over having already analyzed the situation [200].

Another example of updating user models is signaling bystanders when a machine is powered up so that they will know that it has power and can move. A continuous signal (such as a light) is easily ignored. Light or sound can in general be used to warn people about proximity to a hazardous zone. For some machines, it is possible to slow the operation speed when humans enter a hazardous area so that they know their arrival has been observed [365].

As a rule, an automated control system used to maintain safety should provide real-time indication to operators or other appropriate people that it is functioning, and it should also provide information about its internal state (such as the status of the sensors and actuators), its control actions, and its assumptions about the state of the system. Any processing that requires several seconds should provide a status indicator so the operator can distinguish processing from failure [54]. In a system where rapid response by operators is necessary, timing requirements must be placed on the feedback information that the operator uses to make decisions.

The status of safety-critical components or variables should be highlighted in some way and presented unambiguously and completely. If an unsafe condition is detected by the automated system, the operator should be told what anomaly was detected, what action was taken, and the current system configuration. Overrides of potentially safety-critical failures or any clearing of the status data should not be permitted until all of the data has been displayed and perhaps not until the operator has acknowledged seeing it.

For example, a system may have a series of faults that can be overridden safely if they occur singly, but multiple faults could result in a hazard. In this case, the operator should be made aware of all safety-critical faults prior to issuing an override command or resetting a status display. After an emergency stop, Askren and Howard suggest that the operator should be required to complete a restart sequence before any machinery is activated; the automation should not simply go to the next operation after activation [26].

When safety interlocks and alarms have been removed or bypassed for tests, maintenance, or other reasons, the computer (or operator) should verify that the interlocks have been restored prior to resumption of normal operation. While the overlocks and alarms are being overridden, their status should be displayed on the operator's or tester's console [54].

Mode confusion is a type of situation awareness problem that has been implicated in accidents where humans are controlling advanced automation. In the simplest case, mode confusion occurs when the human thinks the hardware and software are operating in one mode but they are actually in a different mode. The human then may make critical control decisions based on their wrong conception of the current operating mode. For example, the human thinks the system is in *test* mode when it is really in *live fire* mode. Automation complexity and the proliferation of control modes in new systems are confusing operators.

Mode confusion most often occurs when automation can change its mode without a direct command from the operator. While the mode change may be annunciated to the operator in some way, the operator may not expect a change at that time and miss the annunciation. The crash of an A320 aircraft while landing at the Mulhouse-Habsheim Airport, France, in 1988 and another while landing at Bangalore, India, in 1990 involved pilot confusion about the current mode of the aircraft. Solutions to this problem are beyond the level of detail in this book. The interested reader is referred to *Engineering a Safer World* [217] and other papers and books on this topic.

Feedback to detect faults Besides updating mental models, appropriately presented feedback is also necessary for the operator to monitor for faults and errors. If a human operator must monitor computer decision-making, then the computer must make decisions in a manner and at a rate that the operator can follow [28]. If this constraint is not enforced, the operator will not be able to follow the system's decision sequence to detect disagreements and trace them back through the sequence to determine their origin.

This result has important implications for maintaining operator confidence in the automated system. A loss of confidence may lead to the operator disconnecting the automatic system, perhaps under conditions where that could be hazardous such as during certain critical points in an automatic landing of an airplane. When the operator can observe on the displays that proper corrections are being made by the automated system, he or she is less likely to intervene inappropriately, even in the presence of disturbances that cause large control actions.

Some human factors experts warn against encouraging operators to adopt a management-by-exception strategy, where the operator waits for alarm signals before taking action. This strategy does not allow operators to prevent disturbances by looking for early warnings and trends in the process state. Also, according to experimental studies by Swaanenburg and colleagues, it is not the strategy adopted by operators as their normal supervisory mode [366]. For operators to anticipate undesired events, they need to be continuously informed about changes to the process state to allow monitoring the system progress and dynamic state. A display that provides only an overview and no detailed information about the state may not provide the information necessary for detecting imminent alarm conditions.

The problem of feedback in emergencies is complicated by the fact that disturbances may lead to the failure of sensors, and thus the information available to the operator or to an automated system becomes increasingly unreliable as the disturbance progresses. In the Fukushima nuclear plant shutdown, the designed displays, by which both the operators and the head office were supposed to understand and monitor the operational status of all units, was not available because it had lost its power supply after the tsunami. In addition, most of the main control rooms, which were the only source of information on the plant status, became incapable of providing the plant parameters due to the loss of power.

12.4.2.2 Alarms Alarms play an important role in human operators detecting faults, but alarm design is difficult. The most significant problems lie not in the design of the alarms themselves, but in the overall system alarm design philosophy.

Alarms are used in control systems to alert operators to events or conditions in the system or controlled process, particularly hazardous conditions. The Asiana flight 214 crash of a Boeing 777 at San Francisco Airport in 2013 was, according to the National Transportation Safety Board (NTSB), caused by the flight crew's mismanagement of the airplane's airspeed during final approach. One of the causal factors involved was that the pilots were unaware that the autothrottle was not maintaining the speed set by the pilots. After the accident, Asiana suggested that the aircraft include an audible warning to alert pilots when the throttle changes to a setting in which it is no longer maintaining the set speed. Boeing argued that it was the pilots' responsibility to carefully monitor air speed.

In general, when the automated system detects an anomalous condition or event, the operator should be informed of the current system state, the anomaly that occurred, and any action that might have been taken in response to it [54]. As discussed previously, automatic control can hide failures and errors until they are beyond the operator's ability to recover. This implies that an automated system should monitor for any unusual events or recovery actions and either fail obviously or make any graceful degradation visible to the operator.

Alarms are especially important in systems where operators must respond quickly to low-probability events for which they may not be watching. In complex systems, more devices or conditions may need to be monitored than is practical for humans to do. Automating alarms and warnings allows a large number of variables to be monitored at the same time. However,

> *It has been stated that man is a poor monitor, yet for detecting some situations, man is clearly superior to any automatic monitor. If he does have monitoring difficulty in large transport aircraft, it would appear to arise from the requirement that he monitor*

a large number of systems and perform other duties at the same time. In spite of many laboratory studies showing the parallel processing capabilities of the human, pilots generally perform many of their tasks as single-channel processors, especially when a task is somewhat out of the ordinary. It is not uncommon, for example, to see pilots concentrate on lateral navigation during a difficult intercept maneuver, to the exclusion of airspeed control. [409, p. 1007]

Designing good alarm systems is difficult. The following sections outline a few principles to consider in system design.

Counter alarm fatigue Examples abound of alarm systems that overwhelm the humans responding to them. If many alarms occur that are not critical, humans may simply tune them out. Alarm fatigue is a serious problem.

Lees suggests that the alarm system should have a properly thought-out philosophy that relates:

- the variable alarmed;
- the number, types, and degrees of alarm; and
- the alarm displays and priorities

to factors such as instrument failure and operator confidence, the information load on the operator, the distinction between alarms and statuses, and the action that the operator has to take [206].

The "green board" is a human factors ideal denoting a control panel that produces a signal only when it communicates significant information. This ideal is seldom reached:

At any given time, dozens of enunciator lights may be lit, all of them signaling conditions that are expected and are of no particular consequence to the operation of the plant; these lights can mask the signals from a few indicators that show something is awry. . . . To make a perfect green board is difficult because a condition that is normal at one stage of operation may be a symptom of malfunction at another time. Nevertheless, there is room for much improvement in the signal lights that are designed to alert the operators, but that, like the boy who cried wolf, may be ignored because they so often proclaim there is a problem when none exists. [31, p. 169]

Counter the incredulity response The *incredulity response* is where operators do not believe a major accident is taking place: The operator is more likely to think that a problem with the instruments or alarms has caused a spurious signal. When the operators have been subjected to a substantial number of false alarms, a real one may very well not be believed [230].

To counteract the incredulity response, systems should be designed to keep spurious alarms to a minimum and straightforward checks provided to allow operators to distinguish accidents from faulty instrument performance. In order to issue alarms early enough to avoid drastic countermeasures, the alarm limits must be set close to the desired operating point. This goal is difficult to achieve for some dynamic processes that have fairly wide operating ranges, and it also adds to the problem of spurious alarms. In addition, measurements of variables that indicate the status of the process always contain some statistical and measurement errors.

Provide the information necessary to check alarm validity When response time is not critical, most operators will attempt to check the validity of the alarm [409]. Providing information in a form such that this human validity check can be made quickly and accurately, and not become a source of distraction, increases the probability of the operator acting properly. Validity checks must also be possible on the alarm system itself—for example, quick checks of sensors and indicators such as a simple "press to test" for smoke detectors. Lewis argues that, in some situations, safety actions should be mandated even though the operator may believe that malfunctioning instruments are the cause of the problem [230].

Limit the number of alarms Alarms certainly have an important part to play in alerting an operator to a problem, but too many alarms may have just the opposite effect. The ease with which alarms can be added and the number of parameters that can be checked in automated systems encourages installing them in large numbers in order to produce a feeling of safety.

In too many systems, alarms are not rare. A shift supervisor testified at the TMI hearings that the control room never had less than fifty-two alarms lit [137]. Swaanenburg and colleagues [366] note that having to push an *alarm acknowledge* button one to five times per minute leaves an operator with little time to do anything else. In addition, too many alarms will cause confusion and lack of confidence.

Although the TMI accident is widely believed to have been the result of operator error, the control room design is just as likely a reason for the operator problems, as discussed in previous chapters and in appendix D. Patrick Haggerty, a member of the commission that investigated the accident, concluded: "*What was apparent about the accident was that if a hundred alarms go off in the first 2 minutes and they all look alike, and some of them are important and some are not, then there's something wrong with the control room*" (quoted in [108, p. 52]).

In fact, too many alarms can elicit exactly the wrong response and interfere with operator action to rectify the problems that caused the alarms. This phenomenon is demonstrated by the CHIRP report described in chapter 6 of a pilot who, during an emergency, was so overwhelmed by all the alarms and their verbal assault that he could not talk to his copilot and was tempted to try to eliminate the cacophony first rather than focus on the main problem.

Patterson [299] notes further that a review of the existing situation showed that this pilot's complaints were justified. Some of the aircraft had as many as fifteen auditory warnings, they were not conceived as a set, and there was no internal structure to assist the learning and retention of the warnings. A number of the warnings produced sound levels over 100 dB at the pilot's ear, and virtually all of them came on instantaneously at their full intensity. In addition, if two of the warnings came on simultaneously, they produced a combined sound that made it difficult to identify either of the conditions involved. From his experiments on aural warnings, Patterson [299] concludes that the number of immediate-action warning sounds should not exceed approximately six, and each sound should have a distinct melody and temporal pattern. While many of these alarm problems have been solved for aircraft, they still exist in many other systems. For example, each hospital device manufacturer may build in many alarms that combine with a large number of automated devices on each hospital floor. Nurses in hospitals today may have thousands or even tens of thousands of alarms go off every week with little distinction between their importance.

Distinguish routine alarms from safety-critical alarms Routine alarms should be easily distinguishable from safety-critical alarms. Often, it is important to ensure that the operator cannot clear a safety-critical alert without taking corrective action or performing subsequent actions required to complete an interrupted operation. The Therac-25 human–computer interface allowed operators to proceed with treatment five times after error messages simply by pressing one key on their keyboard. No distinction was made between errors that could be safety critical, such as the one indicated by the *malfunction 54* message that resulted in the accidents, and those that were not.

Prioritize alarms Alarms should be categorized as to which are of the highest priority. The format of the alarm (such as auditory cues or message highlighting) should indicate the degree of urgency. In addition, alarms with more than one mode, or more than one condition that can trigger the alarm for the mode, must clearly indicate which condition is responsible for the alarm display [409]. Again, with the Therac-25, one message—*malfunction 54*—meant that the dosage given was either too low or too high, without providing information to the operator about which had occurred.

Provide as much timing and sequencing information as possible In general, determining the cause of an alarm may be difficult. In complex, tightly coupled plants, the point where the alarm is first triggered may be far away from where the fault actually occurred. Proper decision-making often requires knowledge about the timing and sequencing of events. However, because of system complexity and built-in time delays due to sampling intervals, information about conditions and events is not always timely or even presented in the sequence in which the events actually occurred.

Complex systems are often designed such that monitored variables are sampled at a frequency appropriate for their expected response to events and state changes. Some variables may be sampled every few seconds; for others the intervals may be measured in minutes. Therefore, changes of state variables are not necessarily recorded at the time or in the sequence they occur. In addition, changes that are negated within the sampling period may not be recorded at all. Thus, events become separated from their circumstances, both in sequence and time [52].

At Three Mile Island, more than a hundred different alarm lights were lit on the control board, each signaling a different malfunction but providing little information about sequencing and timing. Brookes [52] claims that it is common for operators to suppress alarms in order to destroy old information whenever they need real-time alarm information for current decisions. So many alarms occurred at TMI that the computer printouts were running hours behind the events and at one point jammed, losing valuable information.

Although the problem of sampling intervals is not entirely solvable (except by not using sampling), as much temporal information as possible about the events and state changes occurring in the system should be provided to operators.

The proliferation of alarms and the problems of false alarms have led to the development of sophisticated disturbance and alarm analysis systems. The problem is most serious in nuclear power plants, and alarm analysis was pioneered by the nuclear energy industry [206]. At the Wylfa Nuclear Power Station, for example, there are two reactors, each with approximately 6,000 fuel channels, 2,700 mixed analog inputs, and 1,900 contacts. Alarm analysis essentially involves identifying and displaying a *prime cause* alarm along with associated *effect* alarms. Determining such a relationship between alarms is difficult: basically a team of engineers studies the various situations that can occur in the plant and the alarms to which these give rise. Incompleteness in this analysis can lead to serious problems.

Perform an alarm analysis and consider the alternatives to alarms While it is easy to suggest that an overall alarm philosophy and design be created, this process is not straightforward to perform. Rasmussen suggests that it may be extremely difficult, if not impossible, for the designers of a plant to conduct an alarm analysis that considers not only all possible failures in the plant and in the instrumentation but also all combinations of failures. If the analysis is to be useful and not mislead the operator, thus causing even more problems than it attempts to solve, it must be comprehensive and include low-likelihood hazardous conditions.

Weiner and Curry [409] ask whether the necessarily complex logic in automated alarm analysis is too complex for operators to perform validity checks and thus leads to overreliance on the system. They also worry that the priorities might not always be appropriate and that operators might not recognize this fact.

Alarm analysis, furthermore, does not solve the potential problem of operators relying on alerting and warning systems as primary rather than back-up devices. After studying thousands of near accidents reported voluntarily by aircraft crews and ground support personnel, one US government report recommended that the altitude alert system (an aural signal) be disabled for all but a few long-distance flights. Investigators found that this signal had caused decreased altitude awareness in the flight crew, resulting in more frequent overshoots—instead of leveling off at 10,000 ft, for example, the craft continues to climb or descend. A study of such overshoots noted that they rarely occur in bad weather, when the crew is most attentive [303, 408].

Finally, Norman contends that alarms have been overused. Instead, he suggests

> *What is needed is continual feedback about the state of the system, in a normal natural way, much in the manner that human participants in a joint problem-solving activity will discuss the issues among themselves. This means designing systems that are informative, yet non-intrusive, so that interactions are done normally and continually, where the amount and form of feedback adapts to the interactive style of the participants and the nature of the problem. We don't know how to do this with automatic devices: Current attempts tend to irritate as much as inform, either failing to present enough information or presenting so much that it becomes an irritant.* [286, p. 143]

12.4.3 Identifying and Designing the Activities and Functions Provided by Humans

Beyond designing the controls and feedback, a critical aspect of system design is determining what activities and functions will be provided by the human operator and the design of the human tasks to reduce errors in performing them.

Related to this design problem is the need to carefully tailor the feedback to provide an appropriate type and amount of information for the specific tasks. This process, in turn, depends on the nature of the task being performed. Tasks impose different psychological demands and require different skills.

Task design for safety-critical functions needs to take account of human characteristics and variability in the design rather than trying to change humans—perhaps losing many of the advantages of humans in the process—or eliminating them.

If designers want to take advantage of human problem-solving ability and reduce the number of human actions that cause accidents, then they need to match the task to the human instead of the other way around. Designers should not assume that the systems they build will be operated by perfect humans who never take shortcuts or break the rules [184]. Instead, they should assume that people will behave as they have in the past, and the designs should be able to withstand, without serious accidents, the sort of behavior that experience and psychology show will occur. In other words, systems should be designed so that they do not need perfect decisions or perfect decision makers.

One simple observation here that was mentioned in chapter 6 is that when computers are used to automate everything that can be automated, the human can be left with a miscellaneous set of tasks that then become more difficult to execute without error. This approach can degrade the work environment for human operators while depriving the system of the benefits of human flexibility, creativity, and discretion [51; 248].

In fact, we now have the tools, including computers and improved human-error models, to create better environments for humans to work in than the manual environments we are replacing. The power of computers provides more design flexibility than has been available previously to tailor our systems to human requirements instead of vice versa. As Green says about aircraft, "*This means that the onus has moved from training the pilot to cope with what is practically achievable to designing a system that matches the human's capabilities*" [121].

Three basic goals in designing to match human characteristics are maintaining alertness, designing for error tolerance, and allocating tasks appropriately.

12.4.3.1 Combatting lack of alertness In routine tasks, alertness tends to degenerate. "*There is not enough 'motivational capital' to go around, to cover the multitude of boring, repetitious tasks on the diligent accomplishment of which all monitoring—and hence safety-engineering—depends*" [129]. Ravetz argues that automation is not the solution: it can only reduce the quantity of tasks, not their quality. Also, it does not eliminate the perhaps even more boring task of monitoring the automation.

As risk awareness increases, accident potential decreases. DeVille [80], a safety specialist for the US Air Force, argues that awareness of the elevated risks associated with a particular operation or procedure provides a better measure of accident potential than does an evaluation of the risk level itself. In a mature system that has been refined through experience, he finds that a distraction or unplanned event has little effect when a person is operating at a high level of awareness. Normally, this level of awareness is typical of the more difficult operations—those that require planning and preparation and usually have high visibility with top management. In his experience, these are performed extremely well and, in the vast majority of cases, are accident free.

When performing the normal or routine elements of a job with standard levels of visibility and interest, the identical unplanned event that causes loss of the same level of awareness leads to a high accident potential. Complacency, according to DeVille [80], moves us rapidly into the danger zone. This happens when people simply do not believe it could or will ever happen to them.

One of the ways to combat complacency and maintain adequate awareness and interest is to provide challenges in the routine phases of a job. A challenge can maintain operator involvement and resistance to distraction. Such challenges may be in the form of new tasks to relieve periods of passive monitoring. These tasks might be aimed at creating and improving skills and qualifications and maintaining mental models of the system and might be found in mechanical and electronic maintenance, in production planning and management, in quality control, and in system development [289]. The duties should be meaningful—not "make work"—and should be directed to the primary task itself [409].

Another way to keep operators challenged is to allow them latitude in deciding how they will accomplish tasks. Most tasks can be accomplished by different strategies with identical efficiency. Human–automation interaction can be designed to enforce one best way to do a job, or it can broaden and enhance the range of options offered to each individual [248].

By leaving the choice of problem-solving strategy up to the individual, monotony and error proneness can be reduced. This approach often has the additional advantage, in terms of safety, of introducing enough flexibility that operators can improvise when a problem cannot be solved by the limited set of behaviors allowed. Many accidents have been avoided when operators jury-rigged devices or procedures to cope with unexpected events and failures.

In chapter 6, an experiment was described that studied the effects of sleep deprivation. Results showed that alertness and performance held up remarkably well on a simulator where the tasks were stimulating, varied, provided good feedback, and required the active involvement of the subjects in most operations. In this experiment, relatively few tasks involved either passive or repetitive action; the designers deliberately resisted the pressures to automate, and thereby remove human error, except to help overcome operator load [16].

12.4.3.2 Designing for error tolerance As argued in chapter 6, what is called human "error" is often a necessary condition for successful problem-solving and decision-making in the control of complex systems, in which adaptation to unfamiliar situations is crucial and cannot be avoided. As Reason has said: "Systematic error and correct performance are two sides of the same coin" [327].

Not only do humans use experimentation to solve problems and learn about the system in order to maintain a correct mental model of its current status, but they also try to get rid of routine decision-making and choice by establishing rules for behavior; errors are an inevitable result of the experimentation necessary to establish and update these rules.

Under manual control, human operators often obtain enough feedback about the results of their actions within a few seconds to correct their own errors [28]. Only when the results of human action are irreversible and therefore have to be reported, result in an accident or incident, or are nonobservable and thus are not corrected in time do we say that an error has occurred.

By this argument, which has been most eloquently presented by Rasmussen, the aim of the designer should be to build *error-tolerant* systems, in which errors are observable within an appropriate time limit and can be reversed before unacceptable consequences develop. In other words, design so that operators can recognize that they have made an error and correct it.

The same argument applies to computer errors: the software and the system should be designed such that computer errors are observable by operators and reversible. In essence, this design criterion is related to the problem of tight coupling—making the system design error tolerant is equivalent to making it more loosely coupled.

Two requirements for designing error-tolerant systems are providing feedback about errors and providing the ability to recover. A further question is what to do with extremely hazardous states. How much dependence should be placed on the operator to detect and recover from extreme hazards?

Providing feedback about errors The key to making errors observable is to provide feedback about them. The feedback might be information about the effects of operator actions or may simply be information about the action that the operator took on the chance that it was inadvertent, such as echoing back operator inputs and requiring confirmation. Note that this requirement is merely the same feedback principle that is applied to the design of process control, but here it is applied to human behavior.

Humans might monitor themselves, or computers might monitor humans, or there may be a combination of both. Humans are often capable of monitoring themselves given an appropriate system design where they are provided with the right type of information and feedback. Rasmussen says that "*if the possibility of operators to monitor their own performance were considered explicitly during task design in an ordinary engineering way of thinking, a large fraction of reported cases [of human errors] would not reach the printed page*" [314, p. 165].

Humans can also monitor each other—one person can perform the procedure while the other checks the actions to make sure each step is accomplished correctly. Combined human and machine monitoring may be more effective than unaided monitoring by a human, but implementing such monitoring effectively may be difficult.

One experiment demonstrating this difficulty involved the use of electronic checklists for pilots [295]. Electronic checklists provide a memory of pending, completed, and skipped steps, and they can guard against pilots perceiving the expected value of a display rather than the actual value. Also, a touch-operated checklist can use direct-manipulation techniques to aid the pilot in switching from one procedure to another without losing track of partially completed checklists and without getting lost in a bulky paper procedures manual.

The electronic checklist can provide feedback that guards against four errors associated with paper checklists: (1) forgetting what the current item is and thereby inadvertently skipping an item, (2) skipping items because of interruptions and distractions, (3) intentionally skipping an item and then forgetting to return to it, and (4) stating that an item has been accomplished when it was not.

In the experiment, one version of the automated checklist, called *manual-sensed*, required the crew to acknowledge the completion of each item manually. The other version, *automatic-sensed*, automatically indicated completed items without requiring pilot acknowledgment. The primary difference between the two checklists was whether the electronic system or the human checked the system state first. The hypothesis was that even though the manual-sensed checklist was more time-consuming, human-monitoring behavior would be less affected by the presence of machine monitoring. A paper checklist was used as a control. Small problems or potential pitfalls, which they called *probes*, were introduced to test if the pilots were continuing to monitor the system state manually when the automatic-sensed checklist was also monitoring it.

Three probes were used. The first was detected by all four crews with the paper checklist, by two of the four crews with the manual-sensed checklist, and by none of the crews with the automatic-sensed checklist. The second was detected by three of the four crews with the paper checklist, by one of the crews with the manual-sensed checklist, and by none of the crews with the automatic-sensed checklist. The third was detected by all twelve crews.

The experimenters concluded from the results and from their observation of crew behavior during the experiment that both forms of electronic checklist encouraged flight crews *not* to conduct their own checks—manual checking was largely replaced by machine checking. There was no evidence that the manual-sensed checklist was any more successful in promoting human checking than the automatic-sensed checklist. What appeared to help maintain pilot monitoring was conducting the procedure using two pilots in the usual challenge–response mode.

This experiment, although only one, suggests that machine monitoring of the system state will largely *replace*, not add to, human monitoring. Designers cannot assume that machine monitoring provides true redundancy to human monitoring [295]. It would have been interesting to see if a different behavior resulted in this experiment if the automated checklist software had merely aided the human in preventing typical errors, such as by reminding the human that an item had been skipped and never finished but did not perform the checks itself. Computer performance of the task was not necessary to eliminate the first three of the four manual checklist errors. It seems reasonable that simply providing automated assistance to the manual checklist process could reduce pilot error without the pilots starting to depend on the computer to perform the function itself.

Designers commonly over-automate simply because the computer allows us the possibility of doing so. Introducing a powerful new technology often leads to overuse until enough information is available to determine just how much and what use is optimal.

One related and very much open design issue has to do with providing the right type of information to operators about the boundaries of safe operation. Humans need to learn where the boundaries are, but to do so they must sometimes step over them. Rasmussen suggests:

> *It appears to be essential that actors maintain "contact" with hazards in such a way that they will be familiar with the boundary to loss of control and will learn to recover. In "safe" systems in which the margins between normal operation and loss of control are made as wide as possible, the odds are that the actors will not be able to sense the boundaries and, frequently, the boundaries will then be more abrupt and irreversible. Will anti-locking car brakes increase safety or give more efficient transport together with more abrupt and irreversible boundaries to loss of control? A basic design question is: how can boundaries of acceptable performance be established that will give feedback to a learning mode in a reversible way, i.e., absorb violations in a mode of graceful degradation of the opportunity for recovery?* [319, p. 10]

Allowing for recovery Besides providing feedback so that humans can detect their own errors, error-tolerant designs must allow for recovery from actions. Operators must have adequate flexibility to cope with undesired system behavior and not be constrained by inadequate control options. There must also be enough time for these recovery actions to be taken.

In chapter 11, incremental control was described. By performing critical actions incrementally rather than as a single step, the human controller can observe the controlled process, get feedback about previous steps, and then modify or abort control actions before significant damage is done. For this approach to error tolerance to work, the operator must be provided with compensating actions for incremental actions that have undesired effects, essentially the equivalent of the ubiquitous "undo" button.

Unforeseen and extreme hazards The concept of error tolerance implies that the solution to the human-error problem is not to replace operator functions by computers, but to *increase* the operators' options for monitoring themselves and recovering from their own errors. But what about extremely hazardous situations?

The usual way to cope with the problem of variability in human actions under hazardous conditions is to issue mandatory emergency procedures for predicted scenarios. The fact that accidents result from a confluence of several unusual conditions that cannot be adequately foreseen, however, implies that this approach will not be successful because the predicted scenarios may not fit the actual situation. This observation is exacerbated by the fact that the predicted scenarios are exactly those that the designers try to eliminate or control.

It is difficult to anticipate everything that can go wrong and to predetermine how to respond. But note that this is exactly what has to be done when writing software to handle emergency situations. Dealing with rare and unforeseen hazards requires human problem-solving—automatic recovery eliminates the flexibility of on-the-spot diagnosis. Computer response under these conditions will be less effective than human response because the computer can only blindly follow the predetermined procedures.

A conclusion that humans will likely handle unexpected hazards better than computers, however, rests on the assumption that error tolerance can be effectively designed into the system and that the operators will have adequate time and feedback to cope with the situation. If these conditions cannot be fulfilled or if the consequences of an accident are so great that any possibility of human inability to deal with the hazard adequately is unacceptable, then the system must be made fail-safe. Under these conditions, however, the designs cannot assume that humans will be able to monitor computer operations adequately. If operators do not have adequate time and feedback to cope with the situation, they most likely will not have the time and feedback necessary to monitor the computer's attempts to cope with it.

12.4.3.3 Task allocation Even if processes can be made error tolerant, there are strong incentives to automate in order to enhance system productivity or capacity. Computers can provide enhanced control possibilities, if only because of their speed. Whether this should be done strays, once again, into the trans-scientific realm, but designers should recognize that these are not simply engineering decisions and that they have far-reaching implications and results.

If humans are to be included in the control loop in any capacity, a general principle agreed on by almost everyone is that manual involvement must be maintained at some level in order to update the operator's mental model of the system. The controversy begins when considering how much and what type of involvement that should be. If the operator simply acts as an unintelligent actuator for computer-generated commands, the computer could equally well issue the command directly—and should. The available evidence seems to confirm that humans are unlikely to provide any additional benefit such as monitoring or error detection in these circumstances.

At the other extreme, where humans maintain primary control responsibilities, computers can be used to maintain or increase operator effectiveness by supporting human skills and motivation. Computers can, for example, instruct or advise operators, mitigate errors, provide sophisticated displays, and assist when task loads are high.

Computers have complicated task allocation decisions by introducing the possibility of their taking over much, if not all, of the operators' functions. The question then becomes: How should tasks be allocated between computers and humans, and who should do this allocation?

General design considerations All evidence seems to point to the conclusion that using computers to make decisions or to simplify the operator's decisions is dangerous unless carefully done. Simplifying too much of the operator's task, as discussed in chapter 6, may simply lead to more errors.

Automating solely on the basis of the relative abilities of humans and computers may not result in an optimal mix. Performance on a task may depend not only on current conditions but also on what the operator has just finished doing [272]. In addition, Morris and colleagues point out that aptitudes, cognitive styles, and attitudes differ among individuals. Human performance also varies within individuals over time: it may improve with practice and may degrade when the human becomes tired. Task demands may change over time, and the quality of human performance may reflect changes in the nature or difficulty of the tasks that must be performed concurrently [272]. All of this complicates task allocation strategies.

Some differences between computers and humans, such as perceptual abilities, are universal. Humans readily impart meaning into what is seen, and they are excellent at perceptual organization. Computers, on the other hand, have a great deal of difficulty analyzing scenes, but they excel at figure rotation and template matching. Morris and colleagues suggest that such differences can be capitalized on by using them dynamically for task allocation as the character of a visual display changes over time [272]. Machine learning has greatly increased the perceptual ability of automation. At the same time, it also greatly complicates the role of automation in safety.

When the total workload requires a high level of automation, some tasks may be easily automated while others may require human support. To reduce risk, operator decision-making and input may be required for hazardous operations, such as target selection and launch decisions in a weapon system. Other tasks might be designed for varying levels of human–computer interaction, depending on the current workload.

Multiple tasks may interact and substantially affect human performance. Tasks can be complementary in that they provide important information about each other, and performing one makes it easier to perform the other. Alternatively, they may be mutually incompatible in that responsibility for all of them degrades performance on each [336]. Obviously, the goal is to allocate a set of complementary tasks to the operators that will adequately update their mental models, keep them alert and aware, and optimize system safety. This obvious goal is very difficult to achieve, however. The literature on performing task allocation is vast and inconclusive.

Failure detection When humans are assigned monitoring and failure or error-detection tasks, then the relationship of human participation in the control activities and human failure-detection performance is important from a safety standpoint. The difference here from error detection as described in design for error tolerance is that the errors and failures here are in the automation, physical process, or environment. Design for error tolerance is used to deal with errors by the human controllers themselves.

As with task allocation, results from a relatively large number of experimental studies are conflicting [336]: some researchers have found that failure and error detection is enhanced if the human continuously controls the system, while others have found the exact opposite.

Different explanations have been advanced for these contradictory results. Ephrath and Young [97] conclude that total workload is the key. In tasks involving low workload, they hypothesize that failure detection is better when the operator is kept in the control loop. On the other hand, if the workload is high, failure detection is decreased.

Rouse suggests that performing control tasks while monitoring for failures is beneficial if it provides cues that directly help to detect failures and if the workload is low enough to allow the human to utilize these cues. Otherwise, the control tasks simply increase workload and decrease the amount of attention that can be devoted to failure detection [336].

Others believe that the important issue is not just the workload but the *type* of workload. Johannsen and Rouse found that pilots reported less depth of planning under autopilot mode in abnormal environmental conditions, presumably because the autopilot was dealing with the abnormal conditions [164]. In contrast, pilots reported more planning under emergency conditions where the autopilot frees the pilots from online control so they can think about other things.

Sugiyama and colleagues conclude from their studies that the important variable is intermittency of visual information—in other words, whether the operator is monitoring a single instrument or many instruments. They found that a human controller needed longer detection time than a human monitor when the intermittency rate was high [359].

The jury is obviously still out here, as it is for so many aspects of the role of humans in complex system designs. Even if, for various reasons, humans must play only the role of supervisor or monitor, studies have shown that a major benefit can be obtained from allowing them first to interact manually with the system for a while [413]. In all cases, human operators need sufficient practice in controlling the system, preferably not just in a simulator but under normal conditions, to maintain skills and self-confidence if they are expected to take over under abnormal conditions.

Making allocation decisions Besides the question of what tasks the human and computer should perform, there is also the question of who does the allocation of tasks and when. Several approaches are possible. In one, the human is in charge and requests help when desired; the computer is assigned only those tasks that the operator chooses not to perform. One drawback here is the extra operator workload in terms of making allocation decisions and issuing commands to the computer. Alternatively, the human might still make the final decision about task allocation, but the computer could make suggestions.

A second approach is to put the computer in charge of task allocation, perhaps giving the human operator an override or input into the process. Various schemes can then be used by the computer to allocate tasks. In one dynamic allocation scheme, a particular task is allocated according to whether the human or computer has the most resources available to perform it. Rouse and Chu experimented with a scheme whereby the computer only performed tasks when the pilot's workload was excessive. They report substantial improvements in system performance as well as high pilot opinion ratings [336].

Unless one of these default options for task allocator is used, Enstrom and Rouse suggest that the human can become confused about who is supposed to be performing a task at a particular time. Avoiding this problem when the computer is in charge requires displaying the results of the computer's inferences in terms of the allocation of tasks at any particular instant while at the same time not overloading the human with status information; this goal appears to be difficult to achieve [336].

Emergency shutdown A final question regarding task allocation arises as to whether a human or an automated system should take over in an emergency when shutdown or stabilization of a process is necessary. This issue was discussed briefly in the previous section for the case where the human operator is controlling the process. Bainbridge says that when the operator is not involved in online control, he or she may not be capable of taking over control in an emergency and that automatic control is appropriate when shutdown is simple and inexpensive [28].

Manual shutdown is only feasible and safe when there is time for the operator to work out what to do. Overlearned responses, perhaps acquired through frequent practice on a simulator, will work for some systems requiring a medium response time. When a large number of separate actions must be made in a limited amount of time, some might be made by the automatic system, while the remainder are made by a highly practiced operator.

Finally, for systems where the operator has very little time to act, then completely automatic response may be necessary. If reliable automatic response is not feasible for these systems, then Bainbridge says the process should not be built if the costs of failure are unacceptable [28].

12.4.4 Design of Displays for Safety

The display has to be designed to provide the type and amount of information to support the human operator's tasks. The appropriate design, of course, will depend on the nature of the task being performed. Tasks impose different psychological demands and require different skills in interpretation and retrieval of information.

The system hazard analysis can support the designers by providing information about hazardous scenarios. For example, one problem with computer displays is called the *keyhole effect*. Computer displays are limited in the amount of information that can be presented on a single screen. Therefore, related information may be on multiple screens, and users need to click through many screens to find the information they want or need. Figure 12.3 shows an example from a typical control room after several alarms are activated.

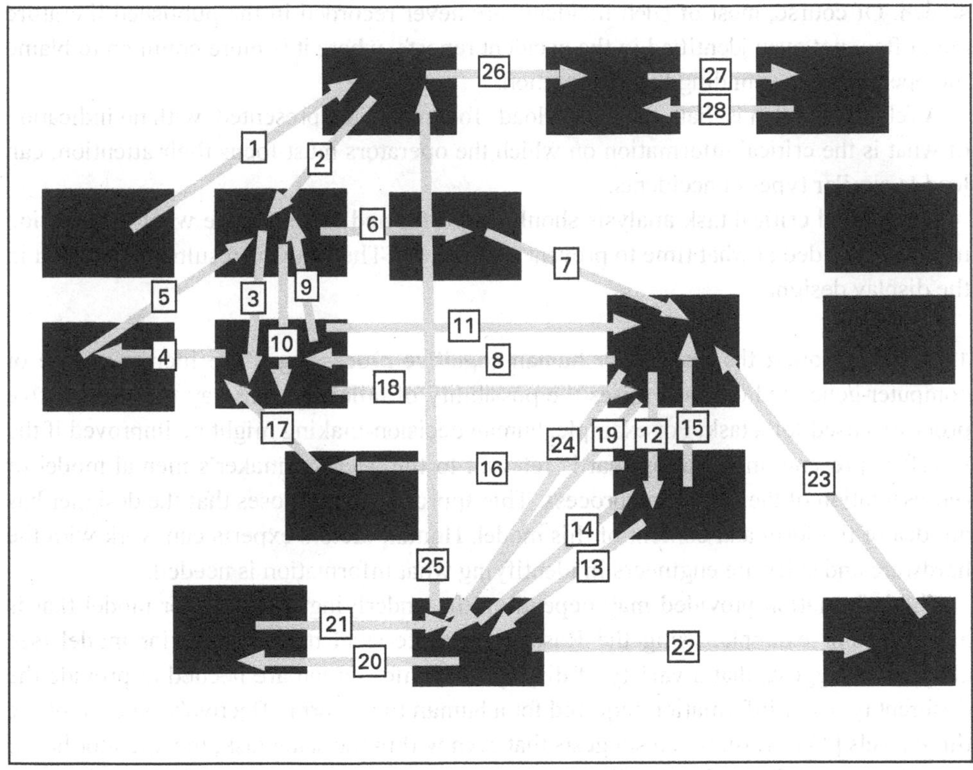

Figure 12.3
A typical screen navigation sequence for a control room operator when several alarms are activated. The operator must spend several minutes navigating among the screens to determine the existing situation in the plant after the alarms sound. (Adapted from User Centered Design Services, Inc. "Control Room Operators and the Keyhole Effect," July 13, 2016. https://mycontrolroom.com/control-room-operators-key-hole-effect/.)

Accidents are occurring because of this problem. In one petroleum refinery accident, for example, described in chapter 8, the operators were unaware that a tank was filling when all the clues in the control room, including the primarily control room display screen, seemed to show that flow of inputs had stopped. The operator would have had to go to a special page to identify the problem, but there was no reason to believe the tank was still filling, and, at the same time, a potentially serious alarm about a different part of the plant activated and the operator switched attention to deal with the alarm. The operator, of course, was blamed for the accident.

As another example, a heart transplant patient was not given the required immunosuppressant medication before his transplant operation. One reason the attending surgeon did not catch this omission was that the Electronic Health Record (EHR) design does not provide clear feedback regarding the status of ordered medications. Determining if the orders are completed required leaving the EHR order screen and moving to an entirely separate screen. There was no reason for the surgical team or the nursing staff in the operating room to suspect that the ordered medication had not been given by the nursing staff in the separate Cardiac Care Unit and thus no reason to navigate to the separate screen [227]

Johannsen and colleagues [165] also have reported rumors of critical incidents that occurred because operators missed important information that did not happen to be on the screen. Of course, most of such incidents are never recorded in the published literature and, often, not even identified in the accident reports, where it is more common to blame the operator for not finding the information.

A related problem is that of data overload. Too much data presented, with no indication of what is the critical information on which the operators must focus their attention, can lead to similar types of accidents.

Hazard and critical task analysis should be performed to determine what information must be provided at what time to prevent an accident. The analysis results can be used in the display design.

12.4.4.1 Tailoring the display for human cognitive processing The increasing use of computer-generated displays raises the possibility of tailoring a display to the cognitive processes used for a task. For example, human decision-making might be improved if the interface presents information that is relevant to the decision-maker's mental model or representation of the controlled process. This approach presupposes that the designer has an idea of the form and content of this model. Human factors experts can work with the hardware and software engineers in identifying what information is needed.

The information provided may depend on the underlying human error model that is assumed. For example, using the Rasmussen three-level human behavior model (see chapter 6) suggests that a variety of displays and information are needed to provide the different types of information required for a human to perform effectively on each of the three levels [336]. Rasmussen suggests that even within the same task, the operator has to move between different levels of behavior and process different types of information; that is, signals, signs, or symbols.

Besides the differences in underlying models of human behavior, the information to be displayed at any time may depend on the level of abstraction at which the operator is thinking about the process [313]. Goodstein discusses how displays may be designed to be compatible with different levels of operator skill using Rasmussen's three levels of behavior [117]. A variety of displays may be needed that provide different types of information to support these various levels.

Bainbridge [28] urges caution, however, in tailoring displays to what is assumed to be the current mode of human cognitive processing:

- An interface may be designed that is ideal for normal conditions but hides the development of abnormal ones—for example, displaying only the data relevant to a particular mode of operation such as start-up, routine operations, or maintenance.
- The use of different skills is partly a function of the operator's experience. With new technology, there is a tendency to reduce the redundancy of information available, but this reduction is likely to cause problems for the novice user [298].
- Interaction under time pressure raises problems. The change between behavior levels (such as between knowledge-based and skill-based behavior) not only is a function of practice but also depends on the uncertainty of the environment: the same task elements may be performed using different skills at different times [28]. The operator may be confused rather than helped by a display based solely on overall skill level.

If not under time stress, the operator may be able to request the information or type of display needed, but this request adds to the workload. Under time stress, operators might not be able to choose or use different displays effectively. Rouse has suggested that the computer might identify what type of skill the operator is using and change the displays [336]. This identification seems difficult and dangerous, however, because it requires information about the current behavior of the operator, which is not easy or perhaps even impossible for the computer or the computer programmer to ascertain.

- Operators might be confused by display changes that are not under their control. They may need time to accommodate to shifts between different display modes, just as they need time to shift between activity modes—such as monitoring and control—even when these are under their control [97]. At the least, a great deal of care is needed to make sure that the different displays are compatible.

12.4.4.2 Ease of interpretation Although it seems intuitively correct that information should be provided to the operator in a form that can be quickly and easily interpreted, this assumption surprisingly may not always be true. If rapid reactions are needed, an easily interpreted display is best; however, some psychological research shows that cognitive processing for meaning leads to better information retention. A display that requires little thought and work on the part of the operator may not support acquisition of the knowledge and thinking skills needed in abnormal conditions [317].

12.4.4.3 Preparing for failure Once operators have learned to work with computer displays, serious problems can arise when these displays fail or are not available during abnormal conditions and emergencies. Humans become dependent on automated systems, although training and practice for equipment failure can reduce this problem. This dependence has two implications for the interface designer.

First, alternative sources of information need to be provided in case the computer-based system fails. For example, manual stations might be provided that allow the operator to manipulate control valves in situations such as the failure of the automatic controls in process plants [206]. It has even been suggested that direct wired displays be used for the main process information and computer displays for quantitative detail [28]. In addition, alternative sources of information provide the operator with a way to detect measuring errors. Displays that provide more detailed or extra data about individual process units, for example, can help in this task.

Second, care should be taken that instrumentation meant to help the operator deal with a malfunction is not disabled by the malfunction itself. This is the familiar common-mode failure problem rearing its head again. As an example of this problem, an engine and pylon came off the wing of a DC-10, severing cables that controlled the leading-edge flaps and also four hydraulic lines, which disabled several warning signals, including a flap mismatch signal and a stall warning light [303]. If the crew had known the slats were retracted and had been warned of a potential stall, they might have been able to save the plane.

12.4.4.4 Displaying critical information in a way easy for humans to process Most systems built today use computer displays, which greatly expands the possibilities for data presentation to the operators. In advanced helicopter systems today, for example, rows of gauges have been replaced by two or three CRT monitors displaying only cautions and warnings or specifically requested information. Designers have learned that the pilot's display menu should be simple and limited to a couple of pages. Other interfaces may include a voice-recognition control system that handles spoken commands and a helmet-mounted display containing the pictorial and digital information needed to fly the aircraft. The design of these systems is driven by pilot workload.

While computer displays allow (and require) many new ways of displaying information to operators, to optimize safety the displays should reflect what is known about how information is used and what kinds of displays are likely to cause human error leading to hazards. The computer displays mediate the human–automation interaction, and even slight changes in the way information is presented can have dramatic effects on performance.

For example, Lees [206] describes several ways in which humans perform diagnostic tasks.

- Respond only to the first alarm generated: this response utilizes a simple rule-based strategy (in Rasmussen's terminology) where the alarm is associated with a particular fault and the operator responds using a rule of thumb. Although incomplete, this strategy may be successful in a large proportion of cases, especially where a particular fault occurs frequently.
- Apply static or dynamic pattern recognition to the control panel displays: static pattern recognition uses instantaneous observation of the displays and matches the pattern to model patterns or templates for different faults. More complex pattern recognition matches the development of the fault over time with dynamic patterns.
- Use some type of mental decision tree, where particular branches of the tree are taken depending on instrument readings.
- Manipulate the controls and observe the effect on the plant.

The diagnostic strategy used has an effect on both display design and on training. Duncan [86] developed training methods for fault diagnosis that reflect different diagnostic strategies and display design. For example, the conventional control panel assists in the recognition of static patterns, but computer consoles do not. Chart recorders aid the recognition of dynamic patterns and instrument faults [206]. When monitoring complicated processes, good operators will scan the displays in order to anticipate undesirable events as much as possible instead of waiting for alarm signals [165]. Conventional rows of recorders and indicators (with normal values aligned horizontally) are appropriate for this purpose.

In contrast, computer consoles require selection of screen, variable identification and requests, and more detailed mental models [165]. The current trend to eliminate more low-tech displays and put everything on computer screens may need to be reexamined, or perhaps the analog displays should be mimicked on the computer screens.

A great deal has been written about how to design computer displays. Only a few highlights are included here about sequential versus parallel presentation, interpretation of displayed information, and layouts.

Sequential versus parallel presentation There is a limit to how much information operators can absorb at any one time, and the display of too much information can be counterproductive. Richardson [328] concludes that there are advantages to systems in which operators have to call for the particular information they think they need. However, such arguments have to be carefully evaluated. A great deal of information can be absorbed relatively easily when it is presented in the form of patterns. Requiring operators to request information specifically, besides the obvious time-delay problems, has serious negative implications in terms of missing important information and reducing the possibility of detecting important patterns. Herry [143] warns of the danger in replacing traditional control panels in which a range of information is presented in a way that may help operators update their mental models of the state of the plant.

In addition, computer displays typically have to be requested and accessed sequentially by the user, a procedure that makes greater memory demands on the operator, negatively affecting difficult decision-making tasks. With conventional instrumentation, all process information is constantly available to the operator, and an overall view of the process state can be obtained by a glance at the console. Detailed readings are needed only if some deviation from normal conditions is found.

The process overview display on a computer console, where only certain variables are displayed and in a digital format, is not equivalent [289]: the overview is more time-consuming and strenuous to read and remember. To obtain additional information about a limited part of the process, the operator has to select consciously among displays.

In a study of electronic interfaces in the process industry, Swaanenburg and colleagues [366] found that most operators considered a computer display more difficult to work with than conventional parallel interfaces, especially with respect to getting an overview of the process state. In addition, operators felt the computer overview displays were of limited use in keeping them updated on task changes; instead, operators tended to rely to a large extent on group displays for their supervisory tasks. Swaanenburg and colleagues conclude that a group display, showing different process variables in reasonable detail—such as measured value, set point, and valve position—clearly provided the type of data operators preferred.

Operators were able to do their jobs using computer consoles, but they tended to feel less confident using them when they were working under pressure during process disturbances. They stressed the advantage of parallel information presentation on multiple screens under these conditions, instead of on a single screen. Keeping track of the progress of a disturbance is very difficult with sequentially presented information [366]; overview displays provided too low a resolution level to be useful. Bainbridge suggests that operators should not have to page between displays to obtain information about abnormal states in the parts of the process other than the one they are currently thinking about; nor should they have to page between displays that provide the information needed for a single decision process as noted previously in the discussion of the keyhole effect [28].

A positive aspect of computer displays is that they are flexible: the information shown can be adapted to the current task and process conditions [165]. Moreover, dialog between humans and computers can be designed in many different ways. At one extreme, the human can select from menus; at the other extreme, the computer can react to commands initiated by the human.

Between these two extremes are mixed-initiative dialogues in which both the human and computer initiate prompts for each other. For controlling dynamic systems, Rouse suggests that command-driven dialogues are best, with a mixed-initiative alternative for special situations, such as when the human has made a mistake [336].

Interpretation of displayed information The format of information is sometimes changed by computer displays in a way that creates problems for operators. An assumption seems to be common that humans are able to process and react to absolute values. Psychological evidence, however, suggests that human operators are more likely to react to change and will tend to process patterns of events, either visual, conceptual, temporal, or strategic [298].

Increasing the amount of internal processing can increase the amount of time it takes for operators to interpret displays, and it can also lead to interpretation errors. Almost everyone is familiar with the increased difficulty of determining how much time is left before some deadline using the digital displays on watches as opposed to watches with analog displays. At the same time, the digital displays make it easier to evaluate absolute time. Determining which activities are most common, most important in terms of safety, and most error prone can help in deciding on the proper format of displays. Digital displays are not necessarily better than their analog counterparts.

Because reaction to displays can be influenced by expectations and assumptions, designs should reflect normal tendencies and expectations. For example, people expect that on a vertically numbered instrument, the higher-value numbers will be at the top, and they expect values to increase clockwise. These expectations may change with culture—Americans expect that moving a light switch upward will turn on the light, while Europeans expect just the opposite. When under stress and in emergency situations, humans tend to revert to stereotype and react in ways that are consistent with their normal expectations even when recent training has been to the contrary.

Icons with a standard interpretation should be used. Researchers have found that icons often pleased system designers but irritated users [165]. Air traffic controllers, for example, found the arrow icons introduced on new displays for directions useless and preferred numbers. Whenever possible, interface designers should mimic the standard displays with which operators have become familiar instead of trying to be creative or unique.

Norman's concept of *semantic distance*—which he divides into semantic directness and articulatory distance—is a variant of this rule. *Semantic directness* requires matching the level of description required in an interface language to the level at which the person thinks of the task. Users must always do some information processing to span the gulf; *semantic distance* reflects how much of the required processing is provided by the system and how much by the user. The more the user must provide, the greater the distance to be bridged [155].

For example, suppose an operator's intent is to control how fast the water level in a tank rises. The operator issues some controlling action and observes the result. If the output shows only the current value, the operator has to observe the value over time and mentally compare the values at different times to determine the rate of change. Displaying the rate of change directly would reduce mental workload.

Either (1) the designer can design an interface that moves toward the user's mental model, making the semantics of the input and output languages match that of the user; or (2) the user can develop competence by building new mental structures to bridge the gulf. The latter requires that the user automate the response sequence and learn to think in the same language as that required by the system [155]. It seems reasonable to assume that minimizing semantic distance and designing to minimize the amount of learning and change required on the part of the user will reduce the potential for human error.

Norman also defines the concept of *articulatory distance*, which has to do with the relationship between the user's intentions and the meanings of expressions. Interface languages should be designed such that the physical form of the vocabulary items is structurally similar to their meanings. The goal of articulatory directness is to couple the perceived form of action and meaning so naturally that the relationships between intentions and actions and between actions and output seem straightforward and obvious. An example is using a moving graphical display rather than a table of numbers.

Layouts Layout is particularly important for the safety-critical information as it needs to be available and salient when needed. The following suggestions for general layout principles should be tailored to give priority to the critical information as identified by the hazard analysis.

Obviously, operator displays and interactions should be clear, concise, and unambiguous. At Three Mile Island, red lights sometimes indicated a satisfactory state of affairs and at other times indicated that something was wrong. Some control panels use the color red to signify three different states: emergency, warning, and normal [151].

In general, the relative position of controls is a stronger cue than shape, which is stronger than color. Labels are the least perceptive [232], and, when used, they should be brief, bold, simple, and clear. Some labels may need to be repeated. The design should make use of color-coding, highlighting (such as blinking), and other attention-demanding devices for safety-critical information. Bilcliffe [37] comments that some designers seem oversold on uniformity, using row upon row of identical switches, lamps, gauges, and meters, or identical lists on digital displays, which invite human errors.

Displays that are used a relatively large fraction of time should be centrally placed, and information that is often used together should be placed near each other or integrated into a single display [336]. For easy identification, a particular system variable should be allocated to a specific area of the computer screen (spatial coding) using windowing or other presentation devices. Johannsen and colleagues [165] suggest that within the window, different types of information can be displayed about the relevant process variables corresponding to different operational tasks.

Warning displays should be simple and brief. Dramatic warning devices, such as flashing lights and loud sounds, are often used to indicate potential problems, but too many attention-grabbing signals can be distracting and have a negative effect, as discussed earlier. In general, designs should be avoided that place undue stress on the operator and cause fatigue, such as glare, inadequate lighting, vibration, or noise.

Many of the design principles described in this chapter identify problems in computer displays, at least in terms of safety. On the positive side, however, a standard principle of human factors design is that an operator is aided by a control panel that mimics the physical layout of the plant. The flexibility and graphics capabilities of computer displays allow computer displays to achieve this goal much better than standard displays. Examples include artificial horizon and moving map displays in aircraft. Computer displays provide the potential for displaying information in a way that greatly helps the operator diagnose malfunctions and their likely effects.

For example, graphical displays allow the status of valves to be shown within the context of piping diagrams and allow the display of the flows of materials. Plots of variables may be shown, highlighting important relationships. The use of graphics in pilot displays of runways in automated approaches, described earlier in this chapter, is another example. Operators can get a much better picture of the state of the plant or system in this way. But remember that the starting point for design should be consideration of the operator's tasks and problems, and a display should evolve as a solution to these [206]. Otherwise, the display may be a solution looking for a problem. An additional danger is that operators may put more faith in advanced displays and not question their contents in practice, even when errors should be obvious.

12.4.4.5 Feedforward assistance and decision aids A final topic in display design is the use of decision aids. Predictor displays show the operator one or more future states of process parameters, as well as their present state or value, through a fast-time simulation of a mathematical model or through some type of analysis that projects forward the progression of a disturbance if nothing is done about it. Such displays can also provide information about the future effects of a particular control action.

The goal of the predictor display is to provide information to operators rather than to tell them what to do. Once again, there are obvious dangers to this approach: it puts a great deal of reliance for safety on the accuracy of the predictor displays and may lull the operators into complacency about the operation of the system.

Predictions can only deal with predetermined sequences, some of which had to be determined even before the system or plant was built and commissioned. Humans can vary their mental models of the system state through feedback about changing conditions or about errors in the original models. But prediction software cannot do this without some special effort and without providing feedback about their performance to the designers. For these and other reasons, predictor displays may be most useful in training environments.

Automated assistance can also be provided in the form of procedural checklists and guides for operations, especially nonroutine operations such as startup, shutdown, and various types of emergency operations that are often associated with accidents.

In addition, decision aids may (1) diagnose underlying causes of events and their importance to safe continued operation, (2) outline alternative actions available and provide additional information such as how changes in important variables affect these actions, and (3) suggest an optimal solution. Procedural checklists have been found to be critical in assisting pilots to respond quickly in an emergency. In many systems, however, if the steps can be prespecified with exactitude, the appropriate response may be to automate the steps if the operators do not have the time to diagnose and respond appropriately.

Even if some human input is required so that the procedures are not totally automated but require some thought, reliance on computer guidance may still reduce operator vigilance. Sometimes, steps that *can* be automated are not, in order to provide the human supervisor with "some room for interpretation in light of his knowledge about the process or objectives that are not shared by the computer" [350]. The question is whether this interpretation will actually be done.

An assumption often made in the design and use of decision aids is that operators will continue to perform an internal simulation using their own mental models and will compare this result with the computer simulation and suggested control actions. This assumption is not realistic if few discrepancies are found over time between the computer simulation and the operator's mental simulation, and the operator then starts to rely on the decision aid. This problem again relates to the fact that humans do not make good monitors: if there are few failures, they will lose vigilance, but if there are many failures, they will lose confidence.

Again, care must be taken to distinguish between providing help and taking over or so simplifying the operator's job that the risk of human error is increased. Automated advice might best be provided only when requested, and serious consideration should be given to designing in ways that keep humans from becoming overly dependent. Operators need to feel that they are in charge of the overall system. Achieving this goal requires that outputs be expressed in terms of advice or suggestions rather than as commands or strict procedures. "By avoiding outputs that are dictatorial, in practice or even only in spirit, the operator can still be innovative when necessary" [337, p. 282].

Another problem with decision aids is that, when following advice, an operator's reactions are slower and less integrated than when the operator generates the sequence of activity [28]. In addition, the operator is not getting practice in making decisions. Without this practice, humans often lack the confidence or skills to intervene when the automation fails or errors are detected.

Finally, decision aids may increase the load on the operator without providing equivalent help. Ephrath reported a study in which system performance got worse with computer aids because the operator made the decision anyway; checking the computer merely added to the operators' workload [28].

12.5 Training and Maintaining Skills

As the role of the operator changes, training must also change. Training for operators working with automated systems has to be more extensive and deeper; the theory of the control system and the control and design models must be taught if the operator is to be expected to detect anomalies and then diagnose and treat them. When software is involved, operators must also be taught the way the software operates in process terms [398].

There was a belief that as more complex software was introduced into systems, the required skill levels for operators would be reduced and money could be saved on training. That belief has proven to be wrong. In fact, required skill levels go up with automated systems, not down, because the operator must have a much better understanding of the process and perhaps the underlying physics.

Additional training requirements include teaching about safety features and training for emergencies.

12.5.1 Teaching about Safety Features

To enhance safety, operators must understand the safety aspects of the design, including the hazards and what has been done to mitigate them. Information about the hazard analysis can highlight for the operator the potential accidents that might occur and how they are controlled so that the operator will understand and appreciate the potential result of removing or overriding controls, changing prescribed procedures, and inattention to safety-critical features or operations [114].

Reviewing past accidents is crucial in preventing future ones. Kletz wrote a series of books on accidents in the chemical process industry that analyze their causes and how similar accidents can be prevented in the future. Similar documentation should be provided for other industries.

For a review of past accidents to be useful, the accidents cannot just be listed and superficially described: they must be carefully analyzed in depth and organized according to their causal factors and appropriate countermeasures.

Maintainers must also be taught the basic safety features of the design so that they do not accidentally undo them during maintenance and also any hazards related to the equipment they are maintaining. Recording these features in the basic documentation is not enough; they must be translated into operational terms and integrated into training programs for both operators and maintainers.

12.5.2 Training for Emergencies

An in-depth understanding of the design and process will help the operator make correct decisions when unexpected events occur and the operator must devise creative solutions to problems that were not anticipated. Simple recognition training will not suffice to develop skills for dealing with unanticipated faults or for choosing corrective actions [85]: operators will have to be trained for general strategies rather than specific responses.

Expecting the operators of complex systems to react to unfamiliar events only by consulting predetermined operational procedures is unrealistic: standard procedures cannot cover all the possibilities, and operators often have to improvise in emergencies. Bainbridge [28] points out the drawbacks of training operators to follow instructions and then putting them in systems to provide intelligence and problem-solving.

In addition to training in the control system interface commands and in the ways that information is obtained, operators need training in systematically looking for information to test hypotheses on possible causes of any disturbance that might arise [366].

It is not enough merely to train operators to react to a single alert or anomaly. Emergency situations usually involve a sequence of events and multiple interdependent failures—where one failure increases the likelihood that another will occur. Procedures for all anticipated situations must be readily understandable, and operators should be exposed to and drilled on all types of alerts and alert combinations to make sure they know how to deal with them.

Refresher courses and practice are important, not only because humans forget with disuse but also to maintain self-confidence and eliminate the fear of intervening. Sten and colleagues [355], in interviews and questionnaires administered to operators in the petroleum industry, found that even experienced operators had a perceived need for training and strongly believed that training in the handling of critical situations was important.

Training on procedures to deal with emergency situations may need to be repeated frequently. The same applies to basic manual control skills in an automated system where the operator is expected to act as backup. In such systems, diagnosis and intervention by an operator will probably be a rare and stressful event. Under stress, humans tend to limit the alternatives they consider and revert to habitual modes of thought and action. Therefore, emergency procedures must be overlearned and continually practiced. Bainbridge [28] notes the irony that the most successful automated systems may provide the operators with the least experience in dealing with alerts and emergencies during normal work and, thus, may need the greatest investment in operator training.

One of the difficulties of training for emergencies is simulating such emergencies and creating the same motivational and stressful atmosphere. While studying the handling of process disturbances in petroleum production, Sten and colleagues [355] found that the startup period for a plant was a crucial time for acquiring high proficiency in handling process deviations and problems. Operators recruited after the startup period did not get sufficient experience in managing crisis situations through daily work alone.

In some systems, operators can get hands-on practice for a short time during each shift or during an operational period, such as during flight for aircraft. If possible, the system should be designed to allow for this practice. In other systems, hands-on control may not be possible, and the only alternative for the type of in-depth training required is to use simulators. For training in emergency response, simulators are obviously required.

Exercises

1. The concept of integrated product teams (IPTs) was created to assist in the communication between the different types of people working on a project. They were not immediately as successful as people expected. What do you think was the reason, and how might those hindrances be alleviated?
2. Your company wants to introduce automation to reduce human operator errors and optimize production. What are three things you would recommend that they do in implementing this goal?
3. Provide an example of a problem created in human–machine automation that arises in the interface. Now provide an example that arises from system or automation design that cannot be totally eliminated by the interface design.
4. You are designing a system where the operators are required to issue a potentially very dangerous command, such as launching a weapon. Describe some design features you might incorporate to prevent an inadvertent or intentional operator action caused by an operator error, a system design error, or a security flaw.
5. Ways to prevent simple connection errors were created to deal with the miles of wires and cables in a modern aircraft. You probably have had to set up a new computer. What are some ways the computer manufacturers use to assist their customers in setting up their new digital equipment? Order the ways you think of in terms of least to most useful.
6. Think of ways to alert the operator that the computer has "failed" and can no longer provide information that can be trusted, perhaps because a physical system component has failed such as a sensor (e.g., a pitot system that senses air pressure on an aircraft). What are some advantages and disadvantages of each potential solution for this goal?
7. Consider the design of displays using Rasmussen's three levels of expert behavior (see also chapter 6). You are designing the display for a specific type of system with which you are familiar. How might it be designed to support the different levels. What are some potential dangers involved in this approach? How might these dangers be mitigated?

8. What problem is being solved by the little spinning dots or hourglasses used in most computer interfaces? What other solutions have you seen used for this problem?
9. Askren and Howard suggest that, after an emergency stop, the operator should be required to complete a restart sequence before any machinery is activated. What do you think is the reason behind this suggestion; that is, what might happen if the automation simply goes to the operation after activation?
10. When digital watches became available, they were instantly very popular, and it was difficult to find an analog watch. Today, analog watches are popular again. The same is true for digital displays of speed in automobiles, which disappeared after one or two seasons (although now they have returned). Why do you think people preferred one or the other? Why do you think the digital displays were introduced in the first place?
11. Take an interface and interaction design for a safety-critical system, perhaps for the Tesla autopilot or some other highly automated automobile, and evaluate it with respect to the principles described in this chapter.
12. Take one of the examples of hazard analysis techniques you performed for the class and a hazardous scenario that was identified that included the behavior of the human operator. How might that hazard be mitigated in the overall system design, including the design of the human interface and interactions.
13. Describe an example of an error-tolerant system you have interacted with where you are able to detect your errors and correct them before bad consequences occurred, or describe how you would create an error-tolerant design for a specific system. This need not be a safety-critical system but could be a text editor or other software tool. What specific characteristics allowed you to do this?
14. As described in the chapter, some researchers have found that error detection is enhanced if the human is continuously controlling the system while others have found the opposite. Consider the different explanation for these contradictory results. Which one seems intuitively correct to you and why? Answer this question in the context of driving a car with autonomous features, such as a Tesla, where you are expected to monitor and intervene if something goes wrong.
15. Increasing the number of aircraft in the skies at any time is limited by human ability to monitor a fixed number of aircraft at any time. Because computers can obviously monitor and control a much larger number, it has been suggested that air traffic control be automated. Technically this is feasible. Why do you think it has not been implemented on a large scale? (Hint: What difficulties might arise in safely handling failures or errors in the automation?)
16. Why might automated systems require more training for human operators than manual ones?

13
Assurance, Assessment, and Certification

The [FAA] administrator was interviewed for a documentary film on the [Paris DC-10] accident. He was asked how he could still consider the infamous baggage door safe, given the door failure proven in the Paris accident and the precursor accident at Windsor, Ontario. The Administrator replied—and not facetiously either—"Of course it is safe, we certified it."
—C. O. Miller,
A Comparison of Military and Civilian Approaches to Aviation Safety

To this point, the book has focused on the most important part in safety engineering, which is how to develop safer systems. Chapter 14 adds information about how to operate and manage them. But there is still the problem of how to convince yourself and others that the system you have developed has eliminated or controlled the most critical hazards.

This chapter describes the current state of safety assurance, assessment, and certification. Here *safety assurance* is defined as providing confidence that hazards have been eliminated or controlled; safety *assessment* provides some type of measurement to predict how many and what type of accidents will occur; and certification is defined as a formal process by an independent authority to approve the use of a safety-critical system. All three are clearly related.

13.1 Assurance of Safety

Safety is a system (emergent) property. What that means is that one cannot just evaluate each component of the system for its safety and then combine these evaluations to determine whether the system as a whole will be safe. Emergent properties are a function of the interactions among the components and not just of individual component behavior (see chapter 3). As system complexity increases, accidents increasingly involve the interactions among components, not just the behavior of the individual components. So, basing assurance arguments on the properties of the components examined individually is not effective for an emergent property such as safety.

Another way of saying this is that the system hazards may be different than the hazards associated with the components. For example, the hazards associated with a valve may involve such things as sharp edges that could cut a person working with the valve. When the valve is used in a braking system, however, the relevant hazards are different and involve the braking system not stopping the vehicle in the time required and the space available. That is, the hazards involve the behavior of the braking system as a whole, as reflected in the interactions among the components of the braking system and interactions with the components of the larger system in which the braking system lies. Evaluating the valve alone can identify that it might have a dangerously sharp edge, but it cannot determine whether it will effectively play its role in stopping the vehicle.

System safety is related to how the components interact and fit together as a whole. Therefore, assuring safety at any point during or after development requires a top-down process that considers the system design and operation as a whole and not just the individual components. Bottom-up processes or, in other words, assuring that individual components operate without hazardous behavior in isolation and then trying to combine those results, cannot be used to assure that hazards are eliminated or controlled in the system as a whole.

At the same time, systems today may have millions of physical parts and tens of millions of lines of software. Top-down assurance of the completed system is usually not practical or even feasible. Some have suggested that it is possible to make arguments using a *safety case*, which is often based on formal logic and documented using some informal notation involving boxes and arrows. Most of the published examples of such arguments are, in fact, logically flawed and thus do not assure safety. For large and complex systems, any such argument is doomed from the beginning, at the least, because it is not practical.

And even if assurance of system safety in such systems were possible, it is unclear how to proceed if the assurance effort is not successful. Making changes in large, complex systems is incredibly difficult. The types of changes that might be possible, such as adding redundancy or monitoring, are very costly and unlikely to be very effective. And, of course, it is almost always too late to start over with a better concept and design.

There is an alternative to after-the-fact assurance, however, which involves building in safety from the beginning and spreading the assurance argument throughout development. This alternative involves

1. using analysis to identify the system safety functional requirements and constraints during concept development;
2. tracing the safety requirements and constraints from the system level to the component level before the components are designed; and
3. ensuring that the safety requirements are satisfied by the design as the design process proceeds, usually involving more analysis to assist in making the lower-level design decisions.

The assurance process is thus spread throughout the development process. At the end, the argument has been made as the design was created. If the design is evaluated throughout the development process, and, in fact the development process is driven by the evaluation results, there is much less chance that at the end one will be confronted by the necessity of an enormously expensive redesign process. If such *safety-by-construction* is used, then the assurance (certification) argument at the end is trivial and simply requires documentation of what was done. A more detailed discussion of this conclusion follows.

13.1.1 Limitations of Traditional Assurance Activities When Used for Safety

Assurance requires convincing ourselves or someone else that the constructed system will have some property of interest. When the property is safety, the goal is to show that the identified hazards have either been eliminated from the system's potential behavior or controlled to an acceptable degree.

In the most common approach to safety assurance, a system is constructed, and then assurance and certification are based on an analysis that the already constructed system will be acceptable in terms of safety. There is an assumption here that effective efforts will be used to make the system safe while it is being constructed, but often the method to be used to do this is either not specified or, if specified, only the results are evaluated. The process looks like that shown in Figure 13.1.

There may be some unit testing during development and even some limited integration testing, but the majority of the integration testing is done at the end of development. Final testing and certification are done once the development is complete.

One of the problems with an emphasis on assurance of the completed design is that if the system is not actually safe and that fact is determined during the assurance activity, then the only solution is to (1) start again, which is usually impossible; (2) patch the completed design the best one can, including adding redundancy or protection systems, which is usually ineffective and costly; or (3) try to compensate for the unsafe aspects during operations. Everything but the first solution is of limited efficacy, and the first is usually impractical.

The after-design assurance approach to safety appears to rest on the assumption that almost all systems are safe when designed, and there is only a need to provide some extra argument about this assumption to convince management, the public, and perhaps a regulatory authority. If some safety flaws are found during assurance, then it is assumed that they will be easy to fix before the system is used. This belief belies all evidence and the known limitations of engineering.

Let's look in more detail in the following sections at the limitations of the typical assurance activities: (1) testing (including simulation and other execution activities), (2) formal mathematical analysis that does not rely on execution, and (3) informal argumentation.

Testing and simulation The problem with testing is that the complexity of our systems today precludes the possibility of exhaustive testing. This argument can be shown to be true simply by evaluating the number of states that the system can potentially reach, which is enormous. The infeasibility of exhaustive testing has always been true for software but is now becoming true for hardware, particularly systems that contain both hardware and software. This fact leads to a common adage in computer science, often attributed to Edsgar Dijkstra, that *testing can only show the presence of errors, not their absence.*

Even if exhaustive testing were possible, it could not assure safety in the operational system. The most important reason is that testing is done against the specified requirements.

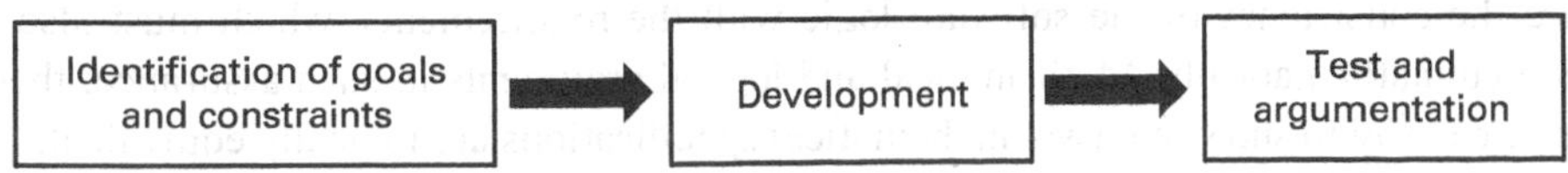

Figure 13.1
Typical development process.

Systems, particularly those containing software, may be able to act in ways that satisfy the specified requirements but are unsafe. Basically, the system may have unintended functions; that is, it may do things beyond what is specified. The only way to prevent this is to somehow ensure that the system satisfies the requirements and does no more than what is specified in the requirements, including unknown behavior that arises because the environment is different than that assumed during development. Demonstration of either of these properties is, for real systems, impossible.

Safety problems almost always stem from system engineering design flaws, including those related to false assumptions about what is the required behavior for the system to be safe and assumptions about the operational environment. Simulation and test equipment can, and probably will, include those same incorrect assumptions such as assumptions about human behavior, unanticipated environmental conditions, or unanticipated uses of the system.

In general, accidents are rare events, usually related to flawed assumptions about system or environmental conditions during development or to changes in the operating system and its environment. Rare events, conditions, interactions, and potential changes that are not anticipated by the designer are equally likely to not be anticipated by the testers.

Moreover, even if safety-critical problems are found during the testing of a complex system, the options are limited. Large projects, particularly those involving complex software, have found that fixing one error often involves introducing new errors. Doing system testing at the end, including testing the human–automation and software–hardware interfaces, is not a cost-effective approach compared to catching errors early, when they can more easily be fixed.

The argument here is not that testing and/or simulation are not useful, only that they are not an adequate way to assure safety and cannot be left for the end of development. Testing and simulation are effective means for achieving other goals and for providing added confidence for a specified set of critical safety requirements. Generating test data or test cases from hazard analysis is an important use for hazard analysis. But testing alone cannot provide the high confidence needed for system safety.

Formal mathematical/logical arguments A second way to create assurance is to use formal mathematical arguments. When the arguments are based on physical principles, such as aerodynamics or fluid dynamics, then strong assurance arguments can be made combining both mathematics and testing, where testing is used to show that the assumptions underlying the mathematics are correct in this particular design. This process, of course, does not preclude accidents when the assumptions about the physical properties are violated or new structures are used for which the physics is not entirely understood.

Starting as far back as sixty years ago, computer scientists have suggested that mathematical arguments can be used to show that software has certain properties. Essentially, the software is treated as a mathematical or logical object. Formal logic is used to demonstrate the consistency of the software logic with the requirements, which must also be specified mathematically. Mathematical and logical arguments in such assurance, therefore, are a way to show that two mathematical specifications are logically equivalent.

Unfortunately, almost all accidents in complex systems stem from unsafe or flawed requirements, so showing consistency between the specified requirements and the software implementation of those requirements does not provide assurance of safety. In addition, such proofs are not applicable to the physical parts of the system or the human components and their interaction with the software, where the hazards commonly lie. Software by itself is an abstraction and cannot cause any physical harm. Only when the software controls something physical does safety come into play. In that case, formal verification of the software alone and not the humans and hardware in concert is not useful for safety.

There are potential benefits from mathematical argumentation about the software properties that can be easily represented in mathematical form. Unfortunately, those properties usually do not include safety, and the formal argumentation techniques do not scale to today's large and complex software.

Informal argumentation A third possibility is to use informal arguments for assurance. Because of the desire to get the system into operation and other cost and political pressures, there is huge pressure for the assurance activity and those informal arguments to be successful, namely, that the argument provides the desired level of confidence. Often assurance processes, because they are late in the process, are cut short when budgets are depleted and delivery dates loom. The same is true for testing or any other activity at the end of development. In addition, if problems are found, going back to the drawing board is just not realistic for most systems nor is abandoning the system. These factors increase the pressure on the informal argumentation. These pressures lead to the appearance of a psychological concept called *heuristic bias* [170] in informal and formal arguments about safety.

Heuristics are simple strategies or mental processes that humans use to quickly form judgments, make decisions, and find solutions to complex problems [171]. Heuristics are important and very useful, but they can also involve biases or systematic errors. One of the most relevant heuristic biases (sometimes called *cognitive biases*) involved in the assurance process is *confirmation bias*.

Confirmation bias is the tendency to interpret new evidence as confirmation of one's existing beliefs or theories. If the goal is to show that something is safe, then people tend to look for data that will support that goal; that is, an argument that the system as designed is safe. We pay more attention to information that satisfies our biases while, at the same time, ignoring information that contradicts the goal. "*Still a man hears what he wants to hear and disregards the rest*" (from "The Boxer" by Simon and Garfunkel).

Experiments have repeatedly shown that people tend to test hypotheses in a one-sided way, by searching for evidence consistent with the hypothesis they hold at a given time [197; 283]. Rather than searching through all the relevant evidence, they ask questions that are phrased so that an affirmative answer supports their hypothesis. A related aspect is the tendency for people to focus on one possibility and ignore alternatives, obviously biasing the conclusions that are reached.

As an example, a person may believe that left-handed people are more creative than right-handed people. Whenever this person encounters a person that is both left-handed and creative, they place greater importance on this "evidence" that supports what they already believe. This person might believe other so-called proof that further backs up this belief while discounting examples that do not support the hypothesis. If our goal is to show that a system is safe, then it is easy to provide evidence or even to prove that it is safe by ignoring evidence to the contrary. Only positive proof is sought.

How can confirmation bias be overcome? One way is to change your goal. For example, look for arguments that the system is *not* safe rather than arguments that it *is* safe. Making an argument for safety is always subject to confirmation bias. Instead, you need to focus on gathering evidence that the system is unsafe. Actively seek out and consider contradictory evidence. This is the standard process used in hazard analysis.

Putting the argument into a graphical format, often containing boxes and arrows, has been suggested as somehow better than simply writing the argument out in natural language [172]. Unfortunately, graphical formats do not eliminate heuristic bias and, in fact, can make it worse because the creators and readers of the argument can be biased by the fact that there is a seeming structure to the argument and therefore assume it must be correct. Structuring informal assurance arguments in some notation made up of boxes and arrows does not make them less problematic; it simply provides false confidence that they are correct and unbiased. In some sense, arguments using informal notations may be *more* dangerous. By hiding the formal logic behind the boxes and arrows, logical fallacies can be difficult to detect.

In addition, such informal arguments about safety are often based on omission: if a hazard cause is not identified, then it is assumed to not exist. This type of argument is similar to the argument that not finding faults during testing means that faults do not exist. But remember, testing can only *find* faults, it cannot show that they do not exist, unless the testing is exhaustive, which is impossible for real systems. The same is true for arguments specified in a graphical notation. It is easy to provide evidence to support any argument if only the evidence that supports it is considered.

What is the alternative to these approaches to safety assurance? The alternative is to assure that the system is safe as it is being constructed. This approach to assurance is similar to what in mathematics is called *proof by construction.* In this approach, hazard analysis is performed during system development to identify scenarios that can lead to hazardous states, and the design is developed using that information (figure 13.2).

Traditional hazard analysis, however, usually comes after a design is essentially complete (see chapters 9 and 10), a process that has some of the same drawbacks as after-the-fact safety assurance. By using more modern hazard analysis techniques, safety can be designed into the system from the earliest stages of a project.

The improved process, which is less resource intensive and more practical, identifies hazards and eliminates them or controls them in the design or operation of the system rather than trying to argue that the system is safe after development (figure 13.2). Essentially, safety is built into the design throughout the system development process, starting from early concept development, and assurance is done in parallel with and integrated into the construction process. Any final assurance process at the end would simply review the documentation of what was done for quality and for compliance with the required standards; the final assurance process then is not an argument that the system is safe, only that the hazards identified by the stakeholders have been properly handled.

That is the best we can do. No complex system is totally safe in all respects and under all conditions. We can, however, create systems that eliminate or control the hazards of most importance to the stakeholders.

Note that testing is used to validate the assumptions underlying the hazard analysis, which is doable if those assumptions are identified during construction of the system.

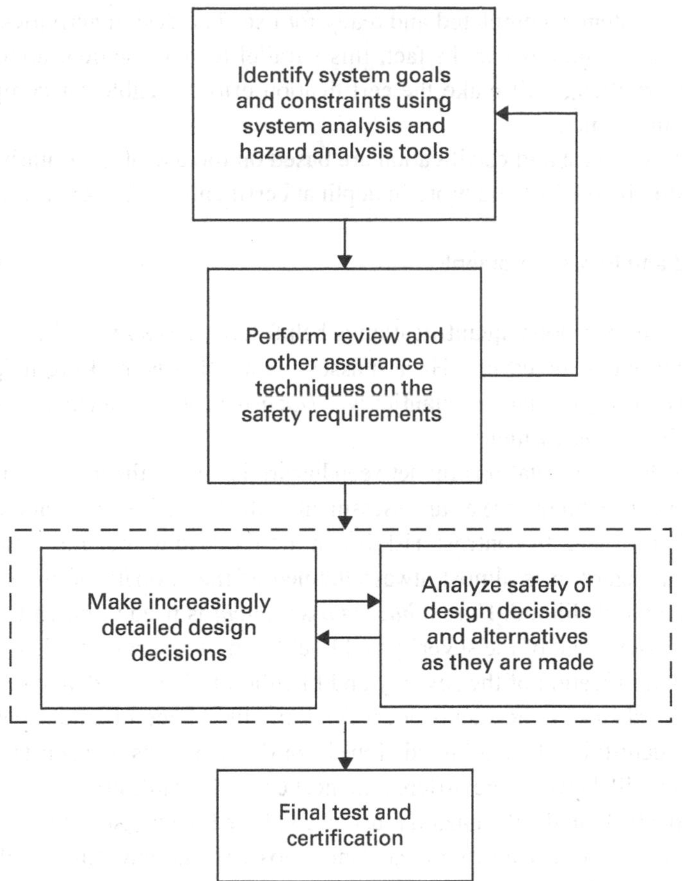

Figure 13.2
An improved process that integrates hazard analysis into the development process.

Thus, testing can play an important role, particularly if the test cases are derived from the hazard analysis, but it cannot be relied on as providing assurance of system safety after the system has been designed and constructed. In fact, unsafe behavior identified during a test (including simulation) should never occur and, if it does, it implies serious flaws in the development process that need to be investigated.

There are several advantages to incorporating safety assurance as the system is developed, including the ability to catch problems earlier and make better design decisions from the beginning of concept and design development. The detailed system and component safety requirements are identified before designs are created, which minimizes the need for backtracking. Problems are caught early when it is still practical to fix them.

It is much too late to try to argue that a system is safe—that is, build a safety case—once the system is completed. Even when trying to increase safety during development, the goal should be to figure out how it will be unsafe and not to talk ourselves into a belief that it is already safe.

But what if you are a government regulatory agency with a responsibility to certify systems for safety? This responsibility does not need to imply that all activities must occur once the system is completed and ready for use. The design activities and analyses can be inspected as they occur. In fact, this parallel to development approach (that is, safety by construction), will make the certification effort feasible for complex systems, which it currently is not.

Sometimes assurance and certification are based on the use of quantitative risk assessment. Therefore, before looking more in depth at certification, let's examine assessment.

13.2 Hazard and Risk Assessment

For some types of decisions, quantification is helpful. *Assessment* can be defined as making a judgment about something. Hazard assessment—that is, making judgments about hazards—may be qualitative or quantitative. *Hazard level* is a term for describing the results of the hazard assessment.

Figure 13.3 shows the relationship between hazard level and the usual definition of risk. The hazard level is defined here as an assessment of the hazard severity and the likelihood of the hazard occurring. In contrast, risk is commonly used to mean an assessment of an accident, not a hazard. It is almost always defined as the severity of an accident or loss combined with its likelihood. Here a *hazard assessment* is used to mean the quantitative or qualitative assessment of the severity and likelihood of a hazard while risk assessment involves the quantification of the severity and likelihood of an accident or loss.

Because hazards lead to accidents, identifying the likelihood of an accident will necessarily involve identifying the likelihood of the hazard. The results of hazard assessment and risk assessment will, however, be different in most cases. Hazards do not necessarily lead to losses or accidents. Usually, the hazard must be combined with a particular state of the system environment for it to lead to an accident or loss. If that environmental state is very unlikely or is prevented in some way, then the hazard level may be high, but the risk may be very low. Risk is also usually affected by the duration of the hazard. The longer the hazard exists, the more likely the environmental conditions that can lead to an accident will occur.

To summarize:

Hazard analysis: identifying hazards and their causal scenarios (causes) at both the system and component level. The analysis is based on an accident causality model that provides assumptions about how and why an accident occurs.

Hazard assessment: making a judgment about hazards; that is, identifying a *hazard level*. Usually, this judgment involves a quantitative or qualitative assessment about the

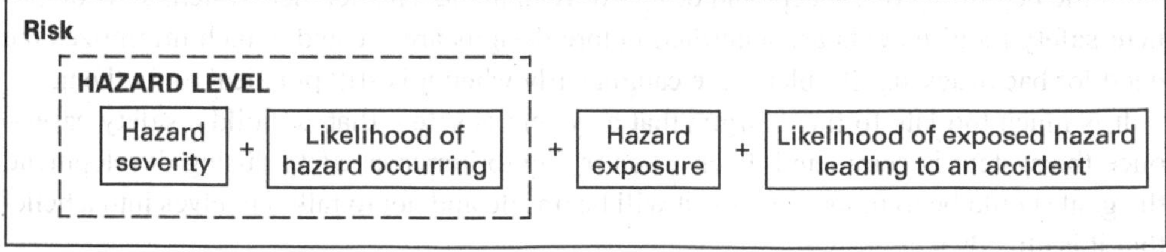

Figure 13.3
The components of risk.

potential severity and likelihood of the hazard but a different type of judgment or assessment is possible.

Risk assessment: making a judgment about a potential accident or loss. Almost always today that judgment involves a quantitative assessment of the severity and likelihood of an accident (versus a hazard), but again, a different type of judgment or assessment is possible.

The major difference here is between analyses and assessments and between qualitative and quantitative assessments. Hazard analysis is the basis for both hazard and risk assessment. Even if quantitative methods are used, qualitative analyses must precede them: hazards and their causal factors must be identified before numerical values can be assigned to them. Thus, the quality of any quantitative analysis depends on how good the qualitative one is.

Another caution to keep in mind: hazards are states, not events, and hazardous system states are not caused simply by system component failures. Therefore, failure analysis is not equivalent to hazard analysis. Failure analysis is primarily used to analyze reliability, while hazard analysis is used to analyze safety.

To obtain the information needed to engineer safety into a system, hazard analysis can usually stop with the qualitative aspects. Knowing the causal factors and using stakeholder identification of relative hazard rankings are adequate for most purposes during the system development phase, when few accurate numerical values are available anyway. But managers often want more, usually, to assist them in decision-making.

13.2.1 Qualitative and Quantitative Hazard and Risk Assessment

A hazard or risk assessment is sometimes used to determine the relative importance and prioritization of the hazards to assist in making decisions about resource allocation early in development. Examples of such decisions are determining which hazards should get the most attention or, indeed, any attention at all and making decisions involving cost and schedule planning.

Assessment may also occur after design and development are complete. The goal in this case is not to guide the development process but to evaluate the final product. The results may be used for internal decision-making or for independent certification to determine the residual risk and whether the system is acceptable for use.

Hazards are usually assessed by combining their individual likelihood and severity. The result is specified in the form of a matrix. Figure 13.4 shows a typical hazard assessment matrix. Many forms of such matrices exist, but most are similar to the example shown in the figure. The categories used are specific to the industry and sometimes the system.

The classic hazard or risk assessment matrix used has two ordinal rating scales: severity and likelihood. Individual hazards are assessed and assigned to a specific box in the matrix. The boxes are usually also associated with the way to deal with hazards in that box, such as the amount of effort to be assigned to eliminating or controlling the hazard or how it should be treated when tradeoffs are required. For boxes in the upper left quadrant, design action may be required to reduce the occurrence of the hazard or respond to its occurrence. For hazards assigned to boxes with low likelihood assessments, no action may be required because an assumption is made that they will not occur. Similarly, if the impact is negligible, no design actions may be required. The accuracy of the hazard

Likelihood		Severity: Catastrophic	Critical	Marginal	Negligible
	Frequent	H1	H2, H7		H13
	Probable	H3, H11		H6, H7	
Likelihood	Occasional	H12	H4, H9		
	Remote			H8, H10	
	Improbable		H5, H9		
	Impossible				

Figure 13.4
An example of a standard hazard level or risk matrix.

assessment results, therefore, is critical in making appropriate decisions during development and operations. Often, colors are used to differentiate between the boxes in the matrix, such as green, amber, and red. Note that likelihood will be different depending on whether the goal is assessing hazard level or risk.

In such a hazard or risk assessment, problems can arise in defining severity and likelihood. While hazard or risk assessment is often thought of as producing quantitative results, in practice it is usually only practical to define it qualitatively; that is, in terms of ordinal rating scales for severity and likelihood. Quantitative values are usually impossible to know, particularly before the system is completed or even designed, depending on when the matrix is created.

Using qualitative scales can give a qualitative scoring that indicates the category or box in which the event falls. A qualitative assessment, although seemingly at least possible, does not allow for sophisticated calculations or subtle differences.

Let's consider severity first.

Severity Severity is usually defined as a set of categories such as:

Catastrophic: may cause death or system loss

Critical: may cause severe injury, severe occupational illness, or major system damage

Marginal: may cause minor injury, minor occupational illness, or minor system damage

Negligible: will not result in injury, occupational illness, or system damage

When determining severity, the specific loss (accident) associated with that hazard is considered. Hazards themselves, which are states of the system, have no associated severity as a state of a system is not usually dangerous itself. Only when the state of the system is combined with a state of the environment can a loss event occur. If the hazard analysis process links hazards to associated losses, the information needed is easily obtained.

Note that these categories are subjective and could potentially be defined in different ways by the stakeholders. For example, what is a severe injury versus a minor one? How much damage or injury is major or minor? Alternatively, or in addition, monetary losses may be associated with the severity categories, although that raises the trans-scientific question of how to determine the monetary value of a human life (see chapter 2).

Other categories might be used. Here are some examples:

A NASA document (NHB 5300.4) lists NASA hazard categories as:

Category 1: loss of life or vehicle (includes loss or injury to the public)

Category 2: loss of mission (includes post-launch abort and launch delay sufficient to cause mission scrub)

Category 3: all others

A Department of Energy standard (DOE 5481.1) for nuclear systems defines three categories of hazard severity:

High: hazards with potential for major onsite or offsite impacts to people or the environment

Moderate: hazards that present considerable potential onsite impacts to people or environment but at most only minor offsite impacts

Low: hazards that present minor onsite and negligible offsite impacts to people or the environment

Severity seems like it would be relatively straightforward to use, but there are problems that arise. The definition of how to define severity itself is one, as discussed earlier. More important is the problem of whether worst-case, credible, or most likely outcomes are considered, or perhaps only predefined common events.

Using the worst case is the most inclusive approach, but concerns may be raised that it is too pessimistic, and everything pretty much will be assigned the highest level of severity. Instead, the worst *credible* outcome might be used. The latter raises the problem of how to define "credible" and can lead to a blurring of the distinction between severity and likelihood, making these two factors not truly independent in the assessment of the hazard level. A third approach is to use the most likely outcome, which again mixes severity and likelihood and reduces their independence.

By using worst-case severity, nearly every hazard can be argued to be potentially catastrophic. The opposite—underestimating severity of hazards—is much more common, however, because there are great pressures for the assessment to be as low as possible.

In many cases, people simply default to assigning severity according to what they think are the most likely outcomes. In aircraft certification, SAE ARP 4761 has an example of a wheel brake failure on landing being assigned "no safety effect" if the brake failure is annunciated to the flight crew (see chapter 10). An assumption is made that if the pilots know about the failure, they will be able to bring the aircraft to a stop safely by, perhaps, steering off the runway or taxiway onto grass. While this is the most likely outcome, it is easy to think of specific situations where the pilots will be unable to prevent an accident even if they know the brakes have failed.

A final possibility, considering only specific predetermined failures or events (called in the nuclear industry a *design basis event*, such as a pipe break or more generally a loss of coolant) can result in the hazard assessment being highly optimistic and often unrealistic due to being too limited in what is considered.

Likelihood Additional problems arise in defining the *likelihood* of a hazard. It is difficult or impossible to determine the likelihood of an event occurring in the future. While likelihood might be defined using historical events, the design of most systems today differs significantly from the same systems in the past: for example, by much more extensive use of software or the use of other new technology and designs. In fact, the usual reason for creating a new system is that existing systems and designs are no longer acceptable.

Note that in hazard assessment, the likelihood of the hazard occurring is used. For risk assessment, the likelihood of the hazard occurring plus the likelihood of it leading to an accident must be determined. Historical data only tells us about the past, but hazard or risk assessment is performed to predict the future. Just because something has not occurred yet does not provide an accurate prediction about the future, particularly when the system or its environment differs from the past.

Even if the design itself does not change in the future, the way the system is used or the environment in which it is used will almost always change over time. The concept of migration toward higher risk over time [320] argues against the applicability of the past as a determinant for the future. And identifying all future changes along with their impacts is essentially impossible.

The example hazard or risk assessment matrix in figure 13.4 categorizes likelihood in terms of frequent, probable, occasional, remote, improbable, and eliminated (or impossible). These categories usually need to be defined more precisely. Here is one common approach used in defense systems:

Frequent: likely to occur frequently

Probable: will occur several times in the system's life

Occasional: likely to occur sometime in the system's life

Remote: unlikely to occur in system's life, but possible

Improbable: extremely unlikely to occur

Impossible: equal to a probability of zero

As the reader can easily see, these definitions are not terribly helpful and simply restate the problem in a different but equally vague form. This same criticism holds for most of the attempts to define qualitative likelihood categories.

Sometimes the qualitative categories are associated with probabilities. An example is the use of probabilistic categories, such as (1) *1.0E-9 and higher*, (2) *between 1.0E-6 to 1.0E-8*, and (3) *1.0E-5 and lower* (see table 13.1 as an example). This probabilistic assignment, however, does not eliminate the question of whether the probabilities can be determined in advance; that is, before extensive operational use of the system.

13.2.2 Limitations of Hazard and Risk Assessment

If the system has not yet been designed or is in the process of being developed and tested, the hazard or risk assessment may be used to determine the amount and type of effort to apply in order to prevent the identified and assessed hazards from occurring. It may also be used to evaluate the effort required with respect to standard design processes mandated by the customer, such as the LOR (Level of Rigor) or the design assurance level to be used in development. There are, of course, serious questions about whether a particular Level of Rigor actually results in measurable differences in safety. This commonly accepted conclusion has never been proven and, at least for software, is most likely incorrect.

Practical limitations Once the categories for severity and likelihood to be used in the matrix are determined, doing the hazard or risk assessment involves assigning the identified hazards to appropriate boxes and thus "assessing" their level. There are many technical and mathematical limitations to assessing anything using such a matrix format, but the more serious limitation is that the hazard assessment is most useful before the system design is complete, but at that time almost nothing is known or knowable about the likelihood of the hazards and, of course, the likelihood of a hazard leading to an accident.

To compensate for this problem, hardware component failures, such as loss of external communication or breaking piston nuts, are often used instead of hazards such as aircraft instability or inadequate separation from terrain. Even a general failure such as "loss of heading information" in an aircraft may or may not be dangerous, depending on how, who, and what is using the heading information and the condition of other parts of the system and environment at the time.

Failures are used because if the system contains standard parts, much is known and documented about their likelihood of failure and failure rates. The obvious problem with this strategy is that reliability level is being assessed rather than hazard level because hazards do not just result from hardware component failures or even from hardware components that have a long history of use. And new components, such as software or new hardware technology, must be omitted from the assessment.

The past is a poor predictor of the future, particularly because the way systems are used and the environment in which they are used will change over time. Accurate prediction about operational behavior is not possible using the hazard level matrix. This common practice of assessing failures rather than hazards makes the resulting assessment dangerously misleading.

Another problem with considering only failures rather than hazards is that *individual* failures are usually considered but not combinations of low-ranked failures. For example, consider the situation in an aircraft where the visual environment degrades at the same time as a loss of altitude information, heading indication, airspeed indication, aircraft health information, or internal communication. Individually, each of these losses may not result in a system hazard or an accident, particularly if it is assumed, as is often the case, that the pilots will react appropriately. When multiple losses occur simultaneously, however, the likelihood of a hazard or of an accident may be significant. Looking at each loss separately in the hazard or risk matrix can lead to a low hazard level or risk assessment due to a low probability of occurrence and low severity level of each of the individual, single-point failures. There is also, of course, usually an assumption of independence of the failures and often a lack of consideration of common failure modes leading to multiple failures at the same time.

It is not surprising that such combinations of failures are not considered given the large number of failures possible in any realistic system—assessing all combinations becomes prohibitively expensive and usually infeasible. However, not considering combinations of failures affects the accuracy of the results, even for a reliability analysis. Major accidents almost always involve multiple failures and other conditions.

Different problems occur when humans are part of the system. One common complication is that assumptions may be made that the flight crew will not only recognize the failure (or hazard) but will also respond appropriately. Ironically, accidents are often blamed on inadequate flight crew or operator behavior while at the same time assuming that they will behave correctly in the hazard or risk assessment.

Clearly, there are many cases where this assumption that the human operator will "save the day" does not hold. The mental model of the system operator plays an important role in accidents. In aircraft, for example, the flight crew must receive, process, and act on numerous sources of feedback about the state of the aircraft in order to interact correctly and safely with the various vehicle and mission systems. Time to perform this decision-making may be very limited. The interaction of control mode displays, pedal and other control position (in a rotorcraft), reference settings for various operating modes, and other visual and proprioceptive feedback can lead to flight crew mode confusion and an accident, particularly when external visual feedback is degraded.

Hazards may only appear improbable if some of the likely factors involved—such as software requirements flaws and complex aspects of human behavior—are not considered. In a military helicopter, Abrecht and his colleagues [5] found many non-failure scenarios that could lead to a hazardous system state but were not considered at all in the official hazard or risk matrix. They later traced these missing scenarios to real losses that had occurred during operational use of the helicopter. They also identified realistic scenarios where the flight crew would not behave appropriately and suggested additional controls to prevent the unsafe behavior as well as important safety requirements for the software. Finally, and perhaps most disturbing, they identified realistic and relatively likely scenarios leading to all the specific hazards that *were* identified but were dismissed as improbable in the official risk assessment.

Similar limitations in the official risk assessment were identified in the software-intensive positioning system for a new naval vessel [4]. Additional risk assessment inadequacies, however, were identified in this system. For example, the likelihood of a loss can vary significantly depending on the external environment in which a hazard occurs. But that factor is not usually considered in the hazard-level matrix. In addition, likelihood and severity may be so entangled—for example, through the external environment—that again they cannot be evaluated along separate and independent dimensions.

The omission of important scenarios and simplifications in the generation of the hazard level matrix will lead to a very inaccurate assessment and perhaps dangerous complacency. At the least, it can lead to misallocation and poor use of resources.

The discussion so far has focused on practicality and the results of actual empirical evaluations. But there are also psychological limitations in the use of the standard hazard assessment matrix.

Psychological limitations Some of the most interesting limitations of hazard or risk assessment stem from what Kahneman and Tversky call heuristic biases [170; 171]. Confirmation bias was described previously in this chapter with respect to assurance, but there are other important biases in risk assessment. Kahneman and Tversky are psychologists who studied how people actually do risk evaluation. It turns out that humans are really terrible at estimating risk, particularly likelihood. Here are a few of the relevant heuristic biases that have been described:

- *Confirmation bias*: People tend to pay more attention to information that supports their views than to evidence that conflicts with them. The result is that people tend to deny uncertainty and vulnerability and overrate estimates that conform to their previous experience or views.
- *Availability heuristic*: People tend to base likelihood judgments of an event on the ease with which instances or occurrences of that or similar events can be brought to mind. While this heuristic may often be a reasonable one to use, it can also lead to systematic bias. For example, psychologists have found that judgments of the risk of various hazards or events will tend to be correlated with how frequently they are mentioned in the news media.
- *Ease of scenario generation*: People will often construct their own simple causal scenarios of how the event could occur, using the difficulty of producing reasons for an event's occurrence as an indicator of the event's likelihood. If no plausible cause or scenario comes to mind easily, an assumption may be made that the event is impossible or highly unlikely.
- *Difficulty in predicting cumulative causes*: People tend to identify simple, dramatic events rather than causes that are chronic or cumulative. Dramatic changes are given a relatively high probability or likelihood whereas a change resulting from a slow shift in social attitudes, for example, is more difficult to imagine and thus is given a lower likelihood.
- *Conjunction fallacy*: An outcome paired with a likely cause is often judged to be more probable than the outcome alone, even though this conclusion violates the laws of probability.
- *Incomplete search for possible causes*: A search is often stopped once one possible cause or explanation for an event has been identified. If that first possible cause is not very compelling, stopping the search at that point leads to nonidentification or underestimation of other more plausible and compelling causes.
- *Defensive avoidance*: Categorizations of risk that conflict with other pressing goals are often rejected or downgraded. Budget and schedule pressures may affect where the hazard is placed in the matrix. The desire for lower categorization of risk can outweigh and suppress objectivity.

One way to overcome these biases is to provide those responsible for creating the matrix with better information about the scenarios that can lead to the loss event, perhaps through use of a structured analysis process that (1) can be used before the design is completed to generate the scenarios, and (2) includes humans, software, and management decision-making in the scenarios and not simply component failures. Such information can be obtained through the use of sophisticated hazard analysis techniques.

If more information is available about potential hazard causal scenarios earlier in development, it may be possible to substitute a different criterion for likelihood in the risk assessment process. Using mitigatibility or controllability rather than likelihood has been proposed to improve the accuracy of hazard assessment [123].

13.2.3 Probabilistic Risk Analysis

Once design and development are complete, the system risk can be evaluated using the actual design. The goal in this case is not to guide the design process but to evaluate the final product. The results may be used internally or for independent certification to decide whether the system is acceptable for use.

Again, risk is generally defined as the combination of severity and likelihood of a loss. It is perhaps unfortunate that risk is defined in terms of one way to compute it because that limits our ability to devise new approaches to assessing and minimizing it. One simple alternative is to define risk as the lack of certainty about an outcome, often the outcome of making a particular choice or taking a particular action. It might also be defined in terms of the potential power of the controls used to prevent losses.

Quantitative probability assessment, if used, is stated in terms of likelihood of occurrence of the hazard or a loss per unit of time, events, population, items, or activity, such as 10^{-7} per year. Note that risk goes beyond just hazard assessment. It starts from hazard level assessment but also should include the exposure of the hazard and the likelihood that the hazard will lead to an accident or loss. Remember, the hazard is defined as only within the system being designed while accidents involve factors outside the boundaries of the system.

Typically, worst-case effects of hazards are all that is needed for hazard assessment, and these are relatively easily determined. Probabilistic risk assessment, however, requires elaborate quantitative analyses. Unfortunately, obtaining probabilistic data about the harmful consequences of some hazards and about human errors is quite difficult and perhaps the most error-prone part of any quantitative risk assessment.

The evaluation of susceptibility to a hazard should consider duration and exposure; that is, how many people are exposed for how long. The warning time, which is the interval between identification of the problem and the occurrence of injury or damage, may also be important.

Probabilistic risk assessments have been performed most frequently on chemical or radiation hazards. The main physical effects of plant or equipment failure arise from escaping gases or liquids catching fire or exploding [150], or from toxicity. To assess the magnitude of an accident, the analyst first needs to determine (1) how much material is likely to escape, (2) what is likely to happen to it over time and distance (the physical consequences), and (3) the effect on people. Chemical engineering models exist to calculate some of these factors for that industry, but the accuracy of the models for some factors, such as gas dispersion, is poor. Considerable simplifying assumptions are needed to make the models manageable because the effects of a release depend on the physical state of the released material, its release rate, natural topography, intervening structure, atmospheric conditions, homogeneity of the gas cloud, ignition sources, and so on. The translation of structural damage into human casualties is problematic. Hope suggests that this translation is so speculative that in practice it can be no more than a statistical assumption, which in any given case may be orders of magnitude wrong [150]. And, of course, the assessment of risk rests on the accuracy of the assessment of the hazard level.

The accuracy of consequence assessment not involving release of material, such as the probability that two planes violating minimum separation standards actually collide, depends on the hazard and system involved. Usually, the number of uncontrolled variables makes prediction very difficult. The probability of human injury, for example, may depend on the time of day or the day of the week, and so on.

At the end of development, risk assessment might be used to make decisions about whether the risk is sufficiently controlled and to make decisions about certification, deployment, and operational use of the system. Quantitative risk assessment of a completed design often is required by certification agencies or is used in public arguments about the safety of a controversial technology such as nuclear power.

Limitations of risk assessment The accuracy and use of probabilistic risk assessment (PRA) are controversial. The arguments in favor are usually based on the ability of the technique to provide input to decision-making, such as decision-making about the certification of plant designs. The major limitations lie in the inaccuracies of the available data, in the consequence models (such as dispersion of heavy gas and explosions) and in the toxicity data for chemical system risk, and the usual lack of any kind of data for other risks.

Perhaps the most important limitations arise when causal scenarios contain events that are not quantifiable. Often these are simply ignored. Many important causal factors—design deficiencies, for example—cannot be easily or reasonably quantified. Because equipment failures are the most easily assessed probabilistically, many probabilistic risk assessments have been criticized for placing more emphasis on these failures than on less easily predicted and quantified factors such as design errors, construction deficiencies, operator actions, maintenance errors, and management deficiencies, which may factor heavily into the actual risk. For example, some probabilistic assessments have emphasized failure probabilities of devices that are in the range of 10^{-7} or 10^{-8} while ignoring installation errors or maintenance errors of those same devices with probabilities in the range of 10^{-2} or 10^{-3} over the same units, such as time.

Quantifying only what can easily be quantified does not provide a realistic estimate of risk. In the space program, where probabilistic risk assessment based on Fault Tree Analysis and Failure Modes and Effects Analysis was used extensively, almost 35 percent of the actual in-flight malfunctions were not identified by the method as credible [242].

Care also has to be taken that quantification does not divert attention away from risk reduction measures. The danger exists that system safety analysts and managers will become so enamored with the statistics that simpler and more meaningful engineering processes are ignored.

The most important question is whether the numbers obtained are credible. Follensbee cites several commercial aircraft accidents caused by events that had been calculated to have a probability of 10^{-9} or less (one was calculated as 10^{-12}) using accepted techniques. In all of these cases, incorrect assumptions about the behavior of the pilots or the equipment led to underestimated failure or risk figures [104]. In several cases, the need for compliance with standard aircraft fail-safe design standards—which might have prevented these accidents—was judged unnecessary based on the calculations.

Some studies have shown widely varying results for parallel assessments performed on the same system by different analysts or analysis teams. The variations in these studies seem to be caused by different initial assumptions, different restrictions of the object under study, the particular failure data used, and different analyses of the hazard consequences [335].

Surprisingly few scientific evaluation studies of PRA have been performed. The most recent was done in 2002 at Riso Laboratory in Denmark [204]. Seven different teams performed a risk assessment for the same chemical facility, an ammonia storage unit. Only storage and loading/unloading were considered. It does not appear that any active control (such as computers) or humans were included, which would have made the assessment more difficult. Even with these limitations in the problem space, differences in assessments of up to four orders of magnitude were found among the teams.

The Union of Concerned Scientists and others have warned against a disproportionate emphasis on meeting predicted quantitative levels of safety rather than on consideration of technical problems and implementation of engineering fixes [9]. The same may be true for human error. Hornick, past president of the Human Factors Society has suggested that "*the general nuclear power community is couching a cavalier attitude towards human factors in the (false?) comfort of risk-assessment statistics*" [151, p. 114].

The accuracy of human error data is also controversial. Chapter 10 briefly described some types of quantitative operator error analysis. In practice, most systems do not provide enough data on human error to be useful for probabilistic modeling. The alternatives are to use laboratory studies or numbers collected over a long time and over many types of systems, but they may not apply to a particular system or to future systems.

The difficulty in extrapolating from laboratory studies stems from the significant differences between the laboratory and industrial settings [312]. Laboratory tasks tend to have (1) a well-defined goal, (2) stable requirements, (3) specific instructions, (4) artificial and low-valued payoffs, and (5) a subject that is controlled by the task. In contrast

> *In "real" tasks only a (sometimes vague) overall performance criterion is given and the detailed goal structure must be inferred by the operator. . . . The task may vary as the demands of the system vary in real time. Operating conditions and the system itself are liable to change. Costs and benefits may have enormous values. There is a hierarchy of performance goals. The operator is usually highly trained, and largely controls the task, being allowed to use what strategies he will. Risk is occurred in ways which can never be simulated in the laboratory.* [312]

The other alternative, using historical data from a large number of systems, has two main drawbacks: (1) the data and tasks from one system may not apply to a different system, and (2) data collection is often biased, incomplete, or inaccurate. Case studies and incidents do not represent all the errors that operators make, merely those that are reported. Unreported errors tend to be those that the operator is able to correct before damage occurs. Monitoring is one way to collect data, but human behavior may be abnormal if the person being monitored is aware of the monitoring.

Human error data also suffers from the difficulty of classifying errors, such as determining whether an error was an operator error or a design error. The variety of classification schemes makes use of data in a different environment unreliable or inapplicable.

Given the relationship between human behavior, system design, and task characteristics, it seems misleading to collect empirical data about human errors without also noting subtle differences in the environment and system design when those errors occurred. There can be wide variation in the environmental situations and physical aspects of the tasks, including stress factors such as noise, temperature, emotional stress, and vibration.

For example, collecting probabilistic data about humans misreading a particular type of dial in poor lighting and then applying the "probability of misreading a dial" to systems with a different dial design and better lighting may be unjustified; noting in the database all conditions under which every error was made is probably impossible; and the number of instances is not large enough for such differences to become unimportant. Errors on particular tasks may also depend on the other tasks the operator is performing at the time. Thus, the context of the task is extremely important, further limiting the situations in which historical data applies.

Besides the problems in collecting it, human numerical error rate data also suffers from various other kinds of uncertainty. Human performance exhibits considerable natural variability based on skill, experience, and personal characteristics. Not only do people differ in their innate capabilities, but the performance of any one individual will vary over time.

Some of these variations are unpredictable, while others seem to be circadian. Performance variations over a 24-hour day arise from fluctuations in the work situation and also from modifications in human capabilities [317]. Historically, safety and productivity are low at night. The fact that we are a diurnal species may explain why many of the major industrial accidents involving human error have occurred at night [103]. Variability in performance may also arise from interactions with an unstable environment, from stress, and from interactions with other workers.

The impossibility of obtaining probability data for unsafe software behavior is so widely accepted that certification in many industries, for example, commercial aircraft, does not allow its use.

Specific methods for assessing human error rates and software and including them in hazard analysis and probabilistic risk assessment have been proposed despite these theoretical and practical problems, but little validation of these methods has been done.

13.3 Certification

Some types of systems are so dangerous or important to the public that they are required to be certified before they can be used. In other instances, there is no legal requirement, but independent groups provide certification for insurance or liability purposes. There are also individual professional certifications to prove competence to engage in an activity, but that usage is not included here. The focus in this section will be on certification of safety-critical systems, not individuals, by the government or some other appropriate authority.

When hazard or risk assessment is used for acceptance analysis (e.g., certification or licensing), more information is usually required. For example, each hazard might need to be documented to show its potential causes, the implemented controls and the tracing of the hazard into the detailed design, and the results of the verification efforts. Basically, this documentation is a description of the potential problem and what has been done about it. The documentation should allow the assessor or reviewer to follow the analyst's reasoning in order to check the correctness of the results [335]. For very critical systems, the informal reasoning may be augmented by formal (and perhaps mathematical) reasoning and argument. The assumptions underlying any formal models or methods used need to be verified but often are not.

Remember, safety is a system property, so certification needs to involve the whole system. It makes little sense to talk about certifying a software or a hardware component being safe or unsafe except when used in a particular system or context. System hazards are different than component hazards as explained throughout this book. Certifying some software as safe and then using that label to argue that it will be safe in a different system is not technically justified. This fact is inconvenient because many people want to certify or assure an individual component's safety level, particularly software, and then apply that level to any system in which it will be used. Many serious losses, however, have occurred when software that was safe in one system was reused in another one [217].

Certification must consider all aspects of the system, including the hardware, software, operators, and environment in which it will be used. Recently, some certification authorities are including management and the safety management system in their evaluations. Potential component failure is only one aspect that must be considered; certification must also consider system design flaws, unsafe interactions among components, human factors arising from the operator or user interacting with the system, and so on.

A final general principle is that certification is not a one-time effort. A system must continue to operate safely throughout its life even when the system itself changes: the users and operators will change their behavior over time, and the environment—including the assumptions about that environment underlying the design and the original hazard analysis—will also change over time.

Any evaluation of certification approaches must consider these principles as well as other aspects, such as those in the legal, political, and moral realm. The first important distinction to consider is between types of certification.

13.3.1 Types of Certification Approaches

Certification methods differ greatly among industries and countries. The approaches commonly used can be broken into two general types (figure 13.5), which determine the type of evidence used in the certification process: (1) prescriptive, or (2) performance-based and goal-setting.

Prescriptive standards In the prescriptive approach, standards or guidelines for product features or development processes are provided that are used to determine whether a system should be certified. Prescriptive standards may be product-based or process-based.

Product-based standards In *product-based* standards, specific design features are required, which may be (a) specific designs, or (b) more general features such as fail-safe design or the use of protection systems. Specific designs and features provide a way to encode and pass on knowledge about past experience and lessons learned from past accidents.

In some industries, practitioners are licensed based on their knowledge of the standards or codes of practice. An example is the existence of electrical codes based on past experience with various designs. Certification then becomes the responsibility of the licensed practitioner, who can lose their license if they do not follow the standards. Licenses may need to be renewed periodically or proof provided that the licensee has kept up with

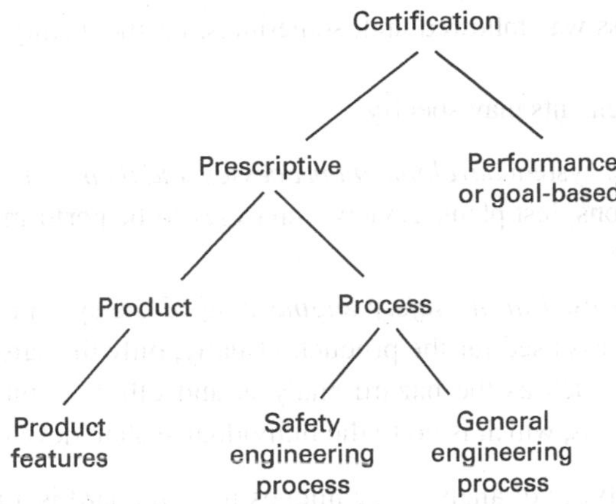

Figure 13.5
Types of certification.

advances in technology. Alternatively, a certification authority or group may attest to the compliance of the product to product standards or codes of practice.

Organizations may also be established that both produce standards and provide certification. These have been particularly useful and successful in industries where government agencies do not require certification but such certification is a company advantage in consumer product sales or in obtaining liability insurance. An example is Underwriter's Laboratory (UL), which was founded in 1894 by a group of insurance underwriters to reduce their risk.

It is difficult to fathom any argument that encoded knowledge in the form of standards about safe design should not be included in any certification effort. Requiring reinvention of this past experience for every project would be prohibitively costly and potentially incomplete and error prone without any clear advantage. Usually even industries that certify systems using a different approach include specific product requirements in their certification standards.

Different industries face different safety problems, and therefore the general approach to safe design may differ among them along with required product design features. For example, commercial aviation has created various types of general fail-safe techniques used to protect against component failures [104]. In addition, specific types of safety equipment are required, such as seat belts, life preservers, emergency exit slides, and oxygen masks in case the cabin becomes depressurized. As an example of more general requirements, nuclear power has traditionally required defense-in-depth and protection systems.

Certification using a prescriptive product approach is usually implemented by inspection and test that the design features provided are effective and implemented properly.

Process-based standards While most systems have specific design features that have been found to be important through many years of experience, complex systems often use additional types of *process-based* standards. Here the standards specify the process to be used in producing the product or system or in operating it (e.g., maintenance or change procedures) rather than specific design features of the product or system itself. Assurance is based on whether the process was followed and, sometimes, on the quality of the process or its artifacts.

The process requirements may specify

- general product or system *development processes and their artifacts*, such as requirements specifications, test plans, reviews, analyses to be performed, and documentation produced; or
- the *process to be used in the safety engineering* of the system and not the general development process used for the product. Usually, only the safety engineering process is specified, such as the hazard analysis and other activities, not the general development process, which is up to the individual system developers.

Requiring specific types of analysis techniques in a standard is a mistake as it means that standards are almost out of date and behind the state of the art. It can take a long time to create a new standard or update an old one, and technology is moving too fast for that to be adequate today. Specifying only a general process or artifacts, for example, identifying the type of results to be produced by a hazard analysis and not the specific way to produce those results (e.g., by using FEMA), ensures that standards are applicable and useful for longer periods of time as technology and tools advance.

Goal-based and performance-based standards Performance-based or goal-setting approaches focus on desired, quantifiable outcomes, rather than required product features or prescriptive processes, techniques, or procedures. The certification authority specifies a threshold of acceptable performance and often, but not always, a means for assuring that the threshold has been met.

Goal-based regulation specifies defined results without specific direction regarding how those results are to be obtained. Basically, the standards set a goal, which may be a probabilistic risk target, and usually it is up to the assurer to decide how to obtain that target. This type of certification is very popular with industry as it provides freedom in selecting how to meet the goals.

An example is a requirement that an aircraft navigation system must be able to estimate its position to within a circle with a radius of ten nautical miles with some specified probability. Another example is for new aircraft in-trail procedure (ITP) equipment "The likelihood that the ITP equipment provides undetected erroneous information about accuracy and integrity levels of own data shall be less than 1E-3 per flight hour" [338]. It is completely up to the applicant to devise their argument that the goal or performance level has been achieved.

13.3.2 National and Industry Practices in Certification

While in the past most certification was based on prescriptive methods (either product or process), there has been growing use of performance-based regulation and assurance by government agencies, starting in the United States during the Reagan administration. The use of performance-based regulation was primarily spearheaded by pressure from those being certified. The American Nuclear Society in 2004, for example, called for the use of risk-informed and performance-based regulations for the nuclear industry, arguing that "*risk-informed regulations use results and insights from probabilistic risk assessments to focus safety resources on the most risk-significant issues, thereby achieving an increase in safety while simultaneously reducing unnecessary regulatory burden produced by deterministic regulations"* [18].

Similar arguments have been made about FAA regulations and procedural handbooks being inflexible and inefficient and rulemaking taking too long. The certification process has moved to performance-based regulations where appropriate, but there are problems (discussed later), particularly with respect to how the performance goals are set and assured.

Certification today in the United States primarily uses prescriptive methods but mixes the two types—product and process—and is facing pressures from industry to introduce more performance-based certification. The defense industry, for example, has traditionally used standards that specify the safety engineering processes to be used during development and operation, but not the tools to implement those processes. Recently, risk assessment has been added and sometimes substituted for certifying hardware along with using Level of Rigor for software.

Commercial aircraft certification is complex but provides an interesting example of the possibilities. The FAA and EASA (European Union Aviation Safety Agency) do not require specific practices for general aircraft certification but issue advisory circulars that recognize acceptable means for showing compliance with safety regulations and airworthiness regulations. Specific product requirements, such as seat belts, were mentioned in the previous section.

The commercial aviation industry has produced the most widely used standard for general airworthiness certification, called SAE ARP 4754, that specifies general aircraft and system development processes. A closely related standard specifies the process for assessment of airworthiness using primarily probabilistic risk assessment for hardware and system functions along with design assurance levels for software (SAE ARP 4761). Table 13.1 shows required probability objectives and assurance levels in SAE ARP 4761. Human factors considerations are handled by separate standards, primarily focused on flight deck displays and controls.

Sometimes certification is a one-time activity that follows the development process and occurs before the product or system is allowed to be marketed or used. More commonly and especially for complex systems, such as aircraft, nuclear power plants, and offshore oil exploration, certification may involve both initial approval as well as oversight of the operational use of the system. Changes to the original system design and certification basis may require recertification activities.

All certification is based on evidence that the certification approach has been followed. Inspection and test may be used if the certification is based on following a product standard. If the certification is based on the process used, engineering artifacts or analyses may be required and reviewed. Performance-based regulation may require a particular type of analysis or may allow any type of reasoning that supports having achieved a particular performance goal.

As an example, the US Department of Defense in MIL-STD-882 has traditionally used a prescriptive process that details the steps that must be taken in the development of safety-critical systems to ensure they are safe. Actual certification is based on the safety assessment report (SAR), which describes the results of prescribed tasks in the standard. The SAR contains the artifacts of the prescribed process, such as a Safety Plan (which must be approved by the DoD at the beginning of the development of the system), a Preliminary

Table 13.1
Typical relation between failure severity, probability requirements, and assurance levels used in aircraft certification

Probability per Flight Hour	Probability Description	Failure Condition Severity	Failure Condition Effect	Development Assurance Level (DAL)
			No safety effect	Level E
1.0E-3	Frequent	Minor	Slight reduction in safety margins; slight increase in crew workload	Level D
1.0E-5	Remote	Major	Significant reduction in safety margins or functional capabilities, significant increase in crew workload	Level C
1.0E-7	Extremely remote	Hazardous	Large reductions in safety margins or functional capabilities; high workload or physical distress to crew such that crew cannot be relied on to perform tasks accurately or completely	Level B
1.0E-9	Extremely improbable	Catastrophic	Failure conditions prevent continued safe flight and landing	Level A

Hazard Analysis, a System Hazard Analysis, a Subsystem Hazard Analysis, an Operating System Hazard Analysis, and so on. The DoD evaluates the quality of the process artifacts provided in the SAR as the basis for approving use of the system. Relatively recently, a probabilistic risk approach has been added to the standard that can be used in lieu of the prescriptive process.

The United Kingdom has developed its own approach to certification. Government oversight of safety in England started after the Flixborough explosion in 1974 (see appendix C) and focuses on the presentation of a *safety case.* The term safety case seems to have arisen from a report by Lord Cullen on the Piper Alpha disaster in the offshore oil and gas industry in 1988 where 167 people died. The Cullen report on the Piper Alpha loss, published in 1990, was scathing in its assessment of the state of safety in the industry [71]. The Cullen report concluded that safety assurance activities in the offshore oil industry

- were too superficial;
- were too restrictive or poorly scoped;
- were too generic;
- were overly mechanistic;
- demonstrated insufficient appreciation of human factors;
- were carried out by managers who lacked key competences;
- were applied by managers who lacked understanding; and
- failed to consider interactions between people, components and systems

The report suggested that regulation should be based around goal setting that would require that stated objectives be met, rather than prescribing the detailed measures to be taken [412]; that is, goal-based rather than prescriptive. In such a regime, responsibility for controlling risks shifted from the government to those who create and manage hazardous systems in the form of self-regulation. This approach has been adopted by the British

Health and Safety Executive and applied widely to industries in that country. While the government sets goals, the operators develop what they consider to be appropriate methods to achieve those goals. It is up to the managers, technical experts, and the operations/maintenance personnel to determine how accidents should be avoided. In general, however, all risks must be reduced such that they are below a specified threshold of acceptability. Problems arise in trying to define and evaluate acceptability.

When performance-based or goal-based certification is used, there are differences in how the performance or goals are to be specified and how the evaluation will be performed. Trans-scientific issues, as defined in chapter 1, can arise in these decisions.

13.3.3 Providing Evidence in Performance-Based Regulation and Safety Cases

The type of evidence required is straightforward with prescriptive regulation, but performance-based regulation requires a more complex argument and evaluation strategy. While the term *safety case* may be used in prescriptive regulation, it is more commonly used in a performance or goal-based regulatory regime.

Risk-based regulation requires a definition of what risk level needs to be attained as nothing is totally risk free. Changing the goal to "acceptably safe" simply changes the problem to defining what is "acceptable" and to whom: To the producer of the system who is paying the cost of making it safe or to the potential victim?

After the 1974 Flixborough explosion, the UK Health and Safety Executive (HSE) introduced the concepts of certification using cost-benefit analysis to attempt to answer that question. In this conception, safety certification involves balancing the benefits from undertaking an activity against protecting those that might be affected by it (see chapter 2). The HSE also instituted the related concept of ALARP or "as low as reasonably practical" and probabilistic risk assessment as the basis for evaluating achievement of the goals [139].

The nuclear power industry was probably the first to use probabilistic risk analysis as a basis for certification. In the United Kingdom, the Nuclear Installations Act of 1965 required covered facilities to create and maintain a safety case containing such analysis in order to obtain a license to operate. Because of the use of standard designs in the nuclear power community and very slow introduction of new technology and innovation in designs, historical failure rates are often determinable.

Other potentially high-risk industries, such as the US nuclear submarine community, have adopted the opposite approach. For example, SUBSAFE does not allow the use of PRA [217]. Instead, they require OQE (Objective Quality Evidence). As defined in chapter 4, OQE may be qualitative or quantitative, but it must be based on observations, measurements, or tests that can be verified. PRA, using that basis, is not allowed.

A second unique aspect of the British approach to safety assurance and required by the HSE is argumentation and approval based on whether risks have been reduced as low as is reasonably practicable (ALARP), as mentioned before. Evaluating ALARP involves an assessment of the risk to be avoided, an assessment of the sacrifice (in money, time, and trouble) involved in taking measures to avoid that risk, and a comparison of the two. The assumed level of risk in any activity or system determines how rigorous, exhaustive, and transparent the risk analysis effort has been. The greater the initial level of risk under consideration, the greater the degree of rigor required to demonstrate that risks have been reduced as far as is reasonably practical [140]. This criterion ignores the question of how the initial level of risk is determined.

The application of ALARP to new systems, where "reasonably practical" has not yet been defined, is questionable. For example, not increasing the accident rate in civil aviation above what it is today does seem like a reasonable goal given the current low rate, but it is not clear how such an evaluation could be performed for new technologies, such as satellite navigation or for new and very different procedures for aviation operations and air traffic control. If an industry is changing and innovating rapidly, historical accident rates are not very useful.

For industries that are not starting at a very low accident level, determining how low is reasonably practical seems philosophically and ethically difficult to determine.

In addition, safety cases are strictly confidential in the United Kingdom; only company officials, regulators, and, in limited circumstances, worker representatives are allowed to see the entire plan. The safety case may contain proprietary information, so that limitation seems reasonable if nonprescriptive standards are used. This type of confidentiality, however, would be unlikely to be acceptable in the United States [354].

While none of these more controversial aspects of assurance and certification need to be present when using a performance- or goal-based certification approach, they are part and parcel of the history and foundation of safety cases and performance-based regulation. For convenience, let's call this risk-based regulation.

The problems associated with safety assurance and safety cases based on an argument that the system will be safe in a given operational context were outlined in the first section of this chapter, including confirmation bias. To reiterate, people will focus on and interpret evidence in a way that confirms the goal they have set for themselves. If the goal is to prove that the system is safe, they will focus on the evidence that shows it is safe and create an argument for safety. If the goal is to show the system is unsafe, the evidence used and the interpretation of available evidence will be quite different. People also tend to interpret ambiguous evidence as supporting their existing position [77].

Engineers always try to build safe systems and to verify to themselves that the system will be safe. The value that is added by system safety engineering is that it takes the opposite goal, namely, to show that the system is unsafe. Otherwise, safety assurance becomes simply a paper exercise that repeats what the engineers are most likely to have already considered.

An example of what can happen is called out in the official Nimrod accident report [128]. In September 2006, a UK Royal Air Force Hawker Siddeley Nimrod aircraft caught on fire during flight and crashed in Kandahar, Afghanistan. All fourteen crew members on board died. The crash, which occurred during a reconnaissance flight, was the biggest single loss of life suffered by the British military since the Falklands War. The report cited the use of safety cases as a major cause and recommended that safety cases be relabeled "risk cases" and that the goal should be "to demonstrate that the major hazards of the installation and the risks to personnel therein have been identified and appropriate controls provided" [128], not to argue that the system is safe. The report also warned against equating quantitative probabilities such as 10^{-6} to "remote." It suggested that such figures give the illusion and comfort of accuracy and a well-honed scientific approach, but, outside the world of structures, numbers are far from exact.

A practical problem with the use of a safety case is that each safety case needs to be evaluated by a certification authority [401]. How can this be done? If every submission is different, it will be difficult to evaluate them in a systematic way, so certification of different systems may occur using different criteria. If certification only occurs within a company in an unregulated industry, will the argument really be evaluated in an independent and unbiased way?

Where will people qualified to do this evaluation be found? Such an evaluation would be extremely time and resource consuming and potentially costly [401]. Companies complain already about the long waits involved in certification. In addition, regulatory agencies are notoriously understaffed and underfunded, but the use of a safety case requires a well-resourced and competent regulator.

Another practical consideration is the role of stakeholders in the certification process. Historically, users of products and systems were expected to evaluate the safety of these products themselves and assume responsibility if they choose to use them. But systems have become so complex that this process is no longer possible. The general public is not capable of evaluating the safety of an aircraft before they fly on it, so they need to trust the regulatory authorities to make this determination. External evaluation of the certification processes by stakeholders, such as airline passengers and pilots or those living in the vicinity of nuclear power plants, is currently possible because the processes are defined in detailed standards.

For example, in the commercial aviation realm, RTCA (Radio Technical Commission for Aeronautics) committees that define certification procedures are open to everyone, including pilots and pilot unions and airline passenger associations. Stakeholders cannot see proprietary company data, but they can help to determine the process for evaluating that data.

In general, consumers, stakeholders, and engineers and scientists outside the company involved have participated in creating the product- and process-based standards and sometimes even the certification process itself, so blind trust in government agencies to ensure our safety is not required.

With goal-based regulation and the use of safety cases, those being certified can determine how the goal will be shown to have been achieved. Each case can be presented using different evidence, and the company itself determines what argument is provided. The details of the safety case cannot be shared outside the government for proprietary reasons. Stakeholders are effectively shut out of the certification process.

While theoretical arguments against safety cases and goal-based certification are interesting, the proof is really "in the pudding." How well have they worked in practice? Unfortunately, careful evaluation and comparison between certification approaches has not been done. Most papers express personal opinions or deal with how to prepare the different types of evidence required, but they do not provide data on whether a particular approach is effective. As a result, there is no real evidence that one type of regulation or certification is better than another.

13.3.4 Designing a Certification Program

There is clearly no one perfect way to certify safety-critical systems. Each approach has limitations, and how they are implemented is critical. Without data about the cost-effectiveness of the various certification approaches, general conclusions are limited to philosophical, logical, and experiential arguments. Culture and history within an industry are also factors in designing certification activities. There are, however, some general practices that seem reasonable and might be considered.

Combine types of certification as appropriate Any viable certification approach will almost certainly contain both types of prescriptive certification. Product-based certification ensures that lessons of the past will be incorporated in the design and operation of new systems. Process-based certification, at the same time, ensures consistency of certification decisions, the handling of new technology and hazards, and relatively efficient and cost-effective implementation of the certification process. Both are capable of including stakeholder inputs and participation.

Most useful approaches to regulatory oversight will combine these two types of certification. In addition, there may be instances where goal- or performance-based certification is appropriate, particularly if there is an accepted and scientifically viable means for evaluating the achievement of the goal.

A common requirement is that the process used, either imposed by standards or selected by the developer, reduces confirmation bias as much as possible. It is easy to fall into the trap of assuming the conclusion—the system is safe—and looking for supporting evidence for the conclusion instead of carrying out a proper hazard analysis.

The certification process chosen should not just be a paperwork or "check the box" exercise. As Haddon-Cave said in the conclusions of the Nimrod accident report, there should not be a culture of "paper safety" at the expense of real safety [128]. At the other extreme, but with the same results, requiring the rigorous use of standard engineering development processes that have nothing to do with safety or safety engineering may delight many developers—who will not have to devote any resources to anything they would not already have done anyway—but provides no contribution to safety. Both extremes will have no impact on operational safety.

Start the process early No matter what type of certification is selected, the process needs to be started early. The analysis done for certification is only useful if it can influence design decisions. That means it should not be done after a design is completed or prepared in isolation from the system engineering effort. If safety cases are created only to argue that what already exists is safe, then the effort will not improve safety and becomes, as has happened in the past, simply a paper exercise to get a system certified and often an exercise in fantasy. One unfortunate result of such an exercise is unjustified complacency by those operating and using the systems. As described below, the US defense system safety standard MIL-STD-882 requires a plan to be submitted at the beginning of development.

Allow flexibility The agency or industry standards can specify the types of results that should be obtained and the factors to be considered—including software, human factors, and operations—but should allow discretion in the way they are obtained. The specific hazard analysis techniques to be used can be included in the certification plan and thus approved or not by the certifier at the beginning of the development process.

Allowing the tailoring of the processes in standards can alleviate some of industry's legitimate concerns. To be most useful, however, qualitative and verifiable quantitative information should be required, not just probabilistic models of the system.

For example, MIL-STD-882, which is used for defense systems, includes submitting a plan early in the development program to the certifying authority outlining the way the system will be certified to be safe. Flexibility is thus allowed in the way the standard is applied. Otherwise, an overly costly and unnecessarily rigid process might be required. The agency, of course, then needs to have personnel who are capable of evaluating the plans.

MIL-STD-882 contains a set of independent tasks, not all of which are appropriate for every project, and not all are required in every plan. The certification authority can approve the company's plan or request modifications. Note that this approach is similar to that used in approval of pharmaceuticals by the FDA. More recently, a risk-based approach has been included in MIL-STD-882 that allows risk assessment to be used in lieu of the specific analysis and management tasks.

A development safety plan should include identifying the accidents (losses) and specific hazards to be considered. In some industries, these losses and hazards are actually defined by the government regulator and are not at the discretion of the designer or operator of the system if the system is highly dangerous to the public. For example, the US government specifies the hazards that must be eliminated or controlled for atomic weapon systems.

MIL-STD-882 does not require the use of particular hazard analysis techniques; the ones to be used are included in the plan that is submitted. Too many standards specify the hazard analysis methods to use and, in essence, over-specify the process. The problem is not only that those methods might not be the best for the particular system being developed, but, more generally, advances in technology and in practices cannot be introduced into standards in less than a decade and sometimes more, by which time even the new analysis techniques may be obsolete. The standards are then always behind the state of the art. Instead, standards should outline general tasks and processes and allow the applicant to tailor the process when appropriate and to produce a plan on how they intend to satisfy the standard.

An integral part of the plan should include how newly identified hazards will be added to the analysis and incorporated into the operational documentation. How will the original assumptions about hazards be re-examined in light of new events and new knowledge?

Finally, the certification deliverables should include the limitations of what was done and the uncertainties as well as the assumptions underlying the analyses and the design procedures used.

Require or at least allow safety by construction Any certification effort should require or at least allow designing safety into the system from the start of development and not simply require arguments at the end. The plan can specify how this goal will be accomplished so that flexibility can be allowed. After-the-fact approaches to completed product assurance, as argued earlier in this chapter, have limited utility.

Require worst-case hazard analysis Any analysis results used for certification should consider worst cases, not just the likely or expected case. It should also include special activities related to identifying hazards and safety-critical requirements or ensuring that hazards are eliminated or controlled in the final design. Simply requiring the use of "good" engineering processes or rigorous development and testing, including software development activities, is not adequate. The processes required and used must include standard safety engineering activities. Reliability analysis should not be substituted for hazard analysis.

Require that the entire system be included as well as the entire system lifetime The certification activity needs to include all factors; that is, it must be comprehensive. It should include not just hardware, software, and operator behavior but also management structure and decision-making. The integrated system must be considered and not just each hazard or component in isolation. Analyzing or treating software in isolation from the system design can result in high-quality but unsafe software.

Certification must consider operations. Updating of certification and lifetime monitoring if deemed important should be included. Too often warning signs of increasing risk over time are ignored by those operating hazardous systems. Oversight can encourage early intervention.

Like assurance, the assumptions underlying the certification decision should be continually monitored during operations and procedures established to accomplish this goal. The system may be working, but not the way it was designed, it may be used differently than assumed in the certification process, or the assumptions underlying the certification process may turn out to be wrong, perhaps because the environment has changed.

Changes to the system and its environment may have been made for all the right reasons, but the drift between the system as designed and the system as it currently exists is rarely if ever analyzed or understood as a whole. Each particular deviation may appear sensible and even helpful to the individuals involved, but risk may be increasing.

Include documentation requirements To make maintaining the certification feasible, the analysis needs to be integrated into system engineering and system documentation so it can be maintained and updated by those operating the system. Certification should not be just a one-time activity but should continue through the lifetime of the system, which includes checking during operations that the assumptions made in the certification decision remain true for the system components and the system environment.

The major problems in updating and maintaining documentation and certification arise in relating the original analysis to the detailed design decisions. When a system design or operating environment is changed, it must be possible to determine what assumptions in the hazard analysis are involved and need to be revisited. Starting a new hazard analysis from scratch is just not feasible in almost all cases. Related to this concern is grandfathering of certification when a product is changed in a seemingly innocuous way: the assumption that risk has not been increased should have to be proven.

Create practical standards and certification practices The process has to be practical and affordable for both the applicant and the certification authority. Otherwise, shortcuts will be taken or resistance created. Many of the activities required in certification today are simply tedious exercises in producing paper analyses and provide little practical benefit in the engineering or operation of safer systems. Most after-the-fact analyses and argumentation fall into this category.

Large amounts of money and effort are often spent on activities that do not improve the safety of the system but are viewed as simply a required exercise to get certified (or insured or protected against potential liability).

The certification process must also be practical to implement with the personnel available in the government certification agency, or the company if external certification is not involved, or by those licensed by the agencies to accept responsibility for approval.

While specific information has to be proprietary, the process must be open to the public both for inputs and review. If the stakeholders are unable to judge the safety of specific systems for themselves, they at least should be able to have a say and oversight into the processes the government or others are using to protect public safety.

13.4 Some General Conclusions

High levels of assurance and safety can only be achieved by creating comprehensive safety engineering programs containing useful activities—not just creating reams of paper that is never used, except perhaps in certification activities or in court cases after accidents occur. Indeed, safety engineering should not just be an afterthought or an isolated activity—it needs to be tightly integrated into the system development process.

Safety must be built into the system from the beginning. It cannot be argued in after the fact or created in isolation from the critical system design decisions.

Exercises

1. Chernobyl had a calculated PRA of 10^{-9} per year (or a mean time between failure of 10,000 years). What do you think went wrong in the analysis? Or did it? (There is a description of what happened at Chernobyl in appendix D).
2. Look up the certification standard(s) for your field or one you are interested in that is regulated. What type of standard is it?
3. Discuss the problems in doing a probabilistic risk assessment for software, where risk is defined as the probability of the software causing or contributing to an accident in a particular system. What are the problems for any complex system (not just software)?
4. A common approach to certification is to start with an analysis to identify the functions that are related to hazards. The system hardware components involved in those functions are provided with probabilistic goals while the software components are required to meet a *design assurance level* (DAL). The DAL is defined in terms of the types of assurance activities (that is, testing, simulation, proof of correctness, and so on) and is usually categorized in terms of categories from low to high. LOR (Level of Rigor) may be defined similarly to the DAL or may involve all software development activities and not just assurance activities. An example is shown in chapter 10. Evaluate this approach.
5. A related approach used to deal with complex systems that include software is *Level of Rigor*. There may be a hazard analysis performed, but if the technique used only considers failures, then software is not included (software has only design errors and does not "fail" in the sense that hardware does). In this approach, the software is developed to a specified *Level of Rigor*, and the overall risk of the system combines a quantitative risk analysis of the hardware with a required LOR for the software development to come up with a risk assessment for the system as a whole. Evaluate this approach.
6. Some industries, particularly the process industry (petrochemicals and nuclear), certify their systems on the basis of the protection system *Safety Integrity Level* (SIL). The process industry puts much reliance for safety on defense and depth and in protection

systems that change a hazardous state into a safe one. What drawbacks and advantages might this approach have for the process industry? For other industries?

7. A SIL might be assigned using probabilistic or qualitative approaches, where qualitative includes a specific product development standard. How the SIL is defined and determined will, of course, determine whether the quality being considered is reliability or safety. How might the SIL be defined to ensure it defines safety and not reliability? Is it possible to identify the SIL for some standard components, software or hardware, and then use that result to determine the safety of the component when it is used in a larger system?
8. In complex systems, an OEM (original equipment manufacturer) usually contracts out the components to subcontractors, for example, in automotive, different companies may produce the propulsion system and the braking system. What implications do these different certification approaches have for the OEMs and subcontractors in these industries?
9. Why do you think that performance- or goal-based standards are more popular with industry than prescriptive standards?

14

Designing a Safety Management System

Management systems must ensure that there is in being a regime which will preserve the first place of safety in the running of the railway. It is not enough to talk in terms of "absolute safety" and of "zero accidents." There must also be proper organisation and management to ensure that actions live up to words.

—British Department of Transport,
Investigation into the Clapham Junction Railway Accident

Most managers do care about safety. The problems usually arise because of misunderstandings about what is required to achieve high safety levels and what the costs really are if safety is done right. Safety need not entail enormous financial or other costs.

—Nancy Leveson,
Engineering a Safer World

Much has been written about the design of safety management systems (SMSs) and safety management in general. While an in-depth treatment of this topic is beyond the scope of an introductory text on system safety engineering, this book would be incomplete without including at least a basic introduction to this important topic. Even the best engineering practices can be negated by poor management decision-making, perhaps caused by misplaced priorities, complacency, inadequate knowledge of system safety engineering, or maybe simply lack of the appropriate information to make better decisions.

Many managers think that safety is a net cost and that the choice is between spending on safety activities and spending on increasing productivity and profit-making activities. Some even think that increasing safety necessarily requires decreasing productivity and profits. If the reader comes away with nothing else from this book, it should be that this conception is wrong.

Paul O'Neill provides an example of how safety and profitability can be mutually reinforcing. When he was hired from the outside to be CEO of Alcoa in 1989, he immediately announced that his primary goal was to make Alcoa the safest company in America and to go for zero injuries. Typical reactions to this announced goal were that the Alcoa board of directors had put a "crazy hippie in charge" and that he was going to destroy the company. In fact, within a year, Alcoa's profits had hit a record high and continued that way until O'Neill retired in the year 2000. At the same time, Alcoa became one of the safest companies in the world [82]. O'Neill clearly understood that safety and productivity are not conflicting and that, in fact, they go hand in hand. He has been quoted as saying "Safety should never be a priority. It should be a prerequisite" [287].

O'Neill was focused on workplace safety, but the same principle applies to system safety. Accidents are expensive. Some catastrophic ones can ruin a company or sometimes even have a negative impact on the future of an entire industry. The nuclear industry is a cogent example, but the same is true for industries with less dire accident consequences. Commercial aviation is an example of an industry that reached its current levels of profitability and success by emphasizing safety. The recent B737 MAX accidents, however, illustrate how devastating even the short-term effects of accidents can be and how difficult it can be to recover. If done effectively, system safety engineering can not only provide an efficient path to increasing profits but also contribute to the long-term sustainability of the company and the industry as a whole.

Of course, there are short-term costs in establishing an effective safety management system and ongoing costs to maintain it. But while establishing an SMS may be expensive, it will not be nearly as expensive as an accident or producing unsafe products, particularly when damage to reputation and future profits are considered.

To reap the gains from investing in system safety, the leaders of an organization need the ability to think beyond very short-term rewards and build a culture and safety management system that values safety as an important component of company success.

The Macondo/*Deepwater Horizon* disaster is an example of the costs of undervaluing safety in achieving organizational goals. In that case, profit and performance pressures implicitly excluded system safety from the core priorities of the companies involved, particularly BP, and led the project to a high-risk state long before the actual catastrophic blowout occurred. Estimations of the direct costs of Macondo to BP alone are at least $65 billion. A narrow focus on potential short-term costs and profits by BP leadership in all the company activities, from upstream exploration to downstream refining, led management to not perceive the potential costs associated with emphasizing short-term profits over longer-term success.

Whether safety is handled well and is cost-effective depends, of course, on how it is done. Even well-intentioned safety activities are frequently done poorly or are limited in their usefulness. Simply investing in safety is not enough.

Another problem is that attention may be focused on only one part of the problem, such as satisfying government regulators, and not on the more important problem of improving safety. Too often, satisfaction of regulations or protection from liability has little to do with preventing accidents. Ironically, activities focused on preventing legal liability may actually lead to increased accidents and, perversely, increased liability. Those making management decisions about system safety need to understand system safety engineering. They need not be experts, but they do need to know the fundamental concepts contained in this book.

Designing a safety management system starts from what is known about the causes of accidents. Obviously, the success of the safety management system must be judged by its ability to eliminate or minimize these causes. Chapter 4 classified accident causes into three levels: (1) mechanisms, (2) conditions, and (3) constraints or systemic factors. The first two, the mechanisms and conditions related to accidents, are mostly covered in standard engineering and science education, such as the nature of chemicals, chemical reactions and energy, how to design electromechanical systems, and how to create correct software. In contrast, the primary focus in system safety engineering is on the level 3 systemic factors, classified in chapter 4 as social dynamics and organizational culture, management practices and organizational structure, operational practices, regulatory and certification policies and practices, and system safety engineering processes.

Chapter 13 considers regulation and certification. Most of the other chapters examine system safety processes and practices, although not their management. This chapter focuses on the rest: social dynamics and organizational culture, management practices and organizational structure, operational practices, and some factors in engineering processes and practices not covered so far: that is, managing safety-critical system development and creating an effective safety information system.

These topics could fill a book or books of their own, so only basic principles are included here. Future books will elaborate on these topics.

14.1 Social Dynamics and Organizational Culture

Social dynamics and organizational culture greatly influence individual decision-making. Edgar Shein, considered to be the "father of organizational culture," defined safety culture as "the values and assumptions in the industry and/or organization used to make safety-related decisions." Figure 14.1 shows Shein's model of the three levels of organizational culture [348].

The essence of culture is the lowest-level set of values and deep cultural assumptions that underlie decision-making: for example, believing that safety does not conflict with productivity and profits and that blame is the enemy of safety. These values and assumptions provide the foundation for creating organizational rules, policies, and practices such as system development processes, maintenance practices, and training. In turn, those

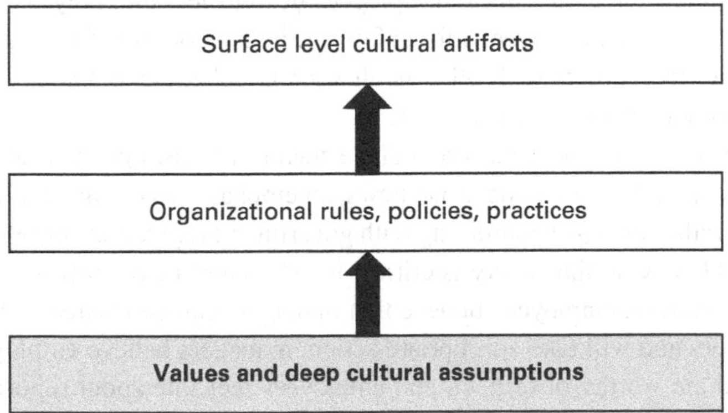

Figure 14.1
Shein's three-level model of organizational culture.

rules, policies, and practices are used in the creation of the organization's cultural artifacts, such as hazard analyses and accident investigation reports. Safety culture is the subset of the organizational or industry culture that reflects safety values.

The middle and top levels are not the culture; they merely reflect the impact of the foundational values on the operation of the organization. While attempting to make changes only at the top two levels may temporarily change behavior and even lower risk over the short term, superficial fixes at the top two levels are likely to be undone or become ineffective over time if they do not address the set of shared values and social norms on the bottom level.

Often historical and environmental factors are key in creating the existing safety culture. For example, William Boeing, when building the commercial aviation industry, was faced with the fact that improved safety would be required to sell aircraft and air travel. In 1955, only 20 percent of US citizens were willing to fly—aircraft crashes were a relatively common occurrence compared with today. In contrast, the nuclear power industry was initiated by passage of the Price-Anderson Act in 1957, whereby the industry agreed to be tightly regulated by the government in exchange for limited liability in case of accidents. In some industries, the safety culture is the result of strong individual leadership, such as Admiral Hyman Rickover in the nuclear navy. Differences in the way safety is handled exist in each of these industries as a result.

The flaws in safety culture noted in chapter 4 as leading to accidents all relate to flawed decision-making. They include such factors as overconfidence and complacency, low priority given to safety, flawed resolution of conflicting goals, and believing that safety conflicts with productivity and profits. The values communicated by BP management at Texas City, *Deepwater Horizon*, and the company's many other costly accidents, were cost and schedule over everything. This type of flawed decision-making needs to be appropriately controlled in an effective safety culture. In the nuclear power accidents described in appendix D, nobody believed the accidents were possible, so they did not take the steps necessary to prevent them or mitigate their consequences.

The safety culture—that is, the values to be used in decision-making—is established by top management. A sincere commitment by management to safety is often cited as the most important factor in achieving it. Employees need to feel that they will be supported if they exhibit a reasonable concern for safety in their work and if they give priority to safety in short-term conflicts with other goals such as schedule and cost. Subcontractors in large projects need the same assurances.

The features of a good safety culture include such things as: openness about safety and the safety goals, a willingness to hear bad news, an emphasis on doing what is necessary to increase safety rather than just complying with government regulations or producing a lot of paperwork, and believing that safety is critical in achieving the organization's goals. In an effective safety culture, employees believe that managers can be trusted to hear their concerns about safety and will take appropriate action, managers believe employees are worth listening to and are worthy of respect, and employees feel safe about reporting their concerns and feel their voice is valued. Safety is enhanced when it is a shared responsibility and employees are considered to be part of the solution and not just the problem. At the same time, in a good safety culture, responsibility is not simply placed on the workforce to keep themselves and others safe. Appropriate responsibilities need to be assigned to individuals at all levels of the organization.

Management establishes the value system under which decisions are made in an organization. The first step, therefore, in improving the safety culture is for management to establish and communicate what is expected in the way of safety-related decision-making and behavior.

The organizational values or culture is communicated in two basic ways. The first is modeling the behavior desired and demonstrating that safety is a high priority, and the second is communicating the expected behavior.

14.1.1 Modeling Desired Behavior

A manager's open concern for safety in everyday dealings with personnel can have a major impact on the reception given to safety. The most important is through actions and demonstrated commitment to safety by

- setting the goals and values to be used in decision-making and establishing and communicating what is expected in safety-related decision-making and behavior;
- personal involvement in safety-related activities;
- setting priorities and following through on them;
- setting up appropriate organizational structures;
- appointing high-ranking, respected leaders to safety-related roles and responsibilities
- providing adequate resources for safety efforts to be effective;
- assigning the best employees to safety-related activities, not just those who are nonessential;
- rewarding employees for safety efforts and establishing appropriate incentive structures;
- responding to initiatives by others;
- supporting employees who exhibit reasonable concern for safety;
- following the safety philosophy in one's own decision-making and expecting the same from everyone;
- minimizing blame by focusing on "what" and "why," not "who";
- listening to and acting on inputs.

An important goal in creating a strong safety culture is establishing what has been called a "Just Culture." Here the highest priority in reacting to incidents or feedback about unsafe behavior is placed on identifying the systemic factors that led to the events and fixing them rather than punishing individuals. Blame is the enemy of safety. In a blame-oriented culture, mistakes are hidden and important lessons are not learned, leading to the same and more serious events in the future.

Sometimes the Just Culture concept is poorly implemented. Simply doing away with punishing people for mistakes accomplishes nothing by itself. Focus needs to be placed on identifying why people behaved the way they did and changing the conditions that led to the undesired behavior.

14.1.2 Documenting Values and Policies

Beyond modeling and rewarding desired behavior, management must communicate in written policy the type of safety culture they want; that is, the desired values for making decisions.

Most companies have extensive safety policy documents that detail exactly how safety should be handled. While these are important, a shorter statement of the philosophical principles and values is a good way to communicate the expected cultural principles of the organization. This philosophical statement should define the relationship of safety to other organizational goals.

Some general cultural principles are applicable to all organizations, while the applicability of others will depend on whether the organization is developing or operating safety-critical systems or whether they are providing safety-related services. Avoid meaningless slogans such as "our goal is zero accidents" or "be mindful of weak signals." The latter was the stated corporate safety philosophy of BP prior to the Texas City explosion.

Instead, remember that the goal is to state the values that are to be used in decision-making, particularly when there are conflicts between goals, at least in the short term, and tradeoffs are required. Some examples of organizational safety philosophy are:

1. All injuries and accidents are preventable.
2. Increasing quality and safety leads to decreasing cost and schedule and, in the long term, to increased profits. Preventing accidents is good business.
3. Safety and productivity go hand in hand. Improving safety management leads to improving other quality and performance factors. Maximum business performance requires safety.
4. Incidents and accidents are an important window into systems that are not operating safely and should trigger comprehensive causal analysis and improvement actions instead of blame and punishment.
5. Safety information must be surfaced without fear.
6. Effective communication and the sharing of information is essential to preventing losses.
7. Safety commitment, openness, and honesty is valued and rewarded in the organization. Each employee will be evaluated on his or her performance and contribution to our safety efforts.

A short list will be more effective than a long one.

A written statement of the safety philosophy and more detailed policy statements are a start, but they are not enough. Employees quickly identify when the written policy differs from the actual behavior of managers. To be successful, there needs to be real commitment by those at the top, not just sloganeering and making statements that are belied by behavior.

The major ideas in creating an effective safety culture are summarized in the following box.

Tips for how management can improve safety culture:

- Set the goals and values to be used in decision-making; establish and communicate what is expected in safety-related decision-making and behavior.
- Support employees who exhibit reasonable concern for safety in their work.
- Create a short, written safety philosophy and more detailed safety policy.
- Ensure that safety philosophy and policy have wide buy-in from managers and employees.
- Follow the safety philosophy in your decision-making and expect the same from everyone.
- Emphasize building safety in and not assurance or after-the-fact assessment.
- Perform assessment with the goal of providing evidence that the design is not safe, not providing an argument that it is safe.
- Require Objective Quality Evidence for assurance or certification. Avoid focusing on compliance and paperwork.
- Demonstrate commitment to safety by
    - personal involvement
    - setting priorities and following through on them
    - setting up appropriate organizational structures
    - appointing high-ranking, respected leaders to safety-related roles and responsibilities
    - providing adequate resources for safety efforts to be effective
    - assigning the best employees to safety-related activities, not just those who are nonessential or expendable
    - rewarding employees for safety efforts
    - responding to initiatives by others
- Minimize blame; focus on "why" not "who."
- Engineer the incentive structure to encourage desirable safety-related behavior.
- Listen and be willing to hear bad news. Follow through.

14.2 Organizational Structure

Systems thinking teaches that the organizational structure influences or drives the behavior of the humans who operate in that organization. If we want people to make good decisions and act in a safe manner, then we need to construct the organization in a way that allows and encourages them to do so.

The organizational control structure must ensure that hazards are eliminated or, if that is not possible, controlled and mitigated. It also must support positive safety culture values as communicated in the documented safety philosophy.

The overall goal is to design a control structure that eliminates or reduces losses. Satisfying this goal requires a clear definition of expectations, responsibilities, authority, and accountability for safety-related tasks at all levels of the control structure. In addition, to operate effectively, the structure requires appropriate feedback and coordination between entities. There should be leading indicators to signal when the controls are becoming ineffective because of internal or external changes. Together, the entire control structure must enforce the safety constraints on behavior through physical design, processes and procedures, and social interactions and culture.

There will probably be significant differences between a safety control structure for development and one for operations. If you are a regulated industry, government regulatory agencies will need to be included in your safety control structure design considerations. Other external groups may be important to consider such as the courts, insurance companies, and user groups.

Start from the types of systemic factors leading to accidents identified in chapter 4: (1) ill-defined and diffused responsibility, authority, and accountability; (2) inappropriate placement of system safety responsibility within the organization; and (3) limited communication channels and poor information flow.

14.2.1 Assigning Responsibility, Authority, and Accountability

A common feature of the accidents described in the appendices and, indeed, most major accidents is ill-defined and defused safety responsibility, authority, and accountability.

Leadership is the key to achieving high levels of safety. This means that leadership of the organizational safety functions should not just be a rotational assignment for training future leaders and managers. Organizations that have few losses appoint leaders for their safety functions who are passionate about safety and the role they play in preventing losses. There should be a career path within the organization that allows those who are committed to safety to rise in the management organization.

As with any effective management system, there must be responsibility, authority, and accountability assignments. A belief that "everyone is responsible for their own and others' safety as well as the safety of the products" leads to excessive accidents. If everyone is responsible for safety, then nobody is.

Responsibility for safety lies at every level of the organizational structure although appropriate responsibilities will differ in each of them. In some organizations, usually those that have a large number of accidents and incidents, a belief is prevalent that responsibility for safety should be pushed down in the control structure. The argument is used that the lower levels have more information about how their parts of the system actually work.

The problem with this argument is that the lower levels also lack perspective: Although they have detailed information about their specific parts of the system, they do not have the equivalent visibility and information about other parts of the system. They cannot anticipate or control the behavior of other components in the larger system. Responsibility for coordinating decision-making necessarily requires more visibility about the larger system state than is possible at lower levels of the management hierarchy. A decision that seems perfectly safe at one level can be dangerous for the organization or system as a whole. The capsizing of the *Herald of Free Enterprise* ferry provides an example. Figure 8.5 shows the complex interactions involved in the decision-making about ferry operations and how many of those factors were beyond the scope and control of the individual decision-makers who were held responsible for the loss.

The Macondo blowout was largely caused by the poor assignment and understanding of the responsibilities and accountability of the three major companies involved and the lack of control by BP over the safety-related activities of its subcontractors. The *Deepwater Horizon* accident report found that the individual contractors had different cultures and management structures, leading to conflicts of interest, confusion, lack of coordination, and flawed decision-making. BP, as the lead company, should have been responsible for ensuring overall system safety.

The same principle applies, of course, when all the relevant parties are within one organization. The *Columbia* Space Shuttle accident report noted that management of safety at NASA involved "confused lines of responsibility, authority, and accountability that almost defies explanation" [112, p. 186].

Management responsibility for safety goes all the way up to the board of directors. The BP board of directors did not have a member responsible for assessing and verifying the performance of BP's major accident hazard prevention programs. Nobody on the BP board of directors had either refinery operational experience or a professional background in offshore drilling relevant to the risks that were being taken at Texas City or Macondo [386]. The same lack of expertise and misunderstanding of their responsibility for safety was true for the Boeing board of directors in the B737 MAX accidents.

Many aerospace accidents, and those in other industries, have occurred after the organization transitioned from oversight to "insight" [215]. Contractors have a conflict of interest with respect to safety and their own goals and cannot be assigned the responsibility that is properly that of the contracting agency or company.

Beyond lacking perspective, lower organizational levels tend to have a short-term focus rather than a longer-term one: The higher organizational levels have a broader focus on system goals as opposed to the goals of their own subcomponent of the system. The higher organizational levels need to control the behavior and interactions among the lower-level components and ensure that safety constraints are being enforced by those beneath them. It is frequently the case that project managers skip safety activities when behind schedule or over budget, which usually ends up costing more time and resources to fix when unacceptable flaws are found late in the process. A long-term focus is the responsibility of the levels above the individual project level.

Each level must provide oversight of the level below by ensuring that they are using appropriate procedures, that they are carrying them out correctly, and that the efforts are effective. There also usually needs to be a management focal point (or points) with responsibility for ensuring that the safety management system is designed and working properly at each level of the management system and as a whole.

Tips for assigning responsibility:

- Appoint leaders who are passionate about safety.
- Create career paths for those committed to preventing losses.
- Provide a clear definition of expectations, responsibilities, authority, and accountability at all levels of the safety control structure.
- Do not make everyone responsible for safety, and do not push responsibility downward without appropriate decision-making and oversight at the higher levels of the organization.
- Establish appropriate safety responsibilities for contractors and oversee their activities.
- Assign someone to be responsible for ensuring that the SMS itself is designed and working properly.

14.2.2 Location of System Safety Activities

Where should safety activities be in the organization? There are some basic principles that need to be considered in deciding where to place system safety activities and how to design and manage feedback and control change.

First, there needs to be a group with enterprise-level responsibilities such as ensuring that required activities are taking place and that they are effective. This group also provides leadership and coordination. Although safety activities will permeate every part of the development and operation of large organizations, a common methodology and approach will strengthen the individual activities. The leader of this group needs to be able to communicate directly with top management and provide input to all types of management decision-making. This implies that the person in this position must report to someone with influence and be seen as having the support of senior management.

Below this top-level safety management, there may be activities at all levels of the organization, with the higher levels having broader responsibilities and each successive lower level having more focused responsibilities appropriate to the level at which they operate: for example, at the business unit, program and project level of a development organization. The same is true for organizations that operate safety-critical systems. A common mistake is to make safety a staff function that has little impact on line operations.

There is no one or even several correct designs for a safety control structure. An effective design will depend on the industry, organizational management style, and so on. Some responsibilities that should be included in the safety control structure for development and for operations are listed at the end of this chapter. How they are assigned will depend on individual factors.

Second, separating system safety engineering from system engineering is a mistake in product development organizations; it can lead to higher costs in achieving safety goals because safety concerns must be addressed starting early in the development process. Skipping safety activities when behind schedule or over budget usually ends up costing more time and resources when unacceptable flaws are found and must be fixed late in the process. Safety engineering efforts should not be focused on after-the-fact assurance. Engineers need safety information at the time they are making safety-related decisions.

Service organizations have different concerns, but there should be some safety engineering component within all departments or groups where decision-making about safety takes place. At the same time, there are reasons for also having independent oversight and decision-making authority outside the primary decision-making group.

For all types of organizations, safety needs to be designed into the workplace, and decisions need to be made about the appropriate place for this responsibility in the organization.

Accidents are often associated with a fractured, organizationally dispersed safety staff [166] and with ill-defined responsibility and authority for safety. Problems arise especially when responsibility is divided across organizational boundaries. There should be at least one person in the organization with overall responsibility for safety.

Interface problems are particularly likely in the building of complex systems that are composites of subsystems and components designed and manufactured by more than one contractor. A central organization is needed to handle the interface issues and to ensure that each subgroup does not assume that another group is taking care of safety. As a simple example, if one subsystem contains ordnance or flammable materials, a central safety organization must make the existence of such materials known to the designers of other subsystems that could provide sources of ignition [245].

With different contractors all making decisions influencing safety, some person or group is required to integrate the information and make sure it is available to all decision makers. A large organizational distance between decision makers and those with technical awareness and competence is, of course, a common problem in engineering organizations, but poor decision-making can have especially disastrous results when safety is involved.

Some general principles for allocating safety responsibilities in the organizational structure are:

- Decision makers need direct links to those who can provide safety information. If critical information has to float up a chain of command, it may be lost or modified either deliberately, usually because of schedule or budget pressures, or inadvertently. Direct communication channels provide more chance that information is received in a timely manner and without being filtered by groups with potentially conflicting interests. Some decision makers may also need fast access to information.
- Direct communication channels to most parts of the organization are required. Safety may be involved in almost all organizational activities.
- Safety must influence decision-making, which means that decision makers must have access to necessary safety information at the time safety-related decisions need to be made.

Safety organizations are frequently separated from engineering using the argument that they need independence. Sometimes they are put into low-status groups that lack influence over the most important safety-related decisions. Their voices need to be heard. The BP managers on board the *Deepwater Horizon* were required to exercise judgment without adequate training, information, procedures, resources, and support to do their job effectively. Safety-critical decisions were being made without a safety engineer on the rig to provide input to these decisions. Well-managed projects assign responsibility to people who specialize in system safety to assist managers in making better decisions.

Safety also should not be placed in assurance organizations, although they play a role in achieving overall system safety. Putting system safety engineering activities such as hazard analysis within the assurance group establishes the expectation that system safety is an after-the-fact or auditing activity only. In fact, the most important aspects of system safety involve core engineering activities such as building safety into the basic design and proactively eliminating or mitigating hazards. By treating safety as an assurance activity only, safety concerns are guaranteed to come too late in the process to have an impact on the critical design decisions. Necessary information may not be available to the engineers when they are making decisions, and instead potential safety problems may be raised at reviews, when doing something about the poor decisions is costly and likely to be resisted. Safety engineering activities must be a central part of the engineering design organization.

Concerns about independence results from a basic dilemma: either the system safety engineers work closely with the design engineers and lose their independence, or the safety efforts remain an independent assurance effort but safety becomes divorced from the engineering and design efforts. The solution to this dilemma, which other groups use, is to

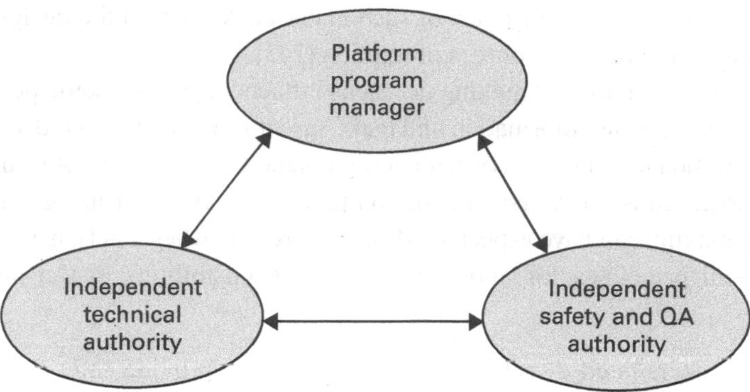

Figure 14.2
The SUBSAFE three-legged stool concept.

separate the safety engineering and the safety assurance efforts, placing safety engineering within the engineering organization and the safety assurance function within the assurance groups. Programmatic decision-making is separated from safety-related decision-making.

The SUBSAFE program for safety in US nuclear submarines uses a unique design to satisfy these requirements. They describe their structure as a separation of powers or a "three-legged stool" (figure 14.2).

Program managers in SUBSAFE can select only from a set of acceptable alternatives created by the Independent Technical Authority (ITA). Technical authority is defined as a process that establishes and assures adherence to technical standards and policy. The ITA, which is essentially doing system engineering, provides a range of technically acceptable alternatives with risk and value assessments. Risk may be specified quantitatively or qualitatively, but, as explained in chapter 4, any assessment of risk must use objective quality evidence, and therefore probabilistic risk assessment or likelihood estimates that cannot be tested or verified are not allowed.

ITA responsibilities (and accountability) in this management structure include:

- setting and enforcing technical standards
- maintaining subject matter expertise
- assuring safe and reliable operations
- ensuring effective and efficient systems engineering
- making unbiased independent technical decisions
- providing stewardship of technical and engineering capabilities

The third leg of the stool is the compliance verification organization. It is equal in authority to the program managers and the ITA.

14.2.3 Communication, Coordination, and Information Flow

An examination of the organizational factors leading to accidents often turns up communication problems. As noted in chapter 4, not only goals and policies but the reasons for decisions, procedures, and choices need to be communicated downward in order to avoid undesirable modifications by lower levels and to allow detection and correction of

Tips for where to locate activities in the control structure:

- Create a high-level group with enterprise-level responsibilities, such as ensuring required activities are taking place and are effective and for providing leadership and coordination.
  - This group ensures a common methodology and approach is being used by everyone.
  - They must have a direct path to top management and provide input to management decision-making.
  - They must report to someone with influence and be seen as having the support of senior management.
- Assign responsibilities so that safety is not just a staff function but has direct impact and influence on line operations.
- Integrate the safety function into the system engineering function. Do not separate system safety and system engineering in product development organizations.
- Include system safety expertise in all departments or groups where safety-related decision-making occurs.
- Provide independent oversight in addition to integrated functions.
- Ensure decision makers have direct links to those who can provide safety analysis and information.
- Create direct communication channels between any parts of the organization where there are safety-related activities and safety decisions need to be made.
- Consider implementing the SUBSAFE separation of powers structure. At the least, ensure that technical decisions are independent of programmatic ones.

misunderstanding and misinterpretation. In the feedback channel, the feedback from operational experience and communication of technical uncertainties and safety issues up the chain of command is crucial for proper decision-making.

Feedback and dissemination of information to those who need it to perform their safety management roles is an important factor to consider in the design of the safety control structure. Feedback is critical in managing any activity. Having access to too much information can be as bad as having too little when it overwhelms the ability to identify the most important information for effective decision-making. In addition, required information for improved safety-related decision-making must be handy; that is, available when needed. Because it is so important in reducing losses, the safety information system is discussed separately later in this chapter.

Communication is also important in coordinating activities and responses to events. People with overlapping responsibilities need communication channels and ways to coordinate their activities to ensure that the safety constraints are enforced. For example, safety-motivated changes in one subsystem may affect another subsystem and the system as a whole. Safety activities must not end up fragmented and uncoordinated.

Interactions must be defined not just between hierarchical components but also between different parts or types of systems at the same level, such as between development and manufacturing or between development and operations. Figure 14.3 shows an example of the types of safety-related information that needs to be communicated between system developers and system operators.

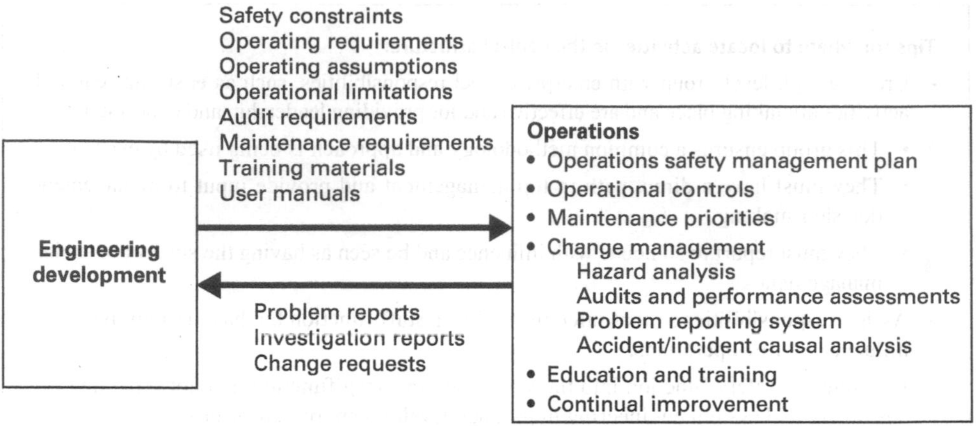

Figure 14.3
Necessary information channels between development and operations to support safety activities in both.

Similar descriptions of the information that needs to be communicated among all the components of the safety control structure must be identified and documented.

One very effective means for communication and coordination devised by the defense department is *working groups*. Defense department projects can span many years, be extremely large and complex, and often involve a large number of participants who are geographically and organizationally distributed. Coordination and communication can become a major problem in such projects. Part of the solution is provided by a hierarchical structure where the high-level management of the project may lie in the Defense Department, but there is a prime contractor that provides the system engineering and coordination among the subcontractors. For such a structure to work, communication becomes critical. Working groups have been successful in such coordination and communication efforts and can be adapted for less complex projects.

A safety working group provides an interface between two hierarchical components of the safety control structure or between two or more components at the same level. Members of these groups are responsible for coordinating safety efforts, reporting the status of unresolved safety issues, and sharing information about independent safety efforts. There may be working groups at each level of the control structure, with their members and responsibilities depending on the level at which they operate.

A corporate working group may be composed of the safety managers for different divisions or programs, while, at the lower levels, working groups may be composed of representatives from groups designing different parts of a product. When there are special types of safety problems that require new solutions, a safety group may be created to share experiences and approaches, such as a software safety working group. A government agency might create a safety working group for a large project composed of representatives from the agency offices and the prime contractor. A prime contractor may create a safety working group composed of the safety managers for all the subcontractors. And it continues similarly down the control structure.

Tips for designing communication and coordination:

- Ensure information necessary for decision-making involving safety is available to decision makers.
- Provide communication channels and a way to coordinate activities among those with overlapping safety responsibilities.
- Make sure safety activities are not fragmented and uncoordinated.
- Define required interactions and ensure that necessary information flow among them is defined and used.
- Identify and document necessary safety-related communication channels among all components of the safety control structure.
- Ensure that feedback and coordination channels exist and are working.
- Consider establishing safety working groups.

14.3 Management of Safety-Critical System Development

Most of the previous chapters in this book describe the activities needed to enhance system safety in a cost-effective way. But little has been said about the overall processes to be used or how to manage safety-critical system development.

Each project is different: system safety engineering processes and tasks need to be tailored to the criticality of the potential hazards; the particular organizational culture, structure, and personnel; the particular industry and application; and so on.

In this section, the standard system engineering V-model is assumed. Changes will be needed for other system development processes. Processes that do not allow the system safety engineering activities to be performed at a reasonable place in the overall process or exclude some of these processes completely should, however, be avoided. Saving time upfront often just means that more time and resources are required in later development and during operations and that fixes that would have been simple are enormously expensive and sometimes impossible.

Figure 14.4 shows the standard system engineering V-model. As is emphasized throughout this book, starting early is the cheapest and most effective way to optimize safety. At the same time, different activities will be required in different phases of the system life cycle.

System safety engineering tasks differ in the various phases of a project, although many merely begin in different phases but then continue, perhaps in slightly different forms, throughout the system life cycle. While they are assigned to various steps here, they could be performed at other times. One overall goal is to start them as early in the process as possible. The tasks listed below will not all be appropriate for every project, and the list is not complete.

Concept development

- Prepare the system safety program plan.
- Identify overall system safety goals for the specific project.

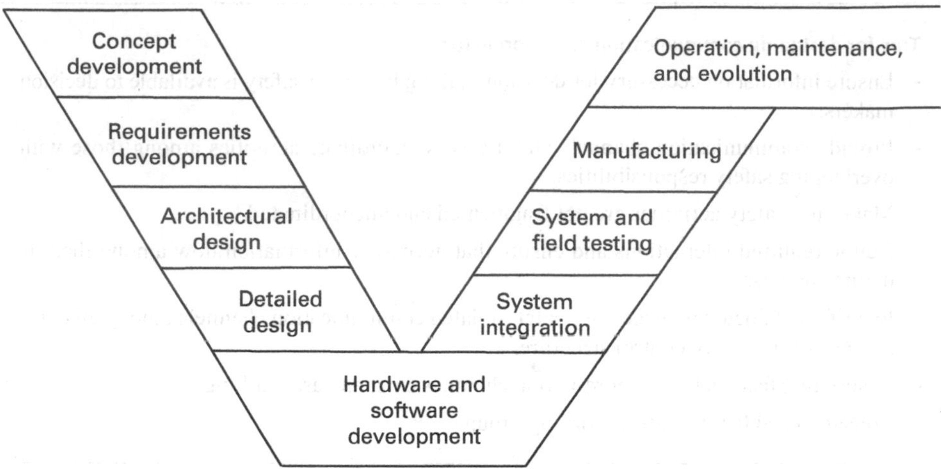

Figure 14.4
The standard system engineering V-model.

- Work with stakeholders to identify and prioritize potential losses.
- Identify system-level hazards, and trace them to the losses. Identify high-level system safety requirements and constraints.
- Translate system hazards into constraints and requirements for the system design and operation.
- Create the development safety information and documentation database, such as the hazard log and tracking system.
- Document the assumptions underlying the safety goals and the operating environment.
- Review lessons learned from similar systems, including past accidents.
- Assist in developing a concept of operations (ConOps) document that includes safety considerations during operations.
- Start the process of identifying test requirements and operational safety requirements. This process will continue through all the development steps.

Requirements engineering

- Refine system-level hazards. Translate them to system-level safety requirements and constraints.
- Add additional test requirements.
- Start the process of developing training requirements, and continue this process throughout development.

System architecture development

- Refine system-level safety requirements and constraints
- Use hazard analysis to provide safety information for the system architecture development.

- Trace system hazards to components and generate component safety requirements and constraints.
- Work with system engineers to perform trade studies on alternative architectures and to eliminate or control of hazards in the architecture selection process.
- Identify critical interface requirements.
- Add test requirements.

System design and development

- Assist in eliminating or mitigating hazards in the system detailed design.
- Add test requirements.

System integration

- Assist in resolving safety-related problems arising in system integration.

System test and evaluation

- Identify hazards in operational testing, and assist in resolving them; for example, hazards in flight test.
- Evaluate operational test anomalies and test results.
- Ensure safety-related information is incorporated into user and training documentation.

Manufacturing

- Establish manufacturing safety requirements and constraints.
- Investigate workplace accidents and incidents.

Operation, maintenance, and evolution

- Create a safety management system.
- Create an operational safety management plan.
- Investigate accidents and incidents.
- Perform safety audits.
- Generate operational safety requirements and constraints.
- Create and monitor leading indicators.
- Review change proposals for safety.
- Perform hazard analysis on operational processes and procedures, particularly changes.
- Monitor operational assumptions.

The box that follows provides some general tips for management of safety-critical development.

14.4 Management of Operational Processes and Practices

Hazard analysis and resolution does not stop with the end of development and the system going into use. There may have been hazardous scenarios omitted during design, or assumptions may have been made about the operating environment that turn out to be

Tips for managing safety-critical development:

- Start safety efforts early.
- Spend time on planning upfront. It will save much more in time and resources than trying to fix problems identified later.
- Design safety into the system; do not wait to add it to a completed design.
- Do not focus on generating a lot of paperwork and documentation. It does not make a system safer; it simply wastes resources.
- Do not rely on compliance with standards to prevent accidents. More than compliance is required.
- Document and validate assumptions and design rationale. Establish traceability between safety requirements and constraints, hazard analysis, identified causal scenarios, and design features to eliminate or control hazards.
- Minimize unnecessary complexity and features that increase complexity.
- Use system engineering processes that integrate the design of hardware, software, and human factors.
- Place high priority on communication about safety in large development projects. Use working groups and other communication channels and ensure they are effective.
- Integrate safety engineers and activities into the main engineering design process.

flawed. Designers may have decided that some causal scenarios would best be handled or perhaps can only be handled during operations. And, of course, the world changes over time, and the designers may not have predicted the way the system would be operated or the changes in its environment. In fact, most of the same or new safety engineering activities are needed after the handoff between developers and operators or users. Additional activities, such as investigating incidents and accidents, are important in obtaining information about changes that are needed to operate safely.

One critical requirement to allow safe operations is recording of the assumptions about operations in the hazard analysis and design rationale along with links between the hazards and the specific design features used to control them. Without such documentation, the cost of reanalysis when contemplating changes during operations can be enormous and, in some cases, impractical. For example, if a change is desired or required to a complex system during operations, the operator will need to know if it will impact a design feature purposely used to eliminate or mitigate an identified hazard. Without appropriate documentation, the developer might not know either. Most of the required information should be documented in the original hazard analysis and design rationale, but it needs to be pulled out and provided in a form that makes it easy to locate and use during operations.

Figure 14.3 shows the types of information that needs to be provided in the handoff between developer and operator. There will also be important information that must go from the operators back to the developers to fix problems identified during operation of the system. Of course, developers need to get as much information as possible from potential operators during development. The best results ensue when operators work closely with developers. Operators should create an operational safety management plan to ensure that risks are contained during operations.

Some important functions that need to take place during operations involve providing a shared and accurate perception of current risk, feedback and learning from events, creating and updating operational procedures, training and contingency management, managing change, and maintenance.

14.4.1 Providing a Shared and Accurate Perception of Risk

Chapter 4 and the appendices provide many examples of accidents resulting from complacency and inaccurate perception of risk. While not an easy problem to overcome, its solution is basic to improving risk-related decision-making. Most people want to make good decisions about risk but simply don't have accurate information and usually have competing pressures. Providing education and accurate information are key here.

Education must include information about not only the hazards and safety constraints enforced by the controls but also about priorities and how decisions are to be made. The safety philosophy statement, discussed earlier, provides information about the safety values to be used in decision-making. In addition, everyone needs to know the risks they are taking in the decisions they make. Often, poor decision-making arises from having an incorrect assessment of the risk being assumed. Using the nonprobabilistic definition of risk provided earlier, this means that the decision makers must know how their decisions will impact the designed controls in the safety control structure.

Telling managers and employees to be "mindful of weak signals" (a common suggestion in the High Reliability Organization, or HRO, literature) simply creates a pretext for blame after a loss event occurs and hindsight provides the clarity that transforms a weak signal (that is, noise) into a strong signal. Instead, everyone must be educated about the hazards associated with the operation of a system and how to recognize them if we expect them to recognize the precursors to an accident. People need to know what to look for, not just be told to look for an undefined something.

System safety engineering should provide this information. Usually, more detailed information is required than simply a standard risk matrix. While the risk matrix, if it is accurate, provides information about the level of risk, it does not provide the detailed information to act appropriately in lowering risk. More sophisticated tools for understanding the risks involved are needed to assist in decision-making.

One problem is that informal risk assessment may change as time passes without a loss. The actual risk has probably not decreased, but our perception of it has. That leads to a change in priorities and in the resolution of conflicting goals. Indeed, the Space Shuttle losses can be partly explained in this way, particularly the *Columbia* accident [216]. Unfortunately, a strange dynamic can arise where success at preventing accidents can actually lead to behavior and decision-making that paradoxically increases risk. A circular dynamic occurs where safety efforts are successfully employed, the feeling grows that accidents cannot occur, which leads to reduction in the safety efforts, an accident, and then increased controls for a while until the system drifts back to an unsafe state and complacency again increases, and so on. This dynamic is a good example of circular causality, described in chapter 7 as a limitation of linear causality models.

The complacency factor is so common that safety management systems need to include ways to deal with it. SUBSAFE, the very successful US nuclear submarine safety program, puts major emphasis on fighting this tendency and has been particularly successful

Tips for maintaining accurate risk perception:

- Provide information beyond the standard risk matrix including specific existing hazards and ways to manage them.
- Protect against informal downgrading of risk when accidents do not occur.
- Focus on identifying and implementing ways to fight ignorance, arrogance, and complacency.

in limiting it. They continually are putting effort into combating what they describe as their three most difficult challenges:

1. Ignorance: not knowing.
2. Arrogance: pride, self-importance, conceit, or assumption of intellectual superiority and presumption of knowledge that is not supported by fact.
3. Complacency: satisfaction with one's accomplishments accompanied by a lack of awareness of actual dangers or deficiencies.

Much of the design and focus of the SUBSAFE program is aimed at combating these three challenges, and the design of any SMS should include steps to provide vigilance against them.

One way to combat this erosion of risk perception is to provide ways to maintain accurate risk assessments in the minds of the system controllers. The better information controllers have, the more accurate will be their risk assessments and therefore the better their decisions. Accident analysis and various kinds of audits should involve a careful investigation of the accuracy of risk perception in the mental models of those controlling high-risk projects. Unearthing the reason for this migration during an accident or incident causal analysis when unrealistic risk perception was involved can help to identify ways to design the safety control structure to prevent it or detect it when it occurs.

14.4.2 Feedback and Learning from Events

Good decision-making and preventing accidents require information flow and properly functioning feedback channels. Cultural problems are a common reason for problems in feedback. There are three general ways to implement feedback channels: audits and performance assessments, accident/incident causal analysis, and reporting systems. All of these activities contribute to establishing a learning culture.

Audits and performance assessments There are many types of audits and performance assessments, but those whose goal is to evaluate the state of safety start from the safety constraints and the assumptions in the design of the safety controls. Audits should involve not just the products and processes but also the safety management system itself and the effectiveness of the controls designed to ensure that losses are prevented. At least some part of a safety audit and performance assessment should be focused on the operation of the safety management system as it was designed and assumed would operate.

The entire safety control structure must be audited and not just the lower levels. In the SUBSAFE program, even the top admirals are subject to a SUBSAFE audit. Not only does this ensure that the program is operating as assumed, but it also provides a positive cultural component to the audit in that all employees see that even the leaders are expected to follow the SUBSAFE rules and that top management are willing to accept and resolve audit findings just like any other member of the SUBSAFE community. People at lower levels of the SUBSAFE safety control structure participate in the performance assessment of those above them, providing a visible sign of the commitment of the entire program to safety and the importance of everyone in preventing losses. Accepting being audited and implementing improvements as a result—that is, leading by example—is a powerful way for leaders to convey their commitment to safety and to its improvement.

Participatory and nonpunitive audits can be very effective. The goal of such an audit is that it be a constructive learning experience and not a judgmental process. It should be viewed as a chance to improve safety rather than a way to evaluate employees. Instead of the usual observation-only audit process by outside audit companies, experts from other parts of the organization not directly being audited should make up the audit team. Various stakeholders may play a role in the audit, and even individuals from the group being audited may be on the audit team. The goal should be to create an attitude that this is a chance to improve our practices and provide a learning activity for everyone involved, including the auditors.

The audit itself should be designed to allow continuous communication with those being audited so as to obtain a full understanding of any identified problems and potential solutions. Unlike the usual rules for outside audits, in a participatory audit, immediate feedback should be provided and solutions discussed. Doing this will reinforce the understanding that the goal is to improve safety and not to punish or evaluate those involved. It also provides an opportunity to solve problems when a knowledgeable team is on the spot and not after the usual written report is provided months after the audit actually occurs. And, of course, it serves as an effective feedback channel as well as a way to communicate that management is committed to the organization's safety goals.

An additional goal of safety audits and performance assessments is to measure the level of safety knowledge and training that actually exists, not what managers think exists or what exists in the training programs and user manuals. The audits can provide important feedback about potential improvement of the training and education activities. In keeping with the nonpunitive audit philosophy, knowledge assessments should not be used in a negative way that is viewed as punishment by those being assessed. Instead, the goal should be to improve training and education efforts.

Incident and accident investigation Incident and accident investigation provide clear evidence that the SMS is not working as designed or expected. The investigation procedures must be embedded in an organizational structure that allows exploitation of the results. All systemic factors involved must be identified and not just the symptoms or technical factors as demonstrated in chapter 8. Was the incident a result of wrong assumptions about operations during system development, an inadequate fix of a previously identified hazard, or a new hazard or new causal scenario of a known hazard that resulted from a change in the way the system is being used or in the environment? Did it result from slow migration to a state of high risk?

Assigning blame should not be the goal of the investigation; rather the goal should be to find out why the safety control structure was not effective in preventing the loss or near loss. That means that the entire safety control structure must be investigated to identify each component's potential contribution to the adverse events.

Managers should not be responsible for investigating incidents that occur in their chain of command: investigators and causal analysts must be managerial and financially independent from those in the immediate management structure involved. Using trained teams with independent budgets and with high-level and independent management should be considered.

Investigation requires more than just writing a report. There must be assignment of responsibility for ensuring that appropriate measures are taken to strengthen any aspects of the safety management system that contributed to the events. Then there should be follow-up to ensure that the fixes were effective. Too often, fixes are made, but there is no attempt to determine whether the fixes were successful in improving safety management until another incident or accident occurs.

Finally, the findings should be used as input to future audits and performance assessments. If there is a reoccurrence of the same factors that led to past incidents and accidents, there needs to be an investigation of why those factors were never corrected or why they reoccurred even if they were removed for a while. If fixes are not effective in removing the causes of incidents, then an investigation of the process of creating recommendations and responding to them is warranted to identify any weaknesses in these processes in the organization and to improve them. That is, not only must the factors involved in the incident be corrected but also the process that led to inadequate fixes being implemented after previous incidents or accidents.

Reporting systems Reporting systems are critical. Often, after an accident, it is found that the same events occurred multiple times in the past but were never reported or, if reported, were never corrected. These events may involve near misses. For example, several aircraft may not fly an airport approach correctly, but because it did not result in an accident, they may not report it, even if a reporting system exists. Not until an accident occurs is action taken.

A common finding in accident reports is that a reporting system existed but was not used, as discussed in chapter 8. These reports then usually include recommendations to train people on how to use the reporting system and require that they use it. This recommendation assumes that the problem is with the potential reporter and not with the design of the reporting system itself.

Examination of the reporting system may find that it is difficult or awkward to use and that information reported appears to go into a black hole. People may believe that there is no point in going through the official reporting system because the organization will not do anything anyway about the factors reported. Often, those finding problems bypass the reporting system and simply go directly to a person or group they believe may be able to solve the problems. While this might be effective and efficient in the short term, it may lead to the same problems occurring in the future because the systemic causes are not eliminated. In some cases, reporting is inhibited by a fear that the information reported will be used against the reporter. Anonymous reporting systems can be helpful here, but establishing a Just Culture and assuring employees that management cares about safety will be even more effective. An anonymous reporting system should be a little-used backup to detect when safety culture problems exist in the organization.

Another factor to consider is that often events are not reported by front-line operators because they identified the problem before it created a perception of risk, or they may have perceived it as only their own error. Most people are not trained to recognize risk when it is created as part of a normal job process. People accept the flaws in design as "normal" and perhaps already known, so rather than reporting them, they just work with or around them. A related factor is that people will generally not report hazardous events when those events do not meet the criteria for required reporting. Near misses may fall in the latter category.

In general, reporting needs to be encouraged. This can be accomplished by maximizing accessibility, minimizing anxiety, and acting on the information obtained. Reporting forms or channels should be easily and ubiquitously available and not cumbersome to fill in or use. To minimize anxiety, there should be a written policy on what the reporting process is; the consequences of reporting; and the rights, privileges, protections, and obligations of those doing the reporting as well as those following up on the reports. Without a written policy, ambiguity exists, and people will disclose less. Alternatively, ambiguity about who is responsible for following up may lead to everyone assuming that someone else will take care of it.

Finally, encouraging reporting involves providing feedback and rewarding the efforts. Immediately after a report is created, the person who provides the information should be informed that the report was received, assured that it will be investigated, and thanked for their input. A second crucial component is providing feedback later about the results of any investigation and any steps that were taken as a result to prevent a reoccurrence. Reporters should not feel like their concerns are being ignored.

Learning and continual improvement Because SMS designs are rarely perfect from the beginning, and the world changes over time, there must be an effective process in place to ensure that the organization is continually learning from safety incidents and improving the SMS itself as well as the workplace, the products, and the services.

If losses are to be reduced over time and companies are not going to engage in constant firefighting, a process for continual improvement needs to be implemented. Identified flaws must not only be fixed (that is, symptom removal), but the larger operational and development safety management systems must be improved, as well as the process that allowed the flaws to be introduced in thc first place. The overall goal is to change the culture from a *fixing orientation*—identifying and eliminating deviations that are symptoms of deeper problems—to a *learning orientation* where systemic causes are included in the search for the source of safety problems.

To accomplish this goal, a feedback control loop is needed to track and assess the effectiveness of the development and operational safety management systems and its controls. Were hazards overlooked or incorrectly assessed as unlikely or not serious? Were some potential failures or design errors not included in the hazard analysis? Were identified hazards inappropriately accepted rather than being fixed? Were the designed controls ineffective? If so, why?

When numerical risk assessment techniques are used, operational experience can provide insight into the accuracy of the models and probabilities used. In various studies of the DC-10 aircraft, the chance of engine power loss with resulting slat damage during takeoff was estimated to be less than one in a billion fights. However, this highly improbable event

Tips for creating a learning culture:

- Design and ensure the continued efficacy of audits, performance assessments, and reporting systems.
- Audit the safety control structure itself (including all levels) and the effectiveness of the designed controls.
- Design audits so they are constructive learning experiences and not a judgmental process.
    - Include as participants members of the groups that are being audited.
    - Use audits as a way to improve safety and as a learning activity and not to evaluate employees.
- Take advantage of audits to evaluate the effectiveness of training and education activities and use them to provide feedback (knowledge assessment) to be used for improving training activities.
- Create effective system-level accident/incident causal analysis procedures that focus on *why* and not *who.*
- Create incident and investigation procedures that identify systemic factors and not just the symptoms of the deeper problems.
- Embed the investigation procedures in an organizational structure that allows exploitation of the results. Assigning blame or finding a root cause should not be the goal.
- Create an accident investigation process that is managerially and financially independent from those in the immediate management structure involved. Consider using highly trained teams with independent budgets and high-level management. Follow up on recommendations to determine whether they were effective and, if not, then why.
- Ensure that reporting systems are easy to use and available and that anonymous reporting channels exist.
    - Encourage reporting and train people to know when it should be used.
    - Provide a written policy.
    - Maximize accessibility, minimize anxiety, and act on information obtained.
    - Provide feedback to those using the reporting channels. Reporters need to feel that their concerns are not being ignored.
- Create a learning culture using feedback loops and the checking during operations of assumptions made in the safety analysis. Make everyone responsible for coming up with solutions, and show that their input matters.
- Assign responsibility to someone in the safety control structure to ensure that learning and improvement is occurring.

occurred four times in DC-10s in the first few years of operation without raising alarm bells before it led to an accident and changes were made. The probability of a certain type of uncontrolled reaction was estimated to be zero (impossible) during design of the Shell Moerdijk chemical plant although it had happened twice before in other Shell plants [87]. Even one event should have warned someone that the models used might be incorrect and that such an event was not impossible.

Experimentation is an important part of the learning process, and trying new ideas and approaches to improving safety should be encouraged but also evaluated carefully to ensure that improvement actually results.

Finally, responsibility must be assigned to someone in the safety management system control structure to ensure that learning is taking place.

14.4.3 Creating and Updating Operating Procedures

Procedures are important. Expecting operators to make correct real-time decisions in case of any event or set of events is unrealistic. Pilots are provided with extensive lists of what to do under various circumstances, particularly emergencies.

At the same time, operators must know when the procedures provided should be abandoned and creative solutions to problems produced. The ability to do this requires extensive training, especially with today's complex and highly automated systems. Inevitably operators will make the wrong decision to either follow the procedures provided when it appears that the procedures are wrong or to incorrectly deviate from procedures when, in fact, they were right. Either way they will be blamed. Jury-rigging saved the day at the Browns Ferry nuclear power plant, while at Three Mile Island the operators were blamed for following the incorrect procedures and training that had been provided. One of the problems noted in chapter 4 was that the TMI operators were trained to be button pushers and did not have the extensive knowledge needed to evaluate the situation correctly. Sometimes after accidents shine a light on procedures, the procedures are found to be flawed or impractical, but operators may still be blamed for not following them.

When operators do not follow the written procedures, the response should not be blame and punishment but instead an attempt to understand why they felt it was safe and why they felt the need to deviate from the procedures. Including operators in the problem-solving effort will not only improve the solutions to any identified problems but improve the safety culture and workplace morale as well.

We leave humans in systems and do not automate everything because humans are adaptable and flexible in the face of unexpected events. Humans are able to adapt both the goals and the means to achieve them and to use problem-solving and creativity to cope with unusual and unforeseen situations. Human error, as noted previously, is the inevitable side effect of this flexibility and adaptability. The goal should be to create a work environment where such flexibility and adaptability is encouraged but those deviating from procedures understand the risks and make an informed decision when doing so. Sometimes the decisions will be wrong, as is true for all decisions. Management needs to accept that and use such times to learn how to improve decision-making in the future.

Because the time to come up with the right response in an emergency may not be available, contingency procedures need to be created and frequently practiced. The story of the *Bankston* mudboat in chapter 4 demonstrated the value of doing this. While chaos surrounded the life rafts on the *Deepwater Horizon* rig, and workers were jumping into the ocean to escape the fire and explosions, the *Bankston* crew calmly executed the emergency procedures they had reviewed and practiced that morning as they always did before any hazardous activity: they first protected the mudboat and its crew from the events that were occurring on the drilling rig and then manned the *Bankston* lifeboats to pull rig personnel out of the sea.

Workers often deviate from the procedures provided even in non-emergency situations. Chapter 6 explains this behavior in terms of operators having to work in the system that actually exists rather than the ideal one existing in the minds of those who design the procedures (see figure 6.4). When procedures are not updated to reflect actual working conditions,

Tips for creating and updating operational procedures:

- Explain the rationale behind the procedures.
- Keep procedures up to date by periodically revisiting them and monitoring when changes in the system or environment make updating necessary.
- Monitor whether procedures are being followed. If not, identify why people feel the need to change them. Evaluate the gap between specified procedures and actual behavior. Encourage input from those deviating from procedures to explain why and participation in any evaluation and correction activities.
- Create procedures through a partnership between operators, system designers, system safety engineers, and human factors experts. Perform an independent review of the resulting procedures along with testing or other types of evaluation.

operators and their supervisors learn not to rely on them. When processes change and following written procedures does not allow efficient operations, workers will create their own procedures, which may not adequately address hazards. At Texas City, BP management allowed operators and supervisors to alter, edit, add, and remove procedural steps without following the required management of change procedures to assess the safety impact of those changes. Following procedures does not guarantee safety, as argued in chapter 8. Instead, safety comes from people being skilled in judging when and how to apply them [76].

After accidents, operators are often *prima facie* assigned blame for not following the procedures. Some accident investigation techniques start by determining whether the operators were noncompliant with the procedures. If not, they are assumed to be the cause of the accident. Especially in workplace safety, efforts to prevent accidents frequently consist of forcing operators to follow procedures. But operators may need to adapt procedures to deal with a changing system and to satisfy multiple goals, such as safety, efficiency, and productivity. Instead of forcing compliance, the best approach is to (1) monitor and identify the gap that exists between specified procedures and actual practice, (2) understand why the gap exists, and (3) update and rewrite the procedures accordingly.

Accidents have occurred because operators were asked to create the procedures they were to use on the assumption that they would have the best understanding about what needed to be done and how. Accidents resulted because the operators had a limited view of the rationale behind the design of the equipment and the hazards involved in its operation. Creating procedures needs to involve a partnership between operators, system designers, system safety engineers, and human factors experts. Appropriate independent evaluation of the resulting procedures from a safety standpoint should also be performed to ensure that tradeoffs in the creating process were made correctly.

14.4.4 Training and Contingency Management

Everyone in the safety management system, not just the lower-level controllers of the physical systems, must understand their roles and responsibilities with respect to safety and why the system—including the organizational aspects of the safety control structure—was designed the way it was. If employees understand the intent of the SMS and commit to it, they are more likely to comply with that intention rather than simply follow rules when it is convenient to do so. Training is not enough; education is required.

Education must include not only information about the hazards and safety constraints enforced by the controls but also about priorities and how decisions are to be made. The safety philosophy statement, discussed earlier, provides information about the safety values to be used in decision-making. In addition, everyone needs to know the risks they are taking in the decisions they make. Often, poor decision-making arises from having an incorrect assessment of the risk being assumed. Using the nonprobabilistic definition of risk provided previously, this means that the decision makers must know how their decisions will impact the designed controls in the safety management system.

As discussed earlier in this chapter, telling employees to be "mindful of weak signals" simply creates a pretext for blame after a loss event occurs. Instead, if we expect people to recognize the precursors of a loss, everyone must be trained on the hazards associated with the operation of the system and how to recognize them. People need to know what to look for, not just be told to look for an undefined something.

Training should also include "why" as well as "what." Understanding the rationale behind the safety rules they are asked to follow will help reduce (1) complacency, (2) what appears to be reckless behavior but to the person made perfect sense, and (3) unintended changes leading to hazards. The rationale includes understanding why previous accidents occurred and what changes were made to try to prevent a reoccurrence. After the *Columbia* Space Shuttle loss, the accident investigation committee was surprised to find that few NASA engineers working in the Space Shuttle program had read the previous *Challenger* accident report [112]. As noted in appendix B, the same causal factors were involved.

With increasing automation and people interacting with it, understanding the hazards and controls involved may require more training than was necessary in the past when the hazards were more obvious and intuitive. People who interact with complex systems need to learn more than just the procedures to follow: they must also have an in-depth understanding of the controlled physical process as well as the logic used in any automated controller they may be supervising or with which they may be interacting. Human operators need to know [217]:

- The system hazards and the reasons behind safety-critical procedures and operational rules.
- The potential result of removing or overriding controls, changing prescribed procedures, and inattention to safety-critical features and operations: past accidents and their causes should be reviewed and understood.
- How to interpret feedback: training needs to include different combinations of alerts and sequences of events, not just single events.
- How to think flexibly when solving problems: controllers need to be provided with the opportunity to practice problem-solving involving safety.
- General strategies rather than specific responses: controllers need to develop skills for dealing with unanticipated events.
- How to test hypotheses in an appropriate way: to update mental models, human controllers often use hypothesis testing to understand the current system states and to update their process models. Such hypothesis testing is common with computers and automated systems where documentation is usually so poor and hard to use that experimentation is often the only way to understand the automation design and behavior. Hypothesis testing,

Tips for training and education:

- Educate; don't just train.
- Make sure everyone understands their roles and responsibilities, why the system was designed the way it was, information about hazards and safety constraints enforced by the controls, and the risks they are taking in the decisions they make.
- Include "why" and not just "what."
- Include in everyone's training:
  - The system hazards and the reasons behind safety-critical procedures and operational rules that are related to their jobs.
  - The potential result of removing or overriding controls, changing prescribed procedures, and inattention to safety-critical features and operations.
- Make sure the causes of past accidents in the same or similar systems are well disseminated and understood.
- Give people the opportunity to practice problem-solving involving safety.
- Teach general strategies rather than specific responses so that controllers can develop skills for dealing with unanticipated events.
- Teach operators how to test hypotheses in an appropriate way and provide ways for them to learn in this way.
- Include overlearning and continual practice for emergency procedures.
- Review contingency procedures before any safety-critical activity.
- Provide continuous training and not just a one-time event when hired.
- Teach about recent events and trends.
- Consider having managers provide at least part of the safety training.

however, can lead to losses. Designers need to provide operators with the ability to test hypotheses safely and controllers must be educated on how to do so.

- Emergency procedures must be overlearned and continually practiced. Controllers need to be taught about operating limits and specific actions to take in case they are exceeded. Requiring operators to make good decisions under stress and without full information simply ensures they will be blamed after a loss occurs.

Training should not be a one-time event for employees but should be continual throughout their employment, if only as a reminder of their responsibilities and the system hazards. Learning about recent events and trends should be part of this training. Assessing the effectiveness of the training, perhaps through regular audits and performance assessments, can be useful in implementing an effective improvement and learning process. Incident and accident investigation results are an important source of information about training effectiveness.

Some companies have adopted a practice where managers provide safety training. Training experts help manage group dynamics and curriculum development, but the training is provided by project or group leaders. By learning to teach the materials, supervisors, and managers are more likely to absorb and practice the key principles. In addition, it has been found that employees pay more attention to messages delivered by their boss than by a trainer or safety expert.

14.4.5 Managing Change

Most accidents occur after some type of change. This fact is not surprising as continuing to do what one has done without any changes in behavior or the environment should theoretically result in the same consequences. Systems will always change and evolve over time, as will the environment in which the system operates. Adaptation and change are an inherent part of any system and are required for an organization to thrive. Because changes are necessary and inevitable, processes must be created to ensure that safety is not degrading. The problem is not change, but *unsafe* change. The SMS, therefore, must have carefully designed controls to prevent unsafe changes and to detect them if they occur despite any efforts to prevent them. The goal is to allow change as long as it does not violate the safety constraints.

Common types of changes that must be controlled are:

1. Physical changes: the equipment may degrade or not be maintained properly.
2. Human changes: human behavior and priorities usually change over time for many reasons or may involve limited-time workarounds.
3. Organizational changes: change is a constant in most organizations, including changes in the safety controls and safety management system.
4. The physical or social environment changes in which the system operates or with which it interacts. These may be permanent or of limited duration.

Controls need to be established to reduce the risk associated with all these types of changes. The controls may be in the physical system itself or in the safety management system. Because operational safety depends on the accuracy of the assumptions and models underlying the design and hazard analysis processes, the operational system should be monitored to ensure that:

1. The system is constructed, operated, and maintained in the manner assumed by the designers.
2. The models and assumptions used during initial decision-making and design are correct.
3. The models and assumptions are not violated by changes in the system operation, such as workarounds or unauthorized changes in procedures, or by changes in the environment.

Changes may be planned (intended) or unplanned (unintended). Each of these types of changes needs to be handled differently.

Planned changes If the changes are planned, a strong and well-designed management of change policy that is enforced and followed should ensure the changes are safe. In many accidents, management of change (MoC) procedures existed, but they were neither effective nor enforced. One example in the 2014 Shell Moerdijk explosion was the switch to a new catalyst without testing it and assuming that previous catalyst properties still held. A second example was the removal of parts of the work instructions for the unit that exploded—again without assessment—because the instructions were not considered critical. Specifically, critical requirements regarding nitrogen flow were removed during periodic updates of the work instructions in an attempt to limit their content to information that was believed essential and to focus on what was incorrectly thought to be the most important from a safety and operational view. Other information was omitted from the work instructions because, over time, understanding of the most appropriate procedures related to the unit involved changed [87]. All of these changes should have been carefully reviewed by safety personnel.

Careful examination of the MoC policies and the enforcement mechanisms along with feedback about whether they are being followed can assist in improving the policies and enforcement mechanisms and compliance with them.

Unplanned changes Unplanned changes provide a more difficult challenge. There needs to be a way for the company to detect unplanned changes that affect safety and to respond to them as well as to institute procedures to prevent them from occurring if possible.

Changes may occur slowly over time, and their impact may not be obvious. One example is the changes in the work instructions at Shell Moerdijk. As the work instructions were amended before each major maintenance action (called a turnaround in chemical plants), important information was omitted—in some cases intentionally and in others unintentionally. Examples include the nitrogen flow requirements mentioned earlier and the required heating rate for the reactor. Changes do not appear to have been reviewed by experts, but if they had been, then the review process was flawed. Another example is the crash of a cargo aircraft while landing at Birmingham Airport [246]. One factor implicated was the general increase in night cargo operations at airports. A CAST (Causal Analysis based on Systems Theory) of this accident found that historical assumptions about airport operations and procedures created to enhance safety needed to be revisited in the light of changes to airline operations over time. CAST [222] is an accident analysis process based on the approach described in chapter 8.

Changes may be known and planned in one system component but appear as unplanned and unknown changes to another component of the system. The change in composition of the catalyst was known by the catalyst manufacturer but not by Shell Moerdijk. Clearly communication is an important factor here.

Leading indicators are commonly used in some industries to identify when the system is migrating toward a state of higher risk. At Shell Moerdijk, as well as most chemical plants, the leading indicators used may be common throughout the whole industry, in this case the number of leaks. Their selection seems to be predicated primarily on ease of collection and lack of having any alternatives for creating more effective ones.

A new approach to identifying leading indicators has been created, called *assumption-based leading safety indicators* [218]. Briefly, the idea is that certain assumptions are made during system development that are used to design safety into a system. When, over time, those assumptions no longer hold, then the organization is likely to migrate to a state of higher risk. Leading indicators, then, can be identified by checking the original safety-related assumptions during operations to make sure that they are still true. The assumptions underlying the design, of course, must be identified and documented for this approach to work.

The use of assumption-based leading safety indicators involves both shaping actions and hedging actions [81]. Shaping actions are those built into the original design of the system to prevent identified hazards based on the assumptions of the hazard analysis and resolution process. Some of these may be actions taken during operations, including the design of operational procedures. Hedging or contingency actions are actions taken in preparation for the possibility that an assumption will fail. In addition, signposts are

Tips for managing and controlling change:

- Design controls and MoC policy to prevent unsafe changes and detect if they occur.
- Evaluate all planned changes, including temporary ones, for their potential impact on safety.
- Assign responsibility for ensuring that MoC procedures are enforced and are being followed. If they are not, find out why and fix the problems.
- Create documentation and procedures that minimize the cost of performing the MoC procedures.
- Create ways to identify unplanned changes that could be unsafe and to respond to these changes.
  - Create assumption-based leading indicators and a risk management program that effectively monitors and responds when potentially unsafe changes are identified.
  - Record assumptions during design and development of the organizational structure, the SMS, the products, and the workplace.
  - Create shaping and hedging actions and a leading indicator checking program, including audits and performance checking as well as signposts.
  - Implement leading indicators to signal when controls are becoming ineffective.
- Ensure that decision makers have information about the current level of risk and the state of the designed safety controls.
- Assign responsibility to respond when feedback shows that the application of the MoC procedures does not match the true level of risk.
- Remain vigilant against the degradation of the safety control structure and the safety culture over time and any increase in complacency.

specific events, planned or unplanned, when specific points in the unfolding future where changes in the current safety controls (shaping and hedging actions) may be necessary or advisable. In essence, they involve planning for monitoring and responding to particular events that may lead to changes in the assumptions underlying the safety controls. For example, new construction or known future changes in the system or in the environment may trigger a planned response, including the triggering of the MoC procedures.

Changes need to be evaluated with respect to their impact on safety. The cost of such evaluations will depend on the scope of the change, the quality of documentation, how the original hazard analysis was performed, and the procedures required in performing the MoC analysis. It is helpful if the analysis process includes traceability from identified hazards to their causal scenarios and to the design features (shaping and hedging actions) that are used to prevent them. Without such traceability, the cost of reanalysis after a change may be impractical and therefore skipped.

In fact, MoC procedures are often skipped in practice, resulting in losses that need never have occurred. There must be responsibility assigned for enforcement of MoC procedures and feedback channels to determine whether they are being followed and, if not, why. If the cause of skipping them is in the MoC change procedures themselves—too onerous, too difficult, too time-consuming, and so on—then the MoC procedures may need to be changed.

One less obvious type of change is the re-evaluation of risk downward over time when no losses occur. Under pressure, people start to violate their own rules and justify it by arguing that safety will not be affected. The processes in the safety control structure need to interrupt this risk re-evaluation before safety margins are eroded. In addition, there needs to be an alerting function to a responsible person when behavior is contrary to the true level of risk. This change can be short-circuited by providing information that allows accurate risk assessment by decision makers. Another strategy is to allow flexibility in how safety goals are achieved so that ignoring rules is less frequently necessary.

14.4.6 Maintenance

Maintenance is critical for safety. Most of the accidents described in the appendices involved poor maintenance practices, as do a lot of accidents in general. One of the reasons systems migrate toward higher states of risk is that physical processes degrade unless they are properly maintained. Human behavior also changes over time, but maintenance in this case involves different types of processes, usually including updating training and education.

Obviously, money and trained personnel are required for proper hardware maintenance to be performed. Preventative maintenance is also important: waiting until equipment breaks down greatly increases risk. At the other extreme, sometimes maintenance activities are delayed to save time and money. Maintenance records are important as is proper oversight and responsibility for ensuring that maintenance requirements are being satisfied.

Reliance on redundancy or on backup and safety equipment is often unjustified when the redundant components are not properly maintained. For example, the blowout preventer on *Deepwater Horizon* was neither maintained properly nor its batteries replaced when required. At the same time, redundancy can be easily defeated if the same person

Tips for maintenance:

- Identify safety-critical equipment.
- Provide proper resources and training.
- Ensure that failure and maintenance information is being recorded accurately.
- Assign oversight responsibilities to ensure that proper maintenance is being performed and that an accurate view of the state of the equipment is provided to decision makers.
- Use preventative maintenance rather than running critical equipment until failure.
- Before returning to potentially dangerous operation:
    - Test the alarms and other equipment to ensure it has been reactivated.
    - Ensure the maintained equipment is operating correctly.
- Provide independent maintenance of redundant components when relying on redundancy for safety.
- Perform trend analysis to warn when risk may be increasing and about any changes in maintenance procedures that are increasing risk.
- Do not startup or perform any safety-related activities when safety-critical equipment is not operational.
- Create ways to ensure that safety interlocks are reset after maintenance activities.
- Create a maintenance information system that contains up-to-date information about maintenance activities—including instructions and maintenance activities along with any relevant trend analysis. This information should be easy to retrieve and should provide an accurate view of the current state of the plant equipment.

maintains all or some of the redundant components. Aircraft have been lost when a maintenance person installed new components, making the same mistake on each of them.

Sometimes alarms do not sound because they were not maintained. Safety-critical equipment must be identified so that it can receive proper attention. The results of the hazard analysis and the documented design rationale can assist in identifying this equipment.

Operating the plant especially during particularly hazardous activities such as startup, when safety-critical equipment such as critical alarms is out of order, must be prevented. In many accidents, decision makers knew that critical equipment was out of order, but that did not deter them from performing potentially dangerous activities.

When critical equipment undergoes maintenance, testing and other means of assurance should be used to make sure the maintenance tasks were performed properly and that the equipment is again operational. A common error when maintenance is completed is to not restart alarms that had to be turned off in order to perform the maintenance.

Trend analysis can provide important information about maintenance needs. Changes in maintenance procedures were not investigated after failures of the *Challenger* O-rings increased over time. In addition, up-to-date documentation, written procedures, and training are necessary. Automation can assist in these activities and provide a way for management to understand the state of the physical equipment and make decisions related to it. It can also be used to provide an overview of the physical state of the system and any risk related to it.

Finally, remember that software needs to be maintained too.

14.5 Creating an Effective Safety Information System

A comprehensive and usable safety information system (SIS) is key to having a successful safety management system. Good decision-making requires accurate and up-to-date information about the safety of the system. After studying organizations and accidents, Kjellan concluded that an effective safety information system ranked second only to top management concern about safety in discriminating between safe and unsafe companies matched on other variables [178].

Control of any activity requires information. Documenting and tracking hazards and their resolution are basic requirements for any effective safety program. All hazards need to be recorded, not just the most critical; otherwise, there is no record that a particular condition has already been evaluated. The complete hazard log and audit trail will show what was done and why decisions were made. Because hazard information may become very large, ways to prioritize and summarize the most significant information are usually necessary and useful for reviews and management oversight.

The SIS can provide the information necessary to detect trends, changes, and other precursors to an accident; to evaluate the effectiveness of the safety controls; to compare models and risk assessments with actual behavior; and to learn from events and improve the SMS. After major accidents, it is often found that the information to prevent the loss existed but was not used or was not available to those involved. Often, lots of information is collected only because it is required for government reports and not necessarily because it is considered a necessity for the operation of an effective SMS.

To determine what should be in the SIS, the responsibilities of those in the safety control structure and the feedback they need to fulfill those responsibilities can be used; that is, the information required to make appropriate decisions. In general, the SIS contains, at a minimum:

- the safety management plan (for both development and operations);
- the status of all safety-related activities for the system;
- the safety constraints and assumptions underlying the design, including operational limitations;
- the results of the hazard analyses (hazard logs) along with tracking and status information on all known hazards;
- the results of performance audits and assessments;
- incident and accident investigation reports and corrective actions taken;
- lessons learned and historical information; and
- results of trend analysis.

Information may be collected by companies or by industries. The sharing of information within an industry can add to the general knowledge about hazards and about effective and ineffective control measures. Within a company, an information system can provide valuable feedback about the hazard analysis process and about the need for additional controls or modifications.

An information system must consider not only accidents but also incidents or near misses. Examination and understanding of near misses can warn of an impending accident and also provide important information about what conditions need to be controlled. Civil aviation and nuclear power are industries where near misses are reported widely.

No matter how the information is collected, understanding its limitations is important. To be most useful, the information must be accurate and timely, and it must be disseminated to the appropriate people in a useful form. Three factors are involved: collection, analysis, and dissemination.

Collection Just having a safety information system does not guarantee that it will be useful or used. Data may be distorted by the way it is collected. Common problems are systematic filtering, suppression, and unreliability. Data collected for accident reports tends to focus on proximal events and actors but not the systemic factors involved such as management problems or organizational deficiencies.

Studies have shown that data from near-miss reporting by operators is filtered and primarily points to technical failure as the cause of the events [151; 180]. Reports by management are similarly filtered and primarily point to operator error as the cause of incidents and accidents [140; 166]. Accident reports are usually made after an event, often by witnesses who were confused and disturbed by it [57; 69]. They may not have been paying much attention until the accident itself and thus find it difficult to recreate preceding events. Even general safety inspections tend to identify limited categories of conditions, especially when checklists are used.

Data collection tends to be more reliable for accidents that are similar to those that have occurred in the past than for new types of systems where past experience on hazards and causal factors is more limited. Software errors and computer problems are often omitted or inadequately described because of lack of knowledge, lack of accepted and consistent categorizations for such errors, or simply not considering them seriously as a causal factor. An effective safety information system must have mechanisms to encourage the inclusion of systemic factors.

Some common deficiencies include not recording the information necessary (a) to detect trends, changes, and other precursors to an accident; (b) to evaluate the effectiveness of the controls used to prevent accidents; (c) to compare risk assessments of those in the industry with actual behavior; and (d) to learn from events and improve their safety management practices.

Experience has shown that it is difficult to maintain reliable reporting of incidents over an extended period, and therefore this type of data is not very useful for estimating probabilities of events or their potential consequences. In some industries, up to three-quarters of accidents may be unreported [57; 69]. Those making reports must be convinced that the information will be used for constructive improvements in safety and not as a basis for criticism or disciplinary action [126; 150]. Sometimes, information is involved that a company wants to keep secret or that those making the reports may worry will be used against them in job performance evaluations. Potential legal liability may also come into play.

Automated monitoring can greatly improve the accuracy and completeness of data collection. Instrumentation can provide information on deviations and trends for selected variables and alarms. The great success of black box aircraft monitoring systems lies in the fact that the parameters to be recorded are standard, correct flying behavior is well-defined, and an authority exists to analyze the data. Other industries do not have these characteristics. Such monitoring systems can be useful in situations where the data to be recorded can be determined a priori, the instruments can be maintained in working order, and the recording device and results can be protected against the effects of an accident, such as a fire or explosion.

Analysis Simply collecting the information, of course, is not enough: it needs to be analyzed and summarized. Problems may arise from the difficulty in systematizing and consolidating a large mass of data into a form useful for learning. Raw quantitative data can be misleading.

Hazard analysis tools can help not only to identify what data needs to be collected but to provide guidance on the importance of the events that are occurring. Data to validate the hazard analysis results and to identify causal factors that were thought to be eliminated or controlled should be part of the information collection and analysis process.

The biggest problem in analysis is that while it is easy to design data collection channels, finding the time and manpower to analyze the collected data may be difficult or impractical. Digital technology is exacerbating the problems as it allows for collecting large amounts of data. Airlines today are particularly suffering from this type of data overload due to large-scale automated aircraft and flight information collection. As a result, the safety information system may contain only summary statistical data that can be easily processed by a computer and not the information about specific hazards, trends, and changes over time that is needed to learn from events before major losses occur.

Dissemination and use Disseminating information in a useful form may be the most difficult part of an SIS. Data is not the same as information and needs to be processed and

Tips for designing a safety information system:

- Accurate and timely feedback and data are important.
- The SIS should provide the information necessary to detect trends, changes, and other precursors to an accident; to evaluate the effectiveness of the safety controls; to compare models and risk assessments with actual behavior; and to learn from events and improve the SMS.
- Use the defined responsibilities of those in the safety control structure to identify the information they need to keep their process models accurate enough for good decision-making.
- Understand the limitations of your collected data.
- To be most useful, information must be accurate and timely, and it must be disseminated to the appropriate people in a useful form.
- Find ways to collect data that minimize distortion of the data (filtering, suppression, and unreliability).
- Keep detailed information on actual safety-related incidents, and ensure that information gets to the people who need it.
- Create ways to improve the comprehension and reliability of data collection.
- Try to include statistical significance on numeric data if possible.
- Use STPA to identify what data needs to be collected, and provide guidance on the importance of the events that are occurring.
- Collect data to validate the hazard analysis results and to identify factors that were thought to be eliminated or mitigated.
- Ensure that data is analyzed and not just collected.
- Present data in a way that people can learn from it and apply it to their daily jobs.
- Keep data up to date.
- Document design rationale and intent, and provide traceability to assist in the change management process.
- Tailor the information provided and the presentation format to the needs of those receiving it.
- Providing too much data, particularly raw data, can be as dangerous as providing too little.
- Adapt the presentation of information to the cognitive styles of the users, and integrate it into the environment in which safety-related decisions are made.
- Use the hazard analysis and safety management system design processes to determine what information is needed, when it is needed, and how it will be used.

presented in a form that people can learn from, apply to their daily activities, and use throughout the system life cycle. It must also be updated in a timely manner.

Useful information about design and procedures must include rationale and intent as well as traceability to assist in the change process. Hazard analysis results comprising hundreds or even thousands of pages of analysis, perhaps in a tabular or graphical form, is not going to be very helpful when an event occurs during operations, and it is important to know whether it is critical or if it resulted from a previously identified causal scenario.

The method for presenting information should be adaptable to the needs and cognitive styles of the users and should be integrated into the environment in which safety-related decisions are made. Simply having information stored in a safety information system does not mean that companies have the structures and processes needed to benefit from it. Again, the original hazard analysis process will provide a great deal of useful input about what information is needed. Providing too much information can be as dangerous as providing too little. To assist in dissemination, information should be integrated into the environment in which safety-related decisions are made, such as computer-assisted design tools, planning, scheduling, and resource allocation systems [179].

Accidents are often repeated, and accident/incident files are one of the most important sources for hazard analysis and control. When conducting an accident investigation, the safety information system and any possible impact on the events need to be examined. Was the information needed to prevent the loss not collected or stored? Was it lost in the collection or analysis process? Was it not available or easily retrieved in the daily activities of those involved?

14.6 Summary

There is no single correct design for a safety management system: alternative organizational structures can be effective with responsibilities distributed in different ways. The culture of the industry and the organization will play a role in what is practical and effective. In general, effective safety management requires

- commitment and leadership at all levels;
- a strong corporate safety culture;
- a clearly articulated safety vision, values, and procedures, shared among stakeholders;
- a safety control structure with appropriate assignment of responsibility, authority, and accountability;
- feedback channels that provide an accurate view of the state of safety at all levels of the safety control structure;
- integration of safety into development and line operations (not just a separate and independent group or a separate subculture);
- individuals with appropriate knowledge, skills, and ability;
- stakeholders with partnership roles and responsibilities;
- a designated process for resolving tensions between safety priorities and other priorities;
- risk awareness and communication channels for disseminating safety information;
- controls on system migration toward higher risk;
- an effective and usable safety information system;
- continual improvement and learning; and
- education, training, and capability development.

General responsibilities for a safety management system The list below (from [217]) can be used in designing a new safety management system, in efforts to evaluate and improve an existing safety management system, and in identifying inadequate controls and management contributions to accidents or incidents. It is not meant to be exhaustive and will need to be supplemented for specific industries and safety programs.

This list only contains general responsibilities and does not indicate how they should be assigned. Appropriate assignment of the responsibilities to specific people and places in the organization will depend on the management structure of each organization. Each general responsibility may be separated into multiple individual responsibilities and assigned throughout the safety control structure, with one group actually implementing the responsibilities and others above them supervising, leading, or overseeing the activity. Of course, each responsibility assumes the need for associated authority and accountability, as well as the controls, feedback, and communication channels necessary to implement the responsibility.

General management

- Provide leadership, oversight, and management of safety at all levels of the organization.
- Create a corporate or organizational safety policy. Establish criteria for evaluating safety-critical decisions and implementing safety controls. Establish distribution channels for the policy. Establish feedback channels to determine whether employees understand it, are following it, and whether it is effective. Update the policy as needed.
- Establish corporate or organizational safety standards and then implement, update, and enforce them. Set minimum requirements for safety engineering in development and operations and oversee the implementation of those requirements, including any contractor activities. Set minimum physical and operational standards for hazardous operations.
- Establish incident and accident investigation standards and ensure recommendations are implemented and effective. Use feedback to improve the standards.
- Establish management of change requirements for evaluating all changes for their impact on safety, including changes in the safety control structure. Audit the safety control structure for unplanned changes and migration toward states of higher risk.
- Create and monitor the organizational safety control structure. Assign responsibility, authority, and accountability for safety.
- Establish working groups.
- Establish robust and reliable communication channels to ensure accurate management risk awareness of the development system design and the state of the operating process. These channels should include contractor activities.
- Provide physical and personnel resources for safety-related activities. Ensure that those performing safety-critical activities have the appropriate skills, knowledge, and physical resources.
- Create an easy-to-use problem reporting system and then monitor it for needed changes and improvements.
- Establish safety education and training for all employees, and establish feedback channels to determine whether it is effective along with processes for continual improvement. The education should include reminders of past accidents and causes and input from lessons learned and trouble reports. Assessment of effectiveness may include information obtained from knowledge assessments during audits.

- Establish organizational and management structures to ensure that safety-related technical decision-making is independent from programmatic considerations, including cost and schedule.
- Establish defined, transparent, and explicit resolution procedures for conflicts between safety-related technical decisions and programmatic considerations. Ensure that the conflict resolution procedures are being used and are effective.
- Ensure that managers who are making safety-related decisions are fully informed and skilled. Establish mechanisms to allow and encourage all employees (including front-line operators) and contractors to contribute to safety-related decision-making.
- Establish an assessment and improvement process for safety-related decision-making.
- Create and update the organizational safety information system.
- Create and update safety management plans.
- Establish communication channels, resolution processes, and adjudication procedures for employees and contractors to surface complaints and concerns about the safety of the system or parts of the safety control structure that are not functioning appropriately. Evaluate the need for anonymity in reporting concerns.

Development

- Implement special training for developers and development managers in safety-guided design and other necessary skills. Update this training as events occur and more is learned from experience. Create feedback, assessment, and improvement processes for the training.
- Create and maintain the hazard log. Establish and maintain documentation and tracking of hazards and their status.
- Establish working groups.
- Design safety into the system using system hazards and safety constraints. Iterate and refine the design and the safety constraints as the design process proceeds. Ensure the system design includes consideration of how to eliminate or reduce contextual factors that cause or contribute to unsafe operator behavior that, in turn, contributes to system hazards. Distraction, fatigue, and so on are risk factors resulting from design that is dependent on humans performing in a way the designer imagined they would rather than behaving as normal humans would in such situations.
- Document operational assumptions, safety constraints, safety-related design features, operating assumptions, safety-related operational limitations, training and operating instructions, audits and performance assessment requirements, operational procedures, and safety verification and analysis results. Document both what and why, including tracing between safety constraints and the design features to enforce them.
- Perform high-quality and comprehensive hazard analyses to be available and usable when safety-related decisions need to be made, starting with early decision-making and continuing through the system's life. Ensure that the hazard analysis results are communicated in a timely manner to those who need them. Establish a communication structure that allows communication downward, upward, and sideways (i.e.,

among those building subsystems). Ensure that hazard analyses are updated as the design evolves and test experience is acquired.

- Train engineers and managers to use the results of hazard analyses in their decision-making.
- Maintain and use hazard logs and hazard analyses as experience is acquired. Ensure communication of safety-related requirements and constraints to everyone involved in development.
- Gather lessons learned in operations (including accident and incident reports) and use them to improve the development processes. Use operating experience to identify flaws in the development safety controls and implement improvements.

Operations

- Create an operations safety management plan
- Develop special training for operators and operations management to create needed skills and update this training as events occur and more is learned from experience. Create feedback, assessment, and improvement processes for this training. Train employees to perform their jobs safely, understand proper use of safety equipment, and respond appropriately in an emergency.
- Establish working groups.
- Maintain and use hazard logs and hazard analyses during operations as experience is acquired.
- Ensure that all emergency equipment and safety devices are operable at all times during hazardous operations. Before safety-critical, nonroutine, potentially hazardous operations are started, inspect all safety equipment to ensure it is operational, including the testing of alarms.
- Perform an in-depth investigation of any operational anomalies, including hazardous conditions (such as water in a tank that will contain chemicals that react to water) or events. Determine why they occurred before any potentially dangerous operations are started or restarted. Provide the training necessary to do this type of investigation and proper feedback channels to management.
- Create management of change procedures, and ensure they are being followed. These procedures should include hazard analyses on all proposed changes and approval of all changes related to safety-critical operations. Create and enforce policies about disabling safety-critical equipment.
- Perform safety audits, performance assessments, and inspections using the hazard analysis results as the preconditions for operations and maintenance. Collect data to ensure safety policies and procedures are being followed and that education and training about safety is effective. Establish feedback channels for leading indicators of increasing risk.
- Use the hazard analysis and documentation created during development and passed to operations to identify leading indicators of migration toward states of higher risk. Establish feedback channels to detect the leading indicators and respond appropriately.

- Establish communication channels from operations to development to pass back information about operational experience.
- Perform in-depth incident and accident investigations, including all systemic factors. Assign responsibility for implementing all recommendations. Follow up to determine whether recommendations were fully implemented and effective.
- Perform independent checks of safety-critical activities to ensure they have been done properly.
- Prioritize maintenance for identified safety-critical items. Enforce maintenance schedules.
- Create and cnforce policies about disabling safety-critical equipment and making changes to the physical system.
- Create and execute special procedures for the startup of operations in a previously shutdown unit or after maintenance activities.
- Investigate and reduce the frequency of spurious alarms.
- Clearly mark malfunctioning alarms and gauges. In general, establish procedures for communicating information about all current malfunctioning equipment to operators and ensure they are being followed. Eliminate all barriers to reporting malfunctioning equipment.
- Define and communicate safe operating limits for all safety-critical equipment and alarm procedures. Ensure that operators are aware of these limits. Assure that operators are rewarded for following the limits and emergency procedures, even when it turns out no emergency existed. Provide for tuning the operating limits and alarm procedures over time as required.
- Ensure that spare safety-critical items are in stock or can be acquired quickly.
- Establish communication channels to plant management about all events and activities that are safety related. Ensure management has the information and risk awareness they need to make safe decisions about operations.
- Ensure emergency equipment and response is available and operable to treat injured workers.
- Establish communication channels to the community to provide information about hazards, and necessary contingency actions and emergency response requirements.

Exercises

1. Evaluate your company's SMS (or another company you may be familiar with) with respect to the principles described in this chapter.
2. Some regulatory agencies or international organizations have guidelines or standards for a safety management system. Select one and compare it to the principles presented in this chapter.
3. What are some effective tools that might help improve risk perception?

4. Examine a safety information system from a company or from a government agency. Evaluate it with respect to collective, analysis, and dissemination of information.
5. Under what conditions do you think procedures are necessary? When might they be harmful? Suggest some reasons why even responsible people might deviate from specified procedures? Provide an example for each of those reasons.
6. What are some ways that a company might identify unplanned changes that affect safety? Create some leading indicators and associate how they might be implemented.

Epilogue: Looking Forward

I may not have gone where I intended to go, but I think I ended up where I needed to be.
—Douglas Adams,
The Long Dark Tea-Time of the Soul

Looking back at what I wrote in the epilogue of my first book on safety engineering, *Safeware*, thirty years ago, I am struck at how little I want to change in the epilogue. With a few small exceptions, my views have not changed drastically.

Safety is a complex, socio-technological problem for which there will be no simple solutions. Unlike the happy endings in fairy tales, the simple, cheap, effective, and magic potion for our problems does not exist. Our solutions will require difficult and costly procedures that in turn require special knowledge and experience on the part of system developers and maintainers. We have no other choice if we want to deal successfully with advanced technology.

In this book, I have tried to describe the problem and the tools and approaches at hand for solving it, particularly in the complex, software-intensive systems that are so common today. The overall approach involves anticipating hazards and preventing accidents before they occur. It requires establishing appropriate managerial and organizational structures and applying safety-enhancing techniques throughout the entire system lifetime—development, maintenance, and evolution.

Some themes are ubiquitous throughout this book:

- Our most effective tool in making things safer is simplicity and building systems that are intellectually manageable.
- Safety and reliability are different and must not be confused. A reliable system is not necessarily safe, and vice versa.
- Building safety into a system will be much more effective than adding protection devices onto a completed design.
- The earlier safety is considered in the development process, the better will be the results.
- To make progress, we must stop oversimplifying accidents and recognize their complex, multifactorial nature. Linear causality models are too simple to describe the losses that are occurring today.

- Fixing only symptoms while leaving the systemic causes intact will not prevent the repetition of most accidents. Concentrating on only one or a few aspects of the problem will probably not have the desired effect. In particular, concentrating only on technical issues and ignoring organizational, social, cultural, and managerial and organizational deficiencies will not result in effective safety programs.
- Simply replacing humans with automation will not solve the safety problem. Human error is integrally related to human flexibility and creativity. We should be working on ways to augment human abilities using automation rather than on ways to replace humans.
- Safety engineering is a subdiscipline of system engineering. This implies that those doing safety engineering must have subject matter expertise in the systems on which they are working. In addition, engineering projects need to be organized to enhance the interaction of the safety specialists with all the other design engineering staff throughout development: hardware, software, human factors, system engineers, application experts, and users. Isolating safety experts by placing them outside of engineering design activities leads to unnecessary losses.
- The safety of system components, including software, can only be evaluated in the context of the system within which they operate. Safety efforts that deal with components in isolation will not be effective.
- Just because the events leading to an accident are not foreseen does not mean the accident is not preventable. The hazard is usually known and can be eliminated or reduced significantly.
- Many of the decisions involved in safety engineering have trans-scientific aspects. Engineers have a duty, however, to clarify the risks for decision makers and to make sure that complacency or other factors or pressures do not interfere with risks being given due consideration in decision-making. Complacency is, in fact, perhaps the most important risk factor.
- We must learn from the past so that we do not repeat the same mistakes in the future.

Much progress in system safety is currently being made, and new editions of this book will surely be required in the future. I have attempted to describe what is currently known. Hopefully, this book will encourage and enable others to build on what is known and to develop new and better approaches to reducing accidents.

Appendix A
Medical Devices: The Therac-25

Between June 1985 and January 1987, a computer-controlled radiation therapy machine, called the Therac-25, massively overdosed six people. These accidents have been described as the worst in the thirty-five-year history of medical accelerators [326].

There is no official accident report on these events. What is included here has been gleaned from lawsuits and depositions, government records, and copies of correspondence and other material obtained from the US Food and Drug Administration (FDA), which regulates these devices.

A.1 Background

Medical linear accelerators (linacs) accelerate electrons to create high-energy beams that can destroy tumors with minimal impact on the surrounding healthy tissue. Relatively shallow tissue is treated with the accelerated electrons; to reach deeper tissue, the electron beam is converted into X-ray photons.

In the early 1970s, Atomic Energy of Canada Limited (AECL) and a French company called CGR went into business together building linear accelerators.

The products of this cooperation were (1) the Therac-6, a 6 million electron volt (MeV) accelerator capable of producing X-rays only, and later (2) the Therac-20, a 20 MeV, dual-mode (X-rays or electrons) accelerator. Both were versions of older CGR machines, the Neptune and Sagittaire, respectively, which were augmented with computer control using a DEC PDP-11 minicomputer.

We know that some of the old Therac-6 software routines were reused in the Therac-20 and that CGR developed the initial software. Software functionality was limited in both machines: the computer merely added convenience to the existing hardware, which was capable of standing alone. Industry-standard hardware safety features and interlocks in the underlying machines were retained.

The business relationship between AECL and CGR faltered after the Therac-20 effort. Citing competitive pressures, the two companies did not renew their cooperative agreement when scheduled in 1981. In the mid-1970s, AECL had developed a radical new *double pass* concept for electron acceleration. A double-pass accelerator needs much less space to develop comparable energy levels because it folds the long physical mechanism required to accelerate the electrons. It is also more economical to produce. Using this double-pass concept, AECL designed the Therac-25, a dual-mode linear accelerator that can deliver either photons at 25 MeV or electrons at various energy levels.

Compared with the Therac-20, the Therac-25 was notably more compact, more versatile, and arguably easier to use. The higher energy takes advantage of the phenomenon of *depth dose*: as the energy increases, the depth in the body at which maximum dose buildup occurs also increases, sparing the tissue above the target area. Economic advantages also come into play for the customer because only one machine is required for both treatment modalities (electrons and photons).

Several features of the Therac-25 are important in understanding the accidents. First, like the Therac-6 and the Therac-20, the Therac-25 is controlled by a PDP-11 computer. However, AECL designed the Therac-25 to take advantage of computer control from the outset; it did not build on a stand-alone machine. The Therac-6 and Therac-20 had been designed around machines that already had histories of clinical use without computer control.

In addition, the Therac-25 software had more responsibility for maintaining safety than the software in the previous machines. The Therac-20 had independent protective circuits for monitoring the electron-beam scanning plus mechanical interlocks for policing the machine and ensuring safe operation. In contrast, the Therac-25 relied more on software for these functions. AECL took advantage of the computer's abilities to control and monitor the hardware and decided not to duplicate all the existing hardware safety mechanisms and interlocks. This omission was an important factor in the accidents.

Some software for the machines was interrelated or reused. In a letter to a Therac-25 user, the AECL quality assurance manager said: "*The same Therac-6 package was used by the AECL software people when they started the Therac-25 software. The Therac-20 and Therac-25 software programs were done independently starting from a common base*" [229]. The reuse of Therac-6 design features or modules may explain some of the problematic aspects of the Therac-25 software design and is another important causal factor.

The quality assurance manager was apparently unaware that some Therac-20 routines were also used in the Therac-25; this was discovered after a bug related to one of the Therac-25 accidents was found in the Therac-20 software. AECL produced the first hardwired prototype of the Therac-25 in 1976, and the completely computer-controlled commercial version was available in late 1982.

The Therac-25 turntable design plays an important role in the accidents. The upper turntable (see figure A.1) rotates accessory equipment into the beam path to produce two therapeutic modes: electron mode and photon mode. A third position (called the field light position) involves no beam at all, but rather is used to facilitate correct positioning of the patient. Because the accessories appropriate to each mode are physically attached to the turntable, proper operation of the Therac-25 is heavily dependent on the turntable position, which is monitored by three microswitches.

The raw, highly concentrated accelerator beam is dangerous to living tissue. In electron therapy, the computer controls the beam energy (from 5 to 25 MeV in the Therac-25) and current, while scanning magnets are used to spread the beam to a safe, therapeutic concentration. These scanning magnets were mounted on the turntable and moved into proper position by the computer. Similarly, an ion chamber to measure electrons was mounted on the turntable and also moved into position by the computer.

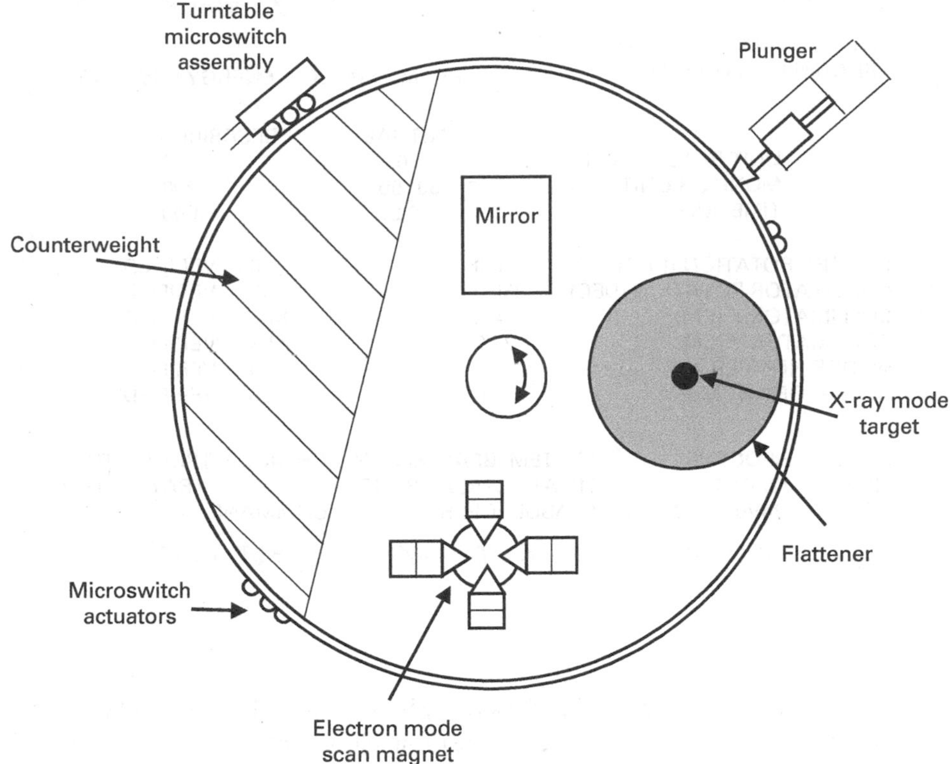

Figure A.1
Upper turntable assembly.

In addition, operator-mounted electron trimmers could be used to shape the beam if necessary. For X-ray (or photon) therapy, only one energy level was available: 25 MeV. Much greater electron-beam current is required for X-ray mode—some 100 times greater than that for electron therapy [326]—to produce comparable output. Such a high dose-rate capability was required because a *beam flattener* was used to produce a uniform treatment field. This flattener, which resembled an inverted ice cream cone, was a very efficient attenuator; thus, to get a reasonable treatment dose rate out of the flattener, a very high input dose rate was required. If the machine produces a photon beam with the beam flattener not in position, a high output dose to the patient results, which is what happened.

This turntable position is the basic hazard of dual-mode machines: if the turntable is in the wrong position, the beam flattener will not be in place. In the Therac-25, the computer was responsible for positioning the turntable (and for checking the turntable position) so that a target, flattening filter, and X-ray ion chamber are directly in the beam path. With the target in place, electron bombardment produces X-rays. The X-ray beam is shaped by the flattening filter and measured by the X-ray ion chamber.

No accelerator beam was expected in the third or field light turntable position. A stainless-steel mirror was placed in the beam path and a light simulated the beam. This let the operator see precisely where the beam would strike the patient and make necessary adjustments before treatment started. There was no ion chamber in place at this turntable position because no beam was anticipated.

```
PATIENT NAME      : TEST
TREATMENT MODE : FIX                        BEAM TYPE: X    ENERGY (MeV) : 25

                                              ACTUAL        PRESCRIBED
          UNIT RATE/MINUTE                       0                 200
          MONITOR UNITS                       50 50                200
          TIME (MIN)                           0.27               1.00

GANTRY ROTATION (DEG)          0.0                          0    VERIFIED
COLLIMATOR ROTATION (DEG)    359.2                        359    VERIFIED
COLLIMATOR X (CM)             14.2                       14.3    VERIFIED
COLLIMATOR Y (CM)             27.2                       27.3    VERIFIED
WEDGE NUMBER                     1                          1    VERIFIED
ACCESSORY NUMBER                 0                          0    VERIFIED

DATE    : 84-OCT-26     SYSTEM :BEAM READY    OP.MODE :TREAT AUTO
TIME    : 12:55: 8      TREAT   :TREAT PAUSE             XRAY  173777
OPR ID  :T24V02-RO3     REASON :OPERATOR      COMMAND
```

Figure A.2
Operator interface screen layout.

Traditionally, electromechanical interlocks have been used on these types of equipment to ensure safety—in this case, to ensure that the turntable and attached equipment were in the correct position when treatment was started. In the Therac-25, software checks were substituted for many of the traditional hardware interlocks.

The operator interface The description of the operator interface here applies to the version of the software used during the accidents. Changes made as a result of an FDA recall are described later.

The Therac-25 operator controlled the machine through a DEC VT100 terminal. In the general case, the operator positioned the patient on the treatment table, manually set the treatment field sizes and gantry rotation, and attached accessories to the machine. Leaving the treatment room, the operator returned to the console to enter the patient identification, treatment prescription (including mode or beam type, energy level, dose, dose rate, and time), field sizing, gantry rotation, and accessory data. The system then compared the manually set values with those entered at the console. If they matched, a *verified* message was displayed, and treatment was permitted. If they did not match, treatment was not allowed to proceed until the mismatch was corrected.

Figure A.2 shows the screen layout.

When the system was first built, operators complained that it took too long to enter the treatment plan. In response, AECL modified the software before the first unit was installed: Instead of reentering the data at the keyboard, operators could simply use a carriage return to copy the treatment site data [266]. A quick series of carriage returns would thus complete the data entry. This modification was to figure in several of the accidents.

The Therac-25 could shut down in two ways after it detected an error condition. One was a *treatment suspend*, which required a complete machine reset to restart. The other, not so serious, was a *treatment pause*, which only required a single key command to restart the machine. If a *treatment pause* occurred, the operator could press the **P** key to "proceed" and resume treatment quickly and conveniently. The previous treatment parameters remained in effect, and no reset was required. This feature could be invoked a maximum of five times before the machine automatically suspended treatment and required the operator to perform a system reset.

Error messages provided to the operator were cryptic, and some merely consisted of the word MALFUNCTION followed by a number from 1 to 64 denoting an analog/digital channel number. According to an FDA memorandum written after one accident:

> *The operator's manual supplied with the machine does not explain nor even address the malfunction codes. The Maintance [sic] Manual lists the various malfunction numbers but gives no explanation. The materials provided give \underline{no} indication that these malfunctions could place a patient at risk.*
>
> *The program does not advise the operator if a situation exists wherein the ion chambers used to monitor the patient are saturated, thus are beyond the measurement limits of the instrument. This software package does not appear to contain a safety system to prevent parameters being entered and intermixed that would result in excessive radiation being delivered to the patient under treatment.*

An operator involved in one of the accidents testified that she had become insensitive to machine malfunctions. Malfunction messages were commonplace, and most did not involve patient safety. Service technicians would fix the problems, or the hospital physicist would realign the machine and make it operable again. She said: "*It was not out of the ordinary for something to stop the machine. . . . It would often give a low dose rate in which you would turn the machine back on. . . . They would give messages of low dose rate, V-tilt, H-tilt, and other things; I can't remember all the reasons it would stop, but there was a lot of them.*"

A radiation therapist at another clinic reported that an average of forty dose-rate malfunctions, attributed to underdoses, occurred on some days. The operator further testified that during instruction she had been taught that there were "so many safety mechanisms" that she understood it was virtually impossible to overdose a patient.

Hazard analysis In March 1983, AECL performed a safety analysis on the Therac-25. This analysis was in the form of a fault tree and apparently excluded the software. According to the final report, the analysis made several assumptions about the computer and its software:

1. Programming errors have been reduced by extensive testing on a hardware simulator and under field conditions on teletherapy units. Any residual software errors are not included in the analysis.
2. Program software does not degrade due to wear, fatigue, or reproduction process.
3. Computer execution errors are caused by faulty hardware components and by "soft" (random) errors induced by alpha particles and electromagnetic noise.

The fault tree resulting from this analysis does appear to include computer failure, although apparently, judging from the basic assumptions, it considers computer hardware failures only. For example, in one OR gate leading to the event of getting the wrong energy, a box contains "Computer selects wrong energy," and a probability of 10^{-11} is assigned to this event. For "Computer selects wrong mode," a probability of $4x10^{-9}$ is given. The report provided no justification of either number.

A.2 Events

Eleven Therac-25s were installed: five in the United States and six in Canada. Six accidents occurred between 1985 and 1987, when the machine was finally recalled to make extensive design changes.

A.2.1 Kennestone Regional Oncology Center, June 1985

Details of this accident in Marietta, Georgia, are sketchy because it was never investigated. There was no admission that the injury was caused by the Therac-25 until long after the occurrence, despite claims by the patient that she had been injured during treatment, the obvious and severe radiation burns the patient suffered, and the suspicions of the radiation physicist involved.

After undergoing a lumpectomy to remove a malignant breast tumor, a sixty-one-year-old woman was receiving follow-up radiation treatment to nearby lymph nodes on a Therac-25 at the Kennestone facility in Marietta. The Therac-25 had been operating at Kennestone for about six months; other Therac-25s had been operating, apparently without incident, since 1983.

On June 3, 1985, the patient was set up for a 10 MeV electron treatment to the clavicle area. When the machine turned on, she felt a "tremendous force of heat . . . this red-hot sensation." When the technician came in, the patient said, "You burned me." The technician replied that that was impossible. Although there were no marks on the patient at the time, the treatment area felt "warm to the touch."

It is unclear exactly when AECL learned about this incident. Tim Still, the Kennestone physicist, said that he contacted AECL to ask if the Therac-25 could operate in electron mode without scanning to spread the beam. Three days later, the engineers at AECL called the physicist back to explain that improper scanning was not possible.

In an August 19, 1986, letter from AECL to the FDA, the AECL quality assurance manager said, "*In March of 1986 AECL received a lawsuit from the patient involved . . . This incident was never reported to AECL prior to this date, although some rather odd questions had been posed by Tim Still, the hospital physicist.*" The physicist at a hospital in Tyler, Texas, where a later accident occurred, reported: "*According to Tim Still, the patient filed suit in October 1985 listing the hospital, manufacturer and service organization responsible for the machine. AECL was notified informally about the suit by the hospital, and AECL received official notification of a law suit in November 1985.*"

Because of the lawsuit (filed November 13, 1985), some AECL administrators must have known about the Marietta accident—although no investigation occurred at this time. FDA memos point to the lack of a mechanism in AECL to follow up on reports of suspected accidents [229].

The patient went home, but shortly afterward she developed a reddening and swelling in the center of the treatment area. Her pain had increased to the point that her shoulder "froze," and she experienced spasms. She was admitted to a hospital in Atlanta, but her oncologists continued to send her to Kennestone for Therac-25 treatments. Clinical explanation was sought for the reddening of the skin, which at first her oncologist attributed to her disease or to normal treatment reaction.

About two weeks later, the Kennestone physicist noticed that the patient had a matching reddening on her back as though a burn had gone right through her body, and the swollen area had begun to slough off layers of skin. Her shoulder was immobile, and she was apparently in great pain. It was now obvious that she had a radiation burn, but the hospital and her doctors could provide no satisfactory explanation.

The Kennestone physicist later estimated that the patient received one or two doses of radiation in the 15,000 to 20,000 rad (radiation absorbed dose) range. He did not believe her injury could have been caused by less than 8,000 rads. To understand the magnitude of this, consider that typical single therapeutic doses are in the 200 rad range. Doses of 1,000 rads can be fatal if delivered to the whole body; in fact, 500 rads is the accepted figure for whole-body radiation that will cause death in 50 percent of the cases. The consequences of an overdose to a smaller part of the body depend on the tissue's radiosensitivity. The director of radiation oncology at the Kennestone facility explained their confusion about the accident as due to the fact that they had never seen an overtreatment of that magnitude before [343].

Eventually, the patient's breast had to be removed because of the radiation burns. Her shoulder and arm were paralyzed, and she was in constant pain. She had suffered a serious radiation burn, but the manufacturer and operators of the machine refused to believe that it could have been caused by the Therac-25. The treatment prescription printout feature of the computer was disabled at the time of the accident, so there was no hardcopy of the treatment data. The lawsuit was eventually settled out of court.

From what can be determined, the accident was not reported to the FDA until after further accidents in 1986. The reporting requirements for medical device incidents at that time applied only to equipment manufacturers and importers, not users. The regulations required that manufacturers and importers report deaths, serious injuries, or malfunctions that could result in those consequences, but healthcare professionals and institutions were not required to report incidents to manufacturers.

The comptroller general of the US Government Accounting Office (GAO), in testimony before Congress on November 6, 1989, expressed great concern about the viability of the incident-reporting regulations in preventing or spotting medical device problems. According to a 1990 GAO study, the FDA knew of less than 1 percent of deaths, serious injuries, or equipment malfunctions that occurred in hospitals [44].

The law was amended in 1990 to require healthcare facilities to report incidents to the manufacturer and to the FDA. At this point, the other Therac-25 users were also unaware that anything untoward had occurred and did not learn about any problems with the machine until after subsequent accidents. Even then, most of their information came through personal communication among themselves and not from the manufacturer or the two government regulatory agencies involved (Canada and the United States).

A.2.2 Ontario Cancer Foundation, July 1985

The second in this series of accidents occurred about seven weeks after the Kennestone patient was overdosed. At that time, the Therac-25 at the Ontario Cancer Foundation in Hamilton, Ontario (Canada), had been in use for more than six months.

On July 26, 1985, a forty-year-old patient came to the clinic for her twenty-fourth Therac-25 treatment for carcinoma of the cervix. The operator activated the machine, but it shut down after 5 seconds with an HTILT error message. The Therac-25's console display read NO DOSE and indicated a TREATMENT PAUSE.

Because the machine did not suspend and the control display indicated that no dose was delivered to the patient, the operator went ahead with a second attempt at treatment by pressing the ***P*** key (the *proceed* command), expecting the machine to deliver the proper dose this time. This was standard operating procedure, and Therac-25 operators had become accustomed to frequent malfunctions that had no untoward consequences for the patient. Again, the machine shut down in the same manner.

The operator repeated this process four times after the original attempt—the display showing NO DOSE delivered to the patient each time. After the fifth pause, the machine went into treatment suspend, and a hospital service technician was called. The technician found nothing wrong with the machine. According to a Therac-25 operator, this scenario also was not unusual.

After the treatment, the patient complained of a burning sensation, described as an "electric tingling shock" to the treatment area in her hip. Six other patients were treated later that day without incident. She came back for further treatment on July 29 and complained of burning, hip pain, and excessive swelling in the region of treatment. The patient was hospitalized for the condition on July 30, and the machine was taken out of service.

AECL was informed of the apparent radiation injury and sent a service engineer to investigate. The US FDA, the then Canadian Radiation Protection Bureau (RPB), and Therac-25 users were informed that there was a problem, although the users claim that they were never informed that a patient injury had occurred. Users were told that they should visually confirm the proper turntable alignment until further notice (which occurred three months later).

The patient died on November 3, 1985, of an extremely virulent cancer. An autopsy revealed the cause of death as the cancer, but it was noted that had she not died, a total hip replacement would have been necessary as a result of the radiation overexposure. An AECL technician later estimated the patient had received between 13,000 and 17,000 rads.

Manufacturer's response AECL could not reproduce the malfunction that had occurred but suspected a transient failure in the microswitch used to determine the turntable position. During the investigation of the accident, AECL hardwired the error conditions they assumed were necessary for the malfunction and, as a result, found some turntable positioning design weaknesses and potential mechanical problems.

The computer sensed and controlled turntable position by reading a 3-bit signal about the status of three microswitches in the turntable switch assembly. Essentially, AECL determined that a 1-bit error in the microswitch codes (which could be caused by a single open-circuit fault on the switch lines) could produce an ambiguous position message to the computer. The problem was exacerbated by the design of the mechanism that extends a plunger to lock the turntable when it is in one of the three cardinal positions: The plunger could be extended when the turntable was way out of position, thus giving a second false position indication.

AECL devised a method to indicate turntable position that tolerated a 1-bit error so that the code would still unambiguously reveal correct position with any one microswitch failure. In addition, AECL altered the software so that the computer checked for "in transit" status of the switches to keep further track of the switch operation and turntable position and to give additional assurance that the switches were working and the turntable was moving.

As a result of these improvements, AECL claimed in its report and correspondence with hospitals that "analysis of the hazard rate of the new solution indicates an improvement over the old system by at least *5 orders of magnitude* [emphasis added]." However, in its final incident report to the FDA, AECL concluded that it "cannot be firm on the exact cause of the accident but can only suspect," which underscored its inability to determine the cause of the accident with any certainty despite their optimistic estimate of the likelihood reduction.

The AECL quality assurance manager testified that they could not reproduce the switch malfunction and that testing of the microswitch was "inconclusive." The similarity of the errant behavior and the patient injuries in this accident and a later one in Yakima, Washington, provides good reason to believe that the Hamilton overdose was probably related to software error rather than to a microswitch failure.

Government and user response The Hamilton accident resulted in a voluntary recall by AECL, and the FDA termed it a Class II recall. Class II means "*a situation in which the use of, or exposure to, a violative product may cause temporary or medically reversible adverse health consequences or where the probability of serious adverse health consequences is remote.*" The FDA audited AECL's subsequent modifications, and after the modifications were made, the users were told they could return to normal operating procedures.

As a result of the Hamilton accident, the head of advanced X-ray systems in the Canadian RPB, Gordon Symonds, wrote a report that analyzed the design and performance characteristics of the Therac-25 with respect to radiation safety. Besides citing the flawed microswitch, the report faulted both hardware and software components of the Therac-25's design. It concluded with a list of four modifications to the Therac-25 necessary for compliance with Canada's Radiation Emitting Devices (RED) Act. The RED law, enacted in 1971, gives government officials power to ensure the safety of radiation-emitting devices.

The modifications specified in the Symonds report included redesigning the microswitch and changing the way the computer handled malfunction conditions. In particular, treatment was to be terminated in the event of a dose-rate malfunction, giving a treatment *suspend.* This change would have removed the option to proceed simply by pressing the **P** key. The report also made recommendations regarding collimator test procedures and message and command formats.

A November 8, 1985, letter, signed by the director of the Canadian RPB, asked that AECL make changes to the Therac-25 based on the Symond's report "to be in compliance with the RED act." Although, as noted earlier, AECL did make the microswitch changes, they did not comply with the directive to change the malfunction pause behavior into treatment suspends, instead reducing the maximum number of retries from five to three. According to Symonds, the deficiencies outlined in the RPB letter of November 8 were still pending when the next accident happened five months later.

Immediately after the Hamilton accident, the Ontario Cancer Foundation hired an independent consultant to investigate. He concluded in a September 1985 report that an independent system (beside the computer) was needed to verify the turntable position and suggested the use of a potentiometer.

The RPB wrote a letter to AECL in November 1985 requesting that AECL install such an independent interlock on the Therac-25. Also, in January 1986, AECL received a letter from the attorney representing the Hamilton clinic. The letter said that there had been continuing problems with the turntable, including four incidents at Hamilton, and requested the installation of an independent system (potentiometer) to verify the turntable position. AECL did not comply: no independent interlock was installed by AECL on the Therac-25s at this time. The Hamilton Clinic, however, decided to install one themselves on their machine.

A.2.3 Yakima Valley Memorial Hospital, December 1985

In this accident, as in the Kennestone overdose, machine malfunction was not acknowledged until after later accidents were understood. The Therac-25 at Yakima, Washington, had been modified by AECL in September 1985 in response to the overdose at Hamilton.

During December 1985, a woman treated with the Therac-25 developed erythema (excessive reddening of the skin) in a parallel striped pattern on her right hip. Despite this, she continued to be treated by the Therac-25, as the cause of her reaction was not determined to be abnormal until January 1986. On January 6, her treatments were completed.

The staff monitored the skin reaction closely and attempted to find possible causes. The open slots in the blocking trays in the Therac-25 could have produced such a striped pattern. By the time the skin reaction was determined to be abnormal, however, the blocking trays had been discarded, so the blocking arrangement and tray striping orientation could not be reproduced. A reaction to chemotherapy was ruled out because that should have produced reactions at the other treatment sites and would not have produced stripes.

When the doctors discovered that the woman slept with a heating pad, they thought maybe the burn pattern had been caused by the parallel wires that deliver the heat in such pads. The staff X-rayed the heating pad but discovered that the wire pattern did not correspond to the erythema pattern on the patient's hip.

The hospital staff sent a letter to AECL on January 31, and they also spoke on the phone with the AECL technical support supervisor. On February 24, the AECL technical support supervisor sent a written response to the director of radiation therapy at Yakima saying, "*After careful consideration we are of the opinion that this damage could not have been produced by any malfunction of the Therac-25 or by any operator error.*" The letter goes on to support this opinion by listing two pages of technical reasons why an overdose by the Therac-25 was impossible, along with the additional argument that there have "apparently been no other instances of similar damage to this or other patients." The letter ends: "*In closing, I wish to advise that this matter has been brought to the attention of our Hazards Committee as is normal practice.*"

The hospital staff eventually ascribed the patient's skin reaction to "cause unknown." In a report written on this first Yakima incident after another Yakima overdose a year later, the medical physicist involved wrote:

At that time, we did not believe that [the patient] was overdosed because the manufacturer had installed additional hardware and software safety devices to the accelerator.

In a letter from the manufacturer dated 16-Sep-85, it is stated that "Analysis of the hazard rate resulting from these modifications indicates an improvement of at least five orders of magnitude"! With such an improvement in safety 10,000,000% we did not believe that there could have been any accelerator malfunction. These modifications to the accelerator were completed on 5,6-Sep-85.

Even with fairly sophisticated physics support, the hospital staff, as users, did not have the ability to investigate the possibility of machine malfunction further. They were not aware of any other incidents and, in fact, were told that there had been none, so there was no reason for them to pursue the matter. No further investigation of this incident was done by the manufacturer or by any government agencies—who, in fact, did not know about it.

About a year later, in February 1987, after a second Yakima overdose led the hospital staff to suspect that this first injury had been due to a Therac-25 fault, the staff investigated and found that the first overdose victim had a chronic skin ulcer, tissue necrosis (death) under the skin, and was in continual pain. The damage was surgically repaired, skin grafts were made, and the symptoms were relieved. The patient is alive today with minor disability and some scarring related to the overdose. The hospital staff concluded that the dose accidentally delivered in the first accident must have been much lower than in the second, as the reaction was significantly less intense and necrosis did not develop until six or eight months after exposure. Some other factors related to the place on the body where the overdose occurred also kept her from having more significant problems.

A.2.4 East Texas Cancer Center, March 1986

More is known about the Tyler, Texas, accidents than the others because of the diligence of the Tyler hospital physicist, Fritz Hager, without whose efforts the understanding of the software problems may have been delayed even further.

The Therac-25 had been at the East Texas Cancer Center (ETCC) for two years before the first serious accident, and more than five hundred patients had been treated. On March 21, 1986, a male patient came into ETCC for his ninth treatment on the Therac-25, one of a series prescribed as a follow-up to the removal of a tumor from his back. This treatment was to be a 22 MeV electron beam treatment of 180 rads on the upper back and a little to the left of his spine, for a total of 6,000 rads over six and a half weeks. He was taken into the treatment room and placed face down on the treatment table. The operator then left the treatment room, closed the door, and sat at the control terminal.

The operator had held this job for some time, and her typing efficiency had increased with experience. She could quickly enter prescription data and change it conveniently with the Therac-25's editing features. She entered the patient's prescription data quickly, then noticed that she had typed *x* (for X-ray) when she had intended to type *e* (for electron) mode. This was a common mistake as most of the treatments involved X-rays, and she had gotten used to typing *x*.

The mistake was easy to fix; she merely used the ↑ key to edit the mode entry. Because the other parameters she had entered were correct, she hit the return key several times and left their values unchanged. She reached the bottom of the screen, where it was indicated that the parameters had been VERIFIED and the terminal displayed BEAM READY, as expected. She hit the one-key command, **B** for *beam on*, to begin the treatment. After a moment, the machine shut down, and the console displayed the message MALFUNCTION 54. The machine also displayed a TREATMENT PAUSE, indicating a problem of low priority.

The sheet on the side of the machine explained that this malfunction was a *dose input 2* error. The ETCC did not have any other information available in its instruction manual or other Therac-25 documentation to explain the meaning of MALFUNCTION 54.

An AECL technician later testified that *dose input 2* meant that a dose had been delivered that was either too high or too low. The messages had been expected to be used only during internal company development.

The machine showed a substantial underdose on its dose monitor display—6 monitor units delivered whereas the operator had requested 202 monitor units. She was accustomed to the quirks of the machine, which would frequently stop or delay treatment; in the past, the only consequences had been inconvenience. She immediately took the normal action when the machine merely paused, which was to hit the ***P*** key to proceed with the treatment. The machine promptly shut down with the same MALFUNCTION 54 error and the same underdose shown by the dosimetry.

The operator was isolated from the patient because the machine apparatus was inside a shielded room of its own. The only way that the operator could be alerted to patient difficulty was through audio and video monitors. On this day, the video display was unplugged, and the audio monitor was broken.

After the first attempt to treat him, the patient said that he felt as if he had received an electric shock or that someone had poured hot coffee on his back. He felt a thump and heat and heard a buzzing sound from the equipment. Because this was his ninth treatment, he knew that this was not normal. He began to get up from the treatment table to go for help. It was at this moment that the operator hit the **P** key to proceed with the treatment. The patient said that he felt like his arm was being shocked by electricity and that his hand was leaving his body. He went to the treatment room door and pounded on it. The operator was shocked and immediately opened the door for him. He appeared visibly shaken and upset.

The patient was immediately examined by a physician, who observed intense reddening of the treatment area but suspected nothing more serious than electric shock. The patient was discharged and sent home with instructions to return if he suffered any further reactions. The hospital physicist was called in, and he found the machine calibration within specifications. The meaning of the malfunction message was not understood. The machine was then used to treat patients for the rest of the day.

In actuality, but unknown to anyone at that time, the patient had received a massive overdose, concentrated in the center of the treatment location. After-the-fact simulations of the accident revealed possible doses of 16,500 to 25,000 rads in less than 1 second over an area of about 1 cm.

Over the weeks following the accident, the patient continued to have pain in his neck and shoulder. He lost the function of his left arm and had periodic bouts of nausea and vomiting. He was eventually hospitalized for radiation-induced myelitis of the cervical cord causing paralysis of his left arm and both legs, left vocal cord paralysis (which left him unable to speak), neurogenic bowel and bladder, and paralysis of the left diaphragm. He also had a lesion on his left lung and recurrent herpes simplex skin infections. He died from complications of the overdose five months after the accident.

User and manufacturer response The Therac-25 was shut down for testing the day after this accident. One local AECL engineer and one from the home office in Canada came to

ETCC to investigate. They spent a day running the machine through tests but could not reproduce a Malfunction 54. The AECL engineer from the home office reportedly explained that it was not possible for the Therac-25 to overdose a patient. The ETCC physicist claims that he asked AECL at this time if there were any other reports of radiation overexposure and that AECL personnel (including the quality assurance manager) told him that AECL knew of no accidents involving radiation overexposure by the Therac-25.

This denial seems odd because AECL was surely at least aware of the Hamilton accident that had occurred seven months before and the Yakima accident, and, even by their account, learned of the Georgia lawsuit around this time (which had been filed four months earlier).

The AECL engineers then suggested that an electrical problem might have caused the problem. The electric shock theory was checked out thoroughly by an independent engineering firm. The final report indicated that there was no electrical grounding problem in the machine, and it did not appear capable of giving a patient an electrical shock. The ETCC physicist checked the calibration of the Therac-25 and found it to be satisfactory. He put the machine back into service on April 7, 1986, convinced that it was performing properly.

A.2.5 East Texas Cancer Center, April 1986

Three weeks later, on April 11, 1986, another male patient was scheduled to receive an electron treatment at ETCC for skin cancer on the side of his face. The prescription was for 10 MeV. The same technician who had treated the first Tyler accident victim prepared this patient for treatment.

Much of what follows is from the operator's deposition. As with her former patient, she entered the prescription data and then noticed an error in the mode. Once again, she used the edit ↑ key to change the mode from X-ray to electron. After she finished editing, she pressed the RETURN key several times to place the cursor on the bottom of the screen. She saw the BEAM READY message displayed and turned the beam on.

Within a few seconds the machine shut down, making a loud audible noise via the (now working) intercom. The display showed MALFUNCTION 54 again. The operator rushed into the treatment room, hearing her patient moaning for help. He began to remove the tape that had held his head in position and said something was wrong. She asked him what he felt, and he replied, "fire" on the side of his face. She immediately went to the hospital physicist and told him that another patient appeared to have been burned. Asked by the physicist to describe what had happened, the patient explained that something had hit him on the side of the face, he saw a flash of light, and he heard a sizzling sound reminiscent of frying eggs. He was very agitated and asked, "What happened to me, what happened to me?"

This patient died from the overdose on May 1, 1986, three weeks after the accident. He had disorientation, which progressed to coma, fever to 104°F, and neurological damage. An autopsy showed an acute high-dose radiation injury to the right temporal lobe of the brain and the brain stem.

User and manufacturer response After this second Tyler accident, the ETCC physicist immediately took the machine out of service and called AECL to alert them to this second apparent overexposure. The physicist then began a careful investigation of his own. He worked with the operator, who remembered exactly what she had done on this occasion. After a great deal of effort, they were eventually able to elicit the MALFUNCTION 54 message.

They determined that data entry speed during editing was the key factor in producing the error condition: If the prescription data was edited at a fast pace—as is natural for someone who has repeated the procedure a large number of times—the overdose occurred. It took some practice before the physicist could repeat the procedure rapidly enough to elicit the MALFUNCTION 54 message at will.

The next day, an engineer from AECL called and said that he could not reproduce the error. After the ETCC physicist explained that the procedure had to be performed quite rapidly, AECL could finally produce a similar malfunction on its own machine. Two days after the accident, AECL said it had measured the dosage (at the center of the field) to be 25,000 rads. An AECL engineer explained that the frying sound heard by the patients was the ion chambers being saturated.

In one lawsuit that resulted from the Tyler accidents, the AECL quality control manager testified that a "cursor up" problem had been found in the service (maintenance) mode at other clinics in February or March of 1985, and also in the summer of 1985.

Both times, AECL thought that the software problems had been fixed. There is no way to determine whether there is any relationship between these problems and the Tyler accidents.

Related Therac-20 problems The software for both the Therac-25 and Therac-20 evolved from the Therac-6 software. Additional functions had to be added because the Therac-20 and Therac-25 operate in both X-ray and electron mode, while the Therac-6 has only X-ray mode. CGR modified the software for the Therac-20 to handle the dual modes.

When the Therac-25 development began, AECL engineers adapted the software from the Therac-6, but they also borrowed software routines from the Therac-20 to handle electron mode, which was allowed under their cooperative agreements.

After the second Tyler, Texas, accident, a physicist at the University of Chicago Joint Center for Radiation Therapy heard about the Therac-25 software problem and decided to find out whether the same thing could happen with the Therac-20. At first, the physicist was unable to reproduce the error on his machine, but two months later he found the link.

The Therac-20 at the University of Chicago is used to teach students in a radiation therapy school conducted by the center. The center's physicist, Frank Borger, noticed that whenever a new class of students started using the Therac-20, fuses and breakers on the machine tripped, shutting down the unit. These failures, which had been occurring ever since the school had acquired the machine, might happen three times a week while new students operated the machine and then disappear for months. Borger determined that new students make many different types of mistakes and use "creative methods" for editing parameters on the console. Through experimentation, he found that certain editing sequences correlated with blown fuses and determined that the same computer bug, as in the Therac-25 software, was responsible. The physicist notified the FDA, which notified Therac-20 users [177].

The software error is just a nuisance on the Therac-20 because this machine has independent hardware protective circuits for monitoring the electron beam scanning. The protective circuits do not allow the beam to turn on, so there is no danger of radiation exposure to a patient. While the Therac-20 relies on mechanical interlocks for monitoring the machine, the Therac-25 relies largely on software.

The software "bug" A lesson to be learned from the Therac-25 story is that focusing on particular software bugs is not the way to make a safe system. Virtually all complex software can be made to behave in an unexpected fashion under some conditions. The basic mistakes here involved poor software engineering practices and building a machine that relies on the software for safe operation.

Furthermore, the particular coding error is not as important as the general unsafe design of the software overall. Examining the part of the code blamed for the Tyler accidents is instructive, however, in demonstrating the overall software design flaws. First, the software design is described and then the errors believed to be involved in the Tyler accidents and perhaps others.

Therac-25 software development and design AECL claims proprietary rights to its software design. However, from voluminous documentation regarding the accidents, the repairs, and the eventual design changes, we can build a rough picture of it.

The software is responsible for monitoring the machine status, accepting input about the treatment desired, and setting the machine up for this treatment. It turns the beam on in response to an operator command (assuming that certain operational checks on the status of the physical machine are satisfied) and also turns the beam off when treatment is completed, when an operator commands it, or when a malfunction is detected. The operator can print out hardcopy versions of the CRT display or machine setup parameters.

The treatment unit has an interlock system designed to remove power to the unit when there is a hardware malfunction. The computer monitors this interlock system and provides diagnostic messages. Depending on the fault, the computer either prevents a treatment from being started or, if the treatment is in progress, creates a pause or a suspension of the treatment.

There are two basic operational modes: treatment mode and service mode. Treatment mode controls the normal treatment process. In service mode, the unit can be operated with some of the operational and treatment interlocks bypassed, and additional operational commands and characteristics may be selected. Service mode is entered only through the use of a password at the service keyboard.

It is clear from the AECL documentation that the software allows concurrent access to shared memory, that there is no real synchronization aside from data that are stored in shared variables, and that the *test* and *set* for such variables are not indivisible operations. Race conditions resulting from this implementation of multitasking played an important part in the accidents.

Details about the specific errors are irrelevant here. The flaws in software involved in accidents almost always stem from inadequate identification of safety requirements and poor overall system engineering. In this case, there are also questions about the training of the software programmer, who may not have had a standard computer science education and, in addition, had serious oral communication problems.

The government and user response The FDA does not approve each new medical device on the market: all medical devices go through a classification process that determines the level of FDA approval necessary. Medical accelerators follow a procedure called *premarket notification* before commercial distribution.

In this process, the firm must establish that the product is substantially equivalent in safety and effectiveness to a product already on the market. If that cannot be done to the FDA's satisfaction, a premarket approval is required. For the Therac-25, the FDA required only a premarket notification. After the Therac-25 accidents, new procedures for approval of software-controlled devices were adopted.

The agency is basically reactive to problems and requires manufacturers to report serious ones. Once a problem is identified in a radiation-emitting product, the FDA is responsible for approving the corrective action plan (CAP). The first reports of the Tyler incidents came to the FDA from the State of Texas Health Department, and this triggered FDA action. The FDA investigation was well underway when AECL produced a medical device report to discuss the details of the radiation overexposures at Tyler. The FDA declared the Therac-25 defective under the Radiation Control for Health and Safety Act and ordered the firm to notify all purchasers, investigate the problem, determine a solution, and submit a corrective action plan for FDA approval.

The final CAP consisted of more than twenty changes to the system hardware and software, plus modifications to the system documentation and manuals. Some of these changes were unrelated to the specific accidents but were improvements to the general safety of the machine. The full CAP implementation, including an extensive safety analysis, was not complete until more than two years after the Tyler accidents.

AECL made its accident report to the FDA on April 15, 1986. On that same date, AECL sent out a letter to each Therac-25 user recommending a temporary "fix" to the machine that would allow continued clinical use. The letter, shown here in its complete form, stated:

> *SUBJECT: CHANGE IN OPERATING PROCEDURES FOR THE THERAC 25 LINEAR ACCELERATOR*
>
> *Effective immediately, and until further notice, the key used for moving the cursor back through the prescription sequence (i.e., cursor 'UP' inscribed with an upward pointing arrow) must not be used for editing or any other purpose.*
>
> *To avoid accidental use of this key, the key cap must be removed and the switch contacts fixed in the open position with electrical tape or other insulating material. For assistance with the latter you should contact your local AECL service representative.*
>
> *Disabling this key means that if any prescription data entered is incorrect then a 'R' reset command must be used and the whole prescription reentered. For those users of the Multiport option it also means that editing of dose rate, dose and time will not be possible between ports.*

On May 2, 1986, the FDA declared the Therac-25 defective, demanded a CAP, and required renotification of all the Therac-25 customers. In the letter from the FDA to AECL, the director of compliance, Center for Devices and Radiological Health, wrote:

> *We have reviewed [AECL's] April 15 letter to purchasers and have concluded that it does not satisfy the requirements for notification to purchasers of a defect in an electronic product. Specifically, it does not describe the defect nor the hazards associated with it. The letter does not provide any reason for disabling the cursor key and the tone is not commensurate with the urgency for doing so. In fact, the letter implies the inconvenience to operators outweighs the need to disable the key. We request that you immediately renotify purchasers.*

AECL promptly made a new notice to users and also requested an extension to produce a CAP. The FDA granted this request. About this time, the Therac-25 users created a users group and held their first meeting at the annual conference of the American Association of Physicists in Medicine. At the meeting, users discussed the Tyler accident and heard an AECL representative present the company's plans for responding to it. AECL promised to send a letter to all users detailing the CAP.

Several users described additional hardware safety features that they had added to their own machines to provide additional protection. An interlock that checked gun current values, which the Vancouver clinic had previously added to their Therac-25, was labeled as redundant by AECL; the users disagreed. There were further discussions of poor design and other problems that caused a 10 to 30 percent underdosing in both modes.

The meeting notes said: *"There was a general complaint by all users present about the lack of information propagation. The users were not happy about receiving incomplete information. The AECL representative countered by stating that AECL does not wish to spread rumors and that AECL has no policy to 'keep things quiet.' The consensus among the users was that an improvement was necessary."*

After the first users group meeting, there were two users group newsletters. The first, dated fall 1986, contained letters from Tim Still, the Kennestone physicist, who complained about what he considered to be eight major problems he had experienced with the Therac-25. These problems included poor screen-refresh subroutines that leave trash and erroneous information on the operator console and some tape-loading problems on startup that he discovered involved the use of "phantom tables" to trigger the interlock system in the event of a load failure instead of using a checksum. He asked the question, "Is programming safety relying too much on the software interlock routines?"

The second users group newsletter, in December 1986, further discussed the implications of the phantom table problem. AECL produced its first CAP on June 13, 1986. The FDA asked for changes and additional information about the software, including a software test plan.

AECL responded on September 26 with several documents describing the software and its modifications but no test plan. They explained how the Therac-25 software evolved from the Therac-6 software and stated that "no single test plan and report exists for the software since both hardware and software were tested and exercised separately and together over many years." AECL concluded that the current CAP improved "machine safety by many orders of magnitude and virtually eliminates the possibility of lethal doses as delivered in the Tyler incident."

An FDA internal memo dated October 20 commented on these AECL submissions, raising several concerns:

> *Unfortunately, the AECL response also seems to point out an apparent lack of documentation on software specifications and a software test plan. . . . concerns include the question of previous knowledge of problems by AECL, the apparent paucity of software quality assurance at the manufacturing facility, and possible warnings and information dissemination to others of the generic type problems.*
>
> *. . . . As mentioned in my first review, there is some confusion on whether the manufacturer should have been aware of the software problems prior to the ARO's [Accidental Radiation Overdoses] in Texas. AECL had received official notification of a law suit in November 1985 from a patient claiming accidental over-exposure from a Therac-25 in Marietta, Georgia. . . . If knowledge of these software deficiencies were known beforehand, what would be the FDA's posture in this case?*

. . . . The materials submitted by the manufacturer have not been in sufficient detail and clarity to ensure an adequate software quality assurance program currently exists. For example, a response has not been provided with respect to the software part of the CAP to the CDRH's [FDA Center for Devices and Radiological Health] request for documentation on the revised requirements and specifications for the new software. In addition, an analysis has not been provided, as requested, on the interaction with other portions of the software to demonstrate the corrected software does not adversely affect other software functions.

The July 23 letter from the CDRH requested a documented test plan including several specific pieces of information identified in the letter. This request has been ignored up to this point by the manufacturer. Considering the ramifications of the current software problem, changes in software QA attitudes are needed at AECL.

AECL also planned to retain the malfunction codes, but the FDA required better warnings for the operators. Furthermore, AECL had not planned on any quality assurance testing to ensure exact copying of software, but the FDA insisted on it. The FDA further requested assurances that rigorous testing would become a standard part of AECL's software modification procedures. "*We also expressed our concern that you did not intend to perform the protocol to future modifications to software. We believe that the rigorous testing must be performed each time a modification is made in order to ensure the modification does not adversely affect the safety of the system.*"

Finally, AECL was asked to draw up an installation test plan to ensure that both hardware and software changes perform as designed when installed. AECL submitted CAP Revision 2 and supporting documentation on December 22, 1986. They changed the CAP to have dose malfunctions suspend treatment and included a plan for meaningful error messages and highlighted dose error messages. They also expanded their diagrams of software modifications and expanded their test plan to cover hardware and software.

A.2.6 Yakima Valley Memorial Hospital, January 1987

On Saturday, January 17, 1987, the second patient of the day was to be treated for a carcinoma. This patient was to receive two film verification exposures of 4 and 3 rads plus a 79-rad photon treatment (for a total exposure of 86 rads.)

Film was placed under the patient, and 4 rads were administered. After the machine paused to open the collimator jaws further, the second exposure of 3 rads was administered. The machine paused again.

The operator entered the treatment room to remove the film and verify the patient's precise position. He used the hand control in the treatment room to rotate the turntable to the field light position, which allowed him to check the alignment of the machine with respect to the patient's body in order to verify proper beam position. He then either pressed the {\em set} button on the hand control or left the room and typed a set command at the console to return the turntable to the proper position for treatment; there is some confusion as to exactly what transpired.

When he left the room, he forgot to remove the film from underneath the patient. The console displayed BEAM READY, and the operator hit the ***B*** key to turn the beam on.

The beam came on, but the console displayed no dose or dose rate. After 5 or 6 seconds, the unit shut down with a pause and displayed a message. The message "may have disappeared quickly"; the operator was unclear on this point.

However, because the machine merely paused, he was able to push the ***P*** key to proceed with treatment. The machine paused again, this time displaying FLATNESS on the reason line. The operator heard the patient say something over the intercom but could not understand him. He went into the room to speak with the patient, who reported "feeling a burning sensation" in the chest. The console displayed only the total dose of the two film exposures (7 rads) and nothing more.

Later in the day, the patient developed a skin burn over the entire treatment area. Four days later, the redness developed a striped pattern matching the slots in the blocking tray. The striped pattern was similar to the burn a year earlier at this same hospital, which had first been ascribed to a heating pad and later officially labeled by the hospital as "cause unknown."

AECL began an investigation, and users were told to confirm the turntable position visually before turning on the beam. All tests run by the AECL engineers indicated that the machine was working perfectly. From the information that had been gathered to that point, it was suspected that the electron beam had come on when the turntable was in the field light position. But the investigators could not reproduce the fault condition.

On the following Thursday, AECL sent in an engineer from Ottawa to investigate. The hospital physicist had, in the meantime, run some tests himself. He placed a film in the Therac-25's beam and then ran two exposures of X-ray parameters with the turntable in field light position. The film appeared to match the film that was left (by mistake) under the patient during the accident.

After a week of checking the hardware, AECL determined that the "incorrect machine operation was probably not caused by hardware alone." After checking the software, AECL engineers discovered a flaw that could explain the erroneous behavior. The software flaws explaining this accident are completely different from those associated with the Tyler accidents.

There is no way to determine what particular software design errors were related to the Kennestone, Hamilton, and first Yakima accidents. Additional unknown software flaws could have been responsible for them. There is speculation, however, that the Hamilton accident was the same as this second Yakima overdose. In a report of a conference call on January 26, 1987, between the AECL quality assurance manager and Ed Miller of the FDA discussing the Yakima accident, Miller notes: "*This situation probably occurred in the Hamilton, Ontario accident a couple of years ago. It was not discovered at that time and the cause was attributed to intermittent interlock failure. The subsequent recall of the multiple microswitch logic network did not really solve the problem*" [266].

Preliminary dose measurements by AECL indicated that the dose delivered under these conditions—that is, when the turntable is in the field light position—is on the order of 4,000 to 5,000 rads. After two attempts, the patient could have received 8,000 to 10,000 instead of the 86 rads prescribed.

AECL again called users on January 26—nine days after the accident—and gave them detailed instructions on how to avoid this problem. In an FDA internal report on the accident, the AECL quality assurance manager investigating the problem is quoted as saying that the software and hardware changes to be retrofitted following the Tyler accident nine months earlier, *but which had not yet been installed,* would have prevented the Yakima accident.

The patient died in April from complications related to the overdose. He had a terminal form of cancer, but a lawsuit was initiated by his survivors alleging that he died sooner than he would have and endured unnecessary pain and suffering due to the radiation overdose. The suit, like all the others, was settled out of court.

Manufacturer, government, and user response On February 3, 1987, after interaction with the FDA and others, including the users group, AECL announced to its customers:

1. a new software release to correct both the Tyler and Yakima software problems
2. a hardware single-pulse shutdown circuit
3. a turntable potentiometer to independently monitor turntable position
4. a hardware turntable interlock circuit

The second item, a hardware single-pulse shutdown circuit, essentially acts as a hardware interlock to prevent overdosing by detecting an unsafe level of radiation and halting beam output after one pulse of high energy and current. This interlock effectively provides an independent way to protect against a wide range of potential hardware failures and software errors. The third item, a turntable potentiometer, was the safety device recommended by several groups after the Hamilton accident.

After the second Yakima accident, the FDA became concerned that the use of the Therac-25 during the CAP process, even with AECL's interim operating instructions, involved too much risk to patients. The FDA concluded that the accidents demonstrated that the software alone could not be relied on to assure safe operation of the machine. In a February 18, 1987, internal FDA memorandum, the Director of the Division of Radiological Products wrote:

> *It is impossible for CDRH to find all potential failure modes and conditions of the software. AECL has indicated the "simple software fix" will correct the turntable position problem displayed at Yakima. We have not yet had the opportunity to evaluate that modification. Even if it does, based upon past history, I am not convinced that there are not other software glitches that could result in serious injury. . . . We are in the position of saying that the proposed CAP can reasonably be expected to correct the deficiencies for which they were developed (Tyler). We cannot say that we are reasonable [sic] confident about the safety of the entire system to prevent or minimize exposure from other fault conditions.*

On February 6, 1987, Ed Miller of the FDA called Pavel Dvorak of Health and Welfare Canada to advise him that the FDA would recommend that all Therac-25s be shut down until permanent modifications could be made. According to Miller's notes on the phone call, Dvorak agreed and indicated that Health and Welfare would coordinate their actions with the FDA [266].

AECL responded on April 13 with an update on the Therac-25 CAP status and a schedule of the nine action items pressed by the users at a users group meeting in March. This unique and highly productive meeting provided an unusual opportunity to involve the users in the CAP evaluation process. It brought together all concerned parties in one place and at one time so that a course of action could be decided on and approved as quickly as possible. The attendees included representatives from

1. the manufacturer (AECL);
2. all users, including their technical and legal staffs;
3. the FDA and the Canadian Bureau of Radiation and Medical Devices;
4. the Canadian Atomic Energy Control Board;
5. the Province of Ontario;
6. the Radiation Regulations Committee of the Canadian Association of Physicists.

According to Gordon Symonds, from the Canadian BRMD, this meeting was very important to the resolution of the problems; the regulators, users, and manufacturer arrived at a consensus in one day.

At this second user's meeting, the participants carefully reviewed all the six known major Therac-25 accidents to that date and discussed the elements of the CAP along with possible additional modifications. They came up with a prioritized list of modifications they wanted included in the CAP and expressed concerns about the lack of independent evaluation of the software and the lack of a hardcopy audit trail to assist in diagnosing faults.

The AECL representative, who was the quality assurance manager, responded that tests had been done on the CAP changes, but that the tests were not documented and that independent evaluation of the software "might not be possible."

He claimed that two outside experts had reviewed the software, but he could not provide their names. In response to user requests for a hard copy audit trail and access to source code, he explained that memory limitations would not permit including such options and that source code would not be made available to users.

On May 1, AECL issued CAP Revision 4 as a result of the FDA comments and the user's meeting input. The FDA response on May 26 approved the CAP subject to submission of the final test plan results and an independent safety analysis, distribution of the draft revised manual to customers, and completion of the CAP by June 30, 1987. The FDA concluded by rating this a Class I recall: a recall in which there is a reasonable probability that the use of, or exposure to, a violative product will cause serious adverse health consequences or death [45].

AECL sent more supporting documentation to the FDA on June 5, 1987, including the CAP test plan, a draft operator's manual, and the draft of the new safety analysis. This time the analysis included the software in the fault trees but used a "generic failure rate" of 10^{-4} for software events. This number was justified as being based on the historical performance of the Therac-25 software. The 10^{-4} number is widely used for software "failure" without any justification. As discussed in chapter 13, such numbers for software make no technical sense.

The final report on the safety analysis states that many of the fault trees had a computer malfunction as a causative event, and the outcome for quantification was therefore dependent on the failure rate chosen for the software.

Assuming that all software errors are equally likely seems rather strange. A close inspection of the code was also conducted during this safety analysis to "obtain more information on which to base decisions." An outside consultant performed the inspection, which included a detailed examination of the implementation of each function, a search for coding errors, and a qualitative assessment of the software's reliability. No information is provided in the final safety report about whether any particular methodology or tools were used in the software inspection or whether someone just read the code looking for errors.

AECL planned a fifth revision of the CAP to include the testing and final safety analysis results. Referring to the test plan at this, the final stage of the CAP process, an FDA reviewer said,

> *Amazingly, the test data presented to show that the software changes to handle the edit problems in the Therac-25 are appropriate prove the exact opposite result. A review of the data table in the test results indicates that the final beam type and energy (edit change) has no effect on the initial beam type and energy. I can only assume that either the fix is not right or the data was entered incorrectly. The manufacturer should be admonished for this error. Where is the QC [Quality Control] review for the test program? AECL must: (1) clarify this situation, (2) change the test protocol to prevent this type of error from occurring, and (3) set up appropriate QC control on data review.*

A further FDA memo indicated:

> *[The AECL quality assurance manager] could not give an explanation and will check into the circumstances. He subsequently called back and verified that the technician completed the form incorrectly. Correct operation was witnessed by himself and others. They will repeat and send us the correct data sheet.*

At the American Association of Physicists in Medicine meeting in July 1987, a third user's meeting was held. The AECL representative described the status of the latest CAP and explained that the FDA had given verbal approval and that he expected full implementation by the end of August 1987. He went on to review and comment on the prioritized concerns of the last meeting. Three of the user-requested hardware changes had been included in the CAP. Changes to tape load error messages and checksums on the load data would wait until after the CAP was done. Software documentation was described as a lower-priority task that needed definition and would not be available to the FDA in any form for over a year.

On July 6, 1987, AECL sent a letter to all users to update them on the FDA's verbal approval of the CAP and to delineate how AECL would proceed. Finally, on July 21, 1987, AECL issued the final and fifth CAP revision.

The major features of the final CAP are:

- All interruptions related to the dosimetry system will go to a treatment suspend, not a treatment pause. Operators will not be allowed to restart the machine without reentering all parameters.
- A software single-pulse shutdown will be added.
- An independent hardware single-pulse shutdown will be added.
- Monitoring logic for turntable position will be improved to ensure that the turntable is in one of the three legal positions.
- A potentiometer will be added to the turntable. The output is used to monitor exact turntable location and provide a visible position signal to the operator.

- Interlocking with the 270-degree bending magnet will be added to ensure that the target and beam flattener are in position if the X-ray mode is selected.
- Beam-on will be prevented if the turntable is in the field light or any intermediate position.
- Cryptic malfunction messages will be replaced with meaningful messages and highlighted dose-rate messages.
- Editing keys will be limited to *cursor up*, *backspace*, and *return*. All other keys will be inoperative.
- A motion-enable footswitch—a type of deadman switch—will be added. The operator will be required to hold this switch closed during movement of certain parts of the machine to prevent unwanted motions when the operator is not in control.
- Twenty-three other changes will be made to the software to improve its operation and reliability, including disabling of unused keys, changing the operation of the *set* and *reset* commands, preventing copying of the control program on site, changing the way various detected hardware faults are handled, eliminating errors in the software that were detected during the review process, adding several additional software interlocks, disallowing changes in the service mode while a treatment is in progress, and adding meaningful error messages.
- The known software problems associated with the Tyler and Yakima accidents will be fixed.
- The manuals will be fixed to reflect the changes.

Figure A.3 shows a typical Therac-25 installation after the CAP changes were made.

Ed Miller, the director of the Division of Standards Enforcement, Center for Devices and Radiological Health at the FDA, wrote in 1987: "*FDA has performed extensive review of the Therac-25 software and hardware safety systems. We cannot say with absolute certainty that all software problems that might result in improper dose have been found and eliminated. However, we are confident that the hardware and software safety features recently added will prevent future catastrophic consequences of failure*" [266]. No Therac-25 accidents have been reported since the final corrective action plan was implemented. There have been, however, many accidents involving radiation therapy with other machines since that time.

A.3 Some Causal Factors

Many lessons can be learned from this series of accidents. A few are considered here. The reader is encouraged to make their own independent list of the causal factors involved. In addition, considering what additional questions might have been pursued in the investigation is instructive.

A.3.1 Overconfidence in Software

There still is the widespread belief that software does not "fail," unlike the hardware devices it is replacing. In many safety-critical systems today—including medical devices, aircraft, nuclear power plants, and weapon systems—standard hardware backups, interlocks, and

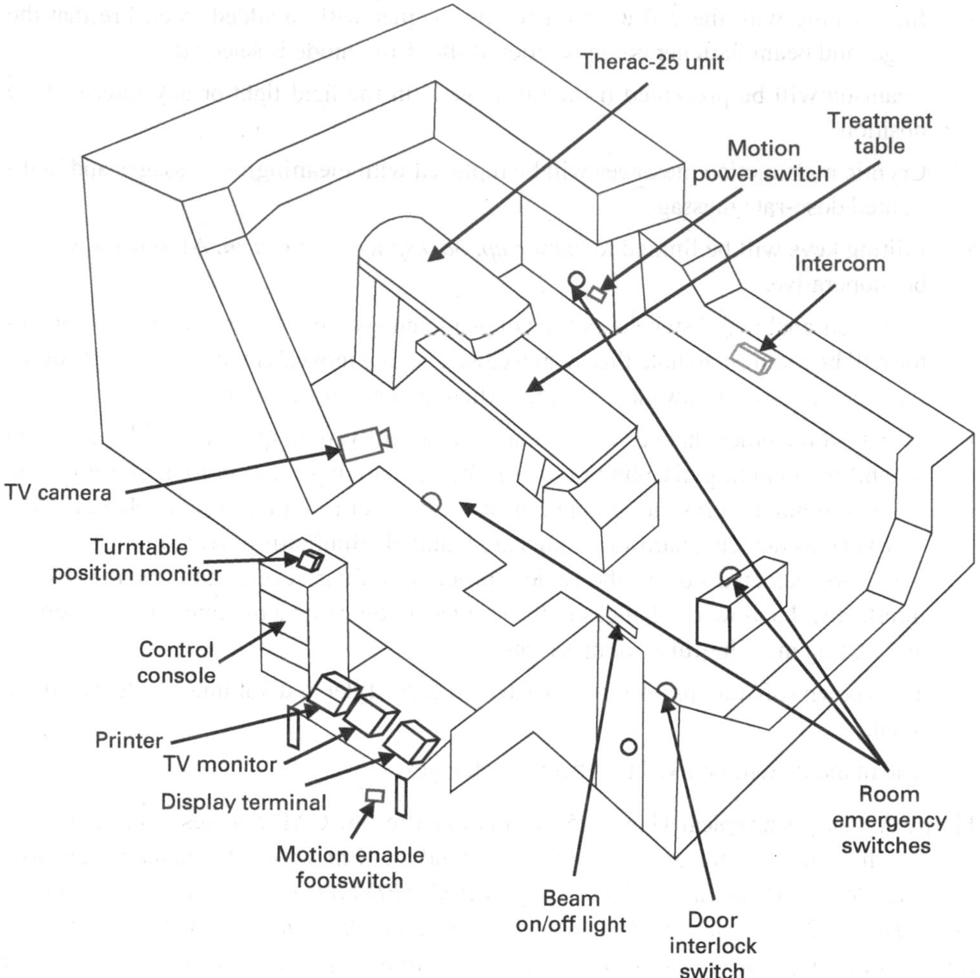

Figure A.3
A typical Therac-25 facility.

other protection devices are being replaced by software. Where hardware protection is still used, it is often controlled by software.

In the Therac-25, even though nearly full responsibility for safety depended on the operation of the software, the first safety analysis on the machine did not include software. When problems started occurring, it was assumed that hardware had caused them, and the investigation concentrated on the hardware. This phenomenon still occurs today although it is more likely to take the form of blaming all problems on the human operators and assuming the software had no impact on the losses.

Eliminating the hardware protection mechanisms that were common before software took over the protection functions is the result of undue confidence being placed on software. Engineers do not purposely create designs where a hardware single point of failure could lead to a catastrophe. The same needs to be true for the software in our critical systems. Protection against software errors can and should be built into both the system and the software itself.

A.3.2 Confusing Reliability with Safety

The Therac-25 software was highly reliable. It worked tens of thousands of times before overdosing anyone, and occurrences of erroneous behavior were few and far between. AECL assumed that its software was safe because it was reliable, and this led to dangerous complacency.

When systems were primarily electromechanical, nearly exhaustive testing was possible, and design errors could be eliminated before operational use. What was left during operations was random wear-out failures. Safety then could be assumed to be effectively approximated by reliability.

Software is very different; it is design abstracted from its physical realization. While the hardware on which the software is executed may fail, the design itself does not fail. In fact, software by itself is not safe or unsafe—safety depends on the *context* in which the software is used. Much of the Therac-25 software had been used on an earlier version of this machine, called the Therac-20. The same flaws that killed people with the Therac-25 were not dangerous because of the design of the Therac-20 hardware that the software was controlling.

A related widespread misunderstanding is that safety of software is enhanced by assuring that the software satisfies the requirements. Almost all software-related accidents have involved requirements flaws, not coding or implementation errors. Perhaps this misguided reliance on assurance techniques such as testing and simulation stems from the fact that we have lots of solutions proposed to ensure that software satisfies its specified requirements, but few to identify the safety-critical software requirements.

Not much has changed in forty years in this regard. Nearly all standards for safety-critical software focus on implementation assurance. If we truly want to reduce software-related accidents, we have to focus less on assurance and more on identifying the safety-critical requirements and building safety into these machines from the beginning of development. Safety cannot be assured in if it is not already there; it has to be designed in from the beginning.

A.3.3 Lack of Defensive Design

The Therac-25 software did not contain self-checks or other error-detection and error-handling features that would have detected the inconsistencies and coding errors. There were no independent checks that the machine was operating correctly. Patient reactions were the only real indications of the seriousness of the problems with the Therac-25. There seemed to be so much confidence in the software that standard hardware safety and defensive design features (see chapter 11) were omitted.

Such verification cannot be assigned to human operators without providing them with some means of detecting errors: the Therac-25 software "lied" to the operators, and the machine itself was not capable of detecting that a massive overdose had occurred. The ion chambers on the Therac-25 could not handle the high density of ionization from the unscanned electron beam at high beam current; they thus became saturated and gave an indication of a low dosage. Engineers need to design for the worst case. We will make little progress if human operators are blamed for accidents when the problems are really in the machine design.

In the Therac-25, audit trails were limited because of a lack of memory in computers at that time. However, today larger memories are available, and audit trails and other design techniques must be given high priority in making tradeoff decisions. The problem today is usually the opposite one: so much data can be and therefore is collected, for instance during aircraft operations, that it is difficult to detect problems until after a serious event has occurred. The problem today is not in collecting data but in identifying trends and behaviors of the hardware, software, and operators that are increasing risk *before* an accident occurs. Simply collecting data alone will not help here. We need sophisticated modeling and analysis tools to analyze that data. Data, whether "big" or not, is not the same as information.

A.3.4 Not Identifying and Eliminating the Systemic Causal Factors

One of the lessons to be learned from the Therac-25 experiences is that focusing on particular software design errors is not the way to make a system safe. Virtually all complex software can be made to behave in an unexpected fashion under some conditions: There will always be another software bug. Just as engineers would not rely on a design with a hardware single point of failure that could lead to catastrophe, they should not do so if that single point of failure is software.

The Therac-20 contained the same software error implicated in the Tyler deaths, but this machine included hardware interlocks that mitigated the consequences of the error. Protection against software errors can and should be built into both the system and the software itself. We cannot eliminate all software errors, but we can often protect against their worst effects, and we can recognize their likelihood in our decision-making.

One of the serious mistakes that led to the multiple Therac-25 accidents was the tendency to believe that the cause of an accident had been determined (e.g., a microswitch failure in the case of Hamilton) without adequate evidence to come to this conclusion and without looking at all possible contributing factors. Without a thorough investigation, it is not possible to determine whether a sensor provided the wrong information, the software provided an incorrect command, or the actuator had a transient failure and did the wrong thing on its own. In the case of the Hamilton accident, a transient microswitch failure was assumed to be the cause even though the engineers were unable to reproduce the failure or to find anything wrong with the microswitch.

In general, it is a mistake to patch just one causal factor (such as the software) and assume that future accidents will be eliminated. Accidents are unlikely to occur in exactly the same way again. If we patch only the symptoms and ignore the deeper underlying causes, or if we fix only the specific cause of one accident, we are unlikely to have much effect on future accidents. The series of accidents involving the Therac-25 is a good example of exactly this problem: fixing each individual software flaw as it was found did not solve the safety problems of the device and more people were injured than was necessary. Admittedly, this machine was designed a long time ago, but too many similar case studies can be written about accidents happening today.

A.3.5 Complacency

Often it takes an accident to alert people to the dangers involved in technology. A medical physicist wrote in 1987 about the Therac-25 accidents:

> *In the past decade or two, the medical accelerator "industry" has become perhaps a little complacent about safety. We have assumed that the manufacturers have all kinds of safety design experience since they've been in the business a long time. We know that*

there are many safety codes, guides, and regulations to guide them and we have been reassured by the hitherto excellent record of these machines. Except for a few incidents in the 1960's (e.g., at Hammersmith, Hamburg) the use of medical accelerators has been remarkably free of serious radiation accidents until now. Perhaps, though we have been spoiled by this success. [326]

Accidents involving medical devices, including radiation therapy, continue today. And such complacency is rampant in all fields.

A.3.6 Unrealistic Risk Assessments

The Therac-25 had a probabilistic risk assessment, including an update after one of the early accidents. The first analysis ignored software. Later changes treated it superficially by assuming that all software errors were equally likely. These probabilistic risk assessments generated complacency and undue confidence in the machine and in the results of the risk assessment themselves. When the first Yakima accident was reported to AECL, the company did not investigate. Their evidence for their belief that the radiation burn could not have been caused by their machine included a probabilistic risk assessment showing that safety had increased by five orders of magnitude as a result of the microswitch fix.

The belief that safety had been increased by such a large amount seems hard to justify. Perhaps it was based on the probability of failure of the microswitch (typically 10^{-5}) AND-ed with the other interlocks. The problem with all such analyses is that they typically make many independence assumptions and exclude aspects of the system—in this case, software—that are difficult, if not impossible, to quantify but that may have a larger impact on safety than the quantifiable factors that are included. There are few if any scientific evaluations of the correctness of such assessments. In our enthusiasm to provide measurements, we should not attempt to measure the unmeasurable.

Specific software flaws leading to a serious loss, of course, cannot be assessed probabilistically. Even if they could, it would require so much information about the flaw that it could be fixed instead of justified away as "will never or rarely occur." I have participated in accident investigations or read accident reports involving software for forty years. In every case, there had been a risk assessment that was used to convince decision makers that the accident could not occur, usually in the exact way that the accident(s) did occur. Even after a loss or significant event, engineers and management often ignored the fact that the risk assessment was obviously wrong and continued to believe in it until the next event and sometimes even after that until a major loss occurred. For some reason, people seem to believe in a calculated number more than actual experience.

If accurate probabilities are not determinable (as is true for systems that contain software), then making them up or simply ignoring the software or other unquantifiable factors in the system risk assessment is not the answer. A better solution is to design for the worst case (instead of assuming that only the average case will occur) and creating better decision-making tools that do not require unsupportable risk assessments.

A.3.7 Inadequate Investigation or Follow-Up on Accident Reports

Every company building safety-critical systems should have audit trails and incident analysis procedures that are applied whenever any hint of a problem is found that might lead to an accident. The first phone call by Tim Still should have led to an extensive investigation of the events at Kennestone. Certainly, learning about the first lawsuit should have triggered an immediate response.

More generally, superficial accident/incident analyses, often placing all or most of the blame on the operators, leads to patching symptoms but not to understanding the deeper underlying and systemic causes of a loss or near loss. This process leads to further accidents that could and should have been prevented. We need to look at the role of the entire system in the accident in order to make progress in safety.

A.3.8 Inadequate System and Software Engineering Practices

The Therac-25 software was created in 1974 and used what are today considered obsolete software engineering practices. There are, however, factors in the accidents related to software engineering that are still common today. Too often, practices that may be acceptable for website or productivity tools development are used and even promoted for safety-critical software.

Some basic software engineering principles that apparently were violated in the case of the Therac-25 and often still today include the following:

- A qualitative system hazard analysis, that includes the software, needs to be performed and the information obtained used to create safety-related requirements specifications and to design safety into the system and software from the beginning. After-the-fact testing and other assurance methods are not useful for safety.
- Software specifications and documentation should not be an afterthought or done after the software development begins.
- Rigorous company software practices and standards should be established.
- Designs should be kept as simple as possible.
- Ways to detect errors and get information about them, such as software audit trails, should be designed into the software from the beginning.
- Computer displays and the presentation of information to the operators, such as error messages, along with user manuals and other documentation need to be carefully designed by human factors experts.

A.3.9 Software Reuse

Important lessons about software reuse can be found in these accidents. A naive assumption is often made that reusing software or using commercial off-the-shelf (COTS) software will increase safety because the software will have been exercised extensively.

The Therac-25 was an improvement of an earlier machine by the same company called the Therac-20, and much of the software was reused. Today overconfidence in reuse is still rampant. A false assumption may be made that reusing software or using COTS software increases safety because the software has been exercised extensively. As stated earlier, software is only safe or unsafe within a specific context. It is not possible to determine safety by looking at the software alone. Reusing software that was safe in one system does not mean it will be safe when used in a different system. Safety is a quality of the system in which the software is used; it is *not* a quality of the software itself. The belief sometimes built into practice or even government standards that reused software is safe or safer is not justified.

A.3.10 Safe versus Friendly User Interfaces

While the interface equipment and principles used for the Therac-25 are obsolete, there are still potential issues even with today's more sophisticated interface tools. Sometimes making the interface easy to use conflicts with safety. For example, eliminating multiple data entry and assuming that operators would check the values carefully before pressing the return key was unrealistic for the Therac-25 and for most systems. The author has been involved in reviews of many newer safety-critical system interfaces and has been surprised by how many included unsafe features. One example is allowing operators to turn off alarms with no indication showing that the alarms have been disabled. One general design principle is that actions to get into or maintain a safe state should be easy to do. Actions that can lead to an unsafe state (hazard) should be hard to do (see chapter 12).

Relying on operators to detect errors and recover before an accident is not realistic, particularly when the operator is not provided with the support to perform this function. Some of the radiation therapy accidents since the Therac-25 overdoses have involved operators not being able to see or react to error messages. The accidents were then blamed on the operators rather than the poor machine and interface design. The same has occurred in other industries, where operators are blamed for accidents that are primarily the result of flawed engineering.

A.3.11 User and Government Oversight and Standards

While the original response of the FDA to the Therac-25 accidents, after they were understood, was impressive, later follow-through and current regulation of medical device software is weak. The same is true for other industries. The FDA puts more emphasis on reporting adverse events in linear accelerators, for example, than on preventing them in the first place. One problem is the difficulty and time required to update standards. In general, standards should never include specific techniques (such as FMEA for medical devices) or the standards will become out-of-date almost immediately without any possibility of updating for perhaps a decade.

Standards can have the undesirable effect of limiting the safety efforts and investment of companies that feel their legal and moral responsibilities are fulfilled if they follow the standards. As the standards often represent the input of commercial companies, there are often conflicts of interest involved in producing effective standards.

One lesson from studying accidents, like those involved with the Therac-25, is that accidents are seldom simple. They usually involve a complex web of interacting events with multiple contributing technical, human, organizational, and regulatory factors. We are not learning enough today from the events nor focusing enough on preventing them.

Note that the input and pressure from the users group was critical in getting the machine fixed and in getting later radiation therapy equipment built in a safer way. An important lesson for users in other industries is that the customers have more power to effect important changes than government oversight agencies. They need to learn how to use their power effectively.

Appendix B
Space: The *Challenger* and *Columbia* Space Shuttle Losses

The *Challenger* and *Columbia* Space Shuttle accidents are combined in this appendix. While the proximate physical causes of the two accidents are different—an O-ring failure in *Challenger* and damage to the thermal tiles in *Columbia*—the accidents stem from the same organizational, management, and cultural flaws.

Safety was an important concern in the Apollo program, but it received special emphasis after three astronauts were killed in a fire on the Apollo launch pad in 1967. Jerome Lederer, a renowned aircraft safety expert, created what was considered at the time to be a world-class system safety program. Over time, that program declined for a variety of reasons that provide important lessons to everyone attempting to improve safety in their organizations. The *Columbia* Accident Investigation Board (CAIB) report describes safety engineering at NASA at the time of the *Columbia* accident as "the vestiges of a once robust safety program" [112, p. 177].

The *Challenger* Space Shuttle was lost on January 28, 1986, five years after the first Space Shuttle flight. After the *Challenger* loss, there was an attempt to strengthen the safety program, but that attempt did not last. The same external pressures, inadequate responses to those pressures, flaws in the safety culture, dysfunctional organizational safety structure, and ineffective safety engineering practices contributed to increasing states of risk. The *Columbia* Space Shuttle loss seventeen years after the *Challenger* accident was due to almost identical organizational, management, and cultural factors. Either the attempts to improve the NASA safety program after the *Challenger* accident were not successful, or they were effective for a while but the same internal factors and outside pressures led to the fixes degrading over time.

What were those pressures? In the case of the Space Shuttle, political and other factors contributed to the adoption of a vulnerable design during the original approval process. Unachievable promises were made with respect to performance in order to keep the manned space flight program alive after Apollo and the demise of the Cold War. While these performance goals even then seemed unrealistic, the success of the Apollo program and the *can-do* culture that arose during it—marked by tenacity in the face of seemingly impossible challenges—contributed to the belief that these unrealistic goals could be achieved if only enough effort were expended. Performance pressures and program survival fears gradually led to an erosion of the rigorous processes and procedures of the

Apollo program as well as the substitution of dedicated NASA staff with contractors who had dual loyalties [216]. The Rogers Commission report on the *Challenger* accident concluded:

> *The unrelenting pressure to meet the demands of an accelerating flight schedule might have been adequately handled by NASA if it had insisted upon the exactingly thorough procedures that were its hallmark during the Apollo program. An extensive and redundant safety program comprising interdependent safety, reliability, and quality assurance functions existed during and after the lunar program to discover any potential safety problems. Between that period and 1986, however, the program became ineffective. This loss of effectiveness seriously degraded the checks and balances essential for maintaining flight safety.* [332, p. 152]

B.1 Background

The Space Shuttle was part of a larger Space Transportation System (STS) concept that arose in the 1960s when Apollo was in development. The concept originally included a manned Mars expedition, a space station in lunar orbit, and an Earth-orbiting station serviced by a reusable ferry or Space Shuttle. Funding for such a large effort, on the order of that provided for Apollo, never materialized.

After many failed attempts, and finally agreeing to what would turn out to be unrealistic compromises, NASA gained approval from the Nixon Administration to develop, on a fixed budget, only the transport vehicle. Because the administration did not approve a low-Earth-orbit station, NASA had to create a mission for the vehicle, which became a space launch.

To satisfy the government's requirement that the system be economically justifiable, the vehicle had to capture essentially all space launch business, both launching and servicing satellites, which included the launch of large satellites for the Air Force. To perform these tasks, design compromises were required that increased risk over what was necessary.

NASA also had to make promises about performance (number of launches per year) and cost per launch that were unrealistic. An important factor in both accidents was the pressure exerted on NASA by an unrealistic flight schedule with inadequate resources and by commitments to customers. The nation's reliance on the Shuttle as its principal space launch capability created relentless pressure on NASA to increase the flight rate to the originally promised twenty-four missions per year. Neither the estimated per-pound payload cost nor the predicted number of flights per year was ever achieved. Over its lifetime, the Shuttle averaged four flights per year, and costs were higher than anticipated.

Even given these constraints, NASA designed and developed a remarkably capable and resilient vehicle, consisting of an orbiter with three main engines, two solid rocket boosters (SRBs), and an external tank. To reduce costs, the booster was to be reusable. Solid propellant boosters were selected because of lower costs and ease of refurbishing them for reuse after they landed in the ocean.

Unfortunately, the Shuttle never met any of its original requirements for cost, ease and speed of turnaround between flights, maintainability, or, regrettably, safety.

Contracts were awarded in 1972: Rockwell was responsible for the design and development of the orbiter, Martin Marietta was assigned the development and fabrication of the external tank, Morton Thiokol was awarded the contract for the SRBs, and Rocketdyne (a division of Rockwell) was selected to develop the orbiter main engines.

Three NASA field centers were given responsibility for the program. Johnson Space Center in Houston, Texas, was to manage the orbiter, while Marshall Space Flight Center in Huntsville, Alabama, was assigned responsibility for the orbiter's main engines, the external tank, and the SRBs. Kennedy Space Center in Merritt Island, Florida, would assemble the Shuttle components, check them out, and conduct launches. The management structure in NASA is fairly complex, but basically there are four management levels, with Level I being the highest.

Because of continual budget cuts and constraints, the original fleet of five Shuttle orbiters was reduced to four. The orbital test flight program began in early 1981, and, after the landing of STS4 in July 1982, NASA declared that the Shuttle was "operational." Including the initial orbital tests, the Shuttle flew twenty-four successful missions over a fifty-seven-month period before the *Challenger* accident.

Budget pressures during operations added to the performance pressures. Budget cuts occurred during the life of the Shuttle: for example, cuts amounted to a 40 percent reduction in purchasing power over the decade before the *Columbia* loss. At the same time, the budget was occasionally raided by NASA itself to make up for overruns in the International Space Station program.

The later budget cuts came at a time when the Shuttle was aging and costs were actually increasing. The infrastructure, much of which dated back to the Apollo era, was falling apart before the *Columbia* accident. In the last fifteen years of the Shuttle program, uncertainty about how long the Shuttle would fly added to the pressures to delay safety upgrades and improvements to the Shuttle program infrastructure.

An important factor in both the *Challenger* and *Columbia* accidents was the pressures exerted on NASA by an unrealistic flight schedule with inadequate resources and by commitments to customers. The nation's reliance on the Shuttle as its principal space launch capability created relentless pressure on NASA to increase the flight rate. The attempt to build up to twenty-four missions per year brought problems such as compression of training schedules, a lack of spare parts, the focusing of resources on near-term projects, and a dilution of the human and material resources that could be applied to any particular flight. Customer commitments may occasionally have obscured engineering concerns.

The Rogers Commission report on the *Challenger* accident notes that at the same time the flight rate was increasing, a variety of factors reduced the number of skilled personnel available to deal with it: retirements, hiring freezes, transfers to other programs like the Space Station, and changing to a single contractor for operations support. A significant number of employees elected not to change to that contractor, which resulted in the need to hire and qualify new personnel.

Budget cuts without concomitant cuts in goals led to trying to do too much with too little. NASA's response to its budget cuts was to defer upgrades and to attempt to increase productivity and efficiency rather than eliminate any major programs. By 2001, an experienced observer of the space program described the Shuttle workforce as "The Few, the Tired" [112].

B.2 Events

While the events differed between the two accidents, the major causes of those events did not.

Challenger The Rogers Commission concluded that the loss was caused by a failure in the joint between the two lower segments of the right solid rocket motor. The failure was the result of the destruction of the seals that are intended to prevent hot gases from leaking through the joint during the propellant burn of the rocket motor. Although the exact cause of the O-ring failure cannot be determined with certainty, the Commission suggested: "*A likely cause of the O-ring erosion appears to have been the increased leak check pressure that caused hazardous blow holes in the putty. Such holes at booster ignition provide a ready path for combustion gases directly to the O-ring. The blow holes were known to be created by the higher pressure used in the leak check*" [332, p. 156].

The solid rocket motor is assembled from four large cylindrical segments—each over 25 ft long, 12 ft in diameter, and containing more than 100 tons of fuel; the resulting four joints are called field joints. These joints are sealed by two rubber O-rings that are installed when the motor is assembled (see figure B.1).

The primary O-ring and its backup, the secondary O-ring, were designed to seal a tiny gap in the joint that is created by pressure at ignition. The O-ring seal must be pressure actuated very early in the solid rocket motor ignition. If this actuation is delayed, the rocket's combustion gases may blow by the O-ring and damage or destroy the seals.

O-ring resiliency is directly related to temperature. If the temperature is low, resiliency is reduced, and the O-ring is very slow to return to its normal rounded shape. If the O-ring remains in its compressed position too long, then the gap between the O-ring and the upstream channel wall may not be sealed in time to prevent joint failure. A second O-ring was used to provide backup if the primary O-ring did not seal the joint.

The Rogers Commission found that the Space Shuttle's SRB problem began with a faulty design of the aft left field joint of the right solid rocket motor and increased as both NASA and the Thiokol management first did not recognize the problem, then later did not fix it after it was recognized, and finally treated it as an acceptable risk.

To help in understanding the events, the terminology used by NASA to classify parts analyzed in the FMEA is important. Criticality 1 means that failure of the part could cause the loss of life or vehicle. Criticality 1R means that the component contains redundant hardware. More than 700 pieces of the Shuttle hardware were listed as Criticality 1 [253].

Leon Ray (an engineer at Marshall) first concluded after tests in 1977 and 1978 that rotation of the SRB field joint under pressure caused the loss of the secondary O-ring as a

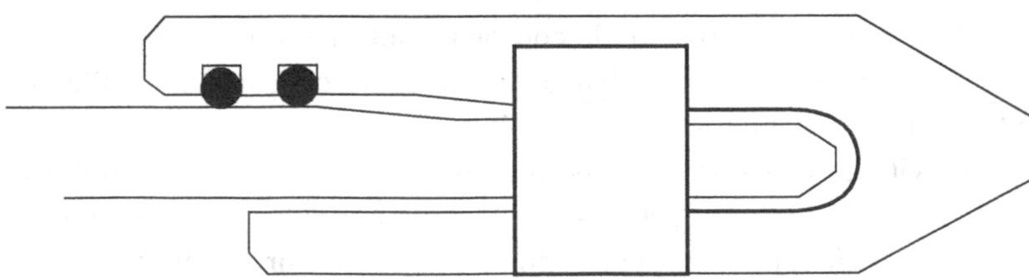

Figure B.1
The Shuttle SRB joint with two O-rings.

backup seal. Nevertheless, in November 1980, the SRB joint was classified on the NASA Shuttle Critical Items List (CIL) as Criticality 1R. The use of R, representing redundancy, signifies that NASA believed that the secondary O-ring would pressurize and seal if the primary O-ring failed to do so. The 1980 CIL does express doubt about the secondary O-ring successfully sealing if the primary should fail under certain conditions—when the motor case pressure reaches or exceeds 40 percent of maximum expected operating pressure—but the joint was assigned a 1R classification from November 1980 through the flight of STS-5 in November 1982.

O-ring anomalies were found during the initial flights, but the O-ring erosion problem was not reported in the Marshall problem assessment system and given a tracking number, as were other flight anomalies. After tests in May 1982, Marshall management finally accepted the conclusion that the secondary O-ring was no longer functional after the joints rotated under 40 percent of the SRB maximum operating pressure, and the criticality was changed to Criticality 1 on December 17, 1982.

Testimony at the Rogers Commission hearings supported the conclusion that, despite the change in criticality from 1R to 1, NASA and Thiokol management still considered the joint to be a redundant seal in all but a few exceptional cases. The disagreement went to a referee for testing, which was not concluded until after the *Challenger* accident.

Because the design was now rated Criticality 1, a waiver was required to allow it to be used. However, most of the problem reporting paperwork tracking the O-ring erosion problem that was generated by Thiokol and Marshall still listed the SRB field joint seal as Criticality 1R long after the status had been changed to Criticality 1. The Rogers Commission suggested that this misrepresentation of criticality may have led some managers to believe—wrongly—that redundancy existed. The problem assessment system operated by Rockwell contractors at Marshall still listed the field joint as 1R on March 7, 1986, more than five weeks after the accident. The commission concluded that, as a result, informed decision-making by key managers was impossible.

Before SRS flight 41-B in February 1984, the O-ring erosion and blow-by problem occurred infrequently. STS 41-B had extensive erosion, but Thiokol filed a problem report concluding that the secondary O-ring was an adequate backup according to the tests they had made. Some engineers at Marshall and Thiokol disagreed with the Thiokol numbers, but approval to fly 41-C was given. After more erosion and blow-by were found on 41-C, NASA asked Thiokol for a formal review of the SRB joint-sealing procedures. Thiokol was asked to identify the cause of the erosion, determine whether it was acceptable, define any necessary changes, and reevaluate the putty then being used. Asbestos-filled putty was used to pack the space in the joint and prevent O-ring damage from the heat of combustion gases. Problems with the putty can lead to the burning of both O-rings and, thus, to an explosion.

A letter in 1984 from Thiokol suggested that the chance of O-ring erosion had increased because of a change (higher pressure) made after STS-9 in the procedures used to check for leaks in the joint seal. The increased air pressure forced through the joint during the O-ring leak check was creating more putty blow holes, allowing more focused jets on the primary O-ring, and thereby increasing the frequency of erosion. The flight experiences supported this conclusion. During the first nine flights (before flight 41-B in January 1984), when the leak check pressure was lower, only one field joint anomaly had been found. When the leak-test pressure was increased for STS 41-B and later flights, over half the Shuttle missions experienced field joint O-ring blow-by erosion of some kind. The same experience was found with the nozzle O-ring.

Thiokol and NASA witnesses at the Rogers Commission hearings agreed that they were aware that the increase in leak-test blow holes in the putty could contribute to O-ring erosion. Nevertheless, Thiokol recommended, and NASA accepted, the increased pressure to ensure that the joint passed the integrity tests. Thiokol did establish plans for putty tests to determine how the putty was affected by the leak check after the 41-C experience, but their progress in completing the tests was slow.

On January 24, 1985, STS 51-C was launched at the coldest temperature to that date—53 degrees. O-ring erosion occurred in both solid boosters, and blow-by was more significant than had been experienced before. The problem assessment report (which was started to track field joint erosion after 41-B) described the O-ring anomaly after 51-C as "as bad or worse than previously experienced. . . . Design changes are pending test results" [332, p. 136]. The design changes being considered included modifying the O-rings and adding grease around them to fill the void left by the putty blow holes created by the leak tests.

Thiokol presented an analysis of the problem on February 8, 1985, that noted a concern that the seal could be lost, but it concluded that the risk of primary O-ring damage should be accepted. Risk acceptance was justified, in part, on the assumption that the secondary O-ring would work even with erosion. During the flight readiness assessment at Marshall for 51-D, Thiokol mentioned temperature, for the first time, as a factor in O-ring erosion and blow-by. Thiokol concluded that "low temperature enhanced probability of blow-by—51-C experienced worst-case temperature change in Florida history" [332, p. 136].

The joint seal problem occurred in each of the next four Shuttle flights. Thiokol conducted O-ring resiliency tests in response to the extensive O-ring problems found on flight 51-C in January and found that the key variable was temperature.

STS 51-B was launched on April 29, 1985. Inspection of the O-rings after the flight revealed a more serious problem than they had previously experienced: the primary nozzle O-ring never sealed at all, and the secondary O-ring was eroded. The erosion was greater than that predicted as the maximum possible by the model they were using. As a result, the Marshall Problem Assessment Committee (a Level III committee) placed a launch constraint on the Shuttle system that applied to flight 51-F and all subsequent flights (including the fatal 51-L *Challenger* flight). Thiokol officials testified at the Rogers Commission hearings that they were unaware of this launch constraint, but Thiokol letters referenced the report containing the constraint.

A launch constraint means that the problem has to be addressed during the flight readiness cycle to ensure that NASA is staying within the "test experience base"—that is, the suspect element is operated only within the parameters for which it has worked successfully before. Although launch constraints are required to be reported to Level II management, this reporting was not done. After the launch constraint was imposed, it was waived for each shuttle flight thereafter, including the fatal *Challenger* flight, 51-L. The Rogers Commission was told that two entries on the O-ring erosion nozzle problem report had erroneously stated that the O-ring erosion problem had been resolved or closed.

After the 51-B erosion problems became known, Roger Boisjoly (an engineer at Thiokol) wrote a memo on July 31, 1985, recommending that a team be set up to solve the O-ring problem. The memo concluded that a catastrophe could occur if immediate action was not taken. On August 19, Thiokol and Marshall program managers briefed NASA Headquarters on the seal erosion problem and concluded that the O-ring seal was a critical matter but that it was safe to fly. An accelerated pace was recommended for the effort to eliminate seal erosion. Thiokol's vice president for engineering, noting that "the result of a leak at any of the joints would be catastrophic," announced on August 20 the establishment of a Thiokol task force to recommend short- and long-term solutions [332].

Early in October, Thiokol management received two separate memos from the O-ring task force describing administrative delays and lack of cooperation. One was written by Roger Boisjoly and warned about lack of management support for the O-ring team's efforts. The other memo started with the word "HELP!" and complained about the seal task force being "constantly delayed by every possible means"; the memo ended with the words "This is a red flag" [332].

Shuttle flight 61-A was launched on October 30, 1985, and experienced both nozzle O-ring erosion and field joint O-ring blow-by. These anomalies were not mentioned at the Flight Readiness Review for the next flight. Flight 61-B was launched on November 26, 1985; it also sustained nozzle O-ring erosion and blow-by.

In December, R. V. Ebeling (manager of Thiokol's solid rocket motor ignition system) became so concerned about the seriousness of the O-ring problem that he told the other members of the seal task force that he believed Thiokol should not ship any more motors until the problem was fixed.

On December 10, Thiokol wrote a memo to NASA suggesting that the O-ring problem be closed with respect to the problem-tracking process. Thiokol officials testified at the Rogers Commission hearings that this was in response to a request from the director of engineering at Marshall. Having the problem listed as closed meant that it would not be involved in the flight readiness reviews, although it could still be worked on. The letter was still in the review cycle at the time of flight 51-L (*Challenger*), but entries were placed on all Marshall problem reports (a closed-loop tracking system) indicating that the problem was considered closed. The commission heard testimony that these entries were "in error."

Flight 51-L was originally scheduled for July 1985, but by January 1985 the flight had been postponed to late November to accommodate changes in payloads. The launch was later delayed further and finally rescheduled for January 22, 1986. The launch was further delayed three times and scrubbed once. The first of these last three postponements was announced on December 23, 1985, in order to accommodate a slip in the launch date of mission 61-L. On January 22, the launch was changed from January 23 to January 25 because of problems caused by the late launch of flight 61-C.

The third postponement occurred on January 25: because of an unacceptable weather forecast for the Kennedy area, the launch was rescheduled for January 27. But the January 27 launch countdown had to be halted when the crew entry and exit hatch handle jammed. By the time the hatch handle was fixed, the crosswinds at Kennedy exceeded the maximum allowable for a return-to-launch abort, and the launch was canceled at 12:26 p.m. and rescheduled for January 28.

At 2 p.m. on February 27, the mission management team met again. At that time, the weather was expected to clear, but temperatures were to be in the low twenties overnight, and a temperature of 26 degrees was predicted at the intended time of launch (9:38 a.m.).

Robert Ebeling, at the Thiokol plant, heard about the predicted low temperatures and called a meeting at 2:30 p.m. with Roger Boisjoly and the other Thiokol engineers. The engineers expressed great concern about launching in such low temperatures because this was far below previous experience and below the temperatures for which the SRBs had been qualified.

A teleconference was set up for 5:45 p.m. with the Thiokol engineering people in Utah, the Thiokol people at Marshall, and the Thiokol representative at Kennedy. NASA participants were all from Marshall, including the Marshall deputy director of science and engineering, the Marshall shuttle project manager, and the Marshall deputy shuttle project manager.

Concerns about the effect of low temperature on the O-rings and joint seal were presented by Thiokol, along with an opinion that the launch should be delayed. A recommendation was also made that Arnold Aldrich, the program manager at Johnson (Level II), be informed about the concerns, but this was never done. The Rogers Commission report criticized the propensity at Marshall to contain potentially serious problems and to attempt to resolve them internally rather than communicate them forward. It also noted that the project managers for the various elements of the Shuttle program felt more accountable to their center management than to the Shuttle program organization. Vital program information frequently bypassed the NSTS (National Space Transportation System) program manager.

The Marshall and Thiokol personnel decided to schedule a second teleconference later in the day, after data could be faxed to the various locations. The Thiokol charts and written data were faxed to Kennedy, and the second phase of the teleconference started at 8:45 p.m. with additional personnel involved. The history of the O-ring erosion and blow-by in the SRB joints of previous flights was presented along with the test results at Thiokol. Bob Lund, Thiokol vice president of engineering, presented the conclusion of the Thiokol engineers that the O-ring temperature must be at or greater than 53 degrees at launch, which was the previous lowest launch temperature. Boisjoly testified at the Commission hearings that no pro-launch statement was made by anyone in the room.

During this conference, the NASA Marshall deputy director of science and engineering is reported to have said that he was appalled by Thiokol's recommendation. The manager of the SRB project at Marshall suggested that Thiokol was using different launch criteria than in the previous twenty-four flights and that, under those criteria, they would not be able to launch until April. After protesting that Thiokol was changing the launch criteria on the night before a scheduled mission, he reportedly exclaimed "You can't do that." The NASA personnel challenged the Thiokol conclusions and recommendations, stressing that the secondary O-ring would seal if the primary one did not. Some Thiokol engineers testified that usually in reviews, the engineers were required to prove that they were ready to launch. In this discussion, however, they felt that the roles had been reversed and that the engineers were required to prove that the launch would *not* be successful.

The teleconference was recessed at about 10:30 p.m. for an off-the-air caucus of Thiokol personnel. This recess lasted for about 30 minutes. During this time, the Thiokol manager of the Space Booster Project at Kennedy continued to argue against the launch. He testified that he said, "If we are wrong and something goes wrong on this flight, I wouldn't want to have to be the person to stand up in front of a board of inquiry and say that I went ahead and told them to go ahead and fly this thing outside what the motor was qualified to" [332, p. 95].

At Thiokol in Utah, about ten engineers participated in the discussion, and very strong objections to the launch were voiced. After the discussion between Thiokol management and engineers was completed, a final management review began. The Thiokol senior vice president said that a management decision had to be made, turned to the Thiokol vice president of science and engineering, and asked him to take off his engineering hat and put on his management hat.

The teleconference was reconvened at 11:00 p.m., at which time Thiokol management stated that they had reassessed the problem and were withdrawing their opposition to the launch—the temperature was a concern, but the data was inconclusive. They concluded that (1) there was a substantial margin to erode the primary O-ring by a factor of three times the previous worst case, and (2) the "harder" O-rings would take longer to seat, but if the primary seal did not seat, the secondary seal would. They therefore recommended that the launch proceed.

NASA Level I and II management and the launch director for 51-L were never told about the initial Thiokol concerns and opposition to the launch—all the discussions and conferences had included only Marshall and Thiokol personnel. The Rogers Commission concluded that the Thiokol management reversed its position and recommended the launch of 51-L, at the urging of Marshall and contrary to the views of its engineers, in order to accommodate a major customer.

There had been heavy rain since *Challenger* had been rolled out to the launch pad, approximately 7 in compared with the 2.5 in that would have been normal for that season. The Rogers Commission report notes that water may have gotten into the joints, and at the time of launch, it was cold enough that water present in the joint would freeze. Tests showed that ice in the joint could inhibit proper secondary seal performance.

Ice accumulated on the launch pad during the night. At the weather briefing for the Shuttle crew, the temperature and ice on the pad were discussed, but the crew was not informed of any concern about the effects of low temperature. After an ice inspection, the launch was delayed to allow more time for the ice to melt. After consultation with senior advisors and without having been told about the O-ring discussions at Marshall or about the doubts of the Thiokol engineers concerning flight safety at those low temperatures, shuttle director Jesse Moore gave the permission to launch. The ambient temperature at launch was thirty-six degrees measured at ground level, fifteen degrees colder than that of any previous launch.

The flight began at 11:39 a.m. on January 28, 1986. Analysis of photographs taken immediately after ignition showed a puff of black smoke coming from the after section of the right SRB. As *Challenger* cleared the tower, gas was already blowing by the rings, although the gap was temporarily plugged by burning rubber and putty. *Challenger* experienced the worst vibrations of any flight to date as it was buffeted by gusts of wind for almost 30 seconds. At 58.7 seconds after ignition, a small flame like a blowtorch appeared at the side of the SRB, unnoticed by anyone on the ground or in the shuttle. It began to burn through the main fuel tank as well as one of the struts that held the rocket to the tank. Less than 14 seconds later, the strut gave way, and the pointed nose of the SRB swiveled inward to pierce the fuel tank.

Seventy-three seconds after the flight had begun, it ended with the hydrogen and oxygen propellants igniting in a huge ball of flame that destroyed the external tank and exposed the orbiter to severe aerodynamic loads that caused complete structural breakup. All seven crew members died. The two SRBs flew out of the fireball and were destroyed by the Air Force range safety officer 110 seconds after launch. It was the worst accident in the history of manned spaceflight and the first time any American astronauts were lost during a mission.

Columbia The loss of the *Columbia* Space Shuttle occurred on February 1, 2003, seventeen years after the *Challenger* accident. At 81.7 seconds after launch, when the Shuttle was at about 65,600 ft and traveling at Mach 2.46 (1,650 mph), a large piece (and at least two smaller pieces) of hand-crafted insulating foam came off an area where the Orbiter attaches to the External Tank. At 81.9 seconds, the large piece struck the reinforced carbon-carbon (RCC) leading edge of *Columbia*'s left wing. This event was not detected by the crew on board or seen by ground support teams until the next day, during detailed reviews of all launch camera photography and videos. This foam strike had no apparent effect on the daily conduct of the 16-day mission, which met all its objectives.

The shedding of external tank foam had a long history. Damage caused by debris has occurred on every Space Shuttle flight, and most missions had insulating foam shed during ascent. Although engineers made numerous changes in foam design and application in the twenty-five years that the external tank was in production, the problem of foam-shedding was not solved, nor had the orbiter's ability to tolerate impacts from foam or other debris been significantly improved.

After images during the fateful *Columbia* launch were examined, attempts were made to image the thermal tiles while the Shuttle was in orbit. There is controversy about whether a potential rescue could have been successfully carried out, but that controversy is omitted from this discussion.

When *Columbia* reentered Earth's atmosphere, the damage allowed hot atmospheric gases to penetrate the heat shield and destroy the internal wing structure, which caused the spacecraft to become unstable and break apart.

The de-orbit burn to slow *Columbia* down for reentry into Earth's atmosphere was normal, and the flight profile throughout reentry was standard. Time during reentry is measured in seconds from "Entry Interface," an arbitrarily determined altitude of 400,000 ft where the orbiter begins to experience the effects of Earth's atmosphere. Entry Interface for STS-107 occurred at 8:44:09 a.m. on February 1. Unknown to the crew or ground personnel, because the data is recorded and stored in the orbiter instead of being transmitted to Mission Control at Johnson Space Center, the first abnormal indication occurred 270 seconds after the Entry Interface point.

The breach on the orbiter left wing, caused by the foam strike on ascent, was of sufficient size to allow superheated air to penetrate the cavity behind the RCC panel. The breach widened, destroying the insulation protecting the wing's leading edge support structure, and the superheated air eventually melted the thin aluminum wing spar. Once in the interior, the superheated air began to destroy the left wing. This destructive process was carefully reconstructed from the recordings of hundreds of sensors inside the wing, and from analyses of the reactions of the flight control systems to the changes in aerodynamic forces.

By the time *Columbia* passed over the coast of California in the predawn hours of February 1, at Entry Interface plus 555 seconds, amateur videos show that pieces of the orbiter were shedding. The orbiter was captured on videotape during most of its quick transit over the western United States. The board correlated the events seen in these videos to sensor readings recorded during reentry. Analysis indicates that the orbiter continued to fly its preplanned flight profile, although, still unknown to anyone on the ground or aboard *Columbia*, her control systems were working furiously to maintain that flight profile. Finally, over Texas, just southwest of Dallas-Fort Worth, the increasing aerodynamic forces the orbiter experienced in the denser levels of the atmosphere overcame the catastrophically damaged left wing, causing the Orbiter to fall out of control at speeds in excess of 10,000 mph.

B.3 Causal Factors in the *Challenger* and *Columbia* Losses

Understanding enough about the *Challenger* and *Columbia* accidents to prevent future ones requires not only determining what went wrong at the time of the losses but also

1. why the high standards of the Apollo program deteriorated over time and allowed the conditions cited by the Rogers Commission as the root causes of the *Challenger* loss; and
2. why the fixes instituted after the *Challenger* loss became ineffective over time.

In other words, it must be determined why the manned space program has a tendency to migrate to states of such high-risk and poor decision-making processes that an accident becomes almost inevitable. The answers to these questions are important as the space program is far from alone in this respect.

All engineering efforts take place within a political, social, and historical context that has a major impact on the technical and operational decision-making. Understanding the context in which decisions were made in the manned space program helps in explaining why bright and experienced engineers made what turned out to be poor decisions and what might be changed to prevent similar accidents in the future.

In the case of the Space Shuttle, political and other factors contributed to the adoption of a vulnerable design during the original approval process. Unachievable promises were made with respect to performance in order to keep the manned space flight program alive after Apollo and the demise of the Cold War. While these performance goals even then seemed unrealistic, the success of the Apollo program and the *can-do* culture that arose during it contributed to the belief that these unrealistic goals could be achieved if only enough effort was expended.

The Rogers Commission study of the Space Shuttle *Challenger* accident concluded that the root cause of the accident was an accumulation of organizational problems [332]. The commission was critical of management complacency, bureaucratic interactions, disregard for safety, and flaws in the decision-making process. It cited various communication and management errors that affected the critical launch decision on January 28, 1986, including a lack of problem-reporting requirements; inadequate trend analysis; misrepresentation of criticality; lack of adequate resources devoted to safety; lack of safety personnel involvement in important discussions and decisions; and inadequate authority, responsibility, and independence of the safety organization. Each of these is examined in more detail next.

Despite a sincere effort to fix these problems after the *Challenger* loss, seventeen years later, almost identical management and organizational factors and cultural flaws were cited in the *Columbia* Accident Investigation Board (CAIB) report. These are not two isolated cases. In most of the major accidents in the past twenty-five years—in all industries, not just aerospace—technical information on how to prevent the accident was known and often even implemented. But in each case, the potential engineering and technical solutions were negated by organizational or managerial flaws.

Space exploration is inherently risky. There are just too many unknowns and requirements to push the technological envelope to be able to reduce the risk level to that of other aerospace endeavors such as commercial aircraft. At the same time, the known and preventable risks can and should be managed effectively.

Flawed safety culture and decision-making Risk and occasional failure have always been recognized as an inherent part of space exploration, but the way the inherent risk was handled at NASA changed over time. In the early days of NASA and during the Apollo era, the belief was prevalent that risk and failure were normal aspects of space flight. At the same time, the engineers did everything they could to reduce it [256]. People were expected to speak up if they had concerns, and risks were debated vigorously. *What if* analysis was a critical part of any design and review procedure.

At some time between those early days and the *Challenger* accident, the culture changed drastically. The Rogers Commission Report includes a chapter titled the "Silent Safety Program." Those on the Thiokol task force appointed to investigate problems that had been occurring with the O-rings on flights prior to the catastrophic *Challenger* flight complained about lack of management support and cooperation for the O-ring team's efforts.

A *culture* can be defined as a shared set of norms and values. It includes the way we look at and interpret the world and events around us (our mental model) and the way we take action in a social context. *Safety culture* is that subset of an organizational or industry culture that reflects the general attitude and approaches to safety and risk management. Trying to change culture and the behavior resulting from it without changing the environment in which it is embedded is doomed to failure. Superficial fixes that do not address the set of shared values and social norms, as well as deeper underlying assumptions, are likely to be undone over time [348].

Perhaps this partially explains why the changes at NASA after the *Challenger* accident intended to fix the safety culture, like the safety activities themselves, were slowly dismantled or became ineffective. Both the *Challenger* accident report and the CAIB report, for example, note that system safety was "silent" and ineffective at NASA despite attempts to fix this problem after the *Challenger* loss.

Understanding the pressures and other influences that have twice contributed to a drift toward an ineffective NASA safety culture is important in creating an organizational infrastructure and environment that will resist pressures against applying good safety engineering practices and procedures in the future. The same types of pressures exist in almost every industry, with slightly different forms.

A partial explanation for the change in the NASA safety culture, noted in both the *Challenger* and *Columbia* accident reports, is that schedule and launch pressures in the Shuttle program created a mindset that dismissed all concerns, leading to overconfidence and complacency. This type of culture has been described as a *culture of denial* where risk assessment is unrealistic and credible risks and warnings are dismissed without appropriate investigation.

Managers begin to listen only to those who provide confirming evidence that supports what they want to hear. Neither Thiokol nor NASA expected the rubber O-rings sealing the joints to be touched by hot gases during motor ignition, much less to be partially burned. However, as tests and then flights confirmed damage to the sealing rings, the reaction by both NASA and Thiokol was to change the definition of "acceptable." At no time did management either recommend a redesign of the joint or call for the Shuttle's grounding until the problem was solved, most likely because of external pressures.

The Rogers Commission concluded that the Space Shuttle's problems began with a faulty design of the joint and later, when the problem was finally identified but no fix was found, treated it as an acceptable risk.

The same behavior contributed to the *Columbia* accident. The CAIB report notes that at the time of the *Columbia* loss, "managers created huge barriers against dissenting opinions by stating preconceived conclusions based on subjective knowledge and experience, rather than solid data" [112, p.192]. An indication of the prevailing culture at the time of the *Columbia* accident can be found in the reluctance of the debris assessment team—created after the launch of *Columbia* to assess the damage caused by the foam hitting the wing—to adequately express their concerns. The exact same thing happened with the O-ring problem. NASA engineers told the CAIB that "by raising contrary points of view about Shuttle mission safety, they would be singled out for possible ridicule by their peers and managers" [112, p. 169].

In an interview shortly after he became center director at the NASA Kennedy Space Center after the *Columbia* loss, Jim Kennedy suggested that the most important cultural issue the Shuttle program faced was establishing a feeling of openness and honesty with all employees where everybody's voice is valued [224]. Statements during the *Columbia* accident investigation and anonymous messages posted on the NASA Watch website document a lack of trust leading to a reluctance of NASA employees to speak up.

At the same time, a critical observation in the CAIB report focused on the managers' claims that they did not hear any concerns expressed by the engineers. The report concluded that not hearing the concerns was due in part to the managers not asking or listening. Managers created barriers against dissenting opinions by stating preconceived conclusions based on subjective knowledge and experience rather than on solid data. In the extreme, they listened only to those who told them what they wanted to hear.

Just one indication of the atmosphere existing at that time were statements in the 1995 Kraft report, which presented the findings of a NASA review of Space Shuttle management. The report dismissed concerns about Shuttle safety by labeling those who made them as being partners in an unnecessary "safety shield conspiracy" [195]. This accusation of those expressing safety concerns as being part of a "conspiracy" is a powerful demonstration of the attitude toward system safety at the time and the change from the Apollo era when dissent was encouraged and rewarded.

Why would intelligent, highly educated, and highly motivated engineers engage in such poor decision-making processes and act in a way that seems irrational in retrospect? One view of culture provides an explanation. Social anthropologists conceive of culture as an ongoing, proactive process of reality construction [270]. In this conception of culture, organizations are socially constructed realities that rest as much in the heads of members as in sets of rules and regulations. Organizations are sustained by belief systems that emphasize the importance of rationality. Morgan calls this the *myth of rationality* [270]. It helps in understanding why, as in both the *Challenger* and *Columbia* accidents, leaders often appear to ignore what seems obvious in retrospect. The myth of rationality "*helps us to see certain patterns of action as legitimate, credible, and normal, and hence to avoid the wrangling and debate that would arise if we were to recognize the basic uncertainty and ambiguity underlying many of our values and actions*" [270, pp. 134–135].

In both the *Challenger* and *Columbia* accidents, the decision makers saw their actions as rational at the time. Understanding and preventing poor decision-making under conditions of uncertainty requires providing environments and tools that help to stretch our belief systems and overcome the constraints of our current mental models; that is, to see patterns that we do not necessarily *want* to see [270; 217].

This phenomenon led to overconfidence and complacency, discounting risk, overreliance on redundancy, and unrealistic risk assessment.

Overconfidence and complacency The Shuttle was mischaracterized by the 1995 Kraft Report as "a mature and reliable system . . . about as safe as today's technology will provide" [195]. NASA believed that it could turn increased responsibilities for Shuttle operations over to a single prime contractor and reduce its direct involvement in ensuring safe Shuttle operations. NASA could then use the "mature" Shuttle to carry out operational missions without continually focusing engineering attention on understanding the mission-by-mission anomalies inherent in a developmental vehicle.

Discounting risk The *Challenger* accident report noted several significant trends in the Shuttle flight readiness reviews. A flight readiness review (FRR) examines tests, demonstrations, analyses, and audits that determine the overall system readiness for a safe and successful flight or launch and for subsequent flight operations. The FRR gives teams responsible for various elements of a NASA flight mission an opportunity to ensure that technical questions raised at earlier reviews have been adequately dealt with and to raise concerns about anything else that might affect mission success.

First, O-ring erosion was not considered early in the program when it first occurred. Then, when the problem grew worse after STS 41-B, there was an early acceptance of the phenomenon without much analysis or research. Later flight readiness reviews gave the problem only a cursory review and often dismissed it as within *acceptable* or *allowable* limits because it was within the database of prior experience.

Morton Thiokol did not accept the implication of tests early in the program that the design had a serious and unanticipated flaw. NASA did not accept the judgment of its engineers that the design was unacceptable, and, as the joint problems grew in number and severity, NASA minimized them in management briefings and reports. Thiokol's stated position was that the condition is not desirable, but it is acceptable. Both groups ignored warnings and over-relied on redundancy. Both Thiokol and NASA Marshall, in fact, continued to rely on the redundancy of the secondary O-ring long after NASA had officially declared that the seal was a nonredundant single point of failure.

Neither Thiokol nor NASA expected the rubber O-rings sealing the joints to be touched by hot gases during motor ignition, much less to be partially burned. However, as tests and then flights confirmed damage to the sealing rings, the reaction by both NASA and Thiokol was to increase the amount of damage considered acceptable. At no time did management either recommend a redesign of the joint or call for the Shuttle's grounding until the problem was solved.

NASA and Thiokol accepted escalating risk apparently because they "got away with it last time." As *Challenger* Commissioner Richard Feynman observed, the decision-making was *"a kind of Russian roulette. . . . [The Shuttle] flies [with O-ring erosion] and nothing happens. Then it is suggested, therefore, that the risk is no longer so high for the next flights. We can lower our standards a little bit because we got away with it last time"* [332, p. 148]. Every time an incident occurred that was a narrow escape, it confirmed for many the belief that NASA was a tough, can-do organization with high intact standards that precluded accidents [216].

The exact same phenomenon occurred with the foam shedding, which had occurred during the life of the Shuttle but had never, prior to the *Columbia* loss, caused serious damage. To be fair, there were a large number of serious hazards existing in the Shuttle program. Hindsight shines a large spotlight on what later turns out to be most relevant. But without a properly funded, designed, staffed, and managed safety program in a culture that understood its importance, identifying the most important hazards in time to fix them makes avoiding serious losses much less likely.

Overreliance on redundancy A common aspect of the type of complacency and overconfidence seen in the manned space program is related to the use of redundancy to increase reliability. One of the rationales used in deciding to go ahead with the disastrous *Challenger* flight despite engineering warnings was that there was a substantial safety margin (a factor of three) in the O-rings over the previous worst case of Shuttle O-ring erosion. Moreover, even if the primary O-ring did not seal, it was assumed that the second, redundant one would. During the accident, the failure of the primary O-ring caused conditions that led to the failure of the backup O-ring.

The design changes necessary to incorporate a second O-ring, in fact, contributed to the loss of the primary O-ring. The design of the Shuttle solid rocket booster (SRB) was based on the US Air Force Titan III, one of the most reliable ever produced. Significant design changes were made in an attempt to increase that reliability further, including changes in the placement of the O-rings. A second O-ring was added to the Shuttle solid rocket motor design to provide backup: if the primary O-ring did not seal, then the secondary one was supposed to pressurize and seal the joint.

In order to accommodate the two O-rings, part of the Shuttle joint was designed to be longer than in the Titan. The longer length made the joint more susceptible to bending under combustion pressure, which led to the failure of the primary and backup O-rings.

In this case, and in a large number of other cases [213], the use of redundancy requires design choices that in fact defeat the redundancy at the same time that the redundancy is creating unjustified confidence and complacency (see chapter 11 on design for safety).

Unrealistic risk assessment A NASA study report in 1999 concluded that the Space Shuttle Program was using previous success as a justification for accepting increased risk [2000]. The practice continued despite this and other alarm signals. William Readdy, head of the NASA Manned Space Program, for example, in 2001 wrote that "the safety of the Space Shuttle has been dramatically improved by reducing risk by more than a factor of five" [112, p. 101]. Note that a similar statement was made for the Therac-25 overdoses described in appendix A, with equally poor justification. It is difficult to imagine where this number came from as safety upgrades and improvements had been deferred while, at the same time, the infrastructure continued to erode. The unrealistic risk assessment was also reflected in the 1995 Kraft report, which concluded that "the Shuttle is a mature and reliable system, about as safe as today's technology will provide." A recommendation of the Kraft report was that NASA should "restructure and reduce overall safety, reliability, and quality assurance elements" [195].

The *Columbia* accident report identified a perception that NASA had overreacted to the Rogers Commission recommendations after the *Challenger* accident, for example, believing that the many layers of safety inspections involved in preparing a Shuttle for flight had created a bloated and costly safety program. Reliance on past success became a substitute for sound engineering practices and for accepting increasing risk. Either the decision makers did not have or they did not use inputs from system safety engineering. "*Program management made erroneous assumptions about the robustness of a system based on prior success rather than on dependable engineering data and rigorous testing*" [112, p. 184].

Many analysts have faulted NASA for missing the implications of the *Challenger* O-ring trend data. One sociologist, Diane Vaughan, went so far as to suggest that the risks had become seen as "normal" [393]. In fact, the engineers and scientists at NASA were tracking thousands of potential risk factors. There were more than 20,000 critical items on the Space Shuttle—that is, items whose failure could lead to the loss of the Shuttle—and at the time of the *Columbia* loss more than 3,000 waivers existed. Waivers were used to allow Shuttle flights to take place even though official safety requirements were not met.

It was not a case that some anomalies, such as O-ring blow-by and shedding foam, had come to be perceived as normal, but that the risk had come to be seen as *acceptable* without adequate data to support that conclusion. In addition, as described later, safety engineering lost funding and the ability to track and evaluate the hazards.

Edward Tufte, an expert in data presentation, faulted the PowerPoint slides used by the Morton Thiokol engineers to present their argument for not going ahead with the *Challenger* launch [378].

While Tufte's insights into the display of data are instructive, it is important to recognize that both the Vaughan and the Tufte critiques are easier to do in retrospect: it is easy to see what is important in hindsight. It is much more difficult to achieve this goal before the critical data has been identified after the accident. Decisions need to be evaluated in the context of the information available at the time the decision is made along with the organizational factors influencing the interpretation of the data and the decision-making process itself.

Risk assessment is extremely difficult for complex, technically advanced systems such as the Space Shuttle. When this engineering reality is coupled with the social and political pressures existing at the time, the emergence of a culture of denial and overoptimistic risk assessment is not surprising.

Shuttle launches are anything but routine, so new interpretations of old data or of new data will always be needed; that is, risk assessment for systems with new technology is a continual and iterative task that requires adjustment on the basis of experience. At the same time, it is important to understand the conditions at NASA that prevented an accurate analysis of the data and the risk and the types of safety culture flaws that contributed to the unrealistic risk assessment.

Confusion between reliability and safety A common factor in flawed risk assessment in the space program and in many other industries is the substitution of reliability engineering and assessment for system safety and hazard analysis. Safety is a system property and needs to be handled from a system perspective. NASA, however, treated safety primarily at the component level, with a focus on component reliability.

For example, the CAIB report notes that there was no one office or person responsible for developing an integrated risk analysis above the subsystem level that would provide a comprehensive picture of total program hazards and risks. Failure Modes and Effects Analysis (FMEA), a bottom-up reliability engineering technique, was the primary analysis method used. Hazard analyses (versus reliability analyses) were performed but rarely used.

As performance pressures increased, the traditional NASA safety standards were watered down by Shuttle program management, who controlled the Shuttle safety program [216]. For example, the Shuttle standard for hazard analyses (NSTS 22254, *Methodology for Conduct of Space Shuttle Program Hazard Analyses*) was changed to specify that hazards be revisited only when there was a new design or when the design was changed: there was no process for updating the hazard analyses when anomalies occurred or even for determining whether an anomaly was related to a known hazard. The hazard of foam shedding was known and identified long before the *Columbia* loss.

NASA delegated safety oversight to its operations contractor USA. USA delegated hazard analysis to Boeing, but as of 2001, "the Shuttle program no longer required Boeing to conduct integrated hazard analyses" [112, p. 188]. Instead, Boeing performed analysis only on the failure of individual components and elements and was not required to consider the Shuttle as a whole; that is, system hazard analysis was not being performed. The CAIB report notes, "*Since the FMEA/CIL process is designed for bottom-up analysis at the component level, it cannot effectively support the kind of 'top-down' hazard analysis that is needed . . . to identify potentially harmful interactions between systems*" [112, p. 188]. Foam from the external tank hitting the forward edge of the orbiter wing was one such interaction.

Assuming risk decreases over time NASA had a perception that less safety, reliability, and quality assurance activity would be required during routine Shuttle operations. Therefore, after the successful completion of the orbital test phase and the declaration of the Shuttle as operational, several safety, reliability, and quality assurance organizations were reorganized and reduced in size.

The chief engineer at NASA headquarters had overall responsibility for safety, reliability, and quality assurance. To carry out this responsibility, he had a staff of twenty people, only two of whom spent 10 percent and 25 percent of their time, respectively, on these issues. Moreover, some safety panels, which were providing safety review, went out of existence or were merged.

> *The unrelenting pressure to meet the demands of an accelerating flight schedule might have been adequately handled by NASA if it had insisted upon the exactingly thorough procedures that were its hallmark during the Apollo program. An extensive and redundant safety program comprising interdependent safety, reliability, and quality assurance functions existed during and after the lunar program to discover any potential safety problems. Between that period and 1986, however, the program became ineffective. This loss*

of effectiveness seriously degraded the checks and balances essential for maintaining flight safety. [332, p. 152]

While safety efforts increased after the *Challenger* accident, they slowly started to degrade again. The manned space program was operating with an increasingly constrained budget. One of the results was workforce reductions. There was little margin in the budget to deal with unexpected technical problems or make shuttle improvements. Safety standards were weakened and became nonmandatory. The priority of safety in the engineering efforts quickly fell back to the pre-*Challenger* level.

Flawed organization structures and practices Organizational change experts have long argued that structure drives behavior. Much of the dysfunctional behavior related to both accidents can be traced to flaws in the NASA organizational safety structure, including poorly defined responsibility and authority, inadequate independence of safety-related decision-making, low influence and prestige leading to insufficient impact, and limited communication channels and information flow.

Poorly defined responsibility, authority, and accountability Both Shuttle accident reports criticized the lack of independence of the safety organization. After the *Challenger* loss, a new independent safety office was established at NASA Headquarters, as recommended in the Rogers Commission report. This group was supposed to provide broad oversight, but its authority was limited, and reporting relationships from the NASA Centers were vague. In essence, the new group was never given the authority necessary to implement their responsibilities effectively, and, most critically, nobody seems to have been assigned accountability. The later *Columbia* accident report noted in 2003 that the management of safety at NASA involved "confused lines of responsibility, authority, and accountability in a manner that almost defies explanation" [112, p. 186].

The problems were exacerbated by the fact that the project manager also had authority over the safety standards applied on the project. NASA safety standards were not mandatory. In essence, they functioned more like guidelines than standards. Each program decided what standards were applied and could tailor them in any way they wanted. While there are advantages to being able to tailor standards, the authority to do so must be placed in the right hands and oversight provided.

In addition, putting the system safety engineering (e.g., hazard analysis) within the assurance group established the expectation that system safety was an after-the-fact or auditing activity only. In fact, the most important aspects of system safety involve core engineering activities such as building safety into the basic design and proactively eliminating or mitigating hazards. By treating safety as an assurance activity only, safety concerns are guaranteed to come too late in the process to have an impact on the critical design decisions. Necessary information may not be available to the engineers when they are making decisions, and instead potential safety problems are raised at reviews, when doing something about the poor decisions is costly and likely to be resisted.

This problem results from a basic dilemma: either the system safety engineers work closely with the design engineers and lose their independence, or the safety efforts remain an independent assurance effort but safety becomes divorced from the engineering and design efforts. The solution to this dilemma, which other groups use, is to separate the safety engineering and the safety assurance efforts, placing safety engineering within the engineering organization and the safety assurance function within the assurance groups. Programmatic decision-making is separated from safety-related decision-making. This design is essentially the "three-legged stool" used in the US Navy's SUBSAFE program and described in chapter 14.

NASA attempted to implement a similar design after the *Columbia* loss by creating an Independent Technical Authority within engineering that is responsible for bringing a disciplined, systematic approach to identifying, analyzing, and controlling hazards. The design of this independent authority is already undergoing changes, with the result unclear at this time. As stated earlier, attempting to change organizational structure without changing culture is difficult.

As contracting of Shuttle engineering outside NASA increased over time, safety oversight by NASA civil servants diminished and basic system safety activities were delegated to contractors. According to the *Columbia* accident report, the operating assumption that NASA could turn over increased responsibility for Shuttle safety and reduce its direct involvement was based on the 1995 Kraft report, which was created to recommend ways to significantly decrease total operating costs while maintaining system safety [195].

The 1995 Kraft report concluded that the Shuttle was a mature and reliable system and that therefore NASA could change to a new mode of management involving outsourcing operations to a private contractor with reduced NASA oversight. A single NASA contractor was given responsibility for Shuttle safety (as well as reliability and quality assurance), while NASA was to maintain "insight" into safety and quality assurance through reviews and metrics. In fact, increased reliance on contracting necessitates more effective communication and more extensive safety oversight processes, not less. The Kraft report is the same report mentioned earlier that suggested that those expressing concern about safety were part of a "safety shield conspiracy" [195].

Many aerospace accidents (and those in other industries) have occurred after the organization transitioned from oversight to "insight" [215]. Contractors have a conflict of interest with respect to safety and their own goals and cannot be assigned the responsibility that is properly that of the contracting agency. In addition, years of workforce reductions and outsourcing had "*culled from NASA's workforce the layers of experience and hands-on systems knowledge that once provided a capacity for safety oversight*" [112, p. 181].

In 1999, three years before the *Columbia* loss and four years after the Kraft report, a team composed of NASA employees, contractors, and Department of Defense experts was created to review and make recommendations about NASA practices, by considering Space Shuttle anomalies and civilian and military aerospace experience. The Space Shuttle Independent Assessment (SIAT) report noted that the program had successfully transitioned to a slimmed-down, contractor-run operation. It warned, however, that significant problems needed to be addressed, including the reduction in resources and staff for critical processes and procedures and replacement of NASA personnel with contract personnel, inappropriate acceptance of risk, a false sense of safety, and erosion of the risk management process by the desire to reduce costs,

The SIAT report suggested that the Space Shuttle Program must rigorously guard against the NASA management tendency to accept risk solely because of prior success and to consider the Shuttle to be an operational vehicle. It noted such problems as high stress level among the workforce, inadequate communication of problems and concerns upward, problems in communication from supervisors downward to workers regarding priorities and changing work environments, and poor communication of requirements and coordination of changes across organizations.

Other specific SIAT findings included concerns such as: moving from NASA oversight to insight; increasing implementation of self-inspection; reducing Safety and Mission Assurance functions and personnel; managing risk by relying on system redundancy and abort modes; and the use of only rudimentary trending and qualitative risk assessment techniques. The report concluded that oversight processes of considerable value, including safety and mission assurance and quality assurance, had been diluted or removed from the program. The report recommended that NASA Safety and Mission Assurance should be restored to the process in its previous role of an independent oversight body and not be simply a "safety auditor."

Inadequate independence of safety-related decision-making The Rogers Commission recommendations centered on an underlying theme: the lack of independent safety oversight at NASA. Without independence, the commission believed, the slate of safety failures that contributed to the *Challenger* accident—such as the undue influence of schedule pressures and the flawed flight readiness process—would not be corrected. In response, NASA created a Headquarters Office of Safety, Reliability, and Quality Assurance, as noted in the previous section, but it did not have the intended result of providing strong safety leadership. While theoretically such an office was a worthy goal, the implementation, for cultural reasons, instead made things worse or at least did not eliminate the problems. Unfortunately, it contributed to a lack of influence and prestige.

Independence does not mean independence from engineering decision-making but from program management concerns such as budget, schedule, and the types of analysis and safety activities that are to be performed.

In retrospect, it is possible to see that the efforts made to improve the safety program after the *Challenger* accident were mostly ineffectual, and significant organizational factors were no better at the time of the *Columbia* accident seventeen years later. The organizational structures at NASA centers placed the safety, reliability, and quality assurance offices under the supervision of the very organizations and activities they were to check. In addition, they were dependent on the NASA management structure to get information and recommendations about safety problems and for implementing suggested changes. But these structures and communication lines were flawed.

After the *Columbia* accident, the CAIB report noted that "*NASA does not have a truly independent safety function with the authority to halt the progress of a critical mission element*" [112, p. 180]. In essence, the project manager "purchased" safety from the quality assurance organization. The amount of system safety applied was limited to what and how much the project manager wanted and could afford. "*The Program now decides on its own how much safety and engineering oversight it needs*" [112, p. 181]. Safety review panels and procedures existed *within* individual NASA programs, including the Shuttle program. Under various types of pressures, including budget and schedule constraints, however, the independent safety reviews and communication channels within the Shuttle program degraded over time and were taken over by the Shuttle Program office.

Independence of engineering decision-making also decreased over time. While in the Apollo and early Shuttle programs the engineering organization had a great deal of independence from the program manager, the Shuttle engineering organization gradually lost its authority and became subservient to the project managers, who again were driven by schedule and budget concerns. In the Shuttle program, all aspects of system safety are in the mission assurance organization. This means that the same group doing the system safety engineering is also doing the system safety assurance—effectively eliminating an independent assurance activity.

Lack of influence and prestige of safety engineering in NASA The Rogers Commission report on the *Challenger* accident observed that the safety program had become "silent" and undervalued. In the testimony to the Rogers Commission, NASA's safety staff, curiously, was never mentioned. No one thought to invite a safety representative to the hearings or to the critical January 27 teleconference between Marshall and Thiokol. No representative of safety was on the mission management team that made key decisions during the countdown on January 28.

The *Columbia* accident report concludes that, once again, seventeen years later, system safety engineers were not involved in the important safety-related decisions although they were ostensibly added to the mission management team after the *Challenger* loss. The isolation of system safety from the mainstream design engineers added to the problem:

> *Structure and process places Shuttle safety programs in the unenviable position of having to choose between rubber-stamping engineering analyses, technical errors, and Shuttle program decisions, or trying to carry the day during a committee meeting in which the other side always has more information and analytical ability.*
>
> . . . *We expected to find the [Safety and Mission Assurance] organization deeply engaged at every level of Shuttle management, but that was not the case.* [112, p. 187]

After the *Challenger* accident, as noted previously, system safety was placed at NASA Headquarters in a separate organization that included mission assurance and other quality assurance programs. For a short period thereafter, this safety group had some influence, but it quickly reverted to a position of even less influence and prestige than before the *Challenger* loss.

Placing system safety in the quality assurance organization, often one of the lower prestige groups in the engineering pecking order, separated it from mainstream engineering and limited its influence on engineering decisions. System safety engineering, for all practical purposes, began to disappear or became irrelevant to the engineering and operations organizations.

Note that the problem here is different from that before *Challenger* where system safety became silent because it was considered to be less important in an operational program. After *Challenger*, the attempt to solve the problem of the lack of independence of system safety oversight, by separating it from the system engineers, quickly led to loss of its credibility and influence and was ineffective in providing lasting independence.

Safety was originally identified as a separate responsibility by the Air Force during the ballistic missile programs of the 1950s to solve exactly the problems seen here—to make sure that safety is given due consideration in decision-making involving conflicting pressures and that safety issues are visible at all levels of decision-making. Having an effective safety program cannot prevent errors in judgment in balancing conflicting requirements of safety and schedule or cost, but it can at least make sure that decisions are informed and that safety is given due consideration. However, to be effective, the system safety engineers must have the prestige necessary to have the influence on decision-making that safety requires.

The CAIB report addresses this issue when it says that: "*Organizations that successfully operate high-risk technologies have a major characteristic in common: they place a premium on safety and reliability by structuring their programs so that technical and safety engineering organizations own the process of determining, maintaining, and waiving technical requirements with a voice that is equal to yet independent of program managers, who are governed by cost, schedule, and mission-accomplishment goals*" [112, p. 187].

Both accident reports note that system safety engineers were often stigmatized, ignored, and sometimes actively ostracized. The *Columbia* investigation report concludes: "*Safety and mission assurance personnel have been eliminated [and] careers in safety have lost organizational prestige*" [112, p. 181]. Losing prestige created a vicious circle of lowered prestige leading to stigma, which limits influence and leads to further lowered prestige and influence and lowered quality due to many of the most qualified engineers not wanting to be part of the group.

At the same time, the organizational responsibility for system safety was not adequately integrated and available to the decision-making management levels. In the absence of a structured process to integrate safety-related analyses and conformance to specifications, information about safety issues was several interfaces removed from the people involved in the decisions on schedules and launch.

Limited communication channels and poor information flow Proper and safe engineering decision-making depends not only on a lack of complacency (that is, the desire and willingness to examine problems) but also on the communication and information structure that provides the information required to do so successfully. For a complex and technically challenging system like the Shuttle with multiple NASA centers and contractors all making decisions influencing safety, some person or group is required to integrate the information and make sure it is available for all decision makers.

Both the Rogers Commission and the CAIB found serious deficiencies in communication and oversight. The Rogers Commission report noted miscommunication of technical uncertainties and failure to use information from past near misses. Relevant concerns were not being reported to management. For example, the top levels of NASA management responsible for the launch of *Challenger* never heard about the concerns raised by the Morton Thiokol engineers on the eve of the launch nor did they know about the degree of concern raised by the erosion of the O-rings in prior flights. The Rogers Commission noted that memoranda and analyses raising concerns about performance and safety issues were subject to many delays in transmittal up the organizational chain and could be edited or stopped from further transmittal by some individual or group along the chain [419]. Information about safety issues was several interfaces removed from the people involved in the decisions about schedules and launch.

The *Challenger* Commission report found miscommunication of technical uncertainties and failure to use information from past near misses. Relevant concerns were not being reported to management. NASA Level I and II management responsible for the launch of 51-L never heard about the concerns raised by the Morton Thiokol engineers about the detrimental effects of cold temperatures on the performance of the SRB joint seal, nor did they know about the degree of concern raised by the erosion of the joint seals in prior flights. In 1985, when temperature became a major concern after STS 51-C and the launch constraint was imposed, higher management levels were never informed.

The problems did not seem to have been fixed after the *Challenger* loss. A report written before the *Columbia* accident notes a "general failure to communicate requirements and changes across organizations" [258]. Later, the CAIB found that "*organizational barriers . . . prevented effective communication of critical safety information and stifled professional differences of opinion . . .* [It was] *difficult for minority and dissenting opinions to percolate up through the agency's hierarchy*" [112].

In general, memoranda and analyses expressing concerns about performance and safety were subject to many delays in transmission up the organizational chain, as well as being subject to numerous stages of editing and potential vetoes on further transmittals.

Superficial or ineffective technical activities A final set of causal factors is related to ineffective technical activities including superficial safety efforts, inadequate hazard analysis, ineffective risk control, and inadequate information collection and dissemination.

As the CAIB investigated the *Columbia* accident, it expected to find a vigorous safety organization, process, and culture at NASA, bearing little resemblance to what the Rogers Commission identified as the ineffective "silent safety" system in which budget cuts resulted in a lack of resources, personnel, independence, and authority. They wrote that NASA's initial briefings to the CAIB on its safety programs described a risk-averse philosophy that empowered any employee to stop an operation at the mere glimmer of a problem. Unfortunately, NASA's views of its safety culture in those briefings did not reflect reality. Shuttle Program safety personnel did not adequately assess anomalies and frequently accepted critical risks without qualitative or quantitative support.

Childs coined a term *cosmetic system safety*, which can be characterized as superficial and perfunctory bookkeeping: hazard logs may be meticulously kept, with each item supporting and justifying the decisions made by project managers and engineers [63]. The CAIB report noted that "*over time, slowly and unintentionally, independent checks and balances intended to increase safety have been eroded in favor of detailed processes that produce massive amounts of data and unwarranted consensus, but little effective communication*" [112, p. 180].

Over time, system safety activities were watered down and made nonmandatory. The project managers had authority over the safety standards that were applied and could tailor them in any way they wanted. Under various types of pressures, including budget and schedule constraints, the independent safety reviews and communication channels within the Shuttle program degraded over time and were taken over by the Shuttle Program Office.

Other more specific problems were evident, including inadequate hazard analyses, incomplete investigations when deviations from expected performance occurred, and not evaluating changes.

Inadequate hazard analysis The CAIB report noted that: "*A large number of hazards reports contained subjective and qualitative judgments, such as 'believed' and 'based on experience from previous flights' this hazard is an accepted risk*" [112, p. 180].

The hazard report on debris shedding (the proximate event that led to the loss of the *Columbia*) was closed as an accepted risk and was not updated as a result of the continuing occurrences [112; 216]. The process laid out in the Shuttle standards allowed hazards to be closed when a mitigation was *planned*, not when the mitigation was actually implemented or determined to be successful.

As noted earlier, reliability engineering and a focus on component failure characterized the Shuttle program with correspondingly less emphasis on hazard analysis. The result was inadequate identification and understanding of hazards that did not involve individual component failure.

Incomplete investigation when deviations from expected performance occurred The Shuttle standard for hazard analyses (NSTS 22254, "Methodology for Conduct of Space Shuttle Program Hazard Analyses") specified that hazards be revisited only when there was a new design or when the design was changed. There was no process for updating the hazard analyses when anomalies occurred or even for determining whether an anomaly was related to a known hazard.

Not evaluating changes Changes frequently precede accidents. At no time was a trend analysis conducted to observe and perhaps trace to a root cause the increase in O-ring anomalies that had started in January 1984. Several changes at that time might have been the cause of the problems and might have been detected in time to avoid the accident if adequate attention had been paid to the problem. These changes to SRB procedures at Kennedy included discontinuation of onsite O-ring inspections; an increase in the leak check pressure on the field joint, which sometimes blew holes through the protective putty; a change in the type of putty used; changes in the patterns for positioning the putty; increased reuse of motor segment casings; and a change in the government contractor who managed the SRB assembly.

Safety information system deficiencies The quality of an organization's safety information system has been found to be the second most important factor distinguishing organizations that have high accident rates [178]. The first is a sincere commitment and priority assigned to safety by management.

Good decision-making about risk is dependent on having appropriate information. Without it, decisions are often made on the basis of past success and unrealistic risk assessment, as was the case for the Shuttle. Lots of data was collected and stored in multiple databases, but there was no convenient way to integrate and use the data for management, engineering, or safety decisions [10].

Creating and sustaining a successful safety information system requires a culture that values the sharing of knowledge learned from experience. Several reports have found that such a learning culture is not widespread at NASA and that the information systems are inadequate to meet the requirements for effective risk management and decision-making [10; 11; 119; 258; 112].

Sharing information across NASA centers is sometimes problematic, and getting information from the various types of lessons-learned databases situated at different NASA centers and facilities ranges from difficult to impossible. Necessary data is not collected, and what is collected is often filtered and inaccurate or tucked away in multiple databases without a convenient way to integrate the information to assist in management, engineering, and safety decisions; methods are lacking for the analysis and summarization of causal data; and information is not provided to decision makers in a way that is meaningful and useful to them. In lieu of such a comprehensive information system, past success and unrealistic risk assessment are used as the basis for decision-making.

As a specific example in the *Challenger* accident, the ineffectiveness of the added O-ring was actually known. An engineer at NASA Marshall Space Flight Center concluded after tests in 1977 and 1978 that the second O-ring was ineffective as a backup seal. Nevertheless, in November 1980, the SRB joint design was classified as redundant until November 1982. Its classification was changed to nonredundant on December 17, 1982, after tests showed the secondary O-ring was no longer functional after the joints rotated under 40 percent of the SRB maximum operating pressure. Why that information did not get to those making the *Challenger* launch decision or was discounted by them is unclear, but communication and information system flaws likely contributed.

After 1983, problem reporting requirements were reduced, and management lost insight into flight safety problems, flight schedule problems, and problem trends. This change represented a breakdown in what had been considered in the system safety business as an outstanding hazard analysis and follow-up program [264]. The new problem-reporting requirements resulted in critical information—such as anomalous events, the status of safety-critical items, criticality levels of components, and launch constraints—getting to the proper levels of management and engineering.

Summary The investigation of accidents creates a window into an organization and the opportunity to examine and fix unsafe elements. The repetition of the same factors in the *Columbia* accident implies that NASA was unsuccessful in permanently eliminating those factors after the *Challenger* loss. These are not problems limited to NASA and the Shuttle program. Rasmussen [320] wrote about the common pattern of organizations migrating to higher states of risk over time. This migration can be due to a variety of reasons such as external pressures and inadequate responses to them, flaws in the safety culture, dysfunctional organizational safety management, and inadequate safety engineering practices. This migration is very likely to continue unless changes are made and safeguards are put into place to prevent that migration in the future.

Appendix C
Petrochemicals: Seveso, Flixborough, Bhopal, Texas City, and *Deepwater Horizon*

C.1 Safety in the Chemical Process Industry

The chemical industry is relatively young, having started with the synthesis of the first organic compound (urea) by the German chemist Fredrich Wohler in 1828. This research changed fundamental thought about the nature of chemical reactions, and by 1920 the chemical industry had become an important factor in industrial development in the West. After 1940, the rate of growth was tremendous; today the chemical process industry involves the manufacture of tens of thousands of chemical substances throughout the world [40].

In the chemical and petrochemical industries, the three major hazards are fire, explosion, and toxic release. Toxic release occurs when a loss of containment allows a dangerous material to escape from a plant in sufficient quantity that its inherent nature (such as toxicity) or subsequent behavior (such as mixing with air and igniting) leads to casualties, property damage, or environmental harm. A *vapor cloud* is a bubble of explosive or toxic gas that drifts in the atmosphere and may explode if it comes upon a source of ignition. Vapor cloud explosions thus usually refer to those explosions that occur in the open air. In recent years, increased environmental concerns have added such hazards as thermal radiation (flares), noise, asphyxiation, and chronic environmental pollution to the list of hazards considered [292].

Of the three traditional hazards (fire, explosion, and toxic release), fire is the most common, explosions are less common but cause greater losses, and toxic release is relatively rare but has been the cause of the largest losses. Since loss of containment is a precursor for all of these hazards, much of the emphasis in loss prevention is on avoiding an escape of explosive or toxic materials through leaks, ruptures, explosions, and so on.

Risk is influenced by several factors [206]:

- *Size of inventory*: as plants have grown in size and output, so has the amount of materials being processed and stored.
- *Energy*: energy is required for a hazardous material to explode inside the plant or to disperse in the form of a flammable or toxic vapor cloud. In most cases, this energy is found in the material itself, either in its basic state or in a potential chemical reaction.
- *Time*: time is involved in both the rate of release and the warning time available to take emergency countermeasures and to reduce the number of people exposed.

- *Intensity/distance relationship*: the area over which a hazard may cause injury or damage will vary. In general, fire has a relatively short range, explosion a greater range, and toxic release a potentially unlimited range, in terms of both distance and time.
- *Exposure*: the number of people or the amount of property exposed to the hazard will obviously affect the amount of loss.

The three major hazards related to the chemical industry have remained virtually unchanged in their nature for many years. Design and operating procedures to eliminate or control these hazards have evolved and been incorporated into codes and standards. This approach sufficed before World War II because the industry operated on a relatively small scale and development was slow enough to allow learning by experience. After World War II, however, the chemical and petrochemical industries began to grow in complexity, size, and new technology at a tremendous rate. The potential for major hazards grew at a corresponding rate. Plants have increased in size, typically by a factor of ten, and they are often single-stream in order to realize economies of scale. Huge pieces of equipment and inventory are now found in these plants, and the processes contain greater stored energy than previously.

The operation of chemical plants also has increased in difficulty, and startup and shutdown have become complex and expensive. Process operating conditions, such as pressure and temperature, have increased, exacerbating the problems of process control and plant construction. Profitability concerns have led to plants being operated under extreme conditions and close to their limits of safety, using relatively sophisticated protection systems.

At the same time, the reliability of large plants has often decreased because of such factors as an increase in maintenance difficulty as layouts become more compact and equipment more sophisticated; increased interdependence and coupling in the plant; and increased economies in capital costs in aspects such as construction and duplication of equipment [206]. Moreover, chemical and petrochemical plants are becoming less isolated from urban environments, and growing quantities of chemicals are being transported over long distances.

The effect of these changes has been to increase the consequences of accidents, to increase environmental concerns such as pollution and noise, to make the control of hazards more difficult, and to reduce the opportunity to learn by trial and error. Toxic exposures, such as the accidental release of methyl isocyanate at Bhopal, have caused losses of unprecedented scale. The blowout that occurred in 2010 when the BP Macondo well was being capped using the *Deepwater Horizon* oil rig in the Gulf of Mexico led to the worst environmental disaster in the history of the United States.

At the same time, the social context has been changing. In the past, safety efforts in these industries were primarily voluntary and based on self-interest and economic considerations. However, pollution has become of increasing concern to the public and to government. Major accidents have drawn enormous publicity and have generated political pressure for legislation to prevent similar accidents in the future. Most of this legislation requires a qualitative hazard analysis, including identification of the most serious hazards and their contributing factors as well as modeling of the most significant accident potentials [335]. The accident at Bhopal—where supposedly a hazard analysis was done—demonstrates that hazard analysis by itself is not enough, however; management practices may be even more important.

Applying hazard analysis in the chemical process industry has special complications compared to other industries that make the modeling of accidents and event sequences especially difficult [335]. The chemical industry has a large number and variety of processes: The number of chemicals is large (and growing daily), each chemical may be produced in a variety of ways, and many of the reactions are not well understood. Thus, experience gained in one study may not be applicable in another.

Although the chemical industry does use some standard reliability and hazard analysis techniques, the unique aspects of the application have led to the development of industry-specific techniques. The hazardous features of many chemicals, for example, are well understood and have been cataloged. Indexes such as the *Dow Chemical Company Fire and Explosion Index Hazard Classification Guide* (usually called the *Dow Index*) and the *Mond Fire, Explosion, and Toxicity Index* (the *Mond Index*) were originally used to select fire protection methods, but they have been expanded to allow for more general hazard identification and evaluation.

Another technique developed for and used primarily in the chemical and petrochemical industries, is called HAZOP (HAZards and Operability Analysis). This technique, described in chapter 10, is a systematic approach for examining each item in a plant to determine the causes and consequences of deviations from normal plant operating conditions. The information about hazards obtained in this study is used to make changes in design, operating, and maintenance procedures.

Hazard analyses on chemical plants are often done late in the design process or on existing plants or designs where the only alternative for controlling hazards without costly design changes or retrofits is to add on protective devices. Along with the introduction of new technology and new plants, however, has come more attention in the early design stages to finding ways of avoiding hazards. Hazard analyses performed early in the design process can be used to help build simpler, cheaper, and safer plants by avoiding the use of hazardous materials, using less of them, or using them at lower temperatures or pressures. Thus, there are some attempts to move the chemical industry from a downstream approach to safety toward a more upstream, system safety approach.

These new trends are especially relevant as computers replace equipment that is well understood and for which standards and codes have been developed through extensive experience. Automated, computer-based control and safety devices are now widely used in the chemical and petrochemical industries. Building inherently safe plants may be even more relevant for this new technology.

C.2 Seveso

The release of a chemical cloud containing the highly toxic chemical dioxin from a plant in Northern Italy in 1976 ranks high on the list of worst man-made environmental disasters. Although it occurred a relatively long time ago, it created concern about public safety that led to improved industrial safety regulations in Europe.

A detailed description of this accident can be found in Lagadec [200]. Most of the information included here is summarized from that description.

C.2.1 Background

The Icmesa Chemical Company is located in Meda, northern Italy, a town of 17,000 residents about 15 miles from Milan. The adjoining community of Seveso was more severely affected by the accident, and thus has given its name to it. The Icmesa plant is owned by the Swiss company Givaudan, which is a subsidiary of the Swiss pharmaceutical giant Hoffman–LaRoche.

The chemical being produced at the time of the accident was trichlorophenol, which is used to make the bactericide hexachlorophene and an herbicide. When the factory was first established, Icmesa had told the authorities that it was to be used for the manufacture of pharmaceutical products. Local law required notification of the mayor and other parties 15 days before the start of any new production. But when the plant was later modified for the manufacture of trichlorophenol, no notification was given to the mayor or any other government regulatory authorities, nor were any of the required certificates obtained [200].

Lawsuits after the accident claimed that some public officials had known since 1972 that the chemical was being produced at Meda but that they had ignored what was happening there. Because no indication of the manufacturing of trichlorophenol had been given, the housing plan of Icmesa was approved in 1973 without inspection of the factory. In 1972, the mayor had requested a report on atmospheric pollution following concerns about the function of the plant, but Icmesa made no mention in the report of the manufacture of the chemical [200].

There have been charges that Givaudan chose Italy in which to manufacture Trichlorophenol because of the absence of restrictive regulations and the weakness of the controls on production of dangerous substances compared to those in Switzerland. In addition, salaries were low, and the trade unions were understanding [200]. The company claims that fewer restrictions in terms of safety and the environment played no part in the siting decision.

The trichlorophenol is not itself dangerous, but during processing, a hydrocarbon called tetrachlorodibenzodioxine (TCDD), or dioxin for short, can be produced as an unwanted byproduct [206]. Normally, dioxin is created only in trace amounts in the reaction that produces trichlorophenol. An accidental increase of temperature and pressure in the reactor, however, can result in large amounts of it. An estimated 2 kg of dioxin was released in this accident.

Dioxin is one of the most poisonous substances known: 500 times more toxic than strychnine and 10,000 times more than cyanide [200]. It can enter the body by ingestion, inhalation, or skin contact. A major symptom of dioxin poisoning is an acne-like skin condition called chloracne, which involves cysts, boils, and inflammation of the sebaceous glands of the skin. A mild case of chloracne usually clears up within a year, but a severe case can last several years [206]. Other effects include skin burns and rashes; gastrointestinal lesions; reduction of sexual potency and libido; and damage to the liver, kidneys, urinary system, thyroid, and nervous system. Nervous system changes may express themselves as memory problems, personality changes, sleep problems, emotional instability, and so on. The chemical is remarkably stable and can be eliminated only in negligible quantities: It accumulates in the liver, nerves, and fatty tissue. The initial appearance of dioxin poisoning symptoms is usually delayed until several days after exposure, when skin lesions (the most obvious symptom) appear.

Dioxin seems to have an unusual ability to interfere with the metabolic processes and is fatal to laboratory animals. No experiments have been done on primates because of the toxicity. Some evidence suggests that dioxin is immunosuppressant, carcinogenic, mutagenic, and teratogenic (causing fetal malformations).

C.2.2 Safety Features

Process control in this plant was manual, and therefore operators had to be present to operate the controls. An automatic system was in the process of being installed. Temperature was critical: The optimal temperature for the reaction was 170 degrees, and the heating device could not cause it to go above 190 degrees, which was well below the critical temperature of 230 degrees. No high-temperature alarm or automatic shutdown switch was installed.

The cooling system was also operated manually, so again operators needed to be present to open the valves of this subsystem. An automatic relief valve was installed on the reactor, but it was designed not to control a possible exothermic reaction but rather as protection for an operation at the start of the reaction [200]. A sudden increase in temperature during the reaction was not anticipated.

Changes had been made in the production process to save money but at the increased risk of exothermic reaction and dioxin forming. These changes allowed a reduction of staff, time, energy, and other costs, but they differed from the original patented Givaudan process [200]. Apparently, no review of the safety features of the new process took place when the changes were made.

C.2.3 Events

At 7 p.m. on July 9, 1976, the workers at the Icmesa plant were told to start a new reaction and distillation cycle, 10 hours later than normal. This cycle usually lasted for 15 hours. The night workers left at 6 a.m., and the plant was unattended for the weekend. When the night shift workers left, they cut off all energy supply to the system, leaving the reactor to cool off.

At 12:37 p.m. on Saturday, the relief valve lifted after a sudden increase in temperature and pressure. An estimated 2 kg of dioxin were released. An Icmesa technician later told the investigating commission that the reactor temperature at the time of the accident was between 450 degrees and 500 degrees and that conditions for the formation of a significant amount of dioxin had developed. The reasons for the sudden increase in temperature and pressure are, according to Givaudan and Icmesa official statements, unknown and unexplainable [200].

A toxic cloud drifted over part of the town. Heavy rain fell and brought the toxic material down to the ground. Some children reported seeing the cloud, but then it disappeared. A plant manager happened to be in the vicinity when the release occurred and took steps to stop it. He notified the man who was taking the place of the man in charge of production (who was on vacation). The commission inquiring into the accident could find no other actions that were taken that day.

On Sunday, the first effects of the accident were noticed: vegetation was burned; animals became ill; and about twenty children had sores on their arms, red spots on their faces, burns on their bodies, high fever, and intestinal problems. An Icmesa engineer sent samples of the burned vegetation to the Givaudan laboratories in Switzerland for analysis. The police launched an inquiry and were told by the company that a cloud of herbicide had spread over the area around the factory—no mention was made of dioxin. The technical director of Givaudan said in a deposition that he had heard of similar dioxin accidents, and he did think of that possibility. But he thought a very high concentration of dioxin would be located only near the relief valve and small concentrations elsewhere. "I could not think at that time that the dioxin could have expanded over a very large area" (translated in [200, p. 50]).

On Monday, the factory was open, and normal work resumed. Icmesa sent a letter to the local health authorities that confirmed that an incident had occurred at the factory, but, again, they mentioned only herbicides: "Not being able to evaluate the nature of the substances carried by those vapors and their exact effects, we have intervened with neighbors asking them not to consume garden products, knowing that the final product is also used in herbicides" [200; 257].

Tuesday, July 13: Health authorities sent a letter to the mayors of Meda and Seveso assuring them that there was no danger to people living in the surrounding areas.

Wednesday, July 14: Analyses at the Givaudan Laboratories in Switzerland confirmed that dioxin was present. Near the factory, the deaths of a large number of animals were reported.

Thursday, July 15: Serious cases of poisoning began to be reported among the population. The mayors announced that precautions should be taken by the residents, such as not eating vegetables grown in area gardens. The mayors met with the Icmesa plant owners, who made no mention of dioxin.

Friday, July 16: Fifteen children, four of whom were in grave condition, were admitted to the hospital, but nobody knew what treatment to give. The residents called for a strike and insisted that the authorities give them accurate information. The Italian authorities took samples for analysis.

Saturday, July 17: The mayors of Seveso and Meda added extra instructions for the residents around the factory, now ordering the burning of polluted garden vegetables and the killing and burning of affected animals. The director of the provincial chemical laboratory established that there could have been a release of dioxin.

Sunday, July 18: The mayor of Meda ordered the factory closed.

Monday, July 19: Five more children were hospitalized while the director of the provincial chemical laboratory learned during a visit to Givaudan that the industrial owners knew that a cloud of dioxin had been released during the accident.

Tuesday, July 20: The local health director returned from Switzerland and informed the mayors. Animals died within a 3 km radius outside the area originally considered endangered.

Wednesday, July 21: More meetings of local authorities were held to decide what to do. Additional protective measures were announced that included not eating meat from animals within the area, closing some establishments, and medical checks of residents.

Friday, July 23: A large meeting of medical experts took place at the police commissioner's office in Milan. They concluded that it was not necessary to take any civil defense actions. The university representatives at the meeting unanimously agreed that further measures were not necessary or urgent. The director general of the Health Service confirmed on the television news that everything was under control.

At the same time, the director of Hoffmann–LaRoche's medical center in Basel declared that the situation was very serious, that draconian measures were necessary, and that 20 cm of the ground surface had to be removed, the factory buried, and the houses destroyed.

Saturday, July 24: The regional health director castigated the Hoffmann–LaRoche medical director:

> *This person was dumped on us; nobody expected him, and nobody expected such severe statements. To my knowledge, he is not an official representative of the company and I shall today request to know on whose behalf he speaks. I have made clear to him the seriousness of what he says. I have the impression that this person is bluffing. And this person will have to answer for his statements.* [200]

Later that afternoon the official position changed, and a decision was made to evacuate 179 people who lived in an area of 2 square miles. They were told not to eat any produce or meat from the area (which, except for the immediate neighbors of the factory, they had eaten for the previous two weeks). The contaminated area was not closed completely, however; a road passing through it continued to be used [206]. The mayors left the meeting and found the populace in an uproar.

July 25–30: The first evacuation of 250 people took place, and the army enclosed the evacuated area with barbed wire. They worked with their bare hands, and no special provisions were taken [200]. Eventually they were given rubber boots. Amazingly, the evacuated residents were allowed to take their clothes, food, and other potentially contaminated objects.

New contaminated areas were detected and subsequently evacuated and cordoned off. Deaths of chickens, rabbits, and dogs began in an area several miles from the company and with a population of 15,000. Women and children were removed from this area, and people were asked not to procreate during the following months. A British expert arrived and suggested that the estimated 2 kg of dioxin released was greatly understated; he warned that there could have been 130 kg.

By the end of July, about 250 cases of skin infection had been diagnosed, fifty people had been admitted to a hospital, 600 people had been told to evacuate their homes, and 2,000 people had been given blood tests. There was fear of outside contamination from people who came from the evacuated area: Some hotels refused to provide them with accommodation [206] and, ironically, Switzerland temporarily closed its border to Italian food products and took air samples along the border in order to detect any contamination that might reach Swiss territory. Some businesses refused to take shipments of furniture coming from the Seveso area. In Italy, an argument ensued about whether therapeutic abortions should be allowed for pregnant women exposed to the dioxin.

In early August, the authorities found that the contaminated area was five times as large as originally thought. This larger area contained forty factories. The mayor of a local town, who had been one of the first people to visit the contaminated area, was found to have an excessive number of white blood cells. Controversy reigned about what to do, which resulted in delays in taking any action. Eventually, decontamination of the area was started.

By the beginning of the next year (1977), cases of chloracne were still being discovered in areas not thought to be contaminated, and many births of malformed children occurred in the second quarter of that year. Whether the percentage of birth defects was greater than normal depends on whose figures are used and which areas are included. More than a year after the release, the first 120 people (out of a total of 800 evacuees) were allowed to return to their homes. Others were readmitted to the area later. A later report found that cancer rates had increased in the area.

C.2.4 Some Causal Factors

As with most serious accidents, extraneous factors combined either to mitigate or to increase the consequences. In this case, a heavy rain brought the toxic cloud down to earth, and later heavy rains raised fears that the dioxin would be further dispersed. The fact that it was a weekend made it difficult to contact authorities immediately to warn the public and take emergency action. Being a weekend also meant that operators had gone home, and nobody was around to stop the release. This latter factor was not just random chance but reflects the management decision to leave the reactor unattended during a cycle. By chance, an employee was in the area, saw the toxic cloud, and was able to take action to avert an even worse accident.

Other factors were not so random.

Complacency and discounting risk The possibility of high temperature and high pressure in the reactor was discounted, and no preparation was taken for this event. After the accident, management at Givaudan and at Icmesa argued that foreseeing such an event was impossible and cited a lack of scientific information. Lagadec points out, however, that the scientific literature between 1971 and 1974 includes descriptions of other accidents where dioxin was formed during the production of trichlorophenol [200]. Milnes had pointed out in 1971 the conditions under which an out-of-control exothermic reaction could develop rapidly and generate temperatures up to 410 degrees, causing the release of large quantities of gaseous products [267]. The commission investigating the Seveso accident concluded that it was improbable that the technical directors of Givaudan and Icmesa could have been unaware of this possibility. The directors confirmed to the commission that they knew before the accident of Milnes's results [200].

During a press conference in Basel, the director of Givaudan indicated that he knew in advance of the risk of toxic products at the Icmesa factory, but that he had never imagined that such a disaster could occur. He explained that this was the reason no emergency plans had been worked out with the local authorities [200].

The workers were told to start a new cycle of reaction and distillation on Friday evening, even though it could not have been finished by the time the workers would leave for the weekend and the process control and cooling systems were manual. The director of Givaudan testified to the commission of enquiry that the manual controls were adequate. The commission concluded: *"This logic renders the responsibility even heavier because it is quite evident that if the merely manual controls were considered adequate, the continuous presence of people who are capable of applying them is an absolute necessity"* (translated in [200, p. 44]).

Uncommunicated and unreviewed changes Changes were made both in the chemical being manufactured and in the production process. These changes violated local regulations and the original patented process. These changes did not receive an adequate review.

Inadequate training The staff was not qualified to deal with these products and was not aware of the risks connected with the manufacture of trichlorophenol. The risks involved not only a major release to the surrounding community but small releases within the plant. On several occasions, production residues had escaped from containers or pipes [200].

Competing priorities Local authorities were never told about the chemical being manufactured, so they never established proper emergency preparation. At the release, local authorities reportedly had difficulty in getting information from the company about the chemicals involved in the release and about the appropriate countermeasures [200]. Doctors at first did not know what treatment to prescribe. Perrow suggests that plant officials tried to avoid a panic simply by not informing the public about what had happened [303]. Lagadec points out that there was silence at the start of the affair, later denials to reassure others and themselves that dioxin was not dangerous, and refusal by both the manufacturers and the local officials to follow up seriously on the effects of the poison. Note the similarity to Bhopal and other serious chemical plant accidents.

Superficial and ineffective safety measures Inadequate alarms and interlocks were installed on the reactor. There was no warning of high temperature or other type of signaling or shutdown equipment, perhaps because a sudden increase in temperature was not anticipated. In addition, the reactor was operated while no operator was present to work the primarily manual controls.

Analysis of the released material could not be done on short notice locally. Samples had to be sent to Givaudan in Switzerland to determine the composition of the materials released. The government agencies, not knowing that a dangerous chemical was being produced, might have had justification for this deficiency, but the company did not.

Finally, the lesson was learned that a pressure relief valve on a plant handling highly toxic substances should not discharge to the atmosphere but instead into a closed system.

C.3 Flixborough

Like Seveso, this accident was important in raising concern about major accidents, this time in the United Kingdom. The accident was investigated by a British Court of Inquiry, and much of the following information comes from this report [79]. The court's investigation was severely limited in its viewpoint and focused primarily on narrow technical issues. Other committees and reports of the British government in the wake of the accident, however, considered additional factors. The accident itself raised the general awareness of *major hazards*—where an accident could threaten the lives of thousands of people—and focused attention on the inadequacy of the existing precautions and controls.

The Advisory Committee on Major Hazards was established shortly after the accident to consider the safety problems associated with large-scale industrial facilities, other than nuclear installations, that conduct potentially hazardous operations. The reports that resulted were highly influential in establishing new research and regulatory initiatives. A Health and Safety Commission was established to propose draft regulations governing the operations of installations where hazardous materials are handled in large quantities. As a result, the Health and Safety at Work Act was passed on July 31, 1974, and a Health and Safety Executive was created to enforce the new regulations.

C.3.1 Background

The explosion occurred in 1974 at the Nypro Ltd. chemical works at Flixborough, a small rural community near Scunthorp, 160 miles north of London. The plant had been built in 1938 as a subsidiary of Fisons Ltd., to manufacture fertilizer. In 1964, it was passed to Nypro to produce caprolactum, an intermediary product in the manufacture of nylon.

Nypro was reorganized in 1967 with the participation of Dutch State Mines, the British National Coal Board, and Fisons Ltd. In August, the company started producing caprolactum from phenol at a rate of 20,000 tons per year. In 1972, the caprolactum capacity was increased to 70,000 tons a year by adding a new unit that used a process based on cyclohexane. The use of cyclohexane introduced a new dimension into the safety problem. When the changes were made, local authorities were not notified.

Nypro was under financial pressure. Instead of the 70,000 tons it was supposed to be producing, output was only 47,000 tons per year at the time of the accident, and the owners were losing money in the operation. They had requested the government's price commission to authorize a 48-percent increase in the price of caprolactum, but this request was refused.

Nypro was the only manufacturer of caprolactum in Great Britain at that time. The factory supplied it to two fiber manufacturers, Courtauld and British Enkalon. These companies were in direct competition with the other big nylon manufacturers, ICI and Dupont, who held patents on a process for the manufacture of caprolactum that most experts agreed was safer.

Cyclohexane has many properties similar to gasoline; it is highly inflammable and dilutes quickly in the air and in the temperature near a hot spot. To make caprolactum, cyclohexane is oxidized by passing it through a set of six reactors. Air and catalysts act on the heated cyclohexane, and the desired product is distilled out. In this process, large quantities of cyclohexane have to be circulated through the plant under pressure and at a temperature of 155 degrees. Loss of containment can release flashed liquid, which produces a large flammable vapor cloud. During the oxidation process, compressed air is injected into the reactors, creating an exothermic reaction. In contrast, the process for making caprolactum involving the hydrogenation of phenol, which Nypro could not use because the patent was held by its customers' competitors, is not exothermic.

On the day of the accident, Nypro had in stock 330,000 gallons (gal) of cyclohexane, 66,000 gal of naphtha, 11,000 gal of methyl benzene, 26,400 gal of benzene, and 450 gal of gasoline. The Petroleum Act of 1928 put the licensing and control of these potentially dangerous substances under the control of the local government. At the time of the explosion, the only licenses that had been issued authorized only 7,000 gal of naphtha and 1,500 gal of gasoline. Thus, the facility stocked more than 400,000 gal of dangerous products while being licensed for only 8,700 gal.

C.3.2 Events

At the beginning of 1974, the maintenance engineer at the Flixborough plant left for personal reasons and by June 1974 (when the plant was destroyed by the explosion) had not been replaced. None of the other engineers had special competence in mechanical engineering.

The plant had a series of six reactors, each slightly lower than the one before. Gravity caused cyclohexane to flow from reactor 1 to reactor 6 through short 28-in-diameter connecting pipes. To allow for expansion, each 28-in pipe contained a bellows (see figure 4.4 in chapter 4).

A small escape of cyclohexane from reactor 5 was discovered on the morning of March 27, 1974. An investigation found a vertical crack in the outer casing of the reactor, which indicated that the internal casing was also defective. The production engineer called the director for that zone of the plant, and they agreed that the installation would have to be closed down, depressurized, and cooled while a complete inspection took place.

The following morning (March 28), the director inspected the crack and found it was about 6.5 ft long. This situation was serious, and the morning was spent deciding what to do. The director wanted to restart production as soon as possible, so a temporary fix was proposed: reactor 5 would be shut down for inspection, and oxidation would continue using the remaining five reactors by building a bypass to link reactors 4 and 6. Once the bypass was in place, the factory would go back on stream.

Nobody at the meeting considered the difficult technical problems involved in constructing a bypass: possible design problems and alternatives were not discussed. Nobody (with perhaps the exception of the area engineer) thought there was a need to inspect the other reactors to find out if any of them had a similar defect that had not yet developed to the point where it could cause an escape. The main goal of the meeting was to restart the oxidation process with a minimum of delay.

The original crack in reactor 5 is now thought to have been caused by the corrosion resulting from past sprinkling of small escapes of cyclohexane with water. Plant cooling water, which contained nitrates, was used as it was convenient and available. The water had penetrated the insulation, and when it evaporated had deposited nitrates on the steel lining. Nitrate-induced cracking was well known to metallurgists at the time but not to other engineers [191].

The openings to be connected between reactors 4 and 6 had a diameter of 28 in, but the largest pipe available on site that could be used for the bypass had a diameter of 20 in. Because the two flanges were at different heights, the connection had to be built in the form of a dogleg of three lengths of pipe welded together and bolted to the flanges at each end.

After the modifications, the plant was started up again without anyone trying to understand the cause of the crack in reactor 5 or to make sure that the other reactors were in good condition. No calculation was made to check whether the bypass could take the load. No reference was made to British standards or to the guidelines published by the manufacturer of the bellows—both of which were violated by the bypass assembly. No piping layout was made besides a full-scale chalk drawing on the workshop floor. And no safety pressure test, either of the piping or of the whole unit, was made before the bypass was installed—such a test would almost certainly have caused the rupture of the bypass assembly. The tests that were made were for leaks and not for the strength of the assembly. No means was used to support the piping from underneath or to prevent lateral movement. The scaffolding constructed was meant as a support during assembly to prevent the weight of the assembly from pulling out the bellows, and it was inadequate for operating conditions. The maintenance manager had given his assistant a sketch of supports for the assembly, but these supports were never erected, and the maintenance manager never insisted that they be installed.

The bypass was completed on the evening of March 29, after 2 days of work. Once in place, it was tested for leaks. A leak was found, so the plant was depressurized. They forgot to mark the leak, however, and had to pressurize the plant again, find the leak, and repair it after depressurizing. Pressurizing a plant is not a simple task—it can take several hours and require thousands of steps. Once the leak was repaired, the plant was repressurized to test again for leaks, but none were found. The pressure was increased for further tests, the plant was depressurized again, and finally startup procedures were carried out.

The bypass assembly was set up by April 1 and appeared to work, but an unusually large use of nitrogen was detected and investigated. The repair held up from April 1 to May 29, even though the bellows were subjected to forces for which they were not designed, and the assembly was not held in place from above nor adequately secured from below. The bypass assembly was not checked closely during this period, but it was looked at in passing by a large number of people. Some observed that under pressure the pipe seemed to lift slightly off the support pipes, but no one noticed anything amiss. The report states, "It must therefore be taken that albeit there may have been some displacement of the assembly during the period, it cannot have been great enough to attract attention" [79].

The plant was shut down only for short periods twice in May. The 4 days preceding the accident, however, were filled with problems. On May 29, a valve was found to be leaking, and the plant was shut down to repair the leak. On the morning of June 1, startup began after repairing the leak and performing escape tests. At 4 a.m., a new leak occurred, others were discovered, and the process was stopped again. They later determined that the leaks had "righted themselves," and at about 5 a.m. operations were restarted.

At this point, pressure went up at an abnormal rate, requiring substantial venting. Shortly thereafter, the process was stopped again because of another leak. This time repairs could not be carried out immediately because the necessary spark-proof tools were not available—they were locked in a shed for the weekend. The process was restarted at 7 a.m. on Saturday morning and continued until 3 p.m. Temperature and pressure problems began again, the pressures being high enough to be disquieting without being alarming.

The precise sequence of events at this point is complex and uncertain because the explosion killed everyone in the control room and destroyed all the instruments. The report is careful to point out that no operator errors were involved. A crucial feature of the situation was that the reactors were subjected to a slightly higher than normal pressure. Pressure would normally be controlled by venting, but this procedure involved the loss of considerable quantities of nitrogen. A number of anomalies were unexplained at this point, especially the fast rise in pressure and the excessive consumption of nitrogen, both of which, according to the report, appear to be independent of the condition of the bypass pipe.

Shortly after the final warmup started, they discovered that they did not have enough nitrogen to begin oxidation and additional supplies would not arrive until after midnight. Although all records of what exactly happened were destroyed in the explosion and ensuing fires, the need to conserve nitrogen would tend to inhibit their use of venting to reduce pressure [206]. Several theories are advanced in the report, but none can be proven.

During the late afternoon, something happened that resulted in the escape of large quantities of cyclohexane. Most people now believe that this event was the rupture of the 20 in bypass pipe, perhaps accompanied by a fire in a nearby 8 in pipe. Through the two 28 in openings from reactors 4 and 6, hot cyclohexane escaped under pressure in massive quantities. Some evidence points to the possibility of two explosions rather than one.

Within 30 seconds, the cyclohexane formed a vapor cloud 700 ft in diameter and 350 ft in height that was composed of 40 to 50 tons of the chemical. A wind of 15 mph drove the mixture 300 ft to the discharge tower of the hydrogen unit, which acted as a source of ignition.

At about 4:53 p.m., a massive unconfined vapor cloud explosion occurred, estimated to be equivalent to the force of 15 to 45 tons of TNT. The explosion was heard 30 miles away and devastated the 60-acre factory site. All buildings within a radius of a third of a mile were destroyed, and more than 2,450 houses were damaged in the vicinity. Windows were shattered in houses up to 8 miles away.

The blast from the explosion shattered the windows of the control room and caused the roof to collapse, killing everyone inside. Of the 28 people who died in the explosion, 18 of them were in the control room. Some of the victims were killed by flying glass while others were crushed by the roof. Seventy-two people were present at the site when the explosion occurred; on a regular working day there would have been 550 present and presumably a great many more deaths and injuries. Off site, 53 people were injured, and hundreds more suffered minor injuries that were not officially recorded.

The flames from the fires on the site rose 250 to 350 ft high and burned for many days. Even after 10 days, the fires were still hindering rescue work at the site. All the fire extinguishing facilities in the plant were immediately destroyed, and it took two and a half days for the firefighters to get to the main sources of the fire.

The material damages from this accident were estimated at \$60 million: more than \$48 million for the reconstruction of the factory, \$10 million for interruption of operations, and \$2 million for third-party liability.

C.3.3 Some Causal Factors

Several conditions combined to mitigate the potential consequences of this accident. First, the site was rural and far from any major population centers. Casualties might have been much greater in a more densely populated area. Second, the wind was light. If it had been stronger, a larger plume might have been created and the explosion could have come later, extending the damage. Daylight and good weather increased the effectiveness of the fire brigades. Finally, the explosion occurred on a weekend when there was a skeleton crew at the plant, so casualties were lower than they might have been.

Other causal factors again show the pattern to be found in most serious accidents.

Complacency and lack of forethought

The report by the Court of Inquiry concludes that nobody among those in charge of the design or the construction of the factory foresaw the possibility of a major accident. No disaster plan had been drawn up.

Although large quantities of unlicensed fluids were stored at the plant, notifying the authorities about them would have made no difference because licensing practices at the time would have allowed the increased storage without requiring additional constraints [206]. However, the accident did reveal the need for better methods of notifying local planning authorities about major hazards and the need for greater guidance in safety matters for these local authorities and for the installations themselves.

Although other vapor cloud explosions had occurred elsewhere in the world, this one was unique in that it provided British industry, the British public, and the rescue squads with the first direct experience of the consequences of such an event. They could no longer ignore the danger [200].

Unheeded warnings The crack in one reactor did not lead them to inspect the other ones. In addition, the bypass pipe had been noted to move slightly during operation, and there had been unexplained anomalies in pressure, temperature, and hydrogen consumption. Yet no thorough investigation was done.

In a broader scope, the chief inspector of factories had been warning, first in 1967 and later from 1970 to 1972, of the problems arising from the use of large quantities of dangerous materials [137].

Conflicting priorities The decisions leading to the accident reflect a conflict of priorities between safety and production. The report said:

> *We entirely absolve all persons from any suggestion that their desire to resume production caused them knowingly to embark on a hazardous course in disregard of the safety of those operating the Works. We have no doubt, however, that it was this desire which led them to overlook the fact that it was potentially hazardous to resume production without examining the remaining reactors and ascertaining the cause of the failure of the fifth reactor. We have equally no doubt that the failure to appreciate that the connection of Reactor No. 4 to Reactor No. 6 involved engineering problems was largely due to the same desire. . . . The design and the construction of the whole link should not have been carried out in a hurry as it was in this case. There should have been time to consider what problems would arise and how they could be suitably dealt with.* [79]

The report recommended that more attention be devoted to the cost-effectiveness of designing continuous process plants so that they could be repaired without shutdown, thus avoiding the need for pragmatic management decisions under competing priorities.

Although the storage of such flammable materials as naphtha, benzene, and gasoline was not involved in the initial explosion, these substances made a major contribution to the heat, flames, and smoke when they ignited, and they also hindered rescue operations. Kletz observes that before the Flixborough explosion, little conscious thought was given to the possibility of increasing safety by reducing inventories of hazardous materials in process and in storage. Up to then, engineers usually designed a plant and accepted whatever inventory was called for by the design. Hazards were instead controlled by adding protective equipment such as trips, alarms, fire protection, and firefighting. In fact, a great deal can be done to reduce inventories without reducing production. As a result of Flixborough, many companies have reduced their stocks of hazardous intermediates, and there has been renewed interest in intensification (see chapter 11) [191].

Ineffective organizational structure The installation did not have a sufficient complement of qualified and experienced people. There was no works manager and no adequately qualified mechanical engineer on site. As a result, some engineers had been asked to assume responsibilities for which they were not qualified. In addition, the engineers who were there were overworked.

With the departure of the maintenance engineer, there was no mechanical engineer on site having sufficient qualifications to deal with complex or novel engineering problems or having the status and authority to enforce necessary safety measures. The maintenance engineer's duties were assigned temporarily to a subordinate whose qualifications were not sufficient for the job. The personnel that remained at the plant did not seem to realize that they lacked important knowledge to carry out their assigned duties.

The problems were exacerbated by the fact that the director and technical director were both chemical engineers with no training in mechanics. The company had recognized that the structure of the engineering section was weak, and a consultant from the National Coal Board had been called on in 1974 for advice about reorganization.

There was also no engineering organization independent of production-line management that was responsible for assessing the overall system and ensuring that proper controls were exercised. The role of safety officer at the Flixborough plant was poorly defined. He considered himself responsible to the personnel director although he had direct access to the director general. Kletz has suggested that the accident demonstrated the need for higher status of safety advisors. "*In highly technical industries like the process industries, it is not sufficient to employ as safety advisor only an elderly foreman who sees his job as taking statements from men who fall off a bicycle*" [182, p. 109].

Superficial safety activities Anomalies were not carefully reviewed, and the stress testing and safety analysis of the effects of the bypass were insufficient. No safety organization was responsible for such analyses and reviews. More specifically:

- Nobody was familiar with the British standard then in existence for piping in petrochemical plants, which specifically prohibited the type of bypass pipe that had been installed between the reactors.
- No calculations were made that took into account the forces arising from the dogleg shape of the connecting pipe between reactors 4 and 6. No drawing of the bypass pipe was made other than in chalk on the workshop floor. No calculations were done to check whether the bellows would withstand the forces caused by the dogleg-shaped pipe connection. No pressure testing was carried out either on the pipe or on the complete assembly before it was fitted. The pressure testing done on the plant was not up to the safety valve pressure. In addition, the test was pneumatic, not hydraulic.
- The scaffolding to support the pipe was intended only for use during construction. It was not suitable as a permanent support for the bypass assembly during normal operation.
- The 20-in bypass pipe was not constructed and installed in accordance with the current standards and codes of practice. The bellows manufacturer produced a guide for their use that made it clear that two such bellows should not be used out of line in the same pipe without adequate support for the pipe.

The explosion occurred during plant startup. The report suggests that special attention should be given to factors that necessitate the shutdown of a chemical plant so as to minimize the number of shutdown–startup sequences and to reduce the frequency of critical management decisions. After a defect was discovered, the plant was restarted, but the remaining reactors were not examined nor was the cause of the failure of the fifth reactor determined.

Unreviewed changes The process was changed, and the capacity was tripled, but safety questions were not rethought. The court's report on the accident recommended that existing regulations for modifying steam boilers be extended to apply to pressure systems containing hazardous materials.

C.4 Bhopal

The Bhopal accident, considered the worst industrial accident in history, has been widely investigated in newspapers, scientific articles, and books. Several good references exist for further information and are listed in the bibliography. A book by Bogard [40] and an article by Ayres and Rohatgi [27] are especially helpful in understanding this accident, and much of the information below comes from these sources.

C.4.1 Background

The accident at Bhopal, ten years after Flixborough, involved the release of methyl isocyanate (MIC), a highly reactive, toxic, volatile, flammable, and unstable chemical. This substance is unusually dangerous, both to store and to handle. Union Carbide is one of two major US producers; Bayer makes the chemical in Europe. Most of Union Carbide's production of MIC is at its plant in Institute, West Virginia, which began to produce it in 1967. Production in India by Union Carbide India Ltd., began on a small scale in 1977.

Demand for MIC dropped sharply after 1981, leading to reductions in production and pressure on the company to cut costs: The Bhopal plant was operating at less than half its capacity when the accident occurred. Union Carbide put pressure on its Indian subsidiary to reduce losses, but no specific details were given about how this was to be done [27].

In response, the maintenance and operating personnel were cut in half [39]. As the plant lost money, many of its skilled workers left for more secure jobs and either were not replaced or were replaced by unskilled workers. Maintenance procedures were severely cut back, and the shift-relieving system was suspended: if no replacement showed up at the end of a shift, the following shift went unmanned. "*Despite the minimal training of many of these workers in how to handle nonroutine emergencies at the plant, no consideration was seriously given to the idea of shutting the facility down*" [39, p. 29].

C.4.2 Safety Features

MIC is used in the production of a number of pesticides and in the production of polyurethanes (which are used in plastics, varnishes, and foams). It is highly volatile, and the vapor is heavier than air. One of the major hazards in handling MIC is contact with water, which results in large amounts of heat. The concentrated gas burns any moist part of the body, including the throat, nasal passages, eyes, and lungs. The hazardous and toxic nature of MIC was described in the Union Carbide operating manual, along with the chemical's ability to cause fatal pulmonary edema.

Union Carbide specified requirements for operating procedures, storage facilities, and safeguards to reduce the hazards of producing MIC. The basic design features at Bhopal are shown in figure 4.5. The chemical was to be stored in underground tanks encased in concrete. The Bhopal facility used three double-walled, stainless-steel tanks, each with a capacity of 60 tons. The operating manual specified that the tanks were never to contain more than half their maximum volume, or a standby tank was to be available to which some of the chemical could be transferred in case of trouble. The Bhopal tanks were interconnected so that the MIC in one tank could be bled into another tank. As specified in the operating manual, the tanks were embedded in concrete.

The chemical was also to be stored in an inert atmosphere of nitrogen gas at 2 to 10 psi over atmospheric pressure. Regular, scheduled inspection and cleaning of valves and piping was specified as imperative, and storage time was limited to twelve months maximum. If staff were doing sampling, testing, or maintenance at a time when there was a possibility of a leak or a spill, the Union Carbide operating manual specified that they were to use protective rubber suits and airbreathing equipment.

In order to limit its reactivity, MIC was supposed to be maintained at a temperature near 0°C. A refrigeration unit was provided for this purpose. In addition to the refrigeration system and standby tank, the plant had several backup protection systems and lines of defense [27]:

- A vent gas scrubber was designed to neutralize any escaping gas with caustic soda. The scrubber was capable of neutralizing about 8 tons of MIC per hour at full capacity.
- The flare tower was supposed to burn off any escaping gas missed by the scrubber; the toxic gases would be burned high in the air, making them harmless.
- Small amounts of gas missed by the scrubber and the flare tower were to be knocked down by a water curtain that reached 40 to 50 ft above the ground. The water jets could reach as high as 115 ft, but only if operated individually.
- In case of an uncontrolled leak, a siren was installed to warn the workers and the surrounding community.

C.4.3 Events

The Indian government required the Bhopal plant to be operated completely by Indians [252]. At first, Union Carbide flew plant personnel to West Virginia for intensive training and had teams of US engineers make regular onsite safety inspections. But by 1982, financial pressures had led Union Carbide to give up direct supervision of safety at the plant, even though it retained general financial and technical control [252]. No American advisors were resident at Bhopal after 1982.

Several Indian staff who were trained in the United States resigned and were replaced by less experienced technicians. When the plant was first built, the operators and technicians had the equivalent of two years of college education in chemistry or chemical engineering. In addition, Union Carbide provided them with six months of training. When the plant began to lose money, educational standards and staffing levels were reportedly reduced [27].

In 1983, the chemical engineer managing the MIC plant resigned because he disapproved of falling safety standards and was replaced by an electrical engineer. A significant number of operating staff had been turned over in the previous three years, and the new employees lacked safety training [233]. Morale at the plant was low, and management and labor problems followed the financial losses. "*There was widespread belief among employees that the management had taken drastic and imprudent measures to cut costs and that attention to the details that ensure safe operation were absent*" [233].

These are just some examples of the unsafe conditions that were permitted:

- At the time of the accident, it was estimated that the chloroform contamination of the MIC was four to five times higher than specified in the operating manual, but no corrective action was taken [419].

- The MIC tanks were not leak-tight to a required pressure test.
- Workers at Bhopal regularly did not wear safety equipment, such as gloves or masks—even after suffering adverse symptoms like chest pains and vomiting—because of high temperatures in the plant. There was no air conditioning [252].
- Inspections and safety audits at the plant were few and superficial [27].

Prior warnings and events presaging the accident were ignored. The Bhopal plant had six minor accidents between 1981 and 1984, several of which involved MIC. One worker was killed in 1981, but official inquiries required by law were often shelved or tended to minimize the government's or the company's role [40]. A leak similar to the one involved in the events described below had occurred the year before [27]. Within the government, at least one person tried to bring up questions of hazards to those inside and in the immediate area of the plant. He was forced to resign [40].

The local press had published articles criticizing the safety practices at the plant in detail, including one that virtually predicted the accident [27]. Local authorities appear to have done nothing in response. The articles, titled "Please Save Our City" and "Bhopal Sitting on the Mouth of a Volcano," were part of Rajukman Kesmani's crusade to get the authorities to recognize the potential danger. He had even posted warnings: "Poison gas. Thousands of workers and millions of citizens are in danger" [252]. The crusade was unsuccessful, although it was mentioned in debates in the state legislature [199].

Five months before the accident, local Union Carbide India management decided to shut down the refrigeration system. Refrigeration is important in MIC storage in order to control dangerous chemical reactions (exothermic degradation). The most commonly advanced explanation for why the refrigeration system was turned off is cost cutting. The local management claimed that the unit was too small and never worked satisfactorily. There is disagreement about whether Union Carbide in the United States approved this measure.

Because the chemical had a higher temperature without the refrigeration than that allowed by the alarm system, the set point of the temperature alarms on the MIC tanks was raised from 11°C to 20°C, and logging of tank temperatures was halted.

Six weeks before the accident, production of MIC had been halted because of an oversupply of the pesticides the chemical was used to make. Workers performed routine maintenance while production was stopped.

At about 10:30 p.m. on December 2, 1984, a relatively new worker was assigned to wash out some pipes and filters, which were clogged. The pipes being cleaned were connected to the MIC tanks by a relief valve vent header, which was normally closed. The worker properly closed the valves to isolate the tanks from the pipes and filters being washed, but nobody inserted a required safety disk (called a slip blind) to back up the valves in case they leaked. The maintenance sheet contained no instruction to insert this disk, although there was a note to the night shift to wash the pipe [27].

The worker who had been assigned this task did not check to see whether the pipe was properly isolated because, he said, it was not his job to do so. Reportedly, he knew that the valves leaked, but the safety disks were the job of the maintenance department. The pipe-washing operation should have been supervised by the second shift supervisor, but that position had been eliminated in the cost cutting.

At the time of the accident, tank 610 (from which the MIC escaped) contained 40 to 50 tons out of the total capacity of 60 tons, which violated the safety requirements: The tanks were not to be more than half-filled with MIC, and a spare tank was to be available to take any excess. The adjacent tank, 611, was thought to contain 15 tons on the basis of shipping records, but actually contained nearer to 21 tons. Tank 619, the spare tank, contained less than one ton, although the level gauge showed that it was about 20 percent full [27]. Many of the gauges were not working properly or were improperly set. This fact may reflect design flaws in the equipment, although operation and maintenance errors also played a part.

When the night shift came on duty at 11 p.m., the first sign of trouble was detected. The pressure gauge indicated that pressure was rising (10 psi instead of the recommended 2 to 3 psi) but was at the upper end of the normal range. The temperature in the tank was above 20°C. Both instruments were ignored because they were believed to be inaccurate.

The instrumentation for detecting pressure and temperature levels of the chemicals at the plant was faulty and unreliable; workers did not trust the information it provided. They were told to use eye irritation as the first sign of exposure in the absence of reliable detection mechanisms to gauge threshold levels accurately. The plant had few automatic shutoffs or alarm systems that might have detected and stopped the gas leak before it spread beyond the facility.

The leak of liquid from an overhead line was first discovered around 11:30 p.m., after some workers noticed slight eye irritation. Leaky valves were common—small leaks occurred from time and time and were not considered to be significant [27]. The workers looked for the leak, and one saw a continuous drip on the outside of the MIC unit. He reported it to the MIC supervisor, but the shift supervisor did not consider it urgent and decided to postpone an investigation until after the tea break.

At 12:40 p.m. on December 3, the control room operator noticed that the pressure gauge on tank 610 was approaching 40 psi and the temperature was at the top of the scale (25°C). At about 12:45 a.m., loud rumbling noises were heard from the tank. The concrete around the tank cracked.

The temperature in the tank rose to 400°C, causing an increase in pressure that ruptured the relief valve. The pressurized gas escaped in a fountain from the top of the vent stack and continued to escape until 2:30 a.m. The MIC was vented from the stack 108 ft above the ground, well above the height of the water curtain. A safety audit two years earlier had recommended increasing the capacity of the water curtain system, but this had not been done. Eventually 50,000 lbs of MIC gas would escape.

The operator turned off the water-washing line when he first heard the loud noises at 12:45 a.m. and turned on the vent scrubber system, but the flow meter showed no circulation of caustic soda. He was unsure whether the meter was working. To verify that the flow had started, the operator would have had to check the pump visually, which he refused to do unless accompanied by the supervisor; the supervisor declined to go with him.

The vent scrubber was not kept operational because it was presumed not to be necessary when production was suspended. By the time it was turned on, it was too late to help. Allegedly, there was not enough caustic soda to neutralize the MIC, anyway [199]. The vent scrubber was designed to neutralize only small quantities of gas at relatively low pressure and temperature. The pressure of the escaping gas during the accident exceeded the scrubber's design by two and a half times, and the temperature was at least 100°C more than the scrubber could handle. There is speculation that had the vent scrubber been in operation, the reaction of MIC with caustic soda would have caused further release of heat and still greater pressures [27]. If this had happened, much of the MIC would have ultimately escaped anyway.

The operator never opened the valve connecting tank 610 to the spare tank 619 because the level gauge erroneously showed it to be partially full. The assistant plant manager was called at home at 1 a.m. and ordered the vent flare turned on. He was told that it was not operable. At the time of the accident, the flare tower was out of service for maintenance: A section of the pipe connecting it to the tank was being repaired. In any case, the design of the flare tower was inadequate to handle the 50 tons of MIC that escaped during the accident. Ayres and Rohatgi [27] claim that if it had been turned on, the result would have been a violent explosion.

The plant manager learned of the leak at 1:45 a.m., when he was called by the city magistrate. When the leak of MIC was serious enough to cause physical discomfort to the workers, they panicked and fled, ignoring the four buses that were intended to be used to evacuate employees and nearby residents. A system of walkie-talkies, kept for such emergencies, was never used. The MIC supervisor could not find his oxygen mask and ran to the boundary fence where he broke his leg attempting to climb over it. The control room supervisor stayed in the control room until the next afternoon, when he emerged unharmed.

The toxic gas warning siren was not activated until 12:50 a.m., when MIC was seen escaping from the vent stack. It was turned off after only 5 minutes, which was Union Carbide policy. It remained off until turned on again at 2:30 a.m. The police were not notified, and when they called between 1 and 2 a.m., they were given no useful information.

A large number of squatters had settled in the vacant areas surrounding the plant. Such settlements of desperately poor migrant workers, in search of any form of employment, are common in India. They also take advantage of whatever water and electricity are available. The plant was located by the railway station, bus station, hospitals, and so on; settlements are also common near such facilities. The railway station in India is a community center.

No information had been provided to the public about protective measures in case of an emergency or other information on hazards. The siren sounded so many times a week—estimates range from twenty to thirty—that there was no way for people living near the plant to know what the siren signified; the emergency signal was identical to that used for other purposes, including practice drills. If the people had known to stay home, close their eyes, and breathe through a wet cloth, the deaths could have been prevented.

Eventually the army came and tried to help by transporting people out of the area and to medical facilities. This help was delayed by the fact that nobody at the plant notified authorities about the release.

Agnees Chisti, a journalist who was in a small hotel visiting the city, described the events outside the plant after the release:

> *I was trying to get one of the foreign stations on the radio at about 2:30 a.m. when I felt some choking in my throat and I thought it was some ordinary case of bad throat or something. I tried to get some cough syrup, but then I realized that it was something much more than that. There was a burning sensation in the eye. I somehow opened the*

door of my room. At first I thought it was a hotel problem. I went out. I was running for some open space where I could get relief but I could get relief nowhere. . . .

When I came out, I saw hordes of people moving towards some direction. I was new to Bhopal City; I didn't know all the routes. . . . Anyway, I walked. And that was a ghastly experience. I saw ladies, almost undressed, straight out of bed in petticoats, children clinging to their breasts, all wailing, weeping, some of them vomiting, some falling down. I now presume falling down dead. . . . they were falling dead, family members were leaving their own family members behind and running for safety. . . .

Fifteen months later, the area resonates with the coughs and groans of perpetually sick passengers. Deaths due to the aftereffects of exposure to gas are even now not infrequent although medical records no longer mention "MIC poisoning" as the cause of the ailments as was done before. [66, p. 7–8]

The result was the worst industrial accident in history. The weather and wind direction contributed to the consequences. In addition, because the release took place in the middle of the night, most people were asleep, and it was difficult to see what was happening.

Infants and children under the age of twelve as well as elderly people were the most likely to die from exposure to the chemical. The victims were primarily local residents but also included eighty-seven railway employees and the superintendent of the Bhopal railway station. Large numbers of animals also died. The exact number of fatalities and injuries is a matter of controversy. Estimates of deaths range from 1,750 up to 10,000. Many deaths occurred among refugees who escaped from Bhopal into the surrounding areas, where no systematic count was possible, and deaths continued over a long period and thus were not attributed directly to the accident.

By 2 days after the accident, Prime Minister Gandhi said that 20,000 victims had received treatment at various hospitals in the city. Eventually there were 232,691 victim claims for compensation. A house-to-house survey of about 25,000 families found that one out of twenty-five families affected by the gas leakage experienced blindness or partial blindness of one of its members [66].

Nobody knows for sure what caused the runaway reaction. Experiments attempting to reproduce the accident conditions determined that there must have been at least 45 tons of MIC, one-half to one ton of water, and one to one and a half tons of chloroform in the tank for the reaction to have occurred.

The most likely cause was water getting into the tank, but a possibility exists that another contaminant entered the tank through the nitrogen line or the scrubber and catalyzed the reaction [27]. One widely accepted hypothesis is that a small amount of water got into the storage tank through leaky valves during or after the cleaning of the pipes and filters earlier in the day. The cleaning itself had been necessary because two of four valves that should have been open to allow water flow were clogged. The extra pressure from the clogged pipes is an alternative hypothesis for the leak of water into the tank [252]. A further hypothesis implicates slow chemical reactions of the MIC with the stainless-steel walls of the storage tank itself [40]. Union Carbide claimed that the amount of water in the tank may have been very large (up to 240 gal of water) and suggested the possibility of deliberate sabotage.

C.4.4 Some Causal Factors

A pattern is clearly emerging, with most of the systemic factors mentioned in chapter 4 involved with the accidents described so far. Bhopal is no exception.

Discounting risk Union Carbide went into large-scale production of MIC without having performed adequate research on the stability of the chemical. After the accident, when there was serious concern about the other MIC tanks at the plant, neither Union Carbide nor Bayer knew of an effective inhibitor for the type of reaction that occurred [27].

The accident came as a complete surprise to almost everyone, including the Union Carbide scientists and risk assessors. "*The belief that such a catastrophe could not happen with such modern technology, that so many safety systems could not fail simultaneously, that the Bhopal plant was a 'model' facility, etc., directed necessary attention away from the overall production process and its possible worst consequences and fostered a general atmosphere of safety*" [39, p. 24]. In addition, contrary to official government policy, the plant was located in a highly populated area.

Complacency Not having accidents in the past led to complacency. Union Carbide and the Indian government pointed to the relatively minor nature of chemical accidents at the plant to support their refusal to install backup safety equipment or to move the plant away from populated areas: "*The extrapolation from the past safety record of the plant, which in reality was not all that good but was presented as such, was in essence an excuse or alibi for failing to deal squarely with the potential, however small, of a catastrophic accident. The assurance that past performance is an adequate guide in the assessment of hazards is itself an incentive for maintaining the status quo*" [39, p. 19].

Ignoring warning signs The management and the state government ignored the risk and warning signs before the accident and then made the consequences of the leak worse by repeated denials of the urgency and magnitude of the disaster [199].

According to Bogard [40], the significance of early warning signs was muted by several factors:

(1) the reluctance of the company to become involved in the day-to-day affairs of the plant's safety and maintenance routines, despite the control it exercised over high-level decisions and budgetary matters
(2) the cost of new equipment and training
(3) the importance of pesticides to the region's farmers

Poor safety engineering and operations Designers did not anticipate an MIC release of anywhere near the magnitude that occurred or prepare for one. Emergency equipment was inadequate for the job, and the plant was not designed to cope with a major leak. Emergency training and procedures were also inadequate, such as the policy of turning off the warning siren after 5 minutes. Ayres and Rohatgi suggest that those who framed such a policy never conceived of an accident of this magnitude [27].

There was systematic analysis and training for serious events to some extent, but it was not comprehensive. For the most part, workers did not have adequate training on what to do about nonroutine events and conditions. Emergency procedures and drills were held, but there was no systematic analysis of hazardous conditions that could lead to a major accident [419]. The procedures and training, accordingly, did not sensitize plant personnel to the "importance of various seemingly minor deficiencies that could combine to produce a major disaster" [419, p. 446].

Ayres and Rohatgi note that none of the parties—plant designers, operators, company management, or responsible public officials—anticipated the possibility of an accident the size of the one that happened [27]. However, all the significant components had occurred repeatedly, including corroded pipes, pump failures, leaky valves, staff failures to follow procedures specified in the operating manual, systems down for maintenance, and even the major leaks of MIC and phosgene, which had happened previously one or more times. The adverse weather conditions (an atmospheric inversion) were unusual but by no means unprecedented. The only new factor was the combination of all of these things at one point in time.

Low priority assigned to safety All levels of management were lax in enforcing safety-related policies [27]. The Indian government inspection agencies were understaffed and uninformed about the safety problems related to MIC. "Safety was given a low priority by all the parties involved" [199].

Inadequate training was provided for emergencies. Even the supervisors forgot what they had been taught when the emergency occurred. There appeared to be no real evacuation plans. The scope of the disaster was thus greatly increased because of total unpreparedness [252].

Flawed resolution of conflicts between safety and other goals The safety equipment was inadequate, and much of what was there was shut down to save money. Cutbacks in staff, training, and maintenance to reduce costs undoubtedly also contributed to the accident.

Company policy was to turn off the siren and not to report minor leaks or accidents. The typical excuse for such secrecy is that the company does not want to create "unnecessary" alarm and fear [27]. Even the local Bhopal community was uninformed about what was going on in the plant. An effective public information program would have been relatively inexpensive. Even simply sending a sound truck through the streets telling people to breathe through wet cloths would have saved thousands of lives [27].

Ineffective organizational structure Union Carbide had transferred legal responsibility for safety inspections to overseas operators even though it retained general management, technical, and financial control [252], creating divided responsibilities.

The organizational structure involved a complicated relationship between Union Carbide and Union Carbide India, which was mostly under the supervision of Indian managers carrying out general directives from the parent company. The directives issued from the parent company and the Indian government were not implemented in the actual practices and everyday operations at the plant, including the behavior of plant operators, managers, and government bureaucrats [199].

Superficial safety activities Leaks were routine occurrences and the reasons for them were seldom investigated: problems were either fixed without further examination or ignored. While the initial design of the plant included many safety features, there was an absence of an ongoing review and audit process. The scenario assumptions for the protection systems were obviously incomplete, and, as noted earlier, inspections by outside authorities were few and superficial [27].

A safety audit in 1982 by a team from Union Carbide noted several safety problems at the plant, including several involved in the accident. There is debate about whether the information was fully shared with the Union Carbide India subsidiary and about who was responsible for making sure changes were made. However, no follow-up seems to have occurred to make sure that the problems had been corrected. The report noted such things as filter-cleaning operations without using slip blinds, leaking valves, the possibility of contaminating the tank with material from the vent gas scrubber, and bad pressure gauges. The audit recommended raising the capacity of the water curtain and pointed out that the alarm at the flare tower from where the MIC leaked was nonoperational, and thus any leakage could go unnoticed for a long time. According to the Bhopal manager, all the improvements called for in the report had been taken care of, but that was obviously untrue [27].

Ineffective risk control Many of the safety devices were inadequately designed. Ayres and Rohatgi describe several improvements that could have been incorporated into the system design, some of which are on the Bayer plant in Germany. Moreover, MIC was an intermediary that was convenient but not necessary to store. Since Bhopal, many companies have reduced their stocks of hazardous intermediates [191].

C.5 The Texas City Refinery Explosion

The Texas City Refinery explosion occurred on March 23, 2005, when a hydrocarbon vapor cloud was ignited and violently exploded at the isomerization (ISOM) process unit at BP's Texas City refinery in Texas City, Texas. Fifteen workers were killed, 180 others injured, and the refinery itself was severely damaged.

Texas City was the second largest oil refinery in the state and the third largest in the United States, with an input capacity of 437,000 barrels or about 10 million gal of gasoline per day. It also produces jet fuels, diesel fuels, and chemical feed stocks. Its 1,200-acre site contains twenty-nine oil refining units and four chemical units. The ISOM unit involved in the explosion was used to convert low-octane hydrocarbons, through various chemical processes, into hydrocarbons with higher octane ratings that could then be blended into unleaded gas.

BP's own accident investigation report stated that the direct cause of the explosion was heavier-than-air hydrocarbon vapors combusting after coming in contact with an ignition source, probably a running vehicle engine. Both the US Chemical Safety Board [383] and a panel headed by former secretary of state James Baker produced independent investigation reports [29]. All these reports plus other sources were used in writing this appendix.

In a press release issued on August 17, 2005, the company noted that the Texas City explosion was the worst tragedy in BP's recent history and that it would "do everything possible to ensure nothing like it happens again." After the March explosion, however, other safety incidents involving deaths and serious injuries occurred at the Texas City plant related to the same deficiencies. In addition, the causes of BP's better-known *Deepwater Horizon* loss in 2010 were similar to those at Texas City five years earlier. In fact, these same deficiencies had been identified after accidents in BP facilities even prior to the Texas City explosion. None of these potential learning events seem to have resulted in effective fixes.

After the Texas City explosion, BP was charged by the government with criminal violations of federal environmental laws and named in numerous suits from the victims' families. The Occupational Safety and Health Administration (OSHA) later gave BP a record fine, citing more than 700 safety violations and claiming that BP had failed to implement safety improvements following the accident.

C.5.1 Background

The refinery employed approximately 1,800 BP workers at the time of the incident. In addition, approximately 800 contractor workers were onsite supporting turnaround work. In refining, a *turnaround* is a major maintenance event where a process unit is taken out of service and opened up for cleaning, repair, and enhancement. Process units have turnarounds on a typical regular cycle of three to seven years. The final part of the turnaround activities, restarting the process unit, is considered to be particularly hazardous. BP had acquired the refinery as part of its merger with Amoco in 1999. The refinery had had five managers in the six years since BP inherited it.

At the time of the explosion, a turnaround was in progress that had started on February 21, 2005, to do work on the raffinate splitter. The raffinate splitter, which was a 170 ft (50 m) tall tower, is one component of the ISOM site. This raffinate splitter was used to separate out lighter hydrocarbon components from the top of the tower, which condensed and were then pumped to the light raffinate storage tank. The heavier components were recovered lower down in the splitter, then pumped to a heavy raffinate storage tank (figure C.1).

The hydrocarbons in the explosion came from liquid overflow from the blowdown stack during startup activities after the turnaround. Overfilling and overheating of the liquids in the blowdown stack led to hydrocarbon release following the operation of the raffinate splitter overpressure system during the startup procedure.

Most of the deaths in the accident occurred to workers in some trailers placed near the ISOM tower. Plans were made late in 2004 to accommodate contractors in nine single trailers and a double-wide trailer adjacent to the ISOM process unit. A management of

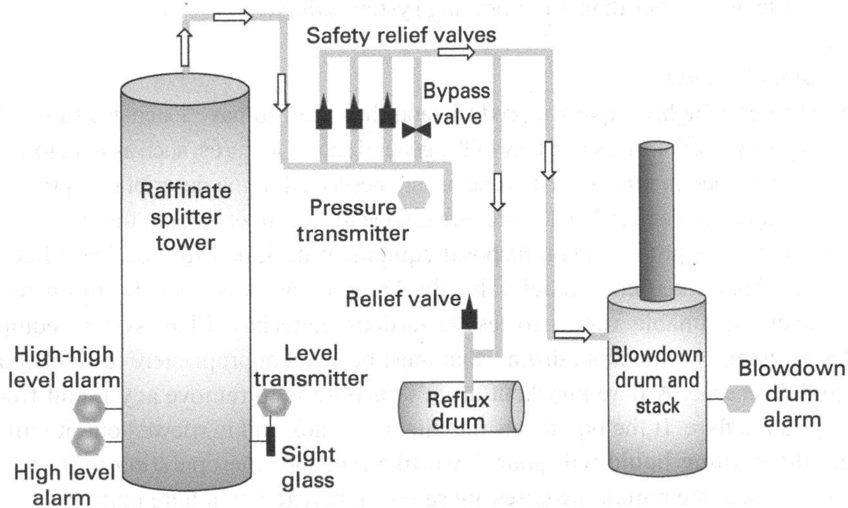

Figure C.1
The basic components involved in the accident. The raffinate splitter tower overfilled, releasing flammable hydrocarbons into the environment through the blowdown stack (from the CSB report [385]).

change (MoC) procedure was used to analyze the addition of the trailers, which found that the double-wide trailer would be less than 350 ft (110 m) from the ISOM plant and therefore had the potential to suffer severe damage in the event of an explosion. Although several action items were created during this assessment that were supposed to be closed before the MoC was approved and the trailers were used, contractors were allowed to use the double-wide trailer anyway starting in November 2004. The other nine trailers arrived on site at the start of 2005, but they were never included in the MoC procedures, so their exposure risk was never assessed.

The refinery was originally built in 1934 and had not been well maintained for several years. Three months before the explosion (January 2005) a consulting firm had examined the plant and found numerous safety issues, including "broken alarms, thinned pipe, chunks of concrete falling, bolts dropping, cigarettes falling 60 ft (18 m), and staff being overcome with fumes" [27] As a member of the Baker Panel that investigated the accident for the Chemical Safety Board (CSB), we heard repeatedly from witnesses that the workers, before the accident, had expressed fear to their families that something bad would happen because of the dangerous conditions allowed to persist in the plant [29].

The Baker Panel reported that BP did not adequately follow the Department of Energy's published safety recommendations. The panel report also suggested that cost cutting, production pressure from BP executives, and failure to invest in improvements led to a progressive deterioration of safety at the refinery [29].

Starting in 2002 (three years before the explosion), BP commissioned a series of audits and studies that revealed serious safety problems at the Texas City refinery, including a lack of necessary preventive maintenance and training. While some additional investments were made after the audits, the investments did not address the core problems at the refinery. In 2004, BP executives challenged their refineries to cut yet another 25 percent from their budgets for the following year. In addition, any safety improvements between 2002 and 2005 were largely focused on personal safety—such as slips, trips, falls, and vehicle accidents—rather than on improving system safety.

C.5.2 Safety Features

Refineries process highly explosive products and thus need to have extensive safety design features to prevent serious losses. Many of the general safety features, such as flare towers and relief valves included in process plant designs, are described in the description in this appendix of the Bhopal accident. A few features are unique to oil refineries or to this refinery.

A flare system is process plant disposal equipment designed to receive and combust waste gases from emergency relief valve discharge or process vent. In an oil refinery, flares convert flammable vapors to less hazardous materials. Flare system equipment includes a vessel, or "knockout drum," that must be sized appropriately to safely contain any liquid discharge. A flare knockout drum's purpose is to remove any liquid from the gas going to the flare. If the liquid gets into the flare stack and is spewed out, it will catch fire, and the burning liquid will pour down like rain into the operating unit. After the liquid is removed, the remaining gases are safely combusted by a flare burner.

Process unit startup is a significantly more hazardous period compared to normal oil refinery operations. BP's Texas City policies and procedures acknowledged that process unit startup is especially hazardous, and that these critical periods experience unexpected and unusual situations. This conclusion is widely accepted in the process industries. As such, BP's process safety guidelines recommended that "supplementary assistance" be provided during unit startups and shutdowns, such as experienced supervisors, operating specialists, or technically trained personnel. One basis for this policy was a previous explosion at Texas City during startup. In 1996, Amoco analyzed data from fifteen years of operations and concluded that incidents during startups were ten times more likely than during normal operation. Despite these guidelines and increased risks, BP did not have supplementary assistance personnel actively involved in the ISOM startup.

Because of the known hazards during startup of a process unit, BP used a checklist that had to be completed, reviewed, and signed off by management before the introduction of hydrocarbons into the raffinate splitter tower. The purpose of the review was to ensure that complete and thorough technical checks were carried out and that all nonessential personnel were clear during the startup operation. The steps in the safety review, as listed in the startup procedure, included completing maintenance work; performing required safety reviews; checking equipment; and ensuring that utilities, control valves, and other equipment were functioning and correctly aligned. Once completed, the startup safety review was to be signed off by refinery operations and safety managers, authorizing the startup work.

As in most process plants, defense in depth was the primary safety design practice. The plant had a variety of alarms to warn operators about dangerous conditions, and pressure relief systems as well as sensors to inform operators about the state of the contents of the various tanks and other equipment that could not be visually checked. Procedures were used to protect against dangerous practices, such as operating trucks around the ISOM where a spark might ignite a vapor cloud.

To protect the raffinate splitter tower from overpressure, three parallel safety relief valves were located in the overheard vapor line 148 ft (45 m) below the top of the tower. The outlet of the relief valves was piped to a disposal header collection system that discharged into a blowdown drum fitted with a vent stack.

The practice at Texas City of discharging hydrocarbon liquids to a sewer system was, in fact, unsafe; industry guidelines recommended against discharging flammable liquids that evaporate readily into a sewer.

A liquid level, normally water, was maintained in the bottom of the blowdown drum. The height of this level was controlled by a gooseneck seal leg piped to a closed drain. A level sight glass was available to monitor the water level and a high-level alarm was set to activate when the liquid level in the drum was close to flowing over the top of the gooseneck seal leg.

While liquid raffinate discharged out the top of the blowdown stack, it also flowed into the process sewer system and into a diversion box and oil/water separator. Both were fitted with high- and high-high level alarms.

C.5.3 Proximate Events

The explosion occurred on the morning of March 23, 2005, during the startup process for the raffinate splitter tower after the maintenance turnaround. During the turnaround, the raffinate splitter had been drained, purged, and steamed out to remove hydrocarbons. At the time of the incident, most of the scheduled maintenance tasks had been completed on the raffinate splitter section, but the Penex reactor, in a separate section of the unit, was waiting for the delivery of a gasket. When BP decided to start up the raffinate splitter section, three contractor crews were still working inside an area of the refinery containing batteries with their related utilities and services of the ISOM unit: one crew was waiting to install the gasket on the Penex reactor, the second was removing some asbestos, and the third was painting equipment inside the unit. Employees from all three crews were injured as a result of the explosion and fire.

During the startup, operations personnel pumped flammable liquid hydrocarbons into the tower for over 3 hours without any liquid being removed, which was contrary to startup procedure instructions. Critical alarms and control instrumentation provided false indications and did not alert the operators about the high level in the tower. Consequently, unknown to the operations crew, the 170 ft (52 m) tall tower was overfilled, and liquid overflowed into the overhead pipe at the top of the tower.

The overhead pipe ran down the side of the tower to pressure relief valves located 148 ft (45 m) below. As the pipe filled with liquid, the pressure at the bottom rose rapidly from about 21 lbs per square inch (psi) to about 64 psi. The three pressure relief valves opened for 6 minutes, discharging a large quantity of flammable liquid to a blowdown drum with a vent stack open to the atmosphere. The blowdown drum and stack overfilled with flammable liquid, which led to a geyser-like release out the 113 ft (34 m) tall stack. This blowdown system was an antiquated and unsafe design; it was originally installed in the 1950s and had never been connected to a flare system to safely contain liquids and combust flammable vapors released from the process.

The released volatile liquid evaporated as it fell to the ground and formed a flammable vapor cloud. The most likely source of ignition for the vapor cloud was backfire from an idling diesel pickup truck located about 25 ft (7.6 m) from the blowdown drum, contrary to safety rules. The fifteen employees killed in the explosion were contractors working in and around temporary trailers that had been previously sited by BP as close as 121 ft (37 m) from the blowdown drum.

This essential safety checklist procedure was not completed prior to the startup work. While the procedure had been applied to unit startups after turnarounds for two years prior to this accident, the process safety coordinator responsible for an area of the refinery that includes the ISOM was unfamiliar with its required use and therefore, it was not conducted. Higher-level management, such as the Texas City operations manager and process safety manager, were required to sign off on the process safety review checklists and authorize the startup. However, none of the safety review procedural steps were taken for the ISOM startup.

During pre-startup equipment checks, key splitter tower instrumentation and equipment were identified as malfunctioning but were not repaired. During the ISOM turnaround, operations personnel reported to the turnaround supervisors that the splitter level transmitter and level sight glass needed repair. The raffinate splitter tower level transmitter was reported by operations personnel as providing inaccurate readings and needing calibration. The tower sight glass was reported as dirty on the inside of the glass so that the tower level could not be visually determined. This equipment could not be repaired when the unit was operating, as the block valves needed to isolate the level transmitter and the sight glass from the process were leaking. A revised work order was issued that added replacement of the isolation block valves to the turnaround job list. The level transmitter was not repaired because BP supervisors determined that there was too little time to complete the job in the existing turnaround schedule. The supervisors planned to repair the level transmitter after the startup.

Both during unit shutdown and the equipment checks in the days preceding raffinate splitter startup, operations personnel found that a pressure control valve was inoperable. The malfunctioning control valve was reported to a frontline supervisor; however, no work order was written, and no repairs were made prior to startup. This same frontline supervisor signed off on the startup procedure that all control valves had been tested and were operational prior to startup.

A functionality check of all alarms and instruments was also required prior to startup, but these checks were not completed. On March 22, instrument technicians had begun checking the critical alarms when a supervisor told them that the unit was starting up and there was no time for additional checks. While some alarms were tested, most were not prior to startup. The supervisor initialed on the startup procedure that those checks had been completed.

Other key safety preparations listed in the startup procedures were omitted or ineffectively carried out. BP guidelines state that unit startup requires a thorough review of startup procedures by operators and supervisors; however, this review was not performed. The procedure also called for adequate staffing for the startup and that any unsafe conditions be corrected. Both of these steps were initialed as being completed by a unit trainee at the direction of his supervisor.

Serious issues with safety-critical equipment that had not been resolved prior to the startup included the inoperative pressure control valve, the defective high-level alarm in the splitter tower, and the defective sight glass used to indicate fluid levels at the base of the splitter. In addition, a critical splitter tower level transmitter had not been calibrated. All this defective equipment figured into the accident causal scenario.

The startup process began on March 23 with the night lead operator starting the filling of the splitter tower. The level transmitter was designed to indicate the raffinate level within a 5 ft (1.5 m) span from the bottom of the splitter tower to a 9 ft (2.7 m) level. It was common practice to fill up to an indicated level of 99 percent even though the procedural requirement was stated as 50 percent. Operations personnel explained that additional liquid level was needed in the tower because in past startups the level would typically drop significantly. To avoid losing the liquid contents of the tower and potentially damaging equipment, board operators typically operated the tower level well above 50 percent.

After the startup was begun, it was stopped, and the raffinate section was shut down to be restarted during the next shift. Starting, but then stopping, the unit was unusual and not covered in the startup procedures, which only addressed one continuous startup. The night lead operator did not use the startup procedure or record completed steps for the process of filling the raffinate section equipment, which left no record of the startup steps completed for the operators on the next shift.

The day supervisor arrived late for work and did not have a handover with the night shift. During the morning meeting, it was decided that the heavy raffinate splitter storage tanks were nearly full, and, therefore, the second day supervisor was told that the startup procedure should not continue, but this information was not passed on.

The startup procedure resumed just before 9:30 a.m. under instructions from the other day supervisor. Before recommencing the tower refill and circulation process, heavy raffinate was drained from the bottom of the tower via the level control valve into the heavy storage tank. The startup procedure required the level control valve to be placed in "automatic" and set at 50 percent to establish heavy raffinate flow to storage.

The day board operator, however, acting on what he believed were the unit's verbal startup instructions and his understanding of the need to maintain a higher level in the tower to protect downstream equipment, closed the level control valve and set it in "manual mode" (rather than the required "automatic mode") with a 50 percent flow rate to establish heavy raffinate flow to storage.

The day board operator said that, from his experience, when the splitter tower bottoms pumps were started and associated equipment filled, the tower level dropped. Operations personnel stated that if the level was maintained at only 50 percent, a drop in liquid level could result in losing heavy raffinate flow from the bottom of the tower, and that loss of flow from the tower bottom's pump to the furnace would shut down the furnace and the startup process.

Note that the day board operator had no written procedure to follow with completed steps initialed to indicate the exact stage of the startup. He observed a 97 percent level when he started circulation and thought that this level was normal; he said he did not recall observing a startup where the level was as low as 50 percent. At 10:10 a.m., 20,000 bpd of raffinate feed was being pumped into the tower and 4,100 bpd was erroneously indicated as leaving the tower through the level control valve. The day board operator said he was aware that the level control valve was shut. He believed he was instructed not to send heavy raffinate product to storage to protect downstream equipment, and that is why he closed the tower level control valve.

The circulation process was restarted just before 10 a.m., and raffinate was once again fed into the tower, even though the level was already too high. Because the level control valve was shut, there was no circulation out of the tower (i.e., no heavy raffinate being transferred to the storage tank), and the splitter tower began to fill up. The defective level transmitter showed the level at less than 100 percent and because the external sight glass was dirty, a visual check to verify the level in the splitter tower was not possible. The sight glass had been nonfunctional for several years.

Burners in the furnace were turned on to pre-heat raffinate going into the tower and to heat the raffinate in the tower bottom. There was still no flow of heavy raffinate from the splitter tower to the storage tank because the level control valve remained closed. The hydrocarbon liquid level in the splitter tower rose to 67 ft (20 m) although the defective level transmitter still indicated the liquid level was 8.65 ft (2.64 m) or 93 percent of the maximum safe level of 9 ft (2.7 m).

As the unit was being heated, the day supervisor, an experienced ISOM operator, left the plant at 10:47 a.m. due to a family emergency. The second day supervisor was devoting most of his attention to the final stages of the ARU (Amine Regeneration Unit) startup; he had very little ISOM experience and, therefore, did not get involved in the ISOM startup. No experienced supervisor or ISOM technical expert was assigned to the raffinate section startup after the day supervisor left, although BP's safety procedures required such oversight.

Just before midday, with heat increasing in the tower, the actual fluid level had risen to 98 ft (30 m). Pressure started to build in the system as the remaining nitrogen in the tower and associated pipework became compressed with the increasing volume of raffinate. The operations crew thought that the pressure rise was a result of overheating in the tower bottoms as this was a known startup issue, so the pressure was released.

By 12:42 p.m., the furnaces had been turned down, and the level control valve was finally opened, draining heavy raffinate from the splitter tower. The operators believed the defective level transmitter reading, which was now down to 7.9 ft (2.4 m) or 78 percent of the maximum safe level. In reality, the fluid level in the 179 ft (52 m) tall splitter tower had now reached 158 ft (48 m). The day board operator thought the level indication was accurate and believed it was normal to see the level drop as the tower heated up.

Although the raffinate flow into and out of the tower were now matching (as the heated raffinate was now leaving the bottom of the tower), heat from this outflow was being transferred via a heat exchanger back into the liquid flowing into the tower from the feed pipe, raising the average temperature inside the column close to the liquid's boiling point. The liquid, already close to the top of the tower and continuing to expand due to the heat, finally entered the overhead vapor line and flowed into the relief valve system.

Pressure built up in the system as fluid filled the pipework running to the safety relief valves and the condenser. At 1:13 p.m., the set pressure of the three pressure relief valves was exceeded, and they opened as the hydrostatic head pressure of the raffinate built to over 60 psi above atmospheric pressure. With the relief valves fully open, over 196,000 L (52,000 gal) of heated raffinate passed directly into the collection header over a 6-minute period before closing, as pressure in the system dropped to their closing or blowdown pressure of 37.3 psi above atmospheric pressure.

Investigating this pressure spike, the day board operator fully opened the level control valve to the heavy raffinate storage tank and shut off the gas fueling the furnace, but the raffinate feed into the splitter tower was not shut off. Hot raffinate flowed into the blowdown drum and stack, and as it filled, some of the fluid started to flow into the ISOM unit sewer system via a 6 in (15 cm) pipeline at the base of the blowdown drum. The high-level alarm for the blowdown drum didn't sound. Eventually the amount of material and pressure in the tower overhead line decreased, which caused the safety relief valves to close after an estimated 51,000 gal (196,500 L) of flammable liquid flowed from the valves into the collection header.

As the blowdown drum and stack filled up, hot raffinate shot out of the top of the stack and into the air, forming a 20 ft (6 m) geyser. The hot raffinate rained down on the ground, ran down the side of the blowdown drum and stack, and pooled at the base of the unit. A radio call was received in the control room that hot hydrocarbons were overflowing from the stack. In response, the board operator and lead operator used the automated control system to shut the flow of fuel to the heater (which they accomplished 5 seconds before the explosion), while the other operators left the satellite control room and ran toward an adjacent road to redirect traffic away from the blowdown drum, as required in BP's emergency procedures.

Hundreds of alarms registered in the automated control system at 1:20:04 p.m., including the high-level alarm on the blowdown drum. ISOM operations personnel stated they did not have sufficient time to assess the situation and sound the emergency alarm before the explosion.

A diesel pick-up truck, with its engine left idling, had been parked about 25 ft (8 m) from the blowdown stack, despite regulations against doing so. The vapor cloud reached the vehicle, and hydrocarbon fumes were drawn into the engine's air intake, causing the engine to race. Nearby workers frantically tried to shut down the engine, without success.

The expanding vapor cloud forced the workers who were trying to shut down the overspeeding truck engine to retreat.

The cloud continued to spread across the ISOM plant, across the pipe-rack to the west and into the trailer area unimpeded. No emergency alarm sounded, and at approximately 1:20 p.m., the vapor cloud was ignited by a backfire seen from the overheating truck engine by nearby witnesses, producing a massive explosion that was heard for miles. The blast pressure wave struck the contractor trailers, completely destroying or severely damaging many of them. The explosion sent debris flying, instantly killing fifteen people in and around the trailers and severely injuring 180 others. The pressure wave was so powerful it blew out windows off-site up to three-quarters of a mile (1.2 km) away. An area estimated to be 200,000 ft^2 (19,000 m^2) of the refinery was badly burned by the subsequent fire that followed the violent explosion, damaging millions of dollars of refinery equipment. The entire ISOM unit was severely damaged by the explosion and subsequent fire, and as a result, remained out of operation for more than two years.

C.5.4 Causal Factors

Because of the serious nature of this accident, several investigation reports were written, from which the following analysis was derived: several BP internal reports [372; 268], the US Chemical Safety Board Investigation Report [383], and the Baker Panel Report [29]. The latter was an independent group that provided an analysis of the nontechnical aspects of the accident. The causal factors identified in these reports include almost every systemic causal factor identified in chapter 4. The same factors had been identified in previous BP accident and audit reports and had never been fixed. They would play a part in the later BP Macondo/*Deepwater Horizon* loss.

C.5.4.1 Technical deficiencies Many technical deficiencies contributed to this accident. Examples of such design flaws include:

- the use of a blowout drum that was not large enough and used outdated technology
- inoperative alarms and level sensors in the ISOM process unit
- inadequate safety devices, such as automated shutdowns or safety interlocks that would be triggered by a high-level of liquid in the tower, instead depending on correct and timely action by operators and following procedures (which, in fact, were inadequate)
- a control board design that did not provide important information and automated control system that hindered the ability of the operations personnel to determine if the tower was overfilling
- critical alarms and control instrumentation that did not provide accurate information to the operators
- an emergency relief system design that did not address the potential for a large liquid release in the event the raffinate splitter tower was overfilled
- trailers placed too close to a process unit handling highly hazardous material

As examples of how these contributed to the events, the tower level indicator showed that the tower level was declining when it was actually filling, the redundant high-level alarm did not activate, and the tower was not equipped with any other level indications or automatic safety devices. In addition, the control board display did not provide adequate information about the imbalance of flows in and out of the tower to alert the operators to the dangerously high level.

The technical deficiencies alone, however, are not enough to understand why the events occurred. We need to understand why the technical deficiencies existed, for example: Were they not previously identified and fixed? Were the engineering processes used in the design of the plant not adequate? Did the design degrade over time or risk increase as changes occurred in the plant and in operations? Were these one-time events or had they occurred previously and nothing effective was done in response? Did operational practices contribute? And did the company culture, management decision-making, and organizational design establish the conditions such that the flawed design could lead to loss of life? The answer to all of these questions is "yes." In fact, as is commonly the case, the technical deficiencies could have easily been avoided or fixed but were not identified or were identified but not fixed due to cultural and management deficiencies.

C.5.4.2 Culture and decision-making The organizational culture can lead to poor decision-making and, as a result, accidents. BP and the Texas City plant reflected management disregard for process risks. Cultural problems at Texas City and in BP as a whole reflected complacency about serious risks and low priority placed on process safety versus short-term profits. Together they created a situation where the plant migrated to a state of higher risk over time. In addition, management did not welcome or listen to reports of safety problems and ignored warning signs of increasing risk. Perhaps not surprisingly, BP managers and white-collar workers generally had a more positive view of the process safety culture at their plants than blue-collar workers and maintenance technicians.

Ignoring hazards Numerous internal surveys, studies, and audits identified deep-seated safety problems at Texas City, but the response of BP managers at all levels was typically "too little, too late."

Some of the most important identified and ignored major hazards include: (1) the blowdown drum had not been replaced by a more modern and safer flare system despite previous serious incident reports and even industry policies that required it to be replaced; (2) occupied trailers were placed dangerously close to the running process despite never completing a risk assessment of the siting according to BP and industry standards; (3) nonessential personnel were not evacuated during the startup despite knowing that a startup was one of the most dangerous activities at the plant; (4) the startup was authorized despite having inadequate staff, malfunctioning instruments and equipment, and without completing a company required pre-startup safety review (PSSR); and (5) during the startup, no qualified supervision was present, and procedures were not followed as had been the practice for some time. In summary, BP did not effectively assess and control the risks of major hazards at Texas City or even follow their own specified procedures.

Previous company audits and assessments had concluded that Texas City had serious deficiencies in identifying and controlling major risks. For example, a company safety audit in 2003 found that "the practice of risk identification and the development of mitigation plans are not evident across the site." A similar 2004 safety assessment concluded that "no formal system exists for identifying high level risks" [372]. The 2005 Telos assessment found that "critical events, (breaches, failures, or breakdowns of a critical control measure) are generally not attended to" [372]. Previous events resulting in fatalities and property damage did not lead to any attempts to fix the factors identified as leading to those events. Identified process safety action items were allowed to increase without closure as well as equipment integrity problems without appropriate action. None of these things were perceived as requiring serious intervention.

In addition, Texas City management did not perform a proper hazard analysis, such as identifying all credible overpressure scenarios, fully documenting overpressure events, or calculating relief flow rates for all the potential overpressure scenarios of the raffinate splitter. Without this information, Texas City management could not ensure that the blowdown disposal system was safely designed or operated safely. When hazard analysis was performed, inappropriate techniques were used such as "What If" or "5 Why's." A trailer siting checklist was never completed.

BP did perform a major accident risk (MAR) assessment, but the results did not provide a warning of the existing high risk of an accident. A MAR assessment theoretically addresses the risk of major accidents with the potential to cause harm to people or the environment and to have a serious adverse impact on BP's reputation. The methodology requires the identification of a representative range of hypothetical events that could lead to a major accident, quantification of the hypothetical likelihood of these events, quantification of the possible physical effects resulting from these events and an assessment of their consequences, and an evaluation of options to mitigate the likelihood and/or consequences of the considered events.

However, MAR assessments are not intended to provide a comprehensive evaluation of hazards at a site, and the use of these assessments has numerous limitations. It is not a replacement for a proper hazard analysis. For example, MAR assessments use event frequencies that are based on industry experience and that reflect average design and operation. For this reason, the MAR assessment process does not reflect instances in which unit conditions are better or worse than the industry average. In this case, conditions at Texas City appeared to be worse than the industry average and even other BP refineries, and, therefore, the MAR assessments performed were overly optimistic. In addition, the MAR assessment process does not account for operation of a facility outside reasonably anticipated parameters, but the Texas City refinery was operated outside both industry practices and BP written standards.

Confusing process safety with personal safety BP's approach to safety largely focused on personal safety rather than on addressing major hazards. Process safety at Texas City and in BP as a whole was not emphasized. Instead, focus, measurement, and rewards were placed on reductions in injury rates and days away from work. The low personal injury rate at Texas City was relied on as an indicator of overall safety at the plant and therefore provided a false sense of confidence about process and plant safety. The behavior-based personal safety program at Texas City did not typically examine safety systems, management activities, or any process safety–related activities. Industry benchmarking was used that did not include any specific process safety metrics.

BP mistakenly interpreted improving personal injury rates as an indication of acceptable process safety performance at its US refineries. Reliance on data such as personal injury rates, combined with an inadequate process safety understanding, created a false sense of confidence that BP was properly addressing process safety risks. Process safety performance was deteriorating, but the use of personal injury rates kept BP from detecting and intervening early enough to prevent a major accident. Something similar occurred in the later BP *Deepwater Horizon* loss. BP should have known that injury data was not effective in assessing process risk because it was cited as a cause in the report on three serious incidents that occurred in BP's petrochemical complex in Grangemouth, Scotland, in 2000. The lessons from this prior accident report were not learned or acted on.

Safety campaigns, goals, and rewards were focused on improving personal safety metrics and worker behaviors rather than on process safety and management safety systems. Compliance with many safety policies and procedures was deficient at all levels of the refinery and the company, but the company focused on individual worker's compliance with safety rules.

It is not that management was not warned about the deficiencies in their safety approach. A safety culture assessment of Texas City was performed by consultants before the accident. The assessment concluded that workers perceived the managers as "too worried about seat belts" and too little about the danger of catastrophic accidents. "*[Individual safety] was more closely managed because it 'counted' for or against managers on their current watch (along with budgets) and that it was more acceptable to avoid costs related to [equipment] integrity management because the consequences might occur later, on someone else's watch*" [372].

The consultants also noted that concern about equipment conditions was not only expressed by BP workers but was strongly expressed by senior members of the contracting companies, who pointed out many specific hazards in the work environment that would not be found at other area plants. The consultants concluded that the tolerance of these kinds of risks contributes to the tolerance of risks they were seeing in individual behavior.

Internal and external process safety audits, the occurrence of major accidents at the plant including three fatalities in 2004, and mechanical integrity reviews led to management focusing even more on personal safety and worker compliance with safety rules. Physical and management deficiencies were not corrected. During the visit of a high-level BP executive to the refinery in July 2004, the site reported a best-ever recordable injury rate. There was no mention of the deteriorating Texas City process safety performance [29].

The results of another safety report in March 2005, right before the explosion, concluded that the site had gotten off to a good start in 2005 with safety performance that "may be the best ever," adding that Texas City had had "the best profitability ever in its history last year" with over $1 billion in profit—more than any other refinery in the BP system" [29].

The lack of focus on process safety can be traced to BP compensation policies. Personnel and management compensation at BP put very little weight on safety, and what was included was primarily personal safety (e.g., days away from work, recordable injuries, and vehicle accidents). Process safety metrics were not included.

Despite the stark findings of safety failures reflected in site performance and assessment reports, BP management persisted in believing that the personal injury rates reflected total safety performance. Focus on injury rates at all levels of BP management helped mask severe shortcomings in process safety.

Ignoring warning signs In 1995, five workers were killed in a Pennzoil refinery when two storage tanks exploded, engulfing a trailer. The conclusion was that trailers should not be located near hazardous materials. BP ignored the warnings; they believed that because the trailer where most of the deaths happened was empty most of the year, the risk was low.

The problem was not simply ignoring what was happening in other companies. In the years prior to the incident, eight serious releases of flammable material from the ISOM blowdown stack had occurred, and most ISOM startups experienced high liquid levels in the splitter tower. Neither Amoco nor BP investigated these events.

In 2004, the year before the Texas City explosion, two of the three major accidents were process safety related. In one of these, which occurred in September 2004, three workers in a Texas City unit were burned, two fatally, with hot water and steam during the opening of a pipe flange. The workers believed the pump system was isolated and were unaware that a check valve was concealing hazardous energy from a leaking discharge block valve. The absence of a bleed valve between the check and block valve did not allow the workers to verify whether the piping was isolated, depressurized, and safe to open.

Two 2004 process safety–related accidents involved fatalities: a major hydrogen fire on July 28, 2005, and another serious accident on August 10, 2005. In March 2004, a furnace outlet pipe ruptured and resulted in a fire that caused $30 million in damage. It was found to be the result of a piping problem that had not been repaired.

Taken as a whole, the accidents revealed a serious decline in process safety and management system performance at the BP Texas City refinery, but almost nothing was done to improve process safety at Texas City in response.

While it is possible to dismiss a lack of alarm from past accidents and incidents as simply the result of hindsight bias, audits, reports, and direct warnings that are ignored do not suffer from the same potential bias. BP management had many direct warnings. Many of the process safety deficiencies causal to the ISOM accident had previously been identified by BP and outside auditors but were never fixed.

In a 2001 presentation, BP Texas City managers stated that the site required significant improvement in performance, or a worker would be killed in the next three to four years. The presentation asserted, however, that unsafe acts were the cause of 90 percent of the injuries at the refinery and called for increased worker participation in the behavioral safety program.

Beginning in 2002, BP Group and Texas City managers received numerous warning signals about a possible major catastrophe at Texas City. In particular, managers received warnings about serious deficiencies regarding the mechanical integrity of aging equipment, process safety, and the negative safety impacts of budget cuts and production pressures.

A 2002 study included serious concerns about the potential for a major site incident at Texas City. The study observed that the Texas City refinery infrastructure and equipment were "in complete decline" and that the refinery needed to focus on improving operational basics such as reliability, integrity, and maintenance management. The leadership culture at Texas City was described in the study as "can do" accompanied by a "can't finish" approach to making needed changes. The study recommended a major overhaul of the basics, including addressing issues such as vulnerability and cultural change. Recommendations included increases in maintenance spending, improving the competency of operators and supervisors, and establishing a "reliability culture." The latter probably came from the fact that BP management was very influenced by the HRO (High Reliability Organization) literature and followed all their teachings.

This study found high levels of overtime and absenteeism resulting from BP's reduced staffing levels and called for applying Management of Change safety reviews to people and organizational changes. The study concluded that personal safety performance at Texas City refinery was excellent, but there were deficiencies with process safety elements such as mechanical integrity, training, leadership, and MoC.

The serious safety problems found in the 2002 study were not adequately corrected, and many played a role in the 2005 disaster.

A 2003 audit identified many of the same deficiencies. The team summarized its conclusions as: "The incentives used in this workplace may encourage hiding mistakes," and "We work under pressures that lead us to miss or ignore early indicators of potential problems" [29].

In August 2004, the Texas City Process Safety Manager gave a presentation to the plant managers that identified serious problems with process safety performance. The presentation showed that Texas City 2004 year-to-date accounted for $136 million, or over 90 percent, of the total BP refining process safety losses, and over five years, accounted for 45 percent of total process safety refining losses. The presentation noted that process safety management was easy to ignore because although the incidents were high consequence, they were infrequent.

The 2005 Texas City Health, Safety, and Security (HSSE) Business Plan warned that the refinery likely would "kill someone in the next 12–18 months." This fear of a fatality was also expressed in the 2004 Texas City Culture Assessment, which concluded "There is an exceptional degree of fear of catastrophic incidents at Texas City." The same fear was expressed in early 2005 by the HSSE manager: "I truly believe that we are on the verge of something bigger happening," referring to a catastrophic incident [29].

Another key safety concern in the 2005 HSSE Business Plan was that the site was not reporting all incidents due to fear of consequences. The plan's 2005 PSM (Process Safety Management) key concerns included mechanical integrity, inspection of equipment including safety critical instruments, and competency levels for operators and supervisors. Deficiencies in all these areas contributed to the ISOM incident.

As a final example, BP and the UK Health and Safety Executive (HSE) concluded from their investigations of the accidents at BP's Grangemouth facility that preventing major accidents requires a specific focus on process safety. BP group leaders communicated the lessons to the business units but did not ensure that needed changes were made.

Tolerating serious deviations from safe operating procedures BP management had long tolerated deviations from safe operating procedures. In the ISOM explosion, the startup was initiated with inadequate staffing, with malfunctioning equipment, and without review, without their own required checklist, without following required procedures, and without evaluating whether nonessential personnel should be in the area. All of this was known to management, who signed off on it.

Cost cutting at the expense of safety Cost cutting and production pressure from BP executives resulted in a lack of needed safety measures and progressive deterioration of safety at the refinery. Each year, budgets were cut. After the merger with Amoco, there was a 25 percent budget cut target. One refinery business unit leader considered the 25 percent reduction to be unsafe because it came on top of years of budget cuts in the 1990s;

he refused to fully implement the target. BP's cost reduction strategy was to "aggressively drive costs out of the system at an accelerated pace relative to other refiners" [29]. The strategy indicated that the cuts would be accomplished through cutting staff, outsourcing, and eliminating unnecessary turnaround costs. Items cut included turnarounds; safety committee meetings; the central training organization; fire drills; maintenance, engineering, supervision, and inspection staff; plant maintenance; and training courses. Safety and maintenance expenditures were a significant portion of the cuts. The refinery's capital expenditures to maintain safe plant operation and to comply with HSE legal requirements were cut $33 million, or 45 percent, from 1998 to 1999.

During 2003, the plant was continually pressured to reduce expenditures. In 2004, BP executives challenged their refineries to cut yet another 25 percent from their budgets for the following year. In 2005, BP executives directed Texas City to cut capital expenditures by an additional 25 percent despite three major accidents and fatalities at the refinery in 2004.

A 2004 culture assessment included statements like: "Production pressure impact managers where it appears as though they must compromise safety"; "Production and budget compliance gets recognized and rewarded before anything else at Texas City"; and "The pressure for production, time pressure, and understaffing are the major cause of accidents at Texas City" [29].

Audits and studies before the explosion revealed serious safety problems at Texas City, including a lack of preventative maintenance and training, both of which contributed to the loss. While some steps were taken to address the findings, the identified core problems were never fixed, and continual budget cutting by BP management made effective efforts impossible.

Not Investing in safety, safety equipment, and maintenance At Texas City, there was too much focus on short-term cost reduction and not enough on longer-term issues like plant reliability and investment for the future. The Baker Panel report concluded that safety was unofficially sacrificed to cost reductions, and cost pressures inhibited staff from asking the right questions; eventually staff stopped asking [29].

A lot of the refinery infrastructure and process equipment at Texas City was in poor condition. This was not simply a failing of BP but also AMOCO, who owned the refinery before BP. Recommendations for replacing equipment by newer and safer alternatives were rejected due to cost pressures. Instead of preventative maintenance, a policy of run-to-failure for process equipment existed at Texas City.

Neither Amoco nor BP replaced blowdown drums and atmospheric stacks, even though a series of incidents warned that this equipment was unsafe. In 1992, OSHA cited a similar blowdown drum and stack as unsafe, but the citation was withdrawn as part of a settlement agreement, and therefore the drum was not connected to a flare as recommended. Amoco, and later BP, had safety standards requiring that blowdown stacks be replaced with equipment such as a flare when major modifications were made. In 1997, a major modification replaced the ISOM blowdown drum and stack with similar equipment, but Amoco did not connect it to a flare. In 2002, BP engineers proposed connecting the ISOM blowdown system to a flare, but a less expensive option was chosen.

BP continued the use of outdated technology, the outdated blowdown drum and stack technology, when replacement with the safer flare option had been a feasible alternative for years. At the same time, BP put off spending on preventative maintenance of safety-critical equipment. During the fifteen years prior to the March 2005 incident, a number of proposals were made to remove blowdown stacks that vent directly to the atmosphere at the Texas City refinery, but none were implemented, primarily due to cost considerations.

The 2004 culture assessment suggested that the poor equipment conditions were made worse in the view of many people by a lack of resources for inspection, auditing, training, and staffing for anything besides "normal operating conditions."

Allowing migration to states of higher risk Rasmussen has suggested that it is common for organizations to migrate to states of higher risk, resulting from pressures for greater productivity and profits in competitive industries [320]. Texas City is a perfect example of this phenomenon.

Reliance was placed on redundancy, but the actual redundancy deteriorated over time. For example, during the emergency, redundant alarms did not sound because they were not functional or not properly calibrated. The same was true for other safety-critical equipment. None of the instruments that indicated raffinate splitter tower water or pressure level were working properly on March 23. BP instrument technicians described the unit instrumentation as being run-down and in disrepair.

The technology used in the plant was outdated but not upgraded due to cost considerations. Maintenance was inadequate, including preventative maintenance on safety-critical systems.

Maintenance was a particular problem, including problems in documentation and inadequate or missing written procedures for testing and maintaining the instruments in the ISOM unit. Raffinate splitter instruments were not included in the list of critical instruments, which led to ineffective maintenance.

Automated maintenance management software was used, but it did not provide an effective feedback mechanism for maintenance technicians to effectively report and track details on repairs performed, future work required, or observations of equipment conditions. The maintenance management software did not include trending reports that would alert maintenance planners to troublesome instruments or equipment that required frequent repair, such as the high-level alarms on the raffinate splitter and blowdown drum.

The Texas City SAP work order process did not include verification that work had been completed. According to interviews, BP maintenance personnel were authorized to close a job order even if the work had not been completed.

During the startup resulting in the 2005 explosion, several instruments in the ISOM raffinate splitter section failed, likely due to inadequate maintenance and testing, contributing to the incident. The mechanical integrity program did not incorporate the necessary training, tools, and oversight. The equipment data sheets were not kept up to date so that incorrect data, such as a wrong specific gravity value for the process hydrocarbons, resulted in miscalibration of critical instruments. Instruments with a history of problems, such as the tower high-level alarm, were not tracked to ensure proper corrective action and avoid breakdown maintenance. Site practices did not ensure that instruments were tested and repaired before unit startup. Appropriate methods and procedures were not used for testing instrument functionality as with the blowdown drum high-level alarm.

Mechanical integrity deficiencies resulted in the raffinate splitter tower being started up without a properly calibrated tower-level transmitter, functioning tower high-level alarm, level-sight glass, manual vent valve, and high-level alarm on the blowdown drum.

BP management were aware or should have been aware of the maintenance deficiencies. A 2003 Maintenance Gap Assessment found that most areas had insufficient resources to conduct the required root cause failure analysis for equipment problems, and that most action items were "not implemented because of budget constraints" [29]. Two major accidents in 2004 and 2005 occurred in part because needed maintenance was identified, but not performed during turnarounds. The 2004 safety audit report noted that while Texas City's mechanical availability worsened from 2003 to 2004, loss of containment accident increased 52 percent, from 399 to 607 per year.

The 2003 audit concluded that the condition of the infrastructure and assets was poor. Management was told that budgets were not large enough to address identified risks but responded that only the money on hand would be spent, rather than increasing the budget. The audit found that "HSE actions items are allowed to become past due and remain in that status without intervention" [29].

The poor condition of the refinery, inadequate training, inadequate process safety management, and the need for leadership to set an example and eliminate exceptions to health and safety standards were all findings of the audit and causally related to the ISOM incident.

One key finding of the 2003 audit was that "we have created an environment where people justify putting off repairs to the future" [29]. The BP investigation team also found an "intimidation to meet schedule and budget" when the discovery of an unsafe pipe conflicted with the schedule to start up a unit.

A BP executive recommended a follow-up inquiry about why the Texas City refinery performance had deteriorated. Analysis concluded that "the current integrity and reliability issues at TCR [Texas City Refinery] are clearly linked to the reduction in maintenance spending over the last decade" [29].

More generally, the working environment at Texas City had eroded over time to one that was characterized by resistance to change and by lack of trust, motivation, and a sense of purpose. This environment, coupled with unclear expectations about supervisory and management behaviors, led to rules not being followed consistently, a lack of rigor, and individuals feeling disempowered from suggesting or initiating improvements.

Management who did not want to hear about safety problems The cultural assessments at Texas City all agreed that personnel were not encouraged to report safety problems, and some feared retaliation for doing so. For example, the 2003 audit concluded that "bad news is not encouraged" [29]. Employees reported that management did not encourage reporting of complaints about safety and sometimes were actively discouraged from reporting safety problems upward. The Baker Panel heard from the spouses of those killed that they were afraid to go to work but did not feel free to complain about unsafe working conditions [29].

C.5.4.3 Organization and management of operations Not only were there general cultural and decision-making problems at Texas City, but the organization and management structures were poorly designed. the CSB, in their report on the Texas City explosion, cited serious concerns about: the effectiveness of the safety management system at the refinery, the effectiveness of BP's corporate safety oversight of its refining facilities, and a corporate safety culture that tolerated serious and longstanding deviations from good safety practice [385].

Diffusion of responsibility and authority, lack of corporate oversight and leadership The accident reports found a lack of corporate oversight of both safety culture and major accident prevention programs. Management responsibilities for process safety at all levels above the refinery level were unclear and mostly unspecified. There was a belief that safety decisions should be made at the lowest possible level with little or no responsibilities assigned to the higher management levels. Management did not set or consistently enforce process safety priorities. They did not articulate a clear message on the importance of process safety and match that message both with policies and actions.

The Baker Panel found that BP tended to have a short-term focus [29]. Its decentralized management system and entrepreneurial culture delegated substantial discretion to US refinery plant managers without clearly defining process safety expectations, responsibilities, or accountabilities. In addition, while accountability is a core concept in BP's management framework for driving desired conduct, BP did not effectively hold executive management and refining line managers and supervisors, both at the corporate level and at the refinery level, accountable for process safety performance at its five US refineries. As just one example, the closure rate for safety-related action items had fallen after this metric was removed from the 2003 formula for calculating bonuses.

One problem noted in the accident reports is that BP did not have a designated, high-ranking leader for process safety in its refining business. In fact, responsibility, authority, and accountability for process safety was missing at the top and mid-management levels of the company. As one example, a common responsibility for a board of directors is to provide oversight of safety, but the BP board of directors did not have a member responsible for assessing and verifying the performance of BP's major accident hazard prevention programs. What oversight management provided focused on personal safety and not process safety.

BP's emphasis on improving personal injury rates created a false sense of confidence that process safety was being addressed. Accidents before the 2005 explosion did not seem to provide an alarm that process safety was not being handled, nor did several audits and process safety assessments at Texas City. For example, a BP internal audit group in London reviewed internal audit reports in 2003 and 2004 and found a number of serious safety deficiencies common throughout the organization, including:

- widespread tolerance of noncompliance with basic HSE rules
- poor implementation of HSE management systems, reducing the effectiveness and efficiency of activities to manage HSE risks and deliver sustainable performance
- lack of leadership competence and understanding to effectively manage all aspects of HSE
- insufficient monitoring of key HSE processes to provide management visibility and confidence in management's ability to deliver and implement necessary interventions

Poor process safety management performance was noted in mechanical integrity, training, process safety information, and management of change.

Lack of leadership and accountability were the most common problems. The report stated that there was a lack of leadership focus on closing action items from audits and other safety reviews, as well as a backlog of maintenance items. The report also found that leadership was not "reinforcing expectations through their own behaviors" [29]. Risk assessment was "often incomplete," business units did not understand or address major hazards, and competency in risk and hazard assessment was poor.

Following the Grangemouth incidents, BP mobilized a task force to undertake a review of all operating units and functions at that refinery. One of the task force's findings was that BP was too focused on short-term cost reduction and not focused enough on longer-term investment for the future. The task force also found that health and safety was unofficially sacrificed to cost reductions. The Baker Panel found that this problem also applied to the US refineries. As noted earlier, BP's response to audit reports warning of serious process safety deficiencies was to cut budgets and focus on short-term profits [29].

The Baker Panel report noted "striking" similarities between the lessons of Grangemouth and the events of the Texas City explosion, including the lack of management leadership, accountability, and resources; poor understanding of and a lack of focus on process safety, coupled with inadequate performance measurement indicators; and untimely completion of corrective actions from audits, peer reviews, and past incident investigations [29]. Effective responses to these problems were not taken, and many of the same problems occurred in the later *Deepwater Horizon* explosion and oil spill. In fact, the CSB Texas City accident report found that a number of managers, including executive leadership, had little awareness or understanding of the lessons from the previous Grangemouth accident report [385].

Amoco managed safety under the direction of a senior vice president. It had a large corporate HSE organization that included a process safety group that reported to a senior vice president managing the oil sector. In the wake of the merger, the Amoco centralized safety structure was dismantled. Many HSE functions were decentralized and responsibility for them delegated to the business segments. Amoco engineering specifications were no longer issued or updated. Voluntary groups, such as the Process Safety Committees of Practice, replaced the formal corporate organization. Process safety functions were largely decentralized and split into different parts of the corporation, resulting in a loss of focus on process safety. These changes to the safety organization created cost savings but led to a diminished process safety management function that no longer reported to senior refinery executive leadership. The Baker Panel concluded that BP's organizational framework produced "a number of weak process safety voices" that were unable to influence strategic decision-making in BP's US refineries, including Texas City [29].

The company's system for assuring process safety performance used a bottom-up reporting system that originated with each business unit, such as a refinery. As information was reported up, however, data was aggregated. By the time information was formally reported at the refining and marketing level, refinery-specific performance data was no longer presented separately. And, as noted, the data collected focused on personal safety, not process safety. As a result, the Baker Panel concluded that a substantial gulf existed between the actual performance of BP's process safety management systems and the company's perception of that performance [29].

In general, significant deficiencies existed in BP corporate systems for measuring process safety performance, investigating incidents and near misses, auditing system performance, addressing previously identified process safety–related action items, and ensuring sufficient management and board of director's oversight. Most of these deficiencies were not new but were identified in previous BP accident reports and in numerous audit reports before the Texas City explosion. Effective action was not taken in response.

Inadequate management of change Accidents often occur after some change, either in the system itself and its procedures and policies, the behavior of the operators, or a change in the environment. As a result, most organizations have MoC procedures specified. These procedures generally assess the risk associated with the change and specify actions to eliminate or mitigate any risks identified. The refinery required that an MoC be initiated for any change that involved process chemicals, process or equipment technology, equipment piping, process control and instrumentation, operating procedures, safe operating limits, relief and safety systems, personnel and staffing, organization, outsourcing, and occupied buildings.

While BP had an MoC policy, it was mostly not followed nor completed before safety-critical operations. The placement of the trailers is an example. A management of change procedure was used to analyze the addition of one of the trailers, which found that the placement near the ISOM plant had the potential to result in severe damage in the event of an explosion. Several action items were created during this assessment, but they were not closed before the MoC was approved and the trailer used. The other nine trailers were never included in the MoC, so their risk was never assessed.

More generally, changes made to processes, equipment, procedures, buildings, and personnel at the refinery were not systematically reviewed to ensure that an adequate margin of safety was maintained.

The 2004 safety audit report at BP headquarters in London found that management of change was poorly defined. In particular, there was a lack of clarity on what constitutes a change in areas such as temporary modifications and organizational change. In addition, the report concluded that MoC systems lacked staff competency, rigorous application, and monitoring. These problems were not rectified before the Texas City explosion.

Safety-critical operations with inoperable protection equipment and personnel in potentially dangerous areas During the startup, there was a lack of supervisory oversight and technically trained personnel. This omission was contrary to BP safety guidelines. Some examples are:

- An extra board operator was not assigned to assist, despite a staffing assessment that recommended an additional board operator for all ISOM startups.
- The process unit was started despite previously reported malfunctions of the raffinate splitter tower's level indicator, the level sight glass, and a pressure control valve, all of which were required for safe operation.
- A relief valve system safety study had not been completed before the startup. In fact, the size of the blowdown drum was insufficient to contain the liquid set to it by the pressure relief valves.
- None of the instruments that indicated raffinate splitter tower water or pressure level were working properly on March 23. They were either not functional or not properly calibrated.
- The pre-startup safety review policy to ensure that nonessential personnel were removed from the areas in and around process units during startups was not performed.

BP instrument technicians described the unit instrumentation as being run-down and in disrepair. Instruments were not calibrated, there were worn and misaligned components, the sight-level glass was not cleaned, and there was a damaged level displacer ("float").

Inadequate training of operators and oversight of critical operations BP had cut the budget for training and reduced staffing: the central training department staff had been cut from twenty-eight to eight, and simulators were unavailable for operators to practice handling abnormal situations, including infrequent and high-hazard operations such as startups and unit upsets.

The reports on the accident noted that the operator training was inadequate, particularly for startup hazardous and abnormal operations. The hazards of unit startup, including tower overfill scenarios, were not adequately covered in operator training. BP admitted after the accident that they had not informed employees of known fire and explosion risks [374].

The operators were provided with outdated and ineffective work procedures that did not address recurring problems that had previously occurred during startup. The operators thought that the procedures could be altered or did not have to be followed during the startup process. In the previous five years, most of the nineteen startups had deviated from written procedures.

BP's Texas City policies and procedures acknowledged that process unit startup is especially hazardous, and as such, BP's process safety guidelines recommended that "supplementary assistance" be provided, such as experienced supervisors, operating specialists, or technically trained personnel during unit startups and shutdowns. These were not provided.

The 2004 culture assessment noted: "The quantity and quality of training at Texas City is inadequate . . . compromising other protection-critical competence" [29].

Inadequate training of managers and supervision of critical operations Not only were the workers poorly trained, but so were the managers as evidenced by the surprisingly poor practices demonstrated during the startup, which was commenced without completing the safety checklists and fixing reported malfunctioning and failed safety-critical equipment. Management did not want to change the schedule and so decided to delay the repair of the level transmitter (a key component in the startup and the accident) until after the startup. Other critical alarms and control instrumentation had been reported as malfunctioning before the startup but were not repaired before allowing the startup to begin.

The frontline supervisor signed off on the startup procedure that all control valves had been tested and were operational prior to startup when this statement was not true. Required checks were not completed although supervisors initialed on the startup checklist that they had been. Specified startup procedures were not followed. Other startup checklists were required to be signed to authorize, but this never occurred.

The lack of supervisory oversight and technically trained personnel during the startup, an especially hazardous period, was an omission contrary to BP safety guidelines. An extra board operator was not assigned to assist, despite a staffing assessment that recommended an additional board operator for all ISOM startups. Startup began before all the employees were safely removed from the site, as they were required to be. A supervisor had a family emergency during the startup and had to leave the plant, but nobody was assigned to replace him.

During the startup, the size of the blowdown drum was insufficient to contain the liquid sent to it by the pressure relief valves. The blowdown drum overfilled, and the stack vented flammable liquid to the atmosphere, which fell to the ground and formed the vapor cloud that ignited. A relief valve system safety study had not been completed.

The problems were not just local to Texas City managers but permeated all levels of the company. BP did not effectively define the level of process safety knowledge or competency required of executive management, line management above the refinery level, and refinery managers. More specifically, the company did not ensure that its US refinery personnel and contractors had sufficient process safety knowledge and competence. The Baker Panel concluded that process safety education and training needs to be more rigorous, comprehensive, and integrated. One problem noted was that most of BP's US refineries over-relied on computer-based training, which contributed to inadequate process safety training of refinery employees and management [29].

Outdated work procedures and lack of operating discipline Work procedures were commonly violated, but that was required to get work done. An example was that the idling vehicle that likely provided the ignition source for the vapor cloud was in a location where it was not allowed. Most of those injured or killed from the resulting explosion did not need to be in the area surrounding the ISOM during the hazardous period of unit startup.

In fact, the work environment encouraged operations personnel to deviate from procedures. Such deviations were not unique actions by an incompetent crew but rather were established work practices. They were needed to protect the unit equipment and to complete the startup in a timely and efficient manner.

The work procedures that were provided were outdated (which BP admitted after the accident) [374] and did not address recurring operational problems during startup, leading operators to believe that procedures could be altered or did not have to be following during the startup process. In the previous five years, most of the nineteen startups had deviated from written procedures. Nothing was done in response to these deviations.

For example, operators biased the tower level on the high side, likely to avoid the possibility of losing the level and damaging the furnace. This bias of running the tower on the high side occurred in nearly all of the last nineteen startups. The operators' actions show that they were aware of the tower's low-level risks but not the high-level risks. Without awareness of the risks of a high level, the operators often conducted startup without knowing the actual amount of material in the tower. Once the feed rises above the span of the transmitter, operators have no ready means to determine how much liquid is in the tower, making overfilling the tower much more likely.

Pressure excursions outside the procedure's alarm set-point also occurred in fourteen of the nineteen startups. Since the beginning of 2003, two startups prior to the March 23 incident had pressures above the relief valve set-points, which likely lifted the relief valves and discharged hydrocarbon vapor to the blowdown drum and stack. Neither of these two startup incidents was investigated.

Finally, the procedures lacked sufficient instructions to the operator to safely and successfully start up the unit. They did not explain the safety implications of deviations, did not include instructions on halting unit operations in the middle of the startup process, as was done, nor did they provide instructions for recommencing startup several hours later by a different operating crew. The hazards and safety implications of a partial startup/shutdown/re-startup were not addressed and specific steps to initiate such activities were not provided. Because the startup procedures did not account for this situation, the operators did not use them and left no record of the changes for the next shift.

In addition, management did not ensure that the startup procedure was regularly updated, even though the startup process had evolved and changed over time with modifications to the unit's equipment, design, and purpose. When procedures are not updated or do not reflect actual practice, operators and supervisors learn not to rely on procedures for accurate instructions. Other major accident investigations reveal that workers frequently develop work practices to adjust to real conditions not addressed in the formal procedures. If there have been so many process changes since the written procedures were last updated that they are no longer correct, workers will create their own unofficial procedures that may not adequately address safety issues.

The startup procedure that was provided did not address the critical events the unit experienced during previous startups, such as dramatic swings in tower liquid level, which could severely damage equipment and delay startup. Specific instructions were not included, such as the unusual stopping and resumption of the ISOM startup (which occurred in this case) or the routing of products to different storage tanks. Tower pressure alarm set-points were frequently exceeded, yet the procedure did not address all the reasons this might happen and the steps operators should take in response.

BP management allowed operators and supervisors to alter, edit, add, and remove procedural steps without conducting MoCs to assess risk impact due to these changes. They were allowed to write "not applicable" (N/A) for any step and continue the startup using alternative methods. Procedural workarounds became accepted as normal.

Despite all these deficiencies, Texas City managers certified the procedures annually as up-to-date and complete.

Worker fatigue On the day of the incident, the day board operator was likely fatigued, having worked 12-hour shifts for 29 consecutive days and generally sleeping 5 to 6 hours per 24-hour period. The night lead operator, who filled the tower from a satellite control room, had worked 33 consecutive days. The day lead operator, who was training two new operators, dealing with contractors, and working to get a replacement part to finish the ISOM turnaround work, had been on duty for 37 consecutive days. Another experienced outside operator, who was helping the day lead operator, had worked 31 consecutive days. All of these individuals were working 12-hour shifts. Operations and maintenance workers sometimes worked high rates of overtime. At the same time, BP did not have a human fatigue-prevention policy.

The CSB issued a recommendation after the Texas City accident for BP to develop a guideline for understanding, recognizing, and dealing with fatigue during shift work [385]. The American Petroleum Institute (API) has a recommended practice that includes work on rotating shifts, for the maximum number of overtime hours, and the number of days to be worked without interruption [2].

Lessons learned generally not captured or acted on Reporting requirements focused on personal safety at BP. In the 2004 Texas City Culture Assessment, many employees reported "feeling blamed when they had gotten hurt or they felt investigations were too quick to stop at operator error as the root cause." There was a "culture of casual compliance." The Baker Panel report notes that BP had not instituted effective root cause analysis procedures to identify systemic causal factors that may contribute to future accidents [29]. Corrective actions, when taken, addressed only immediate or superficial causes but not the true root causes. The Panel suggested that BP had an incomplete picture of process safety performance at its US refineries because their process safety management system likely results in underreporting of incidents and near misses.

The CSB found evidence to document eight serious ISOM blowdown drum incidents from 1994 to 2004 [385]. These incidents were early warnings of the serious hazards of the ISOM and other blowdown systems' design and operational problems, but they were not effectively reported or investigated by BP or earlier by Amoco. Only three of the eight incidents involving the ISOM blowdown drum were even investigated.

One of the techniques used in BP for root cause analysis was 5 Whys, which suggests that will be only one root cause and one linear path to an accident. This type of guidance, provided in BP safety documents, can lead an investigation team to omit important systemic causes.

During the March 2005 ISOM startup, the operating deviations were more serious in degree but similar in kind to past startups. Yet the operating envelope program designed to capture and report excursions from safe operating limits was not fully functional and did not capture high distillation tower level events in the ISOM to alert managers to the deviations.

One way of identifying hazardous conditions or behaviors is through external audits. The Baker Panel found that BP had not implemented an effective process safety audit system for its US refineries and expressed concerns about auditor qualifications, audit scope, reliance on internal auditors, and the limited review of audit findings [29]. The panel also expressed concerns that the principal focus of the audits was on compliance and verifying that required management systems were in place to satisfy legal requirements. It did not appear that BP used the audits to ensure that the management systems were delivering the desired safety performance or to assess a site's performance against industry best practices. The problem of not following through on audit findings was especially apparent with overdue mechanical integrity inspection and testing.

Results of operating experiences, process safety analyses, audits, near misses, and accident investigations were not consistently used to improve process operations and process safety management practices. Many of the deficiencies were not new but were identifiable or even identified in lessons learned from previous accidents. For example, BP did not fully and comprehensively implement across BP's US refineries the lessons from previous serious accidents, including the process accidents that occurred in BP's facility in Grangemouth, Scotland in 2000. These incidents included a large process unit fire and two serious upsets. The lessons learned from Grangemouth were repeated in the Texas City explosion: a lack of leadership and accountability, insufficient awareness of process safety, inadequate performance measurement, a safety program too focused on personal safety, and a failure to complete corrective actions. Although the incidents occurred five years apart at different sites in different countries, many of the underlying deficiencies identified after the incidents appear to be the same, especially as they relate to evaluating process safety performance and then taking corrective actions.

Internal company audits had also identified poor company processes to disseminate lessons learned. But Texas City, as well as BP as a whole, lacked a reporting and learning culture. Reporting bad news was not encouraged and at times was actively discouraged. Texas City managers did not effectively investigate incidents nor take appropriate corrective action. They did not encourage the reporting of incidents or create an atmosphere of trust and prompt response to reports. As an example, in the five years prior to the 2005 disaster, over three-quarters of the raffinate splitter tower startups' levels ran above the range of the level transmitter and in nearly half, the level was out of range for more than one hour. These operating deviations were not reported by operations personnel or reviewed in the computerized history by Texas City managers. During the March 2005 ISOM startup, the operating deviations were more serious in degree but similar in kind to past startups. Yet the operating envelope program designed to capture and report excursions from safe operating limits was not fully functional and did not capture high distillation tower level events in the ISOM to alert managers to the deviations. BP did not monitor recurring instrument malfunctions, such as the high-level switch on the raffinate splitter tower and the blowdown drum, for trends or failure analysis.

Learning from prior events requires a usable safety information system. BP had not implemented one. Such a system must ensure that incidents are recorded in a centralized record keeping system and are available for other safety management system activities. Process hazard analysis and incident trending should also be included. The lack of historical data on the ISOM system incidents prevented BP managers from applying lessons learned to determine whether the system was safe and to understand the serious nature of the problem from the repeated release events. The problems with incident investigation and learning from events were not isolated to the ISOM unit.

Limited communication channels and poor information flow Supervisors and operators did not communicate critical information regarding the startup during the shift turnover. In fact, BP did not have a shift turnover communication requirement for its operations staff. The startup procedure should have provided information on the progress of the startup by the night shift, but it was not filled out and did not provide instructions for a noncontinual startup. The day board operator, therefore, had no precise information about what steps the night crew had completed and what the day shift was to do. Day Supervisor A did not distribute nor review the applicable startup procedure with the crew, despite being required to do so in the procedure.

As an example of the communication problems, on the morning of March 23, the raffinate tower startup began with a series of miscommunications. The early morning shift directors' meeting discussed the raffinate startup, and Day Supervisor B, who lacked ISOM experience, was told that startup could not proceed because the storage tanks that received raffinate from the splitter tower were believed to be full. The shift director stated in post-incident interviews that the meeting ended with the understanding that the raffinate section would not be started. This decision was consistent with a March 22 storage tank area logbook entry that stated the heavy raffinate tank was filling up. The instruction to not start the raffinate section was not, however, communicated to the ISOM operations personnel. In fact, Day Supervisor A told the operations crew that the raffinate splitter would be started.

As another example, prior to recommencing the startup, a miscommunication occurred regarding how feed and product would be routed into and out of the unit. The day board operator believed he was instructed not to send heavy raffinate product to storage and therefore closed the tower level control valve. However, the outside operators believed they were instructed not to send the light raffinate product to storage and manually changed the valve positions so that light raffinate would flow into the heavy raffinate product line. Moreover, no feed or product-routing instructions were entered into the startup procedure or the unit logbook. The CSB determined that the miscommunication likely concerned whether the light or heavy raffinate tanks were full and unavailable to receive additional liquid [385].

In general, there was poor communication at the local level within the corporation as a whole. BP did not have a policy or put emphasis on effective communication.

Superficial safety efforts Cosmetic system safety is a term invented by Childs to describe a common tendency for people to simply go through the motions when performing safety engineering activities [63]. BP epitomized this concept and promoted a checklist mentality. Personnel completed paperwork and checked off the safety policy and procedural requirements even when those requirements had not been met.

No formal system existed for identifying high-level risks. Critical events were generally not investigated. Audits were performed, but corrective actions to identified problems were limited and ineffective. Lessons learned were not captured or acted on.

Rather than ensuring actual control of major hazards, BP Texas City managers relied on an ineffective compliance-based system that emphasized completing paperwork. These general problems were identified by BP in the 2004 culture assessment where Texas City was said to be at "high risk" for a "check the box" mentality. This included going through the motions of checking boxes and inattention to the risk after the check-off. "Critical events, (breaches, failures or breakdowns of a critical control measure) are generally not attended to" [29].

While all of BP's US refineries had active programs to analyze process hazards, the system as a whole did not ensure adequate identification and rigorous analysis of those hazards. The Baker Panel noted that the extent and recurring nature of this deficiency was not isolated, but systemic [29].

Specifically, hazard analyses for the ISOM unit were poor, particularly pertaining to the risks of fire and explosion. For example:

- The consequences of high level and pressure in the raffinate splitter tower and high level in the blowdown drum and stack were not adequately identified. Overfilling the tower resulting in overpressurization of the safety relief valves and liquid overflow to the blowdown drum and stack was not identified.
- High heat-up rates or blocked outlets were not identified as potential causes of high pressure.
- The sizing of the blowdown drum for containment of a potential liquid release from the ISOM was not evaluated. The safeguards listed for the blowdown drum and stack to protect against the hazard of overflow, such as the steam-driven pump-out pump and high-level alarm, were insufficient to protect against the hazards. No recommendations were made by the hazard analysis team to provide additional safeguards.

Inadequate regulatory oversight The problems were not simply within BP. Government oversight was lacking. The CSB report found that OSHA, which was the supervisory authority for refineries, had not carried out planned inspections at the refinery and did not enforce safety rules, though there were many warning signs [385]. After the explosion, OSHA found 301 violations of requirements and imposed a fine of $21 million.

One reason for the inadequate inspections is that only a limited number of OSHA inspectors received the specialized training and experience necessary for complex investigations in refineries.

C.6 Macondo/*Deepwater Horizon*

In a press release issued on August 17, 2005, BP noted that the Texas City explosion was the worst tragedy in BP's recent history and that it would "do everything possible to ensure nothing like it happens again" [29]. The company also said it would act promptly to deal with the accident report recommendations. Yet five years later, an accident with the same causal factors occurred in a BP-managed offshore oil rig.

On April 20, 2010, an explosion of the BP-operated *Deepwater Horizon* oil drilling rig led to the one of the largest environmental disasters in American history and the largest marine oil spill in the history of the petroleum industry. Eleven workers were never found and are believed to have died in the explosion. Ninety-four crew members were rescued by boat or helicopter, with seventeen of those workers treated for injuries. Five million barrels (nearly 210 million gal) of oil were released into the Gulf of Mexico. This estimate is around twenty times the amount spilled in the 1989 Exxon Valdez disaster, which until that time held the record for the largest spill in US waters [31].

The *Deepwater Horizon* was a semisubmersible, mobile, floating, and dynamically positioned oil drilling rig that could operate in waters up to 10,000 ft (3,000 m) deep. The well and its location were named Macondo. The accident is typically referred to as either the *Deepwater Horizon* loss, after the name of the drilling rig involved or, more commonly within the oil and petroleum industry, as the *Macondo* accident.

The well owner and operator was BP, the rig owner was Transocean, and the major contractor involved was Halliburton. As described later, many of the same causal factors in this loss occurred in the BP owned and operated Texas City Refinery and apparently were never rectified. The same factors had been cited in previous BP accidents even prior to the Texas City explosion and at least one between the Texas City and Macondo tragedies, but clearly had not been satisfactorily rectified.

The fire, which was fed by 700,000 gal of oil on board and a continuous flow of hydrocarbons from the well, continued for 36 hours until the rig sank on April 22, 2010 [30]. Hydrocarbons continued to flow from the reservoir for 87 days, although reports in early 2021 indicated that the well site was still leaking.

A massive response followed to protect beaches, wetlands, and estuaries from the spreading oil using skimmer ships, floating booms, controlled burns, and oil dispersant. The months-long spill and the negative effects from the response and cleanup activities, caused extensive damage to marine and wildlife habitats as well as to the fishing and tourism activities in the Gulf of Mexico. This appendix focuses only on the blowout itself and its immediate aftermath and not on the long period of cleanup that followed.

Legal and financial penalties were extensive, primarily for BP. In November 2012, BP settled federal criminal charges with the US Department of Justice by pleading guilty to eleven counts of manslaughter, two misdemeanors, and a felony count of lying to the US Congress. BP also agreed to four years of government monitoring of its safety practices and ethics and was temporarily banned from new contracts with the US government. Record-setting fines and other payments were assessed at over $4.5 billion. In September 2014, a US District Court ruled that BP was primarily responsible for the oil spill because of its "gross negligence and reckless and willful misconduct" [374] and was fined $20.8 billion in fines, the largest corporate settlement in US history. Transocean's and Halliburton's actions were described as "negligent," but most of the blame was apportioned to BP. As of 2018, cleanup costs, charges, and penalties had cost BP more than $65 billion.

Criminal charges were also brought against individual BP managers. Two BP site managers were charged with manslaughter for operating negligently in their supervision of key safety tests before the explosion and for not alerting onshore engineers of problems in the drilling operation. BP's vice president for exploration in the Gulf was charged with obstructing Congress by misrepresenting the rate that the oil was flowing out of the well. A Halliburton manager was charged with instructing two employees to delete data related to Halliburton's cementing job on the oil well. None of the charges against individuals resulted in any jail time—just probation and fines—and no charges were brought against any upper-level executives. Two individuals were acquitted.

After the accidents, attempts were made to strengthen government oversight of offshore drilling, but many were undone by the Trump administration that followed.

Accidents always have technical and managerial causes. On the technical side, the accident involved a well integrity failure, followed by a loss of hydrostatic control of the well that resulted in the release of pressurized oil and gas into the rig. The blowout preventer (BOP) at the seabed was unable to seal the well, and the blowout became an uncontrollable disaster.

On the managerial side, the accident involved a series of regulatory and corporate culture flaws and actions motivated by the lucrative offshore drilling business. The Macondo project started in a high state of risk at the time of its conception, and that continued until its catastrophic end.

C.6.1 Background

Major accidents were not new to BP. Before Macondo, for example, three serious accidents at a BP plant in Grangemouth, Scotland, occurred in 2000. The Baker Panel Texas City investigation report noted that the causes cited in the Grangemouth accidents were repeated at Texas City [29].

A year after the Texas City explosion, on March 2, 2006, an oil spill was discovered on BP's exploration pipeline in western Prudhoe Bay, Alaska. During the 5-day leak, around 260,000 gal of oil poured into the bay and the delicate tundra environment from a corroded pipe, making it the largest oil spill recorded on the North Slope [36]. For many years, warnings from BP employees and state inspectors about corrosion of the pipeline were ignored. Upon analysis after the loss, the pipes were found to have been poorly maintained and inspected. Three years earlier, in November 2003, a gas line ruptured on a BP oil platform in the North Sea, flooding the platform with methane. BP admitted to breaking the law by allowing the pipes to corrode but apparently did not change their inspection and maintenance procedures as a result.

A pipeline is inspected by running an electronic drone—called a "smart pig"—through it. This device was supposed to have been used continuously in the BP Alaska pipeline, though it had not been used for fourteen years. A 2005 BP report stated that the company's corrosion-fighting process was based on a specific budget instead of local demands. BP managers did not expect to have corrosion problems in those lines and therefore did not inspect them as required [36].

None of this was just business as usual in the oil industry. In the five years between Texas City and Macondo, BP admitted to breaking US environment and safety laws and committing outright fraud [374]. The company paid $373 million in fines to avoid prosecution. In that period, BP refineries accumulated 760 "egregious, willful" safety violations handed out by OSHA, while the company with the next largest number had only 8.

Many of the problems stem from high-level management decisions. BP was founded in 1909 as the Anglo-Persian Oil Company (APOC) following the discovery of oil in Iran by an Englishman. In 1992, BP reported losses of $811 million and was on the verge of bankruptcy. An aggressive growth strategy was pursued consisting of significantly reducing the alternative energy division and pursuing numerous mergers and acquisitions. Mergers occurred with rivals Amoco in 1998 and ARCO in 2000.

In 2007, BP's organizational structure was transformed by removing several layers of management and slashing personnel at headquarters. Decision-making authority and responsibility were pushed down in the organizational structure, including decision-making related to safety. Employee compensation was tied to site performance. Because each site manager managed their "asset" autonomously and was compensated for its performance, there was little incentive to share best practices in risk management among the various BP exploration sites. There were also downsides to a system in which a centralized body had little oversight over the setting of performance targets, particularly in an industry where risk management and safety were essential to the long-term success of the company [367].

C.6.2 Safety Features

The *Deepwater Horizon* was a dynamically positioned, semisubmersible mobile offshore drilling unit that entered service in April 2001 and went to work for BP in the Gulf of Mexico the same year. Managing such a complex system is very challenging [275]. The large rig floated on massive pontoons, while the oil derrick rose over twenty stories above the top deck. In the bridge on the main deck, two people monitored the satellite-guided dynamic positioning system, controlling powerful thrusters that kept the 33,000-ton *Deepwater Horizon* centered over the well even in high seas and strong currents. Figure C.2 shows the basic components involved.

The Presidential Commission Report on the *Deepwater Horizon* Accident concluded:

> *On the day of the accident, no effective safeguards were in place to eliminate or minimize the consequences of a process safety incident. The safeguards (or barriers) intended to prevent such a disaster were not properly constructed, tested, or maintained, or they had been removed. The management systems intended to ensure the required functionality, availability, and reliability of these safety critical barriers were*

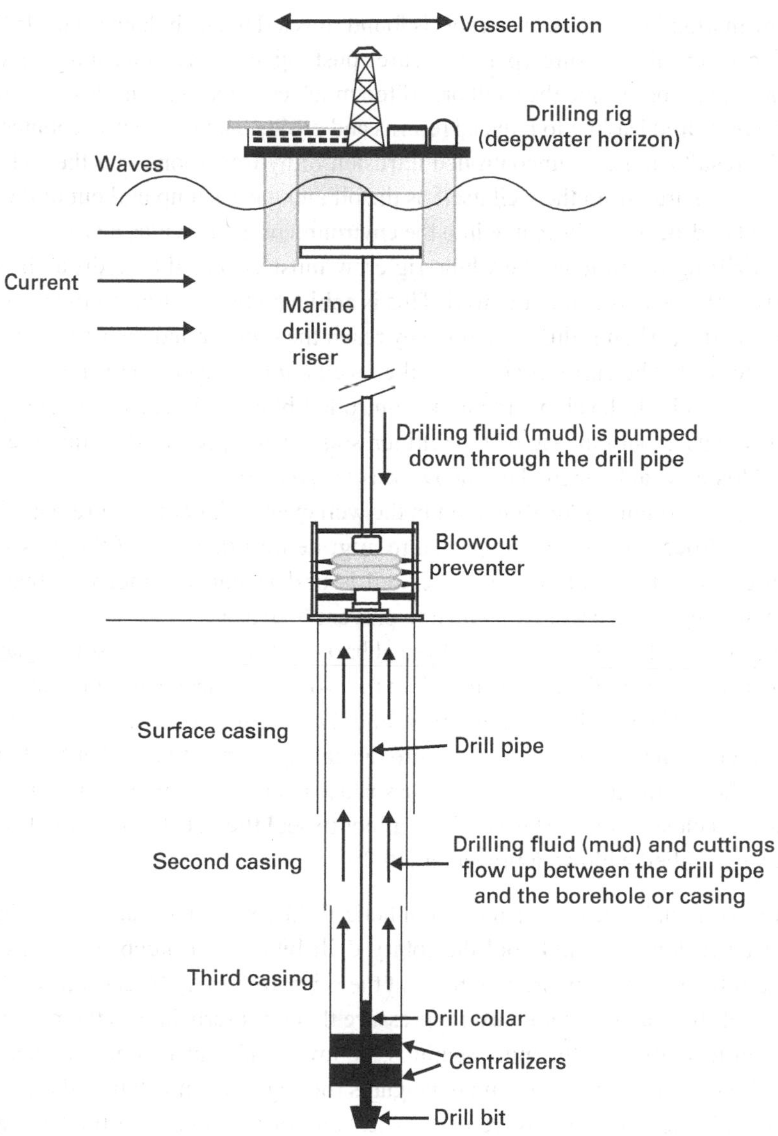

Figure C.2
Some basic components of offshore drilling.

> *inadequate. Ultimately, the barriers meant to prevent, mitigate, or control a blowout failed on the day of the accident.* [275]

To understand this conclusion, it is necessary to know what the safeguards are on such deepwater wells and how they operate.

The goal in deepwater drilling is to find reservoirs of oil and gas from the Middle Miocene era trapped in a porous rock formation at temperatures exceeding 200 degrees. The principal challenge is to drill a path to large, high-pressure reservoirs of oil and gas in a manner that simultaneously controls the enormous pressure and avoids fracturing the geologic formation in which the reservoir is found. It is a delicate balance. The drillers must balance the reservoir pressure (pore pressure) pushing hydrocarbons into the well with counter-pressure from inside the wellbore. If too much counter-pressure is used, the formation can be fractured leading to reduced returns and profits. But if too little counter-pressure is used, the result can be an uncontrolled intrusion of hydrocarbons into the well (called a *kick*), and a discharge from the well itself as the oil and gas rush up and out of the well. An uncontrolled hydrocarbon discharge into the environment is known as a *blowout*.

During drilling operations, the whole rig crew must ensure that hydrocarbons do not migrate from the reservoir into the well. This is achieved by monitoring the well and containing any hydrocarbon influxes before they reach the pipe, called the riser, that connects the rig to the well. The entire process is called *well control*. Well control actions must be initiated before a kick develops into an uncontrolled blowout because gas that travels up the riser is released onto the rig where ignition sources are present, leading to explosions and fires. This is what happened on the *Deepwater Horizon*.

As long as the column of drilling mud in the well exerts higher pressure than that of the formation, hydrocarbons are not expected to migrate into the well. If the pressure of the formation exceeds the mud pressure, the well is underbalanced, meaning that the mud column is no longer sufficient on its own to prevent hydrocarbon.

The Macondo well had a variety of physical barriers (layers of defense) in place to contain the hydrocarbons in the reservoir and, in the worst-case scenarios, prevent a blowout. The well controls the weight of a column of fluid that fills the hole being drilled (mud) and the riser, steel casing, cement used to secure the casing, cement placed at the bottom of a well to seal the hydrocarbon bearing zone, a surface cement plug placed in a shallow location in the well close to the surface of the seafloor to seal the hole being drilled, and a final line of defense called a blowout preventer (BOP).

Mud Mud is a sophisticated blend of synthetic fluids, polymers, and weighting agents that are used to lubricate and cool the rotary drill bits and to keep the well's pressure above that of the reservoir in order to prevent the migration of hydrocarbons into the well. The weight of the column of mud exerts pressure that counterbalances the pressure in the hydrocarbon formation. If the mud weight is too low, fluids such as oil and gas can enter the well, causing a kick. But if the mud weight is too high, it can fracture the surrounding rock, potentially leading to "lost returns"—leakage of the mud into the formation. Rig crews monitor and adjust the weight (density) of the drilling mud as the well is being drilled. This task requires special equipment and sophisticated interpretation of data.

Drillers pump mud down through the drill pipe, where it flows out through holes in the drill bit and then circulates back to the rig through the space between the drill pipe and the sides of the well (the annulus). The mud carries back to the surface bits of rock called cuttings that the drill bit has removed from the bottom of the well. When the mud returns to the rig at the surface, the cuttings are sieved out and the mud is sent back down the drill string. For well control, the drilling mud must exceed the pressure of the formation.

Casing As the well deepens, the crew lines its walls with a series of steel tubes called *casing*. The casing creates a foundation for continued drilling by reinforcing upper portions of the hole as drilling progresses. After installing a casing string, the crews drill farther, sending each successive string of casing down through the prior ones, so the well's diameter becomes progressively smaller as it gets deeper. A completed deepwater well typically telescopes down from a starting casing diameter of 3 ft or more at the well-head to a diameter of 10 in or less at the bottom.

Casing strings also help to control pressures. First, they protect more fragile sections of the well structure outside the casing from the pressure of the mud inside. Second, they prevent high-pressure fluids (like hydrocarbons) outside the casing from entering the wellbore and flowing up the well.

Cement When the original drilling process is complete, cement is used to seal the space between the outside of the pipe and the drill hole wall (the annular space). This space can become a path for escaping gases. The cement must displace all the drilling mud so that the cement remaining in the well does not become contaminated and lose its sealing capacity. If mud channels remain after the cement is pumped, they can become a flow path for gases or liquids from the formation.

The cement slurry must be formulated so that it sets and cures properly under wellbore conditions. Additives, like nitrogen to foam the cement, must be carefully planned to achieve the desired isolation. The design of the cement for Macondo has been largely criticized. The choice of foamed cement, lighter than required for the conditions of the well, apparently contributed to the failure of the cement barrier during the blowout.

A cement slurry must be tested before it is used in a cement job. Because the pressure and temperature at the bottom of a well can significantly alter the strength and curing rate of a given cement slurry—and because storing cement on a rig can alter its chemical composition over time—companies like Halliburton normally fly cement samples from the rig back to a laboratory shortly before pumping a job to make sure the cement will work under the conditions in the well. The laboratory conducts a number of tests to evaluate the slurry's viscosity and flow characteristics, the rate at which it will cure, and its eventual compressive strength.

When testing a slurry that will be foamed with nitrogen (as this one was), the lab also evaluates the stability of the cement that results. A stable foam slurry will retain its bubbles and overall density long enough to allow the cement to cure. The result is hardened cement that has tiny, evenly dispersed, and unconnected nitrogen bubbles throughout. If the foam does not remain stable up until the time the cement cures, the small nitrogen bubbles may coalesce into larger ones, rendering the hardened cement porous and permeable. If the instability is particularly severe, the nitrogen can "break out" of the cement, with unpredictable consequences.

One other consideration is crucial. For the cementing operation to be successful, the casing string has to be properly centralized in the well so that all the areas around the drill pipe are properly sealed by the cement. When a casing string hangs in the center of the wellbore, cement pumped down the casing will flow evenly back up the annulus, displacing any mud and debris that were previously in that space and leaving a clean column of cement. If the casing is not centered, the cement will flow preferentially up the path of least resistance—the larger spaces in the annulus—and slowly or not at all in the narrower annular space. That can leave behind channels of drilling mud that can severely compromise a primary cement job by creating paths and gaps through which pressurized hydrocarbons can flow. Centralizers are used to ensure that the casing is in the proper position. BP's original design called for sixteen or more centralizers to be placed along the long string. In the cementing process that preceded the blowout, only six were used, apparently because only six were available and it would take several days to get more.

Cementing problems are a significant risk factor leading to blowouts. Even following best practices, a cement crew can never be certain how a cement job at the bottom of the well is proceeding as it is pumped. Cement does its work literally miles away from the rig floor, and the crew has no direct way to see where it is, whether it is contaminated, or whether it has sealed off the well. To gauge progress, the crew must instead rely on subtle, indirect indicators like pressure and volume: they know how much cement and mud they have sent down the well and how hard the pumps are working to push it. The crew can use these readings to check whether each barrel of cement pumped into the well displaces an equal volume of drilling mud—producing "full returns." They can also check for pressure spikes to confirm that "wiper plugs" (used to separate the cement from the surrounding drilling mud) have landed on time as expected at the bottom of the well. And they can look for "lift pressure"—a steady increase in pump pressure signifying that the cement has turned the corner at the bottom of the well and is being pushed up into the annular space against gravity.

While these indicators suggest generally that the job has gone as planned, they say little specific about the location and quality of the cement at the bottom of the well. None of them can take the place of pressure testing and cement evaluation logging, which was skipped during the cementing operation at Macondo.

On the day of the accident, the well drilling had been completed and was being temporarily plugged (called temporary abandonment) to change rigs and transition from drilling to production. Later, when a smaller rig was installed at the location, production could start by crews punching holes through the casing and surrounding cement to allow hydrocarbons to flow into the well.

Blowout preventer One other piece of equipment plays an important layer of protection in well control and in the events that occurred. The BOP is located at the top of the well on the seabed. It consists of a stack of enormous valves that rig crews use both as a drilling tool and as an emergency safety device. Once it is put in place, everything needed in the well—drilling pipe, bits, casing, and mud—passes through the BOP.

In case of a blowout, the BOP is designed to shut in the well; that is, to prevent hydrocarbons from exiting the well after they have left the reservoir. The *Deepwater Horizon*'s blowout preventer had several features that could be used to seal the well. The top two were large, donut-shaped rubber elements called *annular preventers* that encircled drill pipe or casing inside the BOP. When squeezed shut, they sealed off the annular space around the drill pipe. The BOP also contained five sets of metal rams. The "blind shear ram" was designed to cut through drill pipe inside the BOP to seal off the well in emergency situations. The BOP rams could be activated manually by drillers on the rig, by a remotely operated unmanned vehicle, or by an automated emergency *deadman* system when extreme conditions occur, such as disconnection of the BOP from the platform. A casing shear ram was designed to cut through casing, and three sets of pipe rams were in place to close off the space around the drill pipe.

Each ram is activated separately. If a kick evolves beyond the point where the driller can safely shut in the well with an annular seal or a pipe ram, he or she can cut the drill pipe, activating the blind shear ram. Considered as the last resource in a well control emergency, BOPs are usually designed with a great deal of redundancy to ensure high reliability. None of this redundancy was effective in this accident.

C.6.3 Proximate Events

The physical cause of the blowout was the failure of a cement barrier, allowing hydrocarbons to flow up the wellbore, through the riser and onto the rig, resulting in the blowout. The following events led to this physical failure. The events and causal factors in the next section were taken from the official Presidential Report on the *Deepwater Horizon* accident [275], the US Chemical Safety Board report on the accident [384], and a research report on the loss [367].

In March 2008, BP paid $34 million for an exclusive lease to drill what was later named the Macondo well. Transocean's drilling rig, *Marianas*, arrived at the location and began drilling the well in November 2009. The *Marianas* drilled for 34 days, reaching a depth of over 9,000 ft. It then had to stop drilling and move offsite to avoid Hurricane Ida about a month later. In January 2010, Transocean's *Deepwater Horizon* rig arrived at the site to continue the operation. Its first task was to lower the giant BOP onto the wellhead that the *Marianas* had left behind. In February, drilling of the Macondo well resumed.

In March, Halliburton personnel sent BP the results of a foam stability test it ran in February on the cement blend it planned to use at Macondo. The data indicated that the cement slurry design was unstable. Halliburton personnel did not comment on the evidence of the cement slurry's instability, and there is no evidence that BP examined the foam stability data in the report at all [275].

After numerous instances indicating fractures in the geologic formation in the March to early April period, BP decided to stop drilling at 18,360 ft, short of the 20,200 ft initially planned. BP informed its lease partners Anadarko and MOEX that well integrity and safety issues required the rig to stop drilling further. Based on logging data, BP concluded that it had drilled into a hydrocarbon reservoir of sufficient size (at least 50 million barrels) and sufficient pressure that it would be economically worthwhile to install a final production casing string that BP would eventually use to recover the oil and gas.

After the exploration phase, BP (like most operators) was to give the job of "completing the well" to a smaller and less costly rig, which would install the hydrocarbon collection and production equipment. To make way for the new rig, the *Deepwater Horizon* would have to remove its riser and blowout preventer from the wellhead. Before it could do those things, the crew had to cement the well for temporary abandonment.

On April 13, just before pumping the final cement job, Halliburton personnel ran a second set of tests on the now slightly altered cement blend they plan to use at Macondo. The foam stability test showed that the cement slurry would be unstable. There is no evidence that Halliburton ever reported these results to BP. Instead, they ran another test on April 18, but the results were not available and were not sent to BP until after the blowout.

On April 14–15, BP engineers selected a "long string" production casing, which would provide a single continuous wall of steel between the wellhead on the seafloor, and the oil and gas zone at the bottom of the well. The other option considered, a "liner," would result in a more complex—and theoretically more leak-prone—system over the life of the well. The liner, however, would have been easier to cement into place at Macondo.

A Halliburton engineer informed BP engineers that computer simulations suggested that the Macondo production casing would need more than six centralizers (used to keep the casing string centered) to avoid channeling in the cement. BP engineers ordered fifteen additional centralizers—the most BP could transport immediately in a helicopter.

When the helicopter delivered the fifteen centralizers to the rig, the BP engineers decided the centralizers were the wrong kind and did not use them. It would take at least 48 hours to get more centralizers delivered, and the BP engineers decided to just go with the six centralizers that were on hand.

The long string production casing was installed. Installation involved use of a "float collar," a simple arrangement of two flapper (float) valves, spaced one after the other through which the mud in the well could flow. In preparation for cementing, the crew attempted to "convert" the float valves. They performed a check to determine whether the float valves were closed and holding. They concluded that they were although the results included anomalous readings of pressure. Circulation pressure was lower than predicted, but the crew decided that the pressure gauge was broken.

The Presidential Commission reported that it is not possible to determine whether the float valves contributed to the blowout, but what is certain is that the BP team did not take the time to consider whether and to what extent the anomalous readings may have indicated other problems or may have increased the risk of the upcoming cement job [275].

Before doing a cement job on a well, common industry practice is to circulate the drilling mud through the well, bringing the mud at the bottom all the way up to the drilling rig. This procedure, known as "bottoms up," lets workers check the mud to see if it is absorbing any gas leaking into it. If so, they can clean the gas out of the mud before putting it back down into the well to maintain the pressure. The American Petroleum Institute says it is "common cementing best practice" to circulate the mud at least once [275].

Circulating all the mud in a well of 18,360 ft, as this one was, takes 6 to 12 hours. But mud circulation on this well was done for just 30 minutes on April 19, not nearly long enough to bring mud to the surface.

One of the final tasks in the temporary abandonment procedures was to cement in place the steel pipe that ran into the reservoir. The cement would fill the space between the outside of the pipe and the rock, preventing any gas from flowing up the sides.

At the time of the accident, they had just finished the task of cementing the well. A three-man Schlumberger team was scheduled to fly out to the rig later in the day to perform a series of tests to examine the new cement seal. A decision was made to send the Schlumberger team home without doing the test, thus saving time and the $128,000 fee. Instead, BP and Halliburton declared the cement job a success.

The rest of the day was devoted to various tests on the well in preparation for temporary abandonment. BP's Macondo team had made numerous changes to the temporary abandonment procedures in the two weeks leading up to the day of the accident. There is no evidence that these changes went through any sort of formal risk assessment or management of change process. On the day of the accident, an operations note was sent to the Macondo team listing the temporary abandonment procedure to be used for the well. It was the first time the BP leaders on the rig had seen the procedure they would use that day. BP first shared the procedures with the rig crew at a midday meeting. After the initial cementing job, the basic sequence was:

1. Perform a positive-pressure test to test the integrity of the production casing.
2. Run the drill pipe into the well to 3,300 ft below the mud line.

3. Displace the mud in the well with seawater, lifting the mud above the BOP and into the riser.
4. Perform a negative-pressure test to assess the integrity of the well and bottom-hole cement job to ensure outside fluids (such as hydrocarbons) are not leaking into the well.
5. Displace the mud in the riser with seawater.
6. Set the surface cement plug.
7. Set the lockdown sleeve.

The crew would never get beyond step 5 before the blowout.

The first step in the procedure listed above was to test well integrity to make sure there were no leaks in the well. The positive-pressure test, among other things, evaluates the ability of the casing in the well to hold under pressure. This test is required by government regulations.

To perform the test at Macondo, the *Deepwater Horizon*'s crew first closed off the well below the BOP by shutting the blind shear ram (there was no drill pipe in the well at the time). Then, much like pumping air into a bike tire to check for leaks, the rig crew pumped fluids into the well (through pipes running from the rig to the BOP) to generate pressure and then checked to see if it would hold. The pressure inside the well remained steady during both tests, showing there were no leaks in the production casing through which fluids could pass from inside the well to the outside.

The next test, the negative-pressure test, was not as successful. This test checks the integrity of the bottomhole cement job. Instead of pumping pressure into the wellbore to see if fluids leak out, the crew removes pressure from inside the well to see if fluids, such as hydrocarbons, leak in, past, or through the bottomhole cement job. In so doing, the crew simulates the effect of removing the mud in the wellbore and the riser (and the pressure exerted by that mud) during temporary abandonment. If the casing and primary cement have been designed and installed properly, they will prevent hydrocarbons from intruding even when that "overbalancing" pressure is removed. A negative-pressure test is successful if there is no flow out of the well for a sustained period and if there is no pressure build-up inside the well when it is closed at the surface. At the Macondo well, the negative-pressure test was the only test performed that would have checked the integrity of the bottomhole cement job.

To conduct a proper negative test at Macondo, BP would have to isolate the well from the effect of the 5,000 ft plus column of drilling mud in the riser and a further 3,300 ft column of drilling mud below the seafloor. Those heavy columns of mud exerted much more pressure on the well than the seawater that would replace them after temporary abandonment. Once this pressure was removed, the downward force of the column of fluids in the well would be less than the pressure of the hydrocarbons in the reservoir, so the well would be in what is called an "underbalanced" state. It was therefore critical to test and confirm the ability of the well (including the primary cement job) to withstand the underbalance. If the test showed that hydrocarbons would leak into the well once it was underbalanced, BP would need to diagnose and fix the problem (perhaps remediating the cement job) before moving on, a process that could take many days.

During the positive-pressure test, the drill crew increased the pressure inside the steel casing and seal assembly to be sure they were intact. The negative-pressure test, in contrast, reduced the pressure inside the well in order to simulate its state after the *Deepwater Horizon* had packed up and moved on. If pressure increased inside the well during the negative-pressure test, or if fluids flowed up from the well, that would indicate a well integrity problem—a leak of fluids into the well. Such a leak would be a worrisome sign that somewhere the casing and cement had been breached—in which case remedial work would be needed to reestablish the well's integrity.

At noon the crew began to run drill pipe into the well in preparation for the negative-pressure test later that evening. Some BP and Transocean executives arrived for a 24-hour tour and the VIP's escorted tour began around 4 p.m. The visiting executives were told that, despite all the troubles they had had with this well, it was one of the top-performing wells in all the BP floater fleet from the standpoint of safety. They had not had a single "lost-time accident" in seven years of drilling.

At 5 p.m., the rig crew began the negative-pressure test. After bleeding pressure from the well, the crew closed it off to check whether the pressure within the drill pipe would remain steady. Oversimplifying slightly, the volume of mud sent into the well should equal the volume of mud returning from the well. An increase in volume is a powerful indicator that something is flowing into the well.

A second, similar indicator is to check that the rate of flow of liquids coming from the well should equal the rate of flow of fluid pumped into the well. Like the first indicator, if flow out of the well is greater than flow into the well, it indicates that a kick may be underway.

In addition to these two primary parameters, the crew can perform visual flow checks.

As a final check, the driller and mudlogger also monitor drill-pipe pressure, but it is a more ambiguous kick indicator than the other parameters because there can be many reasons for a change in drill pressure. If drill pipe pressure decreases while the pump rate remains constant, that may indicate that hydrocarbons have entered the wellbore and are moving up the well past the sides of the drill pipe. The lighter-weight hydrocarbons exert less downward pressure, meaning the pumps do not need to work as hard to push fluids into the well. If drill pipe pressure increases while the pump rate remains constant, that may indicate that heavier mud is being pushed up from below (perhaps by hydrocarbons) and displacing lighter fluids in the well adjacent to the drill pipe. Unexplained changes in drill pipe pressure may not always indicate a kick, but when observed should be investigated. The crew should shut down the pumps and monitor the well to confirm it is static; if they are unable to do so, they should shut in the well until the source of the readings can be determined.

At the beginning of the negative-pressure test, the crew pumped in a spacer liquid, which separates the heavy drilling mud from the seawater, followed by pumping seawater down the drill pipe to push (displace) the mud from below the mud line to above the BOP.

The spacer that BP chose to use during the negative-pressure test was unusual. BP wanted to use some materials that were already on the rig in order to avoid having to dispose of them onshore as hazardous waste according to US government regulations. The regulations have an exception that allows companies to dump water-based drilling fluids overboard if they have been circulated down through a well. At BP's direction, the materials were combined to create an unusually large volume of space that had never previously been used by anyone on the rig or by BP as a spacer, nor had they been thoroughly tested for that purpose [275].

Once the crew had displaced the mud to a level above the BOP, they shut an annular preventer in the BOP, isolating the well from the downward pressure exerted by the heavy mud and spacer in the riser. The crew could now perform the negative-pressure test using the drill pipe: the test involved opening the top of the drill pipe on the rig, bleeding the drill pipe pressure to zero, and then watching for flow. The crew opened the drill pipe at the rig to bleed off any pressure that had built up in the well during the mud-displacement process. The crew tried to bleed the pressure down to zero but could not get it below 266 psi. When the drill pipe was closed, the pressure jumped back up to 1,262 psi.

As the crew conducted the test, the drill shack grew crowded with the night shift arriving as well as the visiting BP and Transocean VIPs. They thought maybe the spacer was leaking down past the annular preventer, out of the riser and into the well. The annular preventer was closed more tightly to stop the leak. They attempted the negative-pressure test two more times, and still the pressure went down but then increased.

One of the workers said he had seen this before and explained it away as the "bladder effect." Another negative test was done. This time, the crew members were able to get the pressure down to zero on a different pipe, called the "kill line," but not for the drill pipe, which continued to show elevated pressure. Whether for these reasons or some other, the men in the shack decided that no flow from the open kill line equaled a successful negative-pressure test even though the direct tests were unsuccessful. Without reconciling the different results from the tests, they declared the test had confirmed the well's integrity. They decided to get on with the rest of the temporary abandonment process.

After concluding that the negative-pressure test was successful, the drilling crew prepared to set a cement plug deep into the well, about 3,000 ft below the top of the well. They reopened the blowout preventor and began pumping seawater down the drill pipe to displace the mud and spacer from the riser (the pipe that connects the rig to the well assembly on the seafloor below, see figure C.2). When the spacer appeared up at the surface, they stopped pumping because the fluid had to be tested to make sure it was clean enough to dump into the Gulf, now that it had journeyed down into the well and back. During this time, clues appeared that there was a problem on the two data displays they had available, but the clues were not noticed. The crew may have been distracted by other matters that had arisen. Mounting evidence of a kick was ignored.

By 9:15 p.m., the crew began discharging the spacer overboard. Suddenly, around 9:45, mud came shooting out, followed quickly by several explosions and fires. After the first explosion, crew members on the bridge attempted to engage the rig's emergency disconnect system (EDS). The EDS should have closed the blind shear ram, severed the drill pipe, sealed the well, and disconnected the rig from the BOP. None of that happened. Alarms started sounding. The power went out, plunging everything on the rig into darkness.

A standby "mudboat" vessel, the *Bankston,* which had been waiting nearby to take on the mud, was told to back off for safety. A mayday call went out, and nearby private boats began to come to the rescue as did the Coast Guard. People tried to get to the lifeboats, but it was dark and smokey and hard to move around and to carry injured crew members. Some people jumped off the rig into the water after seeing the chaos that surrounded the lifeboats.

The mudboat crew, who had prepared for an emergency in a drill that morning, sprang into action doing they had practiced. They immediately detached the mud transfer hose connecting them to the rig. After they had protected themselves, they launched the ship's fast rescue craft and started pulling people out of the water. Lifeboats from the rig made their way to the *Bankston*. The captain of the *Bankston* instructed them to go to the starboard side, which was sheltered from the rig.

A Coast Guard helicopter arrived to oversee the medical evacuation of the injured crew members. Workboats arrived and started spraying water on the fire. Private boats and the coast guard arrived to help look for survivors.

C.6.4 Causal Factors

While the details surrounding this loss were very different than Texas City, the causal factors were very similar. The loss of life and the subsequent pollution of the Gulf of Mexico were the result of complacency and safety culture flaws, poor risk management, a disregard for maintenance of critical equipment, last-minute changes to plans without using appropriate management of change procedures, failure to observe and respond to critical indicators, and a poorly designed safety management system. In this case, there was also inadequate oversight by government agencies who were tasked to protect the environment and oversee offshore oil exploration.

C.6.4.1 Inadequate regulatory oversight Offshore oil and gas exploration were regulated at the time of the accident by the Minerals Management Service (MMS) of the US Department of the Interior. They had a difficult job. Energy independence required increasing domestic production, but previous accidents, such as the Santa Barbara spill in 1969, had increased mandates for environmental protection. A third pressure came from the goal of revenue generation—the expansion of offshore oil and gas production brought in billions of dollars in federal revenues from offshore leases. The incentive to promote offshore drilling conflicted with the mandate to ensure safe drilling and environmental protection. Revenue collection became the dominant objective [275].

Increasing revenues, both for the companies and for the government, meant that offshore drilling needed to move into much deeper waters and the greater risks that entailed. Those increased risks were not matched by greater and more sophisticated regulatory oversight. Industry resisted such oversight, and there was not much political support to overcome that opposition.

The oil and gas industry promised to invest in drilling safety and oil-spill containment technology and contingency response planning in case of an accident, but it never did. The Macondo accident resulted from decades of inadequate regulation, insufficient investment, and incomplete planning [275].

Even if they had wanted to provide strict regulation, the MMS lacked resources, technical training, and experience in petroleum engineering. Over time, those resources fell increasingly short. Industry relied on more demanding and complex technology. The government did not have the expertise to regulate this technology nor the ability to maintain up-to-date regulations. MMS responsibilities increased at the same time as budgets decreased. Offshore fires, explosions, and blowouts increased as a result.

Industry practices also changed. Specialized service contractors, such as Halliburton and Transocean, contracted to BP. When the lessee directly regulated by the government is itself not performing many of the activities critical to well safety, that separation of functions poses heightened challenges for the regulator. But there was no apparent effort by MMS to respond to those challenges by making the service companies more accountable.

Leaders and others at the MMS did not have sufficient expertise and experience in the field to effectively audit and inspect offshore. Funding for training was inadequate. Low salaries, compared to workers in the industry, meant that the MMS had difficulty recruiting the employees it needed to create and enforce regulations. The agency's ethical culture declined, and gifts from oil and gas companies were accepted in some offices.

With respect to the Macondo accident, the deficiencies in regulation led to the following:

- The mandated frequency of pressure tests of BOPs was halved when industry contended they were more reliable than the regulations recognized. Soon afterward, a series of third-party technical studies raised the possibility of high failure rates for the blowout preventers' control systems, annular rams, and blind-shear rams under certain deepwater conditions and due to changes in the configuration and strength of drill pipe used by industry. Two studies commissioned by MMS found that many rig operators, by not testing blowout preventers, were basing their representations on the assumption that the tool would work "on information not necessarily consistent with the equipment in use" [275, p. 74]. Yet, the MMS never revised its blowout preventer regulations nor added verification as an independent inspection item in light of this new information.
- There was no requirement for a negative-pressure test nor a protocol for performing them. Nor were there detailed requirements governing the cementing of a well and testing the cement used for well stability.
- The MMS approved BP's request to set its temporary abandonment plug 3,300 ft below the mud line. In this case, there was an MMS regulation that applied and stated that cement plugs should normally be installed no more than 1,000 ft beneath the mud line. BP requested an exemption, which was approved in less than 90 minutes. It is not clear what steps were taken to ensure that this proposed procedure would provide an adequate level of safety. In general, the MMS granted exemptions from regulatory requirements on a routine basis, without studying each case in detail [our report?].
- Before the Macondo accident, leaseholders and operators were not required to implement a documented safety management program for their well operations; they were asked only to voluntarily adopt such a program [386].
- Many operators were exempted from the requirement that they file plans to deal with major oil spills. It also specifically allowed BP to drill the Macondo well without a detailed environmental analysis.
- The BOP on *Deepwater Horizon* was not certified at the time of the accident. The BOP is supposed to be recertified every five years, but the last time it had been certified was in 2000, thus it was at least five years overdue. The recertification process entails complete disassembly of the BOP on the surface, which can take up to three months or longer and generally requires time in dry dock. As a result, industry common practice is to perform major maintenance on a complicated BOP control system during the shipyard time of a mobile offshore drilling unit in its five-year interval inspection period [275]. The *Deepwater Horizon* never stopped working since its commission, and neither did its BOP. Industry associations and manufacturers require a comprehensive inspection of the BOP every three to five years [275].

- The MMS approved testing of the BOP at lower pressures than required by their regulations. Though testing at lower pressures is also in accord with industry practice to avoid unnecessary wear or damage of the tool while in operation, most tests did not establish the ability of the equipment to perform during blowout conditions with large volumes of gas moving at high speed and high pressure [275].
- MMS well control training provided to BP and Transocean covered initial kick response during drilling operations, but did not include kick detection and indicators, nor emergency response to full-scale blowouts.
- MMS regulations did not cover the use of long-string production casings, require the use of casing centralizers, address the possibility of cement contamination, require BP to conduct or report cement slurry tests nor specify any criteria for test results, address the use of foamed cement or any other specialized cementing technology or even require BP to inform MMS of its use, specify practical indicators of an inadequate cement job, or require measures that would facilitate containment or capping in the event of a blowout.

The need for Coast Guard oversight also increased in the 1990s as industry drilled in deeper waters farther offshore and used more ambitious floating drilling and production systems. But, like the MMS, it faced severe budgetary restraints. As a result, it did not update its marine-safety rules to reflect the industry's new technology, and it shifted much of its responsibility for fixed platform safety to the MMS in 2002.

C.6.4.2 Inadequate industry self-regulation In the United States, the American Petroleum Institute produces standards, recommended practices, specifications, codes, technical publications, reports, and studies that cover the industry and are utilized around the world [1]. In conjunction with API's quality programs, many of these standards form the basis of API certification programs. And the US Department of the Interior has historically adopted those recommended practices and standards, developed by technical experts within API, as formal agency regulations.

The Presidential Commission on the *Deepwater Horizon* Accident concluded that API's ability to serve as a reliable standard-setter for drilling safety is compromised by its role as the industry's principal lobbyist and public policy advocate. Because regulations would make oil and gas industry operations potentially more costly, API regularly resists agency rulemakings that government regulators believe would make those operations safer, and API favors rulemaking that promotes industry autonomy from government oversight.

According to statements made by industry officials to the commission, API's safety and technical standards were a major casualty of this conflicted role. As described by one representative, API-proposed safety standards have increasingly failed to reflect best industry practices and have instead expressed the lowest common denominator—in other words, a standard that almost all operators could readily achieve. Most of the safety standards were really guidelines and suggested practices, but not requirements. Because, moreover, the Interior Department has in turn relied on API in developing its own regulatory safety standards, API's shortfalls have undermined the entire federal regulatory system.

The commission report found that the inadequacies of the resulting federal standards were evident in the decisions that led to the Macondo well blowout. Federal authorities lacked regulations covering some of the most critical decisions made on the *Deepwater Horizon* that affected the safety of the Macondo well.

In addition, for years, API led the effort to persuade the MMS not to adopt regulations about safety and environmental management systems. It instead argued for voluntary, recommended safety practices. Such requirements for safety and environmental management systems are enforced in other countries.

C.6.4.3 Inadequate international oversight Oil and gas exploration and production is a global enterprise. The *Deepwater Horizon* oil rig was built in South Korea. It was operated by a Swiss company under contract to a British oil firm. Primary responsibility for safety and other inspections rested not with the US government but with the Republic of the Marshall Islands—a tiny, impoverished nation in the Pacific Ocean. And the Marshall Islands, a maze of tiny atolls—some smaller than the ill-fated oil rig—outsourced many of its responsibilities to private companies.

In understanding what went wrong in the Macondo accident, this international patchwork of divided authority and sometimes conflicting priorities played a critical role. Under international law, offshore oil rigs like the *Deepwater Horizon* are treated as ships, not real estate. Even though the Macondo well was located in US-protected waters, oil companies are allowed to register the rigs that operate there in foreign countries, reducing the US government's role in inspecting and enforcing safety and other standards.

Critics suggest that this registry process allows oil drillers and producers to shop for jurisdictions that offer them more favorable—and less demanding—terms. Some experts say foreign registration also permitted a confusing command structure and understaffing of the platform. Different types of rigs are classified differently, and the Marshall Islands assigned the *Deepwater Horizon* to a category that permitted lower staffing levels. The rig may not even have met the lower staffing standards.

The Marshall Islands allowed its owner, Transocean, to place an oil-drilling expert above a licensed sea captain in making decisions on the day of the explosion. This dual command structure emphasized oil production over safety and created confusion that may have delayed an effective response to the growing crisis aboard the *Deepwater Horizon*. More rigorous coast guard inspection procedures apply to the relatively small number of oil rigs registered in the United States. A foreign vessel will be reviewed by the Coast Guard, but the inspection is relatively cursory, relying on inspection reports prepared by outside firms that have been paid directly by the owners of the vessel.

C.6.4.4 Technical deficiencies There were certainly a large number of technical deficiencies involved in this accident, such as the cement, the deadman system, and the blowout preventer. The important information, however, is not just that lots of things failed but *why* those failures occurred. With so much redundancy and layers of defense, how could everything fail at the same time? These types of problems seem to occur in many of the process industry accidents, including those described in these appendices, such as Bhopal, Chernobyl, Fukushima, and Texas City.

The immediate cause of the Macondo blowout was a failure to contain hydrocarbon pressures in the well. Three physical barriers could have contained those pressures: the cement at the bottom of the well, the mud in the riser, and the blowout preventer. Each of these barriers had redundancy or additional protection (such as testing). But mistakes and misunderstanding about and sometimes disregard for risk compromised each of those potential barriers, depriving the crew of safeguards until the blowout became inevitable and, in the end, uncontrollable. The design was filled with common cause failure modes, such as inadequate maintenance and common power sources, but these either were never identified or never eliminated.

One important question is why the kick was not detected by the systems in place on the rig, including the people who were responsible for detecting a kick. One explanation is that simultaneous activities were being performed that made kick detection more difficult. The crew was required to perform concurrent activities during the displacement of the mud that eliminated their ability to monitor three of the four kick indicators. In addition, difficult and critical manual calculations, such as calculating the net flow from the well, were required because the crew was not provided with automation to perform them.

The design of the instrumentation that was provided was flawed. The displays contributed to the problems by making fluctuations in data hard to detect. Historical data from previous hours of operation was not directly visible on the displays; it had to be checked manually. The setting and resetting of alarm thresholds in the monitoring systems was done manually and was not performed systematically. There were not enough cameras installed to monitor flow from the wellbore at critical points, such as the overboard line. Operators had to physically measure volume levels with handmade levels and visually confirm the direction of the flow.

BP and Halliburton were able to gather and transmit real-time data from the well. BP even allocated a large room in its Houston Headquarters to monitor the data from the well in the Gulf of Mexico. At the time of the blowout, however, no one was assisting the monitoring of the well onshore. BP had no plans to monitor the data it was getting, arguing that such monitoring "tended to disempower personnel on the rig" [367]. However, the only personnel on the rig monitoring the well at the time of the kick was a Transocean tool pusher. The rig computers in the driller's room operated intermittently and had outdated software.

The rig also needed more sophisticated, automated alarms to alert the driller and mudlogger when anomalies arose. These individuals sit for 12 hours at a time in front of data displays. Given the potential consequences, reliance should not be placed on a system that requires the right person to be looking at the right data at the right time and then understand its significance in spite of simultaneous activities and other monitoring responsibilities. Sometimes detecting problems from the displays required interpreting ambiguous or subtle display changes.

The data sensors themselves had several deficiencies including:

- There was incomplete coverage of the sensors, such as flow meters located after critical line branches.
- Sensors were unreliable, inaccurate, and imprecise. The crew had to perform basic hand measurements to obtain pit volumes, for example. In addition, some critical sensors were affected by disturbances unrelated to the state of the well, such as crane activity.
- None of the monitoring systems had automated alarms or thresholds. The crew had to adjust them manually and constantly.
- Simple well monitoring calculations were not automated and had to be done by hand.

C.6.4.5 Culture and decision-making Almost by definition, bad decision-making usually precedes accidents. Most often, poor decisions arise from complacency and overconfidence. Much of this unjustified confidence stemmed from overreliance on redundancy, ineffective and inaccurate risk assessment, ignoring audits and warning signs, and confusing process safety with personal safety. The complacency that results leads to problems such as tolerating serious deviations from operating procedure; prioritizing cost and schedule over safety; not investing in safety, safety equipment, and maintenance; and allowing migration to states of higher risk.

C.6.4.5.1 Sources of complacency and overconfidence Complacency and overconfidence are widespread in this industry. The MMS and the industry assumed that technological advances had made the equipment they used, such as the BOP, remarkably reliable. This assumption was not necessarily true and could be undermined by the way the technology was used. Other factors also contributed to unjustified complacency. For example, the decision to use a long-string casing instead of a line increased the difficulty of obtaining a reliable primary cement job. The long string decision should have led BP and Halliburton to be on heightened alert for signs of primary cement failure. Even though the well site manager was aware of the lack of expertise of his crew, he did not request expert assistance from onshore personnel to run and interpret the negative-pressure test.

Overreliance on redundancy A common assumption in engineering is that redundancy leads to ultra-high equipment and procedural reliability and to overall system safety. Such redundancy may be in the form of multiple copies of one type of protection or to backups and layers of defense. Chapter 11 discusses the general flaws in this belief.

As is common in the process industry, lots of redundancy was included in the design. Poor decision-making, however, can invalidate the assumption of multiple lines of defense. For example, severely underbalancing the well while displacing the mud from the riser left reliance on the high-risk bottomhole cement as the exclusive barrier in the wellbore. The negative-pressure test was used to test the bottomhole cement, but when the results indicated that that cement was not holding, the results of the test were ignored.

Much of the complacency about intermediate lines of defense arose from the belief that the BOP would be a reliable last line of defense if everything else failed. This belief was unjustified. The blind shear ram, which is supposed to cut the drill pipe and seal off the well, did not work at Macondo. A report by Det Norske Veritas examined fifteen thousand wells drilled off North America and in the North Sea from 1980 to 2006. It found eleven cases where crews on deepwater rigs had lost control of their wells and then activated blowout preventers. In only six of those cases were the wells brought under control, leading the report to conclude that in actual practice, blowout preventers used by deepwater rigs had a failure rate of 45 percent. Was an in-depth hazard analysis ever done on the design of BOPs?

Because BOPs play such a critical role in preventing a disaster, the designs incorporate a lot of redundancy. In this case, the *Deepwater Horizon* BOP could be activated in five different ways:

1. direct activation of the BOP by pressing a button on a control panel on the rig;
2. activation of the emergency disconnect system (EDS) by rig personnel;

3. direct subsea activation of the BOP by a remotely operated vehicle (ROV);
4. activation by the automatic mode function (AMF) or deadman system due to emergency conditions or initiation by an ROV; or
5. activation by the autoshear function if the rig moves off location without initiating the proper disconnection sequence or if initiated by an ROV.

All of this redundancy provided great confidence that, no matter what, they could always find a way to activate their last line of defense. However, none of the five activation modes were successful. A forensic analysis of the BOP after the blowout showed that it failed because it could not cut the pipe to shut in the well. This analysis found that the embedded activation systems lacked sufficient charge, that a previous modification had interchanged some connections of the blind rams preventing their activation robotically, and that the BOP had some leaks and lacked maintenance and recertification; these were all potential contributing factors to the BOP's inability to cut through the drill pipe.

The first two ways to activate the BOP required communication with the rig, but during the emergency, power, communication, and hydraulics connections between the rig and the BOP were cut. In that case, the AMF (deadman) should have been effective, but that failed too. The AMF uses two redundant control systems, the yellow pod and the blue pod, to initiate closure of the blind shear ram. Blind shear rams are a part of the BOP that can shear a drill pipe and seal a wellbore. On the day of the accident, however, only one of the two pods was functioning.

The blue pod was miswired, causing a critical battery to drain, rendering the pod inoperable. A critical solenoid valve in the yellow pod had also been miswired. A solenoid is a valve that opens and closes as a result of an electrically initiated magnetic switching device to control the flow of liquid or gas. In fact, a common factor in accidents relying on redundancy arises when wiring or maintenance is done by the same person or by multiple people making a common mistake. In this instance, the wires within the solenoid all had similar connectors that lacked differentiation to ensure proper wiring. The miswired solenoid valve in the yellow and blue pods would not have passed the manufacturer's factory acceptance testing procedures. However, Transocean had made extensive modifications to the BOP before the blowout and had wired it wrong, apparently without adequate testing of the modified system.

The redundant coils were designed to work in parallel to open the solenoid valve, but the miswiring caused them to oppose one another. In fact, even if both coils had been successfully energized during the blowout (as they were supposed to be), the solenoid valve would have remained closed and unable to initiate closure of the blind shear ram because of the wiring error [386]. During the blowout, a drained battery likely rendered one of these coils inoperable, which serendipitously allowed the other coil to activate alone and initiate closure of the blind shear ram. This lucky failure did not, however, solve the problems given all the other design errors and failures that occurred. For example, there was a significant leak in a key hydraulic system.

A confidential report for Transocean in 2000 on the risk of the *Deepwater Horizon* BOP control system found a large number of single points of failure for the blind shear ram, including a final shuttle valve, which supplies the hydraulics to the blind shear ram. If this valve jammed or leaked, the ram's blades would not budge.

The blind shear ram in the *Deepwater Horizon* BOP did not meet the manufacturer's published design shearing capabilities for the diameter and strength of drill pipe used during all of the *Deepwater Horizon* drilling operations except for the drill pipe used on April 20; thus, for an extended time during the drilling process, the *Deepwater Horizon* BOP could not have reliably sheared the drill pipe during an emergency [386].

Though the emergency valves are often referred to as a fail-safe backup to undersea drilling, they have several weaknesses, particularly if crews do not activate them correctly. The emergency shutoff valve is designed to slam two pieces of metal into the drilling pipe with enough force to cut the pipe and seal the well. However, it isn't always strong enough to work. The devices may be incapable of snapping the joints that connect lengths of drill pipe, which are thicker and stronger than the pipe itself. The crew on the *Deepwater Horizon* had not ensured that the drill pipe was across the blind shear ram and not a pipe joint. In an emergency, there is no time for a driller to make sure the ram's blades are clear of a drill pipe joint, which make up almost 10 percent of the drill pipe's length.

Additionally, as wells have been drilled increasingly deeper, the piping inserted down the shaft has become so strong that the devices may not be capable of shutting off the flow even when the drill pipe is under the blind shear ram. The worst offshore oil spill in history was linked to a blowout preventer that activated but did not completely shut off the well. The Ixtox 1 well off Mexico's Yucatán Peninsula had a blowout in 1979, which was one of the largest spills on record. When workers activated the device, it did not have the strength to cut through the pipe, according to an MMS report. A line carrying hydraulic fluids to the BOP also burst.

In 1990, a blind shear ram could not stop a major blowout on a rig off Texas. It cut the pipe, but investigators found that the sealing mechanism was damaged. And in 1997, a blind shear ram was unable to slice through a thick joint connecting two sections of drill pipe during a blowout of a deep oil and gas well off the Louisiana coast.

Over time, layers of redundancy designed into a system can be inadvertently or even intentionally eliminated. In 2004, BP opted to remove a layer of redundancy from the BOP. It asked Transocean to replace one of the blowout preventer's secondary rams with a "test ram"—a device that would save BP money by reducing the time it took to conduct certain well tests. In a joint letter, BP and Transocean executives confirmed that BP was aware that the change would reduce the built-in redundancy and increase risk. By mistake, Transocean also connected one of the emergency activation mechanisms to the test ram. This modification prevented the blind shear rams from being activated robotically and thus delayed activation.

Problems not only existed in the design of the BOP but also in its testing. For the BOP to effectively contain a kick or blowout, the following safety requirements must be met:

- Power supply and hydraulic pressure from the rig and from the backup systems embedded in the BOP must feed its rams at all times.
- The BOP must be tested at a pressure such that well containment can be guaranteed.
- The blind shear activation modes must be tested and operational under expected blowout conditions (pressure, temperature, power supply, and signal communications).

These requirements were not met at Macondo.

Consider the first requirement. The BOP installed at Macondo was found to have insufficient battery charge to activate the rams, and some investigations report that the BOP activation systems were all connected to the same energy supply at the rig, without any backup.

The second and third requirements have to do with testing. BP and Transocean, with the MMS authorization, never tested the BOP to ensure it could withstand a blowout under Macondo's conditions. The *Deepwater Horizon* was due to be completely recertified five years before, but that was never done, perhaps because that would have taken it out of service for several months. The last maintenance checks had been done in 2005, although the BOP could not be examined thoroughly at that time because it was operating on the seafloor and was therefore inaccessible. The checks that were done at this time found significant problems with the BOP. The control panels on the rig that operate the blowout preventer acted strangely, giving unusual pressure readings and flashing unexplained alarm signals. A critical piece of equipment, the "hot line" that connects the rig to the blowout preventer, was leaking badly.

At the time of the incident, neither recommended industry practices nor US regulations required testing of the redundant components of the AMF/deadman. The industry defended the practice of not testing the BOP under blowout conditions in order to avoid damaging the equipment unnecessarily, but studies published by the MMS before the Macondo accident show that more than half of the BOPs in service were not able to control a blowout, so accurate testing was recommended and described as essential when installing the BOP to ensure well control. But the MMS had never required testing of the deadman, the autoshear, or the underwater robots. The agency did not even require rigs to have these backup systems in place at all, although it did send out a safety alert encouraging their use.

In summary, the BOP was recognized as critical for safety and was designed with redundant components to ensure high reliability. What was not accounted for was common-mode failures and design errors, which resulted from inadequate engineering and underestimation of the risk for such failures. Testing was inadequate to identify these problems.

Confidence in the BOP also did not account for decision-making based on profitability and other social and time pressures that would result in inadequate maintenance of the BOP and decisions to put off replacing the BOP batteries.

Missing, ineffective, and inaccurate risk assessment Inaccurate risk assessment is a major contributor to complacency. BP, Halliburton, and Transocean did not perform a formal risk assessment to identify or address risks of an accident in the design, cementing, and temporary abandonment procedures. Decisions were made without fully appreciating how essential the decisions were to well safety and with limited assessment of their risks. As a result, officials made a series of decisions that saved BP, Halliburton, and Transocean time and money—at least in the short term—but without full appreciation of the associated risks.

BP's project development practices did require a relatively robust risk analysis and mitigation or peer review during the planning phase of the well, but not during execution of those plans. The decision to perform any formal risk analysis was left to the team's discretion. A management of change process was used during planning of the well but was not required for changes to drilling procedures during operations such as the temporary abandonment of the well [367].

The migration of the well to a state of high risk driven by cost and time demands was most evident during the last 10 days before the blowout, when the transition from drilling to production began. Changes were made at the last minute, such as the number of centralizers, without conducting a formal, disciplined analysis of the impact of these decisions and events. Ad hoc decisions and plan deviations at the last minute were based on personal judgment without adequate information, expert input, or a formal risk assessment. There was no consultation with experts inside or outside BP. The few standards and regulations existing at the time were ignored.

Anomalous readings and events occurred without adequate consideration of what they might mean. The last-minute changes did not lead to a heightened alertness to risk. Instead, everyone assumed that the cement was fine and kept running tests and coming up with various explanations for thc unexpected results until they had convinced themselves their assumption was correct.

While a full peer-reviewed risk analysis is not always practical when critical decisions need to be made in a limited amount of time, some controls on decision-making are needed. There does not seem to have been a safety expert on the rig at the time of the temporary abandonment helping with the decision-making nor communication with any such expert on shore when the decisions and changes were made.

The problems were not only in late changes to plans. The use of a long-string production casing rather than a liner had been in the plan all along. It was not until problems started to occur that the team identified the risks associated with using a long string.

Decisions were made on the basis of convenience or time without appropriate consideration of increased risks. Instead, everyone started from the assumption that a kick could not be occurring. They kept running tests and coming up with various explanations for the unexpected results until they had convinced themselves their assumption was correct.

Even the risk assessments that were done were not rigorous. For the cementing task, a simplified decision tree was used in which complex risks, such as the risk of failed cementing, can be forgotten or ignored on the basis of simple and incomplete indicators such as partial returns or lift pressure [367]. The tool focused on reducing risks of lost returns and annular pressure increase while ignoring overarching risks such as well integrity.

In general, analysis and communication of risks within BP used an approach that combined safety risks with efficiency and other risks in a way that obscured the safety risks. The risks were all combined into one number so that the people above had an unclear view of the safety risk that was being taken. Tradeoffs were primarily influenced by costs and reservoir integrity issues that might reduce later retrieval of oil and gas from the well.

Ignoring warning signs and audits It is very easy not to see warning signs, even when there are lots of them. Hindsight bias usually makes events appear to be warnings that are viewed at the time as random occurrences and not identified as suggesting an increase in risk. Deepwater drilling is complicated and challenging. Some crew members dubbed the Macondo the "well deepwater well from hell," but other wells had also earned that nickname.

Audits and reports to management about the existence of specific problems are harder to ignore and misread. An audit of the *Deepwater Horizon* before the Macondo blowout identified the need for equipment that would increase the well's monitoring quality and accuracy. For example, personnel had to perform basic well monitoring calculations by hand, instead of having automated systems to help monitor the well. Inadequacies were identified in the sensors and instrumentation for detecting kicks. There was no camera installed on the rig to monitor flow on the overboard line.

In spite of having all its certifications and inspections up to date, by the time of the accident, the *Deepwater Horizon* was operating with many maintenance deficiencies. For example, a 2009 BP audit on the rig before it headed to the Macondo well identified 390 repairs that needed immediate attention. They would have required more than 3,500 hours of labor to fix and some downtime ashore. But the *Deepwater Horizon* never stopped working between the audit and the accident [275].

Confusing process safety with personal safety As is common in many companies, both BP and Transocean collected, measured, and rewarded personal safety. Both companies achieved low personal worker injury rates and assumed that these indicated that process safety was also high.

One reason for this common practice is that it is much easier to define and collect metrics for personal safety, such as days missed from work, than for process safety. But the result is that often companies advertise slogans like "safety first" but the metrics they collect only encompass a subset of the risks of operations, in this case drilling. At BP, executives and managers had safety at the top of their key performance indicators but it was measured only in the form of recordable injuries. BP did not appear to have tracked how employee decisions impacted process safety or risk [21; 367].

Given the large number of accidents that BP had in their drilling, refinery, and pipeline operations in the ten years before the Macondo blowout and the repetition of the same causal factors in these accidents, it does not appear that process safety was an important priority for the company.

C.6.4.5.2 Results of complacency and overconfidence Complacency and overconfidence can create the action and inaction that leads to serious accidents. For example, it can lead to tolerance of serious deviations from operating procedures; prioritizing cost and schedule over safety; not investing in safety procedures, equipment, and maintenance; and allowing migration of the organization and systems to states of high risk.

Tolerating serious deviations from operating procedures Complacency led to a tolerance for serious deviations from planned operating procedures. Changes were made without apparent concern about the potential to increase the risk of a blowout. One example is the well site leaders not using the real-time data available to them to find the cause for the anomalous pressure readings during the negative-pressure test. Another is not ordering enough centralizers on time or waiting for them to arrive despite the original well design requiring sixteen centralizers. There did not seem to be any real belief that an accident was possible.

Prioritizing cost and schedule over safety Complacency can lead to prioritizing cost and schedule over safety. In the days leading up to the final cementing process, BP engineers focused heavily on their biggest challenge, which they considered to be the risk of fracturing the formation.

According to the *Deepwater Horizon* Presidential Commission Report, a BP engineer, who later resigned from the company, told the commission that:

BP focused heavily on personnel safety and not on maintaining its facilities. He added that BP was preparing to sell the depleted field, and was running it at minimum cost: "The focus on controlling costs was acute at BP, to the point that it became a distraction. They just go after it with a ferocity that is mind-numbing and terrifying. No one's ever asked to cut corners or take a risk, but it often ends up like that." [275, p. 219]

Transocean charged BP approximately $1 million per day ($500,000 for leasing the rig and around the same in contractors' fees). BP originally estimated that drilling the Macondo well would take 51 days and cost approximately $96 million. By the day of the blowout, the rig had been at the location for 80 days, was 43 days late for its next drilling location, and had far exceeded its original budget. A petroleum engineer for the MMS testified to the Presidential Commission Report that BP had applied to use the *Deepwater Horizon* rig to drill in another field by March 8.

The well team leader had drilling efficiency as his number one priority in his contract for 2010. The design of the temporary abandonment procedure was done by a junior engineer who prioritized costs and efficiency. The following are unsafe decisions that potentially saved time or money:

- making a choice of casing that focused on long-term reward and not in short-term risks
- not waiting for more centralizers of the preferred design
- not waiting for foam stability test results or redesigning the slurry
- not performing a cement bond log, which would have tested the integrity of the cement; a team from Schlumberger, an oil services firm, was on board the rig to do the test, but BP sent the team home without doing it the morning of April 20
- not fully circulating the drilling mud, a 12-hour procedure that could have helped detect the gas pockets that later shot up the well and exploded on the drilling rig
- displacing the mud and testing the cement simultaneously
- using spacer made from combined lost circulation materials to avoid disposal issues
- displacing mud from the riser before setting the surface cement plug
- setting the surface cement plug 3,000 ft below the mudline in seawater
- not installing additional physical barriers during the temporary abandonment procedure
- not performing further well integrity diagnostics in light of troubling and unexplained negative-pressure test results
- bypassing pits and conducting other simultaneous operations during displacement
- delaying maintenance of critical equipment

There is nothing inherently wrong with choosing a less costly or less time-consuming alternative as long as it is proven to be equally safe. The problem here is BP's Macondo team did not have a formal system for ensuring that alternative procedures were in fact equally safe. None of these decisions were subjected to a comprehensive and systematic risk or hazard analysis, peer review, or management of change process. Companies with good safety practices reward employees and contractors who take action when there is a safety concern even though such action costs the company time and money in the short term.

Not investing in safety, safety equipment, and maintenance As was true for the BP Texas City explosion, internal company audits reported missing and inadequate equipment, but the equipment was not updated.

Allowing migration to states of high risk In retrospect, it is clear that the *Deepwater Horizon* was quickly moving to a state of high risk for a major process accident. This migration process is natural for companies operating in a competitive environment [320], so it must be limited by specific actions. In this case, not only were procedures changing without any effective management of change process being applied, but the state of the equipment itself was degrading through inadequate maintenance. The same was true at Texas City and other BP major accidents at the time.

The BP and Transocean maintenance policy deviated and bypassed maintenance regulations and recommended practice for critical equipment. At the time of the accident, the *Deepwater Horizon* rig was operating with numerous maintenance issues, as noted previously.

As operator, BP specified the configuration of the BOP but delegated to Transocean its operation and maintenance since the rig started operating in 2000. The required recertification was therefore long overdue. The recertification process requires complete disassembly on the surface, which can take up to three months or longer and generally requires time in dry dock. According to industry practice, the best time to perform major maintenance on a complicated BOP control system is during its five-year interval inspection period [275]. The *Deepwater Horizon* never stopped working since its commission, and neither did the BOP. The batteries, which were dead at the time of the blowout, should have been inspected and changed before it was lowered to the seabed, but they were not.

Transocean had a philosophy of condition-based maintenance where "the equipment shall define the necessary repair work, if any" [384].This philosophy led to improperly maintained critical equipment, such as the BOP and the rig gas and fire sensors. The maintenance actually performed did not follow MMS, API, and manufacturers recommendations.

Adding to the problems, Transocean's Rig Management System II (RMS) complicated maintenance on board. The RMS delivered duplicated and erroneous maintenance orders, leaving unattended relevant equipment, such as the computers that provided important data for understanding the state of the well, which operated intermittently and had outdated software.

C.6.4.5.3 Organization and management As usual, organizational and management contributions are important in understanding why this accident occurred.

Diffusion of responsibility and authority, and lack of corporate oversight and leadership Problems in the BP safety management system were described in the appendix on the Texas City oil refinery explosion. Briefly, assignment of responsibility for safety at higher levels of management were limited and focused on personal safety.

Just as nobody on the BP board of directors had refinery operational experience, no member of the board of directors had a professional background in offshore drilling relevant to the risks that were being taken at a well like Macondo [386]. Management and the board of directors play an important role in establishing an organization's safety culture.

Problems were exacerbated here by having multiple companies interacting without a clear understanding of the safety responsibility, authority, and accountability for each. As owner of the rig, Transocean had primary responsibility for the rig equipment. But operational responsibility was undefined. Because BP was the lead company on the rig, it should have had primary responsibility for ensuring that proper safety procedures were being followed by its contractors.

BP appeared to take responsibility for risk-related decision-making but did not establish effective communications to ensure that they were getting the information they needed. Responsibility cannot be delegated to contractors; their safety activities need to be managed and overseen by the lead organization. The lead company also establishes the basic safety culture on the rig.

In the Congressional hearings following the blowout, BP testified that the massive Gulf oil spill was caused by the failure of a safety device (the BOP) made by Transocean. It was argued that because BP did not own the rig, the responsibility for the safety of drilling operations belonged to Transocean. In turn, Transocean testified that BP was in charge: BP had prepared the plan and given the go-ahead to fill the well pipe with seawater before the final cement cap was installed, which reduced the downward pressure. Both BP and Transocean claimed that the MMS had approved the plan. Both also said that Halliburton poured the cement to plug the well and did not do it right. Halliburton claimed that it was only following BP's drilling plan and that their work was in accordance with the requirements set by BP and followed accepted industry practices.

While such finger pointing is common after a major accident, the real problem was the lack of coordination and leadership for the overall operations. Believing that everyone is and can be responsible for safety is a widespread belief in many industries. The problem is that where everyone is responsible for safety and for resolving conflicts between safety and other goals, then nobody is responsible for safety.

Unclear or shared responsibility is just as problematic. If responsibility is shared, people will assume that the other responsible people will behave autonomously on procedures in which they share responsibility.

A related problem is visibility. Detection of a kick and therefore the prompt activation of the BOP was the responsibility of the Transocean crew. Verifying the quality of the cement barrier by interpreting the negative-pressure test was done by BP engineers. The problem is that many of the safety responsibilities were not unrelated and may require more visibility into the larger system state than is possible at lower levels in the hierarchy. A decision that seems perfectly safe at one level of the system can be dangerous for the system as a whole. Or decisions made in parallel without adequate communication with others can be unsafe.

Personal responsibility for doing one's job safely is of course important. But overall system safety requires leadership and assignment of responsibility and authority to individuals who can be held accountable. In this case, there appeared to have been confusion about who was accountable for ensuring the adequacy of the cement slurry design, determining the risks attendant to changes in operations, and assessing the competence of personnel assigned to perform the negative-pressure test.

In Congressional testimony, a petroleum engineering professor testified that "*The individual contractors have different cultures and management structures, leading easily to conflicts of interest, confusion, lack of coordination, and severely slowed decision-making*" [275, p. 229]. Chapter 14 describes how to design an effective safety management system to overcome these problems.

The CSB report noted that the complexities of multiparty risk management in the offshore industry led to inadequately defined safety roles and responsibilities between the operator and the drilling rig contractor. Ultimately, while BP and Transocean had corporate polices for risk management, neither company ensured their implementation at Macondo (volume 3, section 4.0–4.5 [386]).

The problems were not simply limited to BP. A safety management and safety culture survey of Transocean personnel was conducted just a few weeks before the accident. It found that 46 percent of crew members surveyed felt that some of the workforce feared reprisals for reporting unsafe situations, and 15 percent felt that there were not always enough people available to carry out work safely [275]. Some Transocean crews complained that the safety manual was unstructured, hard to navigate, and not written with the end user in mind, and that there was a poor distinction between what was required and how this should be achieved. According to the final survey report, Transocean's crews "*don't always know what they don't know. [F]ront line crews are potentially working with a mindset that they believe they are fully aware of all the hazards when it's highly likely that they are not*" [275, p. 224].

Transocean provided minimal internal guidance and unclear expectations of the risk management tools its personnel should use for an offshore operation or facility. The same was true for BP. The more rigorous ones were not applied at the Macondo well. Transocean claimed not to have used the more rigorous ones because US regulations did not require them [386]. Clearly, not being required to do something does not abdicate responsibility for not doing it.

In general, for good decision-making, people need adequate training, information, procedures, resources, and support. Otherwise, they are unlikely to be able to do their jobs effectively. Decisions were made on the rig during the temporary abandonment without any of these things. An additional form of support is providing a safety engineer to help with safety-related decisions. There should have been such a person assigned to and on the rig when the critical tasks were being performed. All responsibility was instead placed on managers and workers with conflicts in their responsibilities. Projects need to assign responsibility for assisting with safety-critical decisions assigned to people who specialize in and are responsible for providing the necessary information to managers so they can make better decisions.

Inadequate management of change BP, like most companies, had a management of change policy. To be effective, however, responsibility needs to be assigned to make sure that it is understood and applied. Numerous changes were made to the temporary abandonment procedures in the two weeks before the blowout. There is no evidence that these changes went through any sort of formal risk assessment or management of change process or review of any kind. There were no controls in place to ensure that key decisions were safe and sound from an engineering perspective.

The decision to perform any formal risk analysis was left to the team's discretion; an MoC was optional and applied mainly to decisions to deviate from well plans approved during the project creation stages and not to drilling procedures such as the temporary abandonment of the well.

Inadequate training The accident reports noted that the rig crew had not been trained adequately to do their assigned work or to respond to such an emergency. People were assigned responsibility for safety-related decision-making without adequate training, information, procedures, resources, and support to do their jobs effectively.

The accident reports also noted that Transocean managers deliberately decided not to train their personnel in the conduct or interpretation of negative-pressure tests. The rig workers were supposed to learn about these procedures through general work experience.

The same was true for BP. The two well site leaders displayed unfamiliarity with negative-pressure test theory and practice. Neither calculated pressures or volumes before running the negative-pressure test. When someone suggested that they had heard something about a "bladder effect" that might explain the results, nobody questioned this conclusion even though no such effect existed.

Outdated work procedures, lack of operating discipline, and poor contingency planning As previously noted, procedures changed frequently. Nobody saw the procedure they would use for temporary abandonment until 11 a.m. that day. BP had no consistent or standardized temporary abandonment procedures across its Gulf of Mexico operations, and the formal written guidance was minimal. For example, the guide available did not specify the location of those barriers or the procedure by which they should be set. This left the Macondo engineers to determine such issues for themselves on an ad hoc basis.

As an example, neither BP nor Transocean had pre-established standard procedures for conducting a negative in or pressure test. There was no standard procedure for running or interpreting the test in either MMS regulations or in documented industry protocols. In fact, the regulations and standards did not require BP to run a negative-pressure test at all. BP and Transocean had no internal procedures for this test and had not formally trained their personnel how to do it. The Macondo managers did not provide their team with specific tests for doing it at Macondo.

Compounding the lack of written procedures for the test, BP did not have any policy, or at least did not enforce any policy, that would have required personnel to contact experts on shore to call for a second opinion about confusing data.

A lack of contingency planning probably contributed to the losses. Given the chaos that ensued in departing the rig, clearly the crew had not been prepared adequately for an emergency like a blowout. Comparison with the behavior on the *Bankston* mudboat is instructive.

While chaos reigned on the oil platform, the *Bankston* crew performed spectacularly well. During the hearings that followed the accident [385], the captain of the *Bankston* explained that whenever they perform a potentially dangerous job, they plan extensively. In the morning review, they always rehearse all the potential contingency actions that may be needed during the day. When the explosion occurred and the *Deepwater Horizon* crew were panicking and jumping into the sea, the *Bankston* crew calmly went through their planned actions including immediately releasing the hose carrying the mud to the ship in order to protect it and then boarding their rescue boats to save those on the platform who were jumping into the sea. Without their calm and preplanned response, more lives might have been lost.

Worker fatigue Drilling for oil is hard, dirty, and dangerous work that combines heavy machinery and volatile hydrocarbons extracted at high pressures. The rig operated continuously. Most workers put in a 12-hour day or night shift, working three straight weeks on and then having three weeks off. Fatigue and time pressures may have contributed to the events.

Poor learning from events BP had many major accidents in the ten years preceding the *Deepwater Horizon* loss, with the resulting accident reports identifying the same or similar causal factors. Apparently, these factors were never fixed. In the previous section on the explosion of the Texas City refinery explosion, it was noted that the same causal factors that had been cited in the three serious accidents at the BP Grangemouth Complex in 2000 were also cited in the investigation of Texas City. It was not only BP's refineries that were having problems. In November 2003, a gas line ruptured on BP Forties Alpha platform in the North Sea, flooding the platform with methane. BP admitted breaking the law by allowing the pipes to corrode. In all of these instances and in the Macondo accident, the same factors were identified as occurred in the Macondo blowout.

BP had an electronic incident reporting system at the time but did not use it to identify system-level causes and to track corrective action items for a delayed kick response at Macondo well a month before the blowout [386]. BP was aware of the weaknesses in its risk assessment process; a 2008 BP internal review found that risk assessment required improvement—that is, stronger major hazard awareness and integration of assessment processes and results.

Halliburton and Transocean also had learning problems. Halliburton prepared cement for the Macondo well that repeatedly failed Halliburton's own laboratory tests. Then, despite those test results, Halliburton managers onshore let its crew and those of Transocean and BP on the *Deepwater Horizon* continue with the cement job, apparently without first ensuring good stability results. Halliburton also was the cementer on a well that suffered a blowout in August 2009 in the Timor Sea off Australia. The *Montara* rig caught fire, and the well leaked tens of thousands of barrels of oil over two and a half months before it was shut down. The leak occurred because the cement seal failed, according to the government report on the accident. The report said it would not be appropriate to criticize Halliburton because the operator "exercised overall control over and responsibility for cementing operations" [275]. The inquiry concluded that "Halliburton was not required or expected to 'value add' by doing more than complying with [the operator's] instructions" [275]. This report demonstrates the common focus on placing blame rather than learning from events and how this can interfere with learning and repetition in the future.

Transocean was cited in the accident reports as having poor dissemination of lessons learned between teams and no record keeping of rig and equipment modifications and maintenance histories. Lessons from an earlier near miss were never adequately communicated to its crew. The same is true for a similar near miss on one of its rigs in the North Sea four months prior to the Macondo blowout. On December 23, 2009, gas entered the riser on that rig while the crew was displacing a well with seawater during a completion operation. As at Macondo, the rig's crew had already run a negative-pressure test on the lone physical barrier between the pay zone and the rig and had declared the test a success. The tested barrier nevertheless failed during displacement, resulting in an influx of hydrocarbons. Mud spewed onto the rig floor—but fortunately the crew was able to shut in the well before a blowout occurred. Nearly one metric ton of oil-based mud ended up in the ocean.

The MMS tried to expand data reporting requirements as part of an effort to track and analyze offshore incidents and to identify safety trends and leading and lagging indicators.

The proposal was abandoned when industry complained about compliance cost and overlap with Coast Guard reporting requirements.

As a result, the United States has historically had no legal requirement that industry track or report instances of uncontrolled hydrocarbon releases or near misses. At the same time, the United States has the highest reported rate of fatalities in offshore oil and gas drilling among its international peers, but it has the lowest reporting of injuries. This suggests that there is significant underreporting of injuries.

Limited communication channels and poor information flow Communication is clearly important for safety, particularly when multiple contractors are involved and various corporate cultures, internal procedures, and decision-making protocols must be integrated. While the engineering team was making decisions about one aspect of the well, well site leaders were making decisions about other aspects at the same time. There was no effective communication of critical information across teams, and apparently no one seems to be analyzing the risks involved in the simultaneous actions at a system level.

No BP personnel contacted anyone onshore to get expert opinion, for example, to discuss their inability to bleed off drill pipe pressure during the negative-pressure test. They did not even seek a second opinion from their managers, both of whom were engineers and were on the rig during the negative-pressure test as part of the VIP visit. Such communication was never made clear as necessary or important for the well site leaders. The engineering manager was not fully aware of the well site leader's report of unsafe operations, frustrated personnel, and lack of clarity in procedures.

The contractors' contribution to risk assessment in Macondo was no better. It appears that they focused only on their tasks without providing important information to the decision-makers at BP, sharing valuable lessons with them, or raising awareness of the imminent danger of the operation.

For example, Transocean did not report any risk analysis or mitigation plan regarding their performance of simultaneous operations during critical steps (such as the negative-pressure test) and omitted sharing with BP their incident in the North Sea. Nor is there evidence that someone from Halliburton performed a comprehensive hazard analysis of the foamed cement design and job for the specific conditions of Macondo.

It is unclear whether Halliburton ever reported the pilot tests on the cement that were done soon after work began and showed the results would be unstable.

Faced with anomalous data, decision makers did not seek advice from others with expertise but instead decided to continue with the operation with incomplete or inaccurate data. None of these individuals on the rig understood when and why they should ask for help.

Appendix D
Nuclear Power: Three Mile Island, Chernobyl, and Fukushima Daiichi

D.1 Background

A brief introduction to nuclear power plants is provided to help the reader understand the accidents. An introduction to how safety is handled for nuclear power is also provided. Then a few of the major accidents are described.

D.1.1 How a Nuclear Power Plant Works

A nuclear reactor generates heat as a result of the splitting apart of an atomic nucleus, most often that of the heavy atom uranium. This process is called nuclear fission. The nucleus at the core of each atom contains two types of particles tightly bound together: protons, which carry a positive charge, and neutrons, which have no charge.

When a free neutron strikes the nucleus of a uranium atom, the nucleus splits apart, producing two smaller radioactive atoms, energy, and free neutrons. Most of the energy is immediately converted to heat. The new free neutrons can now strike other nuclei, producing a chain reaction and continuing the fission process.

Free neutrons can be captured by the atomic nuclei of some elements, such as boron or cadmium, which stops them from continuing the fission process. Thus, elements that are strong absorbers of neutrons can be used to control the rate of fission and to shut off a chain reaction almost instantaneously [173; 202].

Other radioactive materials are products of the fission of uranium. They are almost all unstable and hence radioactive, with half-lives ranging from millionths of a second to hundreds of years. They decay by emitting gamma or beta radiation. While the reactor is working, new fission products are constantly being formed, while those formed earlier are constantly decaying.

The free neutrons produced by nuclear fission are inefficient producers of further energy at the high energy at which they were originally emitted. Essentially, they move too fast and miss the uranium atoms too easily. Therefore, they must be slowed down to induce additional fissions efficiently and to sustain the chain reaction.

A *moderator* is used to slow down the neutrons, but it should not capture them, or the reaction will stop. The particular moderator used is one of the main differences between different types of nuclear power plant. Other differences arise in the layout of the various components, the canning material (the material in which the core is held), the coolant, the pressure under which the coolant operates, and whether the reactor uses natural or enriched uranium as fuel.

Nearly every reactor has the following components, shown in figure D.1 [259]:

- The *core* consists of the fuel elements and moderator.
- The *control rods* contain elements that absorb free neutrons and thus control the speed of the reaction.
- The *coolant* carries away the heat generated in the core. The coolant, by removing heat, also keeps the fuel rods from becoming overheated.
- The *heat exchanger* transfers the heat carried by the coolant to water in a secondary circuit, which boils and creates steam.
- The *turbine(s)* converts steam to rotational energy, which powers a generator that creates electricity.
- The water in a secondary circuit, called *feedwater*, transfers the heat from the steam generators to the turbine.
- A *biological shield* absorbs any neutrons produced in fission that escape from the core.
- A *containment vessel* contains any radioactivity that escapes from the core or coolant system.

In the 1950s, when nuclear power and energy plant designs were first being conceived by the five nuclear powers—the United States, Russia, Britain, France, and Canada—only Canada had no interest in building nuclear weapons. The rest wanted to manufacture plutonium as well as generate electricity [259]. To accomplish both requires different reactors or limits the design options to rather unsatisfactory compromises.

For commercial power generation, the United States chose light water reactors, which are cooled and moderated using ordinary water. This type of reactor tends to be simpler

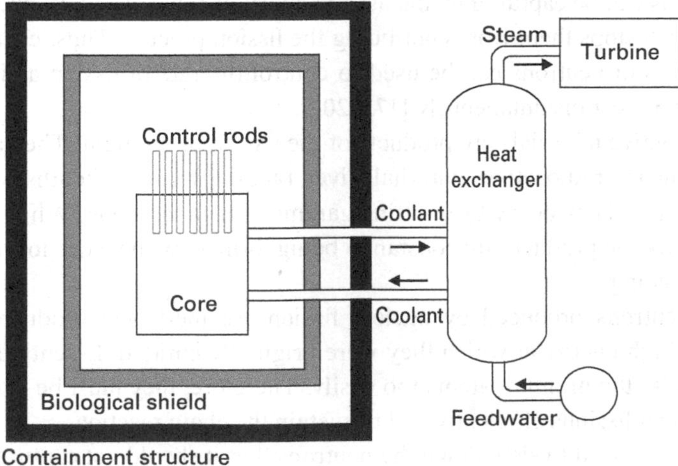

Figure D.1
Generic nuclear power plant components.

and cheaper to build and makes up most of the reactors operating today. Because of the high absorption of neutrons by water, the fuel must be considerably enriched (the uranium content is 3 percent instead of the 0.7 percent of natural uranium).

The United States already had a large fuel-enrichment program, so getting the enriched fuel was not a great problem. The disadvantages of this design are that the fuel cycle is more expensive and the compact size of the core may lead to control problems. On the other hand, the compact size makes for easier export sales, and the need for enriched uranium leads to the possibility of future business in the reprocessing and enrichment of fuel because most countries buying reactors do not have this capability.

The first reactor, built under the direction of Admiral Hyman Rickover, was a power unit for the submarine USS *Nautilus*, launched in 1954. The submarine design influenced the land-based versions, thc first of which was installed at Shippingport, Pennsylvania, and became the first nuclear power plant in the United States.

The reactor core in these reactors is contained in a thick-walled pressure vessel made of welded steel. As stated, the water acts as both a moderator and a coolant. At the top of the core, the hot water (about 320°C) is piped through the side of the pressure vessel to a heat exchanger, in which it is forced through thousands of small pipes immersed in a secondary water circuit. This secondary water is heated by contact. The water in the primary circuit is under high pressure, but the lower pressure of the secondary circuit water allows it to turn to steam when it is heated, and this steam is fed to the steam turbine. The primary coolant water is then pumped back to the reactor to be heated again.

A *pressure equalizer* maintains the coolant at the correct pressure (as the reactor power is changed) by allowing some of the water to evaporate or condense. The control and shutoff rods are generally suspended above the core inside the pressure vessel and are controlled through the lid of this vessel.

This type of reactor is refueled only when it is shut down. To accomplish this, the core is first allowed to cool down, and then a large tank on top of the core is filled with water to provide shielding and also to provide cooling water to remove the fission-produced heat. The lid of the pressure vessel is then removed and the fuel replaced. Because of the time required for this process, the reactor is usually shut down for refueling once per year, and a large percentage of the fuel is replaced at that time.

For safety, the pressure vessel, the core, and the primary coolant circuit are contained in a biological shield of concrete about 7 ft thick. In case a malfunction occurs in the cooling system, an emergency cooling system is provided. Additional safety features of the *defense-in-depth* approach are described later in this appendix.

The light water reactor described is called a pressurized water reactor (PWR). In a boiling water reactor (BWR), the water is allowed to boil within the core, and the resulting steam is piped directly to the turbine. Thus, no secondary coolant circuit is needed. However, something has to be done with the steam in case a malfunction makes the turbine unable to accept it. The reactor is therefore enclosed inside a second vessel, this one made of reinforced concrete, which acts as the biological shield and from which the steam can be channeled into a pool of water under the reactor. The BWR is also refueled when shut down.

The second type of reactor involved in the events described in this appendix is graphite moderated and water cooled. The first American reactor of this type was built at Hanford, Washington, in 1944 for the military production of plutonium. The British also built a graphite-moderated reactor at Windscale at about the same time and for the same purpose. The Soviets chose this type of reactor for civilian use for reasons discussed later.

The Soviet version of this reactor type, called the RBMK (a Russian acronym that roughly translates as "reactor cooled by water and moderated by graphite"), uses low-enriched uranium. Water is boiled directly within the core of the reactor and led off to drive the turbines.

Instead of using a single large pressure vessel to contain both fuel and coolant, the fuel elements are contained in more than 1,600 separate pressure tubes. Water is allowed to boil in the pressure tubes. The resulting steam contains many water droplets (which have to be removed before entering the turbine), so the steam is collected and dried in four huge drums. Each drum contains more than four hundred welds, which must be pressure tight. The graphite temperature is about 700°C under normal conditions in the RBMK reactor, which is above its ignition temperature in air. Therefore, the graphite is blanketed in an inert mixture of helium and nitrogen.

There has been some criticism within the Soviet Union about shortcuts taken in the haste of building the Chernobyl reactors (and others) and thus accusations of substandard construction and quality problems [259].

Although other types of reactors are not involved in these events, they are described briefly for comparison. Graphite-moderated, gas-cooled reactors use enriched uranium and are cooled with carbon dioxide. The carbon dioxide, like the water circulating in a PWR, is used to carry heat to heat exchangers, which change it to steam to drive the turbines. One advantage of this design is that it does not have to be shut down to be refueled. A new design of this type, the advanced gas-cooled reactor (AGR), uses enriched uranium and increases the potential power output of the reactor.

Canada uses heavy-water moderated and cooled reactors called CANDU (CANadian Deuterium Uranium) reactors. The heavy-water moderator is contained in a cylindrical stainless-steel tank (called a calandria), below which is an empty tank into which the moderator can be dumped in case of an emergency. There is also a large building next to the reactor, which is leak-tight and kept empty and under continual vacuum; this structure is to be used to contain any vaporized heavy water and volatile fission products released in a severe reactor accident. CANDU reactors can be refueled without being shut down.

A final type of reactor, called a fast reactor, does not require a moderator but does need a highly enriched fuel and uses liquid metals as coolant. The advantages of this type of reactor are the small size of its core and the possibility of "breeding" more fissile material. Removing heat requires the use of liquid metals, which are circulated through the core by the use of electromagnetic pumps having no moving parts. The liquid metal itself is the only part that moves. These liquid metals react explosively with water, and any accidental leaks of coolant will ignite spontaneously with the moisture in the air. An accidental fire cannot be put out with water but must be smothered with sodium chloride particles.

D.1.2 Safety Features

Nuclear power has the same type of safety engineering problems being faced by other new and potentially dangerous technologies, but, in addition, it has a relatively unique problem with public relations and has had to put a great deal of energy into convincing government regulators and the public that the plants are safe. This requirement, in turn, has resulted in a greater emphasis in some countries on probabilistic methods of risk assessment; the time required for empirical evaluation and measurement of safety is orders of magnitude greater than the pace of technological development [319] and also greater than what the public is willing to accept.

In an effort to promote the development of nuclear power and also partly because the associated hazards were not entirely understood, the industry was exempted from the requirements of full third-party insurance in some countries, for example, by the Price-Anderson Act in the United States [375]. Instead, government regulation and certification were substituted as a means of enforcing safety practices in the industry. The first nuclear power plant designs and sizes were also limited, although this has changed somewhat over the years as confidence in the designs and protection mechanisms has grown. In general, the nuclear industry and its regulators have taken a relatively conservative approach to risk issues.

The nuclear industry uses *defense in depth*, which includes [277]

- a succession of barriers to a propagation of malfunctions, including the fuel itself, the cladding in which the fuel is contained, the closed reactor coolant system, the reactor vessel, any additional biological shield, and the containment building (including elaborate spray and filter systems);
- primary engineered safety features to prevent any adverse consequence in the event of malfunction;
- careful design and construction, involving review and licensing at many stages;
- training and licensing of operating personnel;
- assurance that ultimate safety does not depend on correct personnel conduct in case of an accident; and
- secondary safety measures designed to mitigate the consequences of conceivable accidents.

Licensing is based on an identification of the hazards, design to control these hazards under normal circumstances, and backup or shutdown systems that function in abnormal circumstances to further control the hazards. The backup system designs are based on the use of multiple, independent barriers, a high degree of single element integrity for passive features, and the provision that no single failure of any active component will disable any barrier.

Siting nuclear power plants in remote locations is a form of barrier in which the separation is enforced through isolation. Perrow argues the impracticality of isolation:

> *The ideal spot for a nuclear plant cannot exist. It should be far from any population concentration in case of an accident, but close to one because of transmission economies; it has to be near a large supply of water, but that is also where people like to live; it should be far from any earthquake faults, but these tend to be near coastlines or rivers or other desirable features; it should be far from agricultural activities, but that also puts it far from the places that need its power. The result has been that most of our plants are near population concentrations, but in farming or resort areas just outside of them.* [303, p. 41]

Because of the difficulty in isolating plants, emergency planning has gotten more attention since the Three Mile Island (TMI) accident.

With the nuclear power defense-in-depth approach to safety, an accident requires a disturbance in the process, a protection system that fails, and inadequate or failing physical barriers. These events are assumed to be statistically independent because of differences in their underlying physical principles: a very low calculated probability of an accident can be obtained as a result of this independence assumption [319].

Recovery after a problem has occurred depends on the reliability and availability of the shutdown systems and the physical barriers. A major emphasis in building such systems, then, is on how to increase this reliability, usually through redundancy of some sort.

Probabilistic risk assessment has been proposed for and occasionally used in the nuclear industry. Risk assessments include estimating the reliability of the barriers and protection systems. In Britain, the worst possible accident that could occur, even if all the protection systems and barriers worked within their proper margins of tolerance following a disturbance, is called a *design basis accident* [138]. Any event bigger than a design basis release is an *uncontrolled* release. In the United States, design basis accidents usually refer to a set of disturbances against which the design is evaluated, including foreseeable failures of barriers.

Early risk assessments involved developing scenarios of *credible* or worst-case accident sequences that were still considered within the realm of possibility. These event sequences were called *maximum credible accidents*, and the design was evaluated against them. The probability of the maximum credible accident might be very low, but not so low as to make it impractical to incorporate safeguards. Hammer [131] illustrates this concept with the following simple example, framed in a slightly different context:

> *An aircraft is to be developed to carry a thermal device using a radioactive isotope as the source of energy. The device is nonexplosive; the principal danger is the possibility that a fire could cause dispersion of the radioactive material in airborne products of combustion. The Maximum Credible Accident would be an aircraft crash severe enough to rupture the device and then a fire that would cause isotope dispersion. Since aircraft crashes and fires do occur, it is necessary to provide a container that will not rupture should the Maximum Credible Accident occur.* [131, p. 67]

A possible criticism of the maximum credible accident approach is that the definition of what is credible is subjective. Farmer, in 1967, proposed a more rigorous approach to the assessment of nuclear plant safety that uses probability [99]. He claimed that for any given factory or other industrial installation, the acceptable frequency of accidents that may harm third parties varies inversely with the magnitude of the consequences of those accidents. He suggested that nuclear power plants be required to meet a safety criterion expressed in terms of consequence and probability.

Because empirical measurement of probability is not practical for nuclear plant designs, most approaches to probabilistic risk assessment build a model of accident events and construct an analytical risk assessment based on this model. Such models include the events leading up to the uncontrolled release as well as probability data on factors relating to the potential consequences of the release, such as weather conditions and population density.

The use of probabilistic risk assessment has been quite controversial. The arguments in favor are usually based on the ability of the technique to provide input to decision-making.

Identifying hazards alone does not help in determining how funds should be allocated in reducing them. Comparative probability data can be useful in such decision-making. Decision-making in the certification of plant designs is also eased by the use of such numbers, and probabilistic risk assessment has been advocated for use by government agencies in decision-making about nuclear energy and other potentially hazardous plants.

One trend in Britain is the substitution of *tolerable* risk for *acceptable* risk. If emphasis is placed on probabilistic risk assessment and other quantitative measures, then a need arises to determine how to make decisions based on those numbers, as discussed in chapter 13. The concept of *acceptable* risk has been used in this decision-making process; however, objections have been raised to labeling something with such serious consequences as a nuclear accident as "acceptable." As defined in a publication of the British Health and Safety Executive about risk from nuclear power stations, "*'tolerability' does not mean 'acceptability.' It refers to a willingness to live with a risk so as to secure certain benefits and in the confidence that it is being properly controlled. To tolerate a risk means that we do not regard it as negligible or something we might ignore, but rather as something we need to keep under review and reduce further if and as we can*" [138]. This change in terminology may reflect a change in philosophy, or it may simply be an attempt at better public relations.

Although the introduction of computers into protection systems started only recently, most government regulatory agencies are now being faced with certifying such systems. Emphasis in this certification activity in some countries is placed on the reliability of the operational software and, sometimes, on analytical assessments of this reliability.

D.2 Three Mile Island

The Three Mile Island accident in March 1979 involved a partial meltdown of reactor number 2 of the Three Mile Island Nuclear Generating Station in Dauphin County, Pennsylvania, and a subsequent radiation leak. The accident started when several water pumps stopped working. The Kemeny Commission, which investigated the accident, concluded that "a series of events—compounded by equipment failures, inappropriate procedures, and human errors and ignorance—escalated into the worst crisis yet experienced by the nation's nuclear power industry" [173, p. 81].

D.2.1 Background

Three Mile Island, located 10 miles southeast of Harrisburg, had two nuclear power plants, TMI-1 and TMI-2. Together they had a capacity of 1,700 megawatts—enough electricity to supply 300,000 homes. The two plants were owned jointly by Pennsylvania Electric Company, Jersey Central Power & Light Company, and Metropolitan Edison Company (Met Ed), and were operated by the latter.

At TMI-2, the reactor core held some 100 tons of uranium. The uranium, in the form of uranium oxide, was molded into cylindrical pellets, each about an inch tall and less than a half-inch wide. The pellets were stacked one on top of another inside fuel rods. These thin tubes, each about 12 ft long, were made of Zircaloy-4, a zirconium alloy. This alloy shell—called the *cladding*—transfers heat well and allows most neutrons to pass through. TMI-2's reactor contained 36,816 fuel rods.

TMI-2's reactor had sixty-nine control rods. Control rods contain materials called "poisons" because they are strong absorbers of neutrons and shut off chain reactions. The absorbing materials in TMI's control rods were 80 percent silver, 15 percent indium, and 5 percent cadmium. When the control rods were all inserted in the core, fission was effectively blocked, as explained earlier. Withdrawing them started a chain reaction. By varying the number of and the length to which they were withdrawn, operators could control how much power the plant produced. The control rods were held up by magnetic clamps: in an emergency, the magnetic field is broken and the control rods, responding to gravity, drop immediately into the core to halt fission. This process is called a *scram*.

The primary hazard from nuclear power plants is the potential for the release of radioactive materials produced in the reactor core as the result of fission. These materials are normally contained within the fuel rods. Damage to these fuel rods can release radioactive material into the reactor's cooling water, and this radioactive water might be released to the environment if the other barriers—the reactor coolant system and containment building barriers—are also breached.

A nuclear power plant has three basic safety barriers, each designed to prevent the release of radiation. The first line of defense is the fuel rods themselves, which trap and hold radioactive materials produced in the uranium fuel pellets. The second barrier consists of the reactor vessel and the closed reactor coolant system loop. The TMI-2 reactor vessel, which held the reactor core and its control rods, was a 40 ft high steel tank with walls 8.5 in thick. This tank, in turn, was surrounded by two separated concrete and steel shields, with a total thickness of up to 9.5 ft, which absorbed radiation and neutrons emitted from the reactor core. Finally, all this was set inside the containment building, a 193 ft high, reinforced-concrete structure with walls 4 ft thick.

In the normal operation of a pressurized-water reactor, it is important that the water heated in the core remain below *saturation*—that is, the temperature and pressure combination at which water boils and turns to steam. In an accident, steam formation itself is not a danger, because it too can help cool the fuel rods, although not as effectively as the coolant water. But problems can occur if enough of the core's coolant water boils away that the core becomes uncovered.

An uncovered core can lead to two problems. The first is that the temperature may rise to a point, roughly 2,200°F, where a reaction of water and the cladding could begin to damage the fuel rods and also produce hydrogen. The second problem is that the temperature might rise above the melting point of the uranium fuel, which is about 5,200°F.

Either event poses a potential danger. Damage to the zirconium cladding releases some radioactive materials trapped inside the fuel rods into the core's cooling water. A melting of the fuel itself could release far more radioactive materials. If a significant portion of the fuel should melt, the molten fuel could melt through the reactor vessel itself and release large quantities of radioactive materials into the containment building.

The essential elements of the TMI-2 system during normal operations (figure D.2) included:

- the reactor (fuel and control rods)
- water, which is heated by the fission process going inside the fuel rods to ultimately produce steam to run the turbine; this water, by removing heat, also keeps the fuel rods from becoming overheated

- two steam generators, through which the heated water passes and gives up its heat to convert cooler water in another closed system to steam
- a steam turbine that drives a generator to produce electricity
- pumps to circulate water through the various systems

A pressurizer is a large tank that maintains the reactor water at a pressure high enough to prevent boiling. At TMI-2, the pressurizer tank usually held 800 cubic ft of water and 700 cubic ft of steam above it. The steam pressure was controlled by heating or cooling the water in the pressurizer. The steam pressure, in turn, was used to control the pressure of the water cooling the reactor.

Normally, water to the TMI-2 reactor flowed through a closed system of pipes called the reactor *coolant system* or *primary loop*. The water was pushed through the reactor by four reactor coolant pumps, each powered by a 9,000-horsepower electric motor. In the reactor, the water picked up heat as it flowed around each fuel rod. Then it traveled through 36 in diameter stainless steel pipes, shaped like and called *candy canes*, and into the steam generators.

In the steam generators, a transfer of heat took place. The very hot water from the reactor coolant system traveled down through the steam generators in a series of corrosion-resistant tubes. Meanwhile, water from another closed system—the feedwater or *secondary loop*—was forced into the steam generator.

The feedwater in the steam generators flowed around the tubes that contain the hot water from the reactor coolant system. Some of this heat was transferred to the cooler

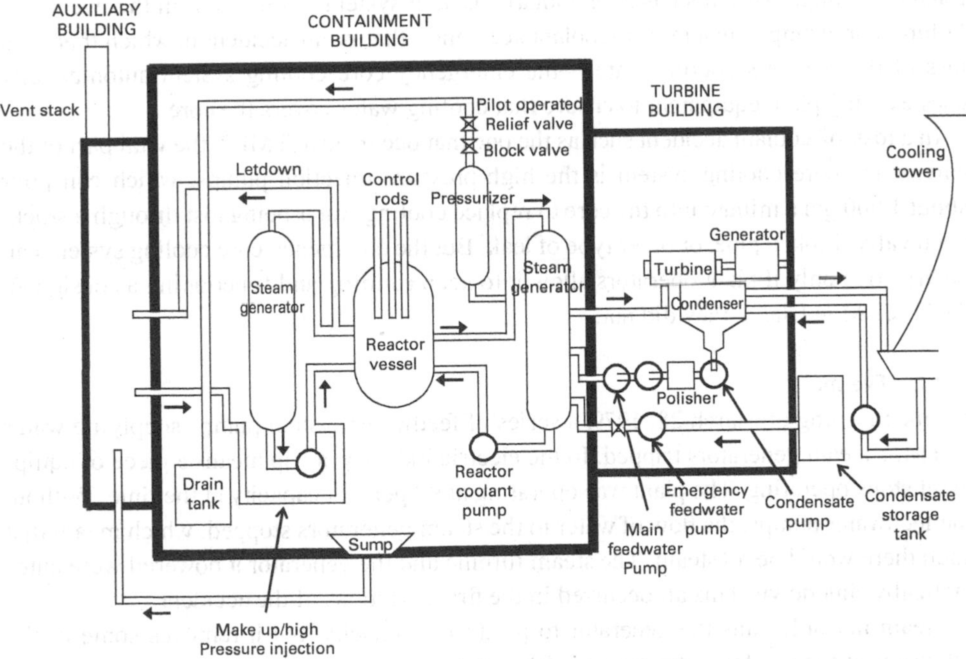

Figure D.2
The Three Mile Island nuclear power plant (Source: John G. Kemeny. *Report of the President's Commission on the Accident at Three Mile Island.* Washington, DC: US Government Accounting Office, 1979 [173]).

feedwater, which boiled and became steam. Just as in a coal- or oil-fired generating plant, the steam was carried from the two steam generators to turn the steam turbine, which ran the electricity-producing generator.

The water from the reactor coolant system, which had now lost some of its heat, was pumped back to the reactor to pass around the fuel rods, pick up more heat, and begin its cycle again.

The water from the feedwater system, which had turned to steam to drive the turbine, passed through devices called condensers. Here, the steam was condensed back to water and was forced back to the steam generators again.

The condenser water was cooled in the cooling towers. The water that cooled the condensers was also in a closed system or loop. It cooled the condensers, picked up heat, and was pumped to the cooling towers, where it cascaded along a series of steps. As it did, it released its heat to the outside air, creating the white vapor plumes that drift skyward from the towers. Then the water was pumped back to the condensers to begin its cooling process over again.

Neither the water that cooled the condensers, the vapor plumes that rose from the cooling towers, nor any of the water that ran through the feedwater system was radioactive under normal conditions. The water that ran through the reactor coolant system was radioactive, of course, because it had been exposed to the radioactive materials in the core.

The turbine, the electric generator that it powered, and most of the feedwater system piping were outside the containment building in other structures. The steam generators, however, which had to be fed by water from both the reactor coolant and feedwater systems, were inside the containment building with the reactor and the pressurizer tank.

A nuclear power facility is designed with many ways to protect against system failure. Each of its major systems has an automatic backup system to replace it in the event of a failure. For example, in a loss-of-coolant accident—that is, an accident in which there is a loss of the reactor's cooling water—the emergency core cooling system automatically uses existing plant equipment to ensure that cooling water covers the core.

In a loss-of-coolant accident such as the one that occurred at TMI-2, the vital part of the emergency core cooling system is the high-pressure injection pumps, which can pour about 1,000 gal a minute into the core to replace cooling water being lost through a stuck-open valve, broken pipe, or other type of leak. But the emergency core cooling system can be effective only if plant operators allow it to keep running and functioning as designed. At Three Mile Island, they did not.

D.2.2 Events

On the morning of March 28, 1979, a series of feedwater system pumps supplying water to TMI's steam generators tripped. In the electric industry, a trip means a piece of equipment stops operating. The plant was operating at 97 percent capacity at the time. Without the feedwater pumps, the flow of water to the steam generators stopped, which meant that soon there would be no steam. The steam turbine and the generator it powered were automatically shut down. This all occurred in the first 2 seconds of the accident.

Steam not only runs the generator to produce electricity, but it removes some of the intense heat the reactor water carries. When the feedwater flow stopped, the temperature of the reactor coolant increased, and the rapidly heating water expanded. The pressurizer level (the level of the water inside the pressurizer tank) rose, and the steam in the top of the tank compressed. Pressure inside the pressurizer built to 2,250 psi, which is 100 psi more than normal. A valve on top of the pressurizer, called a *pilot-operated relief valve* (PORV), is designed to relieve excess pressure. The PORV opened correctly, and steam and water began flowing out of the reactor coolant system through a drain pipe to a tank on the floor of the containment building. Pressure continued to rise, however, and 8 seconds after the first pump tripped, the reactor scrammed—that is, the control rods automatically dropped down into the reactor core to halt its nuclear fission.

The heat generated by fission was essentially zero less than a second later. In any reactor, decaying radioactive materials left from the fission process continue to heat the reactor's coolant water. Although this heat is just 6 percent of that released during fission, it has to be removed to keep the core from overheating. When the pumps that normally supply the steam generator with water shut down, three emergency feedwater pumps started automatically. Fourteen seconds into the emergency, an operator in the control room noticed that the emergency feedwater pumps were running. He did not notice two lights that indicated that a valve was closed on each of the two emergency feedwater lines and thus no water could reach the steam generators. One light was covered by a yellow maintenance tag. Nobody knows why the second light was missed.

With the reactor scrammed and the PORV open, pressure in the reactor coolant system fell. The PORV should have closed 13 seconds into the accident, when the pressure dropped to 2,205 psi, but it stuck open. A light on the control room panel indicated that the electric power that opened the PORV had gone off (the solenoid was deenergized), misleading the operators into assuming that the valve had shut. The PORV would remain open for 2 hours and 22 minutes, draining needed coolant water and starting a loss-of-coolant accident.

In the first 100 minutes of the accident, over one-third of the entire capacity of the reactor coolant system—32,000 gal—escaped through the PORV and out the reactor's let-down system. The let-down system is the means by which water is removed from the reactor coolant system. The make-up *system* adds water. Piping from both runs through the auxiliary building.

If the valve had closed as it was designed to do, or if the control room operators had realized that the value was stuck open and closed a backup valve to stem the flow of coolant water, or if they had simply left on the plant's high-pressure injection pumps, the accident at TMI would have been only a minor event.

Reactor operators are trained to respond quickly in emergencies, with the initial actions ingrained and almost automatic and unthinking. The first alarm in the control room was followed by a cascade of 100 alarms within minutes. The operators reacted quickly as trained to counter the turbine trip and reactor scram. Later, one operator told the commission about his reaction to the incessant alarms, "I would have liked to have thrown away the alarm panel. It wasn't giving us any useful information."

The shift foreman was called back to the control room. He had been overseeing maintenance on one of the plant's polishers—a machine that uses resin beads to remove dissolved minerals from the feedwater system. They were using a mixture of air and water to break up resin that had clogged a pipe that transfers resin from a polisher (demineralizer) to a tank in which the resin is regenerated. Later investigation revealed that a faulty valve in one of the polishers allowed some water to leak into the air-controlled system that opens and closes the polishers' valves. This malfunctioning valve probably triggered the initial pump trip that led to the accident. The same problem of water leaking into the polishers' valve control system had occurred at least twice before at TMI-2. The Kemeny report claims that had Met Ed corrected the earlier polisher problem, the March 28 sequence of events may never have begun [173].

With the PORV stuck open and heat being removed by the steam generators, the pressure and temperature of the reactor coolant system dropped, and the water level in the pressurizer fell. Thirteen seconds into the accident, the operators turned on a water pump. The water in the system was shrinking as it cooled, and therefore more water was needed to fill the system. Forty-eight seconds into the accident, while pressure continued falling, the water in the pressurizer began to rise again because the amount of water being pumped into the system was greater than that being lost through the PORV.

A minute and forty-five seconds into the accident, the steam generators boiled dry because their emergency water lines were blocked. The reactor coolant heated up again, expanded, and helped send the pressurizer level up further.

Two minutes into the incident, with the pressurizer level still rising, pressure in the reactor coolant system dropped sharply. Two large pumps, called high-pressure injection pumps, that are part of the emergency core cooling system, automatically began pouring about 1,000 gal a minute into the system. The level of water in the pressurizer continued to rise, and the operators, conditioned to maintain a certain level in the pressurizer, took this to mean that the system had plenty of water in it. However, the pressure of the reactor coolant system water was falling, and its temperature became constant.

About 2.5 minutes after the high-pressure injection pumps began working, an operator shut down one and reduced the flow of the second to less than 100 gal per minute. The reactor operators had been taught to keep the system from "going solid"—a condition in which the entire reactor and its cooling system, including the pressurizer, are filled with water. A solid system makes controlling the pressure within the reactor coolant system more difficult and can damage the system.

The Kemeny Commission concluded that the falling pressure, coupled with a constant reactor coolant temperature after high-pressure injection came on, should have alerted the operators that a loss-of-coolant accident had occurred and that safety required that they maintain high-pressure injection. Note the hindsight bias in this statement. An operator told the commission, however: "*The rapidly increasing pressurizer level at the onset of the accident led me to believe that the high-pressure injection was excessive, and that we were soon going to have a solid system*" [173, p. 94].

The saturation point was reached 5.5 minutes into the accident. Steam bubbles began forming in the reactor coolant system, displacing the coolant water in the reactor itself. The displaced water moved into the pressurizer, sending its level even higher. Again, this suggested to the operators that there was still plenty of water in the system. They did not realize that water was actually flashing into steam in the reactor. With more water leaving the system than being added, the core was on its way to being uncovered. To respond, the operators began draining off the reactor's cooling water through the let-down system piping.

Eight minutes into the accident, someone discovered that no emergency feedwater was reaching the steam generators. An operator scanned the light on the control panel that indicates whether the emergency feedwater valves are open or closed. One pair of emergency feedwater valves designed to open after the pumps reach full speed were open. But a second pair, called the *twelve-valves*, which are always supposed to be open except during a specific test of the emergency feedwater pumps, were closed. The operator opened the two twelve-valves and water rushed into the steam generators.

The twelve-valves were known to have been closed 2 days earlier, on March 26, as part of a routine test of the emergency feedwater pumps. The Kemeny Commission investigation did not identify a reason why the valves were in a closed position 8 minutes into the accident. They concluded that the most likely explanations were (1) the valves were never reopened after the March 26 test; (2) the valves were reopened after the test, but the control room operators mistakenly closed them during the very first part of the accident; or (3) the valves were closed mistakenly from control points outside the control room after the test. The loss of emergency feedwater for 8 minutes had no significant effect on the outcome of the accident, but it did add to the confusion that distracted the operators as they tried to understand the cause of their primary problem.

The Kemeny Commission noted that during the first 2 hours of the accident, the operators ignored or failed to recognize the significance of several things that should have warned them that they had an open PORV and a loss-of-coolant accident. One of these was the high temperatures at the drain pipe that led from the PORV to the reactor coolant drain tank. One emergency procedure states that a pipe temperature of 200°F indicates an open PORV. Another emergency procedure states that when the drain pipe temperature reaches 130°F, the block valve beneath it should be closed.

The operators testified that the pipe temperature normally registered high because either the PORV or some other valve was leaking slightly. "I have seen, in reviewing logs since the accident, approximately 198 degrees," the shift supervisor told the commission. "But I can remember instances before . . . just over 200 degrees" [173, p. 96]. They therefore dismissed the significance of the temperature readings, which the supervisor recalled as being in the range of 230°F (the top value was recorded as 285°F). He told the commission that he thought the high temperature on the drainpipe represented residual heat: *"Knowing that the relief valve had lifted, the downstream temperature I would expect to be high and that it would take some time for the pipe to cool down below the 200-degree set point"* [173, p. 96].

At 4:11 a.m., an alarm signaled that there was high water in the containment building's sump, an indication of a leak or break in the system. The water, mixed with steam, had come from the open PORV, first falling to the drain tank on the containment building floor. That tank was eventually filled, and the water flowed into the sump. At 4:15 a.m., a rupture disc on the drain tank burst as pressure in the tank rose. This break sent more slightly radioactive water onto the floor and into the sump. From the sump it was pumped to a tank in the nearby auxiliary building.

Five minutes later, instruments showed a higher-than-normal count of the neutrons inside the core, another indication that steam bubbles were present in the core and were forcing cooling water away from the fuel rods. During this time, the temperature and pressure inside the containment building rose rapidly because of the heat and steam escaping through the PORV and drain tank. The operators turned on the cooling equipment and fans inside the containment building. The Kemeny Commission concluded that the fact that they did not know that these conditions resulted from a loss-of-coolant accident indicated a severe deficiency in their training to identify the symptoms of such an accident [173].

At about this time, one of the operators got a call from the auxiliary building saying that an instrument there showed more than 6 ft of water in the containment building sump. When the operator queried the control room computer about this, he got the same answer. He then recommended shutting off the two sump pumps in the containment building. He did not know where the water was coming from and did not want to pump water that might be radioactive outside the containment building. Both sump pumps were stopped at about 4:39 a.m., but as much as 8,000 gal of slightly radioactive water might already have been pumped into the auxiliary building. Only 39 minutes had passed since the start of the accident.

By this time, managers and experts had arrived at TMI-2. The superintendent of technical support testified that what he found was not what he expected. "*I felt we were experiencing a very unusual situation, because I had never seen pressurizer level go high and peg in the high range, and at the same time, pressure being low. They have always performed consistently*" [173, p. 99].

The control room staff agreed. They later described the accident as a combination of events they had never experienced, either in operating the plant or in their training simulation.

Shortly after 5 a.m., the four reactor coolant pumps began vibrating severely. The vibration was a result of pumping steam as well as water, and it was another unrecognized indication that the reactor's water was boiling into steam. The operators were afraid that the violent shaking would damage the pumps or the coolant piping. The control room supervisor and his operators followed what they had been trained to do in this situation: At 5:14 a.m., they shut down two of the pumps and 27 minutes later the other two remaining pumps, stopping the forced flow of cooling water through the core.

By approximately 6 a.m., radiation alarms inside the containment building provided evidence that at least a few of the fuel rod claddings had ruptured from the high gas pressures inside them and allowed radioactive gases within the rods to escape into the coolant water. With coolant continuing to stream out the open PORV and little water being added, the top of the core became uncovered and heated to the point where the zirconium alloy of the fuel rod cladding reacted with steam to produce hydrogen. Some of this hydrogen escaped into the containment building through the open PORV and drain tank, while some remained within the reactor.

At 6:22 a.m., the open block valve was closed, 2 hours and 22 minutes after the PORV had opened. The loss of coolant was stopped, and pressure began to rise, but the damage continued. For some unexplained reason, high-pressure injection to replace the water lost through the PORV and let-down system was not started for almost another hour. Before that time, rising radiation levels were detected in the containment and auxiliary buildings, and evidence indicates that as much as two-thirds of the 12 ft high core was uncovered, with temperatures as high as 3,500 to 4,000°F in parts of the core.

At 6:54 a.m., the operators turned on one of the reactor coolant pumps, but shut it down 19 minutes later because of high vibrations. More radiation alarms went off, and a site emergency was declared. Site emergencies are declared when some event threatens an uncontrolled release of radioactivity to the immediate environment. This declaration initiated a series of actions required by the emergency plan, including notifying state authorities. Later, after more alarms sounded, a general emergency was declared, which indicates an incident that has "potential for serious radiological consequences to the health and safety of the general public."

Four hours after the accident began, the containment building isolated, which is an automatic procedure to help prevent radioactive material from escaping into the environment. Pipes carrying coolant run between the containment building and auxiliary buildings. These pipes close off when the containment building isolates, but the operators can open them. They were opened in this case, and some of this piping leaked radioactive material into the auxiliary building, some of which escaped from there into the atmosphere outside.

In the TMI-2 design, isolation occurs only when pressure in the containment building reaches a certain point. Although large amounts of steam entered the containment building early in the accident through the open PORV, the operators had kept pressure there low by using the building's cooling and ventilation system. The NRC had instituted new criteria for isolation in 1975, but TMI-2 was grandfathered under the old criteria. The Kemeny Commission concluded, however, that the failure to isolate early made little difference in the accident; some of the radioactivity ultimately released into the atmosphere occurred from leaks in the let-down system that continued to carry radioactive water out of the containment building and into the auxiliary building after isolation [173].

At 8:26 a.m., the operators once again turned on the emergency core cooling system's high-pressure injection pumps and maintained a relatively high rate of flow. The core was still uncovered at this time, and evidence indicates that it took until about 10:30 a.m. for the pumps to cover the core again fully.

Many off-site emergency procedures were initiated at this time, which are not detailed here. Activities in the control room became even more difficult as workers had to put on protective face masks with filters to screen out any airborne radioactive particles, making communication among those managing the accident difficult.

At 1:50 p.m. on Wednesday, a "thud" was heard in the control room. It was the sound of a hydrogen explosion inside the containment building, but nobody recognized this until late Thursday. The noise was dismissed at the time as the slamming of a ventilation damper. A pressure spike on a computer strip chart was written off as possible instrument malfunction.

By Friday, the nuclear industry had become deeply involved in the accident, sending experts from around the country. It was also on Friday that concern arose about a hydrogen bubble in the reactor possibly exploding if sufficient oxygen should enter the bubble. Someone in the NRC provided a theory (which turned out to be wrong) that the radiation in the reactor could cause decomposition of water into hydrogen and oxygen, leading to an explosion that might blow the pressure vessel apart. The concern turned out to be unfounded, but it continued through the weekend, with laboratories and scientists outside the NRC providing advice.

The cost of the accident, including cleanup of the buildings and disposal of approximately one million gallons of radioactive water, a substantial amount of radioactive gases, and solid radioactive debris, was estimated between $1 billion and $1.86 billion, not counting the loss of the plant itself.

The public health effects of the accident were negligible; very little radioactive material was released outside the plant. The maximum whole-body dose that could have been received by those outside was about the same dose as a person receives in a year from natural radiation. Plant personnel had slightly greater exposure.

At one point in the crisis, on March 30, the governor of Pennsylvania advised that children and pregnant women living in the vicinity of TMI should leave, and many did so. An evacuation center at Hershey, about 15 miles away, only attracted about two hundred people; the rest made their own arrangements.

The Kemeny Commission report concluded that the major health effect of the accident was found to be mental stress. The public relations consequences for the nuclear industry were much more serious, including a lowering of public confidence in the industry and in its regulatory agencies [173].

D.2.3 Some Systemic Causal Factors

On the positive side, the containment building at TMI remained intact, despite a hydrogen explosion within it, and kept most of the radioactive material from escaping. The building at Chernobyl was destroyed by a hydrogen explosion. Some reactors in the United States also lack strong containment systems, such as the ones producing weapons-grade plutonium for the Department of Energy and early versions of boiling-water reactors.

On the other hand, there were a large number of systemic causal factors that contributed to this accident. While there appeared to be little off-site damage that was incurred, the large number of problems in the operation of this plant that were discovered during the investigation led to important changes in the operation, management, and oversight of nuclear power plants in the United States.

Poor maintenance practices The maintenance force was overworked at the time of the accident and had been reduced in size to save money. There were many shutdowns, and a variety of equipment was out of order.

Review of equipment history for the six months prior to the accident showed that a number of equipment items that were involved in the accident had had a poor maintenance history without adequate corrective action. These included the pressurizer level transmitter, the hydrogen recombiner, pressurizer heaters, make-up pump switches, and the condensate polishers. Despite a history of problems with these polishers, no effective steps were taken to correct the problems. These polishers probably initiated the March 28 sequence of events. In addition, procedures were not changed to ensure that operators would bypass the polishers during maintenance operations to protect the plant from a possible malfunction of the polisher.

Inspection of the valves in the TMI-1 containment building after the accident showed long-term lack of maintenance. Boron stalactites more than a foot long hung from the valves, and stalagmites had built up from the floor.

At the time of the accident, Met Ed had not corrected deficiencies in radiation monitoring equipment pointed out by an NRC audit months before.

Weaknesses in operational procedures and their approval process Some of the key written operating and emergency procedures in use at TMI were inadequate, including the procedures for a loss-of-coolant accident and for pressurizer operation. The Kemeny report notes that the deficiencies in these procedures could cause operator confusion or incorrect action. For example, a 1978 Babcock and Wilcox (B&W) analysis of a certain type of small-break loss-of-coolant accident was misinterpreted by Met Ed. That misinterpretation was incorporated into the loss-of-coolant accident emergency procedures available at the time of the accident [173].

Operating and emergency procedures that had been approved by Met Ed and were in use contained many minor substantive errors, typographical errors, and imprecise or sloppy terminology. Some were inadequate. For example, a 1978 revision in the TMI-2 surveillance procedure for the emergency feedwater block valves violated TMI-2's technical specifications, but no one realized it at the time. The approval of the revision was not done according to Met Ed's own administrative procedures.

In general, plant procedures were deficient. For example:

- Pipe and valve identification practices were significantly below industrial standards. Eight hours into the accident, Met Ed personnel spent 10 minutes trying unsuccessfully to locate three decay heat valves in a high radiation field in the auxiliary building.
- When shifts changed in the control room, there was no systematic check required on the status of the plant and the lineup of valves.
- Although Met Ed procedures required closing the PORV block valve when temperatures in the tailpipe exceeded 130°F, the block valve had not been closed at the time of the accident even though temperatures had been well above 130°F for weeks.
- Performance of surveillance tests was not adequately verified to be sure that the procedures were followed correctly. The emergency feedwater valves that, during the accident, should have been open but were closed may have been left closed during a surveillance test 2 days earlier.
- The iodine filters in the auxiliary and fuel handling buildings were left in continuous use rather than being preserved to filter air in the event of radioactive contamination. Thus, on the day of the accident, their charcoal filtering capacity was partially expended, and they did not perform as designed. The NRC had waived testing requirements to verify the filter effectiveness.
- After the accident, radiological control practices were observed to be deficient. Contaminated and potentially contaminated equipment was found in uncontrolled areas of the auxiliary building.

Operating procedures were not thoroughly reviewed by experts. After the accident, the NRC started to require that plant safety committees review the operating procedures, but this was not standard before the accident [12]. Met Ed had no requirement for an independent (outside of line management) safety assessment of operating procedures.

Inadequate training Many of the operators' actions contributed to the accident sequence. The Kemeny Commission Report contained many instances of hindsight bias where it was stated that operators "should have known that . . ." Given the widespread nature of hindsight bias in accident investigation, it is instructive to examine the common conclusion of operator error in more detail in this accident. It is necessary to look both at the training of the operators and the human factors design of the plant in order to understand why they behaved the way they did.

Most of the operators and others involved in the accident did not fully understand the operating principles of the plant equipment. Their training was severely inadequate.

The lack of depth in this understanding, even by senior reactor operators, left them unprepared to deal with something as confusing as the situation in which they found themselves.

For example, the TMI-2 operators had never been trained for the sequence of the stuck-open PORV, and instructions on how to handle it were not included in their emergency procedures. They also had never had specific training about the danger of saturation conditions in the core, although they were generally familiar with the concept.

Although Met Ed believed that saturation (which could have led to the core being uncovered) had occurred in an incident a year before the accident, the hazards of the problem were not emphasized to the operators. When saturation occurred during the accident, unsurprisingly the operators did not recognize its significance and take corrective action quickly. In addition, the operators were not given adequate information about the temperatures to be expected in the PORV tailpipe after the PORV opened.

In addition, the simulator training of operators at B&W did not prepare operators for multiple-failure accidents. The operators were only trained for an accident in the course of which only *one* thing went wrong. "*They were never trained for a situation in which two independent things could go wrong. And in this particular accident three independent things went wrong*" [174, p. 66].

In fact, the simulator was not programmed to reproduce the conditions that the operators faced during the accident. It was unable to simulate increasing pressurizer level at the same time that reactor coolant pressure was dropping. Nuclear power experts did not think this condition could occur.

The Kemeny report concluded that the training of TMI operators was greatly deficient [173]. While it might have been adequate for the operation of a plant under normal circumstances, insufficient attention was paid to possible serious accidents and to accident sequences involving more than one failure. The content of the operator instructional program did not lead to sufficient understanding of reactor systems.

The person responsible for training at B&W was a witness at the hearings. He was very proud of the last five years of his company's program. When asked what he considered his most important achievement, he replied: "*When I arrived, many courses had been given by engineers. But the engineers don't know how to talk in a way which people can understand. Consequently, the first rule I introduced was that no engineer was authorized to participate in the training of operators*" (quoted in [174, p. 65]).

The Kemeny Commission found that all theory had been taken out of the operator training program and that they were trained to be button pushers. This training was adequate for normal conditions, but the operators had not been prepared for a serious situation [174]. This satisfied NRC requirements and had become standard practice.

Note that lack of understanding of what was happening was not just a problem for the operators: the TMI emergency and engineering personnel also had difficulty in analyzing the events, and the hydrogen bubble issues confused even outside experts. Even after supervisory personnel took charge, significant delays occurred before the full amount of core damage was recognized and stable cooling of the core was achieved. Of the people on duty at the plant when the accident started, none were nuclear engineers, none were even college graduates, and none were trained to handle complex reactor emergencies.

The TMI training program, however, did conform to the NRC standard for training at that time. Moreover, TMI operator candidates had higher scores than the national average on NRC licensing examinations and operating tests.

The Kemeny Commission concluded that NRC standards allowed a shallow level of operator training and prescribed only minimum training requirements. There were no minimal educational requirements for operators and no requirements for checks on psychological fitness or criminal records. In addition, an individual could fail parts of the licensing examination, including sections on emergency procedures and equipment, and still pass the overall examination by getting a passing average score [173].

The NRC also had no criteria for the qualification of those people who conduct the operator training programs and did not regularly conduct in-depth reviews of the training programs.

Understanding the behavior of the operators during the accident also has to be considered in the context of the design of the plant and a lack of attention to human factors.

Lack of attention to human factors The designers paid little attention to the interaction between humans and machines under the rapidly changing and confusing conditions of an accident. The design ignored the needs of operators during a slowly developing small-break accident, perhaps because of a concentration on large-break accidents, which do not allow time for significant operator action.

The TMI-2 control room in general was not adequately designed with the management of an accident in mind: the designers had never systematically evaluated the design to see how well it would serve in emergency conditions. The design problems ranged from the details of the control-board layout and the positioning of controls and displays, to larger-scale problems of system design that lead to human errors in the interpretation of data. "*It is not surprising that the operators made errors, because the design of the control-room details was such as to enhance, not to minimize, errors*" [52, p. 157].

Some examples of control room design and information presentation that confused the operators include:

- The operators were faced with more than one hundred alarms within 10 seconds of the first one. There was no way to suppress the unimportant signals so that the operators could concentrate on the important ones.
- The arrangement of controls and indicators was not well thought out. Some key indicators relevant to the accident were on the back of the control panel. Information was not presented in a clear and sufficiently understandable form. For example, although the pressure and temperature within the reactor coolant system were shown, there was no direct indication that the combination of pressure and temperature meant that the cooling water was turning into steam. Patrick Haggerty (a member of the Kemeny Commission) was appalled by this deficiency:

  It's inconceivable to me that the really serious accident parameters are not grouped together instead of being scattered all over the control room—and even then not very well portrayed. It's inconceivable that somebody hadn't grouped those few indicators that absolutely would have shown the operator what counted—that the core was or was not covered. I just can't believe it. . . . If you take the simplest kind of microcomputer . . . it would probably be competent to provide information and plot out saturation so that it is always visible. (quoted in [234, p. 54])

A B&W official testified that a direct reading of the level of coolant in the core would be difficult to provide and too expensive, and it would create other complications. Although there were several indirect indications, each proved to be faulty or ambiguous.

- Operators were not able to determine the water pressure in the core because engineers had not anticipated this need [146].
- The manual control station of the polisher bypass valve was nearly inaccessible and took great effort, in a physically awkward position, to operate.
- Several instruments went off scale during the course of the events, so the operators lacked some highly significant diagnostic information. These instruments were not designed to follow the course of an accident.
- The computer printer registering alarms was running more than 2.5 hours behind the events and at one point jammed, losing valuable information.

John Kemeny, after a tour of the TMI-1 control room, remarked in the presence of journalists that he did not think that it was a masterpiece of modern technology. "*In fact, I said that it was at least twenty years behind the times. I was heavily criticized for this. And with good reason, because my statement proved to be wrong: we discovered later, in the documents of the NRC, a report written ten years earlier in which an expert had said that the control rooms were, then, twenty years behind the times*" [174, p. 66].

A large number of control room instruments were out of calibration, and many tags were hanging on the instrument panel indicating equipment out of service. One of those tags obscured the emergency feedwater block control valve indicator lights.

While acknowledging the role the operators played in the accident events and suggesting throughout the report that operators should have known better, the Kemeny Commission did refrain from simply blaming them for the accident. The report noted that "*many factors contributed to the action of the operators, such as deficiencies in their training, lack of clarity in their operating procedures, failure of organizations to learn the proper lessons from previous incidents, and deficiencies in the control room*" [174, p. 11].

As discussed in chapter 6, more attention is now being given to human factors by the nuclear power industry.

Ignoring warnings

Similar incidents in other plants demonstrated the incipient problems and the operators' need for clear instructions to deal with events like those that happened. Some of these warnings, including a similar incident at the Davis-Besse nuclear power plant a year before, were detailed in chapter 4. In addition, after an April 1978 incident at TMI, a control room operator had complained to his management about problems with the control room, but no corrective action was taken by the utility [173].

Lack of design for controllability

Both the RBMK reactor at Chernobyl and the TMI PWR reactor are sensitive to perturbations. The B&W reactor at TMI has a once-through cooling system, with a small volume of water compared to other US reactors and an undersized pressurizer [12]. This design makes it respond more to disturbances, and thus it is more difficult for the operators to control.

> *For water-cooled systems, the drive for economy leads to high fuel ratings and therefore to an extremely rapid sequence of accident events given an initiation due, for example, to the loss of coolant. The times available for intervention are such that the reactor has essentially become a fly-by-wire device rather than a piloted one. To give sufficient statistical assurance that such a device will "fly" without crashing, large degrees of redundancy in safety systems have to be provided. The consequent increase in complexity makes it even more likely that the operator, if he intervenes, will do something counterproductive.* [107, p. 22]

Franklin [107] suggests that it is worth giving up a few percent of capital cost-effectiveness for reactors that provide the operator with a half-hour to reflect on the consequences of actions before needing to intervene.

Complacency The Kemeny report points out that a feeling had pervaded the entire US nuclear power industry, and even the government oversight committees, that a major accident could not happen in the United States.

> *After many years of operation of nuclear power plants, with no evidence that any member of the general public had been hurt, the belief that nuclear power plants are sufficiently safe grew into a conviction. One must recognize this to understand why many key steps that could have prevented the accident at TMI were not taken. The Commission is convinced that this attitude must be changed to one that says nuclear power is by its very nature potentially dangerous, and, therefore, one must continually question whether the safeguards already in place are sufficient to prevent major accidents. A comprehensive system is required in which equipment and human beings are treated with equal importance.* [173, p. 9]

As another aspect of complacency and discounting risk, the plant was inadequately designed to cope with the cleanup of a damaged plant [173].

Inadequate safety management The Kemeny report notes significant deficiencies in the management of TMI-2. Shift foremen could not adequately fulfill their supervisory roles because of excessive paperwork not related to supervision. There was no systematic check on the status of the plant and the line-up of valves when shifts changed. Surveillance procedures were not adequately supervised. And there were weaknesses in the program of quality assurance and control. A review of TMI-2's licensee event reports (required by the NRC) disclosed repeated omissions, inadequate failure analyses, and inadequate corrective actions. There was also no group with special responsibility for receiving and acting on potential safety concerns raised by employees. Finally, the report notes that management allowed operation of the plant with a number of poor control room practices.

The GPU Service Corporation (GPUSC) had final responsibility for the design of the plant. However, by its own admission, it lacked the staff or expertise in certain areas. Once construction was complete, GPUSC turned the plant over to Met Ed to run, but Met

Ed did not have sufficient knowledge, expertise, and personnel to operate the plant or to maintain it adequately.

Responsibility for management decisions was divided among the TMI site, Met Ed, and GPU. GPU recognized in early 1977 that integration of operating responsibility into one organization was desirable. An outside management audit, completed in the spring of 1977, recommended clarifying and reevaluating the roles of GPUSC and Met Ed in the design and construction of new facilities, strengthening communications between the two, and establishing minimum standards for the safe operation of GPU's nuclear plants. However, integration of management did not occur until after the accident.

Poor quality assurance Met Ed had a quality assurance plan that met NRC requirements. But the Kemeny report notes that the NRC requirements at that time were inadequate—they did not require that quality assurance programs be applied to the plant as a whole but rather only to systems classified as "safety-related." Neither the PORV nor the condensate polishers were placed in this category. In addition, according to the Kemeny report, the NRC did not require the level of independent review (outside of line management) normally found in the quality assurance programs of safety-critical industries.

Met Ed's implementation of its quality assurance plan was also found to have significant deficiencies by the Kemeny Commission staff and in an NRC post-accident audit. Independent audit of the performance of surveillance procedures was required only every two years. There was no plan for such a review, and, in fact, no review had been made of those TMI-2 procedures that were more than two years old. There were deficiencies in the reporting, analysis, and resolution of problems in safety-related equipment and other events required to be reported to the NRC. Independent assessment of general plant operations was minimal.

Information collection and use The lessons from previous accidents and incidents did not result in new, clear instructions being passed on to the operators. The NRC accumulates an enormous amount of information on the operating experience of plants (2,000 to 3,000 reports a year). However, before the TMI accident, systematic methods were lacking for evaluating these experiences and looking for danger signals of possible generic safety problems. In 1978, the General Accounting Office had criticized the NRC for this failure, but no corrective action had been taken as of the TMI-2 accident.

> *In a number of important cases, General Public Utilities Corporation, Met Ed, and B&W failed to acquire enough information about safety problems, failed to analyze adequately what information they did acquire, or failed to act on that information. Thus, there was a serious lack of communication about several critical safety matters within and among the companies involved in the building and operation of the TMI-2 plant. A similar problem existed in the NRC.* [173, p. 43]

There also seems to have been a lack of closure in the system: important safety issues were raised and studied to some degree but were not carried through to resolution. According to the Kemeny Commission report, the lessons learned from these studies did not reach those people and organizations that most needed to know about them.

Licensing and regulation deficiencies Licensing deficiencies occupied a major place in the Kemeny report. Most of these were fixed after the accident. They are important to understand, however, because many of these same deficiencies can be found in the licensing procedures for other industries, and some of the same mistakes are being repeated with respect to the introduction of computers into the control of nuclear power plants and other hazardous processes and industries.

The Kemeny Commission report concluded that prior to TMI, the NRC paid insufficient attention in licensing reviews to loss-of-coolant accidents of this size, such as might be caused by a stuck-open valve. Instead, the NRC focused most of its attention on large-break loss-of-coolant accidents. Those managing the TMI accident were unprepared for the significant amount of hydrogen generated. TMI illustrates a situation where the NRC emphasis on large breaks did not cover the effects observed in a smaller accident.

The licensing process also concentrated on equipment, assuming that the presence of operators would only improve the situation—they would not be part of the problem. The Kemeny Commission noted a persistent assumption that plants could be made sufficiently safe to be "people-proof." Thus, not enough attention in the licensing process was devoted to the training of operating personnel and to operator procedures.

Moreover, at that time, license applicants were only required to analyze single-failure accidents. They were not required to analyze what happens when two or more systems fail independently of each other (as happened at TMI) or to provide emergency operating procedures for such events.

There was also a sharp delineation between those components in systems that are safety-related and those that are not. Strict reviews and requirements applied to the former; the latter were exempt from most requirements—even though they can have an effect on the safety of the plant. Items not labeled safety-related did not need to be reviewed in the licensing process, were not required to meet NRC design criteria, did not need to be testable, did not require redundancy, and were ordinarily not subject to NRC inspection. "We feel that this sharp either/or definition is inappropriate. Instead, there should be a system of priorities as to how significant various components and systems are for the overall safety of the plant" [173].

The Kemeny Commission also noted that there were no precise criteria as to which components and systems were to be labeled safety-related; the utility made the initial determination subject to NRC approval. For example, at TMI-2, the PORV was not a safety-related item because it had a block valve behind it. On the other hand, the block valve was not safety-related because it had a PORV in front of it.

> *The NRC's reliance upon artificial categories of "safety-related" items has caused it to miss important safety issues and has led the nuclear industry to merely comply with NRC regulations and to equate that compliance with operational safety. Thus, over-emphasis by the NRC process on specific categories of items labeled "safety-related" appears to interfere with the development, throughout the nuclear industry, of a comprehensive safety consciousness, that is, a dynamic day-to-day process for operating safely.* [173, p. 53]

At that time, plants could receive an operating license with several safety issues still unresolved. This placed such a plant in regulatory limbo, with jurisdiction divided between two different offices within the NRC. TMI-2 had this status at the time of the accident, thirteen months after it had received its operating license.

The report noted more generally that there was no identifiable office within the NRC responsible for systems engineering examination of overall plant design and performance, including interaction between major systems, and also no office to examine the interface between machines and humans: "*There seems to be a persistent assumption that plant safety is assured by engineered equipment, and a concomitant neglect of the human beings who could defeat it if they do not have adequate training, operating procedures, information about plant conditions, and manageable monitors and controls. Problems with the control room contributed to the confusion during the TMI accident*" [174].

A further problem noted was the NRC's primary focus on licensing and insufficient attention to the process of assuring nuclear safety in operating plants. The labeling of a problem as *generic* (applying to a number of different nuclear power plants) provided a convenient way to postpone decisions on a difficult problem. Once an issue was labeled generic, the individual plant being licensed was not responsible for resolving the issue prior to licensing. There was a reluctance to apply new safety standards to previously licensed plants.

Finally, the existence of a vast body of NRC regulations tended to focus industry attention narrowly on the meeting of regulations rather than on a systematic concern for safety. For example, because the licensing process concentrated on the consequences of single failures, there was no attempt to prepare operators for accidents in which two systems failed independently of each other.

Emergency preparedness In the approval process for reactor sites, the NRC at that time required licensees to plan for off-site consequences only in the "area of residents"—about a 2-mile radius for TMI. The existence of a state emergency or evacuation plan was also not required. In general, emergency planning had a low priority in the NRC, and the AEC before it, prior to the TMI accident.

The Kemeny Commission report suggests that the reasons for this included the agency's confidence in design reactor safeguards and their desire to avoid raising public concern about the safety of nuclear power. The report also suggests that the attitude fostered by the NRC regulatory approach and by Met Ed at the local level was that radiological accidents having off-site consequences beyond the 2-mile radius were so unlikely as not to be of serious concern [173].

The TMI emergency plan did not require the utility to notify state or local authorities in the event of a radiological accident, and delays occurred in doing this. Met Ed also did not notify its physicians under contract, who would have been responsible for the onsite treatment of injured and contaminated workers, and the emergency medical care training given to these physicians was inadequate. Moreover, the emergency control center for health physics operations and the analytical laboratory to be used in emergencies were located in an area that became uninhabitable in the early hours of the accident, there was a shortage of respirators, and there was an inadequate supply of uncontaminated air [173].

The response to the emergency was characterized by an atmosphere of almost total confusion and a lack of communication at all levels. Almost all the local communities around TMI lacked detailed emergency plans. Many key recommendations were made by people who did not have accurate information, and those who managed the accident were slow to realize the significance and implications of the events that had taken place. "*The fact that too many individuals and organizations were not aware of the dimensions of serious accidents at nuclear power plants accounts for a great deal of the lack of preparedness and the poor quality of the response*" [12; 173].

The Kemeny Commission recommended centralization of emergency planning and response in a single agency at the federal level, with close coordination between it and state and local agencies. This agency would have responsibility both for ensuring that adequate planning takes place and for taking charge of the response to the emergency. They also recommended that siting rules for plants be changed.

D.3 Chernobyl

On April 26, 1986, an accident occurred at the Chernobyl nuclear power plant in the Soviet Union. It is only one of two nuclear energy accidents rated at seven—the maximum level—on the International Nuclear Event Scale. The other was the 2011 Fukushima accident, which is described later in this appendix.

The accident occurred during a safety test meant to measure the ability of the steam turbine to power the emergency feedwater pumps of an RBMK-type nuclear reactor in the event of a simultaneous loss of external power and major coolant leak. The flawed test resulted in steam explosions and melting of the reactor core.

The meltdown and explosions ruptured the reactor core and destroyed the reactor building. This was immediately followed by an open-air reactor core fire that lasted about a week, during which airborne radioactive contaminants were released that were deposited onto other parts of the USSR and Europe. The fire released about the same amount of radioactive material as the initial explosion. In response to the initial accident, a 10 km (6.2 mile) radius exclusion zone was created 36 hours after the accident, from which approximately 49,000 people were evacuated. The exclusion zone was later increased to 30 km (19 miles), and an additional 68,000 people were evacuated.

Following the reactor explosion, which killed two engineers and severely burned two more, a massive emergency operation to put out the fire, stabilize the reactor, and clean up the ejected radioactive material began. During the immediate emergency response, 237 workers were hospitalized, of which 134 exhibited symptoms of acute radiation syndrome (ARS). Among those hospitalized, twenty-eight died within the following three months, all of whom were hospitalized for ARS. The health effects to the general population are uncertain.

In August 1986, representatives from the Soviet Union presented a 382-page report at the International Atomic Energy Agency (IAEA) Expert Conference on the Chernobyl Accident held in Vienna [390]. The report provides an account of the shortcomings in design and reactor operation that led to the accident. Later analyses by others filled in missing pieces, including one by Atomic Energy of Canada Ltd. (AECL) [249].

D.3.1 Background

As described earlier, the Chernobyl reactor design, called the RBMK, is graphite moderated and water cooled. Shields on the side of the RBMK reactor are made of water, sand, and concrete. The bottom and the top both have concrete shields. All the pressure tubes and control rods are attached to this top shield, and the entire reactor and shield complex is inside a confinement building.

The pipes below the reactor core were inside boxes that were connected to a huge pool of water under the entire building. If one of the pipes in the boxes broke, the steam would be forced into the pool, where it and any radioactive particles it contained would be trapped in the water. But all the steam pipes above the core were inside ordinary industrial buildings. Thus, if one of these pipes broke, a release of radioactive steam would occur. The Russians at the time relied on accident prevention and mitigation and provided only partial containment. After TMI, they started to add containment to their reactors.

The reactor is controlled by 211 control rods. The large number of rods is meant to compensate for their relatively slow speed. The effectiveness of a control rod varies with the stage of insertion: a rod is not as marginally effective in the early or late stages of insertion as in the intermediate stages [12]. It takes about 20 seconds for the rods to be fully inserted.

An important aspect of this reactor, with respect to safety, is that it has a *positive void coefficient*. Briefly, the power increases as water flowing through the core decreases—the opposite of most reactors. In a PWR, water is used as a moderator and without it, fissioning stops. However, fuel elements may still melt from the heat generated by decaying fission products. This is what happened in part of the TMI core.

Water in the RBMK is used only for transferring heat. The water does capture some neutrons, however, preventing them from producing further fissions. If the water boils and becomes less dense, the neutrons it would have captured instead go into the graphite, where they cause more fissioning. The absence of water thus increases the reactivity and the rate of fissioning.

To handle this positive void coefficient, the RBMK reactors depend on a complex, computer-run control system [12]. There are four main water circulating pumps: three used in normal operation and a fourth as standby. Emergency feedwater pumps and pressurized accumulators switch on in an emergency between 10 and 20 seconds after the main feedwater stops.

An experiment had been planned while the reactor was being shut down for "medium repair" on April 25, 1986. The purpose of the experiment was to determine whether, after steam had been shut off from the turbo-generator, the inertia of the still rotating generator would be sufficient to generate enough electricity to operate auxiliary motors that were part of the reactor emergency cooling system.

The operation of a nuclear power plant does not rely on the electricity it generates, which is sent out to distribution centers. Lines from other power stations provide electric power to run the essential parts of the plant. The systems requiring off-site electricity include instrumentation, control, and some of the pumps. Other pumps are run by steam generated by the plant. Because of the positive void coefficient in the RBMK, the emergency feedwater pumps must be kept running at all times to circulate water through the reactor.

The Russians design their plants to withstand both an accident and a simultaneous loss of electrical power. Because the reactor is shut down immediately, it cannot generate its own power directly. Normally, electricity would be obtained from the electrical supply to the station or from other reactors at the same site. An extra level of protection is provided, however, in case either of these sources also fail.

The normal backup is to provide diesel engines to drive emergency generators. US plants are required to have emergency batteries for instantaneous response and diesel generators that go from cold start to full power in 10 seconds [12]. The Russians have said that their diesel run-up time is 15 seconds, but the report released in Vienna indicated they need other sources of power for at least 45 seconds. They decided to tap the energy of the spinning turbine—which is so heavy that it takes a while to slow down—to generate electricity for the few seconds before their diesels started.

The experiment was to see how long this electricity would power the main pumps that keep the cooling water flowing over the fuel. The experiment had been done before at other plants with no problems, but they found that the voltage from the generator dropped sharply long before the generator stopped rotating. In the new test, they planned to try to eliminate this problem by using a special system to regulate the magnetic field of the generator. The Vienna report said that the test procedure was not prepared or approved in the proper way and was of low quality. In addition, personnel deviated from the plan, which created the emergency conditions.

The basic plan was to reduce reactor power to less than half its normal output so that all the steam could be put into one turbine. This remaining turbine was then to be disconnected and its spinning energy used to run the main pumps for a short while.

D.3.2 Events

The reactor was operated normally until 1 a.m. on April 25, when the operators started to reduce reactor power in preparation for the test. This operation was done slowly, with the reactor reaching 50 percent power 12 hours later. Only one of the two turbines was needed at this point to take the steam from the reactor, so the second turbine was disconnected at 1:05 p.m. to simulate the loss of off-site power.

At this point, the electrical supply for the four main cooling pumps and two feedwater pumps (which were normally powered from the switched-off generator) were transferred to the operational generator. The emergency cooling system was disconnected from the pipelines at 2 p.m. according to plan so that its power consumption would not affect the test results.

Normally, the test would have continued with the power being reduced to about 30 percent. However, further shutdown of the reactor was delayed because of a request from the grid distribution center at Kiev. The controller at Kiev asked that output be maintained to satisfy an unexpected demand. The continued operation of the reactor for another 9 hours, without its emergency cooling system, was a gross breach of operating regulations, although it did not affect the accident.

The test finally began at 11:10 p.m. The test procedures called for the test to be carried out at a reactor power between 700 and 1000 Mw. To enable the reactor to operate at this low power, the local automatic control system had to be switched off. The global (or average power) control system is meant to be used primarily as a backup to the local control system; the local system is more accurate and smooths out the effects of perturbations. Local control is required during transitions from one operating regime to another [390].

A sudden power reduction causes a quick buildup of xenon in uranium fuel. Xenon is a radioactive gas that readily absorbs neutrons and tends to hasten the reactor shutdown. At this point, sustaining a chain reaction became difficult because many of the neutrons needed for fission were being absorbed by the xenon.

In addition, the core was at such low power that the water in the pressure tubes was not boiling, as it normally does, but was liquid. Liquid water absorbs like xenon. The operators continued inserting control rods in order to continue shutting the plant down. As a result, the reactor power dropped to below 30 Mw. The reactor was almost shut down, and the power was too low for the test.

The operators saw the drastic undershoot and tried to increase power by pulling out control rods. By 1 a.m., they managed to get the reactor back to 200 Mw. This was below the power level specified for the test (which was 700 Mw), but it was as high as they could go because of the xenon and water, and they decided to proceed anyway.

By this time, both the automatic regulator rods (inserted from the bottom of the reactor) and the control rods (inserted from the top of the reactor) had been withdrawn almost to their fullest extent.

Snell describes this state as rather like "driving a car with the accelerator floored and the brakes on—it was abnormal and unstable" [249, p. 15]. Running this type of reactor with all the control rods withdrawn is a very serious error because some of them are needed for emergency shutdown. If they are all pulled out well above the core, it takes too long for them to move back into the high-power part of the reactor in an emergency, and the shutdown is very slow. A minimum of thirty rods is considered minimal for safety, but only six to eight were in the core at this time.

At 1:03 and 1:07 a.m., two more cooling pumps (the normal reserve pumps) were switched on, in addition to the six pumps that were operating. Switching on the reserve pumps was done to ensure that at the end of the experiment (during which four of the pumps were required to be shut down), four pumps would remain operating to provide reliable cooling of the reactor core. Under normal levels of power output, adding the pumps would not have been a problem, but at 200 Mw, the reactor required many manual adjustments to maintain a safe balance of steam and water.

One explanation for the following events is that the operation of these pumps greatly increased the flow of cold water into the reactor, decreased the rate of steam generation, and reduced the level of water in the steam separator below the emergency level. An emergency shutdown would normally have occurred automatically at this point.

The AECL explanation suggests that the operator had to take over manual flow of water returning from the turbine, as the automatic controls were not operating well—the plant was never intended to operate at such low power. Controlling the flow of water is a complex task to carry out manually, and the operator did not succeed in getting the flow correct. The reactor was so unstable it was close to being shut down by the emergency rods.

In either case, because a shutdown would abort the test, the operators short-circuited the emergency protection signals. At 1:23 a.m., the shutdown control values of the second generator were closed, and the reactor continued to operate at a power of 200 Mw. They decided to start the test.

Normally, the reactor would shut down automatically if the remaining turbine was disconnected, as was planned for the test. However, they wanted a chance to repeat the test immediately if it were not done correctly, which they could not do if the reactor was shut down. Therefore, an operator disabled this shutdown signal also. The remaining automatic shutdown signals would go off on abnormal power levels but would not react immediately to the test conditions. The second turbine was then tripped.

At this point, the reactor was operating at 200 Mw and still producing steam. Because both generators were turned off, there was no place for the steam to go, and steam pressure began to rise. At the same time, the flow of water through the reactor began to decrease because four of the eight pumps were working off the generator, which was deprived of steam and running down. Further steam generation resulted, and, since steam absorbs neutrons much less effectively than water does, there was a rapid rise of reactivity. The reactor power increased to more than 530 Mw in less than 3 seconds. At this point (1:23.40 a.m.), the operator pushed the emergency *scram* button, which inserted all control and shutdown rods into the reactor.

By this time, as noted earlier, the automatic control rods were at the bottom of the core (where they had been withdrawn in order to get the reactor power up to 200 Mw), and the manual adjustments by the operators in their attempts to achieve the test conditions had resulted in practically all the other control rods being withdrawn to the top of the core. When the *scram* button was pressed, the rods went down, but the operator saw that they stopped before they reached the correct depth. He then cut off the drive couplings so that they fell into the core under their own weight.

People outside the reactor building reported that they heard two explosions, one after the other, at 1:24 a.m. Hot fragments and sparks flew up above the reactor building, and some of them started a fire on the roof of the turbine room.

Total agreement is lacking about the detailed cause of each of these explosions. There is consensus, however, that the first was caused by the great amount of heat produced by the increased rate of fissioning. This heat production occurred over too short a time for heat transfer to occur. Enormous pressure built up in the vertical pressure pipes, which lifted the concrete slab above the reactor, took off the roof of the building above the concrete slab, and blew the fuel out.

Experts also differ about the cause of the reported second explosion. Some believe it was a hydrogen explosion, while others believe it was caused by an additional fuel excursion [12]. Megaw explains it as resulting from the sharp temperature increase in the reactor core, the rupture of the cooling channels (releasing steam onto the hot graphite moderator and producing water gas), and the chemical reaction between the overheated zirconium canning and water [259].

The initial fires were started by the burning debris that was expelled after the second explosion. The core graphite also caught on fire. As the fission products were expelled from the reactor, they were carried aloft by the heat released from the fire and, because of local weather conditions, were pushed high into the atmosphere, where they were subject to various wind patterns.

The first hint outside the Soviet Union that anything was wrong came from Sweden [259]. At the Forsmark nuclear generating station just north of Stockholm, a worker arriving for the morning shift on Monday, April 28, set off an alarm on a radiation detector. Assuming that the leak was from their own reactor, the operators shut it down and closed the plant. Later tests showed that the radiation was not from Forsmark or any other Swedish reactor.

The Swedes alerted the Americans, who first thought that the radioactivity was the result of a Russian underground weapons test that accidentally leaked to the atmosphere. By the afternoon, Swedish scientists had identified the radioactive components and determined that they must have come from a reactor and not from a weapons test. Radioactive atoms are unstable, continually trying to change themselves into a more stable configuration [259]. The greater the instability, the faster they change. The rate of decay is measured by the *half-life*—decay time for half of the atoms in a particular sample. The radioactive products formed in fission all have different half-lives and decay at different rates. Because in a weapons test they are all formed at the same time, they start to decay at the same time. But in a reactor, they are formed over a long period and start to decay at different times.

From an examination of the proportions of various fission products present in a sample, Swedish scientists were able to determine that the radioactivity came from at least a partial meltdown of a reactor. Using the time that the radioactive cloud first appeared on their coast and known wind speeds and directions, they were able to trace the cloud backwards over Latvia and Minsk to Kiev.

Swedish diplomats asked the Russians for an explanation but were told that there was no information. At 9 p.m., an announcement was made on Soviet television: "*An accident has occurred at the Chernobyl Nuclear Power Plant and one of the reactors has been damaged. Measures are being taken to eliminate the consequences of the accident. Aid is being given to those affected. A Government Commission has been set up*" [259, p. 19].

Little additional information was provided for a while, but this has been true of many nuclear incidents around the world; some countries with more open news organizations have been able to withhold information for shorter times, but rarely has information immediately after the incident been very forthcoming and sometimes not even later.

At 8 a.m. on the morning of Tuesday, April 29, a scientific attaché from the Soviet Embassy in Bonn went to Atomforum, a nongovernment agency representing West Germany's nuclear power industry, and asked if the Germans knew anyone who could advise them about how to put out a graphite fire in a reactor. The Germans gave some advice and suggested they ask the British, who had experience with the graphite fire at Windscale. The Russians also asked the Swedish nuclear authority for advice. The American government offered various types of assistance, but the offer was politely refused. The Russians did invite an American bone marrow specialist, Dr. Robert Gale, to provide medical assistance. By this time, satellite photos of the damaged reactor had appeared in the Western press.

By April 30, other countries became aware that a radioactive cloud from the accident was depositing radioactive materials on their territory. Some countries banned the sale of milk, advised the population against drinking water, and gave out potassium iodide to counteract the effects of iodine on the thyroid.

Thirty-one plant workers and firefighters died from acute radiation sickness and burns. The thousand families living in a workers' town 1 mile from the plant were evacuated 12 hours after the explosion.

The evacuation of nearby Pripyat and seventy-one villages within 18 miles of the plant started the next day [252]. The number of deaths or cancers attributable to the accident that have occurred already or to be expected in the future is unknown and probably unknowable.

Systemic causal factors Even with somewhat limited knowledge about what actually happened, it is clear that the common systemic causes for major accidents existed for this case.

Lack of design for controllability As noted in the section on Three Mile Island, nuclear reactor designs differ greatly in their ease of control and their dependence on added-on control and trip systems that may fail or may be neglected.

The official Russian position was that the operators were to blame for the accident. Scientists from other countries who attended the Vienna meeting did not entirely accept this argument. They agreed that the immediate cause was the operators' actions, but they argued strongly that a prime factor was also the basic design of the reactor (the positive void coefficient), which made the reactor difficult to control [259].

Lord Marshall (chairman of the British Central Electricity Generating Board) said that his government had warned the Soviets nine years earlier of the defects in the RBMK reactor design. "You can make any design safe by having clever enough operators," he said, "but the designers of Chernobyl gave the operators too difficult a task" [259, p. 65].

Lord Marshall claimed that the Soviets chose this design because it would save money. They knew it had defects, but they thought they could compensate. He added, "We are not talking about hindsight here. This was a judgment made in advance."

The head of the Russian delegation replied that he did not know about any British warnings in 1977. "*He declined to compare the quality of the Chernobyl reactors with western ones and said that the accident was caused by a series of awkward and silly mistakes by the operators*" [259, p. 65].

When serious accidents happen, they usually occur in ways not foreseen—which is what makes them serious. But the fact that the exact events leading to an accident are not foreseen does not mean the accident is not preventable. The hazard is usually known, and measures can often be taken to eliminate or reduce it. The exact events at Chernobyl did not have to be predicted to know that the reactor design made the operators' job too difficult: *some* events would occur during the life of the reactor that would require manual control under stress. Note that in this case, the operators were under extra pressure because the test was scheduled to be carried out just before a planned shutdown for routine maintenance. If the test could not be performed successfully this time, then they would have to wait for another year for the next shutdown [249]. In addition, the test procedures had been developed by a station electrical engineer, and the operators thought it was an electrical test. They did not conceive of it as a nuclear test.

Note the similarity with the DC-10 cargo door accident described in chapter 4. Applegate had warned that some events would occur during the life of the DC-10 that would lead to a cabin floor collapse. In both cases, the decisions were made to try to eliminate the preceding events because that would require the fewest tradeoffs with other goals such as lower cost or desired functionality. The necessity to make such tradeoff decisions leads to most of the difficulties in dealing with safety issues.

The positive coefficient of the RBMK reactor makes it very difficult to control, and safety is dependent on the water supply. Supposedly, the British had previously warned the Soviet Union that the RBMK design had serious defects and that the design gave the operators too difficult a job. The RBMK operators had minutes, and maybe only seconds, to react to an emergency.

The Soviets most likely built RBMK reactors, instead of the safer PWR reactor, for several reasons. For one, they wanted to produce plutonium and electricity in the same plants [390], and they did not have the resources of other countries to build separate reactors for these two goals. They also did not have the ability (at the time the RBMK reactors were built) to make the large pressure vessels or steam generators required by the PWR design [12]. The RBMK requires neither of these, and the graphic core comes in easy to construct modules. The Soviets later built PWRs after they had developed the necessary manufacturing capability.

Training deficiencies Operators and other staff were not trained in the technological processes in a nuclear reactor. They had also "lost any feeling for the hazards involved" [390]. In addition, as at TMI, the operators at Chernobyl had no simulator training for the accident sequence that occurred.

Approval of operating procedures The Soviet report said that neither the plant's chief engineer nor the resident representative of the atomic safety committee was consulted in the test design, and the test procedures were not reviewed by safety personnel, nuclear engineers, or physicists.

The test was perceived as an electrical test only and had been conducted uneventfully before. Thus, the operators did not think carefully enough about the effects on the reactor. There is a strong possibility, in fact, that the test was being supervised by representatives of the turbine manufacturer instead of the normal operators.

Complacency The Soviets admitted that the operators (and management) at the Chernobyl plant had become complacent. Chernobyl-4 was a model plant; of all the RBMK type plants, it ran the best. Its operators felt they were an elite crew, and they had become overconfident. The plant had run so well that they began to be too relaxed—they slipped into the dangerous attitude that an accident could never happen [12]. A US observer in Vienna noted that the Soviets were still incredulous that the accident could have happened.

The Chernobyl reactor had no containment structure but rather a *confinement building*, which had fans and filters. In the case of a release of radioactive gases, it was believed that the filters would remove the radioactivity before it was vented outside. But the building was not designed to withstand much overpressure [12].

In Vienna, the Soviets admitted that they were learning many of the lessons the United States had learned after TMI: the need for better training of operators and for simulator training, including that focused on accident sequences; the necessity to have procedures checked by a safety committee before tests are performed; and the danger of complacency. Much of the damage at Chernobyl could have been avoided had the Soviets paid more attention to the TMI investigations [12]. Unfortunately, this is true for most industries and most major accidents. Learning from the errors of others seems to be difficult.

D.4 The Fukushima Daiichi Nuclear Power Plant Accident

Background

On March 11, 2011, a magnitude 9 earthquake, called the Great Tohoku Earthquake, shook northeastern Japan and generated a tsunami. The earthquake and tsunami triggered a serious nuclear accident at the Fukushima Daiichi Nuclear Power Plant, owned by the Tokyo Electric Power Company (TEPCO). The flooding from the tsunami led to three nuclear meltdowns, three hydrogen explosions, the release of an enormous amount of radioactive substances into the air, and the evacuation of approximately 150,000 residents. This was the largest nuclear disaster since the Chernobyl accident in 1986.

Large amounts of water contaminated with radioactive isotopes were also released into the Pacific Ocean during and after the accident. The plant's operator has since built new walls along the coast and has created a 1.5 km long "ice wall" of frozen earth to stop the flow of contaminated water. An ongoing cleanup program to both decontaminate affected areas and decommission the plant has been estimated will take thirty to forty years.

By March 2012, one year after the accident, all but two of Japan's nuclear reactors had been shut down, some of which had been damaged by the quake and tsunami. Authority to restart the others after scheduled maintenance was given to local governments, which all decided against reopening them. The debate about national energy policy was changed overnight. According to the *Japan Times*, *"By shattering the government's long-pitched safety myth about nuclear power, the crisis dramatically raised public awareness about energy use and sparked strong anti-nuclear sentiment"* [414]. The official government investigation report was scathing:

> *Fukushima cannot be regarded as a natural disaster. . . . It was a profoundly man-made disaster—that could and should have been foreseen and prevented. And its effects could have been mitigated by a more effective human response. . . . Governments, regulatory authorities and Tokyo Electric Power [TEPCO] lacked a sense of responsibility to protect people's lives and society. They effectively betrayed the nation's right to be safe from nuclear accidents.* [198]

The report concluded that a culture of complacency about nuclear safety and poor crisis management led to the nuclear disaster. It claimed that the disaster might have been prevented if the power plant had been prepared to handle such a severe accident.

In fact, the Onagawa Nuclear Power Plant, which was owned and operated by a different utility, the Tohoku Electric Power Company, which was closer to the epicenter and experienced a higher tsunami, was able to shut down safely. So too was the Fukushima Daiini Nuclear Power Plant, which was approximately 12 km to the south of Fukushima Daiichi. The Daiini plant, which like Fukushima Daiichi was owned by TEPCO, had incorporated design changes that improved its resistance to flooding.

In addition to complacency, investigations concluded that the Japanese regulatory system engaged in a "network of corruption, collusion, and nepotism" where senior regulators accepted high paying jobs at the companies they once oversaw. Several top energy officials were fired by the Japanese government after the accident. In addition, three TEPCO executives were indicted for negligence resulting in death and injury. A court found all three men not guilty.

Safety features In 1967, when the Fukushima Daiichi Nuclear Power Plant was built, Japan was not capable of designing its own nuclear power plants. TEPCO instead imported nuclear technology from General Electric (GE) to build six separate light water boiling reactors. Fukushima Daiichi was one of the world's twenty-five largest nuclear power stations. It was designed to withstand an earthquake with a peak ground acceleration of 0.18 g and a response spectrum based on the 1952 Kern County earthquake in California. The earthquake design basis for all six reactor units at Fukushima ranged from 0.42 g to 0.46 g. All the units had a containment structure.

The Fukushima reactors were not designed for a large tsunami nor had the reactors been modified when concerns were raised in Japan and by the IAEA about their inadequate design.

As described in the beginning of this appendix, nuclear reactors generate electricity by using the heat of the fission reaction to produce steam, which drives turbines that generate electricity. When the reactor stops operating, the radioactive decay of unstable isotopes in the fuel continues to generate decay heat for a time and requires continued cooling. This decay heat amounts to approximately 6.5 percent of the amount produced by fission at first, then decreases over several days before reaching shutdown levels. Afterwards, spent fuel rods typically require several years in a spent fuel pool before they can be safely transferred to dry cask storage vessels.

In the reactor core, high-pressure systems cycle water between the reactor pressure vessel and heat exchangers. These systems transfer heat to a secondary heat exchanger via the essential service water system, using water pumped out to sea or an onsite cooling tower. Units 2 and 3 at Fukushima had steam turbine driven emergency core cooling systems that could be directly operated by steam produced by decay heat and that could inject water directly into the reactor. Some electrical power was needed to operate valves and monitoring systems.

Unit 1 had a different, entirely passive cooling system, the isolation condenser (IC). It consisted of a series of pipes run from the reactor core to the inside of a large tank of water. When the valves were opened, steam flowed upward to the IC, where the cool water in the tank condenses the steam back to water, which runs under gravity back to the reactor core. Unit 1's IC was operated only intermittently during the emergency in order to maintain the reactor vessel level and to prevent the core from cooling too quickly, which can increase reactor power. As the tsunami engulfed the station, the IC valves were closed and could not be reopened automatically due to the loss of electrical power, but they could have been opened manually. After the accident, TEPCO declared that the cooling systems for Units 1 to 4 were beyond repair.

When a reactor is not producing electricity, its cooling pumps can be powered by other reactor units, the grid, diesel generators, or batteries. Two emergency diesel generators were available for each of Units 1 to 5 and three for Unit 6.

In accordance with GE's original specifications for the construction of the plant, each reactor's emergency diesel generators and DC batteries, crucial components in powering cooling systems after a power loss were located in the basements of the reactor turbine buildings. Mid-level GE engineers expressed concerns, relayed to TEPCO, that this left them vulnerable to flooding. Nothing was done in response.

In the late 1990s, three additional backup diesel generators for Units 2 and 4 were placed in new buildings located higher on the hillside, to comply with new regulatory requirements. All six units were given access to these diesel generators, but the switching stations that sent power from these backup generators to the reactors' cooling systems for Units 1 through 5 were still located in the poorly protected turbine buildings.

All three of the generators added in the late 1990s were fully operational after the tsunami. If the switching stations had been moved inside the reactor buildings or to other flood-proof locations, power would have been provided by these generators to the reactors' cooling systems and thus the catastrophe would have been averted. The switching station for Unit 6 was protected inside the only GE Mark II reactor building and continued to function.

A seawall was used to protect the Fukushima Daiichi plant from a tsunami. The wall had originally been planned to be 98 ft (30 m) above sea level, but when the plant was built, TEPCO leveled the coast to make it easier to bring in equipment. That put the new plant at only 33 ft (10 m) above sea level, a third of the height originally intended.

In another change, TEPCO altered the piping layout for the emergency cooling system. The original plans separated the piping systems for two reactors from each other in the isolation condenser. However, in the actual construction, the two piping systems were connected outside the reactor. These changes were not noted, in violation of regulation. After the tsunami, the isolation condenser should have taken over the function of the cooling pumps by condensing the steam from the pressure vessel into water to be used for cooling the reactor. However, the condenser did not function properly, and TEPCO could not confirm whether a valve was opened.

The Fukushima Daiini Nuclear Power Plant was also struck by the tsunami. However, this power plant had incorporated design changes that improved its resistance to flooding, thereby reducing flood damage. The diesel generators and related electrical distribution equipment were located in the watertight reactor building, and therefore this equipment remained functional. By midnight, power from the electricity grid was being used to power the reactor-cooling pumps. Seawater pumps for cooling were protected from flooding. Although three of four pumps initially failed, they were restored to operation.

Events Understanding this accident requires starting with events that occurred before the actual proximal events associated with the loss of control at the Fukushima Nuclear Power Plant. Information about all the events is taken from [98; 198; 380; 414].

The Japanese government decided to introduce nuclear power in the 1950s. Some local governments expressed interest in hosting nuclear power plants in the hope of obtaining subsidies as well as gaining a major source of employment. But as the local governments and the national governmental bodies responsible for nuclear energy scouted Japan's countryside for sites to put future nuclear power plants, they found that people lived close to many of the potential sites. Assuming they could not convince residents to move, they instead switched to inspiring trust in the residents that the plants were indeed safe.

In reality, however, few discussions on safety took place at the national level until the 1970s. They believed that nuclear energy safety could be assured by maintaining the safety standards established in other countries, from which the nuclear energy technology was being imported. It is reported that, between 1973 and 1974, Kinji Moriyama, then the Japan Atomic Energy Commission's chairman, said that it was only necessary for the commission to consider the construction of nuclear power plants, with no need for input on safety from academics; that is, there was no need to question or doubt that nuclear reactors were completely safe.

In response to the rise of the antinuclear movement in the 1970s, the government Nuclear Safety Commission (NSC) separated from the Atomic Energy Commission in 1978 with the goal of improving safety regulation. The NSC, the commissioners of which were appointed by the prime minister with Diet approval, was assigned the function of double-checking safety regulations, establishing regulatory policies, and issuing recommendations through the prime minister to regulatory bodies.

In case of a nuclear emergency, the NSC was to provide technical advice based on requests made by the prime minister. Specifically, the NSC was to set up a headquarters organization called an Emergency Technical Advisory Body within its secretariat, and a local body of the Emergency Technical Advisory Body (the Nuclear Emergency Response Headquarters or NERHQ) at a local off-site center to which it would dispatch, among others, the NSC commissioners and the advisors for emergency responses. In turn, the NSC was to collect information and perform investigations and analyses, as well as prepare technical advice.

In a reorganization of the central government in January 2001, the Ministry of Economy, Trade and Industry (METI)—the former Ministry of International Trade and Industry (MITI)—took charge of all safety regulations on nuclear power. In this process, the Nuclear and Industrial Safety Agency (NISA) was established as a "special agency" inside the Agency for Natural Resources and Energy (ANRE), an extra-ministerial bureau, to take charge of ensuring energy safety and industrial safety. NISA was also responsible for nuclear emergency response.

At the same time in 2001, the Ministry of Education, Culture, Sports, Science and Technology (MEXT) was formed, with the former Ministry of Education merged with the former Science and Technology Agency (STA), which was established in 1956 originally to take on a central role in Japan's administration of nuclear energy. MEXT served as the authority for radiation protection, including monitoring.

The Nuclear Emergency Response Support Headquarters (NERHQ) was established within MEXT in case of a nuclear emergency. Its responsibility primarily was to provide advice for monitoring conducted at the off-site center, to analyze the monitoring data, and to dispatch disaster medical assistance teams to the site. The National NERHQ, with the support of NISA, was supposed to play a pivotal role in the government's emergency response measures such as taking protective action on behalf of the residents, including their evacuation. To this end, it was also expected, in terms of public relations in an emergency, to provide relevant information to the public via local governments, promptly, accurately, and in an easily understandable, clear-cut manner. An off-site center was designated for each nuclear facility as the base for responding to a nuclear disaster and was supposed to act as the base of the national nuclear emergency preparedness system. During the accident events, none of these groups operated as planned.

When the earthquake struck, units 1, 2, and 3 at Fukushima Daiichi were operating, but units 4, 5, and 6 had been shut down for scheduled inspection and refueling. Even though three of the reactors were shut down, their spent fuel pools still required cooling.

The earthquake produced forces that exceeded the seismic design tolerances for the reactors for continued operation, but were within the design tolerances at units 1, 4, and 6. Immediately after the earthquake, reactors 1, 2, and 3 shut down their fission reactions in a controlled manner by inserting control rods to scram the reactors. As the reactors were now unable to generate power to run their own coolant pumps, emergency diesel generators came online, as designed, to power the electronics and the coolant systems. Electric power was required to operate the pumps that circulated coolant through the reactors' cores. Continued circulation removes residual decay heat, which continues to be produced after fission has stopped.

The emergency power generators operated normally until the tsunami destroyed the generators for all the reactors except Unit 6. The two generators cooling Unit 6 were undamaged and were sufficient to be pressed into service to cool the neighboring Unit 5 along with their own reactors, averting the overheating that the other reactors suffered.

The largest tsunami wave was 43 to 46 ft (13 to 14 m) high and hit approximately 50 minutes after the initial earthquake. The water flooded the plant's ground level, which was 33 ft (10 m) above sea level. The waves flooded the basements of the power plant's turbine buildings, which disabled the emergency diesel generators and caused a loss of power to the circulating pumps. TEPCO then notified authorities of a "first-level emergency."

The switching stations that provided power from the three backup generators located higher on the hillside failed when the building that housed them was flooded. All AC power was lost to units 1 through 4, and all DC power was lost on Units 1 and 2, while some DC power from batteries remained available on Unit 3.

Steam-driven pumps provided cooling water to reactors 2 and 3 and prevented their fuel rods from overheating, as the rods continued to generate decay heat after fission had ceased. Eventually, those pumps stopped working, and the reactors began to overheat. The lack of cooling water led to meltdowns in reactors 1, 2, and 3.

More batteries and mobile generators were dispatched to the site, but were delayed by poor road conditions. The first arrived March 11 at 21:00, almost 6 hours after the tsunami hit. Unsuccessful attempts were made to connect portable generating equipment to power water pumps. The failure was attributed to flooding at the connection point in the Turbine Hall basement and the absence of suitable cables. TEPCO switched its efforts to installing new lines from the grid. One generator at Unit 6 resumed operation on March 17, while external power did not return to Units 5 and 6 until March 20.

As workers struggled to supply power to the reactors' coolant systems and restore power to their control rooms, three hydrogen chemical explosions occurred. The pressurized gas was vented out of the reactor pressure vessel, where it mixed with the ambient air and eventually reached explosive concentration limits in Units 1 and 3.

On March 12, leaking hydrogen mixed with oxygen exploded in Unit 1, destroying the upper part of the building and injuring five people. On March 14, a similar explosion occurred in the Reactor 3 building, blowing off the roof and injuring eleven people. On March 15, the Reactor 4 building filled with hydrogen due to a shared vent pipe with Reactor 3, which led to a third explosion. In each case, the hydrogen–air explosions occurred at the top of each unit, in their upper secondary containment buildings.

Not only was the power supply damaged, but the tsunami also destroyed or washed away vehicles, heavy machinery, oil tanks, and gravel. It destroyed buildings, equipment installations, and other machinery. Seawater from the tsunami inundated the entire building area and even reached the high-pressure operating sections of Units 3 and 4 and a supplemental operations common facility (the Common Pool Building).

After the water retreated, debris from the flooding was scattered all over the plant site, hindering movement. Manhole and ditch covers had disappeared, leaving gaping holes in the ground. In addition, the earthquake lifted, sank, and collapsed building interiors and pathways; access to and within the plant site became extremely difficult. Recovery tasks were further interrupted as workers reacted to the intermittent aftershocks and tsunami.

The amount of damage sustained by the reactor cores in Units 1, 2, and 3 during the accident and the location of molten nuclear fuel within the containment buildings is unknown; TEPCO has revised its estimates several times. Reactor 4 was not operating when the earthquake struck. All fuel rods from Unit 4 had been transferred to the spent fuel pool on an upper floor of the reactor building prior to the tsunami. The explosion on March 15 damaged the fourth-floor rooftop area of Unit 4, creating two large holes in a wall of the outer building.

Units 5 and 6 were also not operating when the earthquake struck. Unlike Unit 4, their fuel rods remained in the reactor. Units 5 and 6 shared a working generator and switchgear during the emergency and were successfully shut down 9 days later on March 20.

The Fukushima Daiichi Nuclear Power Plant (NPP) operators were overwhelmed and confused on learning that the nuclear reactors were successively losing their power sources. The loss of electricity resulted in the sudden loss of their monitoring equipment, such as scales, meters, and the central control room functions. Lighting and communication were also affected. The groups that were supposed to assist them in an emergency were cut off by communication failures.

The decisions and responses to the accident, therefore, had to be made on the spot by operational staff at the site without appropriate tools and manuals. Despite great effort, the Fukushima Daiichi operators were unable to control the overheating reactors after many of the functions, including electricity, required to effectively cool down the reactors had been lost.

The loss of the main control room functions and of lighting and communication systems inhibited or delayed onsite emergency response to a great extent. The lack of access obstructed the delivery of necessities such as alternative water injection using fire trucks, the recovery of the electric supply, and the vent. The operators had to work in the yard where there was rubble and debris, worrying about the aftershock or the recurrence of a tsunami as well as the safety of their families. The working environment further deteriorated with time due to an increase in radiation level and the hydrogen gas explosions in the reactor buildings.

The Safety Parameter Display System, by which both the operators and the head office would have been able to instantly grasp and monitor the operational status of all units, was not available because it had lost its power supply after the tsunami. In addition, most of the main control rooms, which were the only source of information on the plant status, also became incapable of providing the plant parameters due to the loss of the power supply.

The local operator confusion was mirrored in the other parts of the system that were trying to cope with the events. Setting up the Local Nuclear Emergency Response Headquarters (NERHQ) at the off-site center for this accident, which was supposed to lead the emergency response, required a lot of time due to the delays and cancellations of the arrivals of necessary personnel as well as the loss of power.

In addition, the off-site center for both the Fukushima Daiichi NPP and the Fukushima Daiini NPP was located only about 5 km from the Fukushima Daiichi NPP, and was not equipped with air cleaning filters to insulate it from radioactive substances. Those factors ultimately forced the off-site center to relocate its functions.

Even after the establishment of the Local NERHQ, ground communication lines remained disconnected, causing serious problems in the sharing of information, and in liaison and coordination with relevant organizations. The communication problems led to the NERHQ having inaccurate information about the state of the plant. For example, they mistakenly believed for a while that the IC system of Unit 1, used to transfer residual and decay heat, was operating normally. They therefore took no further actions to compensate for the loss.

Residents also did not receive accurate information in the evacuation orders issued by NERHQ, including news about the seriousness of the accident or the expected term of their evacuation. Unaware of the severity of the accident, they thought that they would be away from their homes for only a few days. They headed to the evacuation shelters literally with just the clothes on their backs. Ultimately, however, they were subjected to a long-term evacuation. In addition, the residents were forced to relocate to other evacuation shelters whenever the evacuation zone changed, increasing their stress. Some evacuees unknowingly evacuated to areas that were later found to have high doses of radiation.

Similar communication problems occurred at the national emergency response level. Because of cybersecurity concerns, mobile phones could not be used in the basement of the prime minister's office, where the relevant government officials assembled. With other communication means not available, it was difficult for these officials to gather information on the accident. For example, it was impossible to obtain almost any emergency monitoring results during the initial response stage of this accident because the monitoring posts, which were overly concentrated along the Fukushima Prefectural coastline, became unusable in the wake of the earthquake and tsunami. The Fukushima Prefecture was unable to implement prompt emergency monitoring when most of their monitoring posts were washed away in the tsunami. With communication lines broken by the earthquake, only one of the twenty-four monitoring posts was functioning properly following the disaster.

Communication problems were exacerbated by confusion about responsibilities. For example, Fukushima Prefecture officials were under the impression that the response to the nuclear disaster was mainly to be carried out at the off-site center. The ineffectiveness of the off-site center thus pushed the prefecture's response to the disaster into a state of confusion.

To make things worse, when the nuclear accident happened, a large number of personnel both in the Fukushima Prefecture and in the municipalities were tied up with their response to the earthquake and tsunami disasters.

The emergency radiation medical system was unable to deal with accidents that involved the release of large amounts of radioactive substances over a wide area because of the inappropriate locations of primary radiation emergency hospitals: the hospitals themselves had to be evacuated and were unable to treat any patients. But even if they had been available, there was a lack of decontamination facilities and inadequate or almost nonexistent radiation training of the hospital staff. As a result, some of those who were injured at the Fukushima Daiichi Nuclear Power Plant did not have their injuries treated for 3 days.

The structures in the central government to deal with a nuclear accident were totally ineffective. It has been claimed that they fell into a so-called "elite panic," in which they refused to pass on critical information for fear of inciting panic among the general public. For example, NISA delayed publishing most of the results of monitoring conducted from March 11 to March 15 that was sent from the off-site center.

Many of the problems in the central government response were simply due to poor planning. For example, the NSC used a group email system for mobile phones to summon the advisors for emergency responses and to set up a national Emergency Technical

Advisory Body. However, the group email was not delivered to some advisors, and disruptions in public transportation and telecommunications meant that nearly all of the advisors for emergency responses who were summoned did not convene on March 11.

Causal factors The official government accident report concluded that the accident was the result of collusion between the government, the regulators, and TEPCO, and of a lack of governmental oversight. It concluded that the causes of the accident were foreseeable and that the plant operator, TEPCO, had not met basic safety requirements such as risk assessment, preparing for containing collateral damage, and developing evacuation plans. At a meeting in Vienna three months after the disaster, the International Atomic Energy Agency faulted lax oversight by the Ministry of Economy, Trade, and Industry, saying the ministry faced an inherent conflict of interest as the government agency in charge of both regulating and promoting the nuclear power industry.

TEPCO admitted for the first time on October 12, 2012, that it had failed to take stronger measures to prevent disasters for fear of inviting lawsuits or protests against its nuclear plants. Japanese authorities later admitted to lax standards and poor oversight and to engaging in a pattern of withholding and denying damaging information. Once the events started, those responsible for responding to it did not pass on critical information to the public to avoid creating panic.

While there were physical equipment design flaws that led directly to the events, these mostly arose, as usual, because of a poor safety culture leading to flawed decision-making, ineffective organizational structures and practices, and superficial or ineffective technical activities.

Flawed safety culture and decision-making In March 2012, Prime Minister Yoshihiko Noda said that the government shared the blame for the Fukushima disaster, saying that officials had been blinded by a false belief in the country's "technological infallibility." Risks both before and during the accident were downplayed, and complacency was widespread.

The Japanese nuclear power establishment had what can be characterized as a culture of denial. The basis of this culture was what has been called the Safety Myth, which was a belief that serious accidents could never happen in nuclear power plants in Japan. The belief resulted in an inability to consider such events as a reality, to plan for them, and to respond to the events appropriately.

Most of the players in the Japanese nuclear power industry, including the operators, regulators, nuclear experts, and even the local governments that obtained subsidies for hosting nuclear power plants, relied on the continued operation of the existing reactors. A frequently used term, "Nuclear Village," illustrated the fundamentally close, insular, conservative, and interconnected nature of these players [47; 198]

The unspoken understanding in the Nuclear Village was that the "risk of shutting down existing reactors in order to avoid any potential risk of accidents" outweighed the "risk of accidents occurring as the result of inaction," and therefore the shared Safety Myth prevailed that nothing could go wrong in Japan's nuclear sector [198]. As a result, preparations for severe accidents and for compound nuclear disasters remained inadequate.

According to a survey conducted by the Atomic Energy Society of Japan of its members, the nuclear scientists believed that the light water reactor was such a proven technology that it was no longer the target of research. At the same time, power companies did not welcome research on nuclear safety for fear that the public might think there were safety issues [198].

Historical context also comes into play here. Japan was a nation poor in energy resources, so promotion of nuclear power was important. But local governments and citizens were skeptical after Japan's experience as a target of the nuclear bomb attacks over Hiroshima and Nagasaki during World War II and the entrenched antinuclear sentiment it engendered [47; 198]. The promotion of the Safety Myth became important in promoting and maintaining the nuclear power industry. The myth played an important role in affecting attitudes and societal acceptance of nuclear power technology, as well as in determining the administrative framework of safety regulations.

Promotion of this myth, however, led to inadequate regulation and preparations for an accident. Concern existed that if regulatory oversight was required, it would logically follow that the existing regulations were inadequate and that their application was deficient. The regulators substantially increased the number of items covered in inspections, lengthened the time needed to complete inspections, and increased the time and labor to produce the review documents, substituting paperwork for effective oversight and preparedness. In a "paperwork culture," employees spend all their time proving that the system is safe but little time actually doing the things necessary to make it so. The Japanese nuclear operators eventually shifted to document-based inspections to show that every aspect of the hardware was designed and maintained to be safe, which severely limited true oversight and also kept those working on safety occupied with producing large amounts of paperwork [47]. The micromanagement of hardware, partially due to lack of resources at the regulatory agencies, in lieu of other types of oversight, left the country open for a disaster.

The limitations of the government and TEPCO's efforts become even more obvious when compared to others in the Japanese nuclear community. The Tohoku Electric Power Company, which owned and operated the Onagawa Nuclear Power Plant, had top management that strongly advocated safety and had a culture that prioritized safety above all. While historical data showed tsunamis in the Tohoku region to average 3 m prior to construction, they conducted surveys and research to determine potential tsunami height and constructed their plant 14.7 m above sea level, which was barely above the height of the actual tsunami that they experienced on March 11, 2011. Employees were also prepared for disasters through periodic training sessions against extreme situations, which allowed them to stay calm during the actual disaster and to avoid a catastrophe [380]. In contrast, almost all the emergency response actions by the government and TEPCO were inadequate due to the shared assumption that such accidents could never occur.

Unrealistic risk assessment and over-reliance on redundancy Complacency was created or at least facilitated by unrealistic risk assessment and by reliance on redundancy. The government regulators believed that the probability of a severe accident was so small that they could assume they would never happen [380]. Authorities grossly underestimated the tsunami risks that would follow the magnitude 9.0 earthquake. The 40 ft high tsunami that struck the plant was double the height of the highest wave predicted by officials. The erroneous assumption that the plant's cooling system would function after the tsunami worsened the disaster.

Defense-in-depth is a type of serial redundancy that is the standard design approach in nuclear power. It involves creating multiple, supposedly independent layers of defense. The goal of this design approach is to compensate for potential human and mechanical failures [159]. The problem, as is often the case in complex systems, is that the layers of defense were not independent and thus had single failure modes and causes. In addition, mechanical failures and human errors are not the only causes of accidents.

The first layers of protection usually include design features while emergency response is used in case the design is deficient. Frequently, as in this case, the system designers and operators are so sure that the design features will be effective that they begin to believe that emergency response will never be needed and put less emphasis on it [47].

In particular, TEPCO and the government regulators believed that huge tsunamis would not occur in the near future. It was judged that natural disaster countermeasures were not urgent because such disasters occur once every one hundred years, and the lifespan of the reactor was shorter than that. In 2006, however, the Japanese government overseers set up a study group on flooding, which concluded that no basis existed for assuming that the probability of a tsunami hitting the Fukushima Daiichi Nuclear Power Plant was extremely low [198]. No appropriate response was taken in response to this new information.

When the tsunami hit, the Safety Parameter Display System, by which the Fukushima Daiichi NPP operators and the head office were supposed to be able to monitor the operational status of all units, was not available because it had lost its power supply due to the tsunami. In addition, most of the main control rooms, which were the only source of information on the plant status, became incapable of providing the plant parameters for the same reason [198]

The manuals on emergency operating procedures provided to the operators proved to be unusable because they assumed the ability to monitor the state of the nuclear reactors and were not designed to cope with the loss of all electric power for a long period of time, which is what happened [198]. In fact, nearly all preparation and training did not consider the possibility of the complete and simultaneous loss of AC power at multiple nuclear power units.

As described in the events, most of the protection systems in the plant were also unavailable because they also depended on a common power source, which failed. The government's preparedness system in the event of a nuclear disaster had been constructed on the assumption that the infrastructure, including communication and transportation networks, would function and operate as in ordinary circumstances. No countermeasures were put in place in case they did not. The off-site center, in particular, did not have sufficient logistical and personnel support in place. It was not equipped with air filters to block the penetration of radioactive materials.

The scale, complexity, and speed of the progress of the accident had not been assumed in the nuclear emergency preparedness drills. The government was also facing the huge task of responding to the tremendous damage that had been wrought by the earthquake and tsunami on an extremely large scale [198], also not anticipated.

The nuclear emergency preparedness system of Fukushima Prefecture was based on the assumption that a nuclear disaster, an earthquake, and a tsunami would not occur simultaneously. Their Regional Disaster Prevention Plan noted that an earthquake was not assumed to cause a nuclear emergency because the national government had confirmed the seismic safety of the nuclear power plants.

Nuclear power, in general, emphasizes *design-basis* accidents in preparation for a loss. Briefly, design-basis accidents are those caused by events that are anticipated to occur. Probabilistic analysis is often used to identify the expected cases, and defense-in-depth is then used to handle them. Worst-case analysis is used in other industries.

Ignoring high-consequence, low-probability events Ignoring high-consequence, low-probability events is common in the process industry. In fact, the term accident is commonly replaced with "high-consequence, low probability event," and accidents are then assumed to be low probability. An accident needs to be proven to be "low probability"; it cannot be assumed to be low probability by definition. In fact, talking about "low probability" when it is usually not possible to measure probability in a complex system makes little sense.

Almost everyone in the Japanese nuclear power community, including the government regulators, appear to have believed the Safety Myth and therefore did not prepare for and were unable to respond well to the actual events.

Ignoring warnings Credible warnings, including those from whistleblowing, are sometimes dismissed without taking appropriate action. In the case of Fukushima, in 2000, Kei Sugaoka, a Japanese-American nuclear inspector who had done work for General Electric at Fukushima Daiichi, told the nuclear regulator about a cracked steam dryer that he believed was being concealed. In response, NISA illegally divulged Sugaoka's identity to TEPCO, effectively expelling him from the industry. TEPCO was merely instructed to inspect its reactors by itself. It was allowed to keep operating its reactors for the next two years even though, an investigation ultimately revealed, its executives had actually hidden other, far more serious problems, including cracks in the shrouds that cover reactor cores [291].

Other warnings were also ignored. In October 1991, one of two backup generators for Reactor 1 failed after flooding in the reactor's basement. Seawater used for cooling leaked into the turbine building from a corroded pipe through a door and some holes for cables. One of the two power sources was completely submerged, but its drive mechanism was not affected. An engineer was quoted as saying that he informed his superiors of the possibility that a tsunami could damage the generators. TEPCO installed doors to prevent water from leaking into the generator rooms. The power supply was not cut off by the flooding, and the reactor was stopped for only one day.

An in-house TEPCO report in 2000 recommended safety measures against seawater flooding based on the potential for a 50 ft (15 m) tsunami. TEPCO leadership said the study's technological validity "could not be verified." After the tsunami, a TEPCO report said that the risks discussed in the 2000 report had not been announced because "announcing information about uncertain risks would create anxiety" [64, p. 84].

A study within TEPCO in 2008 identified an immediate need to better protect the facility from flooding by seawater. This study mentioned the possibility of tsunami waves up to 33 ft (10.2 m). Headquarters officials insisted that such a risk was unrealistic and did not take the prediction seriously. An earthquake research center suggested that TEPCO and NISA revise their assumptions for possible tsunami heights upward based on their finding about another earthquake, but this was not seriously considered at the time [64].

The US Nuclear Regulatory Commission warned of a risk of losing emergency power in 1991, and NISA referred to that report in 2004 but took no action to mitigate the risk.

Warnings by government committees about tsunamis higher than the maximum of 18 ft (5.6 m) forecast by TEPCO and government officials were also ignored.

A seismologist named Katsuhiko Ishibashi wrote a 1994 book titled *A Seismologist Warns* criticizing lax building codes, which became a bestseller when an earthquake in Kobe killed thousands shortly after its publication. In 1997, he coined the term "nuclear earthquake disaster," and in 1995 wrote an article for the *International Herald Tribune* warning of a cascade of events much like those that actually occurred during the Fukushima disaster [64].

The IAEA had expressed concern about the ability of Japan's nuclear plants to withstand earthquakes. At a 2008 meeting of the G8's Nuclear Safety and Security Group in Tokyo, an IAEA expert warned that a strong earthquake with a magnitude above 7.0 could pose a serious problem for Japan's nuclear power stations. The region had experienced three earthquakes of magnitude greater than 8 in the past 159 years.

In 2007, Japan received an Integrated Regulatory Review Service (IRRS) of IAEA, a peer review that evaluated legal systems and regulatory organizations. NISA, in particular, received several recommendations and suggestions, such as "NISA should continue to develop its efforts to address the impacts of human and organizational factors on safety in operation," and "NISA should develop a strategic human resources management plan to face future challenges." NISA did not take any concrete measures in response to these warnings nor to any of the others.

Flawed resolution of conflicting goals The conflict of one government organization both promoting nuclear power and overseeing safety in Japan was recognized, and a complex organization was established that supposedly reduced conflicts. However, NISA, as a part of METI, had to play a role in protecting the framework for promoting nuclear power for the sake of its parent organization, which compromised its position as the guardian of nuclear safety.

The NSC also lacked independence from the administrative institutions promoting nuclear power and did not play its expected role of checking regulations administered by NISA. On the contrary, in some cases they received instructions from NISA. The NSC almost never exercised its authority to issue recommendations through the prime minister to regulatory bodies in spite of a number of nuclear accidents and incidents, and it avoided any regulation that appeared to be an obstacle to the promotion and utilization of nuclear power.

As just one concrete example, the comprehensive nuclear emergency preparedness drill conducted by the national government in cooperation with local governments was superficial in nature because it was aimed primarily at not worrying or confusing local residents. It was ineffective as a response to actual accidents.

TEPCO also had conflicting goals. The nuclear power division was under constant pressure to maintain high uptime at its nuclear plants at the expense of safety [380].

Migration toward states of higher risk Changes occurred that increased risk even during construction. An example is the construction of a seawall of only 33 ft (10 m) instead of the planned 98 ft (30 m) above sea level to make it easier to bring in equipment. No evaluation of the risk involved in that change was made then or later when warnings about potential tsunami heights were made.

Risk continued increasing during operations. Until the 1980s, TEPCO was active and enthusiastic in its institution of safety provisions. The workers studied safety design philosophy for nuclear power plants and learned the lessons of past accidents. They were trained in an atmosphere of constant initiatives to improve and upgrade processes. But as the core focus of work at nuclear power plants shifted to maintenance and operational issues, and as the pressure to cut costs increased to cope with the partial liberalization of the electricity market, the practical aspects of safety design started to recede into the background. Repairs and inspections of plants were increasingly outsourced to reduce personnel costs, which lessened awareness of the onsite realities of nuclear power plants [47].

Ineffective organizational structures and practices Japan's government was criticized after the accident as having a "rigid bureaucratic structure, reluctance to send bad news upwards, need to save face, weak development of policy alternatives, eagerness to preserve nuclear power's public acceptance, and politically fragile government" [64]. These, along with TEPCO's very hierarchical management culture, contributed to the way the accident unfolded.

Diffusion of responsibility and authority The complex administrative framework of the national government blurred the responsibilities of each stakeholder involved in safety regulations, such as NISA, NSC, and MEXT, creating coordination problems between them. Ambiguity existed in all the government organizations responsible for nuclear power about who was responsible for what. Spreading responsibility over different ministries contributed to the confusion during the accident about who was responsible for assessing the accident information, giving evacuation orders, utilizing the data that was collected, and communicating risk information.

NISA did not have the ability to provide appropriate supervision of TEPCO partly because METI did not give NISA institutional independence. Reliance on a traditional personal management system based on rotation among jobs also did not ensure that NISA had sufficient nuclear expertise to be effective as a regulatory body.

The NSC, in turn, did not correct incomplete regulations or provide effective regulation policies. Scientists could not sound alarms about the safety of the nuclear power plants. The NSC lacked independence from administrative institutions promoting nuclear power and did not play its expected role of checking regulations administered by NISA. On the contrary, in some cases they received instructions from NISA, an organization that they were supposed to supervise, illustrating the fact that the NSC was seriously ignored as far as actual operations were concerned [198]. The NSC almost never exercised its authority to issue recommendations through the prime minister to regulatory bodies in spite of a number of nuclear accidents and incidents, and it avoided any regulation that appeared to be an obstacle to the promotion and utilization of nuclear power.

TEPCO, which was supposed to be subject to nuclear safety regulatory supervision, strongly pressured regulatory authorities for postponement of regulations and softening of regulatory criteria. It took advantage of its information superiority and its close relationship with NISA's parent organization METI, which was the supervising authority for the electric power business and was promoting nuclear power policies.

NISA was responsible for regulation of all industrial safety, not just nuclear safety, but it did not have sufficient human resources to carry out this charge. It had been occupied with handling various nuclear incidents since its establishment in January 2001 and lacked sufficient personnel capable of addressing mid- to long-term nuclear safety challenges [198]. In addition, periodic personnel transfers for NISA staff members made it difficult to develop specialized technical ability along with expertise and experience with nuclear regulations.

The Nuclear Reactor Regulation Act (Act No. 166 of 1957) required operators of nuclear power plants to assign a lead engineer for each nuclear reactor to oversee safe operations. In reality, however, one engineer was made responsible for several reactors; at the Fukushima Daiichi Nuclear Power Plant, only two engineers were in charge of six reactors. This assignment of lead engineers for the nuclear reactors, along with the work schedule of plant operators, was not sufficient to respond to multiple events occurring at the same time. In addition, the lead engineers had not received special training or education to prepare them for serious accidents.

TEPCO, although it owned as many as seventeen nuclear reactors, had never had a CEO whose expertise was rooted in nuclear technology or engineering, and only a limited number of executives could really appreciate the managerial responsibilities of the nuclear power business. Past CEOs rose up through administrative departments, which were more valued in TEPCO than technical organizations such as the Nuclear Power Division [47].

As a result, the top management could not establish effective leadership and management for nuclear safety, called out in the IAEA Fundamental Safety Principles [160]. They did not understand the details of the highly specialized operations of the Nuclear Power Division and, instead, entrusted matters, such as safety provisions, to underlings familiar with nuclear power. Lack of experience and expertise contributed to an inadequate sense of responsibility of top management as a nuclear operator and to a weak sense of crisis [47; 160].

Lack of independence and low-level status of safety personnel The attempt to separate government regulation of nuclear power from its promotion was, in fact, unsuccessful. The regulatory authorities were, in reality, not independent.

Also, in the framework of the periodic personnel rotation system, the safety regulator position was considered anything but a plum career move. It was expected that nothing would go wrong, and, if anything did, the officials would be blamed. Therefore, these officials were not experts with a strong sense of mission. The limitation of personnel with technical expertise at NISA led to micromanagement of hardware—in particular, document-based inspections—instead of taking a holistic approach to ensuring system safety.

Limited communication channels and poor information flow Communication problems were rampant both before the events and during the crisis. The information Japanese people received about nuclear energy and its alternatives had long been tightly controlled by both TEPCO and the government. Effort was expended in maintaining the safety myth and ensuring that the public did not question the safety of nuclear power.

After the events began, communication and information flow almost totally broke down. NISA, as well as other institutions, refused to pass on critical pieces of information for fear of inciting panic among the general public.

Examples of communication limitations include [380]:

- Experts called in to help manage the crisis were stunned at how little the leaders in the prime minister's office knew about the measures available to them.
- Delays occurred in releasing data on dangerous leaks at the facility. For example, data was mailed to the prefecture government, but not shared with others. Emails

from the government to Fukushima, containing vital information about evacuation and health advisories, were deleted and unread. NISA did not publish immediately most of the results of the monitoring conducted during the period from March 11 to 15 and sent from the off-site center [47; 198; 161]. TEPCO officials were instructed not to use the phrase "core meltdown" at press conferences.

- A lack of trust existed between the major actors, which hindered the response.
- The Japanese government was slow to accept assistance from US nuclear experts. US military aircraft provided information to the Japanese government (METI), but officials did not act on it nor forward it to those who could act on it. As a result, some residents were unnecessarily exposed to radiation.
- The establishment repeatedly played down the risks of the events and suppressed information about the movement of the radioactive plume, so some people were evacuated from more lightly to more heavily contaminated places. Decontamination equipment was slow to be made available and then slow to be utilized.
- Miscommunication with the Fukushima site also blocked a prompt emergency response. Information such as the operational status of emergency equipment was not accurately communicated from the shift team to the NPP Emergency Response Center, which was located on the second floor of a seismic isolation building of the Fukushima Daiichi NPP. For example, they mistakenly believed for a while that the IC system of Unit 1 to transfer residual and decay heat was operating normally, leading to no further actions.
- NISA, the head of the NERHQ, was unable to collect and share information concerning the progression of the accident and the progress of the response, and it could not provide NERHQ with appropriate expert advice. The TEPCO head office could not understand or monitor the operational status of the reactors because the Safety Parameter Display System was unavailable.
- NISA expended an inordinate amount of time reviewing the scope of evacuation and was unable to draft a proposal for the specific designation of evacuation areas in a prompt manner.
- While the off-site center was expected to collect information locally and report it to NISA, it could not do this because of inadequate and failed communication channels. For example, almost any emergency monitoring results were impossible to obtain during the initial response stage because the monitoring posts, which were overly concentrated along the Fukushima Prefectural coastline, became unusable in the wake of the earthquake and tsunami.
- Summoning and convening the NSC advisors to establish an Emergency Technical Advisory Body was delayed. Group email to mobile phones was not delivered, and disruptions in public transportation and telecommunications meant that nearly all of the advisors for emergency responses were unable to meet on March 11 [47].
- Due to information security concerns, mobile phones could not ordinarily be used in the basement of the prime minister's office, where the relevant government officials assembled. With other communication means not available, it was difficult for these officials to gather information on the accident rapidly.

- TEPCO did not disclose facts about the accident's progression to local residents, Japanese citizens, and stakeholders across the world in a timely and appropriate manner. In addition, the information disclosed was not always sufficient. According to the government accident report, if TEPCO had proactively published an alert with regard to the anomalies of reactors, they could have reduced the impact of radioactive substances on local citizens [198].

These types of communication failures were never anticipated, nor were preparations made to overcome them. They led not only to inadequate emergency response by most every responsible group but also a lack of coordination between them.

Inadequate planning and training NISA was not prepared to cope with serious accidents. Plans for responding to a nuclear disaster were found to not function in the case of serious accidents in which radioactive substances were released into the environment at a large scale. Specifically, the off-site center for both the Fukushima Daiichi NPP and the Fukushima Daiini NPP was located only about 5 km from the Fukushima Daiichi NPP and was not equipped with air cleaning filters to insulate it from radioactive substances. Those factors ultimately forced the off-site center to relocate its functions.

NISA was also responsible for the comprehensive nuclear emergency preparedness drills conducted annually by the national government. These drills did not anticipate severe accidents or complex disasters at all and were virtually useless as a measure to increase preparedness for nuclear accidents [198].

Participants in the drills changed every year due to frequent personnel transfers and changes of administration in the government. The various organizations in charge of the drills were therefore required to brief participants from scratch every time the drill was conducted. The time available to brief participants from the government, including politicians as well as bureaucrats, was very limited, but a large amount of time was necessary to be adequately prepared. In practice, the drills were only conducted for a limited scenario [198].

Operator training in response to severe accidents was not provided during normal operations or periodic inspections, leading to a lack of staff experience or training in activating emergency equipment. Basic procedures to take in case of a serious accident were not adequately shared among the operators at the Fukushima Daiichi NPP, in contrast to those at the Fukushima Daiini NPP.

In the end, "Plant workers had no clear instructions on how to respond to such a disaster, causing miscommunication, especially when the disaster destroyed backup generators" [98].

Superficial or ineffective technical activities Planned control and feedback channels did not operate as designed or expected. An inordinate amount of confidence was placed on redundancy and in defense-in-depth that did not actually exist due to lack of independence and single point failure modes.

The Fukushima Nuclear Power Plant was built based on turnkey agreements, by which the overseas manufacturer undertook all operations from plant design to materials procurement, construction, and trial operation, delivering to the customer a plant in a fully operational state. The reactor buildings were designed to be small to keep the reactor compact and economical. The Japanese users were excluded from the reactor design process and had little or no input.

Unit 1 was originally designed to withstand a tornado (common in the United States), which is why the diesel generators were located in the basement of a poorly protected turbine building. In contrast, the Unit 6 diesel generators and cooling systems were protected by a fortified reactor building. The diesel generator for reactor 1 was submerged under seawater by the tsunami, and unplanned interactions and common mode failures led to five of the reactors eventually losing their emergency power supply for cooling the reactors [24; 47].

The cooling systems were designed to be heavily dependent on electricity for high-pressure water injection, depressurizing the reactor, low pressure water injection, the cooling and depressurizing of the reactor containers and removal of decay heat at the final heatsink [198]. When the power failed, these functions became impossible to perform.

References

1. Adams, E. E. "Accident Causation and the Management System." *Professional Safety* 22, no. 10 (October 1976): 26–29.
2. American Petroleum Institute. "Industry Mission," 2017. http://www.api.org/about/industry-mission.
3. American Petroleum Institute. *API RP 75—Recommended Practice for Development of a Safety and Environmental Management Program for Offshore Operations and Facilities*, 3rd Edition, 2004.
4. Abrecht, Blake. "Systems Theoretic Process Analysis Applied to an Off-Shore Supply Vessel Dynamic Positioning System." SM Thesis, Aeronautics and Astronautics Department, Massachusetts Institute of Technology, 2016b.
5. Abrecht, B., D. Arterburn, D. Horney, J. Schneider, B. Abel, and N. Leveson. "A New Approach to Hazard Analysis for Rotorcraft." AHS Technical Specialists' Meeting on the Development, Affordability, and Qualification of Complex Systems, Huntsville AL, February 9–10, 2016a.
6. Ackoff, Russell L. "Towards a System of Systems Concepts." *Management Science* 17, no. 11 (July 1971): 661–671.
7. Ackoff, Russell. "A Lifetime of Systems Thinking." The Systems Thinker. https://thesystemsthinker.com/a-lifetime-of-systems-thinking/.
8. Ackoff, Russell, and Fred Emery. *On Purposeful Systems*. Abingdon, Oxfordshire, UK: Routledge, 2005.
9. Adato, Michelle, James MacKenzie, Robert Pollard, and Ellyn Weiss. *Safety Second: The NRC and America's Nuclear Power Plants.* Union of Concerned Scientists. Bloomington: Indiana University Press, 1987.
10. Aerospace Safety Advisory Panel (ASAP). "The Use of Leading Indicators and Safety Information Systems at NASA." NASA, March 2003a.
11. Aerospace Safety Advisory Panel (ASAP). *Annual Report.* NASA, January 2003b.
12. Ahearne, John F. "Nuclear Power after Chernobyl." *Science* 236 (May 8, 1987): 673–679.
13. Air Force Space Division. *System Safety Handbook for the Acquisition Manager.* SDP 127–1, US Air Force, January 12, 1987.
14. Aitken, A. "Fault Analysis." In *High Risk Safety Technology*, edited by A. E. Green, 67–72. New York: John Wiley & Sons, 1982.
15. Allen, David J. "Digraphs and Fault Trees." *Hazard Prevention*, January/February 1983, 22–25.
16. Allnutt, M. F., D. R. Haslam, M. H. Rejman, and S. Green. "Sustained Performance and Some Effects on the Design and Operation of Complex Systems." In *Human Factors in Hazardous Situations*, edited by D. E. Broadbent, J. Reason, and A. Baddeley, 81–93. Oxford, UK: Clarendon Press, 1990.
17. Amalberti, R. "The Paradoxes of Almost Totally Safe Transportation Systems." *Safety Science* 37 (2001): 109–126.
18. American Nuclear Society. "Risk-Informed and Performance-Based Regulations for Nuclear Power Plants." Position Statement 46, June 2004.
19. Anderson, T., and P. A. Lee. *Fault Tolerance: Principles and Practice*. Englewood Cliffs, NJ: Prentice-Hall, 1981.
20. Andow, P. K., F. P. Lees, and C. P. Murphy. "The Propagation of Faults in Process Plants: A State of the Art Review." In *Proceedings of the 7th International Symposium on Chemical Process Hazards*, 225–237. University of Manchester, Institute of Science and Technology, April 1980.

21. Anonymous. "Blown Balloons." *Aviation Week and Space Technology*, September 20, 1971, 17.

22. Anonymous. "Doomed Journey: Tracking the KAL Tragedy." *Time Magazine* 128, no. 9 (September 1, 1986): 18.

23. Anonymous. "BP Study Blames Managers for 2005 Blast at Texas Refinery." *New York Times*, May 3, 2007.

24. Anonymous. "Design Flaw Fueled Japanese Nuclear Disaster." *Wall Street Journal*, 2011. http://www.wsj.com/articles/SB10001424052702304887904576395580035481822.

25. Arbous, A. G., and J. E. Kerrich. "Accident Statistics and the Concept of Accident-Proneness." *Biometrics* 7, no. 4 (December 1951): 340–432.

26. Askren, William B., and John M. Howard. "Software Safety Lessons Learned from Computer-Aided Industrial Machine Accidents." COMPASS '87, Washington, DC, June 1987.

27. Ayres, Robert U., and Pradeep K. Rohatgi. "Bhopal: Lessons for Technological Decision-Makers." *Technology in Society* 9, no. 1 (1987): 19–45, Pergamon Journals.

28. Bainbridge, Lisanne. "Ironies of Automation." In *New Technology and Human Error*, edited by Jens Rasmussen, Keith Duncan, and Jacques Leplat, 271–283. New York: John Wiley & Sons, 1987.

29. Baker, James, chair. *The Report of the BP U.S. Refineries Independent Safety Review Panel*, January 2007.

30. Banks, W. W., and Cerven, F. "Predictor Displays: The Application of Human Engineering in Process Control Systems." *Hazard Prevention*, January/February 1985, 26–32.

31. Barrett, Richard S. "The Human Equation in Operating a Nuclear Power Plant." In *Accident at Three Mile Island: The Human Dimensions*, edited by David L. Sills, C. P. Wolf, and Vivien B. Shelanski, 161–171, Boulder, CO: Westview Press, 1982.

32. Bartley G., and B. Lingberg. "Certification Concerns of Integrated Modular Avionics (IMA) Systems." AIAA 27th Digital Avionics Systems Conference (DASC), December 2008.

33. Bassen, H., J. Silberberg, F. Houston, W. Knight, C. Christman, and M. Greberman. "Computerized Medical Devices: Usage, Trends, Problems, and Safety Technology." 7th Annual Conference on IEEE Engineering in Medicine and Biology Society (IEEE EMBS), Chicago, September 1986.

34. Bastl, W., and L. Felkel. "Disturbance Analysis Systems." In *Human Detection and Diagnosis of System Failures*, edited by Jens Rasmussen and William B. Rouse, 451–473. New York: Plenum Press, 1981.

35. Bell, B. J., and A. D. Swain. *A Procedure for Conducting a Human Reliability Analysis for Nuclear Power Plants.* Washington, DC: US Nuclear Regulatory Commission, Sandia National Laboratories, NUREG/CR-2254, May 1983.

36. Benner, Ludwig. "Accident Perceptions: Their Implications for Accident Investigations." In *Readings in Accident Investigation: Examples of the Scope, Depth, and Source*, edited by Ted S. Ferry, 3–6. Springfield, IL: Charles C. Thomas Publisher, 1984.

37. Bilcliffe, Denis S. C. "Human Error Causal Factors in Man–Machine Systems." *Hazard Prevention*, January/February 1986, 26–31.

38. Bird, Frank E., and Robert G. Loftus. *Loss Control Management.* Loganville, GA: Institute Press, 1976.

39. Boebert, Earl. Personal communication. 1995.

40. Bogard, William. *The Bhopal Tragedy.* Boulder, CO: Westview Press, 1989.

41. Bond, Peter. *Heroes in Space*. New York: Basil Blackwell, 1987.

42. Borthwick, David. Report of the Montera Commission of Inquiry, Commonwealth of Australia, Canberra, June 17, 2010.

43. Bowman, W. C., G. H Archinoff, V. M. Raina, D. R. Tremaine, and N. G. Leveson. "An Application of Fault Tree Analysis to Safety-Critical Software at Ontario Hydro." *Conference on Probabilistic Safety Assessment and Management (PSAM)*, April 1991.

44. Bowsher, C. A. *Medical Devices: The Public Health at Risk*, GAO Report GAO/T-PEMD-90–2, US Government Accounting Organization, 1990a.

45. Bowsher, C. A. *Medical Device Recalls: Examination of Selected Cases*, GAO Report GAO/PEMD-90–6, US Government Accounting Organization, October, 1990b.

46. Brehmer, Berndt. "Development of Mental Models for Decision in Technological Systems." In *New Technology and Human Error*, edited by Jens Rasmussen, Keith Duncan and Jacques Leplat, 111–120. New York: John Wiley & Sons, 1987.

47. Bricker, Mindy Kay, ed. *The Fukushima Daiichi Nuclear Power Station Disaster: Investigating the Myth and Reality.* The Independent Investigation Commission on the Fukushima Nuclear Accident. Routledge, 2014.

48. Brilliant, Susan, John C. Knight, and Nancy G. Leveson. "The Consistent Comparison Problem in n-version Programming." *IEEE Transactions on Software Engineering*, SE-15(11), November 1989.

49. Brilliant, Susan, John C. Knight, and Nancy G. Leveson. "Analysis of Faults in an n-version Software Experiment." *IEEE Transactions on Software Engineering* SE-16, no. 2 (February 1990): 231–237.

50. Brodeur, Paul. *Outrageous Misconduct: The Asbestos Industry on Trial.* New York: Pantheon Books, 1985.

51. Brodner, P. "Qualification Based Production: The Superior Choice to the 'Unmanned Factory.'" In *Analysis, Design, and Evaluation of Man–Machine Systems, International Federation of Automatic Control*, edited by G. Mancini, G. Johannsen, and L. Martensson, 11–14. New York: Pergamon Press, 1986.

52. Brookes, Malcolm. "Human Factors in the Design and Operation of Reactor Safety Systems." In *Accident at Three Mile Island: The Human Dimensions*, edited by David L. Sills, C. P. Wolf, and Vivien B. Shelanski, 155–160. Boulder, CO: Westview Press, 1982.

53. Brooks, Frederick P. "No Silver Bullet: Essence and Accidents of Software Engineering." *Computer* 20, no. 4 (April 1987): 10–19.

54. Brown, Michael L. *Software Systems Safety Design Guidelines and Recommendations*, NSWCTR 89–33. Dahlgren, VA: Naval Surface Warfare Center, March 1989.

55. Bupp, Irvin C., and Jean-Claude Derian. *Light Water: How the Nuclear Dream Dissolved.* New York: Basic Books, 1978.

56. Calder, J. "Scientific Accident Prevention." *American Labor Legislative Review* 1 (January 1911): 14–24.

57. Cameron, Rondo, and A. J. Willard. *Technology Assessment: A Historical Approach.* Dubuque, IA: Kendall/Hunt, 1985.

58. Carroll, John M. *HCI Models, Theories, and Frameworks: Toward a Multidisciplinary Science.* San Francisco: Morgan Kaufmann Publishers, 2003.

59. Carroll, John S. "Incident Reviews in High-Hazard Industries: Sense Making and Learning under Ambiguity and Accountability." *Industrial and Environmental Crisis Quarterly* 9, no. 2 (1995): 175–197.

60. Checkland, Peter. *Systems Thinking, Systems Practice.* New York: John Wiley & Sons, 1981.

61. Chen, L., and A. Avizienis. "On the Implementation of *N*-Version Programming for Software Fault-Tolerance during Program Execution," *Proceedings of Computer Software and Applications Conference (COMPSAC) '77*, November 1977, 149–155.

62. Chestnut, Harold. "Information Requirements for Systems Understanding." *IEEE Transactions on Systems Science and Cybernetics* SSC-6, no. 1 (January 1970): 3–12.

63. Childs, Charles W. "Cosmetic System Safety." *Hazard Prevention*, May/June 1979.

64. Clarke, Richard A., and R. P. Eddy. *Warnings: Finding Cassandras to Stop Catastrophe.* New York: Harper Collins, 2017.

65. Cross, A. "Fault Trees and Event Trees." In *High Risk Safety Technology*, edited by A. E. Green, 49–65. New York: John Wiley & Sons, 1982.

66. Chisti, Agnees. *Dateline Bhopal.* New Delhi: Concept Publishing Company, 1986.

67. Cooper, J. H., "Accident-Prevention Devices Applied to Machines." *Transactions of the ASME* 12 (1891): 249–264.

68. Cotgrove, Stephen. "Risk, Value Conflict, and Political Legitimacy." In *Dealing with Risk: The Planning, Management and Acceptability of Technological Risk*, edited by Richard F. Griffiths, 122–140. Manchester, UK: Manchester University Press, 1981.

69. Council for Science and Safety. *The Acceptability of Risks.* London: Barry Rose Publishers, 1977.

70. Cox, Anthony. "What's Wrong with Risk Matrices." *Risk Analysis* 28, no. 2 (2008): 497–512.

71. Cullen, William. *The Public Inquiry into the Piper Alpha Disaster*, vols. 1 and 2. Report to Parliament by the secretary of state for energy by command of Her Majesty, November 1990.

72. Cullyer, W. J. Invited presentation at Compass '90, Gaithersburg, Maryland, June 1990.

73. Danaher, James W. "Human Error in ATC System Operations." *Human Factors* 22, no. 5 (May 1980): 535–545.

74. Daniels, J. T., and P. L. Holden. "Quantification of Risk." *Loss Prevention and Safety Promotion in the Process Industries*, G33–G45. Rugby, UK: Institution of Chemical Engineers, 1983.

75. Duke, Boyce W. "Program Manager's Handbook for System Safety and Military Standard 882B," *Hazard Prevention*, March/April 1986, 15–21.

76. Dekker, Sidney. *The Field Guide to Understanding Human Error.* Farnham, UK: Ashgate, 2002.

77. Dekker, Sidney. *Just Culture: Balancing Safety and Accountability.* Farnham, UK: Ashgate, 2007.

78. DeMillo, Richard, Richard J. Lipson, and Alan Perlis. "Social Processes and Proofs of Theorems and Programs," *Communications of the ACM* 22, no. 5 (May 1979): 271–280.

79. Department of Employment. *The Flixborough Disaster: Report of the Court of Inquiry*. London: Her Majesty's Stationery Office, 1975.

80. DeVille, Edsel. "Mishap Potential versus Risk: A Cumulative Viewpoint." *Hazard Prevention*, January/February 1988, 12–15.

81. Dewar, James. *Assumption-Based Planning*. Cambridge: Cambridge University Press, 2002.

82. Duhigg, Charles. *The Power of Habit*. New York: Random House, 2014.

83. Drucker, Peter. "They're Not Employees, They're People." *Harvard Business Review* 80, no. 2 (2002): 70–77.

84. Dulac, Nicholas, Brandon Owens, Nancy Leveson, Betty Barrett, John Carroll, Joel Cutcher-Gershenfeld, Stephen Friedenthal, Joseph Laracy, and Joseph Sussman. *Demonstration of a Powerful New Approach to Risk Analysis for NASA Project Constellation*. MIT CSRL Final Report, March 2007. http://sunnyday.mit.edu/ESMD-Final-Report.pdf.

85. Duncan, K. D. "Reflections on Fault Diagnostic Expertise." In *New Technology and Human Error*, edited by Jens Rasmussen, Keith Duncan, and Jacques Leplat, 53–61. New York: John Wiley & Sons, 1987.

86. Duncan, K. D. "Fault Diagnosis Training for Advanced Continuous Process Installations." In *New Technology and Human Error*, edited by Jens Rasmussen, Keith Duncan, and Jacques Leplat, 209–221. New York: John Wiley & Sons, 1987.

87. Dutch Safety Board. *Explosions MSPO2 Shell Moerdijk*. The Hague, NL: Dutch Safety Board. 2015.

88. Eaton, John, and Charles A. Haas. *Titanic: Triumph and Tragedy*. Wellingborough, Great Britain: P. Stephens Publishers, 1987.

89. Eckhardt, D. E., A. K. Caglayan, J. C. Knight, L. D. Lee, D. F. AcAlllister, M. A. Vouk and J. P. J. Kelly. "An Experimental Evaluation of Software Redundancy as a Strategy for Improving Software Reliability." *IEEE Transactions on Software Engineering* SE-17, no. 7 (July 1991): 692–702.

90. Eckhardt, D. E., and L. Lee. "A Theoretical Basis for the Analysis of Multiversion Software Subject to Coincident Errors." *IEEE Transactions on Software Engineering* SE-11, no. 1 (December 1985): 1511–1516.

91. Eddy, Paul, Elaine Potter, and Bruce Page. *Destination Disaster*. New York: Quadrangle/New York Times Book Company, 1976.

92. Edwards, M. "The Design of an Accident Investigation Procedure." *Applied Ergonomics* 12 (1981): 111–115.

93. Edwards, W. "Dynamic Decision Theory and Probabilistic Information Processing." *Human Factors* 4 (1962): 59–73.

94. Embrey, D. E. "A New Approach to the Evaluation and Quantification of Human Reliability in Systems Assessment." *Third National Reliability Conference*, 1981.

95. Embrey, D. E. "Modeling and Assisting the Operator's Diagnostic Strategies in Accident Sequences." In *Analysis, Design, and Evaluation of Man–Machine Systems*, edited by G. Mancini, G. Johannsen, and L. Martensson, 219–224, International Federation of Automatic Control. New York: Pergamon Press, 1986.

96. Embrey, D. E. "SHERPA: A Systematic Human Error Reduction and Prediction Method." *Contemporary Ergonomics*, 2009, 113–119.

97. Ephrath A. R., and L. R. Young. "Monitoring vs. Man-in the-Loop Detection of Aircraft Control Failures." In *Human Detection and Diagnosis of System Failures*, edited by Jens Rasmussen and William B. Rouse, 143–154. New York: Plenum Press, 1981.

98. Fackler, Martin. "Japan Power Company Admits Failings on Plant Precautions." *New York Times*, October 12, 2012.

99. Farmer, F. R. "Quantification, Experience, and Judgement." In *European Major Hazards*, edited by B. H. Harvey, 51–59. London: Oyez Scientific and Technical Services, 1984.

100. Ferry, Ted S. *Safety Program Administration for Engineers and Managers*. Springfield, IL: Charles C. Thomas Publisher, 1984.

101. Fischhoff, Baruch, Christoph Hohenemser, Roger Kasperson, and Robert Kates. "Handling Hazards." In *Risk and Chance*, edited by Jack Dowie and Paul Lefrere, 161–179. Milton Keynes, England: Open University Press, 1980.

102. Fischhoff, B., P. Slovic, and S. Lichtenstein. "Fault Trees: Sensitivity of Estimated Failure Probabilities to Problem Representation." *Journal of Experimental Psychology: Human Perception and Performance* 4, no, 2 (1978): 330–334.

103. Folkard, S. "Circadian Performance Rhythms: Some Practical and Theoretical Implications." In *Human Factors in Hazardous Situations*, edited by D. E. Broadbent, J. Reason, and A. Baddeley, 95–105. Oxford: Clarendon Press, 1990.

104. Follensbee, R. E. "The Fail-Safe Concept." FAA Seattle Aircraft Certification Office Systems Designated Engineering Representative Workshop, September 14, 1993.

105. France, Megan E. "Engineering for Humans: A New Extension to STPA." PhD Diss., Aeronautics and Astronautics Department, MIT, June 2017.

106. Frank, Nancy. *Crimes against Health and Safety*. New York: Harrow & Heston, 1985.

107. Franklin, Ned. "The Accident at Chernobyl." *The Chemical Engineer*, November 1986, 17–22.

108. Friedlander, Gordon D. "Nuclear Power Plant Safety II." *IEEE Spectrum*, May 1980, 70–75.

109. Frola, F. Ronald, and C. O. Miller. *System Safety in Aircraft Acquisition*. Washington, DC: Logistics Management Institute, January 1984.

110. Fuller, J. G. "We Almost Lost Detroit." In *The Silent Bomb*, edited by Peter Faulkner. New York: Random House, 1977.

111. Fuller, J. G. "Death by Robot." *Omni* 6, no. 6 (March 1984): 45–46 and 97–102.

112. Gehman, Harold, chair. *Columbia Accident Investigation Report*. US GAO, August 2003.

113. Gerhart, Susan L., and Lawrence Yelowitz. "Observations on the Fallibility of Applications of Modern Programming Methodologies." *IEEE Transactions on Software Engineering* 2, no. 3 (September 1976): 195–207.

114. Giffen, Carol S. "Operations Hazard Analysis." *Hazard Prevention*, May/June 1987, 23–25.

115. Gloss, David S., and Miriam Gayle Wardle. *Introduction to Safety Engineering*. New York: John Wiley & Sons, 1984.

116. Goldberg, Jack. "Some Principles and Techniques for Designing Safe Systems." *Software Engineering Notes* 12, no. 3 (July 1987): 17–19.

117. Goodstein, L. P. "Discriminative Display Support for Process Operators." In *Human Detection and Diagnosis of System Failures*, edited by Jens Rasmussen and William B. Rouse, 185–198. New York: Plenum Press, 1981.

118. Gordon, John. "The Epidemiology of Accidents." A presentation to the American Public Health Association Annual Meeting, Boston, MA, November 12, 1948. Later published in the *American Journal of Public Health*, August 29, 2011.

119. Government Accounting Office (GAO). "Survey of NASA's Lessons Learned Process," GAO-01–1015R, September 5, 2001.

120. Granger-Morgan, M. "Choosing and Managing Technology-Induced Risk." *IEEE Spectrum*, December 1981, 53–60.

121. Green, R. "Human Error on the Flight Deck." In *Human Factors in Hazardous Situations*, edited by D. E. Broadbent, J. Reason, and A. Baddeley, 55–64. Oxford: Clarendon Press, 1990.

122. Greenhouse, Steven. "BP Faces Record Fine for '05 Refinery Explosion," October 30, 2009. https://www.nytimes.com/2009/10/30/business/30labor.html.

123. Gregorian, Dro, and Sam M. Yoo. "A System-Theoretic Approach to Risk Assessment of New Development Programs." Master's thesis, System Design and Management Program, MIT, 2021.

124. Griffin, Michael. "System Engineering and the 'Two Cultures' of Engineering." *IEEE Engineering Management Review*, 2007.

125. Griffiths, Richard F. *Dealing with Risk: The Planning, Management and Acceptability of Technological Risk*. Manchester, UK: Manchester University Press, 1981.

126. Guntzburger, Y., and T. C. Pauchant. "Complexity and Ethical Crisis Management. *Journal of Organizational Effectiveness: People and Performance* 1, no. 4 (2014): 378–401.

127. Haddon, William. "The Prevention of Accidents." In *Preventive Medicine*, edited by Duncan W. Clark and Brian MacMahon, 595. Boston: Little Brown and Company, 1967.

128. Haddon-Cave, Charles. *The Nimrod Review*, HC 1025. London: The Stationery Office Limited, October 28, 2009.

129. Hagen, E. W. "Common-Mode/Common-Cause Failure: A Review." *Nuclear Safety* 21, no. 2 (March–April 1980): 184–191, 323–325.

130. Hamlet, Richard. "Testing for Trustworthiness." In *Directions and Implications of Advanced Computing*, edited by J. P Jacky and D. Schuler, 97–104. New York: Ablex Publishing Company, 1989.

131. Hammer, Willie. *Handbook of System and Product Safety*. Englewood Cliffs, NJ: Prentice-Hall, 1972.

132. Hammer, Willie. *Occupational Safety Management and Engineering*. Englewood Cliffs, NJ: Prentice-Hall, 1976.

133. Hammer, Willie. *Product Safety Management and Engineering*. Englewood Cliffs, NJ: Prentice-Hall, 1980.

134. Hansen, Carl M. *Universal Safety Standards*. New York: Universal Safety Standards Publishing Company, 1914.

135. Hansen, Carl M. "Standardization of Safeguards." In *Proceedings of the Fourth Safety Congress*, 1915, 139–146.

136. Harriss, R. C., C. Hohenemser, and R. W. Kates. "The Burden of Technological Hazards." In *Energy Risk Management*, edited by G. T. Goodman and W. D. Rowe, 103–138. New York: Academic Press, 1979.

137. Harvey, B. H. *European Major Hazards*. London: Oyez Scientific and Technical Services, 1984.

138. Health and Safety Executive, *The Tolerability of Risk from Nuclear Power Stations*. London: Her Majesty's Stationery Office, 1988.

139. Health and Safety Executive. *Safety Case Regulations for Offshore Oil Drilling*, 2005.

140. Heiler, Kathryn. "Is the Australian Mining Industry Ready for a Safety Case Regime." *31st International Conference of Safety in Mines Research Institute*. Brisbane, Australia, October 2005.

141. Heinrich, H. W. *Industrial Accident Prevention: A Scientific Approach*. New York: McGraw-Hill, 1931.

142. Heinrich, H. W., Dan Petersen, and Nestor Roos. *Industrial Accident Prevention*. New York: McGraw-Hill, 1980.

143. Herry, N. "Errors in the Execution of Prescribed Instructions: Design of Process Control Work Aids." In *New Technology and Human Error*, edited by Jens Rasmussen, Keith Duncan, and Jacques Leplat, 239–245. New York: John Wiley & Sons, 1987.

144. Hidden, Anthony. *Investigation into the Clapham Junction Railway Accident*. London: Department of Transport, Her Majesty's Stationery Office, 1990.

145. Higgs, J. C. "A High Integrity Software Based Turbine Government System." *Proceedings of the Third IFAC/IFIP Workshop on Safety of Computer Control Systems 1983 (SAFECOMP '83)* 16, no. 18 (September 1983): 207–218.

146. Hirschhorn, Larry. "The Soul of a New Worker." *Working Papers*, January/February 1982, 42–47.

147. Hoagland, Mel. "The Pilot's Role in Automation." In *Proceedings of the ALPA Air Safety Workshop*. Airline Pilots Association, 1982.

148. Hollnagel, Erik. *Safety-I and Safety-II*. CRC Press, 2014.

149. Hollnagel, Erik. *Safety-II in Practice*. Routledge, 2018.

150. Hope, S. "Methodologies for Hazard Analysis and Risk Assessment in the Petroleum Refining and Storage Industry." *Hazard Prevention*, July/August 1983, 24–32.

151. Hornick, Richard J. "Dreams—Design and Destiny." *Human Factors* 29, no. 11 (1987): 111–121.

152. Houck, Oliver A. "Worst Case and the Deepwater Horizon Blowout: There Ought to Be a Law." *Environmental Law Reporter*, 40 ELR 11036, November 2010.

153. Howard, Walter B. "Efficient Time Use to Achieve Safety of Processes, or How Many Angels Can Stand on the Head of a Pin?" In *Loss Prevention and Safety Promotion in the Process Industries*, A11–A19. Rugby, UK: Institution of Chemical Engineers, 1983.

154. Hunns, D. M. "Discussions around a Human Factors Data Base." In *High Risk Safety Technology*, edited by A. E. Green, 181–215. New York: John Wiley & Sons, 1982.

155. Hutchins, Edwin L., James D. Hollan, and Donald A Norman. "Direct Manipulation Interfaces." *Human–Computer Interaction* 1 (1985): 311–338.

156. Inge, J. R. "The Safety Case: Its Development and Use in the United Kingdom." *Equipment Safety Assurance Symposium 2007*, Bristol, UK, 2007.

157. International Maritime Organization. *About IMO*, 2017. http://www.imo.org/en/About/Pages/Default.aspx.

158. Institute of Medicine of the National Academies. "Assessing the Effects of the Gulf of Mexico Oil Spill on Human Health." A Summary of the June 2010 Workshop. The National Academy Press, 2010.

159. International Atomic Energy Agency. *Safety of Nuclear Power Plants: Design and Safety Requirements*, Report, Vienna: IAEA, 2000. http://www-pub.iaea.org/MTCD/publications/PDF/Pub1099_scr.pdf.

160. International Atomic Energy Agency. *Fundamental Safety Principles: Safety Fundamentals*. Report. Vienna: IAEA, 2006. http://www-pub.iaea.org/MTCD/publications/PDF/Pub1273_web.pdf

161. Investigation Committee on the Accident at Fukushima Nuclear Power Stations of Tokyo Electric Power Company. *Interim Report*, 2011. http://www.cas.go.jp/jp/seisaku/icanps/eng/interim-report.html

162. The Investigation Committee on the Accident at Fukushima Nuclear Power Stations of Tokyo Electric Power Company. *Final Report*, 2012. http://www.cas.go.jp/jp/seisaku/icanps/eng/final-report.html.

163. iSixSigma, 2008. https://www.isixsigma.com/?s=5+whys.

164. Johannsen, G., and W. B. Rouse. "Problem Solving Behavior of Pilots in Abnormal and Emergency Situations." 1st European Annual Conference on Human Decision Making and Manual Control, Delft University, Netherlands, 1981, 142–150.

165. Johannsen, G., J. E. Rindorp, and H. Tamura. "Matching User Needs and Technologies of Displays and Graphics." In *Analysis, Design, and Evaluation of Man–Machine Systems, International Federation of Automatic Control*, edited by G. Mancini, G. Johannsen, and L. Martensson, 51–61. New York: Pergamon Press, 1986.

166. Johnson, William G. *MORT Safety Assurance System*. New York: Marcel Dekker, 1980.

167. Jorgans, Joseph. "The Purpose of Software Quality Assurance: A Means to an End." In *Developing, Purchasing, and Using Safe, Effective, and Reliable Medical Software*, 5–9. Arlington, VA: Association for the Advancement of Medical Instrumentation, October 1990.

168. JPL Special Review Board. *Report on the Loss of the Mars Polar Lander and Deep Space 2 Missions*. NASA Jet Propulsion Laboratory, March 22, 2000.

169. Juechter, J. S. "Guarding: The Keystone of System Safety." In *Proceedings of the Fifth International System Safety Conference*, pp. V-B-1–V-B-21. System Safety Society, July 1981.

170. Kahneman, D., and A. Tversky. "On the Psychology of Prediction." *Psychological Review* 80, no. 4 (1973): 237–251.

171. Kahneman, D., P. Slovic, and A. Tversky. *Judgment under Uncertainty: Heuristics and Biases*. New York: Cambridge University Press, 1982.

172. Kelly, T., and Rob Weaver. "The Goal Structuring Notation—A Safety Argument Notation." In *Proceedings of the International Conference on Dependable Systems and Networks*, 2004.

173. Kemeny, John G. *Report of the President's Commission on Three Mile Island (The Need for Change: The Legacy of TMI)*. Washington, DC: US Government Accounting Office, 1979.

174. Kemeny, John G. "Saving American Democracy: The Lessons of Three Mile Island." *Technology Review* 83, no. 7 (June/July 1980): 65–75.

175. Kemp, Emory. "Calamities of Technology." *Science Digest*, July 1986, 50–59.

176. Kinnersly, P. *The Hazards of Work: How to Fight Them*. London: Pluto Press, 1973.

177. Kivel, M. *Radiological Health Bulletin*. Rockville, MD: US Food and Drug Administration, Center for Devices and Radiological Health, December 1986.

178. Kjellan, Urban. "An Evaluation of Safety Information Systems at Six Medium-Sized and Large Firms. *Journal of Occupational Accidents* 3 (1982): 273–288.

179. Kjellan, Urban. "Deviations and the Feedback Control of Incidents." In *New Technology and Human Error*, edited by Jens Rasmussen, Keith Duncan, and Jacques Leplat, 143–156. New York: John Wiley & Sons, 1987.

180. Kjellen, Urban. "A Changing Role of Human Actors in Accident Control—Implications for New Technology Systems." In *New Technology and Human Error*, edited by Jens Rasmussen, Keith Duncan, and Jacques Leplat, 169–175. New York: John Wiley & Sons, 1987.

181. Klein, Gary A., Judith Orasano, R. Calderwood, and Caroline E. Zsambok. *Decision Making in Action: Models and Methods*. Ablex Publishers, 1993.

182. Kletz, Trevor. "The Flixborough Cyclohexane Disaster." In *1975 Loss Prevention Symposium*, vol. 9, 1975.

183. Kletz, Trevor A. "Human Problems with Computer Control." *Plant/Operations Progress* 1, no. 4 (October 1982): 209–211.

184. Kletz, Trevor. *Myths of the Chemical Industry*. Rugby, UK: The Institute of Chemical Engineers, 1984.

185. Kletz, Trevor. "Simpler, Cheaper Plants or Wealth and Safety at Work." In *European Major Hazards*, edited by B. H. Harvey, 33–41. London: Oyez Scientific and Technical Services, 1984.

186. Kletz, Trevor, "Eliminating Potential Process Hazards, Part. I." *Hazard Prevention*, September/October 1985a, 4–15.

187. Kletz, Trevor. "Eliminating Potential Process Hazards, Part II." *Hazard Prevention*, November/December 1985b, 6–11.

188. Kletz, Trevor. *What Went Wrong?: Case Histories from Process Plant Disasters*. Gulf Publishing Company, 1988b.

189. Kletz, Trevor. "Wise after the Event." *Journal of Control and Instrumentation* 20, no. 10 (October 1988): 57–59.

190. Kletz, Trevor. *An Engineer's View of Human Error*. Rugby, UK: Institution of Chemical Engineers, 1990.

191. Kletz, Trevor. "Plants Should Be Friendly." In *Safety and Loss Prevention in the Chemical and Oil Processing Industries*, 423–433. Rugby UK: The Institution of Chemical Engineers, 1990.

192. Kletz, Trevor. *Lessons from Disaster*. Houston, TX: Gulf Publishing Company, 1991.

193. Knight, John C., and Nancy G. Leveson. "An Experimental Evaluation of the Assumption of Independence in Multi-Version Programming." *IEEE Transactions on Software Engineering* SE-12, no. 1 (January 1986): 96–109.

194. Knight, J. C., and Nancy G. Leveson. "A Reply to the Criticisms of the Knight and Leveson Experiment." *ACM Software Engineering Notes*, January 1990.

195. Kraft, Christopher. *Report of the Space Shuttle Management Independent Review*, February 1995. http://www.fas.org/spp/kraft.htm.

196. Kuhn, Thomas S. *The Structure of Scientific Revolution*. Chicago: The University of Chicago Press, 1962.

197. Kunda, Ziva. *Social Cognition: Making Sense of People*. Cambridge, MA: MIT Press, 1999.

198. Kurokawa, K., K. Ishibashi, K. Oshima, H. Sakiyama, M. Sakurai, K. Tanaka, and Y. Yokoyama. "The Official Report of the Fukushima Nuclear Accident Independent Investigation Commission." Executive summary. Tokyo: The National Diet of Japan, 2012. http://warp.da.ndl.go.jp/info:ndljp/pid/3856371/naiic.go.jp/en/report/.

199. Ladd, John. "Bhopal: An Essay on Moral Responsibility and Civic Virtue." *Journal of Social Philosophy* 22, no. 1 (1991): 73–91.

200. Lagadec, Patrick. *Major Technological Risk: An Assessment of Industrial Disasters*. New York: Pergamon Press, 1982.

201. Lagadec, Patrick. *States of Emergency*. London: Butterworth-Heinemann, 1990.

202. Lamarsh, John R. "Safety Considerations in the Design of Light Water Nuclear Power Plants." In *The Three Mile Island Nuclear Accident: Lessons and Implications*, edited by Thomas H. Moss and David L. Sills, 13–19. New York: The Annals of the New York Academy of Sciences, 1981.

203. Landwehr, Carl. *Security and Safety*. Washington, DC: Naval Intelligence Center, 1993.

204. Lauridsen, Kurt, Igor Kozine, Frank Markert, Aniello Amendola, Michalis Christou, and Monica Fiori. *Assessment of Uncertainties in Risk Analysis of Chemical Establishments*, RISO-R-1344(EN). Roskilde, Denmark: Riso National Laboratory, May 2002.

205. Lederer, Jerome. "How Far Have We Come? A Look Back at the Leading Edge of System Safety Eighteen Years Ago." *Hazard Prevention*, May/June 1986, 8–10.

206. Lees, Frank P. *Loss Prevention in the Process Industries*, vols. 1 and 2. London: Butterworths, 1980.

207. Lehtela, M. "Computer-Aided Failure Mode and Effect Analysis of Electronic Circuits." *Microelectronic Reliability* 30, no. 4 (1990): 761–773.

208. Leplat, Jacques. "Occupational Accident Research and Systems Approach." In *New Technology and Human Error*, edited by Jens Rasmussen, Keith Duncan, and Jacques Leplat, 181–191. New York: John Wiley & Sons, 1987a.

209. Leplat, Jacques. "Some Observations on Error Analysis." In *New Technology and Human Error*, edited by Jens Rasmussen, Keith Duncan and Jacques Leplat, 311–316. New York: John Wiley & Sons, 1987b.

210. Lerner, Eric. "Automating U.S. Air Lanes: A Review." *IEEE Spectrum*, November 1982, 46–51.

211. Leveson, Nancy G. "Software Fault Tolerance: The Case for Forward Recovery." In *Proceedings of the AIAA Conference on Computers in Aerospace*, 67–79. Hartford, CT, 1983.

212. Leveson, Nancy G., chair. "Committee for Review of Oversight Mechanisms for Space Shuttle Flight Software Processes." In *An Assessment of Space Shuttle Flight Software Development Processes*. Washington, DC: National Research Council, National Academy Press, 1993.

213. Leveson, Nancy G. *Safeware*. Reading, MA: Addison-Wesley, 1995.

214. Leveson, N. G. "A New Accident Model for Engineering Safer Systems." *Safety Science* 42, no. 4 (2004): 237–270.

215. Leveson, Nancy. "The Role of Software in Spacecraft Accidents." *AIAA Journal of Spacecraft and Rockets* 41, no. 4 (July 2004): 564–575.

216. Leveson, Nancy G. "Technical and Managerial Factors in the NASA *Challenger* and *Columbia* Losses: Looking Forward to the Future in Kleinman." In *Controversies in Science and Technology Volume 2*, edited by Cloud-Hansen, Matta, and Handelsman. New Rochelle, NY: Mary Ann Liebert Press, 2008.

217. Leveson, Nancy. *Engineering a Safer World*. Cambridge, MA: MIT Press, 2012.

218. Leveson, Nancy. "A Systems Approach to Risk Management through Leading Safety Indicators." *Reliability Engineering and System Safety* 136 (April 2015): 17–34.

219. Leveson, N. G. "CAST Analysis of the Shell Moerdijk Accident," 2017. http://sunnyday.mit.edu/shell-moerdijk-cast.pdf.

220. Leveson, Nancy. *Shortcomings of the Bow Tie and Other Safety Tools Based on Linear Causality.* MIT Technical Report, 2019. http://psas.scripts.mit.edu/home/nancys-white-papers.

221. Leveson, Nancy. *An Improved Design Process for Complex, Control-Based Systems using STPA and a Conceptual Architecture.* MIT ESL Technical Report, July 2020.

222. Leveson, Nancy G. *CAST Handbook,* 2019. http://psas.scripts.mit.edu/home/materials/.

223. Leveson, Nancy G., Stephen S. Cha, John Knight, and T. J. Shimeall. "The Use of Self-Checks and Voting in Software Error Detection: An Empirical Study." *IEEE Transactions on Software Engineering* SE-16, no. 4 (April 1990): 432–443.

224. Leveson, Nancy, Joel Cutcher-Gershenfeld, Betty Barrett, Alexander Brown, John Carroll, Nicolas Dulac, Lydia Fraile, and Karen Marais. *Effectively Addressing NASA's Organizational and Safety Culture" Insights from Systems Safety and Engineering Systems.* Engineering Systems Division Symposium, MIT, March 29–31, 2004.

225. Leveson, Nancy, Nicolas Dulac, Betty Barrett, John Carroll, Joel Cutcher-Gershenfeld, and Stephen Friedenthal. *Risk Analysis of NASA Independent Technical Authority.* NASA final report, June 2005. http://sunnyday.mit.edu/ITA-Risk-Analysis.doc.

226. Leveson, Nancy, Nicolas Dulac, David Zipkin, Joel Cutcher-Gershenfeld, John Carroll, and Betty Barrett. "Engineering Resilience into Safety Critical Systems." In *Resilience Engineering,* edited by Erik Hollnagel, David Woods, and Nancy Leveson. Farnham, UK: Ashgate Publishing, 2006.

227. Leveson, Nancy, Aubrey Samost, Sidney Dekker, Stan Finkelstein, and Jai Raman. "A Systems Approach to Analyzing and Preventing Hospital Adverse Events." *Journal of Patient Safety* 16, no. 2 (June 2020): 162–167.

228. Leveson, Nancy G., and John P. Thomas. *STPA Handbook.* 2018, ch. 7. http://psas.scripts.mit.edu/home/.

229. Leveson, Nancy G., and Clark S. Turner. "An Investigation of the Therac-25 Accidents." *IEEE Computer* 26, no. 7 (July 1993): 18–41.

230. Lewis, E. E. *Introduction to Reliability Engineering.* New York: John Wiley & Sons, 1987.

231. Lewycky, Peter. "Notes toward an Understanding of Accident Causes." *Hazard Prevention,* March/April 1987, 6–8.

232. Lihou, David. "Case Studies of Inadequate Instrumentation." *The Chemical Engineer,* November 1986, 41.

233. Lihou, D. A. "Management Styles—The Effects on Loss Prevention." In *Safety and Loss Prevention in the Chemical and Oil Processing Industries,* 423–433. Rugby, UK: The Institution of Chemical Engineers, 1990.

234. Lombardo, Thomas G. "TMI: An Insider's Viewpoint." *IEEE Spectrum,* May 1980, 52–55.

235. Lowe, D. R. T. "The Hazards of Risk Analysis." *Reliability Engineering* 9 (1981): 243–256.

236. Lowe, D. R. T., and C. H. Solomon. "Hazard Identification Procedures." *Loss Prevention and Safety Promotion in the Process Industries,* G8–G24. Rugby, UK: Institution of Chemical Engineers, 1983.

237. Lucas, Deborah A. "Mental Models and New Technology." In *New Technology and Human Error,* edited by Jens Rasmussen, Keith Duncan, and Jacques Leplat, 321–325. New York: John Wiley & Sons, 1987.

238. Lucas, Deborah A. "New Technology and Decision Making." In *New Technology and Human Error,* edited by Rasmussen, Jens, Keith Duncan, and Jacques Leplat, 337–340. New York: John Wiley & Sons, 1987.

239. Lutz, Robyn R. "Analyzing Software Requirements Errors in Safety-Critical, Embedded Systems." In *Proceedings of the IEEE Software Requirements Conference,* January 1992: 126–133

240. Machol, Robert C. "The *Titanic* Coincidence." *Interfaces* 5, no. 5 (May 1975): 53–54.

241. Mackall, Dale A. *Development and Flight Test Experiences with a Flight-Crucial Digital Control System.* NASA Technical Paper 2857, Dryden Flight Research Facility, NASA, November 1988.

242. MacKenzie, James J. "Nuclear Power: A Skeptic's View." *IEEE Technology and Society Magazine,* March 1984, 9–15 and 15–18.

243. Mackley, William B. "Aftermath of Mount Erebus." *Flight Safety Digest,* September 1982, 1–5.

244. Mahon, Thomas. *Report of the Royal Commission to Inquire into the Crash on Mount Erebus, Antarctica of a DC-10 Aircraft Operated by Air New Zealand Limited.* Wellington, New Zealand: P. D. Hasselberg, 1981.

245. Malasky, Sol W. *Safety Second.* Rochelle Park, NJ: 1974.

246. Malmquist, Shem, Nancy Leveson, Gus Larard, Jim Perry, and Darren Straker. "Increasing Learning from Accidents: A Systems Approach Illustrated by the UPS Flight 1354 CFIT Accident," May 2019. http://sunnyday.mit.edu/UPS-CAST-Final.pdf.

247. Manuele, Fred A. "Accident Investigation and Analysis." In *Readings in Accident Investigation: Examples of the Scope, Depth, and Source*, edited by Ted S. Ferry, 201–211. Springfield, IL: Charles C. Thomas Publisher, 1984.
248. Margulies, F. "Flexible Automation: New Options for Men, Economy, and Society." In *Analysis, Design, and Evaluation of Man–Machine Systems, International Federation of Automatic Control*, edited by G. Mancini, G. Johannsen, and L. Martensson, 11–14. New York: Pergamon Press, 1986.
249. Marples, David R. *The Social Impact of the Chernobyl Disaster*. New York: St. Martin's Press, 1988.
250. Marrett, Cora Bagley. "The President's Commission: Its Analysis of the Human Equation." In *Accident at Three Mile Island: The Human Dimensions*, edited by David L. Sills, C. P. Wolf, and Vivien B. Shelanski, 203–214. Boulder, CO: Westview Press, 1982.
251. Marshall, Eliot. "NRC Takes a Second Look at Reactor Design." *Science* 207 (March 28, 1980): 1445–1448.
252. Martin, Mike W., and Roland Schinzinger. *Ethics in Engineering*. New York: McGraw-Hill Book Company, 1989.
253. McConnell, Malcolm. *Challenger: A Major Malfunction*. Garden City, NJ: Doubleday and Company, 1987.
254. McCormick, G. Frank. "When Reach Exceeds Grasp." Unpublished essay.
255. McCormick, Norman J. *Reliability and Risk Analysis*. New York: Academic Press, 1981.
256. McCurdy, Howard. *Inside NASA: High Technology and Organizational Change in the US Space Program*. Baltimore, MD: Johns Hopkins University Press, October 1994.
257. McDonald, Gerald W. "An Introduction to System Safety." National Security Industrial Association Conference, October 1988.
258. McDonald, Harry, chair. *Shuttle Independent Assessment Team (SIAT) Report*. NASA, February 2000.
259. Megaw, James. *How Safe? Three Mile Island, Chernobyl, and Beyond*. Toronto: Stoddard Publishing Company, 1987.
260. Messersmith, Donald H. *Lincoln Memorial Lighting and Midge Study*. Unpublished report prepared for the National Park Service, CX-2000-1-0014.N.p, 1989.
261. Miles, Ralph F., Jr. "Introduction." In *Systems Concepts: Lectures on Contemporary Approaches to Systems*, edited by Ralph F. Miles Jr., 1–12. New York: John F. Wiley & Sons, 1973.
262. Mill, John Stuart. *A System of Logic, Ratiocinative, and Inductive: Being a Connective View of the Principle of Evidence, and Methods of Scientific Inquiry*. London: J. W. Parker, 1843.
263. Miller, C. O. "A Comparison of Military and Civilian Approaches to Aviation Safety. *Hazard Prevention*, May/June 1985, 29–34.
264. Miller, C. O. "The Broader Lesson from the *Challenger*." *Hazard Prevention*, January/February 1987, 5–7.
265. Miller, C. O. "Down with 'Probable Cause,'" International Society of Air Safety Investigators Seminar, Canberra, Australia, November 7, 1991.
266. Miller, Ed. "The Therac-25 Experience." Conference of State Radiation Control Program Directors, 1987.
267. Milnes, M. H. "Formation of Tetracholorodibenzodioxin by Thermal Decomposition of Sodium Trichlorophenate." *Nature* 232 (1971): 395–396.
268. Mogford, J. *Fatal Accident Investigation Report, Isomerization Unit Explosion, Final Report*. Texas City, TX: BP.
269. Moray, N. "The Role of Attention in the Detection of Errors and the Diagnosis of Failures in Man–Machine Systems." In *Human Detection and Diagnosis of System Failures*, edited by Jens Rasmussen and William B. Rouse, 155–170. New York: Plenum Press, 1981.
270. Morgan, Gareth. *Images of Organizations*. Newbury Park, CA: SAGE Publications, 1986
271. Morgan, Kenneth O. *Britain Since 1945: The People's Peace*, 3rd ed. Oxford: Oxford University Press. ISBN 0191587990, 2001, 180.
272. Morris, N. M., W. B. Rouse, and S. L. Ward. "Experimental Evaluation of Adaptive Task Allocation in an Aerial Search Environment." In *Analysis, Design, and Evaluation of Man–Machine Systems, International Federation of Automatic Control*, edited by G. Mancini, G. Johannsen, and L. Martensson, 67–72. New York: Pergamon Press, 1986.
273. Mostert, Noel. *Supership*. New York: Alfred A. Knopf, 1974.
274. Mulhouse Society. *Collection of Appliances and Apparatus for the Prevention of Accidents in Factories*. Mulhouse, Alsace-Lorraine: Society for the Prevention of Accidents in Factories, 1895.
275. National Commission on the BP Deepwater Horizon Oil Spill and Offshore Drilling. *The Gulf Oil Disaster and the Future of Offshore Drilling, Report to the President*, 2001.

276. National Response Team. "The National Response System," 2009. https://www.nrt.org/.

277. National Nuclear Energy Policy Group. *Nuclear Power Issues and Choices*. Cambridge, MA: Ballinger, 1977.

278. National Transportation Safety Board. *Crash During a Nighttime Nonprecision Instrument Landing, UPS Flight 1354, Birmingham, Alabama, August 14, 2013*. Accident Report NTSB/AAR-14/02, 2014.

279. Neumann, Peter G. "Some Computer-Related Disasters and Other Egregious Horrors. *ACM Software Engineering Notes* 11, no. 5 (October 1986): 18–19.

280. Neumann, Peter G. *Computer-Related Risks*. ACM Press, 1994.

281. Newell, A., and H. A. Simon. *Human Problem Solving*. Englewood Cliffs, NJ: Prentice-Hall, 1972.

282. Nichols, T., and P. Armstrong. *Safety or Profit: Industrial Accidents and Conventional Wisdom*. Falling Wall Press, 1973.

283. Nickerson, Raymond S. "Confirmation Bias: A Ubiquitous Phenomenon in Many Guises." *Review of General Psychology* 2, no. 2 (1998): 175–220.

284. Nielsen, Dan. "Use of Cause–Consequence Charts in Practical Systems Analysis." In *Theoretical and Applied Aspects of System Reliability and Safety Assessment*, 849–880. Philadelphia: SIAM, 1975.

285. NOPSA, 2005. http://nopsa.gov.au/safety.asp.

286. Norman, D. A. "The Problem with Automation: Inappropriate Feedback and Interaction, not 'Over-Automation.'" In *Human Factors in Hazardous Situations*, edited by D. E. Broadbent, J. Reason, and A. Baddeley, 137–145. Oxford: Clarendon Press, 1990.

287. O'Neill, Paul. Quoted in "Safety Should Never Be a Priority." Alcumus Epermits, January 12, 2018. https://www.banyardsolutions.co.uk/safety-never-priority/.

288. O'Reilly, C. "Control of Industrial Major Accident Hazards—Off-Site Planning by Local Authorities." In *European Major Hazards*, edited by B. H. Harvey, 107–123. London: Oyez Scientific and Technical Services, 1984.

289. Olsson, G. "Job Design in Complex Man–Machine Systems." In *Analysis, Design, and Evaluation of Man–Machine Systems*, edited by J. Randa, International Federation of Automatic Control, 313–318. New York: Pergamon Press. 1989.

290. O'Neil, Meaghan. "Application of CAST to Hospital Adverse Events." M.S. thesis, System Design and Management, MIT, May 2014.

291. Onishi, N., and K. Belson. "Culture of Complicity Tied to Stricken Nuclear Plan." *New York Times*, April 27, 2011, A1.

292. Ozog, Henry. "Hazard Identification, Analysis, and Control." *Hazard Prevention*, May/June 1985, 11–17.

293. Ozog, Henry, and Lisa M. Bendixen. "Hazard Identification and Quantification." *Hazard Prevention*, September/October 1987, 6–13.

294. Page, Don, Paul Williams, and Dennis Boyd. *Report of the Inquiry into the London Ambulance Service*. London: Communications Directorate, South West Thames Regional Health Authority, February 1993.

295. Palmer, Everett, and Asaf Degani. "Electronic Checklists: Evaluation of Two Levels of Automation." *Sixth Symposium on Aviation Psychology*. Columbus, OH, 1991.

296. Park, William T. *Robot Safety Suggestions*. SRI International, Technical Note No. 159, April 29, 1978.

297. Parnas, David L. "Software Aspects of Strategic Defense Systems." *Communications of the ACM* 28, no. 12 (December 1985): 1326–1335.

298. Patrick, J. "Information at the Human-Machine Interface." In *New Technology and Human Error*, edited by Jens Rasmussen, Keith Duncan, and Jacques Leplat, 341–345. New York: John Wiley & Sons, 1987.

299. Patterson, R. D. "Auditory Warning Sounds in the Work Environment." In *Human Factors in Hazardous Situations*, edited by D. E. Broadbent, J. Reason, and A. Baddeley, 37–44. Oxford: Clarendon Press, 1990.

300. Pavlovich, J. G. *Formal Report of the Investigation of the 30 April 1999 Titan IV B/Centaur TC-14/Milstar-3 (B32) Space Launch Mishap*. US Air Force, 1999.

301. Pereira, Steven J., Grady Lee, and Jeffrey Howard. "A System-Theoretic Hazard Analysis Methodology for a Non-Advocate Safety Assessment of the Ballistic Missile Defense System." AIAA Missile Sciences Conference, Monterey, CA, November 2006.

302. Perrow, Charles. "The President's Commission and the Normal Accident." In *Accident at Three Mile Island: The Human Dimensions*, edited by David L. Sills, C. P. Wolf, and Vivien B. Shelanski, 173–184. Boulder, CO: Westview Press, 1982.

303. Perrow, Charles. *Normal Accidents: Living with High-Risk Technologies*. New York: Basic Books, 1984.

304. Perrow, Charles. "The Habit of Courting Disaster." *The Nation*, October 1986, 346–356.

305. Petersen, Dan. *Techniques of Safety Management*. New York: McGraw-Hill Book Company, 1971.

306. Petroski, Henry. *To Engineer Is Human: The Role of Failure in Successful Design.* New York: Vintage Books, 1992.

307. Petroski, Henry. *Success Through Failure: The Paradox of Design.* Princeton, NJ: Princeton University Press, 2006.

308. Pettit, Ted. A., Sergio R. Concha, and Herb Linn. "Application and Use of Interlock Safety Devices." *Hazard Prevention* 19, no. 6 (November/December 1983): 4–9.

309. Pickering, William. "Systems Engineering at the Jet Propulsion Laboratory. In *Systems Concepts: Lectures on Contemporary Approaches to Systems,* edited by Ralph F. Miles Jr., 125–150. New York: John F. Wiley & Sons, 1973.

310. Poucet, Andre, and Paul DeMeester. "Assessing the Reliability of Complex Systems—An Interactive Approach." *Perspectives in Computing* 4, nos. 2/3 (Summer/Fall 1984): 47–54.

311. Quintanilla, S. Antonio Ruiz, "Social and Organizational Factors." In *New Technology and Human Error,* edited by Jens Rasmussen, Keith Duncan, and Jacques Leplat, 125–128. New York: John Wiley & Sons, 1987.

312. Rasmussen, Jens. "Some Remarks on Mental Load." In *Mental Workload: Its Theory and Measurement,* edited by N. Moray. New York: Plenum Press, 1979.

313. Rasmussen, Jens. "Models of Mental Strategies in Process Plant Diagnosis." In *Human Detection and Diagnosis of System Failures,* edited by Jens Rasmussen and William B. Rouse, 241–258. New York: Plenum Press, 1981.

314. Rasmussen, Jens. "Human Factors in High Risk Technology." In *High Risk Safety Technology,* edited by A. E. Green, 143–169. New York: John Wiley & Sons, 1982.

315. Rasmussen, Jens. "Approaches to the Control of the Effects of Human Error on Chemical Plant Safety." In *Proceedings of the International Symposium on Preventing Major Chemical Accidents,* American Institute of Chemical Engineers, February 1987a.

316. Rasmussen, Jens. "Cognitive Control and Human Error Mechanisms." In *New Technology and Human Error,* edited by Jens Rasmussen, Keith Duncan and Jacques Leplat, 53–61. New York: John Wiley & Sons, 1987a.

317. Rasmussen, Jens. "The Definition of Human Error and a Taxonomy for Technical System Design." In *New Technology and Human Error,* edited by Jens Rasmussen, Keith Duncan, and Jacques Leplat, 23–30. New York: John Wiley & Sons, 1987.

318. Rasmussen Jens. "Reasons, Causes, and Human Error." In *New Technology and Human Error,* edited by Jens Rasmussen, Keith Duncan, and Jacques Leplat, 293–301. New York: John Wiley & Sons, 1987.

319. Rasmussen, Jens. "Human Error and the Problem of Causality in Analysis of Accidents." In *Human Factors in Hazardous Situations,* edited by D. E. Broadbent, J. Reason, and A. Baddeley, 1–12. Oxford: Clarendon Press, 1990.

320. Rasmussen, J. "Risk Management in a Dynamic Society: A Modeling Problem." *Safety Science* 27, no 2/3 (1997): 183–210.

321. Rasmussen, J., Keith Duncan, and Jacques Laplat, eds. *New Technology and Human Error.* New York: John Wiley & Sons, 1987b, 23–30.

322. Rasmussen, Jens, Annelise Mark Pejtersen, and L. P. Goodstein. *Cognitive System Engineering.* New York: John Wiley & Sons, 1994.

323. Rasmussen, Jens, and William B. Rouse. *Human Detection and Diagnosis of System Failures.* New York: Plenum Press, 1981.

324. Rasmussen, Norman C. "Methods of Hazard Analysis and Nuclear Safety Engineering." In *The Three Mile Island Nuclear Accident: Lessons and Implications,* edited by Thomas H. Moss and David L. Sills, 20–36. New York: Annals of the New York Academy of Science, Vol. 365, 1981.

325. Rasmussen, Norman C. "The Application of Probabilistic Risk Assessment Techniques to Energy Technologies." In *Readings in Risk,* edited by Theodore Glickman and Michael Gough, 195–205. New York: Resources for the Future, 1990.

326. Rawlinson, J. A. "Report on the Therac-25." OCTRF/OCI Physicists Meeting, Kingston, Ontario, May 1987.

327. Reason, James. *Human Error.* Cambridge: Cambridge University Press 1990.

328. Richardson, J. G. "Major Hazards Research." In *European Major Hazards,* edited by B. H. Harvey, 79–88. London: Oyez Scientific and Technical Services, 1984.

329. Robert, J. M. "Learning by Exploration." In *Analysis, Design, and Evaluation of Man–Machine Systems,* edited by G. Mancini, G. Johannsen, and L. Martensson, 189–193. New York: Pergamon Press, International Federation of Automatic Control, 1986.

330. Roberts, Verne L. "Defensive Design." *Mechanical Engineering,* September 1984, 88–93.

331. Rogers, William P. *Introduction to System Safety Engineering.* New York: John Wiley & Sons, 1971.

332. Rogers, William P., chair. *Report of the Presidential Commission on the Space Shuttle Challenger Accident.* Washington, DC: US Government Accounting Office, 1986.

333. Roland, Harold E., and Brian Moriarty. *System Safety Engineering and Management.* New York: John Wiley & Sons, 1983.

334. Rose, C. W. "The Contribution of Operating Systems to Reliability and Safety in Real-Time Systems." *Proceedings of IFAC Safecomp '82.* Elmsford, NY: Pergamon, 1982.

335. Rouhiainen, Veikko. *The Quality Assessment of Safety Analysis*, Publication 61. Espoo, Finland: Technical Research Center of Finland, 1990.

336. Rouse, William B. "Human–Computer Interaction in the Control of Dynamic Systems." *Computing Surveys* 13, no. 1 (March 1981): 71–99.

337. Rouse, W. B., and N. M. Morris. "Conceptual Design of a Human Error Tolerant Interface for Complex Engineering Systems." In *Analysis, Design, and Evaluation of Man–Machine Systems*, edited by G. Mancini, G. Johannsen, and L. Martensson, 281–286, International Federation of Automatic Control. New York: Pergamon Press, 1986.

338. RTCA. "Safety, Performance and Interoperability Requirements Document for the In-Trail Procedure in the Oceanic Airspace (ATSA-ITP) Application," DO-312, 2008.

339. Ruckelshaus, William D. "Risk in a Free Society." *Risk Analysis* 4, no. 3 (1984): 157–162.

340. Ruckelshaus, William D. "Risk, Science, and Democracy." in *Readings in Risk*, edited by Theodore S. Glickman and Michael Gough, 105–118. New York: Resources for the Future, 1990.

341. Ryder, E. A. "The Control of Major Hazards—The Advisory Committee's Third and Final Report." In *European Major Hazards*, edited by B. H. Harvey, 5–16. London: Oyez Scientific and Technical Services, 1984.

342. Sagan, Scott D. *The Limits of Safety: Organizations, Accidents, and Nuclear Weapons.* Princeton, NJ: Princeton University Press, 1993.

343. Saltos, R. "Man Killed by Accident with Medical Radiation." *Boston Globe*, June 20, 1986.

344. Samost, Aubrey. "A Systems Approach to Patient Safety: Preventing and Predicting Medical Accidents Using Systems Theory." SM graduate thesis, Engineering Systems Division, MIT, June 2015.

345. Sawyier, Fay. "The Case of the DC-10." In *Professional Responsibility for Harmful Actions*, edited by Martin Card and Larry May, 388–401. Dubuque, IA: Kendall Hunt, 1984.

346. Senge, Peter. *The Fifth Discipline*. Doubleday, 1990.

347. Sheen, Barry. *Herald of Free Enterprise Report.* Marine Accident Investigation Branch, Department of Transport (originally *Report of Court No 8074 Formal Investigation*, HMSO, London), 1987.

348. Shein, Edgar. *Organizational Culture and Leadership.* San Francisco, CA: Jossey-Bass Publishers, 1985.

349. Sheridan, T. B. "Trustworthiness of Command and Control Systems." *IFAC Proceedings* 21, no. 5 (June 1988): 427–431.

350. Sheridan, T. B., L. Charny, M. B. Mendal, and J. B. Roseborough. "Supervisory Control, Mental Models, and Decision Aids." In *Analysis, Design, and Evaluation of Man–Machine Systems*, edited by J. Randa, 175–181. International Federation of Automatic Control. New York: Pergamon Press. 1989.

351. Shore, John. *The Sachertort Algorithm and Other Antidotes to Computer Anxiety.* New York: Penguin Books, 1986.

352. Spectrum Staff. "Too Much, Too Soon: Information Overload." *IEEE Spectrum*, June 1987, 51–55.

353. Spray, Stanley D. "Principle-Based Passive Safety in Nuclear Weapon Systems." In *High Consequence Operations Safety Symposium.* Albuquerque, NM: Sandia National Laboratories, July 13, 1994.

354. Steinzor, Rena. "Lessons from the North Sea: Should 'Safety Cases' Come to America?" *Boston College Symposium on Environmental Affairs Law Review*, 2010.

355. Sten, T., L. Bodsberg, O. Ingstad, and T. Ullebert. "Handling Process Disturbances in Petroleum Production." In *Analysis, Design, and Evaluation of Man–Machine Systems*, edited by J. Randa, International Federation of Automatic Control, 127–131. New York: Pergamon Press. 1989.

356. Sterman, John D. *Business Dynamics: Systems Thinking and Modeling for a Complex World.* New York: McGraw-Hill, 2000.

357. Stieglitz, William I. "Engineering for Safety." *Aeronautical Engineering Review*, February 1948.

358. Sugarman, Robert. "Nuclear Power and the Public Risk." *IEEE Spectrum*, November 1979, 59–111.

359. Sugiyama, S., J. Yuhara, and S. Horiuchi. "The Effects of Participatory Model on the Detection of Dynamic System Failure." In *Analysis, Design, and Evaluation of Man–Machine Systems*, edited by J. Randa, International Federation of Automatic Control, 279–283. New York: Pergamon Press, 1989.

360. Suokas, Jouko. *On the Reliability and Validity of Safety Analysis*, Publication 25. Espoo, Finland: Technical Research Center of Finland, September 1985.

361. Suokas, Juoko. "The Role of Management in Accident Prevention." *First International Congress on Industrial Engineering and Management*, Paris, June 11–13, 1986.

362. Suokas, Juoko. "Evaluation of the Quality and Safety and Risk Analysis in the Chemical Industry." *Risk Analysis* 8, no. 4 (1988): 581–591.

363. Suokas, Juoko. "The Role of Safety Analysis in Accident Prevention." *Accident Analysis and Prevention* 20, no. 1 (1988): 67–85.

364. Suokas, Juoko, and Veikko Rouhiainen. "Quality Control in Safety and Risk Analysis." *Journal of Loss Prevention in Process Industry* 2, no. 2 (April 1989): 67–77.

365. Suominen, J., and T. Malm. "Intelligent Safety Systems Provide Production Adapted Safety Strategies for Occupational Accident Prevention. In *Ergonomics of Hybrid Automated Systems II*, edited by W. Karwowski, 889–896. New York: Elsevier Science Publishers, 1990.

366. Swaanenburg, H. A. C., J. J. Zwaga, and F. Duijunouwer. "The Evaluation of VDU-based Man–Machine Interfaces in Process Industry. In *Analysis, Design, and Evaluation of Man–Machine Systems*, edited by J. Randa, International Federation of Automatic Control, 71–76. New York: Pergamon Press, 1986.

367. Tafur, Maria Fernanda. "The Underestimated Value of Safety in Achieving Organizational Goals: CAST Analysis of the Macondo Accident." SM Thesis, Integrated Design and Management Program, MIT, May 2017.

368. Taylor, D. J., D. E. Morgan, and J. P. Black. "Redundancy in Data Structures: Improving Software Fault Tolerance." *IEEE Transactions on Software Engineering* SE-6, no. 6 (November 1980): 585–594.

369. Taylor, Donald H. "The Role of Human Action in Man—Machine System Errors." In *New Technology and Human Error*, edited by Jens Rasmussen, Keith Duncan and Jacques Leplat, 287–292. New York: John Wiley & Sons, 1987.

370. Taylor, J. R. "Sequential Effects in Failure Mode Analysis." *Theoretical and Applied Aspects of System Reliability and Safety Assessment*, 881–894. Philadelphia: SIAM, 1975.

371. Taylor, J. R. "An Integrated Approach to the Treatment of Design and Specification Errors in Electronic Systems and Software." In *Electronic Components and Systems*, E. Lauger and J. Moltoft. North-Holland Publishing Company, 1982.

372. Telos Group. *Texas City Site Report of Findings*. BP, January 21, 2005.

373. Thomas, John. "Extending and Automating a Systems-Theoretic Hazard Analysis for Requirements Generation and Analysis." PhD diss., Engineering Systems Division, MIT, June 2013.

374. Thomas, Pierre, Lisa A. Jones, Jack Cloherty, and Jason Ryan. "BP's Dismal Safety Record." *NBC News*, May 27, 2010. https://abcnews.go.com/WN/bps-dismal-safety-record/story?id=10763042.

375. Thomson, J. R. *Engineering Safety Assessment: An Introduction*. New York: John Wiley & Sons, 1987.

376. Thygerson, Alton L. *Accidents and Disasters: Causes and Countermeasures*. Englewood Cliffs, NJ: Prentice-Hall, 1977.

377. Transocean. *Macondo Well Incident Report*. Vol. I, 2011.

378. Tufte, Edward. "Power Point Does Rocket Science: Assessing the Quality and Credibility of Technical Reports." Accessed March 15, 2023. https://www.edwardtufte.com/bboard/q-and-a-fetch-msg?msg_id=0001yB.

379. Turner, Barry. *Man-Made Disasters*. London: Wykeham Publications, 1978.

380. Uesako, Daisuke. *STAMP Applied to Fukushima Daiichi Nuclear Disaster and the Safety of Nuclear Power Plants in Japan*. SM thesis, MIT System Design and Management Program, June 2016.

381. US Bureau of Ocean Energy Management, Regulation and Enforcement. *Report Regarding the Causes of the April 20, 2010 Macondo Well Blowout*, Volume II of Joint Investigation Team Report of Investigation. Department of Interior, 2011.

382. US Bureau of Safety and Environmental Enforcement. "30 CFR—Mineral Resources," chapter II. US Department of the Interior.

383. US Chemical Safety and Hazard Investigation Board. *Investigation Report*, Report No. 2005–04-I-TX. Refinery Explosion and Fire, 2006.

384. US Chemical Safety and Hazard Investigation Board. *Investigation Report. Drilling Rig. Explosion and Fire at the Macondo Well*. 2016.

385. US Coast Guard. *Report of Investigation into the Circumstances Surrounding the Explosion, Fire, Sinking and Loss of Eleven Crew Members Aboard the Mobile Offshore Drilling Unit Deepwater Horizon in*

the Gulf of Mexico April 20–22, 2010. Volume I of Joint Investigation Team Report of Investigation, MISLE Activity Number: 3721503. Department of Homeland Security.

386. US Department of Defense. *MIL-STD-882: Department of Defense Standard Practice: System Safety.* Revised edition, May 11, 2012. First published July 1969.
387. US Environmental Protection Agency. "National Oil and Hazardous Substances Pollution Contingency Plan," part 300, 40 CFR Chapter I, Subchapter J Superfund, Emergency Planning, and Community Right-To-Know Programs.
388. US Federal Aviation Administration. "Analysis Techniques." In *System Safety Handbook*, chapter 9, December 30, 2000.
389. US Federal Emergency Management Agency. "National Oil and Hazardous Substances Pollution Contingency Plan (NCP) Overview." Accessed January 5, 2017. https://www.epa.gov/emergency-response/national-oil-and-hazardous-substances-pollution-contingency-plan-ncp-overview.
390. USSR State Committee on the Utilization of Atomic Energy. *The Accident at the Chernobyl Nuclear Power Plant and Its Consequences.* Report presented at the AIAE Experts Meeting, Vienna Austria, August 25–29, 1986.
391. Van de Putte, T. "Purpose and Framework for a Safety Study in the Process Industry." *Hazard Prevention*, January/February 1983, 18–21.
392. Van Horn, David J. "Risk Assessment Techniques for Experimentalists." In *Chemical Process Review*, edited by John M. Hoffman and Daniel C. Maser, 23–29. Washington, DC: American Chemical Society, 1985.
393. Vaughn, Diane. *The Challenger Launch Decision.* Chicago: University of Chicago Press, 1997.
394. Veseley, W. E., F. F. Goldberg, N. H. Roberts, and D. F. Haasl. *Fault Tree Handbook*, NUREG-0492. Washington, DC: US Nuclear Regulatory Commission, 1981.
395. Vicente, Kim J. *A Field Study of Operator Cognitive Monitoring at Pickering Nuclear Generating Station*, Technical Report CEL 9504, Cognitive Engineering Laboratory, University of Toronto, 1995.
396. Vicente, K. J., and J. Rasmussen. "Ecological Interface Design: Theoretical Foundations." *IEEE Transactions on Systems, Man, and Cybernetics* 22, no. 4 (July/August): 1992.
397. Von Bertalanffy, Ludwig. *General Systems Theory: Foundations.* Braziller, NY, 1969.
398. Voysey, Hedley. "Problems of Mingling Men and Machines." *New Scientist*, August 1977, 416–417.
399. Wahlstrom, B., and E. Swaton. *Influence of Organization and Management on Industrial Safety.* International Institute for Applied Systems Analysis, 1991.
400. Warner, Frederick. "Forward: The Foundations of Risk Assessment." In *Dealing with Risk: The Planning, Management and Acceptability of Technological Risk*, edited by Richard F. Griffiths, ix–xxii. Manchester, England: Manchester University Press, 1981.
401. Wassyng, Alan, Tom Maibaum, Mark Lawford, and Hans Bherer. "Software Certification: Is There a Case against Safety Cases?" In *Monterey Workshops 2010*, LNCS 6662, edited by R. Calinescu and E. Jackson, 206–227. Springer-Verlag, 2011.
402. Watt, Kenneth E. G. *The Titanic Effect.* Stamford, CT: Sinauer Associates, 1974.
403. Weaver, W.W. "Pitfalls in Current Design Requirements." *Nuclear Safety* 22, no. 3 (May–June 1981).
404. Weick, Karl E. "Organizational Culture as a Source of High Reliability." *California Management Review* 29, no. 2 (Winter 1987), 112–127.
405. Weil, Vivian. "Browns Ferry Case." In *Professional Responsibility for Harmful Actions.* Edited by Martin Curd and Larry May, 402–411. Dubuque, IA: Kendall Hunt, 1984.
406. Weinberg, Alvin M. "Science and Trans-Science." *Minerva* 10 (1972): 209–222.
407. Weinberg, Gerald. *An Introduction to General Systems Thinking.* New York: John Wiley & Sons, 1975.
408. Weiner, Earl L. "Control Flight into Terrain Accidents: System-Induced Errors." *Human Factors* 22, no. 5 (May 1980): 170–177.
409. Weiner, Earl L., and Renwick E. Curry. "Flight-Deck Automation: Promises and Problems." *Ergonomics* 23, no. 10 (1980): 995–1011.
410. Weiner, Norbert. *Cybernetics: or the Control and Communication in the Animal and in Engineering.* Cambridge, MA: MIT Press, 1965.
411. Whitfield, D., and G. Ord. "Some Human Factors Aspects of Computer Aiding Concepts for ATCOs." *Human Factors* 22, no. 5 (May 1980): 569–580.
412. Whyte, D. "Moving the Goalposts: The Deregulation of Safety in the Post Piper Alpha Offshore Oil Industry," 1997.

413. Wickens, C. D., and C. Kessel. "Failure Detection in Dynamic Systems." In *Human Detection and Diagnosis of System Failures*, edited by Jens Rasmussen and William B. Rouse, 155–170. New York: Plenum Press, 1981.

414. Wikipedia, "Fukushima Nuclear Disaster." Accessed December 31, 2021.

415. Wolf, C. P. "Some Lessons Learned." In *Accident at Three Mile Island: The Human Dimensions*, edited by David L. Sills, C. P. Wolf, and Vivien B. Shelanski, 215–232. Boulder, CO: Westview Press, 1982.

416. Wray, Tony. "The Everyday Risks of Playing Safe." *New Scientist*, September 8, 1988, 61–65.

417. Young, William. "A System-Theoretic Security Analysis Methodology for Assuring Complex Operations Against Cyber Disruptions." PhD Diss., System Design and Management, MIT, 2017.

418. Yourdan, E., and L. Constantine. *Structured Design*. Englewood Cliffs, NJ: Prentice-Hall, 1979.

419. Zebroski, Edwin L. "Sources of Common Cause Failures in Decision Making Involved in Man-Made Catastrophes." In *Risk Assessment in Setting National Priorities*, edited by James J. Bonin and Donald E. Stevenson, 443–454. New York: Plenum Press, 1989.

420. Zsambok, Caroline E., and Gary Klein, eds. *Naturalistic Decision Making*, Hillsdale, NJ: Lawrence Erlbaum Associates, 1997.

Index

A300 Birmingham airport crash, 458
A320 Bangalore crash, 371
Abstraction, 51–52, 60, 68, 69
Accident investigation, 12, 449–450
Active protection, 333. *See also* Passive protection
Adaptation, 55, 457
Alarms. *See* Warning systems
ALARP, ALARA, 44, 421–422
Alcoa, 429–430
Alertness, 148, 169, 370, 377, 378
Analytic decomposition, 56–59, 62
Articulatory distance, 391
Asbestosis, 40–41
Asiana B777 San Francisco crash, 372
Assessing safety, 397
Assumption-based leading indicators. *See* Leading indicators
Assumptions, 117, 143, 253, 300, 308, 452, 467
  accident causality, 179, 181, 188, 192, 283, 404
  design of safety controls, 347, 350, 409, 448, 459, 462, 499, 514, 518
  documenting of, 111, 243, 244, 245, 247, 320, 357, 444, 446, 467
  flawed, 14, 99, 122, 126, 316, 400, 519–520
  about human behavior, 149, 205, 208, 268, 309, 350, 363, 387, 390, 392, 400, 410, 413
  monitoring and updating, 244, 245, 341–342, 425, 426, 445, 457
  operational, 12, 67, 104, 196, 244, 246, 445, 449
  underlying concept of a system, 52, 56, 57, 58, 61, 63
  unrealistic (simplifying), 22, 59, 87, 89, 117, 122–123, 258, 280, 306, 412, 517–518
  validating or verifying, 242, 363, 402, 416
Assuring safety, 137, 244, 318, 319, 331, 333, 397–402, 399–402, 422
Atlas missile, 19
Audits (Auditing), 12, 245, 246, 448–449, 456, 575, 599
Audit trails, 244, 461
Authority limiting, 336
Aviation (aeronautics), xiv, 12–14, 462

B737 MAX accidents, 115, 229, 430, 437
B757 accident (Cali, Columbia), 184, 204
Backup system, 15, 118, 322, 354, 517, 597, 645. *See also* Defense-in-depth; Redundancy; Shutdown system
  Bhopal, 545
  Challenger O-ring, 506–507, 527
  Chernobyl, 634, 635
  Deepwater Horizon (Macondo), 597
  Fukushima Daiichi, 642, 645, 651
  humans as, 170–173, 311, 361, 394
  nuclear industry, 613, 618
  software as, 316, 347, 495
  spurious behavior, 345
  Three Mile Island, 619
Bad Apple Theory, 71, 205–206
Bankston mud boat, 108–109, 356, 453
Barriers, 182, 335, 351
Behaviorism, 9, 189
Bhopal MIC accident, 29, 74, 88, 94, 189, 332, 544–552
  complacency, 84
  government oversight, 115
  isolation and containment, 351
  layers of protection, 118–119, 182, 189–190
  management factors, 17
  precursors, ignoring warnings, 90
  spurious alarms, 169
  unquantifiable factors, 87
Blame. *See* Human error
Boeing 767 incident in Canada, 156
Bow tie analysis, 190, 283–284, 289
Brainstorming, 248, 249, 289, 291
Bravo oil rig, North Sea, 78
Browns Ferry nuclear power plant, 34, 91, 120, 157, 345, 453

CAST (Causal Analysis based on System Theory), 215, 458
  example of, 215–235
Causality, 73–75
  circular, 119, 192–193, 199
  complex systems, 191–193
  legal approach, 76
  linear, 181–190, 216, 322
  multifactorial, 79–81

Causality (cont.)
organizational factors, 78
oversimplification in, 76, 202–205
subjectivity in, 75–76
systemic factors, 80, 81–126, 431, 498
systems theoretic, 322
technical failures, 77–78
Cause-Consequence Analysis (CCA), 281–283
Certification, 319, 404–405, 413, 415–427. *See also* Regulation; Standards
Chain-of-failure-events model of causality, 73, 181–190, 216, 322
Challenger Space Shuttle, 37, 74, 90, 120, 124, 461, 506–512, 513–527. *See also* Space Shuttle
complacency, 84
learning from, 74
redundancy, 124, 345
safety margin, 120
systemic and organizational factors, 80, 83, 94, 95, 101, 102, 196
unrealistic risk assessment, 86
Change. *See* Management of change
Checklists, 250–251, 285, 289–290, 317–318, 379–380, 392, 555, 556
checklist mentality, 117, 462
compared to HAZOP guidewords, 285, 288–289
design checklists, 317
overreliance on, 122
Chemical industry, 16–17, 529–607
Chernobyl, 29, 81, 83, 108, 427, 633–640
management of change, 112
China airlines 747 accident, 33
Circular causality, 119, 447
Clapham Junction railway accident, 93, 151
Codes of practice. *See* Standards
Columbia Space Shuttle, 74, 455, 512–527. *See also* Space Shuttle
systemic and organizational factors, 80, 83, 94, 95, 101, 196, 437
unrealistic risk assessment, 86, 447
Common-cause failure, 119, 120, 121, 276, 274, 276, 279, 346, 347, 387
Common-mode failure, 119, 346, 409
Complacency (overconfidence), 83, 85–99, 107, 131, 138, 196, 377–378, 495–497
automation related, 131–132, 148, 169, 316, 335, 392
Bhopal, 550
Chernobyl, 640, 641
Deepwater Horizon/Macondo, 590, 595, 598, 600, 649
Flixborough, 541
Fukushima Daiichi, 641, 648
handling in SUBSAFE, 447–448
relation to training, 455
risk assessment, 410, 447, 499, 598
role of redundancy, 119–121, 193
Seveso, 536
Space Shuttle program, 83, 84, 513–514, 516–518
Texas City, 561
Therac-25, 498–499
Three Mile Island, 629
Complexity, 30–32, 123
decompositional, 55
definition, 55–56
dynamic, 55, 61
increasing, 30–32
interactive, 55
limitations, 188–190
linear, 181–190
non-linear, 55, 61
Compliance, 9, 207–208, 310, 454, 577
Concurrent engineering, 19
Confirmation bias, 401–402, 411, 422, 424
Constraints, 194, 207, 245, 294
Contingency management, 104, 107–108, 454–456
Bankston mudboat example, 356, 453
Bhopal, 550
Chernobyl, 108
Deepwater Horizon/Macondo, 605
Fukushima Daiichi, 649, 656
Three Mile Island, 632–633
Continual improvement. *See* Learning
Cosmetic system safety, 117, 525, 537, 543, 551–552, 577
Cost-benefit analysis, 421
Coupling, 8, 46, 56, 59, 308, 330–331, 361, 379, 530. *See also* decoupling
Critical Items List, 264
Crystal River nuclear power plant, 157
Culture. *See* Safety culture
Cut sets, 270

Damage reduction, 46–47, 322
Data overload, 386
DC-10 accidents, 76, 79, 80, 126, 149, 202, 345, 451
Decentralization, 33
Decision aids, 334–335, 392–393
Decision making, bottom up, 234, 436
Decomposition. *See* Analytic decomposition
Decoupling, 31, 330–331
Deepwater Horizon (Macondo), 17, 74, 111, 430, 439, 578–607
finger pointing, 100–101
ill-defined responsibility, 100–101, 436–437
inadequate learning from events, 104
inadequate training and contingency management, 107, 108, 356
maintenance, 113
management of change, 111
mixing product safety and workplace safety, 10
oversight and regulation, 115–116
poor operating procedures, 107
redundancy, 121, 345, 460
safety culture, 432, 436–437
schedule and budget pressures, 94
short-term vs. long-term thinking, 430
training, 108
Defense-in-depth, 15, 118, 182, 189–190, 613, 650, 656
Defense systems, xiv, 17–22, 419
Defensive design, 124, 497
Design assurance level (DAL), 409, 419, 427, 427
Design basis event, 407, 614, 642, 651
Design for error tolerance (Rasmussen), 161, 378, 381
Designing safety in, 123, 398, 402, 404, 425
Design precedence, 320–323
Design rationale, 247, 320, 357, 446, 454, 461, 464

Diversity. *See* Redundancy
Domino model, 6, 7, 149, 184–186
 Adams model, 185
 Bird and Loftus model, 185
Dow Index, 262–263

Emergence (emergent property), 50, 60, 62, 63–66, 397
Emergency management. *See* Contingency management
Energy model of causality, 180–181, 262–263, 322
Epidemiological model of causality, 190–191
Error tolerance. *See* Design for error tolerance
Event Tree Analysis (ETA), 277–281
Exposure, 32, 46, 322, 351, 404, 530

Fail-safe. *See* Protection system
Failure
 analysis, 405
 definition, 48–49
 modes (software), 145, 316
 single-point, 274
Failure Modes and Effects Analysis (FMEA), 188, 279–266, 413
Failure Modes and Effects Criticality Analysis (FMECA), 279–268, 274
Fall-back states, 354
Fault Hazard Analysis (FHA), 266–268, 271
Fault Tree Analysis (FTA), 188, 268–277, 413, 477
Ford Pinto gas tank, 37, 44
Feedback control, 64
Feedforward control, 64
Fishbone diagrams, 291
5 Whys, 105, 290–291, 562, 575
Flixborough accident, 77, 93, 102, 202, 332, 420, 537–543
 ignoring warnings, 90
 isolation, 351
 management of change, 110, 112, 357
French weather balloons, 122, 354
Fukushima Daiichi nuclear power plant, 102, 103–104, 372, 640–657
 complacency, 83, 88, 118
 management of change, 110, 112
 operator training and emergency management, 103–104, 109
 oversight and regulation, 114
 procedures, 107
 redundancy (defense-in-depth), 118, 120, 345, 372
Functional FMEA, 264, 266
Functional resonance model, 187–188
Fundamental attribution error (attribution effect, correspondence bias), 76

Guidelines, 35. *See also* Standards

HAZAN, 288
Hazard
 analysis definition, 11, 35, 45, 53, 46, 237, 241–243
 analysis, 244–253, 404
  limitations and criticisms, 257–259
  operational system, 253
  preliminary, 247
  subsystem, 247
 assessment, 404
 control, 46, 321–322
 elimination, 46, 321, 323–333
 level, 404, 406
 severity, 406–408
 reduction, 46, 321
 resolution, 244
 tracking system, 244, 461
Hazard Indices, 262–263
HAZOP, 17, 284–289
Healthcare, 23–24, 106, 479
Health check, 353
Heinrich's triangle (pyramid), 6–7
Herald of Free Enterprise ferry accident, 192, 215–235, 436
Heuristic bias, 401, 410–412
Hierarchical Task Analysis. *See* Human task analysis
Hierarchy (hierarchy theory), 66–68
 hierarchical control loops, 194, 294
High-Reliability Organization (HRO) theory, 234, 436, 447, 455
Hindsight bias, 92, 203–205, 219, 599
HRA (Human Reliability Analysis), 306–309. *See also* Human task analysis
Human-centered design, 11, 362, 363
Human error, 54, 453
 and blame, 7, 33, 77, 209–213
 definition, 157, 176
 human–machine mismatch, 176
 relationship to experimentation, 160–161
 systems view, 208–209
Human error analysis, 303–311
Human factors, 11, 627–628
 human–machine interaction, 147, 173, 175, 360, 364
 human–machine interface, 35, 501
Human task analysis, 303–311
 qualitative, 303–305
  Action Error Analysis (AEA), 304
  Hierarchical Task Analysis, 304, 310
  Operator Task Analysis (OTA), 304
  Work Safety Analysis (WSA), 304
 quantitative, 305–310
  Data Store, 306
  THERP, 306
Human-task mismatch (Rasmussen), 157–158, 311
Hyatt-Regency walkway collapse, 357
Hypothesis testing (operators), 163, 455

Independence, 82, 102, 499, 522–523, 654
Independent Technical Authority (ITA), 440
Industrial hygiene. *See* Workplace safety
Industrial safety. *See* Workplace safety
Influence (prestige) of safety personnel, 82, 102, 438–439, 441, 520, 522–524, 570
Intellectual manageability, 56
Interlock, 7, 337–339
Intrinsically safe design, 323
Ishikawa diagrams. *See* Fishbone diagrams

John-Mansfield Corporation. *See* Asbestosis
Jury-rigging, 157, 453
Just Culture, 433, 450

Keep alive signal, 353
King's Cross underground fire, 91
Knowledge-based behavior, 159
Korean Air Lines plane shot down, 367

Law. *See* Legal system
Law of large numbers, 59
Leading indicators, 244, 458
Learning, 451–453. *See also* Reporting systems
  from experience (events), 12, 104–105, 126, 179, 201–237, 448–453, 640
  from failure, 34
  inhibiting learning, 74
  limitations in learning from experience, 12, 21, 34, 125, 530, 552, 606, 640
  in safety management system, 451–453
  from success vs. from failure, 35, 99
  trial and error, 34
Legal system (role of courts), 40–42, 76
Level of Rigor (LOR), 133, 409, 419, 424, 425, 427
Levels of cognitive control, 158–160
Levels of safety, 117
Licensing. *See* Regulation
Lockin, 335, 336–337
Lockout, 335–336

Macondo. *See* Deepwater Horizon
Maintenance, 112, 460–461, 567–568, 602, 624
Management by exception, 372
Management of change, 110–112, 245, 357, 426, 457–460
  Challenger, 526
  Deepwater Horizon, 111
  Deepwater Horizon (Macondo), 604
  Flixborough, 543
  Fukushima Daiichi, 112, 652–653
  planned changes, 457–458
  Seveso, 112, 536
  Texas City, 454, 571
  unplanned changes, 458–460 (*see also* Leading indicators)
Masking, 33
Mental models (role in safety), 60, 161–163, 195, 365, 382, 448
Migration to states of higher risk (Rasmussen), 55, 408, 458
  in Deepwater Horizon (Macondo) accident, 602
  erosion of risk perception, 448
  in Fukushima Daiichi accident, 652–653
  identifying during audits, 448, 466, 568
  impact on certification, 426
  in Space Shuttle accidents, 513
  in Texas City explosion, 567–568
MIL-STD-882, 419–420, 424, 425. *See also* Standards
Minuteman ICBM, 19, 20
Mode confusion, 359, 371, 410
Model (definition), 253
Mond Index, 262–263
Monitor (monitoring), 123, 164–170, 244, 341–342
Montara oil rig fire, 105
Mount Erebus crash, 167
Multi-version software, 346, 348

Nimrod accident, 257, 422, 424
NORAD incidents, 120, 125, 150, 329, 369
North Anna nuclear power plant, 156
Nuclear power, 14–16, 421, 432, 462, 609–657
Nuclear weapons, 251, 339–341

Objective Quality Evidence (OQE), 88, 421, 435
Occupational safety. *See* workplace safety
O'Neill, Paul, 429–430
Operations (processes and practices), 11, 104, 106, 243, 244, 244, 245, 426, 445–456
Overconfidence. *See* Complacency

Panic button, 352
Passive protection, 333, 356
Performance assessments. *See* Audits
Performance checks, 341
Petrochemicals. *See* Chemical industry
Piper Alpha oil rig accident, 420
Policy, 11, 434–435
Predictor display, 392
Priority, 93
Probabilistic risk assessment, 16, 412–415, 421, 440, 499, 562
  of humans, 305–310, 414
  of software, 415
Procedures, operating, 453–454
  following procedures dilemma, 177, 207–208, 560
  role in accidents, 106, 550, 560, 573–574, 605, 624–625, 640
Process models, 195–196, 366
Productivity. *See* Safety vs. efficiency
Protection system, 16, 118–121, 123, 315, 316, 325, 352–356, 399, 650. *See also* Defense-in-depth
  downstream, 96
  upstream, 96
Proximate cause, 76

Ranger 6 spacecraft, 345
Reasonableness checks, 346
Recovery, backward, 348
Recovery, forward, 348
Redundancy, 343–348, 380
  common-cause failures, 120, 347, 560, 567
  functional, 345
  limitations of, 30, 112, 120, 123, 125, 315, 327, 460–461
  overreliance on, 118–121, 193, 517, 595–598, 649–651, 656
Referred symptoms, 168
Regulation (government oversight), 12, 113–116, 420, 501, 578, 590–593, 630–632, 652–654
  risk-based, 421–422
Reliability
  analysis, 255, 263, 425
  vs. safety, 49, 49–50, 97–98, 336, 409, 497, 519, 526
Reporting systems, 12, 244, 450–451, 574–575, 576
Resilience, 55
Responsibility, 100
Reuse, 98, 138, 333, 500
Risk. *See* Probabilistic risk assessment
  acceptable, 37, 421, 615
  assessment (evaluation), 36, 39, 86–88, 404–415, 405, 517, 562, 598–599, 649–651

definition, 30, 43–44, 320
discounting, 85–86, 516–517, 536
displacement, 121
likelihood, 408
matrix, 405–408
perception, 447, 448, 459, 519
reducing, 35
risk–benefit analysis, 36–38
risk–cost analysis
vs. safety, 43–44
Root cause seduction, 202
Rule-based behavior, 159

Safety
definition, 43–44
vs. risk, 44–45
Safety-by-construction, 398, 402–403, 425
Safety case, 398, 403, 404, 420, 421–423, 424
Safety culture, 82, 429–430, 431–433, 448, 514–516, 561, 568–571, 648–649
complacency and overconfidence (*see* Complacency)
culture of denial, 514, 648
paperwork culture, 116, 117, 257, 424, 432, 577, 648, 649
safety myth, 648
Safety factors, 322, 343
Safety information system, 124–126, 461–465, 526–527, 576, 630
analysis, 463
collection, 462–463
dissemination and use, 463–465
Safety integrity level (SIL), 132
Safety management system, 246, 429–468, 520–522, 568–571, 602–604, 629–630
learning in, 236
organizational structure, 542–543, 551, 435–443
Safety margins, 38, 120, 123, 139, 316, 343
Safety plans, 20, 419, 424, 425, 443, 445, 446, 462, 468
Safety-related system, 117
Safety vs. efficiency (safety vs. productivity), 5–6, 429–430, 542, 551, 565–567, 600–601
Sanity check, 353–354
Search, 254
backward search, 255
bottom-up, 256
forward search, 254–255
top-down, 255–256
Semantic distance, 390
Separation of powers, 440. *See also* SUBSAFE; Three-legged stool
Seveso dioxin release, 93, 110, 112, 531–537
Management of change, 110, 112
Shell Moerdijk, 452, 457–458
SHERPA, 310
Shutdown system, 15, 16
Simulation (simulator), 107, 399
Skill-based behavior, 158
Skill levels, 171
Situation awareness, 172, 359, 370
Sociotechnical system, 54–55, 57, 59, 63, 66, 188, 198, 292, 297
Space industry (astronautics), 23
Space Shuttle, 82–83, 90, 124, 136, 447, 503–527. *See also Challenger* Space Shuttle; *Columbia* Space Shuttle
Specifications, 52
Spurious shutdown, 344–345
STAMP (systems theoretic) model of causality, 179, 193–199, 292
Standards, 35, 113, 243, 247, 317, 318, 416, 501
goal-based or performance-based, 418, 421–423, 424
prescriptive, 416–418, 424
process-based, 417–418, 424
product-based, 416–417, 424
STPA (System-Theoretic Process Analysis), 292–303
compared to FMECA, 299, 300
compared to FTA, 299, 300–302
compared to HAZOP
SUBSAFE, 22–23, 88, 421, 440, 447–448, 448–449
Superficial safety activities. *See* Cosmetic system safety
Swiss cheese model, 186–187
Symptom referral, 33
System
definition, 51–54
environment, 53–54
open vs. closed, 54
properties, 52–54
state, 53
System-of-systems, 53
System Safety, 20–22
System Theory, 54, 59–68, 194. *See also* system thinking
model of causality (STAMP), 193–199, 322
System thinking, 55, 59, 193

Testing for safety, 137, 244, 347, 399, 402–403
Texas City oil refinery explosion, 74, 112, 352, 437, 552–578
ignoring warning signs, 91, 92
management of change, 112
mixing system safety and workplace safety, 10, 98–99, 196
poor operating procedures, 106, 454
safety culture, 103, 112, 432
training deficiencies, 108, 112
Therac-25 radiation overdoses, 90, 136, 362, 375, 473–501
complacency, 89, 498
confusing safety with reliability, 97–98, 497
inadequate government oversight, 501
lack of defensive design, 124, 497–498
overconfidence in software, 495–496
role of redundancy and complexity, 327, 338
unrealistic risk assessment, 499
THERP, 306–308. *See also* Human Task Analysis
Three-legged stool, 440
Three Mile Island (TMI), 37, 78, 87, 117, 155–156, 375, 615–633
alarms, 374, 375
common-cause failure, 345
complacency and overconfidence, 83
government oversight, 114
human-factors, 156, 164, 165, 167, 374, 453
ignoring warning signs, 91

Three Mile Island (TMI) (cont.)
inadequate training and emergency management, 107–108, 109
jury-rigging, 157
maintenance, 94, 113, 367
role of operator error, 151–152
training, 108, 453
Titanic, 85
Titanic Coincidence, 85, 118
Titanic Effect, 86
Titan missile, 19
Traceability, 294, 319, 320, 459, 464
Training, 107–108, 175, 393–395, 453, 454, 454–456
Bhopal, 550–551
Chernobyl, 108, 640
Deepwater Horizon (Macondo), 108, 604–605
Fukushima Daiichi, 108, 649–650, 656
Seveso, 536
Texas City, 108, 572–573
Three Mile Island, 107–108, 625–627
Trans-scientific questions, 38–40, 44, 47, 69, 148, 175, 381, 407, 421, 472
trend analysis, 461

Unintended consequences, 55, 61
Unknown unknowns, 50, 58, 59, 191, 326, 342
Unsafe Control Action (UCA), 295–298
Usability (vs. safety), 362, 367
Use cases, 319
Utilitarianism, 36, 37

Validity checks, 374

Warning signs, 90, 541–542, 564–565, 651
Warning systems, 165–166, 172, 372–376, 537, 560–561, 594, 599–600
alarm analysis systems, 375–376
alarm fatigue, 350, 356, 373
alarm validity checks, 374
false alarms, 148, 153, 165, 169, 373, 375
incredulity response, 373
Warsaw A320 accident, 50, 189
WASH-1400, 277–278
Watchdog timer, 353
Working groups, 442
Workplace safety, 1–2
confusion with process safety, 10, 98–99, 562–563, 600
Worst-case analysis, 29, 45, 248, 268, 295, 407, 425, 651